Communications in Computer and Information Science 1850

Rationale
The CCIS series is devoted to the publication of proceedings of computer science conferences. Its aim is to efficiently disseminate original research results in informatics in printed and electronic form. While the focus is on publication of peer-reviewed full papers presenting mature work, inclusion of reviewed short papers reporting on work in progress is welcome, too. Besides globally relevant meetings with internationally representative program committees guaranteeing a strict peer-reviewing and paper selection process, conferences run by societies or of high regional or national relevance are also considered for publication.

Topics
The topical scope of CCIS spans the entire spectrum of informatics ranging from foundational topics in the theory of computing to information and communications science and technology and a broad variety of interdisciplinary application fields.

Information for Volume Editors and Authors
Publication in CCIS is free of charge. No royalties are paid, however, we offer registered conference participants temporary free access to the online version of the conference proceedings on SpringerLink (http://link.springer.com) by means of an http referrer from the conference website and/or a number of complimentary printed copies, as specified in the official acceptance email of the event.

CCIS proceedings can be published in time for distribution at conferences or as post-proceedings, and delivered in the form of printed books and/or electronically as USBs and/or e-content licenses for accessing proceedings at SpringerLink. Furthermore, CCIS proceedings are included in the CCIS electronic book series hosted in the SpringerLink digital library at http://link.springer.com/bookseries/7899. Conferences publishing in CCIS are allowed to use Online Conference Service (OCS) for managing the whole proceedings lifecycle (from submission and reviewing to preparing for publication) free of charge.

Publication process
The language of publication is exclusively English. Authors publishing in CCIS have to sign the Springer CCIS copyright transfer form, however, they are free to use their material published in CCIS for substantially changed, more elaborate subsequent publications elsewhere. For the preparation of the camera-ready papers/files, authors have to strictly adhere to the Springer CCIS Authors' Instructions and are strongly encouraged to use the CCIS LaTeX style files or templates.

Abstracting/Indexing
CCIS is abstracted/indexed in DBLP, Google Scholar, EI-Compendex, Mathematical Reviews, SCImago, Scopus. CCIS volumes are also submitted for the inclusion in ISI Proceedings.

How to start
To start the evaluation of your proposal for inclusion in the CCIS series, please send an e-mail to ccis@springer.com.

Alberto Abelló · Panos Vassiliadis ·
Oscar Romero · Robert Wrembel ·
Francesca Bugiotti · Johann Gamper ·
Genoveva Vargas Solar · Ester Zumpano
Editors

New Trends in Database and Information Systems

ADBIS 2023 Short Papers, Doctoral Consortium and Workshops:
AIDMA, DOING, K-Gals, MADEISD, PeRS, Barcelona, Spain
September 4–7, 2023, Proceedings

Editors
Alberto Abelló
Universitat Politècnica de Catalunya
Barcelona, Spain

Oscar Romero
Universitat Politècnica de Catalunya
Barcelona, Spain

Francesca Bugiotti
University of Paris-Saclay
Gif-sur-Yvette, France

Genoveva Vargas Solar
CNRS
Villeurbanne Cedex, France

Panos Vassiliadis
University of Ioannina
Ioannina, Greece

Robert Wrembel
Poznan University of Technology
Poznan, Poland

Johann Gamper
Free University of Bozen-Bolzano
Bozen-Bolzano, Italy

Ester Zumpano
University of Calabria
Rende, Italy

ISSN 1865-0929 ISSN 1865-0937 (electronic)
Communications in Computer and Information Science
ISBN 978-3-031-42940-8 ISBN 978-3-031-42941-5 (eBook)
https://doi.org/10.1007/978-3-031-42941-5

This Springer imprint is published by the registered company Springer Nature Switzerland AG
The registered company address is: Gewerbestrasse 11, 6330 Cham, Switzerland

Preface

This CCIS volume includes research papers from the 27th European Conference on Advances in Databases and Information Systems - ADBIS, research papers from workshops accompanying ADBIS, and research papers from the PhD consortium.

The 27th ADBIS conference was held in Barcelona, Spain, on September 4–7, 2023, as a full on-site event. It received significant attention from both the research and industrial communities. A total of 77 papers were submitted to the conference, 14 of which were accepted as long papers, appearing in the LNCS proceedings volume of the conference. The papers were reviewed by an international Program Committee constituted of 96 members.

This CCIS volume includes 25 papers from the main ADBIS Conference, at an acceptance rate of 39% (25/63). These papers cover the following topics: Data Integration, Data Quality, Consistent Data Management, Metadata Management, Index Management, Query Processing, Temporal Graphs, Data Science, and Fairness.

This volume includes also the papers accepted at five workshops which were co-located with ADBIS 2023. Each workshop had its own international program committee, whose members served as the reviewers of the workshop papers included in this volume. The maximum paper acceptance rate at each of these events did not exceed 50%. In total, 63 papers were submitted to these workshops, out of which 29 were selected for presentation at the conference and publication in this volume, giving an overall acceptance rate of 46%.

The following workshops were run at the ADBIS 2023 conference.

AIDMA: 1st Workshop on Advanced AI Techniques for Data Management and Analytics, chaired by Allel Hadjali (Engineer School ENSMA, France), Anton Dignös (Free University of Bozen-Bolzano, Italy), Danae Pla Karidi (Athena Research Center, Greece), Fabio Persia (University of L'Aquila, Italy), George Papastefanatos (Athena Research Center, Greece), Giancarlo Sperlì (University of Naples "Federico II", Italy), Giorgos Giannopoulos (Athena Research Center, Greece), Julien Aligon (Toulouse 1 Capitole University, France), Manolis Terrovitis (Athena Research Center, Greece), Nicolas Labroche (University of Tours, France), Paul Monsarrat (RESTORE, France), Richard Chbeir (University of Pau and the Adour Region, France), Sana Sellami (Aix-Marseille University, France), Seshu Tirupathi (IBM Research Europe), Torben Bach Pedersen (Aalborg University, Denmark), and Vincenzo Moscato (University of Naples "Federico II", Italy). Artificial Intelligence (AI) methods are now well established and have been fully integrated by the data management and analytics (DMA) community as an innovative way to address some of its challenges in different application domains. The AIDMA workshop fully embraced this new trend in data management and analytics. The workshop aimed to gather researchers from Artificial Intelligence, data management, and analytics to address the new challenges in a variety of application domains.

DOING: 4th Workshop on Intelligent Data – from Data to Knowledge, chaired by Mirian Halfeld-Ferrari (Université d'Orléans, France) and Carmem S. Hara (Universidade Federal do Parana, Brazil). Texts are an important source of information and communication in diverse domains. The intelligent, efficient, and secure use of this information requires, in most cases, the transformation of unstructured textual data into data sets with some structure and organized according to an appropriate schema that follows the semantics of an application domain. In this context, the workshop focused on transforming data into information and then into knowledge, focusing on data related to health and environmental domains. The main aim was to gather researchers in NLP (Natural Language Processing), DB (Databases), and AI (Artificial Intelligence) to discuss two main problems: (1) how to extract information from textual data and represent it in knowledge bases and (2) how to propose intelligent methods for handling and maintaining these databases with new forms of requests.

K-Gals: 2nd International Workshop on Knowledge Graphs Analysis on a Large Scale, chaired by Mariella Bonomo (University of Palermo, Italy) and Simona E. Rombo (University of Palermo, Italy). Knowledge graphs are powerful models to represent networks of real-world entities, such as objects, events, situations, and concepts, by illustrating their relationships. This workshop aimed to provide an opportunity to introduce and discuss new methods, theoretical approaches, algorithms, and software tools that are relevant for research on Knowledge Graphs in general and on a large scale in particular. Specific topics included how Knowledge Graphs can be used to represent knowledge, how systems managing Knowledge Graphs work, and which applications may be provided on top of a Knowledge Graph.

MADEISD: 5th Workshop on Modern Approaches in Data Engineering and Information System Design, chaired by Ivan Luković (University of Belgrade, Serbia), Slavica Kordić (University of Novi Sad, Serbia), and Sonja Ristić (University of Novi Sad, Serbia). How to support information management processes in order to produce useful knowledge and tangible business value from collected data is still an open issue in many organizations. One of the central roles in addressing this issue is played by databases and information systems. In this context, the goal of this workshop was to address open questions regarding data engineering and information system design for the development and implementation of effective software services in support of information management and data-driven decision making in various organization systems.

PeRS: 2nd Workshop on Personalization and Recommender Systems, chaired by Marek Grzegorowski (University of Warsaw, Poland), Aleksandra Karpus (Gdańsk University of Technology, Poland), Tommaso di Noia (Politechnic University of Bari, Italy), and Adam Przybyłek (Gdańsk University of Technology, Poland). Recommender systems are present in our everyday lives when we read news, log in to social media, or buy something in an e-shop. Closely related to Recommender Systems is User Modelling since it enables personalization as an essential aspect of novel recommendation techniques. In a broader sense, user representation, personalized search, adaptive educational systems, and intelligent user interfaces are also related. In this context, the aim of this workshop was to extend the state of the art in User Modelling and Recommender Systems by providing a platform where industry practitioners and academic researchers can meet and learn from each other.

This volume also includes student papers accepted at the *Doctoral Consortium* (DC). The ADBIS 2023 DC was a forum for PhD students to present their research projects to the scientific community. DC papers describe the status of a PhD student's research, a comparison with relevant related work, their results, and plans on how to experiment, validate, and consolidate their contribution. The DC allowed PhD students to establish international collaborations with members of the ADBIS community and conference participants. The ADBIS 2023 DC received 22 papers, evaluated through a selective peer-review process by an international program committee. After a thorough evaluation, 10 papers were selected for presentation and 8 of these are included in this volume, giving an overall acceptance rate of 45%. The ADBIS 2023 DC program also included a panel about the meaning of success in scientific careers with accomplished professionals in academia, industry, and entrepreneurship. Finally, recognizing the importance of Diversity and Inclusion (D&I), according to the initiative promoted by the conference, the DC organized a data-driven activity to create inclusion awareness in research together with the D&I conference program. The PhD consortium was chaired by Genoveva Vargas Solar (French Council of Scientific Research, France) and Ester Zumpano (University of Calabria, Italy).

Comprehensive information about ADBIS 2023 and all its accompanying events is available at https://www.essi.upc.edu/dtim/ADBIS2023/index.html. This web page is also available from the official ADBIS portal - http://adbis.eu, which provides up-to-date information on all ADBIS conferences, persons in charge, tutorials and keynotes, publications, and issues related to the ADBIS community.

ADBIS history in a nuthsell. The first ADBIS Conference was held in Saint Petersburg, Russia (1997). Since then, ADBIS has been continuously organized as an annual event. Its previous editions were held in: Poznan, Poland (1998); Maribor, Slovenia (1999); Prague, Czech Republic (2000); Vilnius, Lithuania (2001); Bratislava, Slovakia (2002); Dresden, Germany (2003); Budapest, Hungary (2004); Tallinn, Estonia (2005); Thessaloniki, Grece (2006); Varna, Bulgaria (2007); Pori, Finland (2008); Riga, Latvia (2009); Novi Sad, Serbia (2010); Vienna, Austria (2011); Poznan, Poland (2012); Genoa, Italy (2013); Ohrid, North Macedonia (2014); Poitiers, France (2015); Prague, Czech Republic (2016); Nicosia, Cyprus (2017); Budapest, Hungary (2018); Bled, Slovenia (2019); Lyon, France (2020); Tartu, Estonia (2021); and Turin, Italy (2022).

Acknowledgements. We would like to wholeheartedly thank all participants, authors, PC members, workshop organizers, session chairs, doctoral consortium chairs, workshop chairs, volunteers, and co-organizers for their contributions in making ADBIS

2023 a great success. We would also like to thank the ADBIS Steering Committee and all sponsors.

July 2023

Alberto Abelló
Panos Vassiliadis
Oscar Romero
Robert Wrembel
Francesca Bugiotti
Johann Gamper
Genoveva Vargas-Solar
Ester Zumpano

Organization

General Chairs

Oscar Romero	Universitat Politècnica de Catalunya, BarcelonaTech, Spain
Robert Wrembel	Poznan University of Technology, Poland

Program Committee Chairs

Alberto Abelló	Universitat Politècnica de Catalunya, BarcelonaTech, Spain
Panos Vassiliadis	University of Ioannina, Greece

Workshop Chairs

Francesca Bugiotti	CentraleSupélec, France
Johann Gamper	Free University of Bozen-Bolzano, Italy

Doctoral Consortium Chairs

Genoveva Vargas-Solar	French Council of Scientific Research, France
Ester Zumpano	University of Calabria, Spain

Tutorials Chairs

Patrick Marcel	Université de Tours, France
Boris Novikov	National Research University – Higher School of Economics, Russia

Publicity Chair

Mirjana Ivanović	University of Novi Sad, Serbia

Proceedings and Website Chair

Sergi Nadal	Universitat Politècnica de Catalunya, BarcelonaTech, Spain

Special Issue Chair

Ladjel Bellatreche	University of Poitiers, France

Diversity and Inclusion Chairs

Barbara Catania	Università degli Studi di Genova, Italy
Genoveva Vargas-Solar	French Council of Scientific Research, France

Organizing Committee

Besim Bilalli	Universitat Politècnica de Catalunya, BarcelonaTech, Spain
Petar Jovanovic	Universitat Politècnica de Catalunya, BarcelonaTech, Spain
Anna Queralt	Universitat Politècnica de Catalunya, BarcelonaTech, Spain

Steering Committee Chair

Yannis Manolopoulos	Open University of Cyprus, Cyprus

Members

Andreas Behrend	TH Köln, Germany
Ladjel Bellatreche	ENSMA Poitiers, France
Maria Bielikova	Kempelen Institute of Intelligent Technologies, Slovakia
Barbara Catania	University of Genoa, Italy
Tania Cerquitelli	Politecnico di Torino, Italy
Silvia Chiusano	Politecnico di Torino, Italy
Jérôme Darmont	University of Lyon 2, France

Johann Eder Alpen-Adria	Universität Klagenfurt, Austria
Johann Gamper	Free University of Bozen-Bolzano, Italy
Tomáš Horváth	Eötvös Loránd University, Hungary
Mirjana Ivanović	University of Novi Sad, Serbia
Marite Kirikova	Riga Technical University, Latvia
Manuk Manukyan	Yerevan State University, Armenia
Raimundas Matulevicius	University of Tartu, Estonia
Tadeusz Morzy	Poznan University of Technology, Poland
Kjetil Nørvåg	Norwegian University of Science & Technology, Norway
Boris Novikov	National Research University – Higher School of Economics, Russia
George Papadopoulos	University of Cyprus, Cyprus
Jaroslav Pokorny	Charles University in Prague, Czech Republic
Oscar Romero	Universitat Politècnica de Catalunya, Spain
Sergey Stupnikov	Russian Academy of Sciences, Russia
Bernhard Thalheim	Christian Albrechts University of Kiel, Germany
Goce Trajcevski	Iowa State University, USA
Valentino Vranić	Slovak University of Technology in Bratislava, Slovakia
Tatjana Welzer	University of Maribor, Slovenia
Robert Wrembel	Poznan University of Technology, Poland
Ester Zumpano	University of Calabria, Spain

Program Committee

Cristina D. Aguiar	Universidade de São Paulo, Brazil
Syed M. Fawad Ali	Accenture DACH, Germany
Bernd Amann	LIP6, Sorbonne Université, CNRS, France
Witold Andrzejewski	Poznan University of Technology, Poland
Sylvio Barbon	State University of Londrina, Brazil
Andreas Behrend	University of Bonn, Germany
Khalid Belhajjame	PSL, Université Paris-Dauphine, LAMSADE, France
Ladjel Bellatreche	ENSMA Poitiers, France
Josep L. Berral	Universitat Politècnica de Catalunya, BarcelonaTech, Spain
Maria Bielikova	Kempelen Institute of Intelligent Technologies, Slovakia
Sandro Bimonte	INRAE, France
Paweł Boiński	Poznan University of Technology, Poland

Lars Ailo Bongo	UiT The Arctic University of Norway, Norway
Drazen R. Brdjanin	University of Banja Luka, Bosnia and Herzegovina
Ismael Caballero	Universidad de Castilla - La Mancha, Spain
Damiano Carra	University of Verona, Italy
Silvia Chiusano	Politecnico di Torino, Italy
Isabelle M. Comyn-Wattiau	ESSEC Business School, France
Dante Conti	Universitat Politècnica de Catalunya, BarcelonaTech, Spain
Antonio Corral	University of Almería, Spain
Jerome Darmont	ERIC - Université Lumière Lyon 2, France
Christos Doulkeridis	University of Pireaus, Greece
Johann Eder	Universität Klagenfurt, Austria
Markus Endres	University of Augsburg, Germany
Javier A. Espinosa-Oviedo	University of Lyon, France
Lorena Etcheverry	Universidad de la República, Uruguay
Georgios Evangelidis	University of Macedonia, Greece
Fadila Bentayeb	Université Lumière Lyon 2, France
Flavio Ferrarotti	SCCH, Austria
Flavius Frasincar	Erasmus University Rotterdam, The Netherlands
Johann Gamper	Free University of Bozen-Bolzano, Italy
Matteo Golfarelli	University of Bologna, Italy
Jānis Grabis	Riga Technical University, Latvia
Francesco Guerra	University of Modena e Reggio Emilia, Italy
Alberto Gutierrez-Torre	Barcelona Supercomputing Center, Spain
Tomáš Horváth	Eötvös Loránd University, Hungary
Mirjana Ivanovic	University of Novi Sad, Serbia
Stefan Jablonski	University of Bayreuth, Germany
Pokorny Jaroslav	Charles University, Czech Republic
Petar Jovanovic	Universitat Politècnica de Catalunya, BarcelonaTech, Spain
Alexandros Karakasidis	University of Macedonia, Greece
Zoubida Kedad	University of Versailles, France
Attila Kiss	Eötvös Loránd University, Hungary
Georgia Koloniari	University of Macedonia, Greece
Haridimos Kondylakis	FORTH-ICS, Greece
Julius Köpke	University of Klagenfurt, Austria
Georgia Koutrika	Athena Research Center, Greece
Dejan Lavbič	University of Ljubljana, Slovenia
Audrone Lupeikiene	Vilnius University, Lithuania
Federica Mandreoli	Università di Modena e Reggio Emilia, Italy
Yannis Manolopoulos	Aristotle University of Thessaloniki, Greece

Patrick Marcel	University of Tours, France
Adriana Marotta	Universidad de la República, Uruguay
Miguel A. Martinez-Prieto	University of Valladolid, Spain
Alejandro Maté	University of Alicante, Spain
Jose-Norberto Mazon	Universidad de Alicante, Spain
Sara Migliorini	University of Verona, Italy
Alex Mircoli	Università Politecnica delle Marche, Italy
Angelo Montanari	University of Udine, Italy
Lia Morra	Politecnico di Torino, Italy
Tadeusz Morzy	Poznan University of Technology, Poland
Kjetil Nørvåg	Norwegian University of Science and Technology, Norway
Boris Novikov	National Research University - Higher School of Economics, Russia
Omar Boussaid	Université Lumière Lyon 2, France
George Papastefanatos	Athena Research Center, Greece
Veronika Peralta	University of Tours, France
Giuseppe Polese	University of Salerno, Italy
Elisa Quintarelli	Università degli Studi di Verona, Italy
Milos Radovanovic	University of Novi Sad, Serbia
Franck Ravat	Institut de Recherche en Informatique de Toulouse, France
Stefano Rizzi	University of Bologna, Italy
María del Mar Roldán	University of Málaga, Spain
Gunter Saake	University of Magdeburg, Germany
Kai-Uwe Sattler	TU Ilmenau, Germany
Milos Savic	University of Novi Sad, Serbia
Marcos Sfair	Sunye Universidade Federal do Paraná, Brazil
Spiros Skiadopoulos	University of the Peloponnese, Greece
Claudia Steinberger	Universität Klagenfurt, Austria
Sergey A. Stupnikov	FRC CSC RAS, Russia
Bernhard Thalheim	CAU Kiel, Germany
Goce Trajcevski	Iowa State University, USA
Genoveva Vargas-Solar	French Council of Scientific Research, France
Goran Velinov	Ss. Cyril and Methodius University in Skopje, North Macedonia
José R. R. Viqueira	Universidade de Santiago de Compostela, Spain
Szymon Wilk	Poznan University of Technology, Poland
Ester Zumpano	University of Calabria, Spain

Additional Reviewers

Giorgos Alexiou
Paul Blockhaus
Andrea Brunello
Loredana Caruccio
Thanasis Chantzios
Marco Franceschetti
Luca Geatti
Balasubramanian Gurumurthy
Maude Manouvrier
Barbara Oliboni
Nicolas Ringuet
Vassilis Stamatopoulos

Contents

Data Integration

Data Quality

Metadata Management

Contributions from ADBIS 2023 Workshops and Doctoral Consortium

AIDMA: 1st Workshop on Advanced AI Techniques for Data Management, Analytics

DOING: 4th Workshop on Intelligent Data - From Data to Knowledge

K-Gals: 2nd Workshop on Knowledge Graphs Analysis on a Large Scale

MADEISD: 5th Workshop on Modern Approaches in Data Engineering, Information System Design

PeRS: 2nd Workshop on Personalization, Recommender Systems

Doctoral Consortium

Index Management and Data Reconstruction

Estimating iSAX Parameters for Efficiency

Mihalis Tsoukalos(✉), Nikos Platis, and Costas Vassilakis

Department of Informatics and Telecommunications, University of the Peloponnese, Tripolis, Greece
{mtsoukalos,nplatis,costas}@uop.gr

Abstract. iSAX is considered one of the most efficient indexes for time series. Several parameters affect the construction of an iSAX index: the sliding window size, the threshold value, the number of segments and the maximum cardinality, the last two being related to the SAX representation. In this paper (i) we consider the effect of each parameter on the efficiency of the iSAX index, (ii) we evaluate the importance of each parameter, and, (iii) suggest how to optimize these parameters.

Keywords: Time series · Indexing · SAX · Parameter tuning

1 Introduction

Indexing is often used for time series processing, to speed up query evaluation and subsequence searches required in data mining tasks. One of the most efficient time series indexes is iSAX [6]. Nevertheless, in order to create an iSAX index, four parameters must be provided, and the choice of their values is crucial to the performance and accuracy of the iSAX index.

In this work, we try to alleviate the effort of finding optimal iSAX parameters by (a) presenting a framework that quantifies the quality of an iSAX index, (b) demonstrating the importance of each individual parameter and (c) proposing a heuristic for optimizing their values to enhance the efficiency of iSAX indexes.

2 Preliminaries and Related Work

2.1 Time Series and the SAX Representation

A *time series* $T = (t_1, \ldots, t_n)$ is an ordered list (sequence) of n *data points* $t_1, \ldots, t_n$. The SAX (Symbolic Aggregate Approximation) representation [4] is a symbolic representation of a time series which transforms T into a representation with m elements where $m << n$. This is referred to as *dimensionality reduction.*

The computation of the SAX representation of a time series T is illustrated in Fig. 1(a). In this example, T contains 12 points, pictured as circles. The data value space is partitioned into a number of areas (shown with the different color shades); the number of areas is called the *cardinality* of the SAX representation. Each area is labeled with a binary number, called a *SAX word.* To compute

A. Abelló et al. (Eds.): ADBIS 2023, CCIS 1850, pp. 3–12, 2023.
https://doi.org/10.1007/978-3-031-42941-5_1

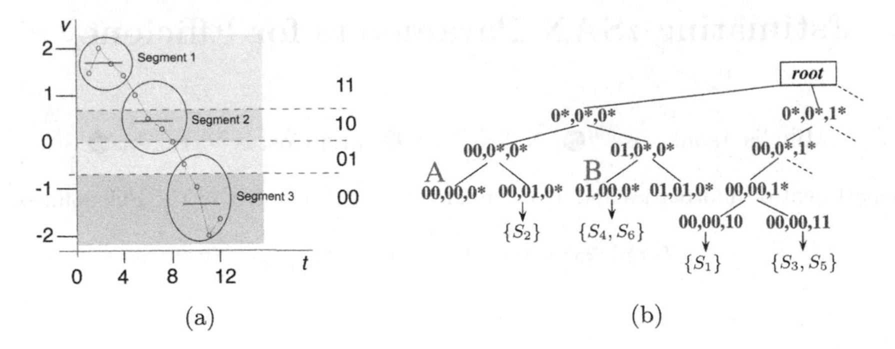

Fig. 1. (a) The SAX representation of a time series T: $SAX(T) = \{\mathbf{11}, \mathbf{10}, \mathbf{00}\}$. (b) Part of an *i*SAX index. The number of segments is 3 and the maximum cardinality is 4. Stars (*) denote that the SAX representations may be promoted in deeper levels.

$SAX(T)$, we divide T into *segments* that contain the same number of data points. In our example, T is divided into the 3 segments of 4 points each. For each segment, we compute the average value (shown with the red lines) of its data points and note the label of the area that this average value falls in; all these labels form $SAX(T)$. In our example, $SAX(T) = \{\mathbf{11}, \mathbf{10}, \mathbf{00}\}$.

Using a different partitioning would lead to a different SAX representation of a time series. Most interestingly, it is possible to create a *hierarchy* of SAX representations by merging consecutive areas of the partitioning. Referring to Fig. 1(a) again, areas **11** and **10** could be merged to an area labeled **1** and areas **01** and **00** to another one labeled **0**. In this case, the SAX representation would be $SAX(T) = \{\mathbf{1}, \mathbf{1}, \mathbf{0}\}$. It can be seen that when merging adjacent areas as described, the SAX representation may be computed from the representation of higher cardinality just by dropping the last digit of each SAX word.

Note that all previous computations are performed using the *normalized* (in $[-1, 1]$) version of a subsequence [2,4].

2.2 The *i*SAX Index

*i*SAX is an index for time series that utilizes the SAX representation. In practical applications, it is usual to have a very long time series T and decompose it in (contiguous) *subsequences* using a *sliding window* of size w. Then, all these subsequences are indexed using *i*SAX and queries are seeking subsequences with requested properties using the index.

The SAX representations of the subsequences are used to construct an unbalanced *tree*, as will be explained below. Subsequences are stored in *terminal nodes* only, whereas *inner nodes* are used for traversing the index. All subsequences indexed in an *i*SAX tree are of the same length, and subdivided into a specific number of segments, thus resulting in SAX representations of the same number of SAX words. For performance reasons, the SAX representations are computed once, using a large *maximum cardinality* value. As shown above, to get SAX representations corresponding to lower cardinalities, we drop the rightmost digits of the SAX words. Also for performance reasons, each terminal node of the tree stores a number of subsequences up to a specified *threshold*.

To construct an *i*SAX index (see Fig. 1(b)), the root node of the tree does not store any subsequences and only contains pointers to its children. Then, the (first-level) children of the root node are labeled with SAX representations of the specified number of segments, corresponding to cardinality 2. For each subsequence to be indexed, its SAX representation corresponding to cardinality 2 and the relevant child of the root node are considered:

- If the number of subsequences already stored in this terminal node is below the threshold value, the subsequence is stored there.
- Otherwise, the node is *split*. First, two new terminal nodes are created, labeled by *promoting* one of the SAX words to a cardinality higher by one level, thus adding the digits **0** and **1**, respectively for the two new nodes, to the relevant SAX word. Then, the subsequences already stored in the original terminal node are distributed to the two new nodes according to their SAX representations (of one level higher cardinality on the promoted SAX word). Last, the two new terminal nodes are linked to the original one, which now becomes an inner node.

If a subsequence is about to be stored in a terminal node that already holds the maximum number of subsequences allowed by the threshold value and cannot be further split because all its words are using the maximum cardinality, we have an *overflow*. In this case, the *i*SAX parameters must be adjusted to increase the capacity of the index and the process starts over.

2.3 Using *i*SAX

Having indexed a set of subsequences using *i*SAX, we may use the index to accelerate queries on the set. In this work, we consider the following two types of queries: (i) given a subsequence S, check if it exists in the set, and (ii) given a subsequence S, find its *nearest neighbor* (NN), based on the Euclidean distance, in the set. For both queries, we compute $SAX(S)$ and use it to traverse the *i*SAX index, down to the terminal node that corresponds to $SAX(S)$.

For the first query, it is obvious that if S exists in the set, it must have been indexed under this terminal node. For example, if $SAX(S) = \{00, 00, 00\}$, we start from the root of the tree and in each level we descend to the leftmost child node of Fig. 1(b), ending down to node A that does not contain any subsequences; this means that S does not exist in the set. For the second query, subsequences similar to S (near neighbors of S) are also expected to have been indexed under this terminal node. For example, if $SAX(S) = \{01, 00, 00\}$, we are going to reach node B containing two candidate subsequences S_4 and S_6, which will be further processed to find the nearest neighbor to S. In the case where $SAX(S)$ with a cardinality of 2 does not match any child of the root node, we say that we have an *NN miss*.

2.4 Related Work

To the best of our knowledge, there is no prior work regarding the definition of quality of *i*SAX indexes. Furthermore, there is no prior work related to the estimation and tuning of all the parameters of an *i*SAX index.

An initial evaluation of the effect of some of the parameters (namely, the number of segments and the threshold value) is provided in [6]. However, several assumptions considered in [6] are completely different than ours, and, additionally, we take into account all the *i*SAX parameters.

Another loosely related work is [1], where the authors try to find the best SAX segment size and cardinality based on the distribution of values of the time series. Their technique uses variable cardinality per SAX word as well as a variable number of SAX words.

3 Overview of *i*SAX Parameters

As discussed above, an *i*SAX index depends on the following parameters:

The sliding window size w**.** The sliding window size depends on the time series, the data, and the problem at hand. For instance, if a time series contains samples on every second and the data should be analyzed per minute, it would be reasonable to consider a sliding window of size 60.

The number of segments s**.** The number of segments determines the (maximum) number of children of the root node of the *i*SAX index. Qualitatively, the number of segments affects the approximation of the SAX representation: using a large number of segments, each subsequence is subdivided into many small segments, therefore its approximation is tighter and fewer subsequences will be stored in each terminal node; this would make the search for a specific subsequence faster. On the other hand, such an index would perform worse for NN search, since only the subsequences stored in a single terminal node are considered as NN candidates (cf. Sect. 2.3), and the ones present therein may not actually be near neighbors overall.

The maximum cardinality value mc**.** The maximum cardinality value is the least important parameter of an *i*SAX index. In our approach, the parameters are chosen in such a way that all subsequences can fit into the *i*SAX index without an overflow.

The threshold value th**.** The threshold value is important as it affects both the capacity and the performance of the *i*SAX index. Using a very small threshold value would hamper the capacity of the index; in order to avoid overflows, a higher maximum cardinality value would be required, but such a setup would create a deeper, less efficient tree. On the other hand, using a very large threshold value would adversely affect the efficiency of the *i*SAX index in a different way: each terminal node would contain too many subsequences, therefore many comparisons would be required to locate a given subsequence or the NNs among them.

4 Evaluating *i*SAX Performance

4.1 Methodology

The fundamental question this paper tries to answer is *whether we can find appropriate parameter values to construct efficient* iSAX *indexes*. An efficient *i*SAX index exhibits two properties: first, it is *fast* when searching for a specific subsequence; and second, it is *accurate* when looking for the NN of a given subsequence—in this case, speed is much less important.

The speed of an *i*SAX index can be quantified by the *number of accesses to subsequences* during search. As a more general measure, we report the *ratio* of number of accesses to subsequences over the length of the full time series, thus having an average of the number of times each subsequence was accessed. Our queries include both subsequences of the same time series, which will certainly be found (we used 2000 of these for each test run), and subsequences (of length equal to the sliding window size) from an external time series, which will most probably not be found in the index (the external time series was of length 2000, therefore $(2000 - w + 1)$ subsequences were queried for each test run).

The accuracy for NN queries can be measured by the *difference* between the distance of the approximate NN returned by the *i*SAX index and the distance of the exact one computed using an exhaustive search. As a summary measure of the differences for all the NN queries that we perform, we report the *Root Mean Square Error* of them, $RMSE = \sqrt{\sum_{i=1}^{k}(approx_i - exact_i)^2/k}$. In this case, it is only meaningful to use subsequences of the external time series.

In order to accelerate the exact NN computations, we use the Matrix Profile (MP) [3,8]: given two time series, MP finds the distance of the NN of each subsequence of the first time series among the ones of the second time series.

We also report the number of NN misses. To include them in the RMSE, we set in this case $approx_i = 2\sqrt{w}$, since this is the largest Euclidean distance between two normalized (in $[-1, 1]$) subsequences of length w.

In real world scenarios, NN queries are much more useful than exact searches: we usually look for similarities in subsequences between different time series without knowing a priori what to expect. This is why we concentrate on evaluating the *i*SAX index for NN queries, without overlooking its speed in the process.

4.2 Test Datasets

We used four main datasets in our experiments, with different characteristics. We refer to them as **ECG**, **EOG**, **EEG** and **Synthetic**. The first three are publicly available very long time series, and from each one we used three continuous parts of 200k, 400k and 500k data points. The fourth dataset was created using a random number generator, and its values follow the uniform distribution; again, variants of 200k, 400k and 500k were generated. Histograms showing the distribution of the values for the datasets with 500k data points are provided in Fig. 2. The distribution of values is similar for all the variants of each dataset.

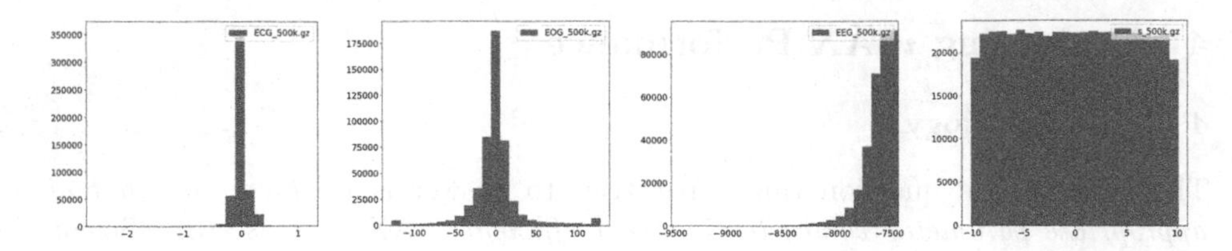

Fig. 2. Histograms of all 500k datasets: ECG, EOG, EEG and Synthetic.

Time series with many similar values (for example, the ECG dataset whose most values are close to 0) tend to be difficult to index using *i*SAX because the SAX representations of many subsequences fall into the same *i*SAX branch; thus the *i*SAX tree is very unbalanced and there is high risk for overflow unless large threshold and/or maximum cardinality values are used. On the contrary, time series whose values distribution is more varied produce more balanced, compact and efficient indexes.

5 Experimental Evaluation

5.1 Experimenting with *i*SAX Parameters

We experimented with various combinations of *i*SAX parameters in order to discover more about their effect on the performance of an *i*SAX index. In order to explore the parameter value space, we applied grid search, using sliding window sizes equal to 120, 300 and 600; number of segments values of 5, 6 and 10; and threshold values of 450, 600 and 1000. The maximum cardinality value was 64 across all experiments, which did not cause any overflows.[1]

Table 1 shows the 5 best and 5 worst RMSE values over the 324 total tests, along with the corresponding datasets and *i*SAX parameters. Figures 3, 4 and 5 show the RMSE, NN misses and accesses over time series length for all datasets and for all values of the parameters tested. Some patterns emerge from these tables and graphs: First, the distribution of the values of the datasets affects the quality of NN queries using *i*SAX. The best results for the RMSE were all achieved with the EEG dataset, whose values follow the normal distribution with a relatively high standard deviation. On the contrary, all the worst ones were recorded with the ECG dataset, whose values are mostly concentrated around 0; as discussed above, this leads to a very unbalanced index. This result is confirmed in Fig. 3. Second, a smaller number of segments leads to better quality NN queries (lower RMSE and fewer NN misses). All the best results used 5 or 6 segments, whereas all the worst ones used 10 segments, for various dataset and parameter values. Third, higher threshold values lead to better quality NN queries (lower RMSE and fewer NN misses), which is expected considering that more subsequences are contained in each terminal node, which constitute the

[1] The source code and test data are available upon request.

Table 1. The 5 **best** and 5 **worst** RMSE values and their parameters.

Dataset	Window	Segments	Threshold	RMSE	Acc/Len	NN Misses
EEG 200k	120	5	1000	0.431	29.016	0
EEG 500k	120	6	1000	0.439	19.333	0
EEG 400k	120	5	1000	0.439	22.037	0
EEG 500k	120	5	1000	0.456	20.776	0
EEG 400k	120	6	1000	0.471	20.170	0
ECG 400k	300	10	1000	5.710	14.165	33
ECG 500k	300	10	450	5.724	14.341	31
ECG 400k	300	10	600	5.817	14.491	34
ECG 400k	300	10	450	5.892	14.571	34
ECG 200k	600	10	1000	6.011	13.369	27

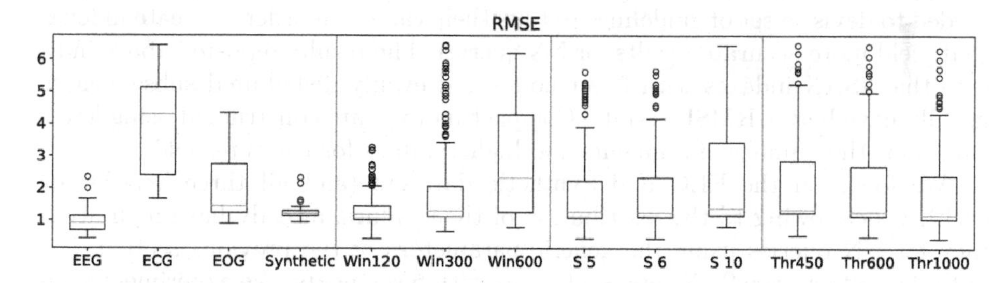

Fig. 3. RMSE for all the values of the parameters tested. The last three plots aggregate results over all test cases.

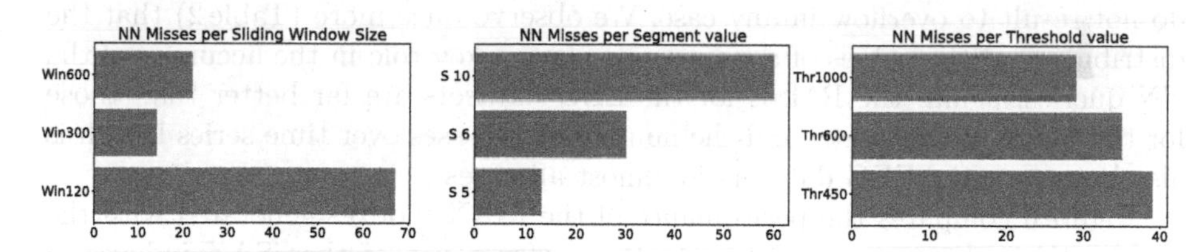

Fig. 4. NN Misses for all the values of the parameters tested, aggregated over all test cases.

candidate sequences among which the nearest neighbor will be selected. However, this should not be taken to an extreme, as this would adversely affect the speed of the index, as discussed in Sect. 3, and this starts being evident even for the relatively low values used (Fig. 5). Last, a faster search, indicated by a lower number of accesses to subsequences, will most probably not be of high quality. This is reasonable since the search for the NN will not be broad enough. As shown in Table 1, the low number of accesses may be due to the dataset itself and the combination of parameter values (low threshold, high number of segments).

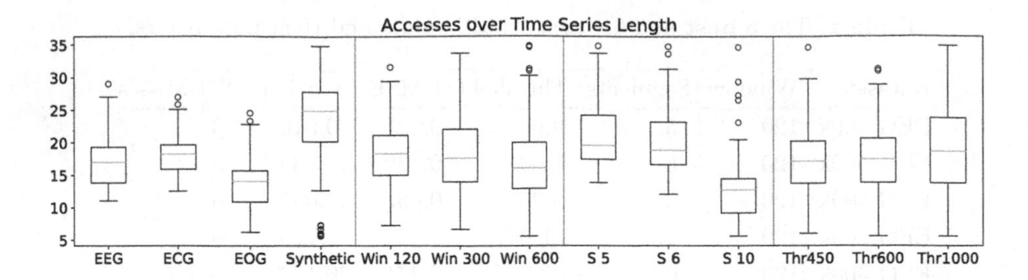

Fig. 5. Number of accesses to subsequences divided by time series length. The last three plots aggregate results over all test cases.

5.2 Optimizing *i*SAX Parameters

Having gained intuition of how each parameter affects the *i*SAX index, we proceeded to devise a set of guidelines to tune their values, in order to create indexes that yield more accurate results for NN queries. The results reported above indicate that *i*SAX indexes with fewer nodes and evenly distributed subsequences usually have better RMSE results. Compact indexes are constructed using lower values for the number of segments and higher values for the threshold.

We focus on the EEG and Synthetic datasets (and all three lengths for each), which, owing to the distribution of their values, already have high accuracy for NN queries using the generic parameters of the previous section. For both datasets we use SAX representations with 5 segments. We experiment with threshold values of 2000 and 2500 to see how the results are affected. A maximum cardinality of 16 for the EEG dataset and 64 for the Synthetic dataset do not result to overflow in any case. We observe once more (Table 2) that the distribution of the values of a time series plays a key role in the accuracy of the NN queries: again, the RMSE for the EEG datasets are far better than those for the Synthetic datasets, and the number of accesses over time series length is also better for the EEG datasets in almost all cases.

Figure 6 compares the performance of the *i*SAX indexes generated with the new, optimized parameter values to the performance of the *i*SAX indexes of Sect. 5.1. As the RMSE of the new indexes is considerably reduced compared to the RMSE of the previous ones, we infer that using parameters that reduce the number of *i*SAX nodes does produce more efficient *i*SAX indexes. As expected, there is a trade off between accuracy and number of accesses to subsequences. These results could be further improved by using different parameters for each dataset, length and sliding window size.

Summarizing, the guidelines proposed for optimizing the parameters of an *i*SAX index are: (i) Use relatively small number of segments and small maximum cardinality values and an appropriate threshold value in order to avoid overflows. (ii) Provided that we have an appropriate segments value, the threshold value plays a key role in the accuracy of the NN results. Increasing it will improve accuracy but deteriorate performance. (iii) Before creating an *i*SAX index, it should be first determined whether its main use would be to search for specific subsequences

Table 2. The results of the optimized parameters for the EEG and Synthetic datasets.

Length	Window	Threshold	EEG		Synthetic	
			RMSE	Acc/Len	RMSE	Acc/Len
200k	120	2000	0.3584	43.398	0.972	41.341
400k	120	2000	0.3602	28.648	0.955	31.436
500k	120	2000	0.3754	25.554	0.946	28.374
200k	120	2500	0.3257	51.345	0.936	46.970
400k	120	2500	0.3439	31.637	0.905	35.257
500k	120	2500	0.3497	28.942	0.889	31.339
200k	300	2000	0.5692	39.650	0.984	47.914
400k	300	2000	0.5113	25.960	0.981	33.493
500k	300	2000	0.5537	23.458	1.027	30.508
200k	300	2500	0.5390	45.615	0.961	52.017
400k	300	2500	0.4637	29.987	0.925	37.717
500k	300	2500	0.5198	26.028	0.986	33.510
200k	600	2000	0.7099	34.666	1.024	45.671
400k	600	2000	0.6925	23.247	1.000	34.681
500k	600	2000	0.7288	20.306	0.998	32.719
200k	600	2500	0.6685	37.746	0.992	49.599
400k	600	2500	0.6652	26.301	0.946	38.045
500k	600	2500	0.6821	23.175	0.946	34.655

or finding the nearest neighbors of given subsequences; subsequently, parameters should be set accordingly, to optimize the performance of the main use. Larger threshold values mean better NN accuracy but slower searching, as they lead to more subsequence accesses. (iv) Large segment values usually lead to degraded performance; however, the length of time series is a factor that must be taken into account. (v) The data and its features should be carefully analyzed. The distribution of the data plays a key role in the performance of *i*SAX.

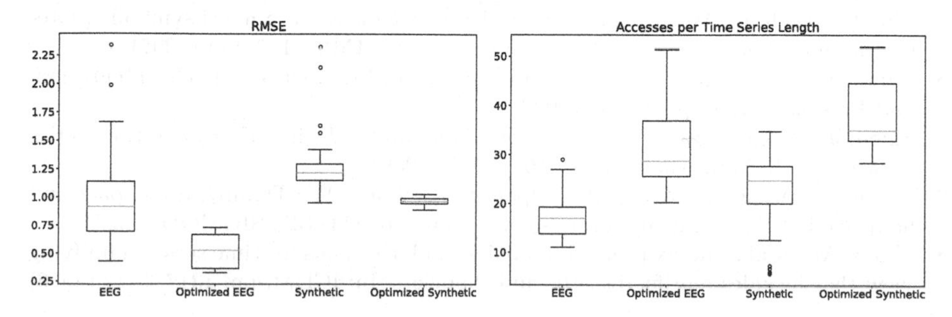

Fig. 6. Comparison of (a) RMSE values and (b) accesses over time series length, between the initial and the optimized parameters for the EEG and Synthetic datasets.

6 Conclusions and Future Work

In this paper, we proposed a methodology for quantifying the performance of an *i*SAX index with regard to both speed and quality of NN queries, taking into account the number of subsequence accesses and the accuracy of the approximate NNs computed using the index. We conducted experiments using four datasets and the analysis of these metrics demonstrated that the parameters used to build an *i*SAX index influence greatly its efficiency. We studied the impact of each parameter and proposed a set of guidelines for finding values that will optimize the performance of *i*SAX.

In the future, we plan to analyze in more depth the effect of the maximum cardinality value, considering also a wider range of values. Other optimizations could arise from taking into account the histogram of the dataset while partitioning the data space [1] and constructing the SAX representation. It would also be interesting to devise ways for automatic estimation of effective parameter values, possibly approaching this as a multicriteria optimization problem. A different direction would be to consider other types of queries such as k-NN queries. Finally, newer versions of *i*SAX [5] as well as Dumpy [7] should be similarly analyzed.

Acknowledgements. This research was partly funded by the SodaSense project (https://sodasense.uop.gr) under grant agreement No. MIS 6001407 (co-financed by Greece and the EU through the European Regional Development Fund).

References

1. Castro, N.C., Azevedo, P.J.: Automatically estimating iSAX parameters. Intell. Data Anal. **19**(3), 581–595 (2015)
2. Keogh, E., Kasetty, S.: On the need for time series data mining benchmarks: a survey and empirical demonstration. In: Proceedings of the Eighth ACM SIGKDD, pp. 102–111 (2002)
3. Law, S.M.: STUMPY: a powerful and scalable python library for time series data mining. J. Open Sour. Softw. **4**(39), 1504 (2019)
4. Lin, J., Keogh, E., Wei, L., Lonardi, S.: Experiencing SAX: a novel symbolic representation of time series. Data Min. Knowl. Discov. **15**(2), 107–144 (2007)
5. Palpanas, T.: Evolution of a data series index. In: Information Search, Integration, and Personalization, pp. 68–83 (2020)
6. Shieh, J., Keogh, E.: iSAX: disk-aware mining and indexing of massive time series datasets. Data Min. Knowl. Disc. **19**, 24–57 (2009)
7. Wang, Z., Wang, Q., Wang, P., Palpanas, T., Wang, W.: Dumpy: a compact and adaptive index for large data series collections, pp. 111–137. SIGMOD (2023)
8. Yeh, C.M., et al.: Matrix profile I: all pairs similarity joins for time series: a unifying view that includes motifs, discords and shapelets. In: ICDM, pp. 1317–1322 (2016)

RILCE: Resource Indexing in Large Connected Environments

Fouad Achkouty[1](✉), Elio Mansour[2], Laurent Gallon[4], Antonio Corral[3], and Richard Chbeir[1]

[1] Univ Pau & Pays Adour, E2S-UPPA, LIUPPA, EA3000, Anglet, France
{fouad.al-achkouty,richard.chbeir}@univ-pau.fr
[2] Scient Analytics, Paris, France
elio.mansour@scient.io
[3] University of Almeria, Almería, Spain
acorral@ual.es
[4] Univ Pau & Pays Adour, E2S-UPPA, LIUPPA, EA3000, Mont de Marsan, France
laurent.gallon@univ-pau.fr

Abstract. The purpose of this study is to present a solution to a particular challenge related to indexing resources in a connected environment. In essence, multiple factors impact significantly resource indexing in connected environments such as uneven spatial distribution of devices, their heterogeneous capacities, varying usage, and querying purposes and priorities, to mention a few. This work proposes a hybrid resource indexing approach able to cope with devices' capacities (especially their storage capacity). To validate our approach, several types of queries have been submitted to test the index's efficiency. Preliminary experiments show promising results and prove the efficiency and usefulness of the proposal.

Keywords: IoT · Resource indexing · Sensing devices

1 Introduction

Connected environments (IoT, Edge Computing, Fog Computing, etc.) have been adopted in many industries linked to the environment, supply chain, and smart cities [3]. According to statistics, connected resources[1] will reach 27 billion devices in 2023 [7]. With the increase of connected resources, several problems related to data storage, routing, networking, privacy, and devices' deployment have emerged.

In this paper, we address the retrieval problem by proposing a hybrid indexing approach called Resource Indexing in Large Connected Environment (RILCE). In RILCE, one node (having enough processing capacities) generates a global index taking into account individual devices capacities and coverage zones of the environment. Then, it distributes to each device a local index providing it with the capability to interact, exchange information and respond to queries directly or indirectly in a fully distributed manner without relying on the initial node (owning the global index).

[1] The terms resource and device will be used interchangeably in the rest of the paper.

A. Abelló et al. (Eds.): ADBIS 2023, CCIS 1850, pp. 13–22, 2023.
https://doi.org/10.1007/978-3-031-42941-5_2

However, many factors interfere with the index creation and querying process. Therefore, the following challenges should be addressed:

- Challenge 1: How to optimize network lifecycle(Reduce network latence and response lag) and querying by taking into consideration the capacity of the devices in each index?
- Challenge 2: How to consider covered and uncovered zones in order to maximize the index coverage?

Our approach aims at addressing the aforementioned challenges as we will see in the next sections.

The rest of the paper is organized as follows. Section 2 reviews related works and compares the different approaches to ours. Section 3 details the main definitions and preliminaries used to ease the understanding of our approach concepts. Section 4 presents our proposal explaining the index generation and query execution algorithms, while Sect. 5 describes the experiments conducted to validate our approach in various cases and discusses the results. Finally, Sect. 6 concludes the paper and pins down several future works.

2 Related Work

The number of connected nodes has expanded in recent years, generating a tremendous volume of data. A connected environment's indexing method aims to quickly locate and obtain a resource that holds the needed data from a collection of linked devices. In what follows, we will present the research studies belonging to the spatial resource indexing. Spatial Indexing uses geographical coordinates such as longitude, latitude, and altitude to represent the environment(e.g., kd-tree, R-tree).

In [10] and [6], the authors present the Geographic Hash Table (GHT) and the DIFS(extension of GHT), tree-based algorithms that uses an indexing tree to retrieve data from devices. Each node is assigned a hash key k representing the geographical coordinates of the node and its value accessible via queries. When a query is issued, It starts with nodes that cover precisely the query range traversing the tree until it reaches a node that covers the entire network.

In [5], a quad-tree index technique is proposed. The GH-indexing, using a divide-and-conqueror approach, builds a tree by encoding the IoT resources into geo-hashes. Starting at the root node, the query will go down the tree, evaluating the child nodes and comparing their minimum bounding rectangles against the region of query to discover matches.

The authors in [4] presented an indexing technique following a modified version of DP-means. The approach groups sensors into different clusters and queries are forwarded to the proper gateway by a discovery service layer. When the discovery service receive a query, it forwards it to the gateway with the shortest distance between its centroid and the query.

In [11, 12] provide an approach that adopts minimum energy principles. The authors propose algorithms named ECH and EGF-tree, where the connected environment is divided into sub-regions based on grid division. In this proposal, sensors report their data to the base station. The authors provide a query aggregation plan: sub-queries return their result to the base station using in-network aggregation and the query results are derived from the sub-queries.

There are also approaches that use mathematical calculations, such as probabilities to index IoT devices. In [8], the authors implement discovery services that are connected via Gaussian mixture models. Then, probabilistic indexing is used to determine which discovery service or gateway is responsible for forwarding a query. Lower-level discovery services become less sensitive to the probability. Query forwarding continues until the requested resource is reached.

In [1], a tree-structured method named BCCF-tree is presented. The BCCF tree has two main layers: the internal node level (nodes with pivots) and the leaf node level (nodes that have containers). This technique creates clusters by measuring the distance r between pivots and then fetching the distance (less than r) between one of the pivots and the centroid of a cluster. When users issue queries, the distance between the pivots and the query shrinks while going down the tree.

Finally, C. Dong et al. [2] propose a method named A-DBSCAN which clusters devices using DBSCAN. This generates clusters with high device density. Therefore, every cluster is then re-clustered using k-means. In this approach, users can provide feedback which is used to improve future iterations.

In Table 1, we present a comparison of the reviewed approaches based on three criteria: i) consideration of device capacities (Yes: the approach considered devices' capacities; No: otherwise); ii) consideration of different query types (Fast: queries sent directly to devices regardless of response availability, Urgent: queries that attempts various routing paths to find a query answer, Default: queries will try all possible paths to reach the target while taking into account node capacities to reduce the complexity); and iii) indexing coverage denotes the type of zones taken into account by the approach (Covered: zones that are covered by at least one device, Both: covered and uncovered zones, Undefined: approaches that do not handle indexing coverage at all).

Table 1. Comparison table of indexing approaches for IoT resources

	Devices Capacities	Types of Queries	Indexing Coverage
GHT/DIFS [6,10]	No	Fast	Covered
GH-Indexing [5]	No	Fast	Undefined
Mod. DP-means [4]	No	Fast	Covered
ECH [11]/EGF-Tree [12]	No	Fast	Covered
DSIS [8]	Yes (capacity is the number of sensors in the WSN)	Urgent	Undefined
BCCF-Tree [1]	Yes	Fast	Undefined
Multi-index [9]	No	Fast	Undefined
A-DBSCAN [2]	No	Fast	Covered
Our approach	Yes	Default	Both

None of the existing approaches fully consider our entire criteria as well as the challenges that we intend to address. We present next our proposal starting with some preliminaries to clarify and formally define the used terminology.

3 Definitions and Assumptions

In this section, we will present preliminaries and assumptions used in our approach.

Definition 1 (Global Index).
A global index gi is a 3-tuple matrix-based structure defined as follows:

$$gi : (D, Z, bRule) \quad where : \tag{1}$$

- *D is the set of devices that constitute the matrix rows*
- *Z is the set of covered and uncovered zones that constitute the matrix columns*
- *bRule is the binary association rule that maps a zone to a set of devices.* ■

Every row in gi will be used to generate the local index of each device. The algorithmic behavior of *bRule* is the main contribution of this study and will be explained in details in the following section. In what follows, uncovered zones will be represented in the matrix with a bar (e.g., $\bar{z}$).

Example 1. Let's assume that a device (PC) located in the monitoring office has the capability to store the entire index of a connected environment (composed of: 4 devices d_1-d_4, 4 covered zones, and 3 uncovered zones). The global index will correspond to the following: $gi_1 = \begin{bmatrix} & z_1 & z_2 & z_3 & z_4 & \bar{z}_5 & \bar{z}_6 & \bar{z}_7 \\ d_1 & 1 & 1 & 1 & 0 & 1 & 0 & 0 \\ d_2 & 0 & 1 & 1 & 1 & 0 & 1 & 0 \\ d_3 & 1 & 0 & 1 & 1 & 0 & 0 & 1 \\ d_4 & 1 & 1 & 1 & 1 & 0 & 0 & 0 \end{bmatrix}$

4 Proposed Approach

We present here the scope of the contribution and the proposed algorithms. In this study, we don't address the physical network communication problems and technologies and assume that messages can be sent between any devices using the physical layer (as it is the case in our motivating scenario). Two types of devices are considered:

- High-Capacity (HC) device: is a device which has the storage capacities to log a big part or the entire environment and deployed devices.
- Low-Capacity (LC) device: is a device that has a limited storage capacities and consequently cannot store enough information to have a visibility of the entire environment.

Users can interact with the environment by sending a query to specific devices or by broadcasting it within a location. When receiving a query, a device can choose to respond to the query or forward it to others. When possible, the query can also be sent to the orchestrator which forwards it following the global index. In order to react to queries effectively and by maximizing query response using limited capacity devices, this paper focuses on the global device index generation. Due to space limitation, the clustering algorithm of zones won't be detailed here.

4.1 Global Index Generation

In order to explain the global index generation algorithm we took an example of an environment having 4 devices and 7 zones in which 4 of these zones are covered (Z1, Z2, Z3, Z4) and 3 of them are uncovered($\bar{Z}5, \bar{Z}6, \bar{Z}7$). For the index generation, two main algorithms are implemented. The index initialization (Algorithm 1) is responsible for initializing the global index (of the entire environment or a part of it) using, in its matrix representation, a set of devices as rows and zones as columns. The device indexing (Algorithm 2) adds the corresponding indexes, creating consequently a device-to-zone relation. Both algorithms are executed on an orchestrator.

In Algorithm 1, we show the pseudo-code of the global index initialization. More concretely, after initializing the index matrix, we sort it following the devices capacities (line 2). We end up by assigning each device to a zone that maximizes its coverage. An illustration is given in Fig. 1 (step 1). When device capacity is between 1 and $\|Z\| + \|\bar{Z}\|$, Algorithm 2 needs to be applied.

Algorithm 1: initializeGlobalMatrix()

```
  Input  : Set of Devices D, Covered Zones Z, Uncovered Zones Z̄
  Output: gi                                    // Initial global index matrix
1 gi ← matrix(||D||, ||Z|| + ||Z̄||);                         // index creation
2 Sort(gi, Asc, D.C);          // index ascending ordering according to capacities
3 foreach d_j ∈ gi.D do
4 |  gi(d_j, gi.Z ⋂_max d_j.s.cz) = 1 // assigning each device to a best covered zone
5 end
6 Return gi
```

After completing the initialization phase, binary values must be added to create a device-to-zone relationship using Algorithm 2. Its inputs are: the initial global index (output of Algorithm 1), α indicating the number of covered zones that should be indexed per row (also taking into consideration the device capacity), β is the number of uncovered zones per row. It also has some local variables: the maximum capacity, and a counter used start-row for optimization purpose (equal to 2 on the first iteration of Algorithm 2 since Algorithm 1 is considered as an iteration). In this study, we aimed to keep α always greater than β to foster covered zones over uncovered ones.

For each iteration, we get the starting row using the *cpt* by increasing the variable start_row that indicates the starting row. In other words, the row of a device having a capacity less than *cpt* will be skipped since it has been already maximized, and no index value can be added (lines 1–4). We note that at the end of each recursive call, we increment *cpt* (lines 30). In lines 5–8, a minimal set of index entries is created to consider all the covered zones. This is done by adding an index (value 1) next to each column in the matrix except for the devices that belong to the column of the final covered zone. For these devices, an index is added at the beginning of the matrix. The *createMinimalCycle* function (line 7) is responsible for doing that. After generating this cycle, all devices in covered zones will be able to interact, which highlights the importance of this step. This iteration can be seen in step 2 of Fig. 1. Since $D1$ is located in $Z1$, and $Z1$ is near the zone $Z2$ in the index, $D1$ adds an index for $Z2$. Since $Z5$ is an uncovered zone, an index for $Z1$ is added to $D4$.

Algorithm 2: generateIndex()

```
   Input           : gi, α, β, max_capacity, cpt
   Output          : gi                                    // final index matrix
   Local Variables: start_row = 0, cpt = 2, max_capacity
 1 for i in 0..||gi.D|| do
 2 |   if (d_i.C<=cpt) then
   |   |   // device capacity less or equal cpt
 3 |   |   start_row++;
 4 end
 5 if (cpt == 2) then
 6 |   α = α − 1
 7 |   createMinimalCycle(gi) // The function will generate the cycle that connects the
   |       zone with each other through common devices
 8 |   α = α − 1
 9 else if (cpt == 3 and β ≠ 0) then
10 |   for i in start_row..||gi.D|| do
11 |   |   gi(i, i + ||Z̄||) = 1 // set 1 on the diagonal of uncovered zones columns.
12 |   end
13 |   addIndexOnRemainingConveredZones(gi) // if the diagonal reaches the end of
   |       the index and there are still devices that hadn't received an index, we add an
   |       index in the covered zone part for these devices following the number of index
   |       in a column
14 |   β = β − 1
15 else if (cpt%2 == 1 and α ≠ 0 ) then
16 |   addIndexOnCoveredZones(gi) // put 1 on covered zones following column sum
17 |   α = α − 1
18 else if (cpt%2 == 0 and β ≠ 0 ) then
19 |   addIndexOnUncoveredZones(gi) // put 1 on uncovered zones following column sum
20 |   β = β − 1
21 else if (α ≠ 0 and cpt < max_capacity) then
22 |   addIndexOnCoveredZones(gi)
23 |   α = α − 1
24 else if (β ≠ 0 and cpt < max_capacity ) then
25 |   addIndexOnUncoveredZones(gi)
26 |   β = β − 1
27 else if (α == 0 and β == 0 or cpt > max_capacity) then
28 |   verifyCycle(gi) // verify that the index has been indexed correctly + adding one
   |       from uncovered zones to covered zones when the sum of indexes are equal(on the
   |       same device)
29 else
30 |   cpt++
31 |   generateIndex()
32 end
33 Return gi
```

After the minimal cycle creation (leading to the decrease of α twice), we can start indexing uncovered zones (if $\beta \neq 0$). A diagonal of the uncovered zones columns is created to ensure we have information about all uncovered zones (lines 10–12). After creating the diagonal, we add indexes in covered zones for the remaining devices in order to foster covered over uncovered zones. The *addIndexOnRemainingConveredZones* function is responsible for doing that (line 13). The result of this step is illustrated in step 3 of Fig. 1. $Z5$, $Z6$, and $Z7$ are indexed by $D1$, $D2$, and $D3$, respectively. We have access to every zone that is uncovered at this stage. An index for $Z2$ is added for $D4$ because it hasn't been indexed yet.

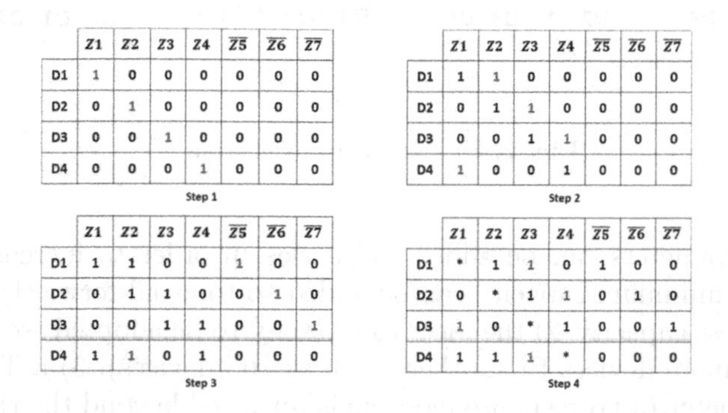

	Z1	Z2	Z3	Z4	$\overline{Z5}$	$\overline{Z6}$	$\overline{Z7}$
D1	1	0	0	0	0	0	0
D2	0	1	0	0	0	0	0
D3	0	0	1	0	0	0	0
D4	0	0	0	1	0	0	0

Step 1

	Z1	Z2	Z3	Z4	$\overline{Z5}$	$\overline{Z6}$	$\overline{Z7}$
D1	1	1	0	0	0	0	0
D2	0	1	1	0	0	0	0
D3	0	0	1	1	0	0	0
D4	1	0	0	1	0	0	0

Step 2

	Z1	Z2	Z3	Z4	$\overline{Z5}$	$\overline{Z6}$	$\overline{Z7}$
D1	1	1	0	0	1	0	0
D2	0	1	1	0	0	1	0
D3	0	0	1	1	0	0	1
D4	1	1	0	1	0	0	0

Step 3

	Z1	Z2	Z3	Z4	$\overline{Z5}$	$\overline{Z6}$	$\overline{Z7}$
D1	*	1	1	0	1	0	0
D2	0	*	1	1	0	1	0
D3	1	0	*	1	0	0	1
D4	1	1	1	*	0	0	0

Step 4

Fig. 1. Algorithm Steps Example

Indexes are added for columns with the fewest indexes to have an even distribution. From lines 15–17, when α is still not equal to 0, the *addIndexOnCoveredZones* function is triggered by adding indexes in covered zones for all the devices that didn't maximize their capacities. The indexes are added so that the sum of indexes is calculated before adding an index, and the column with the lowest number of indexes will be indexed (step 4 of Fig. 1). α is decremented at the end of this step. The same process is repeated if β is not equal to 0 for uncovered zones using the *addIndexOnUncoveredZones* function (lines 18–20). Indexes are added for uncovered zones based on the index summation of each column, and β is decremented by 1. The modulo is used to alternate between covered and uncovered zones. Plus, the same functions are used in lines 21–26 and repeated until α and β reach 0. β will converge before α since it is lesser.

After the algorithm convergence in lines 27–28, we check that all devices have a number of entries for covered zones greater than entries in uncovered zones. When the number of indexes in uncovered zones is greater or equal to those in covered zones on the same device, an entry from the uncovered zone is removed and added for the covered zones using the *verifyCycle* function.

4.2 Query Execution Illustration

After the generation of the global index, it is divided and distributed to the devices, each device receiving its own chunk. In Fig. 2, an example of an 'urgent' query is demonstrated using our motivating scenario. Considering $Z1$ is the source zone initiating the query and $\bar{Z}6$ is the destination zone. $D1$ will broadcast the query to all devices stored inside its index. When the devices receive it, they will also broadcast the query to other known devices until reaching the destination. Since $D2$ knows information about $\bar{Z}6$, it will respond to the user with an uncovered zone response.

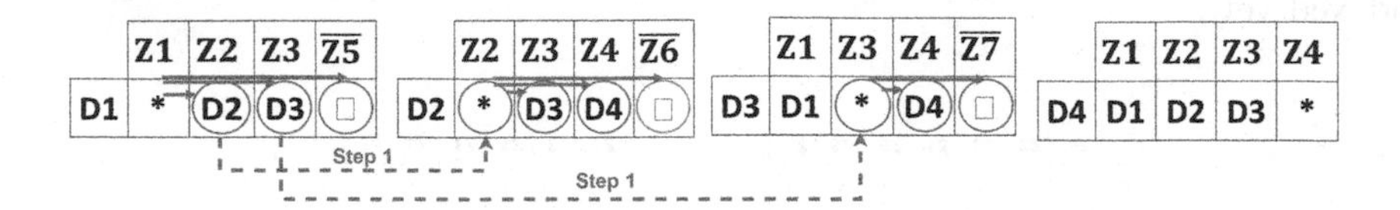

Fig. 2. Urgent query execution

Several parameters can be added to queries in order to increase their efficiency: 1) the minimum capacity can be added so to avoid forward the query to devices with less capacity, 2) the forward strategy to be adopted (sent to devices connected to more devices than others such as HC devices), 3) a Time To Live or TTL parameter (a no-response message is returned beyond the threshold), to mention a few.

5 Experiments

In the following section, we present the set of experiments conducted to validate our approach. The experiments concern index generation and query execution. The tests were conducted on a 16 GB RAM machine with a I7 CPU core on windows 10 operating system. The coverage percentage used is 80%, meaning that 80% of the zones are covered. We also used a 70% overlapping, meaning that 70% of the covered zones will have more than one device, creating a diversified environment: zones with low and high device density while also having uncovered zones. We did not compare our approach to existing ones since none of them can meet all the required criteria in their indexing scheme (Sect. 2). Data and queries of our tests have been simulated using our generator. Run times are averaged over ten runs.

Figure 3 demonstrates the impact of the zone number variation on the matrix creation. The number of devices used is 3000 with $\alpha = 80$ and $\beta = 20$. The index took 11.3 s to be generated for 100 zones, while it took 45.6 s for 400 zones. The higher the number of zones, the more time is needed to create the index. We can see that the increase in time is linear in terms of zones.

In Fig. 4, we fixed the number of zones to 300, with $\alpha = 80$ and $\beta = 20$, and we varied the number of devices. We can note that the generation of the global index took 4 s for 1000 devices, while it took 57 s for 4000 devices. The creation of the index takes longer the more devices there are. However, the behavior of the algorithm is a bit different in that case representing a quadratic complexity.

In Fig. 5, we compared the behavior of α and β on the global index creation. For $\alpha = 10$ and $\beta = 5$, the index creation took 14 s. Moreover, For $\alpha = 70$ and $\beta = 40$, the index creation took 59.76 s. We note that during this experiment, we fixed the number of devices to 3000, the number of zones to 300, and the devices' capacity to 64. The greater the value of α and β, the more time needed to create the index. This is because more iterations are required, plus, as previously mentioned in Algorithm 2, the process converges when $\alpha = \beta = 0$ or when the counter reaches $max(C)$. In addition, when $\alpha \geq 40$, the time is almost the same since the maximum capacity is reached. Therefore, since execution is almost the same for $\alpha + \beta \geq 64$, no execution is done beyond 64 bits.

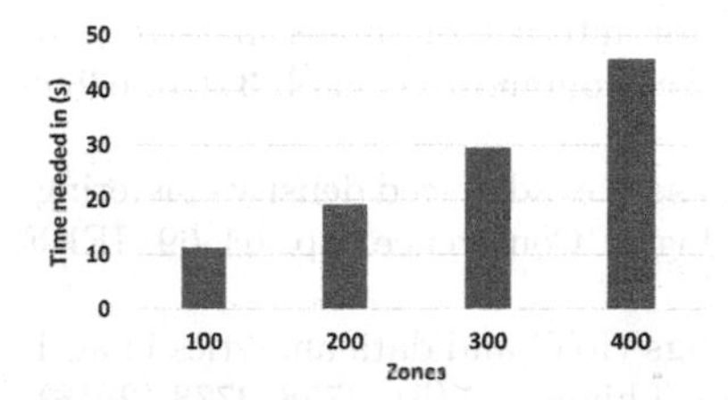

Fig. 3. Impact of zones on matrix creation

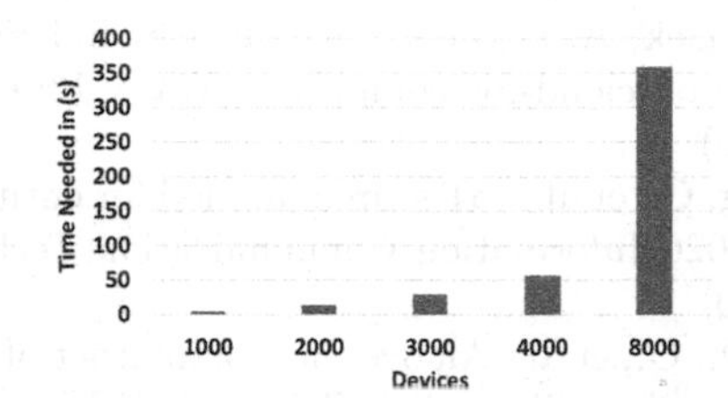

Fig. 4. Impact of devices on matrix creation

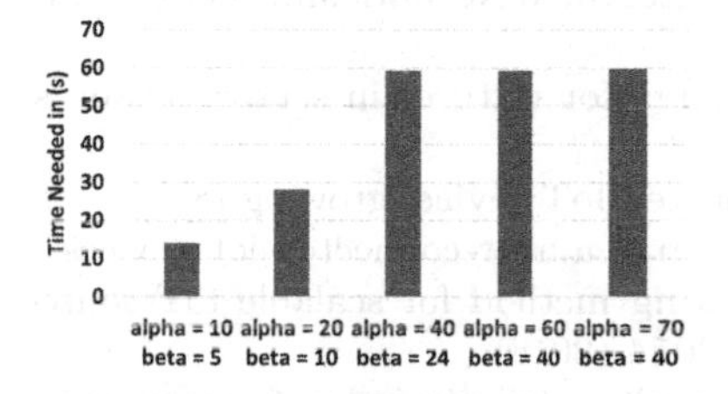

Fig. 5. Impact of α and β on matrix creation

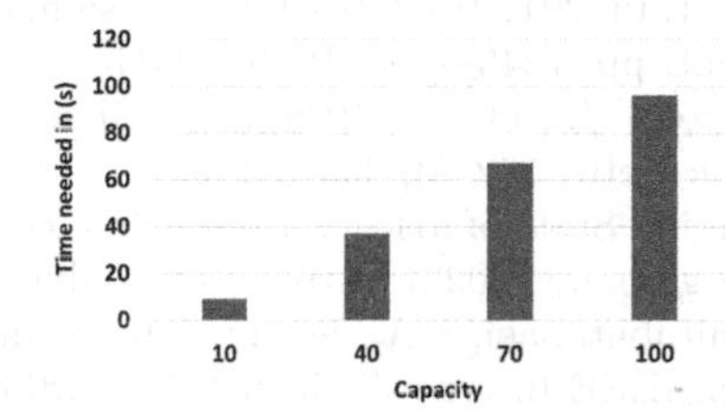

Fig. 6. Impact of capacities on matrix creation

In Fig. 6, we show the impact of the devices' capacities on the global index creation. We fixed the number of devices to 3000 and the number of zones to 300 with $\alpha = 80$ and $\beta = 20$ and increased the devices' capacity from 10 to 100. Our approach takes longer with devices that have bigger capacities as it is capacity dependent while maintaining a linear behavior. After analyzing the different metrics that affect the global index creation, we generated an index of 3000 devices and 300 zones with an $alpha = 80$ and $beta = 20$ and a maximum capacity of 20. We executed 80 queries, and all of these queries returned a response. The average response time of 10 runs was 0.086103 s.

6 Conclusion and Future Works

In this paper, we present a capacity-aware indexing approach that consists of indexing resources in connected environments with devices while using a decentralized architecture. The approach consists of different steps that generate a global index that is distributed over devices to use as a local index. Different evaluations on index creation and query execution were done and came out with good results. In future works, we are going to implement fast queries since most of the approaches are used along with other types of queries that fit user needs. We will also detail the spatial clustering algorithm.

References

1. Benrazek, A.-E., et al.: An efficient indexing for internet of things massive data based on cloud-fog computing. Trans. Emerg. Telecommun. Technol. **31**(3), e3868 (2020)
2. Dong, C., et al.: IoT search method for entity based on advanced density clustering. In: 2020 Information Communication Technologies Conference, pp. 64–69. IEEE (2020)
3. Elijah, O., et al.: An overview of internet of things (IoT) and data analytics in agriculture: benefits and challenges. IEEE Internet Things J. **5**(5), 3758–3773 (2018)
4. Fathy, Y., et al.: A distributed in-network indexing mechanism for the internet of things. In: 2016 IEEE 3rd World Forum on IoT, pp. 585–590 (2016)
5. Fathy, Y., et al.: Distributed spatial indexing for the internet of things data management. In: 2017 IFIP/IEEE Symposium on Integrated Network and Service Management, pp. 1246–1251. IEEE (2017)
6. Greenstein, B., et al.: DIFS: a distributed index for features in sensor networks. Ad Hoc Netw. **1**(2–3), 333–349 (2003)
7. Hasan, M.: State of IoT 2022: number of connected IoT devices growing 18% to 14.4 billion globally (2022). https://iot-analytics.com/number-connected-iot-devices/
8. Hoseinitabatabaei, S.A., et al.: A novel indexing method for scalable IoT source lookup. IEEE Internet Things J. **5**(3), 2037–2054 (2018)
9. Huang, C.-Y., Chang, Y.-J.: An adaptively multi-attribute index framework for big IoT data. Comput. Geosci. **155**, 104841 (2021)
10. Ratnasamy, S., et al.: GHT: a geographic hash table for data-centric storage. In: Proceedings of the 1st ACM International Workshop on Wireless Sensor Networks and Applications, pp. 78–87 (2002)
11. Tang, J., et al.: An energy efficient hierarchical clustering index tree for facilitating time-correlated region queries in wireless sensor network. In: 2013 9th International Wireless Communications and Mobile Computing Conference, pp. 1528–1533. IEEE (2013)
12. Zhou, Z.B., et al.: EGF-tree: an energy-efficient index tree for facilitating multi-region query aggregation in the internet of things. Pers. Ubiquit. Comput. **18**(4), 951–966 (2014)

Query Processing

A Cost-Effective Query Optimizer for Multi-tenant Parallel DBMSs

Mira El Danaoui(✉), Shaoyi Yin, Abdelkader Hameurlain, and Franck Morvan

IRIT Laboratory, Paul Sabatier University, Toulouse, France
{mira.el-danaoui,shaoyi.yin,abdelkader.hameurlain, franck.morvan}@irit.fr

Abstract. For a multi-tenant DBMS, at any point in time, a tenant may submit an SQL query associated with a performance SLO (Service-Level Objective). Any violation in the SLO requires the provider to pay penalties. As such, the DBMS which is in charge of the provider's long-term benefit, should find a fair compromise between the financial costs brought on by resource consumption and the SLO satisfactions that affect the provider's income. Such compromises should be carefully considered due to the instability of multi-tenant query workload. In this paper, we propose an execution plan selection strategy which changes the optimization objective in response to the query workload. This latter information is provided by the queue manager, a new component of our extended optimizer, which controls the selected execution plans' scheduling as well as resource allocation. Experimental results showed that it proved successful in terms of the provider's long-term profit despite variations in the multi-tenant workload.

Keywords: Multi-tenancy · Service Level Objective · Query Optimization · Resource Allocation · Queuing Policy

1 Introduction

Multi-tenancy, the ability of a system to share resources transparently among multiple tenants, is considered as one of the principle features of cloud services [10]. Early research work on database management for supporting multi-tenancy focused on database design for SaaS (Software-as-a-Service) applications [3, 16]. Then, academic and industrial researchers realized the importance of managing multi-tenancy related issues inside the DBMSs [1, 14], like buffer pool sharing [15] and resource management [7, 18]. Regarding resource management in multi-tenant DBMSs, existing works [7, 18] consider resource allocation on the per-tenant level. In this paper, we are interested in the case where all tenants share a single DBMS instance running on a cluster managed by a cloud provider. That is the case in many DBaaS (DataBase-as-a-Service) offers, like Google BigQuery. An interesting approach is that, instead of allocating a certain amount of resources to every tenant, the DBMS assigns resources on a per-query level while considering the tenant's specific needs for each query. At any time, tenants may submit an SQL query and a performance SLO (Service-Level Objective) set between them and

A. Abelló et al. (Eds.): ADBIS 2023, CCIS 1850, pp. 25–34, 2023.
https://doi.org/10.1007/978-3-031-42941-5_3

the service provider. For the same query, different tenants may require distinct performance SLOs and be ready to pay different prices. Any violation in the SLO requires the provider to pay fines. Thus, the DBMS which is in charge of the provider's long-term gain, should find a fair compromise between resource consumption-related financial costs and SLO satisfactions that affect the provider's income. Such compromises should be considered carefully due to multi-tenant query workload's instability.

Indeed, during periods where too many queries arrive instantly from different tenants, the DBMS may need to reject some of them in order to execute others with SLO guarantees. As for periods where there are enough resources to execute all arriving queries, optimization should also be made to improve the economic benefit while meeting SLOs. In commercial DBMSs, there is often a workload manager or an equivalent module to deal with these problems. In the multi-tenancy context, many new policies have been proposed [17, 18] in order to take the SLOs into account. They all queue up the queries before processing them by the DBMS kernel, and monitor their execution without affecting the query optimizer's decisions. However, based on work in [20], we argue that the query optimizer can play an important role in managing the workloads associated with multi-tenant queries. Thus, in this paper we propose to extend the query optimizer to allow query workload variations to directly affect the optimizer's decisions, like the execution plan selection and the resource allocation to the selected plans.

Notably, we extend the query optimizer as follows: (1) we propose a workload-aware execution plan selection strategy (called WAPS) that alters the optimization objective (e.g., minimize the response time, or maximize the economic benefit) from one query to another based on query workload; and (2) we add a new component, the execution plans' queue manager, that guides the selected plans' queueing and resource allocation.

In this paper, Sect. 2 analyzes the related work. Section 3 gives an overview of the extended optimizer architecture and discusses our execution plan selection strategy along with our execution plans' queue management system. Finally, Sect. 4 shows experimental results and Sect. 5 concludes the paper and outlines future work.

2 Related Work

As stated before, the unstable query workload and the need for meeting performance SLOs should be considered by query optimizers of multi-tenant DBMSs. While most of the industrial multi-tenant DBMSs focus on the workload management in a global manner, their query optimizers still aim to minimize the query response time [4, 6]. A previous work [19] proposes to use the multi-objective query optimization (MOQO) in order to make tradeoffs between the response time and the monetary cost. Other works similar to our previous work [20] fix the objective to maximizing the provider's benefit while dealing with the performance SLOs as constraints. In this work, we focus on the provider's long-term economic benefit and believe it is useful if we can alter the objective of the query optimizer whenever needed. For example, when resources are limited, the optimizer can choose the execution plan that minimizes the resource consumption as a way to have more queries executed, or the fastest plan in order to quickly release resources; when there are enough resources, the optimizer may consider factors like economic benefit. To our knowledge, this idea hasn't been developed in the literature.

Given how crucial resource allocation is to query optimization, we propose to consider the multi-tenant query workload and the performance SLOs when allocating resources to the selected execution plans. Prompt execution of a selected plan (especially when it is long-running and non-urgent) may cause the rejection of the next coming ones. Thus, we designed a fine-grained execution plans' queue management system. Many research studies exist in resource allocation that consider deadline constraints in the context of multi-tenancy (e.g., MapReduce jobs scheduling [5, 12]), but we are interested in the ones that consider queue management [2, 9]. Work in [9] introduced the "Earliest Maximal Waiting Time First (EMWTF)" which assigns resources to deadline-constrained queries in order of increasing maximal waiting time. In our paper, deadline-constrained plans are sorted into urgent and non-urgent while this isn't tackled by [9].

3 Overview of the Extended Query Optimizer

In this section, we present an overview of the extended query optimizer's architecture. Then, we introduce a detailed algorithm for the proposed execution plan selection, and present our execution plans' queue management system.

3.1 Architecture Description

The standard optimization problem can be defined as follows: for a query Q, corresponding to a space of execution plans E, and a cost function which assigns a cost for every plan p in E, find the plan that minimizes the execution cost for Q. Thus, the query optimizer consists of: (1) a **Search Space** which represents the virtual set of all possible execution plans for a given query, (2) a **Search Strategy** which is used to look through the search space for an optimal (or close to optimal) execution plan, and (3) a **Cost Model** which assigns a cost for every execution plan in the considered search space [8].

In a multi-tenant RDBMS, the standard query optimizer selects execution plans which might disappoint the tenant or the provider, or both. Firstly, because the performance SLO (Service Level Objective) isn't considered. In fact, all tenants are treated alike no matter how much they are ready to pay. Thus, tenants with large spending won't get better service than those with lower budgets. Secondly, the query optimizer has a fixed objective (e.g., minimize the response time) in spite of the system's current or future workload. For a multi-tenant RDBMS, at any point, multiple tenants would submit many queries instantly, causing the system to be overloaded to execute them all.

Our approach tackles the above aspects and tries to reduce the losses of both the tenant and the provider. Hence, we reformulate the problem of query optimization as follows: Given a query Q, and a performance SLO, corresponding to a space of execution plans E, and a cost model which assigns different costs (e.g., response time, resource consumption and monetary cost) for every plan p in E, based on the system's current workload, the optimizer chooses an execution plan which tends to maximize the provider's long-term profit. Then, instead of directly sending the selected plan to the Executor, we add it into a queue, so that the waiting plans can be rescheduled based on their urgency and other priorities. Queuing the execution plans aims to optimize the global resource allocation in order to reduce SLO violations and increase the provider's long-term profit. Thus, the query plans' queue manager serves as a component of our extended query optimizer, as seen in Fig. 1. We give some further explanations below.

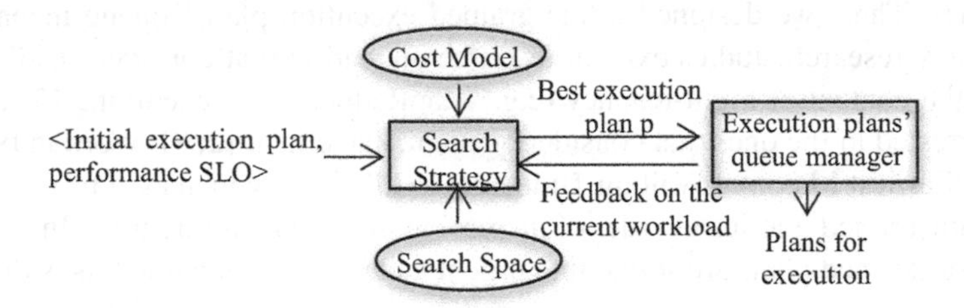

Fig. 1. Extended query optimizer's architecture

Once a query Q is submitted, an initial plan is generated for it. Our extended optimizer takes this plan and the performance SLO as inputs. The latter is a key driving factor for us to make the extension. First, we argue that the performance SLO should have an impact on the execution plan selection. Then, based on the performance SLO, each selected plan will be recognized as urgent or non-urgent. When resources are limited, a prompt execution of non-urgent plans may delay the execution of upcoming urgent plans and may cause them to miss their deadlines. In order to raise the chance to have always resources available for urgent plans without excessive resource reservation, we propose to design a fine-grained queue management system. One advantage of having execution plans' queue manager inside the query optimizer is that, based on the queuing situation, we can deduce the system's current workload and use this information to guide the execution plan selection for incoming queries. In our work, we characterize the system's workload by three statuses: peak, normal and idle time. Peak time is the period when there are insufficient resources to execute all urgent plans. However, normal time is when enough resources exist for all urgent plans and selected non-urgent plans. Idle time is when there are enough resources to execute all of the urgent and non-urgent execution plans (i.e., some resources stay idle). We will describe the proposed Workload-Aware Plan Selection WAPS algorithm in detail below.

3.2 Workload-Aware Execution Plan Selection

As discussed earlier, an execution plan selection is based on two factors: the performance SLO and the workload of the system (peak, normal or idle time). For fairness, we assume that a performance SLO is defined for each query by the tenant and the provider, and it corresponds to a deadline that satisfies the tenant and the provider can meet. In the literation, there are proposals for automatically generating this kind of SLO (e.g., [17, 20]), and we believe that other methods will appear. This work doesn't focus on the detail of these methods, but only considers their result: the performance SLO of a query Q, which is defined by two parameters $QCT_{SLO}(Q)$ and D_Q. $QCT_{SLO}(Q)$ means the expected Query Completion Time of Q agreed between the tenant and the provider, and D_Q means the hard deadline before which the execution of Q must be finished. The values of these two parameters are correlated for ad-hoc queries but independent for other queries, like reporting queries. For ad-hoc queries, the deadline D_Q is deduced by: D_Q = current time + $QCT_{SLO}(Q) * \tau$; $\tau(>1)$ is the tolerance ratio. For reporting queries, the tenant may fix a deadline in advance, e.g., next Monday at 6 a.m. Once the SLO is well established, the tenant can use the provider's services.

Algorithm 1: Workload-Aware Execution Plan Selection Strategy for a Query Q

```
Input: Query Q (Po, TQ, QCTSLO, DQ), E the set of valid plans generated
for Q (abide by the deadline of Q), Is_Peak_Time a Boolean value in-
dicating whether it is peak time
Output: Pbest, the best execution plan for Q
1  if Is_Peak_Time then
2    Pbest = PlanSelection_Peak (Q, E)
3  else
4    Pbest = PlanSelection_NotPeak(Q, E)
5  end if
6  return Pbest
7  Function PlanSelection_Peak(Q, E)
8  PQCT ← a valid plan which minimizes the QCT
9  PRES ← a valid plan which minimizes the number of resources
10 if RES(PQCT) - RES(PRES) <= a*RES(PRES) then
11   Pbest = PQCT
12 else
13   Pbest = PRES
14 end if
15 return Pbest
16 Function PlanSelection_NotPeak (Q, E)
17 PQCT ← a valid plan which minimizes the QCT
18 PEB ← a valid plan which maximizes the provider's profit per second
19 if BFT(PEB) - BFT(PQCT) <= a'*BFT(PEB) then
20   Pbest = PQCT
21 else
22   Pbest = PEB
23 end if
24 return Pbest
```

According to Algorithm 1, once a tenant submits a query Q to the DBMS, an initial plan P_o is generated for it. Our algorithm takes a query Q in the form of a tuple $(P_o, T_Q, QCT_{SLO}, D_Q)$ and generates a plan for Q denoted by P_{best}, where T_Q represents the submission time of Q. The search space is created by generating a finite subset of plans equivalent to P_o. Then, a cost model assigns different costs (response time, resources consumption and monetary cost) for every plan evaluated in the subset. Thus, only valid plans are considered; where a plan is valid if it can meet the deadline. Then, an efficient search strategy is applied to explore the search space for (1) $\boldsymbol{P_{QCT}}$, the execution plan that has the shortest *QCT*, or (2) $\boldsymbol{P_{RES}}$, the plan that consumes the least resources, or (3) $\boldsymbol{P_{EB}}$, the plan that maximizes the provider's economic benefit per time unit. For every submitted query Q, we check the system's workload status, which is updated by the execution plans' queue manager regularly. If it is peak time, the algorithm will choose either P_{RES} as a way to conserve resources for other plans, or P_{QCT} in order to release resources quickly. The second scenario happens when P_{QCT} guarantees a relatively reduced consumption of resources and a shortened completion time *(Line: 10); a* represents a certain percentage. However, if it is normal or idle time, then the algorithm will choose P_{EB} which allows increasing the provider's global profit or P_{QCT} which allows releasing resources quickly. The latter occurs when P_{QCT} ensures a relatively significant economic benefit and a reduced completion time *(Line: 19).*

In this section, we outlined the various workloads the system may have. Then, for every submitted query *Q*, a workload-aware execution plan selection strategy was proposed. In the next section, we present the Execution Plans' Queue Management System.

3.3 Execution Plans' Queue Management System

Once an execution plan is selected for a submitted query *Q*, the execution plans' queue management module (or Queue Manager for short), takes control over its queueing. The latter decides to which queue (urgent or non-urgent queue) the execution plan should be submitted. However, that isn't the only function of the Queue Manager. Given that we are working in a multi-tenant RDBMS environment, then at some point of time, the system may become overloaded. Thus, the optimizer needs to set its priorities clear in order to decide which execution plans will be sent to the Executor first. In actuality, urgent execution plans, are assigned a priority over the non-urgent ones, when insufficient resources exist for plans in both queues. At that point, it becomes the Queue Manager's responsibility to decide which plans are allocated the resources. These multiple functions of the Queue Manager are shown in Fig. 2, and more explained hereafter.

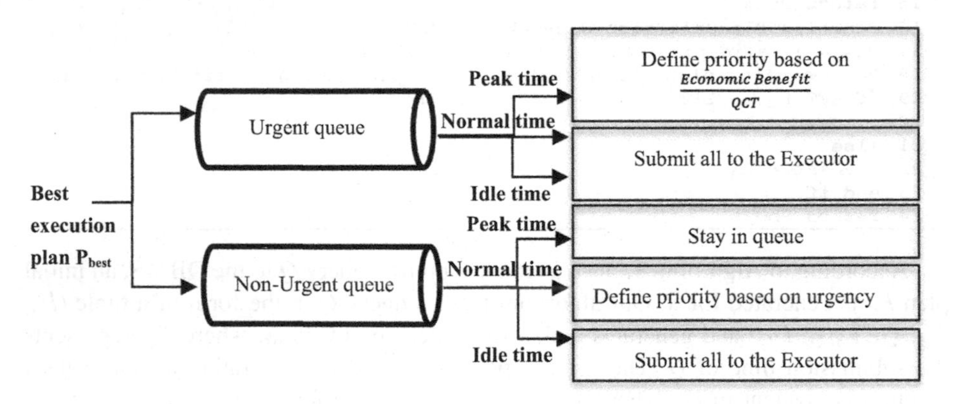

Fig. 2. Control Flow of the Execution Plans' Queue Management System

As shown in Fig. 2, the queue manager submits the selected execution plans to one of the queues. This decision is based on an urgency factor calculated for every selected plan and expressed as a ratio of query completion time to elapsed time. However, as the queues allow the share of resources among them, this makes resource allocation reliant on the three system's workloads: idle, normal and peak time. During idle time, when enough resources exist for all urgent and non-urgent plans, both queues empty. However, during normal time, the urgent queue empties after all urgent execution plans receive resources. Non-urgent plans, on the other hand, get the remaining resources, if any; plans with the highest urgency factor and for which there are enough resources are assigned the resources first. Nonetheless, during peak time, no enough resources exist to execute all urgent execution plans. Thus, a priority score is evaluated for every urgent plan and expressed as the economic benefit per second. Urgent plans with the highest priority score are assigned the resources first. Whereas non-urgent execution plans stay in the queue for later execution. Hence, when the system is constrained, it makes sense that plans with high economic benefit and relatively short *QCT* are prioritized.

4 Experiments and Discussion

In this section, we outline the experiments that are done to test the performance of our extended optimizer.

4.1 Experimental Setting

Within a simulated extended optimizer, we implemented the proposed WAPS algorithm. In fact, a simulation is performed when the real system doesn't exist or when it does but real experiments are time- and money-consuming [11, 13]. The following describes in details the different aspects of our implementation and simulation.

Execution Plan and Workload Generator
In our context, the optimizer takes as input execution plans of queries submitted from different tenants and the corresponding performance SLOs. However, the generation of concrete execution plans isn't the focus of this work and fortunately, the WAPS algorithm need only high-level descriptions of the execution plans. Therefore, every execution plan is simulated as a vector with necessary information (e.g., query id, response time, resource consumption, and the performance SLO), such that the queue manager could be tested and its performance could be measured.

In order to reflect the diversity of query workload, we used a recently published TPC-DS performance results as a basis [http://tiny.cc/databricks_TPCDS]. In the executive summary, we have the estimated response times of 8 different execution plans for each of the 99 queries, which serves as the input of our execution plans generator. For each query, we add 12 random values between the minimum and the maximum response times, in order to simulate a search space with 20 execution plans. The number of consumed nodes for each plan is generated randomly between 1 and max_nb_nodes, where max_nb_nodes is positively correlated to the response time. As for the queries' arrival times, we assume that they follow a Poisson distribution, which is a decent approximation of real-world systems with many separate occurrences that don't reflect one another [2]. Given the set of 99 queries and that multiple tenants can submit the same query, a total of 670 query submissions are generated.

Simulated Environment and SLO Generator
Our referencing TPC-DS performance result was obtained with a parallel relational DBMS running on a shared-nothing architecture with 256 nodes, each consisting of 61 GiB of RAM, 2.112 GHz/8-core Intel processors and 8 1,9 TB (SSD) drives. In our work, we consider the same setting but with *n* nodes, with *n* a variable that will be used

for the performance evaluation. As for generating the performance SLOs, we consider three types of tenants: premium, standard, and basic; premium tenants having the highest willingness to pay, followed by the standard and basic having the lowest. According to the type of the tenant, he is assigned a parameter τ, which is a tolerance rate used to define the SLO. The higher the status of the tenant is, the lower τ is. Execution plans' expected price, monetary cost and economic benefit are evaluated as work in [20] proposed. Using Python *pandas* library, we import our input data (produced by the above generators) as a dataframe. We use the discrete-event simulation method to model the full process from query arrival to the end of its execution.

Comparison Points and Metrics
Experiments were conducted in order to evaluate the performance of the extended optimizer. That is done by validating the proposed Workload-Aware Plans' Selection WAPS algorithm. Thus, we compare it to a strategy that selects only plans that minimize the response time (labeled as RT), or plans that maximize the provider's economic benefit (labeled as EB) or plans that minimize resources' consumption (labeled as RC). Comparisons were made over two metrics: (M1) the Saturation Threshold with regard to the Number of Nodes (labeled as STNN), i.e., the least number of nodes needed to keep the economic benefit positive, under a fixed query arrival rate; the less the saturation threshold is, the better the optimizer can handle overload within limited amount of resources, and (M2) the Saturation Threshold with regard to the query Frequency Factor (labeled as STFF), i.e., the highest frequency factor the system can handle while keeping a positive economic benefit, under a fixed number of nodes. The larger the saturation threshold is, the better the optimizer can handle a heavy load of incoming queries.

In order to evaluate (M1), we begin with a cluster of 100 nodes under a frequency factor of 45 queries/second. Then, we gradually contract the cluster by removing nodes while evaluating the provider's economic benefit per second for every cluster size. However, in order to evaluate (M2), we fix the cluster's size to 35 nodes. We multiply the query arrival time by decreasing and increasing factors while sustaining the consistency of Poisson distribution. This is done to record the economic benefit per second at accelerated and decelerated frequency of arriving queries under a fixed cluster node.

4.2 Results and Discussions

Experimental results are described and explained hereafter. We start by spotting the saturation threshold STNN (metric M1) of all methods under comparison. As shown in Fig. 3(a), with 100 nodes, the four strategies showed a positive gain. However, as the number of nodes decreases gradually, all four strategies started to lose profit. We can spot the STNN of WAPS at 23 nodes, and at 31 nodes for RC and EB and at 34 nodes for RT. Thus, WAPS can continue with fewer resources than RT, EB and RC while still making a better gain. Also, we need to spot the saturation threshold STFF (metric M2) for all four methods. As shown in Fig. 3(b), at a low frequency factor, all four methods recorded a positive gain. However, when queries started to arrive at a faster rate, RT's, EB's and RC's economic gain started to decline at a higher pace relative to WAPS. We can spot the STFF of RT, EB and RC at 120, 230 and 240 queries/second respectively.

While the STFF of WAPS is at 400 queries/second. Thus, our strategy manages queries arriving at almost 3.3 times faster rate than RT and 1.7 times faster than RC and EB.

Findings in Fig. 3 are predictable and arise from: (i) when the system is constrained or overloaded, EB may choose plans that have extended execution times. This leads more plans to be rejected and forces the provider to bear greater penalties, (ii) although RT and RC ensure fast resource liberation and efficient resource consumption respectively, their selected plans may have limited profit.

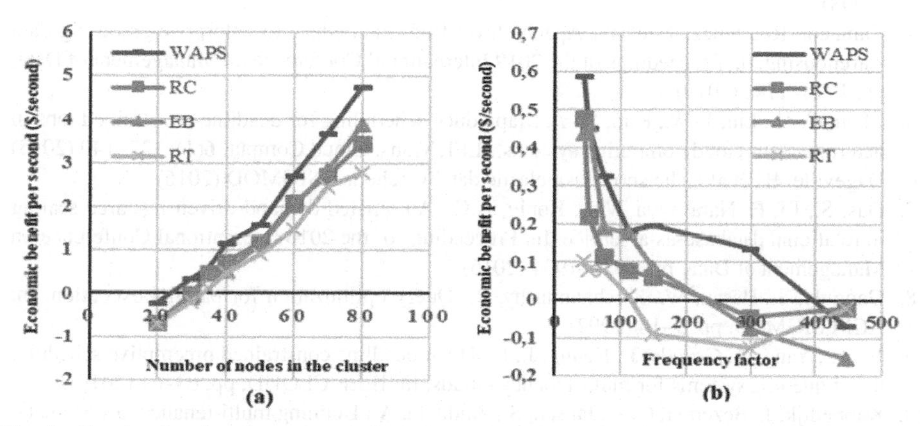

Fig. 3. Graphs showing the economic benefit per second as a function of the number of nodes in the cluster (a) and the frequency factor (b) for WAPS, RC, EB and RT

We also evaluated the performance of the proposed queue management system and compared it to a work in the literature [9]. Results showed that, our queue manager has a great ability to handle more frequent peak times with fewer resources while sustaining positive gain. Due to space limitation, we aren't able to display the graphs.

5 Conclusion

In this paper, we designed an extended query optimizer for multi-tenant parallel DBMS. We proposed an execution plan selection strategy which is based on query workload. The latter information is indicated by the queue manager, a new component of our extended optimizer, which controls the scheduling and resource allocation of the selected plans. We implemented the execution plan selection strategy and compared it to single objective query optimization strategies. Experimental results showed that our strategy tackles all workloads efficiently by ensuring a positive gain within a reduced cluster of nodes and at high query arrival rates. We are aware of the limitations brought on by doing a simulation, but it is crucial to get initial feedbacks on our proposal's efficiency.

In future, we plan to tackle the problem of long-running queries by developing an effective method that gives the optimizer fine control over how to slice long queries without the Executor's intervention. In fact, long-running queries may take over all resources, leading to resource contention, and resulting in major rejection rates and fines.

References

1. Abadi, D., et al.: The seattle report on database research. ACM SIGMOD Rec. **48**(4), 44–53 (2020)
2. Abhaya, V.G., Tari, Z., Zeephongsekul, P., Zomaya, A.Y.: Performance analysis of EDF scheduling in a multi-priority preemptive M/G/1 queue. IEEE TPDS **25**(8), 2149–2158 (2013)
3. Aulbach, S., Grust, T., Jacobs, D., Kemper, A., Rittinger, J.: Multi-tenant databases for software as a service: schema-mapping techniques. In: ACM SIGMOD 2008, pp. 1195–1206 (2008)
4. Camacho-Rodriguez, J., et al.: Apache hive: from mapreduce to enterprise-grade big data warehousing. In: Proceedings of the 2019 International Conference on Management of Data, pp. 1773–1786 (2019)
5. Chen, C.-H., Lin, J.-W., Kuo, S.-Y.: Mapreduce scheduling for deadline-constrained jobs in heterogeneous cloud computing systems. IEEE Trans. Cloud Comput. **6**(1), 127–140 (2015)
6. Dageville, B., et al.: The snowflake elastic data warehouse. SIGMOD (2016)
7. Das, S., Li, F., Narasayya, V.R., Konig, A.C.: Automated demand-driven resource scaling in relational database-as-a-service. In: Proceedings of the 2016 International Conference on Management of Data, pp. 1923–1934 (2016)
8. Ganguly, S., Hasan, W., Krishnamurthy, R.: Query optimization for parallel execution. In: ACM SIGMOD, pp. 9–18 (1992)
9. Jia, R., Yang, Y., Grundy, J., Keung, J., Li, H.: A deadline constrained preemptive scheduler using queuing systems for multi-tenancy clouds. In: IEEE CLOUD, pp. 63–67 (2019)
10. Kabbedijk, J., Bezemer, C.-P., Jansen, S., Zaidman, A.: Defining multi-tenancy: a systematic mapping study on the academic and the industrial perspective. J. Syst. Softw. **100**, 139–148 (2015)
11. Law, A.M., Kelton, W.D., Kelton, W.D.: Simulation Modeling and Analysis, vol.3. Mcgraw-Hill New York (2007)
12. Liu, L., et al.: Preemptive hadoop jobs scheduling under a deadline. In: IEEE SKG 2012, pp. 72–79 (2012)
13. McHaney, R.: Understanding Computer Simulation. Bookboon (2009)
14. Narasayya, V., Chaudhuri, S.: Multi-tenant cloud data services: state-of-the- art, challenges and opportunities. In: Proceedings of the 2022 International Conference on Management of Data, pp. 2465–2473 (2022)
15. Narasayya, V., Menache, I., Singh, M., Li, F., Syamala, M., Chaudhuri, S.: Sharing buffer pool memory in multi-tenant relational database-as-a-service. Proceed. VLDB Endowment **8**(7), 726–737 (2015)
16. Ni, J., Li, G., Zhang, J., Li, L., Feng, J.: Adapt: adaptive database schema design for multi-tenant applications. In: Proceedings of the 21st ACM CIKM, pp. 2199–2203 (2012)
17. Ortiz, J., Lee, B., Balazinska, M., Gehrke, J., Hellerstein, J.L.: SLA orchestrator: reducing the cost of performance SLAs for cloud data analytics. In: USENIX ATC 18, pp. 547–560 (2018)
18. Tan, Z., Babu, S.: Tempo: robust and self-tuning resource management in multi-tenant parallel databases. Proc. VLDB Endow. **9**(10), 720–731 (2016)
19. Trummer, I., Koch, C.: Approximation schemes for many-objective query optimization. In: Proceedings of the 2014 ACM SIGMOD International Conference on Management of Data, pp. 1299–1310 (2014)
20. Yin, S., Hameurlain, A., Morvan, F.: SLA definition for multi-tenant DBMS and its impact on query optimization. IEEE TKDE **30**(11), 2213–2226 (2018)

Video Situation Monitoring to Improve Quality of Life

Hafsa Billah, Abhishek Santra, and Sharma Chakravarthy(✉)

IT Lab and CSE Department, University of Texas at Arlington, Arlington, TX, USA
{uxb7123,abhishek.santra}@mavs.uta.edu, sharmac@cse.uta.edu

Abstract. Video situation monitoring can be used for surveillance, regulatory (e.g., traffic) aspects, societal applications such as improving quality of life, and others. Situations (e.g., inactivity, non-participation in group activities, etc.) in the assisted living or smart home environment are difficult to detect using current customized video processing solutions. These solutions require each action or situation to be predefined and are human capital intensive both for software development and usage, not flexible, and cannot support real-time situation detection. The alternative solution is to manually detect situations using human-in-the-loop which is not practical and scalable.

We propose an alternative to the above approaches which avoids (or minimizes) human-in-the-loop and can be extended to support 'real-time' detection/analysis. Our approach extracts video content *only once* and allows 'ad-hoc' queries (or situations) to be specified and processed without having to develop new software. In this paper, we make a case for developing expressive data models and continuous query processing of extracted video contents leveraging ideas from stream data and complex event processing. We also propose two alternative representations of extracted video contents with different benefits, identify a number of situations relevant to the assisted living domain, and indicate how they can be detected using the proposed approach. *Our main contribution is defining and establishing a novel framework for a problem that is becoming very important as part of big data analytics.*

Keywords: Video content extraction · Continuous queries · Boost quality of life

1 Introduction

Ubiquity of inexpensive devices such as CCTV cameras, dash/personal cams, and others generate large volumes of video data on a day-to-day basis. Video analysis is useful for postmortem situation detection (or forensics) for surveillance, security, or personal interest. Additionally, video analysis can also be used for improving quality of life provided we develop the right framework that is not human capital intensive (as it is now) and can make it work in 'real-time'. It is important to identify whether someone has walking issues (inactive) to provide help timely or whether someone is taking part in group activities. Currently, these situation types are monitored using sensor data, custom solutions, or manual analysis (by watching recorded videos exhaustively) after the fact. Though situation monitoring for **sensor Stream Processing (SP)** and **Complex**

A. Abelló et al. (Eds.): ADBIS 2023, CCIS 1850, pp. 35–45, 2023.
https://doi.org/10.1007/978-3-031-42941-5_4

Event Processing (CEP) have matured, there is no corresponding framework, representation or modeling, and query language capability for applying them to video contents [17] (e.g., individuals, groups, etc.)

With the advent of deep learning algorithms **Video Content Extraction (VCE)** have become more efficient. They can extract a variety of complex information such as feature vectors, bounding boxes, backgrounds to some extent, activity, etc. Traditional SP and CEP frameworks cannot be applied directly to the extracted video contents without appropriately representing them with a data model. Although there exist some SP frameworks (e.g., Oracle Multimedia, Amazon Kinesis [13]) for storing, searching, and video (or images) retrieval based on metadata, they have not pushed the boundaries to represent and query video contents continuously.

In this paper, we postulate that video content analysis can be effectively achieved by synergistically integrating approaches from three domains –VCE, SP, and CEP– for obtaining an end-to-end holistic solution. Contents[1] extracted from each video frame using VCE algorithms are streamed to a continuous query processing system. These contents need to be represented using expressive data models and processed with new operators to detect situations and raise alerts for video situation monitoring. *A key advantage of this approach is that as each component advances, its effect can be leveraged by the other components to produce a system that is more than the sum of its parts.*

The contributions of this paper are:

- **Establishing** the viability of well-defined and proven (low-risk) approaches for video content analysis (Sect. 2).
- **Identifying situations** to be detected for assisted living domain and **advantages and disadvantages** of different representation models for detecting the situations (Sect. 2).
- **Leveraging** tested approaches in a novel way for developing a domain-independent video content analysis (QVC) framework (Sect. 4).
- **Proposing** two representation models for extracted video contents (Sect. 4) and query processing architecture (Sect. 5).
- Approaches to situation detection using new operators of Continuous Query Language for Video Analysis (CQL-VA) or graph analysis techniques (Sect. 6).

The related work and conclusions in this paper are in Sects. 3 and 7, respectively.

2 Challenges

The situation types that can be identified are dependent on the accuracy and completeness of the extracted contents by the latest VCE algorithms. The goal of this paper is to identify situations for assisted living domain considering what actually can be extracted using VCE algorithms. A representative set of such situations is shown in

[1] Our focus is on the efficient processing of extracted contents from VCE algorithms. Real-time video content extraction is a different problem altogether. However, this is a prerequisite for video situation monitoring in real-time using our framework.

Table 1. These situations are used for demonstrating how the proposed representations and framework can be used for detecting them without human-in-the-loop. In our view, situations shown in Table 1 are starting points for automated video situation analysis in the assisted living domain. A combination of different categories of VCE algorithms is required in the pre-processing step to answer the queries in Table 1. The challenge is to represent the different content categories extracted by VCE algorithms, pose queries on the representation models, and compute them efficiently.

As mentioned earlier, the existing SP or relational models donot have support for representing the various extracted content types (e.g., multidimensional feature vectors, bounding boxes, etc.) For example, a traditional relational model would require four columns to represent a bounding box, losing the context of the bounding box as a whole. The number of columns will increase in the case of multi-dimensional feature vectors. Besides, computations (e.g., distance, join, etc.) on the extracted contents becomes more complex (e.g., eight columns need to be joined for bounding boxes). Maintaining the order of the contents is also important to preserve video context, as they cannot be processed in arbitrary order. Hence, an enhanced relational model with a set of primitive operators is required to answer queries in Table 1.

Table 1. Example Situations Relevant for Assisted Living Domain

Query Type	Example Situations	Data Model
Q1	a) Someone being in the **same position** (or **inactive**) for an extended period of time (specified as a parameter)	R++
Q2	a) **Bumping** into the door or other large object, b) A cane/walking stick **falling** from someone's hand	R++
Q3	a) Identifying **isolated** individuals (not part of a group). b) Counting isolated individuals. (To help participate in group activity)	Graph
Q4	a) **Tracking** (in terms of direction) the movement of a **group** (e.g., in the assisted environment or even in sports)	Graph

Though the relational model has support for grouping, ordering, aggregation, etc., analyzing the complex relationships (e.g., groups, community, etc.) between objects is difficult. For example, identifying groups of objects for answering Q3-Q4 in Table 1 requires clustering the same objects across consecutive frames. This would require multiple self-joins if the relational model is used, whereas graph models can represent an object as a node, and clustering algorithms can be applied to the nodes. Besides, the existing graph algorithms can be leveraged to identify more complex situations. The current graph models [17,18] for video analysis cannot answer the situations Q3-Q4 listed in Table 1 as they represent graphs as fixed scene classes or graphs per frame and cannot extract groups or communities. Hence, multiple ways to extract video contents and the use of an appropriate graph representation model are important for detecting certain types of situations. Since both the graph and relational model have their own advantages, the availability of multiple models provides a wider choice for detecting situations (from the same contents).

3 Relevant Work

In Table 2, different categories of video situation analysis framework, their representation techniques, and supported functionalities are shown. The **assisted living solutions** [3,11,12] mostly use sensor data obtained from wearable or IoT devices and classify the events using machine learning techniques. These systems are designed for a specific event and their focus is on optimizing the machine learning algorithms for sensor data or improving the wearable device utility, accuracy, etc. Though fall detection systems handle video data, they are customized solutions as well. For identifying new types of events, a custom device or algorithm is needed to be incorporated into the current solutions. The existing **video content analysis systems** are designed for surveillance or monitoring applications. We have categorized these solutions into Low-level content analysis, and Graph analysis as shown in Table 2.

Table 2. Summary of existing video content analysis literature vs. MavVStream

Category	Systems	Data Model	Supported
Assisted living	Fall detection [3, 12]	Video, Sensor	Falling/not-falling classification using deep learning
	Isolation detection [11]	Sensor	Abnormal/normal behavior classification using deep learning
Low-level content analysis	BilVideo [8], SVQL [10], LVDBMS [4]	Relational	**Relational** (select, project), **Spatial** (bounding box overlap, direction), and **Low-level** (appear, before, ...) operators, fixed event and content database
Graph analysis	VidCep [17]	Knowledge Graph	Pattern (horse riding, ...) matching using bounding box based spatial (overlap,...) and temporal (direction, ...) operators
	Equi-Vocal [18]	Scene Graph	Patterns using spatial bounding box relationship (e.g., front) and object ids
	EDCAR [6]	Scene Graph	Patterns (pick up, set down, ...) using bounding box
QVC	MavVStream	**R++ (relations with vectors and arrable)**	**Window-based relational operators, CQL-VA operators** [5]: compress consecutive tuples (CCT), consecutive join (cJoin), and select with feature vector similarity matching condition, direction, and distance
	MavVGraph	Graphs, MLN	**Graph Algorithms:** clustering, searching a group

4 Proposed Framework

Our proposed QVC framework (shown in Fig. 1) is composed of three modules: A) Video Content Extraction (VCE), B) Extracted content representation (R++ with arrable or Graphs), and C) Processing continuous queries for situation detection using CQL-VA operators (proposed in [5]) or graph analysis (or querying) techniques.

A) Video Content Extraction (VCE): The VCE component of the proposed framework uses state-of-the-art and open-source object detection (YOLO [14]) and tracking (Deepsort [16]) algorithms. VCE extracts **frame id (fid)**: a unique identifier of a frame, **object id (oid)**: a unique identifier of an object across a maximum number of frames, **label**: object class (e.g., car, person, etc.), **object label confidence score, bounding box ([BB])**: object location in a frame, and **feature vector ([FV])**: a multi-dimensional vector representing features of an object. Once all the content attributes are extracted, the raw outputs are post-processed for appropriate formatting and assigning a timestamp (ts) to each frame.

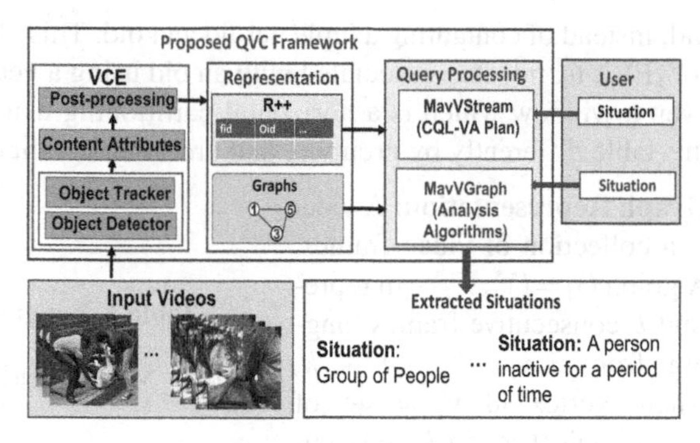

Fig. 1. QVC Framework Components

B) Representation: The two proposed representation models (R++ and graph) are discussed as follows.

R++ Representation: The arrable data model proposed in Aquery [9] represents an ordered set of vectors with a single tuple (see Fig. 2(c)). Moreover, arrable is backward compatible with the relational model and has size and order-preserving properties. Video content can be represented and processed using arrable and its properties effectively. Therefore an enhanced relational model (termed **R++**) with support for arrable is proposed in this paper. The R++ model supports traditional numerical values (e.g., fid, oid, ...), categorical values (e.g., object label, direction, ...), and vectors (e.g., multi-dimensional, ragged to capture [BB], [FV], ...).

A tuple is created in an R++ table for each object detected in a video frame along with its associated attributes (extracted using VCE). An example of an R++ table is shown with five different types of object attributes in Fig. 2(b). Here, fid, oid, and ts are numerical types whereas [BB], and [FV] are vector types. [BB] vector size is four whereas [FV] vector size varies depending on the feature vector type. In Fig. 2, frame 2 contains two objects with oid 1 and 2, and there are two tuples for each object in frame 2. For supporting time-based windows, each frame is associated with an actual timestamp (shown as an integer for convenience). The R2A operator proposed in [5] converts a table to an arrable table (example shown in Fig. 2(c)). This table contains one tuple per

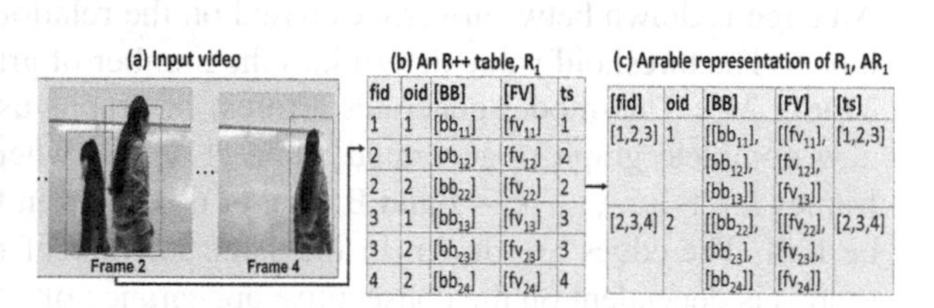

Fig. 2. R++ representation. (a) Input video, (b) R++ table with vectors, (c) R++ table with arrable by grouping on oid and ordering on ts. Bounding box: $[bb_{ik}]$, feature vector: $[fv_{ik}]$ of ith object in k^{th} frame.

oid, instead of containing a tuple per fid and oid. This allows us to represent all the [BB] (or [FV], ts, fid, etc.) associated with an oid using a vector across the entire relation (or even a window which is a horizontal partitioning using ts). It is possible to represent this table differently by grouping and ordering on other scalar attributes.

Graph Representation: A video is a collection of video frames. A graph $G_i = (V_i, E_i)$ can represent k consecutive frames ranging from $i, i+1, ..., i+k$, where vertex id V_i = set of unique object ids in k consecutive frames and edge E_i = relationship (e.g., distance) between object ids. Each vertex v in V_i and edge e in E_i has a list of vertex labels (e.g., bounding box, feature vector, etc.) and edge labels (direction and distance) respectively. Each vertex and edge label has a list of values (maximum length k) computed across k consecutive frames. We are proposing two graph models (M_1 and M_2) in addition to the graph model M_3 used in literature [17,18]. All models generate a forest of graphs.

Table 3. # of graphs in alternate graph models.

Dataset	Video	Length	No. of frames	No. of graphs (M_1)	No. of graphs (M_2)	No. of graphs (M_3)
CAMNET [19]	PRG14	18 min	22393	**23**	1588	21527
	PRG1	24 min	26339	**26**	395	24848
DETRAC [15]	40181	68s	1700	**33**	948	1700
	39851	32s	1420	**20**	668	1414

1. **Model M_1:** This model generates a forest of complete graphs based on the node ratio. Here, the node ratio value is 1- *jaccard index*(A, B), where, A = set of object ids in frame $i, i+1, ..., i+k-1$ and B = set object ids in frame $i+k$. A new graph is generated for frame $i+k$ when the node ratio $>$ a given threshold. The number of frames used in each graph may vary. The same object (or object id) is represented as a single node in a graph for all the frames along with as many labels as needed. An edge is drawn between vertices based on the relationship and its values as edge labels. The threshold value determines the number of graphs generated.
2. **Model M_2:** This model generates a forest of graphs using a different criterion. A new complete graph is generated when $A \not\subset B$, where A = set of object ids in frames $i, i+1, ..., i+k-1$ and B = set of object ids in frame $i+k$ where B cannot be null. The edges are drawn in the same manner of model M_1. The number of graphs is dependent on the consecutive appearance of objects.
3. **Model M_3** [17]**:** This model generates a set of complete graphs, one for each nonempty frame in a video. The number of graphs generated $=$ number of nonempty frames.

In Table 3 the above graph models are compared using videos from CAMNET [19] and DETRAC [15] dataset. M_1 (with node ratio > 0.85) and M_2 generate significantly fewer numbers of graphs than M_3. M_1 generates fewer graphs even for crowded traffic videos in DETRAC. Although the number of labels is likely to be the same in all models (there are also ways to compress them), processing can be more efficient.

C) Processing and Analysis: This is elaborated in the following section.

5 Query Processing Architecture

MavVStream Architecture: MavVStream framework (shown in Fig. 3) is an extension of earlier stream and event processing system MavEStream [7]. It includes basic relational operators along with aggregates, *enhanced* relational representation with support for vectors and arrables, CQL-VA operators such as compress consecutive tuples (CCT), efficient join using consecutive Join (cJoin), CCT with join using similarity match condition (to compare objects for similarity), and conditions to process vector attributes, such as [BB] and [FV] vectors.

This is a client-server architecture where users can submit CQL-VA queries to MavVStream server. A multi-threaded feeder feeds R++ tuples continuously to the query processor in the server. Once the client submits a continuous query, a query plan is generated by the server. Each operator in the query plan object is associated with input and output buffers. The system also has support for both physical and logical windows with flexible window specifications with hop size to support disjoint and tumbling/rolling windows.

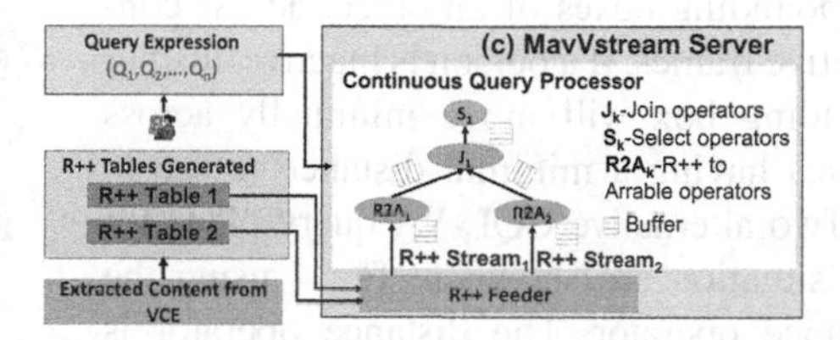

Fig. 3. MavVStream architecture and its modules [7]

MavVGraph Analyzer: A forest of graphs is generated from the extracted video contents for any of the models described in Sect. 4. Processing or analysis is done on the forest of graphs. The choice of the model may be based on the analysis and efficiency considerations. For example, for identifying groups or clusters using model M_3 the MavVGraph analysis extracts groups or clusters from n consecutive frames using the K-means clustering algorithm on object-bounding box centroids. At first, N_c (number of clusters) farthest bounding box centroids from the first nonempty frame are chosen as initial cluster centroids. The objects in a frame are clustered using the cluster centroid computed at a previous frame. The most strictly formed groups are considered to have minimal intra-cluster distance. The clusters (formed in a frame) with minimum sum of squared error (SSE) in d neighboring frames are given as output.

6 CQL-VA Query Processing or Graph Analysis

In this section, we represent how situations shown in Table 1 can be detected with the proposed framework with some preliminary results. Two situations (Q1 and Q3 one from each representation model) were used for the experiments on videos from Mavvid [1] dataset (contains situations addressed in [5] and Table 1). The videos in this dataset were generated by concatenating Youtube videos (e.g., capitol riot, soccer match, etc.) as well as videos generated in ITLAB. Additionally, Q3 experiments were conducted on arbitrary video downloaded from YouTube in the assisted living environment. All the videos in the dataset were pre-processed using the YOLO and deep sort implementation on an NVIDIA Quadro RTX 5000 GPU with 10 GB memory. The query processing experiments were conducted on the same machine with 2 processors (Intel Xeon), 48 cores, and 746 GB of main memory.

Ground Truth (GT) for Evaluation: Acc(vce) is used as the ground truth to be fair. GT is manually determined on VCE output. It is compared with query output.

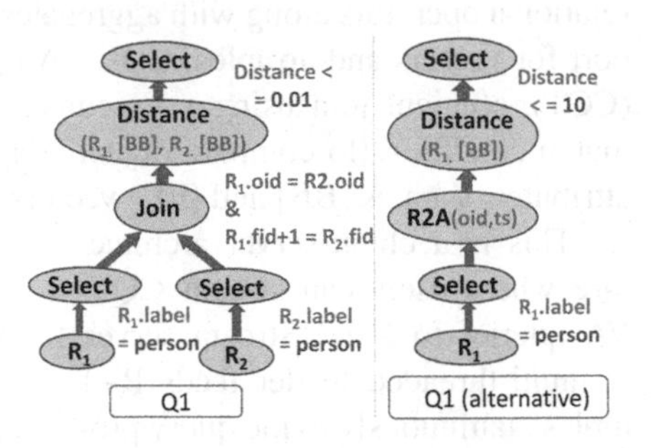

Fig. 4. Alternative query plans for Q1.

Q1 (Someone in the Same Position for an Extended Period of Time): This situation is addressed by computing the distance between the bounding boxes of an object across consecutive frames. If a person is inactive, his/her bounding box will move minimally across frames having a minimal distance (close to 0). Two alternative CQL-VA query plans for this situation are shown in Fig. 4 using the distance operator. The distance operator is designed to compute the Euclidean distance between two bounding boxes columns of R++ tables or the first and last bounding boxes of an arrable. In Q1, a self-join is performed between two R++ tables (R_1 and R_2 are the same tables). This brings together every pair of frames for an object in a window and the distance operator is applied on top of that. The rows with a distance value < a threshold are chosen to identify object inactivity. In Q1 (alternative) the R++ table is converted into an arrable table using the R2A operator. The group by and assuming order attributes for R2A is object id and timestamp respectively. The distance operator was applied on arrable table and rows with a distance < threshold is chosen.

The experiment results for this query are shown in Table 4. Here, a 5 min video from the MavVid dataset was used, with other different types of situation instances present. There were 36 instances of the situation in question and 34 of them were identified correctly. The 2 situation instances were not brought out correctly as there were two objects bounding boxes having distance > 0.01. The total query execution time (T_e) using a 100s disjoint window was 86s. Additionally, we have experimented with the same video with the query plan Q1 (alternative) in Fig. 4. Since, this computes the distance between the first and last frame of an object, only 16 instances of the situation were detected and the accuracy was 44.4% for a window size of 100s and threshold value 10. However, processing this plan took 50s, which is less than the query plan with join. The reason for this is an object can be active in the first 'n' frames of a video and then become inactive or vice

Table 4. Results of Q1 on MavVid dataset. V_{id} 7002 length = 5 min., # of frames = 8939. T_e = query processing time.

V_{id}	Query Plan	GT instances	CQL-VA instances	Acc (vce)	T_e
7002	Q1	36	34	94.4%	86s
	Q1 (alternative)	36	16	44.4%	50s

versa. The distance between the first and last bounding boxes increases in that case. The Q1 (alternative) plan is more efficient and less accurate than the Q1 plan (with join).

Q2 (Bumping Into the Door, Cane Falling From Hand): Object label information and distance between bounding boxes are required to identify these situations. For Q2(a) in Table 1, when a person bumps into the door, the distance (between person and door) will be minimal (close to 0). For computing this query, we need to select the rows with "person" and "door" (or other object types) from the same R++ table. Since the door and person are different objects a self-join needs to be computed where the object id is different. A distance operator can be applied to the joined table, and an R++ output column containing numeric distance values is generated as a result. Finally, the rows with a distance of 0 (or close to 0) can be selected. The query plan is similar to Fig. 4 (Q1), with different join and selection conditions.

Q3, Q4 (Identifying Isolated Individuals, Tracking Groups): Typically isolated individuals are not part of a group (defined as two or more objects in close proximity) in a video. Hence, this situation can be identified by extracting groups using the MavV-Graph clustering algorithm discussed in Sect. 5 and finding out which objects are not part of a group/cluster in a certain video. MavVid and Youtube datasets were used for the experiments (shown in Table 5) of this situation. In this table, three representative frames are shown from each video from MavVid and Youtube datasets. The average number of objects in both the videos on MavVid and Youtube is 2.08 and 2.9 respectively. Hence, two different N_c (number of clusters) values (2 and 3) are chosen for analysis. Acc(vce) is calculated for the whole video, by setting $d = 3$. For video 7002 in the MavVid, the cluster having the minimum SSE value at frame 856 is the output cluster (O_c) among frames 855–857. From the O_c, we can say clusters having more than 2 elements form a group (object id 63 and 64 forms a group), and one element cluster (object id 20) are isolated individuals. When $N_c = 3$, all the output clusters have one element and three isolated individuals are identified. There were 37 isolated individuals (positive situation instances) and 42 individuals (negative instances) part of a group in this video. When $N_c = 2$, 27 (TP) isolated individuals were detected correctly and 74.7% accuracy (TP = 27, FP = 10, TN = 32, FN = 10) was obtained. When $N_c = 3$, 20 isolated individual instances were correctly identified (TP) and 64.5% accuracy (TP = 20, FP = 11, TN = 31, FN = 17) was obtained. The processing time was 11.5s.

Table 5. **Results of Q3** with three representative frames. Clusters are shown in braces (object id as members). fid = Frame id, T_e = query processing time. V_{id} 7002 (MavVid) length = 5 min., # of frames = 8939. V_{id} 8002 (Youtube [2]) length = 5s, # of frames = 200.

V_{id}	N_c	fid	Clusters per frame	SSE	O_c	Q3 Output	Acc (vce)	T_e
7002	2	855	{20,63},{71}	48416.0	{20},	{20}	74.7%	11.5s
		856	{20}, {63,64}	80.1	{63,64}			
		857	{20,32}, {64}	87216.6				
	3	846	{71}, {20,74}, {81}	2628.62	{71},	{71},	64.5%	
		847	{71}, {74}, {20}	0.0	{74},	{74},		
		848	{71}, {74}, {20}	0.0	{20}	{20}		
8002	2	159	{2}, {1,3,8,9}	73221.6	{2},	{2}	100%	0.59s
		160	{2}, {1,3,7,9}	70093.3	{1,3,7,9}			
		161	{2}, {1,3,7,11}	70620.3				
	3	159	{2}, {1}, {3,8,9}	9450.7	{2}, {1},	{2},	85.7%	
		160	{2}, {1}, {3,7,9}	7956.5	{3,7,9}	{1}		
		161	{2}, {1}, {3,7,11}	8007.0				

For video 8002, there was 1 isolated individual (object id 2) and 5 individuals were part of a group. 100% accuracy was obtained for $N_c = 2$. However, for $N_c = 3$ two isolated individuals (object id 2 (TP) and object id 1 (FP)) were identified. Accuracy is 85.7% (TP = 1, FP = 1, TN = 5) in this case. The processing time for this video is 0.59s. It is evident that changing the value of N_c drastically changes the accuracy and has no effect on the processing time. Hence, choosing an appropriate number of clusters is important for answering this query accurately. Q4 can be similarly identified, by computing the direction of each object in a group across frames.

7 Conclusions and Future Work

In this paper, we have identified three components of an automated video situation monitoring framework for improving quality of life. We proposed two representation models, identified a number of situations for the assisted living domain, and showed how they can be detected. Future work will extend the query classes and graph analysis.

Acknowledgments. This work was partly supported by NSF Grants CCF-1955798 and CNS-2120393.

References

1. Mavvid dataset prepared in ITLAB (2023). https://itlab.uta.edu//downloads/mavVid-datasets/MavVid_Merged_v1.zip
2. Video downloaded from youtube representing assisted living situation (2023). https://www.youtube.com/watch?v=YzxTSzaCZMo&t=1191s
3. Amsaprabhaa, M., et al.: Multimodal spatiotemporal skeletal kinematic gait feature fusion for vision-based fall detection. Expert Syst. Appl. **212**, 118681 (2023)
4. Aved, A.J., Hua, K.A.: An informatics-based approach to object tracking for distributed live video computing. Multimedia Tools Appl. **68**(1), 111–133 (2014)
5. Billah, H., Arora, M., Chakravarthy, S.: A continuous video content querying system for situation detection. https://doi.org/10.48550/arxiv.2211.14344

6. Caruccio, L., Polese, G., Tortora, G., Iannone, D.: EDCAR: a knowledge representation framework to enhance automatic video surveillance. Expert Syst. Appl. **131**, 190–207 (2019)
7. Chakravarthy, S., Jiang, Q.: Stream Data Management: A Quality of Service Perspective. Springer (2009)
8. Dönderler, M.E., Şaykol, E., Arslan, U., Ulusoy, Ö., Güdükbay, U.: BilVideo: design and implementation of a video database management system. Multimedia Tools Appl. **27**(1), 79–104 (2005)
9. Lerner, A., Shasha, D.: AQuery: query language for ordered data, optimization techniques, and experiments. In: Proceedings of the 29th International Conference on Very Large Data Bases-Volume 29, pp. 345–356. VLDB Endowment (2003)
10. Lu, C., Liu, M., Wu, Z.: SVQL: a SQL extended query language for video databases. Int. J. Database Theory Appl. **8**(3), 235–248 (2015)
11. Prenkaj, B., et al.: A self-supervised algorithm to detect signs of social isolation in the elderly from daily activity sequences. Artif. Intell. Med. **135**, 102454 (2023)
12. Tohidypour, H.R., Shojaei-Hashemi, A., Nasiopoulos, P., Pourazad, M.T.: A deep learning based human fall detection solution. In: Proceedings of the 15th International Conference on PErvasive Technologies Related to Assistive Environments, pp. 89–92 (2022)
13. Varia, J., Mathew, S., et al.: Overview of amazon web services. Amazon Web Services (2014)
14. Wang, C.Y., Bochkovskiy, A., Liao, H.Y.M.: YOLOv7: trainable bag-of-freebies sets new state-of-the-art for real-time object detectors. arXiv preprint: arXiv:2207.02696 (2022)
15. Wen, L., et al.: UA-DETRAC: a new benchmark and protocol for multi-object detection and tracking. Comput. Vis. Image Underst. **193**, 102907 (2020)
16. Wojke, N., Bewley, A.: Deep cosine metric learning for person re-identification. In: 2018 IEEE WACV, pp. 748–756. IEEE (2018). https://doi.org/10.1109/WACV.2018.00087
17. Yadav, P., Salwala, D., Curry, E.: VID-WIN: fast video event matching with query-aware windowing at the edge for the internet of multimedia things. IEEE Internet Things J. **8**(13), 10367–10389 (2021)
18. Zhang, E., Daum, M., He, D., Balazinska, M., Haynes, B., Krishna, R.: EQUI-VOCAL: synthesizing queries for compositional video events from limited user interactions [technical report]. arXiv preprint: arXiv:2301.00929 (2023)
19. Zhang, S., Staudt, E., Faltemier, T., Roy-Chowdhury, A.K.: A camera network tracking (CamNet) dataset and performance baseline. In: 2015 IEEE WACV, pp. 365–372. IEEE (2015)

Remembering the Forgotten: Clustering, Outlier Detection, and Accuracy Tuning in a Postdiction Pipeline

Anna Baskin[1], Scott Heyman[1], Brian T. Nixon[1(✉)], Constantinos Costa[2], and Panos K. Chrysanthis[1]

[1] Department of Computer Science, University of Pittsburgh, Pittsburgh, USA
{afb39,sth66}@pitt.edu, {nixon.b,panos}@cs.pitt.edu
[2] Rinnoco Ltd, 3047 Limassol, Cyprus
costa.c@rinnoco.com

Abstract. The ever-increasing demand to use and store data in perpetuity is limited by storage cost, which is decreasing slowly compared to computational power's exponential growth. Under these circumstances, the deliberate loss of detail in data as it ages (referred to as data decay) is useful because it allows the cost of storing data to decrease alongside the data's utility. The idea of data postdiction as a data decay method uses machine learning techniques to recover previously deleted values from data storage. This paper proposes and evaluates a new pipeline using clustering, outlier detection, machine learning, and accuracy tuning to implement an effective data postdiction for archiving data. Overall, the goal is to train a machine learning model to estimate database features, allowing for the deletion of entire columns, which can later be reconstructed within some threshold of accuracy using the stored models. We evaluate the effectiveness of our postdiction pipeline in terms of storage reduction and data recovery accuracy using a real healthcare dataset. Our preliminary results show that the order in which outlier detection, clustering, and machine learning methods are applied leads to different trade-offs in terms of storage and recovery accuracy.

Keywords: Data postdiction · Data Decaying · Lossy Compression · Clustering · Outlier Detection

1 Introduction

Enterprises are generating data at unprecedented rates as more services are hosted online and IoT devices are widely deployed. From government, to industry, to entertainment and healthcare, all enterprises are collecting and exploiting this data for business intelligence, and online decision making. For example, the National Institutes of Health (NIH) initiated the *All of Us* research program [6] with the goal of establishing one of the most diverse health databases

A. Baskin, S. Heyman, B. T. Nixon—These authors contributed equally

A. Abelló et al. (Eds.): ADBIS 2023, CCIS 1850, pp. 46–55, 2023.
https://doi.org/10.1007/978-3-031-42941-5_5

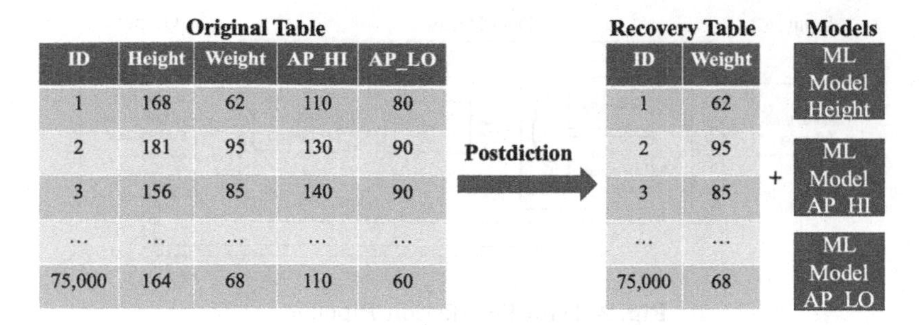

Fig. 1. Example of Basic Postdiction on Medical Data

ever created for precision medicine and disease prediction. Also, telecommunication companies are collecting data that can be used from churn prediction of subscribers, city localization, 5G network optimization and user experience assessment [3]. The demand to support many analytic-oriented processing scenarios using larger collections of data that span over a longer period of time has led to an ever-increasing demand in storing big data in perpetuity.

The continuous storage of big data has the potential to facilitate extensive analysis. Furthermore, the rapid growth of computational power allows such analysis to be performed at faster rates. However, the expenses associated with storing such a vast amount of data is increasingly becoming a constraining factor. The storage cost is decreasing at a much slower rate compared to the exponential growth of computational power. This is not a new challenge and significant research has been done into *lossless compression* as a solution to this problem as well as into *data reduction* techniques [12], e.g., sampling [11], aggregation (OLAP) [8], dimensionality reduction (LDA, PCA) [7], synopsis (sketches) [1] and lossy compression [9]. Only recently the focus was shifted on utilizing *data decay* [4,5], i.e., the deliberate loss of detail in data as it ages, as it allows the cost of storing data to decrease alongside the data's usefulness.

Data postdiction [2,3] is a new way to implement data decay making a statement about the past value of some tuple, which does not exist anymore as it had to be deleted to free up disk space. Data postdiction relies on existing Machine Learning (ML) algorithms to abstract data into compact models that can be stored and queried when necessary [1–3]. This allows for the deletion of entire data columns, as shown in Fig. 1, where one or more columns are saved to be able to recover the postdicted columns using the stored models.

Postdiction currently considers the complete dataset to be decayed similar to compression. In contrast to compression, postdiction can retrieve individual data values without recovering the entire compressed dataset. Similar to lossy compression, postdiction exhibits a *storage-accuracy* trade off. The goal of this work is to improve storage reduction and data recovery accuracy by exploiting data partitioning.

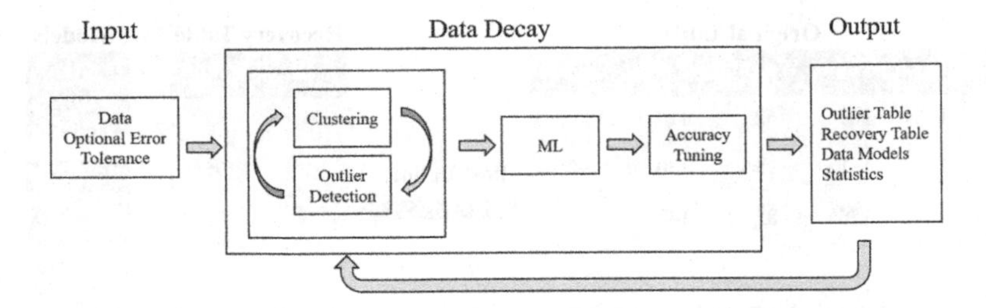

Fig. 2. Data Postdiction Pipeline

This paper hypothesizes that *clustering, outlier detection*, and *accuracy tuning* can be used to implement data partitioning that improves the original postdiction effectiveness. The contribution of this paper is a novel postdiction pipeline that explores these methods to produce the best possible data storage reduction for a specified *error tolerance*, i.e., data recovery accuracy (Sect. 2).

We use our postdiction pipeline to observe the effects of various combinations of clustering, outlier detection using LSTM (Long Short-Term Memory) to generate the machine learning models, with the goal of observing how the three strategies of clustering, outlier detection, and accuracy tuning can be used together to create better postdiction (Sect. 3).

2 Our Approach: Postdiction Pipeline

In our approach, we extend the original postdiction methodology, which uses only machine learning for data decay (Fig. 1), into a postdiction pipeline using clustering, outlier detection, and accuracy tuning, as shown in Fig. 2, to achieve improved postdiction results. The improved results take two forms: (1) the best reduction balance between data storage and recovery error (similar to the original postdiction), and (2) the best reduction in data storage given a specified data recovery accuracy/error tolerance.

Workflow. In the pipeline, input data is passed first into clustering or outlier detection. They can be run in any order: outlier detection before clustering, after clustering, or before and after clustering. Each of the generated clusters are then used to train a machine-learning model. At this stage the pipeline produces a recovery table, the machine learning models and a table of outliers, representing the best achievable storage reduction and recovery accuracy, without any specified error tolerance threshold.

Next, if the error tolerance is provided, accuracy tuning is used to predict each value in the recovery table, pruning any results outside a designated error threshold and placing them in the outlier table alongside the previously detected outliers. At the end, the pipeline outputs the final table of outliers, the recovery table, the machine learning models, and statistics relating to the number of outliers and the recovered accuracy.

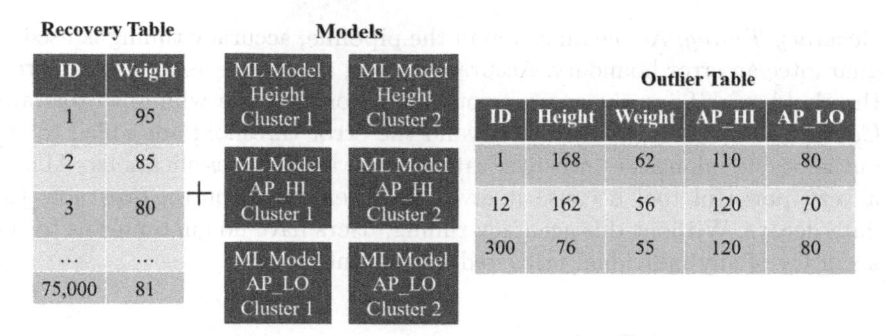

Fig. 3. Output of Our Postdiction Pipeline

Ideally, we would like to decay the maximum number of columns using the minimum number of features. An ideal sample output of the pipeline including the recovery table, machine learning models, and outlier tables is shown in Fig. 3. The example assumes that `Weight` best predicts all other fields/columns of the original table (shown in Fig. 1) and that each field can be partitioned into two clusters, possibly using different clustering methods, each of which could be compacted into a single model using different ML algorithms.

Rationale. The underlying reasons for each of the four component of our postdiction pipeline are:

- *Outlier detection:* Given that a dataset's distribution is unknown a priori, outlier detection is important for postdiction for a couple of reasons. Machine learning models typically struggle to predict outliers, but their very presence in the dataset deteriorates model performance. Therefore, if the outliers are removed and stored before running the machine-learning model in postdiction, users will be able to recover outliers with one hundred percent accuracy and have a better model for recovering the rest of the data. However, storing the outliers is a trade-off from traditional postdiction. On one hand, more storage space is required to maintain the outliers, but storing them means that more data can be recovered accurately.
- *Clustering:* Clustering algorithms group data in such a way that the data in each group are more similar to each other than to data in other groups. Clustering is important for postdiction because it can make machine learning models perform better. Having data that are all similar and close together makes it easier to recover later at a higher accuracy. Furthermore, if the algorithm finds multiple clear and distinct clusters of data, the model will have clear decision thresholds, which will allow for more accurate recovery. Recovering the data, however, will now require storing a model for each cluster, which can increase storage costs.
- *Machine Learning:* Any machine learning algorithm can be used in our pipeline to generate the compact ML models, for example, any type of recurrent neural network or logistic regression. The choice could be determined by the type of the data.

- *Accuracy Tuning:* As the final step in the pipeline, accuracy tuning is used to guarantee an error boundary. Accuracy tuning allows the user to set an error threshold, specifying the exact amount of error they are willing to tolerate. Continuous variable predictions beyond the error threshold are added to the outlier table, alongside incorrect categorical variable classifications. This is a very powerful tool because it gives the user more control over how the data decays. Without this accuracy tuning, users have no guarantee as to the accuracy of any individual recovered data point.

3 Experimental Evaluation

In our experimental evaluation, we compare the results from combinations of outlier detection algorithms, clustering algorithms and postdiction workflows.

3.1 Experimental Setup

Algorithms. We selected a representative sample of outlier detection and clustering algorithms, which worked in fundamentally different ways, as described below, to observe their effectiveness in the postdiction pipeline. These were implemented using the *scikit-learn* library in Python.

- *Outlier Detection:* The methods of outlier detection chosen for experimentation were Z-score and Isolation Forest. Z-score is a basic statistics measurement describing how many standard deviations a single point in the data is above or below the mean of the data. Z-score outlier detection was chosen for its simple and consistent results. Any data point greater than three times the standard deviations from the mean is considered an outlier. Isolation Forest outlier detection was chosen because of its ability to be applied to very large data sets. Rather than profiling all data points to calculate the outliers (as in Z-score), Isolation Forest is based on the principle of detecting only anomalies. This makes it faster than Z-score on large datasets.
- *Clustering:* The clustering methods chosen for experimentation were K-means, Density-based spatial clustering (DBSCAN), and Gaussian Mixture Modelling (GMM). These are some of the de facto methods of clustering, which is one of the main reasons for choosing them. Additionally, all of these methods work in different ways and can lead to different results. This is deliberate in order to maximize the chance of getting results that are not dependent on which clustering algorithm is chosen. Something important to note is that DBSCAN does not take in as input the number of clusters desired for output.
- *Machine Learning:* We experimented with LSTMs as they are capable of processing both single data points and sequences of data. While the dataset used by this paper does not contain sequential data, postdiction was originally concocted as a method of reducing the size of time-series telecommunication data [3]. The LSTM models were built using the *Keras* library with default configuration.

Table 1. Clustering Algorithm Comparisons

Pipeline	Clustering Method	Outliers	Recovered Accuracy
Control - No Clustering / Outlier Detection	N/A	0	18.7425
Clustering Only	K-means	0	30.97082
	DBSCAN	0	20.25868
	Distribution	0	20.07547
Outlier Detection Before Clustering	K-means	14301	47.43767
	DBSCAN	14301	37.87043
	Distribution	14301	38.00366
Outlier Detection After Clustering	K-means	28602	38.10119
	DBSCAN	14301	38.60343
	Distribution	28602	38.00964
Outlier Detection Before & After Clustering	K-means	19294	52.53579
	DBSCAN	23050	48.80418
	Distribution	14738	38.64695
Clustering + Accuracy Tuning	K-means	18036	100.0000
	DBSCAN	40104	100.0000
	Distribution	19557	100.0000
Outlier Detection Before & After Clustering + Tuning	K-means	28064	100.0000
	DBSCAN	39921	100.0000
	Distribution	26273	100.0000

Dataset. The data we used for experimentation is a healthcare dataset [10] containing *age, gender, height, weight, ap_hi, ap_lo, cholesterol, gluc, smoke, alco, active*, and *cardio*. The columns are broken down into two types. *age, height, weight, ap_hi,* and *ap_lo* are continuous and the rest are categorical. The categorical data has two or three options. The table contains 70,000 rows.

3.2 Experiment 1: Cluster and Outlier Detection Impact

For each combination of continuous variables (*age, height, weight, ap_hi, ap_lo*) predicting all other fields, we ran every combination of outlier detection, clustering, and pipeline workflow and output both the number of outliers (representative of the storage costs of the outlier table) and the recovered accuracy. The recovered accuracy is the percentage of the recovered data that is within a guaranteed error threshold (we used 5%) of the original data. Table 1 shows the outcome for height predicting weight.

We compared the results of our postdiction pipeline to the control, which used only machine learning to represent original postdiction with same accuracy of 5%. In all cases, outlier detection and clustering improved the accuracy of original postdiction, though these accuracy benefits must be weighed against the additional storage costs. Thus, the aim was to reduce these additional storage costs by storing the minimal number of outliers while maintaining a high recovered accuracy.

Outlier Detection Algorithm Comparison. We used the default settings for both outlier detection methods and found that they both performed similarly well. In their default settings, Isolation Forest's lower threshold causes it to store more outliers than Z-score, but both methods allow for adjustments using their hyperparameters. We used Z-score outlier detection as the basis of comparison moving forward because of its straight-forward and reproducible methodology and its lower outlier storage costs at the default values.

Clustering Algorithm Comparison. We found that K-means and GMM performed the best on our dataset, with 36% of columns performing best using a K-means clustering algorithm, and 72% of columns performing best using GMM (including ties, which bring the sum to greater than 100%). The differences among columns indicate that there is no ideal single clustering algorithm applicable for the entire dataset, and rather each column must be evaluated to determine the most suitable clustering algorithm.

We concluded that DBSCAN is not an effective algorithm for our dataset because DBSCAN removes points that it considers outliers from the cluster/clusters and saves them as their own cluster. For our dataset, DBSCAN's optimal value of epsilon often leads to one very poor cluster containing only outliers. Therefore, we had to choose a value of epsilon (we used 15) optimized for our dataset to create more meaningful clusters. Even so, DB-Scan only performed best in 4% of the cases, all of which were tied with another algorithm.

Pipeline Comparison. Looking at the control, which had no outlier detection or clustering performed on the data, we observed very low accuracy, but no outliers had to be saved. Conversely, performing outliers detection before training the model and then accuracy tuning after training the model resulted in very high accuracy with many outliers stored. This is another example of the tradeoff between outlier storage and recovered accuracy.

While accuracy tuning can guarantee that 100% of the recovered data satisfies the 5% error threshold, it will always require storing the same or many more outliers as without accuracy tuning. This tradeoff can be further manipulated by setting a higher or lower error tolerance, with a higher error tolerance requiring storing fewer outliers but guaranteeing a lower recovered accuracy.

Finally, we hypothesized that running outlier detection before and after clustering would improve the machine learning models, meaning that the accuracy tuning would need to pick fewer additional outliers. However, we found that the best pipeline configuration for this dataset was to use K-Means Clustering and Accuracy Tuning without Outlier Detection as it had the fewest number of outliers while maintaining the error threshold.

3.3 Experiment 2: Optimizing Multi-Column Storage Performance

While some algorithms/workflows generally performed better, overall the ideal combination of outlier detection, clustering, and workflow was unique to each combination of fields. As mentioned earlier, ideally, we would like to decay the maximum number of columns using the minimum number of features. To that end, we used each continuous variable to predict all the other features in the dataset with every combination of the above variables for each. Then, we chose

Table 2. Minimum Percentage Outliers when Column predicts Row

	age	height	weight	ap_hi	ap_lo
age (16 bits)		**20.33286**	**20.35571**	**20.29857**	**20.34857**
height (16 bits)	**10.45571**		25.76429	**20.40714**	**19.71429**
weight (32 bits)	28.36571	25.76571		37.81571	38.19286
ap_hi (16 bits)	28.26857	36.17571	38.32857		33.55857
ap_lo (16 bits)	31.25143	31.29429	38.18143	31.24857	
cholesterol (2 bits)	25.6028571	25.8171429	25.9528571	26.2242857	26.0557143
gluc (2 bits)	**20.43**	**20.43**	**20.43**	**20.43**	**21.18857**
smoke (1 bit)	**3.131429**	**3.365714**	**3.047143**	**0**	**5.97**
alco(1 bit)	**1.757143**	**1.76**	**1.75**	**0**	**0**
active (1 bit)	**0**	**0**	**0**	**0**	**0**
cardio (1 bit)	**14.37**	**17.43429**	**16.30286**	31.58571	31.54
Total Outliers	28100	33779	33427	33660	35646
Percentage Outliers	40.14%	48.26%	47.75%	48.08%	50.92%

the minimum percentage outliers that can recover 100% of the data and guarantee a 5% error threshold. The minimum percentage outliers are shown in Table 2.

This table shows the percent of data that would have to be saved as the outlier table if that column was used to recover the row based on a 5% error tolerance. For example, looking at the *age* column, it is observed that if *age* was used to recover *height*, then 10.46% of the data would have to be saved as outliers. If *age* were used to recover *weight* 28.37% of the data would need to be saved as outliers, etc. We used bold text to highlight any value we considered promising which stored under a quarter of the original data as outliers.

Next, we calculated the storage saved using two predictors: *age* and *ap_hi*. We used *age* to predict the bold rows (*height, gluc, smoke, alco, active*, and *cardio*) because *age* had the lowest percentage outliers at 40%. We chose *ap_hi* because, while it had a higher percentage of outliers, it predicted two different 16 bit integer fields (*age* and *height*) rather than just one. Finally, we considered whether using *ap_hi* to predict only *age* and *height*, rather than the additional low-bit fields, would save memory by reducing the number of outliers.

We used the size of the fields to calculate the table storage costs. The original table had 70,000 rows, with each row needing 120 bits to store, so the table cost 1,050 kilobytes in storage. To calculate the storage costs of the postdiction output tables, we calculated the costs to store the outliers table plus the recovery table of non-outliers, stored without the columns which can be reconstructed using the saved machine learning models. Overall, the total storage size was:

$$outlier\_table\_size = 120 \times \ num\_outliers$$
$$recovery\_table = (120 - recoverable\_column\_size) \times (70{,}000 - num\_outliers)$$
$$total\_size = \ outlier\_table\_size + \ recovery\_table$$

Postdiction using *age* found 28,100 outliers, and the outlier and recovery tables were 89.03% the size of the original table. Postdiction using *ap_hi* found 33,660 outliers, and the outlier and recovery tables were 83.99% the size of the original table. Finally, postdiction using *ap_hi* to estimate only *age* and *height* found 24,290 outliers, with the outlier and recovery tables at 82.59% the size of the original table. Using *ap_hi* to predict only *age* and *height* therefore performed

Table 3. Data storage size with and without postdiction

Size/Dataset (KB)	Healthcare Dataset	Healthcare Dataset × 10	Healthcare Dataset × 20
Original Table Size	1050	10500	21000
Outlier Table Size	364.35	5081.025	10197.96
Recovery Table Size	502.81	3973.915	7921.496
Total Size	867.16	9054.94	18119.456
Size with Models	1779.16	9966.94	19031.456

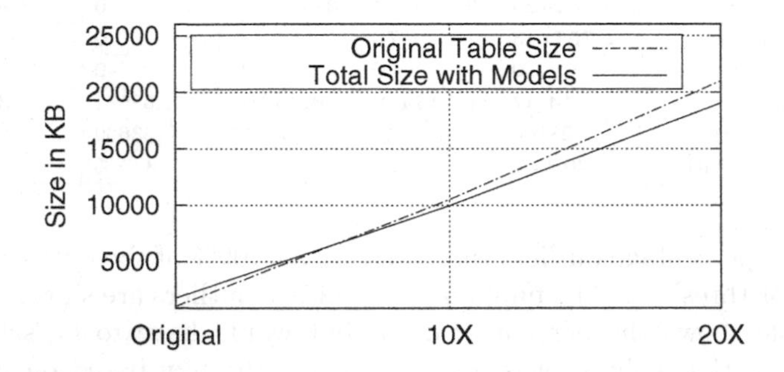

Fig. 4. Data storage size with and without the postdiction pipeline on the healthcare dataset, the dataset copied 10 times, and the dataset copied 20 times

the best. In this case it is better to predict only high-value fields like integers in order to keep outliers low, rather than predicting many binary fields which do not lead to many storage savings in the recovery table.

The above calculations do not take into account the cost of storing the trained LSTM models. LSTMs are robust models with significant storage costs (each model is 228 KB), and a model is created not just for each recoverable field, but for each cluster. Using *ap_hi* to recover age and height requires four LSTM models, totaling 912 KB. Assuming the proportion of outliers stays the same and postdiction can reduce the output table size by 17.41%, the original table would have to be approximately 5,238 KB in order for storage benefits to outweigh the costs of the LSTM. This clearly indicates postdicton's applicability to files of certain size and above (large files/big data). This is also shown in Table 3 where postdiction saved 533.06 KB or 5% on the original dataset expanded 10 times (10,500 KB) and 1968.544 KB or 10% on the original dataset expanded 20 times (21,000 KB). It should be noted that the proportion of outliers remains relatively constant when the dataset is doubled from 10,500 to 21,000 KB and after the crossover point in Fig. 4, the gap between the original size and the total size produced by the postdiction pipeline increases with the size of the dataset, reflecting the gains in savings.

Finally, we can adjust the storage-accuracy tradeoff by changing the error threshold. Selecting an error threshold of 5%, the default of this paper, yielded 24,296 outliers. A higher error threshold yields significant benefits; 10% reduces the number of outliers over 80% to 4,447 outliers. At an error threshold of around 20% there are little benefits for further tolerance, as the number of outliers is now less than 0.1% of the original database.

4 Conclusion

Postdiction is a powerful new concept that gives users the ability to have more control over the storage of their data. In this paper, we proposed the first postdiction pipeline utilizing clustering, outlier detection, and accuracy tuning to increase the accuracy of the original postdiction. We showed how this pipeline could be applied to decay a healthcare dataset and demonstrated the potential that at a large scale, the pipeline will create storage savings.

In this paper, we have only illustrated the potential of our proposed postdiction pipeline and its capabilities as a general postdiction framework. Additional research could observe how our pipeline functions with different and larger amounts of data, including sequential data. Further experimentation could involve different machine-learning, outlier detection, and clustering algorithms, or implement more complex multi-column table reduction using multiple attributes to predict another, or chaining predicted columns together to increase the number of deletable columns. Finally, data postdiction could be combined with traditional methods of lossless compression to increase storage benefits, though this introduces a new trade-off between storage and the inconvenience of decompression for single-value queries.

References

1. Cormode, G., Garofalakis, M., Haas, P.J., Jermaine, C.: Synopses for massive data: samples, histograms, wavelets, sketches. Found. Trends Databases **4**(1–3), 1–294 (2012)
2. Costa, C., Charalampous, A., Konstantinidis, A., Zeinalipour-Yazti, D., Mokbel, M.F.: TBD-DP: telco big data visual analytics with data postdiction. In: 19th IEEE International Conference on Mobile Data Management, pp. 280–281 (2018)
3. Costa, C., Konstantinidis, A., Charalampous, A., Zeinalipour-Yazti, D., Mokbel, M.F.: Continuous decaying of telco big data with data postdiction. GeoInformatica **23**(4), 533–557 (2019)
4. Kersten, M.L.: Big data space fungus. In: CIDR 2015, Seventh Biennial Conference on Innovative Data Systems Research (2015)
5. Kersten, M.L., Sidirourgos, L.: A database system with amnesia. In: CIDR (2017)
6. NIH: All of us research program. https://allofus.nih.gov/
7. Reddy, G.T., et al.: Analysis of dimensionality reduction techniques on big data. IEEE Access **8**, 54776–54788 (2020)
8. Song, J., Guo, C., Wang, Z., Zhang, Y., Yu, G., Pierson, J.M.: HaoLap: A Hadoop based OLAP system for big data. J. Syst. Softw. **102**, 167–181 (2015)
9. Tian, J., et al.: CuSZ: an efficient GPU-based error-bounded lossy compression framework for scientific data. In: ACM International Conference on Parallel Architectures and Compilation Techniques, pp. 3–15 (2020)
10. Ulianova, S.: Cardiovascular disease dataset (2019). https://www.kaggle.com/datasets/sulianova/cardiovascular-disease-dataset
11. Yan, Y., Chen, L.J., Zhang, Z.: Error-bounded sampling for analytics on big sparse data. Proc. VLDB Endow. **7**(13), 1508–1519 (2014)
12. Zhang, R.: Data reduction. In: Liu, L., Özsu, M.T. (eds.) Encyclopedia of Database Systems, pp. 1–6. Springer, Cham (2016)

Advanced Querying Techniques

Diversification of Top-k Geosocial Queries

Hassan Abedi Firouzjaei(✉), Dhruv Gupta, and Kjetil Nørvåg

Department of Computer Science, Norwegian University of Science and Technology (NTNU), Trondheim, Norway
{hassan.abedi,dhruv.gupta,noervaag}@ntnu.no

Abstract. In this work, we investigate the problem of diversifying top-k geosocial queries. To do so, we model the diversification objective as a bi-criteria objective that maximizes both user diversity and geosocial proximity. Due to the intractability of the problem, discovering the ideal results is only possible for limited datasets. Consequently, we introduce two heuristic algorithms to address this challenge. Our experimental findings, based on real-world geosocial datasets, demonstrate that the proposed algorithms surpass existing methods in terms of runtime performance and accuracy.

1 Introduction

The rise of location-based social networks (LBSNs) like Yelp and Foursquare has resulted in an abundance of geosocial data, prompting numerous research efforts to focus on effectively managing and efficiently searching such data. Top-k geosocial queries, a subset of queries performed on geosocial data, have gained significant attention due to their practical applications in location-based advertising [2], activity planning [15,22], and ride-sharing platforms [22]. Despite extensive research on the efficient processing of geosocial queries [2,3,10,12,13,15,16,22], existing work often overlooks result diversity which is an essential aspect of the search for enhancing user satisfaction [21] by summarizing search results [5], addressing query ambiguity, and eliminating redundant results [1].

In this study, we aim to diversify search results for geosocial queries by considering user similarities derived from vertex embedding techniques and user proximities based on their locations. By analyzing social ties on the social network graph, we can capture user similarities and leverage them to diversify query results. For instance, Fig. 1 illustrates how two cliques of users exhibit similarities due to shared vertices (v_4 and v_5). We combine these similarities using a max-sum diversification objective function [4]. However, determining an optimal solution for this bi-criteria objective function is an NP-hard problem. To address this challenge, we propose two algorithms based on well-designed heuristics. Experimental results indicate that these heuristic-based algorithms can diversify search results effectively.

2 Problem Formulation

In this section, we first introduce the preliminary definitions of the concepts that are needed for the formulation of the problem. Then we give a detailed description of the problem we are addressing.

A. Abelló et al. (Eds.): ADBIS 2023, CCIS 1850, pp. 59–70, 2023.
https://doi.org/10.1007/978-3-031-42941-5_6

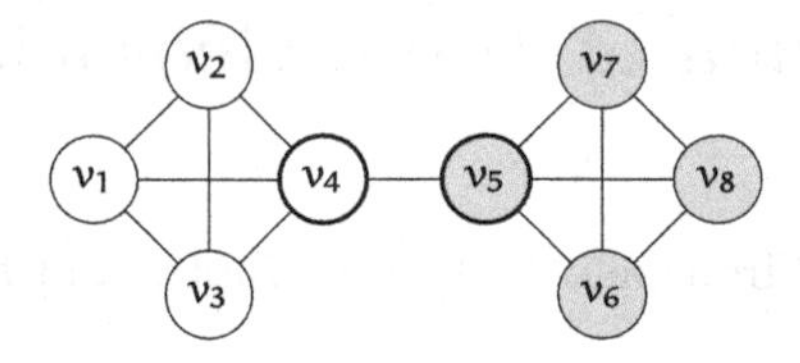

Fig. 1. State-of-the-art vertex embedding techniques can capture a wide range of similarities between users. For example, a typical embedding algorithm such as Node2vec will produce representations where vertices belonging to each of the two cliques are closer to each other with respect to some distance, such as Euclidean. Meanwhile, the pair-wise distance between two vertices from each of the two cliques will normally be larger. In this work, we use this feature to capture user diversity.

2.1 Basic Definitions

A social graph of users, $G = (V, E)$, is an unweighted and undirected graph, where V and E are the sets of vertices and edges, respectively. Each vertex $v \in V$ represents a user and $e_{i,j} \in E$ denotes the edge connecting v_i and v_j. An embedding on vertices of G is a mapping $f : V \rightarrow \mathbb{R}^d$ such that $d \ll |V|$, where f preserves some measure of similarity between vertices such as community membership or vertex roles. A user is represented as a triple $u = (v, l, e)$, where v is the vertex representing the user, l is the current location of the user, and e is the embedding vector of v. The set of all users is U. Given G, the social distance between two users is equal to the length of the shortest path connecting vertices representing them, normalized by the diameter of G. We define the social proximity as $P_{social}(u_i, u_j) = 1 - \frac{sp(u_i.v, u_j.v)}{sp_{max}}$, where sp_{max} is the maximum of the length of the shortest paths in G, which is its diameter, and $sp(u_i.v, u_j.v)$ is the length of the shortest path between $u_i.v$ and $u_j.v$. The geographical distance between two users is defined as $D_{spatial}(u_i, u_j) = \frac{\delta_H(u_i.l, u_j.l)}{d_{max}}$, where $\delta_H(u_i.l, u_j.l)$ is the distance between $u_i.l$ and $u_j.l$ computed using Haversine formula (the distance between objects on the surface of Earth). And d_{max} is the maximum spatial distance between locations of any two pairs of users, used to normalize the value of $D_{spatial}$ between 0 and 1. The geo-social proximity of two users, which is the combined measure of how socially and spatially close they are, is defined as $P(u_i, u_j) = \frac{P_{social}(u_i, u_j)}{1+\alpha \cdot D_{spatial}(u_i, u_j)}$, where variable $\alpha \in [0, 1]$ adjusts the importance given to social proximity and geographical distance in computing $P()$. We measure the social dissimilarity between two users by computing the Euclidean distance between the embedding vectors associated with vertices representing them, normalized by the maximum Euclidean distance between any two embedding vectors present in the database. Although, we chose Euclidean distance, any other distance metric can be utilized for defining and computing user dissimilarity, depending on the application. The user dissimilarity is defined as $D(u_i, u_j) = \frac{\delta(u_i.e, u_j.e)}{\delta_{max}}$, where $\delta(u_i.e, u_j.e)$ is the Euclidean distance between vertex embeddings of u_i and u_j, and δ_{max} is the maximum distance between any two distinct embedding vectors from users in U.

2.2 Problem Definition

Given U and G, a *diversified top-k geosocial (DTkGS) query*, is a quadruple $q = (u_q, k, \alpha, \beta)$, where u_q is the query user, with the goal to find a socially diverse subset $R_q \subseteq U \setminus \{u_q\}$ of k users which are geo-socially close to u_q. Given the query defined as q, we formulate the task of finding R_q as an instance of the max-sum diversification problem [4] where the goal is to find the optimal R_q such that the value of the bi-criteria objective function $\Psi(R_q) = \beta \times \text{Prox}(R_q) + (1-\beta) \times \text{Div}(R_q)$ is maximized. $\text{Prox}(R_q)$ measures the geosocial proximity and $\text{Div}(R_q)$ measures diversity of R_q. The variable $\beta = [0, 1]$ adjusts the trade-off between geo-social proximity and social diversity. Moreover, proximity and diversity of R_q are defined as $\text{Prox}(R_q) = \frac{1}{k} \times \sum_{u \in R_q} P(u_q, u)$ and $\text{Div}(R_q) = \frac{2}{k(k-1)} \times \sum_{u_i, u_j \in R_q} D(u_i, u_j)$.

Lemma 1. *Finding the optimal solution for the diversified top-k geosocial query that maximizes $\Psi(R_q)$ is NP-hard.*

Proof. Suppose $\beta = 0$; then answering the query would be equal to finding the R_q such that it maximizes the $\Psi(R_q) = \text{Div}(R_q)$ which is equivalent to solving the maximum edge weight clique problem under cardinality constraint (MEWC) [9] which is shown to be a particular case of the quadratic knapsack problem that is an NP-complete problem. Consequently, finding the optimal R_q for the query is in NP-Hard [9].

3 Proposed Heuristics

Given the intractability of optimally solving the DTkGS query, we introduce two efficient algorithms employing distinct heuristics to get reasonably accurate results.

3.1 Fetch and Refine

Fetch and Refine (FNR) is our first heuristic (see Algorithm 1). It produces its result in two key steps. In the first step, it constructs a candidate set of size K where $K > k$, and splits the candidate set into two disjoint subsets R_q and S where $|R_q| = k$, and $|S| = K - k$. R_q is initialized with k users with highest geosocial proximities, while S is initialized with the remaining $K - k$ users with highest geosocial proximities. In the second step, which consists of one or more iterations, for every $u^- \in R_q$ and $u^+ \in S$, the algorithm computes a gain value for swapping u^- with u^+ on the score of R_q using the following heuristic:

$$\text{gain}(R_q, u^-, u^+) = \Psi(R_q \setminus \{u^-\} \cup \{u^+\}) - \Psi(R_q). \tag{1}$$

At the end of each iteration, the best pair (u^-, u^+) with the largest positive gain is chosen, and u^- is swapped with u^+. The function `FindBestPair` (in Algorithm 1), given R_q and S, returns the pair u^- and u^+ which maximizes the aforementioned gain, in addition to the actual value of the gain. In Algorithm 1, Δ indicates the gain of the pair returned by `FindBestPair`. This procedure is repeated until no improvement in gain is achieved via swapping. The intuition behind FNR is that finding the globally optimal solution may not be feasible due to the hardness of answering the diversified top-k geosocial query. However, by limiting the size of the search space by only including top-K users with higher geosocial proximities, we can find the locally optimal result set with the highest social diversity among these users. The algorithm's name alludes to the filter-and-refine algorithmic framework, which utilizes one or more filtering steps before producing its results to reduce the query processing cost. The complexity of the FNR algorithm can be conceptualized as follows. It takes $O(|U| \log |U|)$ to compute the geosocial proximities. Moreover, the function `FindBestPair` takes $O(k(K-k))$ to run. The loop in Algorithm 1 lines 10–13 runs a finite number of times; the exact number of times it runs is dependent on the nature of the input data.

Algorithm 1: Fetch and Refine

Input: U, G, u_q, k, α, β, K
Output: R_q

1 $S, R_q, Q \leftarrow$ NewSet(), NewSet(), NewMaxPriorityQueue()
2 **foreach** u in $U \setminus u_q$ **do**
3 | Enqueue(u, $P(u_q, u)$, Q)
4 **while** Size(R_q) + Size(S) $<$ K and not Empty(Q) **do**
5 | **if** Size(R_q) $<$ k **then**
6 | | Add(Dequeue(Q), R_q)
7 | **else**
8 | | Add(Dequeue(Q), S)
9 $u^-, u^+, \Delta \leftarrow$ FindBestPair(R_q, S)
10 **while** $\Delta > 0$ **do**
11 | Remove(u^-, R_q); Remove(u^+, S)
12 | Add(u^+, R_q); Add(u^-, S)
13 | $u^-, u^+, \Delta \leftarrow$ FindBestPair(R_q, S)
14 **return** R_q

3.2 Best Neighbour Search

Best Neighbour Search (BNS) is our second heuristic (see Algorithm 2), which utilizes a local search procedure to find its result. Similar to FNR, BNS works in two major steps. In the first step, it constructs a candidate result-set. Then, in the second step, which consists of one to I_{max} iterations, it identifies the next best neighbouring candidate result set of current R_q and switches to it. Algorithm 2 works similarly to Algorithm 1. The major difference between the two is that FNR only probes the neighbouring solutions involving $K - k$ users in S while BNS globally checks all neighbouring candidate result-sets, and picks the best with respect to the score value. Similar to Algorithm 1, the function `FindBestPair` is used to compute a combined gain to choose the best neighbour at each iteration to move. The complexity of BNS algorithm is computed as follows. Computing the geosocial proximities takes $O(|U|\log|U|)$ to compute, and `FindBestPair` takes $O(k(K-k))$ to run. Furthermore, each iteration of the second loop takes $O(k(K-k))$ for maximum of I_{max} iterations. Thus, the resulting complexity is $O(|U|\log|U| + k(K-k)I_{max})$.

4 Experimental Evaluation

In this section, we describe the details of the experiments to evaluate the performance of the proposed algorithms. To that end, we first describe the data, baselines, setup, and metrics used for the experiments. Then we present the results and discuss their subsequent implications. To enable the reproducibility of our study, we make available the evaluation artefacts here https://github.com/habedi/adbis-2023-paper.

4.1 Data

We used the data of the Gowalla location-based social network from [11], which include the social network of 196K users and the information about 6.4M user check-ins. We extracted the largest connected subgraph of users who had at least one check-in. Furthermore, we extracted the largest connected subgraphs of users who had a check-in in four geographical regions: the USA ($|V| = 45,474$ and $|E| = 215,726$), France ($|V| = 8,449$ and $|E| = 29,347$), Germany ($|V| = 7,018$ and $|E| = 25,822$), and New York ($|V| = 2,187$ and $|E| = 4,958$). We used subgraphs of these regional subgraphs with sizes that correspond to the number of users in the social network in our experiments. For each user, their latest check-in was used as their location.

Algorithm 2: Best Neighbour Search

Input: U, G, u_q, k, α, β, I_{max}
Output: R_q

1 Lines 1 to 3 of Algorithm 1
2 **while** not Empty(Q) **do**
3 | Lines 6 to 8 of Algorithm 1
4 $u^-, u^+, \Delta \leftarrow$ FindBestPair(R_q, S); $c \leftarrow 0$
5 **while** $\Delta > 0$ and $c < I_{max}$ **do**
6 | Lines 5 to 8 of Algorithm 1
7 | $c \leftarrow c + 1$
8 **return** R_q

4.2 Baselines

We compared the performance of our algorithms with the following baselines:

- *Naive Baselines*: we implemented two baselines, JGeoSoc and JUserDiss, which naively pick top-k users based on their respective heuristics. JGeoSoc uses geosocial proximity measure, and JUserDiss utilizes only user dissimilarity to u_q, respectively.
- GMC and GNE are two state-of-the-art diversification methods from [19] that utilize a measure called maximal marginal relevance, which is an improvement on MMR [5], to construct the result-set incrementally. One major difference between the two algorithms is that GNE uses greedy randomized adaptive search, while GMC lacks randomization.
- BSwap from [20], is another state-of-the-art diversification method that, given top-k relevant objects, which in our context are users with higher geosocial proximity, tries to diversify them by swapping the least dissimilar item in the current result-set with the next most relevant item. The algorithm uses the parameter θ to set the threshold on the maximum drop in relevance that it tolerates before terminating. We set this to $\theta = 0.1$ based on [20].

4.3 Settings

All algorithms were implemented in Java. We ran the experiments on a machine with an Intel Core i9 5.3 GHz CPU with 32 GB of RAM running Ubuntu 22.04. The metrics used for evaluation are computed over 50 runs, and their average values are presented. We used Node2vec [8] to compute the embeddings, with dimensionality $d = 16$. Furthermore, we used $I_{max} = 10$, $K = 5 \times k$ (these values are determined by an empirical study of performance with different parameters), and $\alpha = 0.5$ during the experiments. We studied the performance with k ranging from 5 to 35, β from 0 to 1, and $|U|$, i.e., the subgraph size, from 120 to 2000. The default values for k, β, and $|U|$ were 5, 0.5, and 500, respectively.

4.4 Evaluation Metrics

For measuring the quality of results, the difference between the score of the optimal result-set, and the result-set produced by each diversification method was used. The term gap percentage refers to the difference mentioned above when it is scaled to represent a value between 0 and 100. Moreover, other metrics used in this work include average response time for the query and average score.

4.5 Results

Due to the problem's complexity, we can only calculate optimal result-sets for small query parameters. On each dataset, we obtained an optimal result-set for a 120-user subgraph, with parameter $k = 5$ and varying β from 0 to 1 over five runs. The optimal sets were found by searching through every possible k-sized user subset and choosing the one with the highest $\Psi()$ value. We computed the gap percentage between the score of the optimal result-set, and the result-set returned by each method. A smaller gap is better. Figure 2 includes the gap percentage for each method while varying β. Note that as β gets larger, the diversity decreases.

Interestingly, for both naive baselines, JGeoSoc and JUserDiss, the gap between the score of the result-set returned bythese two methods and the optimal result-set is noticeably large. On the other hand, GMC and GNE have small relative gaps. Our proposed method BNS has a gap similar to GMC and GNE. Moreover, BSwap's performance was relatively stable on three of the datasets. On Germany's subgraph, the gap decreased with an increase in value of β. We suspect this is due to the similarity of the embedding vectors in the subgraph. However, the gap is relatively high when β is small, and when β gets larger, the gap gets smaller. This is because FnR picks the subset of K users based on their geosocial proximities. This does not necessarily lead to a better overall score when the weight given to diversity is larger.

We examined the scalability of the methods with respect to the score of the result-set, i.e., $\Psi(R_q)$ and response time, with respect to the value of β (from 0 to 1), k (from 5 to 35), and the number of users associated with vertices of a subgraph, i.e., $|U|$ (ranging from 120 to 2000 users). Due to the overall similarity between the results, in the rest of this paper, we omit the results for the subgraphs of Germany and New York since they are similar to the results for the USA and France subgraphs. Figure 3 shows the average score, for parameters $k = 5$, $|U| = 500$, while increasing β.

Our method, BNS, performed on par or better than GMC and GNE on all datasets. The average score for the method decreases as β becomes larger, which can be due to the disparity between values of $Prox()$ and $Div()$. The score for JUserDiss dropped on all datasets as β got larger. This is because JUserDiss only utilizes social dissimilarity between users. On the other hand, the increase in the average score for JGeoSoc is relatively low. FnR method performed similarly to BSwap regarding the average score. Both methods had a drop in their average score as β got larger. Figure 4 shows the average score for parameters $\beta = 0.5$, $k = 5$, while increasing the size of the users in the subgraph, i.e., $|U|$ from 120 to 2000. The performance of GMC, GNE, and BNS is similar. These three methods performed better on all datasets regardless of the size of the subgraph. Additionally, the increase in the subgraph size seems not to have affected the average scores of the result-sets returned by these methods, which may indicate their stability relative to the input size. On the other hand, JUserDiss performed the worst, and FnR performed better than BSwap.

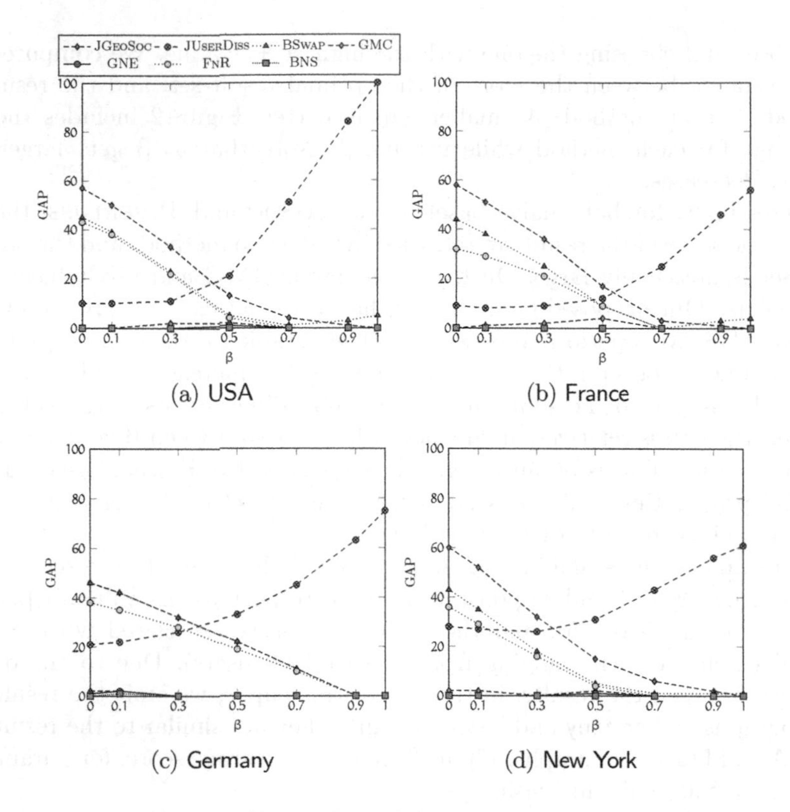

Fig. 2. Gap while varying β

Figure 5 shows the average score for parameters $\beta = 0.5$, and $|U| = 500$, while increasing the value of k from 5 to 20 in steps of 5. With the increase in the result-set size, the average score for all methods except JUserDiss steadily decreases. On the other hand, the average score for JUserDiss increases when k gets larger. It shows that either geosocial proximity or diversity increased by making the result-set larger. BNS performed similarly to GMC and GNE as before. All three methods have performed better than other methods, including FnR. Although FnR did not produce result-sets with high scores as the top three methods, its performance is better than three of the five baselines, including BSwap and JGeoSoc. Finally, Fig. 6 shows the average response time of each method for parameters $k = 5$, $\beta = 0.5$, while varying k. Our methods FnR and BNS consistently performed better than the GMC and GNE as the size grew. The response time for other baselines stayed relatively stable as k got larger.

4.6 Discussion

BNS, GMC, and GNE demonstrated superior result-set scores and stability with increased data size. Their respective strategies for maximizing relevance, integrating randomization with local search, and diversifying initial result-sets

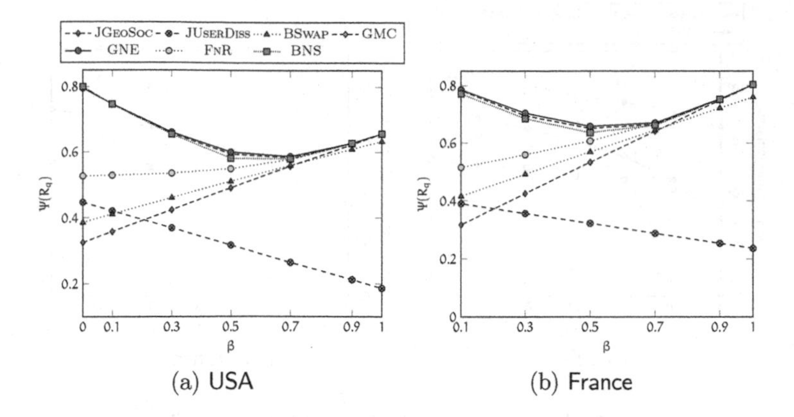

Fig. 3. Average score while varying β

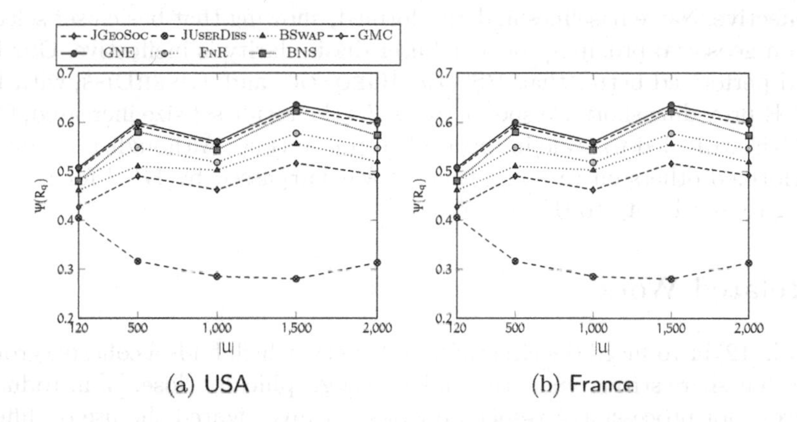

Fig. 4. Average score while varying |U|

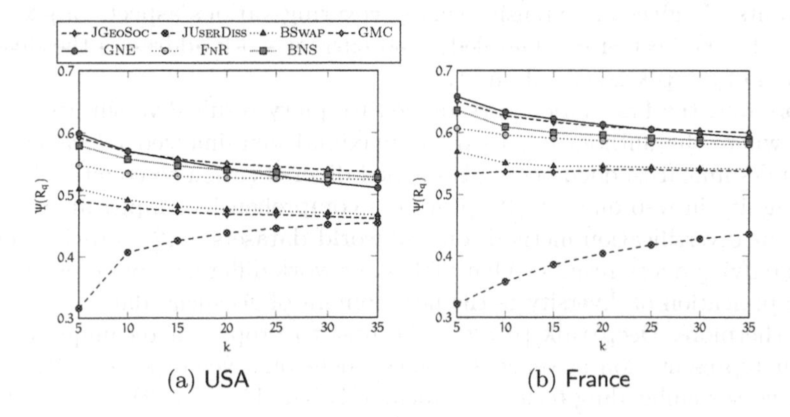

Fig. 5. Average score while varying k

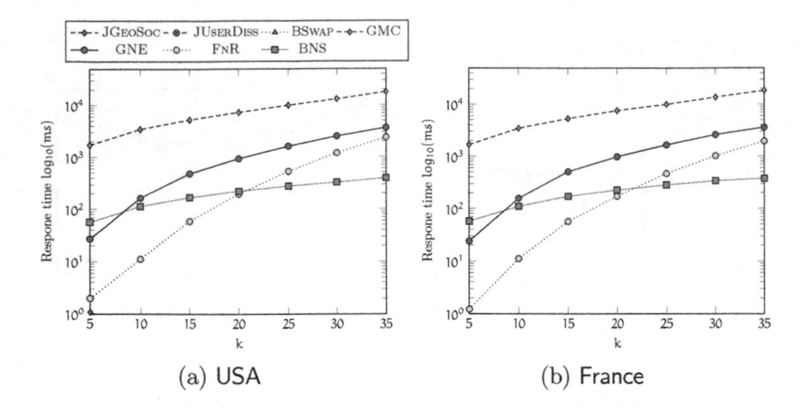

Fig. 6. Average response time while varying k

were effective. Naive baselines underperformed, showing that basic user selection based on geosocial proximity or social user dissimilarity is ineffective. Our FNR method performed better than BSWAP, JGEOSOC, and JUSERDISS, with BNS and FNR providing shorter response times. As the result-set size increased, GMC and GNE struggled to scale in terms of response time. Although FNR initially outperformed others on response time, it was surpassed by BNS for k >= 10, showing its sensitivity to the size of K.

5 Related Work

Liu et al. [12] introduced the circle of friend query, which finds a cohesive group of people that share strong social ties and are geographically close. [3] introduced a framework for processing geo-social queries. [2] investigated the use of different ranking functions to process geo-social queries. [22] proposed novel geo-social queries which could find cohesive user groups in an LBSN under acquaintance constraints. [7] gives an extensive survey covering various aspects of geo-social queries. To the best of our knowledge, no current work addresses the diversity of results of a top-k geosocial query.

Moreover, the first major work related to query result diversification is due to [5], where the authors proposed the maximal marginalized relevance to re-rank and summarize documents. [1] applied diversification to counter the effect of ambiguity in web queries. [19] provide a comprehensive empirical evaluation of various diversification methods on real-world datasets. [6,21] provide surveys on diversifying query results. What makes our work different from previous work is the application of diversity in the new domain of geosocial data.

Furthermore, DeepWalk [14] was the first to propose a technique to learn a latent representation of vertices in low-dimensional vector space. Other well-known vertex embedding techniques include LINE [17], Node2Vec [8], and their variations. Although the similarity measure employed during the learning embeddings of vertices can be chosen arbitrarily, most embedding methods mainly work by learning the immediate neighbourhood of the vertices. A recent deviation from this trend is VERSE [18] which can capture a wider range of similarities, including the roles of vertices.

6 Conclusion

In our research, we utilized vertex embedding for geosocial query result diversification, presenting an optimization problem tackled by two heuristics. These methods showed better overall accuracy and runtime compared to the baselines. Our future plan is to further improve efficiency through parallel computation for the distance between embedding vectors.

References

1. Agrawal, R., Gollapudi, S., Halverson, A., Ieong, S.: Diversifying search results. In: WSDM (2009)
2. Armenatzoglou, N., Ahuja, R., Papadias, D.: Geo-social ranking: functions and query processing. VLDB J. **24**(6), 783–799 (2015)
3. Armenatzoglou, N., Papadopoulos, S., Papadias, D.: A general framework for geo-social query processing. PVLDB **6**(10), 913–924 (2013)
4. Borodin, A., Jain, A., Lee, H.C., Ye, Y.: Max-sum diversification, monotone submodular functions, and dynamic updates. TALG **13**(3), 1–25 (2017)
5. Carbonell, J., Goldstein, J.: The use of MMR, diversity-based reranking for reordering documents and producing summaries. In: SIGIR (1998)
6. Drosou, M., Pitoura, E.: Search result diversification. SIGMOD Rec. **39**(1), 41–47 (2010)
7. D'Ulizia, A., Grifoni, P., Ferri, F.: Query processing of geosocial data in location-based social networks. IJGI **11**(1), 19 (2022)
8. Grover, A., Leskovec, J.: Node2vec: scalable feature learning for networks. In: KDD (2016)
9. Hosseinian, S., Fontes, D.B.M.M., Butenko, S., Nardelli, M.B., Fornari, M., Curtarolo, S.: The maximum edge weight clique problem: formulations and solution approaches. In: Butenko, S., Pardalos, P.M., Shylo, V. (eds.) Optimization Methods and Applications. SOIA, vol. 130, pp. 217–237. Springer, Cham (2017). https://doi.org/10.1007/978-3-319-68640-0_10
10. Huang, C.Y., Chien, P.C., Chen, Y.H.: Exact and heuristic algorithms for some spatial-aware interest group query problems. JIT **21**(4), 1199–1205 (2020)
11. Leskovec, J., Krevl, A.: SNAP datasets: stanford large network dataset collection. http://snap.stanford.edu/data (2014)
12. Liu, W., Sun, W., Chen, C., Huang, Y., Jing, Y., Chen, K.: Circle of friend query in geo-social networks. In: DASFAA (2012)
13. Mouratidis, K., Li, J., Tang, Y., Mamoulis, N.: Joint search by social and spatial proximity. TKDE **27**(3), 781–793 (2015)
14. Perozzi, B., Al-Rfou, R., Skiena, S.: DeepWalk: online learning of social representations. In: KDD (2014)
15. Shen, C.Y., Yang, D.N., Huang, L.H., Lee, W.C., Chen, M.S.: Socio-spatial group queries for impromptu activity planning. TKDE **28**(1), 196–210 (2016)

16. Song, X., et al.: Collective spatial keyword search on activity trajectories. GeoInformatica **24**(1), 61–84 (2019). https://doi.org/10.1007/s10707-019-00358-x
17. Tang, J., Qu, M., Wang, M., Zhang, M., Yan, J., Mei, Q.: Line: large-scale information network embedding. In: WWW (2015)
18. Tsitsulin, A., Mottin, D., Karras, P., Müller, E.: VERSE: versatile graph embeddings from similarity measures. In: WWW (2018)
19. Vieira, M.R., et al.: On query result diversification. In: ICDE (2011)
20. Yu, C., Lakshmanan, L., Amer-Yahia, S.: It takes variety to make a world: diversification in recommender systems. In: EDBT (2009)
21. Zheng, K., Wang, H., Qi, Z., Li, J., Gao, H.: A survey of query result diversification. Knowl. Inf. Syst. **51**(1), 1–36 (2016). https://doi.org/10.1007/s10115-016-0990-4
22. Zhu, Q., Hu, H., Xu, C., Xu, J., Lee, W.-C.: Geo-social group queries with minimum acquaintance constraints. VLDB J. **26**(5), 709–727 (2017). https://doi.org/10.1007/s00778-017-0473-6

Multi-disciplinary Research: Open Science Data Lake

Vincent-Nam Dang[1,2](✉), Nathalie Aussenac-Gilles[1], and Franck Ravat[1,2]

[1] IRIT, CNRS, Université de Toulouse, Toulouse, France
vincent-nam.dang@irit.fr
[2] Université Toulouse Capitole, Toulouse, France

Abstract. Open Science aims to establish an interdisciplinary exchange between researchers through knowledge sharing and open data. However, this interdisciplinary exchange requires exchanges between different research domains and there is currently no simple computerized solution to this problem. Although the data lake adapts well to the constraints of variety and volume offered by the Open Science context, it is necessary to adapt this solution to (1) the accompaniment of data with metadata having a specific metadata model depending on the domain and community of origin, (2) the cohabitation of open and closed data within the same open data management platform, and (3) a wide diversity of pre-existing research data management platforms to deal with. We propose to define the Open Science Data Lake (OSDL) by adapting the Data Lake to this particular context and allowing interoperability with pre-existing research data management platforms. We propose a functional architecture that integrates multi-model metadata management, virtual integration of externally stored (meta)data and security mechanisms to manage the openness of the platforms and data. We propose an open-source and plug-and-play technical architecture that makes adoption as easy as possible. We set up a proof-of-concept experiment to evaluate our solution with different users from the research community and show that OSDL can meet the needs of transparent multidisciplinary data research.

Keywords: Interdisciplinarity · Open Science · Data Lake

1 Introduction

The need for interdisciplinarity in research is growing [1]. This need is expressed through the increasing efforts to implement Open Science (OSci). Data management solutions exist within research communities to handle community-specific data. However, interdisciplinarity brings in new challenges with the management of a wide variety of research data and a need for data openness. The establishment of bridges between communities creates a different context for the design of new solutions. New actors, with their own knowledge and needs, are emerging in relation to intra-community solutions. New contexts also emerge creating additional constraints to ensure that needs are always met. There is a need to

A. Abelló et al. (Eds.): ADBIS 2023, CCIS 1850, pp. 71–81, 2023.
https://doi.org/10.1007/978-3-031-42941-5_7

manage the cohabitation of open and closed data or the management of a wider variety of data and needs around this data, notably with metadata or processing. Specifically in the case of OSci, there are several additional challenges [15]: (1) the need for interoperability with a wide variety of existing data management solutions, (2) data and metadata format issues, (3) a rapid increase in the volume of data generated, both batch data and stream data, or even real-time data, (4) the need for significant time and resources for the implementation of common standards or metadata models. The data lake is a big data analytics solution that addresses the wide variety and volume of data. The data lake has become popular in research data management projects that mix several communities (EOSC with ESCAPE [7], Data Terra with Gaia data project[1], ESA/NASA with MAAP [4], European Commission with Destination Earth [8]). However, Open Big Data is a specific context that brings many additional constraints. We propose a new functional and technical data lake architecture adapted to the OSci context and evaluated by experimentation: the Open Science Data Lake.

In part 2, we explore the different OSci data management platforms and the place of datalakes within them. In part 3, we propose a functional architecture detailing the important additions to transform a multi-zone data lake architecture into an OSDL. In part 4, we propose a plug-and-play and open-source technical architecture. In part 5, we evaluate our solution through a proof of concept evaluated by users and compared to 3 existing data set search platforms.

2 Related Works

Open Science is made up of a large number of data management platforms of all types. More than 3,000 platforms are listed on Re3data[2]. These platforms can be diverse, depending on the type and theme of the data, the volume or the community needs. These platforms can be based on noSQL databases, such as MongoDB [18], domain- or data-type-specific databases [16], data-warehouses[3], catalog-type web applications[4], specific solutions such as Dataverse[5], or many others. However, these solutions all have their limitations: a lack of scalability of interoperability with other platforms, a lack of variety in analyses or the type of data that can be managed, a lack of openness and others reasons.

The need to unify data access points to offer greater richness in data retrieval is growing. For this reason, more and more projects are based on data lakes[6]. This big data analysis solution meets a wide range of analysis and data volume management needs. It can be adapted to all fields and all types of data, whether in physics [3], medicine [11] or biology [13]. Data lake architectures have evolved over time [10,14]. Initially intended as a raw data storage area, other functional

[1] https://www.gaia-data.org/en/.
[2] https://www.re3data.org/.
[3] http://www.biosino.org/bmdc/aboutUs/organization.
[4] https://www.re3data.org/.
[5] https://ada.edu.au/.
[6] https://data.openei.org/data_lakes.

areas have been integrated to meet more needs, including data processing and metadata management. However, these architectures are designed to manage models with a fixed metadata model, in which metadata will be generated during the data life cycle in the data lake. As it stands, managing pre-existing metadata is not part of the data lake context. This is an obstacle to managing the variety present in OSci. In order to move forward with OSci, the FAIR Principles help define the directions in which this information sharing can take place [17]. With regard to the FAIR principles, the data lake lack of mechanism to meet the I3 principle, which concerns the interconnection of metadata. More focused on interoperability [5], the data lake does not functionally possess the mechanisms needed to be interoperable with other platforms. However, this is not a trivial issue. There are over 1600 standards[7] for metadata definition, including models, guidelines or terminology artifacts [12]. These different standards continue to evolve and expand with the adoption of OSci.

3 OSDL: Functional Architecture

The number of asset profiles in OSci is enriched compared with the classic data lake context [9]. There is a whole gradient of data types, from internal data to open data. The opening up of data and platforms creates the presence of users external to the initial context of the platforms. Approaching the problem of OSci as a whole requires to take these assets into account, as well as the large volume and variety of data from OSci. But it is also necessary to integrate the wide variety of pre-existing system assets for data management. Designing

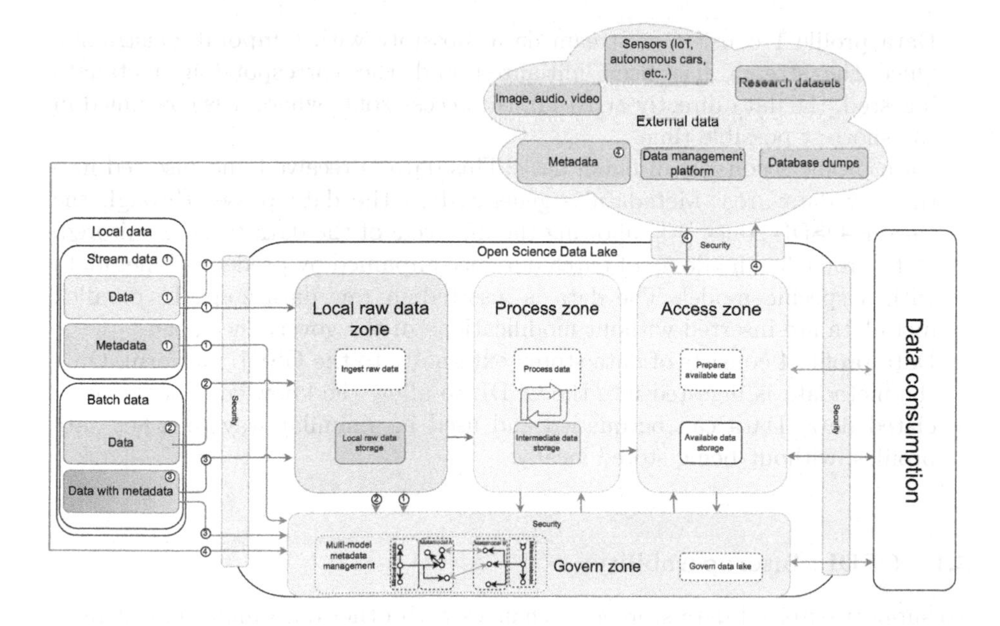

Fig. 1. Functional architecture

[7] https://fairsharing.org/.

an Open Big Data [2] solution requires taking into account 2 major aspects, in addition to the constituents of a Big Data solution. Many data and data management solutions already exist. We need to integrate these pre-existing data and enable interoperability with pre-existing data management platforms in OSci. In addition, the enrichment of the assets to be managed, compared to a usual Big Data context, requires the design of security mechanisms as a core object of the architecture to protect against the associated threats specific to OSci [9]. With regard to the FAIR Principles, we need to address the issue of interoperability. We propose a functional architecture of OSDL (see Fig. 1) where we find the 4 main zones of a multi-zone data lake [6]: the raw data zone ingests the data in the original format, the process zone allows the implementation of treatments on the data, the access zone allows the access and consumption of the processed data and the governance zone contains the metadata as well as the governance mechanisms of the data lake. We observe a new type of storage to be integrated into the OSDL architecture: **external storage**, i.e. external data is stored in existing data management platforms. The volume of OSci data does not allow to copy, store and manage it as local data. This new type of storage requires the ability to manage data and metadata acquisition protocols from data management platforms. Metadata can be used to index large volumes of data. However, it is necessary to integrate the possibility of retrieving metadata only when it is needed, to avoid an explosion in metadata volume. In addition to the two usual profiles (batch and stream data), external storage creates two new data profiles with batch data accompanied by metadata and metadata alone to be ingested. Figure 1 illustrates stream data with orange arrows, batch data with red arrows and metadata with purple arrows.

- Data profile 1 consists of stream data, possibly with temporal constraints. Once the stream has been initialized and the corresponding metadata ingested, the data directly arrives in the access zone, where it is consumed in the shortest possible time.
- Data profile 2 consists of batch data. This data is received and inserted into the raw data area. Metadata is generated as the data passes through the various OSDL zones [10], allowing the life cycle of the data to be monitored.
- Data profile 3 is made up of batch data accompanied by predefined metadata with a specific model. The data is inserted in raw data zone. In parallel, metadata are inserted without modifications in the governance zone.
- Data profile 4 consists of data stored externally to the OSDL platform. Only the metadata is ingested into the OSDL to allow the knowledge of the associated data. Data can be queried and used in a similar way to other data profiles, without being stored locally.

3.1 OSDL: Interoperability

To support external data storage, exchanges with other data management platforms have to be handled. This requires interoperability between platforms and

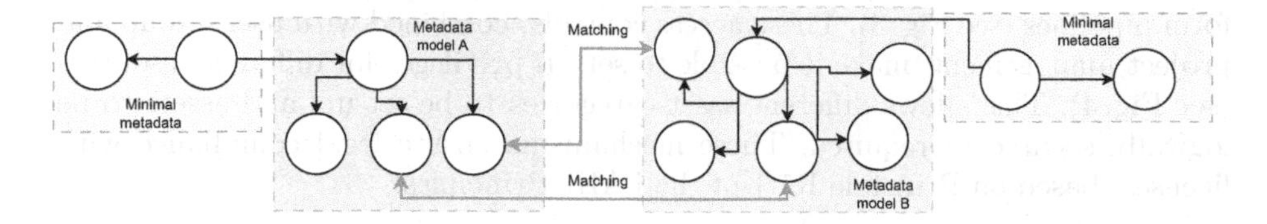

Fig. 2. OSDL metadata management: multi-model with matching

OSDL. We take as our definition of interoperability the one we proposed in a previous article [5]. We aim to enable the exchange of usable information on the different datasets. The data to be exchanged is the dataset metadata, and the useful information is the one about this metadata, the so-called metadata models. For the sake of simplicity, we deal with the 2 layer categories: system layers and process layers. For system layers, we chose to use a REST API to enable communications. In Re3Data.org, the REST API is the most widespread type of API among data management platforms, with almost 45% of platforms having communicated information about their API to Re3Data (interoperability by standardization). In addition to standardization with a large number of platforms, REST API technologies enable simple interfacing with a wide range of existing communication technologies (interoperability by gateway implementation). For process layers, we proposed to adopt multi-model metadata management (Fig. 2). This requires to handle matchings between models. Multi-model management means that external metadata can be stored, but also that these metadata can be used to query external platforms. In this way, metadata can be retrieved when needed, rather than stored locally; and no pivot model is required. We have explored interoperability and matchings more in depth in a former paper [5].

3.2 OSDL: Data Security

To avoid any loss of data, trust or time for researchers, security mechanisms are necessary [9]. Access control to OSDL resources is integrated into all the plat-

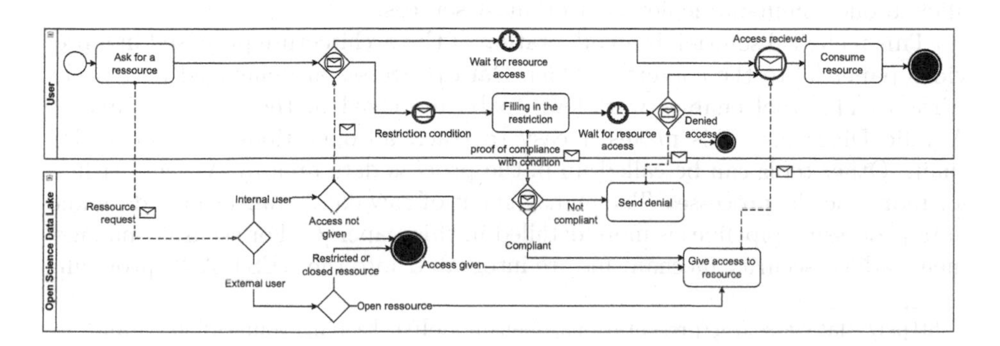

Fig. 3. Access control on OSDL resource process

form pipelines (see Fig. 3). These access controls, combined with user, group and project management, make it possible to set up privileges for different resources (see Fig. 4). This allows different asset categories to be set up, and assets to be logically secured as required. These mechanisms ensure legal compliance with licenses, based on Principle R1.1 of the FAIR Principles.

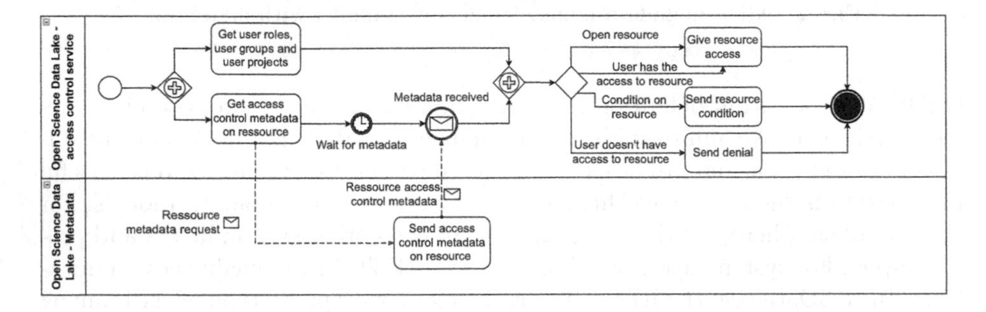

Fig. 4. Access to user privileges and required privileges for a resource

4 OSDL: Technical Architecture

OSDL must also be technically adapted to OSci. For this aspect, this architecture must be an Open Source solution [2]. However, to avoid the durability issues encountered in Dataverse[8], this architecture needs to be modular and easy to be maintained by external developers. In addition, mechanisms must be devised to ensure an adoption as wide and simple as possible. We propose an open-source implementation (Fig. 5) of the OSDL (the code is available in a git Repository[9]). We chose tools by considering the longevity, the openness of code and the use of REST APIs for interacting with them in a concern of simplicity, use, maintenance and interoperability. The entire architecture has been designed as containerized, using Docker containers. Automatic deployment tools have been developed to allow a one command deployment on most servers.

This technical solution is an adaptation of the architecture proposed in a previous paper [6] to the context of OSci. Data processes are managed by Apache Airflow. This tool enables workflows to be managed in the form of Directed Acyclic Diagrams. This makes it easy to track all operations in a processing chain. Other tools can be called up in the process data area by Apache Airflow for more specific processes. The management of raw data, transformed data and data processing pipelines is more detailed in this paper [6]. For security management, added security mechanisms are integrated into the REST API, providing

[8] https://dataverse.org/presentations/open-monolith-keeping-your-codebase-and-your-headaches-small.

[9] http://github.com/vincentnam/docker_datalake.

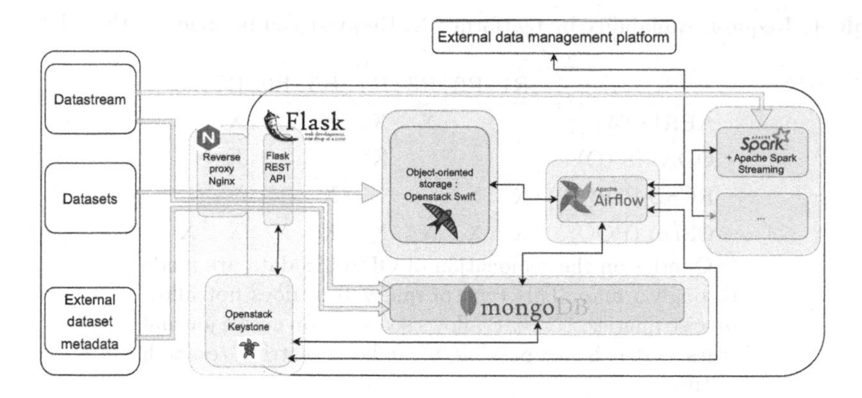

Fig. 5. OSDL: Technical architecture

a single access point to all OSDL resources. User management is implemented with Openstack Keystone, based on legal information in metadata.

Multi-model management is implemented in a MongoDB database. Models and matchings between models are stored independently in different collections. MongoDB allows to take advantage of noSQL flexibility on models and to handle JSON-LD for semantic metadata and linked data without functional redundancy with other data lake services. Moreover, the native interoperability by standardization with REST APIs eliminates pre-processing operations on received messages and reduces the load. The document format allows us to keep the list of matched keys for each model, so that match requests can be simple selections. From a technical point of view, the metadata management tool must be able to store and query metadata. Other needs are met by other data lake services (such as quality assurance pipelines with Airflow). Based on this, MongoDB is not a composite service (like OpenMetadata[10] or Opendatadiscovery[11], which relies on external database services and ElasticSearch) or based on a particular technology (like Apache Atlas[12], which relies on Hadoop). Since the solution meets our needs, this simplicity ensures lower maintenance and development costs. These aspects are essential to ensure that the solution is sustainable and that the problems encountered with monolithic solutions are not transposed to modular solutions. This is a major aspect of the solution's adoption in OSci.

5 Evaluation

We have set up an experimental implementation of a proof of concept of OSDL[13] (see Git repository for an in-depth technical view). The aim is to evaluate the time saved by the user, the ability of OSDL to adapt to user needs, and the

[10] https://open-metadata.org/.
[11] https://opendatadiscovery.org/.
[12] https://atlas.apache.org/.
[13] https://anonymous.4open.science/r/opendatalake_expe-6522.

Table 1. Request availability by platform ; X: Request can be made in this platform.

	R1	R2	R3	R4	R5	R6	R7	R8
AERIS (A)			X	X			X	
ODATIS (O)				X*			X	
RCSB PDB (P)	X	X			X	**		
OSDL (POC)	X	X	X	X	X	X	X	X

*: Queries on the geolocation of ODATIS data are made through a map. This type of query tool does not allow precise queries; **: PDB allows searches on entire journal titles, but it is not possible to make sub-string research in titles.

ability to implement a unified tool for cross-community access to research data with OSDL. We have selected metadata from 3 platforms from different domains and communities with different metadata models (AERIS[14], ODATIS[15], RCSB PDB[16]). Cross-platform communication mechanisms were simulated by integrating metadata into a single database, due to the lack of a method for scripting communications with all platforms. Matchings between models were integrated into a specific collection, and POC queries were sent to our database on a metadata path as well as on all equivalent metadata in the matching. We designed a set of 8 queries on metadata to specify search across multiple attributes, including natural phenomena and protein data (described in Github repository). We set up an experimental scenario with 11 users that were asked to execute the 8 queries on the 4 data retrieval platforms (AERIS, ODATIS, RCSB PDB and the OSDL proof of concept). We selected users so as to approximate the distribution proposed in a study on OSci (cf. Q12[17]) with 3 categories of comfort with open dataset search platforms that we assimilate as equivalent to those in the study: comfortable (≈20%), somewhat comfortable (≈40%) and not comfortable (≈40%). The users have not been trained to use the platforms (in order AERIS, ODATIS, RCSB PDB and finally the POC of OSDL). We measured the time required by each user to perform queries on each platform.

Table 2. Request mean time for each platform

Mean time for request (in second)	AERIS	ODATIS	RCSB PDB	OSDL
Without error	26.74	22.73	31.08	22.96
With error	27.84	21.67	34.32	22.93

[14] https://www.aeris-data.fr/.
[15] http://www.odatis-ocean.fr/.
[16] https://www.rcsb.org/.
[17] https://map.scnat.ch/en/activities/open_data_survey.

We have observed that OSDL enables a greater variety of metadata requests (see Table 1) thanks to the richness provided by multi-model management associated with matchings between these models. To manage the models of the 3 platforms, we had to set up two JSON documents weighing a total of 3.3Kb. This theoretically allows us to retrieve information from almost 200 different platforms present on Re3data having implemented ISO 19115 (the model implemented on the ODATIS platform).

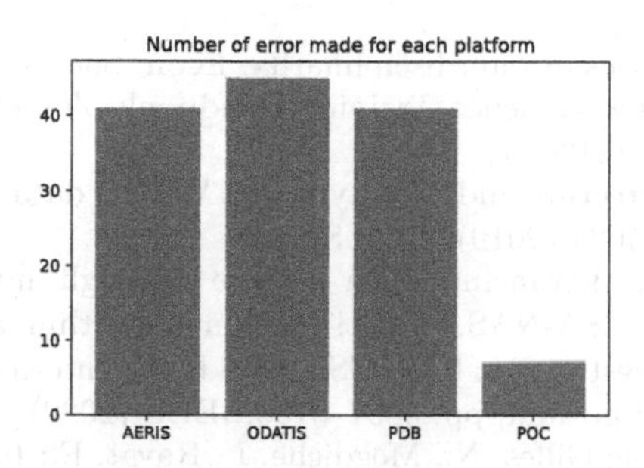

Fig. 6. Number of errors in user querying, for a total of 88 queries per platform

We have found that OSDL provides a data retrieval tool with average usage times at least equivalent to other platforms (see Table 2), while at the same time providing tools that are simpler to use and more user-friendly (see Fig. 6). We managed to integrate data from existing OSci platforms without the need to modify existing platforms. OSDL is interoperable with other pre-existing platforms without other specific requirements than a (meta)data acquisition mechanism.

6 Conclusion

The specificities brought to Big Data by Open Science (OSci) mean that new constraints must be taken into account, with the arrival of new assets. Interoperability and data security are 2 new components to be integrated into the very heart of Open Big Data solution design. We have proposed a data lake architecture adapted to OSci: the Open Science Data Lake (OSDL). Its novel architecture is based on recognized data lake architectures, enabling (i) local data integration by adding (ii) external data storage management for interoperation with existing OSci data management solutions, and (iii) security mechanisms at the very heart of the architecture to guard as far as possible against loss of data, trust or time in the research knowledge creation process. We carried out a POC which we evaluated through an experiment with users from the world of scientific research. This evaluation enabled us to show that OSDL saves time and broadens the scope of data retrieval by researchers. By design, OSDL's allows integration of metadata from other platforms without any additional workloads for the other platforms. With regard to the FAIR principles, our solution meets principles 1 and 3 of metadata Interoperability, which is a necessary but not sufficient step towards data interoperability, and all the layers of interoperability [5]. Further work will focus on adding mechanisms to enable scaling-up through automation of meta-metadata exchanges, by designing of a federation of OSci data management platforms.

References

1. Barry, A., et al.: Logics of interdisciplinarity. Econ. Soc. **37**(1), 20–49 (2008)
2. Bezjak, S., et al.: Open Science Training Handbook. Zenodo (2018). https://doi.org/10.5281/zenodo.1212496
3. Bird, I., et al.: Architecture and prototype of a WLCG data lake for HL-LHC. EPJ Web Confer. **214**, 04024 (2019). EDP Sciences (2019)
4. Bugbee, K., et al.: Advancing open science through innovative data system solutions: the joint ESA-NASA multi-mission algorithm and analysis platform (MAAP)'s data ecosystem. In: IGARSS 2020 - IEEE International Geoscience and Remote Sensing Symposium, pp. 3097–3100. IEEE (2020)
5. Dang, V.N., Aussenac-Gilles, N., Megdiche, I., Ravat, F.: Interoperability of open science metadata: what about the reality? In: Nurcan, S., Opdahl, A.L., Mouratidis, H., Tsohou, A. (eds.) Research Challenges in Information Science: Information Science and the Connected World. RCIS 2023. LNBIP, vol. 476. Springer, Cham (2023). https://doi.org/10.1007/978-3-031-33080-3_28
6. Dang, V.N., Zhao, Y., Megdiche, I., Ravat, F.: A zone-based data lake architecture for IoT, small and big data. In: 25th International Database Engineering & Applications Symposium (IDEAS 2021) (2021)
7. Di Maria, R., Dona, R.: Escape data lake. EPJ Web Confer. **251**, 02056 (2021). EDP Sciences (2021)
8. Juarez, J.D., Schick, M., Puechmaille, D., Stoicescu, M., Saulyak, B.: Destination earth data lake. Tech. rep, Copernicus Meetings (2023)
9. Peisert, S., et al.: Open science cyber risk profile (oscrp), version 1.3.3 (2017). https://doi.org/10.5281/zenodo.7268749
10. Ravat, F., Zhao, Y.: Data lakes: trends and perspectives. In: Hartmann, S., Küng, J., Chakravarthy, S., Anderst-Kotsis, G., Tjoa, A.M., Khalil, I. (eds.) DEXA 2019. LNCS, vol. 11706, pp. 304–313. Springer, Cham (2019). https://doi.org/10.1007/978-3-030-27615-7_23
11. Ren, P., et al.: MHDP: an efficient data lake platform for medical multi-source heterogeneous data. In: Xing, C., Fu, X., Zhang, Y., Zhang, G., Borjigin, C. (eds.) WISA 2021. LNCS, vol. 12999, pp. 727–738. Springer, Cham (2021). https://doi.org/10.1007/978-3-030-87571-8_63
12. Sansone, S.A., et al.: Fairsharing as a community approach to standards, repositories and policies. Nat. Biotechnol. **37**(4), 358–367 (2019)
13. Sarramia, D., Claude, A., Ogereau, F., Mezhoud, J., Mailhot, G.: CEBA: a data lake for data sharing and environmental monitoring. Sensors **22**(7), 2733 (2022)
14. Sawadogo, P., Darmont, J.: On data lake architectures and metadata management. J. Intell. Inf. Syst. **56**, 97–120 (2021)
15. Tanhua, T., et al.: Ocean fair data services. Front. Mar. Sci. **6**, 440 (2019)
16. Wang, Y., et al.: PGG.SV: a whole-genome-sequencing-based structural variant resource and data analysis platform. Nucleic Acids Res. **51**(D1), D1109–D1116 (2023)

17. Wilkinson, M.D., et al.: The fair guiding principles for scientific data management and stewardship. Sci. Data **3**(1), 1–9 (2016)
18. Zhou, C., et al.: GTDB: an integrated resource for glycosyltransferase sequences and annotations. Database **2020**, 219704410 (2020)

Be High on Emotion: Coping with Emotions and Emotional Intelligence when Querying Data

Sandro Bimonte[1], Patrick Marcel[2], and Stefano Rizzi[3](✉)

[1] TSCF, INRAE - University of Clermont Auvergne, Aubière, France
[2] LIFAT - University of Tours, Blois, France
[3] DISI - University of Bologna, Bologna, Italy
stefano.rizzi@unibo.it

Abstract. Emotional Intelligence (EI) is the capacity to use emotions to properly guide our actions. In this paper, we adopt the EI approach to explore the interplay between data, emotions, and actions, thus lying the foundations for an emotional approach to querying. The framework we propose relies on a four-layer model that describes (i) how emotions are connected to each other, (ii) which data may give rise to emotions, (iii) which emotions will be triggered in each user when seeing each piece of data, and (iv) which actions will be done as a consequence. The application scenario we propose for our framework is that of Business Intelligence, specifically, of a set of KPIs connected to the users' goals. To illustrate our proposal, we introduce a working example in the field of e-commerce and use the Datalog syntax to formalize it.

Keywords: Emotional Intelligence · Business Intelligence · KPIs

1 Introduction and Motivation

The emotions raised by our own needs accompany our everyday life and, like it or not, have a strong influence on our choices and decisions. *Emotional Intelligence* (EI) is the capacity to recognize, manage, and use emotions to *properly* guide our own reasoning and actions. In other words, it is the ability to identify the *right* emotions and adopt them to adapt our actions and behavior, according to the context and environments [9]. It has been proved that EI is associated to high levels of job performance, mental health, and decision making. Therefore, during recent years, EI has been increasingly developed in both professional and social contexts.

In the context of Information Systems, how to cope with emotions during exploratory data analyses has not been investigated yet. In this preliminary work, we adopt the EI approach to explore the interplay between data, emotions, and actions, thus lying the foundations for an emotional approach to data querying.

This work has been supported by the French National Research Agency under the IDEX-ISITE project, initiative 16-IDEX-0001 (CAP 20-25), and the project ANR-20-PCPA-0002.

A. Abelló et al. (Eds.): ADBIS 2023, CCIS 1850, pp. 82–91, 2023.
https://doi.org/10.1007/978-3-031-42941-5_8

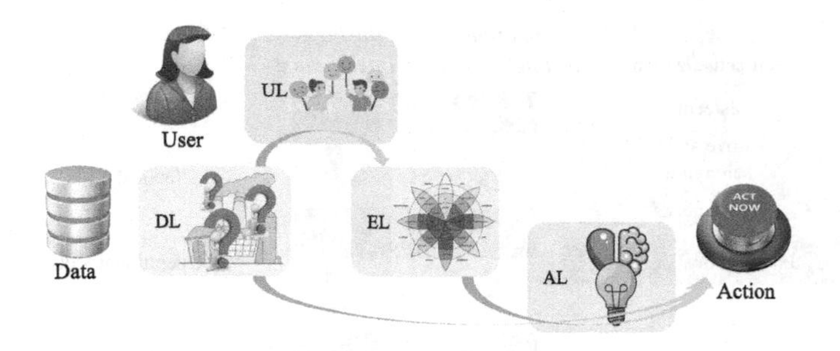

Fig. 1. The AUDE model

Indeed, seeing the results of a query on a dataset may trigger an emotion, which in turn can lead to a decision and/or an action.

Example 1. Consider a user, say Kathy, who has to decide whether to move or not from the city where she lives, say A, to a new one, say B. To make up her mind, Kathy compares the two cities from two points of view: the average salary and the crime rate. Checking some open databases available, Kathy discovers that —contrary to her belief— the average salary in B is higher than the one in A, which *surprises* her. On the other hand, the crime rate in B is quite high, which raises *fear* in her. Combining surprise and fear gives rise to *alarm*, an emotion that implies an active response. The action Kathy can take is to examine additional factors, e.g., the quality of life, before taking her final decision. Importantly, this action comes from a correct recognition of Kathy's emotions. Had surprise not been recognized, Kathy's behaviour would have been exclusively dominated by fear, in which case she would probably have been paralyzed and unable to make a decision.

The framework we propose to keep emotions into account when querying data relies on an *Action-User-Domain-Emotion (AUDE, in Latin "dare") model* that includes four layers: (i) an *emotion layer* (EL) that defines and connects emotions based on a classification drawn from affective science; (ii) a *domain layer* (DL) that characterizes the data that may trigger emotions via a set of queries; (iii) a *user layer* (UL) that connects the two by expressing the emotions of each single user as related to the query results (s)he sees; and (iv) an *action layer* (AL) that maps the emotions triggered by query results into actions consistently with the guidelines of EI. The overall picture is sketched in Fig. 1. In what follows, for the sake of readability, we adopt Datalog rules as a comprehensive way to formalize all the layers (following the syntax of Soufflé: https://souffle-lang.github.io/index.html).

Example 2. Considering again the example above, the EL expresses alarm as the combination of surprise and fear [16]; the DL includes the queries on the average salary and the crime rate in each city; the UL associates user's emotions to query results, e.g., low average salary makes Kathy angry, high crime rate makes her fearful; the AL associates emotions to possible actions, e.g., stand by

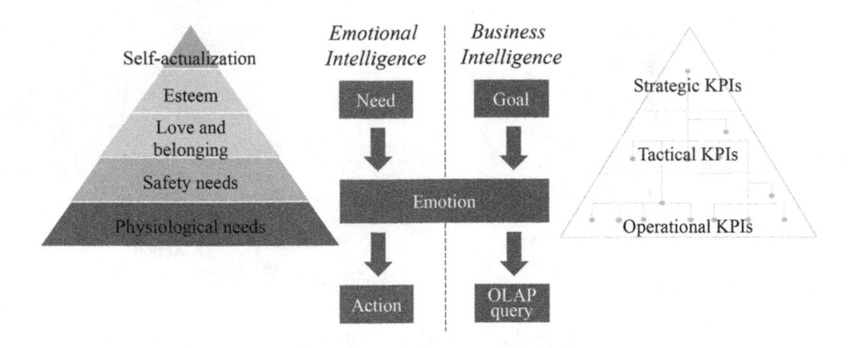

Fig. 2. Emotional intelligence vs. business intelligence

in case of alarm, or pursue your goal in case of excitement. A formalization of what stated above can be as follows:

$$\begin{aligned} evaluateQualityOfLife() &\leftarrow alarm() \\ alarm() &\leftarrow surprise(), fear() \\ surprise() &\leftarrow avgSalary(x, \text{'A'}), avgSalary(y, \text{'B'}), y > x \\ fear() &\leftarrow crimeRate(x, \text{'B'}), x > 70 \\ avgSalary(x, city) &\leftarrow employee(\_, \_, city), x = avg\ s : employee(\_, s, city) \end{aligned}$$

An interesting application scenario we envision for our emotional framework is that of Business Intelligence (BI), which gives computational support to users in exploring and analyzing data. Main citizens of BI systems are Data Warehouses (DWs) and OLAP tools, the latter enabling analyses of huge volumes of data stored in the former according to the multidimensional model. In order to support decision-makers in developing OLAP sessions when exploring data, several approaches for recommending OLAP queries have been devised (e.g., [4,15]). These approaches recommend new queries based on those formulated during the past or current sessions, usually relying on some query similarity metrics and in some cases considering the query interestingness [8]. However, none of these works take into account the complexity of the emotions that may arise during an analysis session.

In this scenario we propose to draw a parallel between the fulfillment of a goal, assessed by a *key performance indicator* (KPI), and the satisfaction of a need. As shown in Fig. 2, needs are often represented in the form of a pyramid [14]; on the other hand, KPIs can be distinguished into *strategic*, *tactical*, and *operational* [11], determining a pyramidal classification that closely resembles the one of needs. Satisfying a goal/need will likely trigger a positive emotion, while the failure of a goal/need may trigger a negative emotion. In turn, depending on the emotion, the user will take an action, which in the BI application will consist of asking a new OLAP query of checking a new KPI. Specifically, we suggest to apply our framework to BI as follows: (i) for the EL, we refer to Plutchik's *wheel of emotions* [16]; (ii) the DL is expressed as a tree of KPIs following the approach proposed in [13]; (iii) the UL specifies how (i.e., through which emotion in the EL) each user will react to the evaluation of each KPI in the DL; (iv) the AL follows the guidelines of EI to map each combination of a KPI in the DL and an emotion in the EL into an action, i.e., a new OLAP query to be formulated or a different KPI to be evaluated.

The paper outline is as follows. Section 2 discusses related work. Section 3 introduces the EI model and showcases it on a BI scenario. Section 4 concludes the paper and outlines research avenues.

2 Related Work

There have been several efforts in the literature to classify emotions in affective science so as to distinguish or contrast one emotion from another. Two main approaches have been pursued to this end. In *dimensional models*, such as the *vector model* and the *circumplex model*, emotions are conceptualized by defining where they lie in two or three dimensions, e.g., *valence* and *arousal* [18]. Conversely, in *discrete emotion theory*, people are thought to have an innate set of basic emotions, which can be distinguished by an individual's facial expression and biological processes, and are cross-culturally recognizable [2,5]. For instance, Ekman identified anger, disgust, fear, happiness, sadness, and surprise based on the corresponding facial expressions [6]. In Plutchik's *wheel of emotions*, eight emotions are distinguished, namely, joy, trust, fear, surprise, sadness, disgust, anger, and anticipation [16]. Emotions can have different degrees (e.g., serenity and ecstasy are, respectively, a mild and an intense form of joy), and have opposites (e.g., sadness is the opposite of joy). The author also theorized twenty-four primary, secondary, and tertiary dyads, i.e., *feelings* composed of two emotions.

As to KPIs, in [13] they are defined as metrics for evaluating goals in the context of BI and decision making. A KPI consists of an aggregate query, a target value to be achieved, and one or more thresholds that discriminate between good and bad performance. In turn, goals are frequently used in the design of BI systems to represent and engineer the users' requirements.

Finally, there has been many works on guiding users analyzing large datasets. Discovery-Driven Exploration (DDE) of data cubes [21], pioneered by Sunita Sarawagi [19,20,22] proposed techniques for interactively browsing interesting cells in a data cube. DDE was essentially motivated by explaining unexpected data (e.g., notable discrepancies) in the result of a cube query, to be explained by generalization (rolling-up, to check whether the discrepancy follows a general trend) [22] or by detailing (drilling-down, to understand what causes the discrepancy) [19]. DDE can be seen as a particular case of Exploratory Data Analysis (EDA), the general notoriously tedious task of interactively analyzing datasets to gain insights [10], that has attracted a lot of attention recently [3,7,12,23]. In any case, all those works assume that the exploration is limited to the use of dedicated primitives (e.g., classical OLAP or SQL operations), and even though user profiles incorporating preferences [1] or intentions [4] may be used, to the best of our knowledge, none of them account for the user's emotions when querying.

3 The AUDE Model

In this section we describe the different layers of our model, with specific reference to the BI application scenario outlined in Sect. 1.

3.1 Modeling Emotions

The EL models human emotions. To this end, in this paper we use Plutchik's model [16] (although any other model could be used instead). Plutchik formulated ten postulates among which there is a small number of basic, primary emotions (*joy*, *trust*, *fear*, *surprise*, *sadness*, *disgust*, *anger*, and *anticipation*); all other emotions are mixed or derivative states occurring as combinations, mixtures, or compounds of the primary emotions. Primary emotions are hypothetical constructs whose properties and characteristics can only be inferred from various kinds of evidence, and they can be conceptualized in terms of pairs of polar opposites. Primary emotions can be of three intensity degrees (mild, basic, and intense); for example, distraction is a mild form of surprise, and rage is an intense form of anger. Twenty-eight secondary emotions are derived by combining each primary emotion with the others. They can be modeled as Boolean queries, for instance:

$$love() \leftarrow trust(), joy() \tag{1}$$

$$curiosity() \leftarrow trust(), surprise() \tag{2}$$

3.2 Modeling Domains

The DL describes parts of the dataset susceptible to trigger emotions. In general, emotions are triggered by events of daily life, and correspond to the satisfaction (or not) of human *needs*. As already mentioned, the seminal work of Maslow [14] represents needs organized in a pyramid. On the bottom of the pyramid there are human basic survival needs, such as eating and sleeping, followed by safety needs, related for instance to health and employment. The two next layers include needs related to love and belonging (e.g., friendship and family) and to esteem (e.g., respect and freedom). On the top level are individual accomplishment needs related to achieving one's full potential, which also includes creative activities. Noticeably, it is recognized that having a gradual bottom-top satisfaction of needs is the only successful way.

In our BI application scenario, The DL contains the definitions of KPIs. A KPI is a numerical metric used to monitor the achievement of a business goal; it consists of a query (typically, an aggregate one), a target value to be achieved, and one or more thresholds that discriminate between good and bad performance [13]. To represent KPIs we adapt the Business Intelligence Model of [13], which connects KPIs (triangles) to goals (ovals); goals can be AND- or OR- decomposed into subgoals. If a goal is a conjunction of subgoals, then they all must succeed for the goal to succeed; if it is defined as a disjunction of subgoals, then at least one of them must succeed. In this work we will assume for simplicity that each (sub)goal is connected to exactly one KPI.

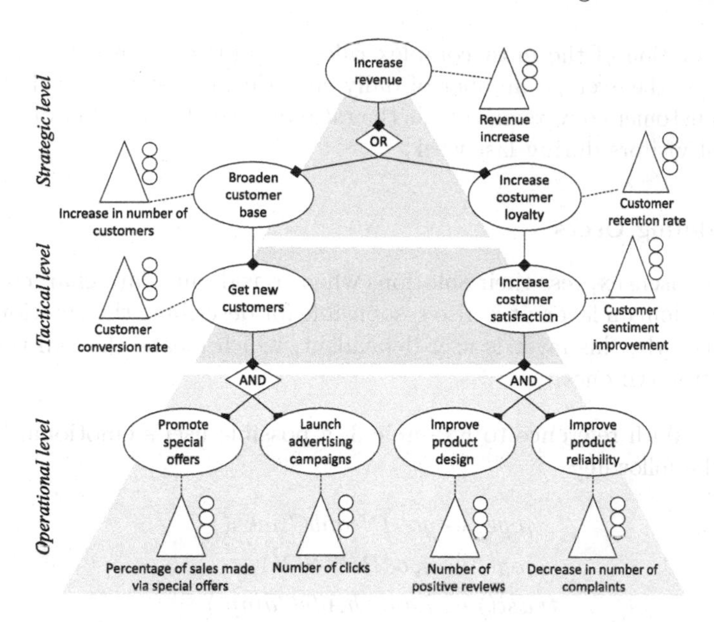

Fig. 3. The Business Intelligence Model for an e-commerce company

Example 3. Consider the Business Intelligence Model for an e-commerce company shown in Fig. 3, focused on the *Increase revenue* strategic goal. This goal can be achieved either by broadening the customer base or by increasing the customers' loyalty. At the tactical level, increasing the customers' loyalty is declined into increasing the customers' satisfaction; this, in turn, is obtained at the operational level by improving the design and reliability of products. Each goal is related to a KPI; for instance, the *customer retention rate* is used to check if the customers' loyalty has increased. Here are some examples of how this Business Intelligence Model can be coded on a simplified database schema that represents sales and clicks for the e-commerce site:

$$SALE(IPaddress, date, time, revenue, specialOfferYN)$$
$$CLICK(IPaddress, date, time)$$

Three goals are expressed as Boolean queries and the associated KPIs are expressed as aggregation queries (the rules for computing *dailyClicks*, *total Visitors*, *percSaleViaSpecOff*, and *totalCustomers* are omitted for brevity):

$$LaunchAdvCamp() \leftarrow numbClicks(x), x > 1000 \quad (3)$$
$$numbClicks(x) \leftarrow x = avg\ c : dailyClicks(date, c), lastWeekDays(date) \quad (4)$$
$$goodNumbClicks() \leftarrow numbClicks(x), x > 2000 \quad (5)$$
$$goodPPVSO() \leftarrow percSaleViaSpecOff(x), x > 0.3 \quad (6)$$
$$GetNewCustomers() \leftarrow custConvRate(x), x > 0.3 \quad (7)$$
$$custConvRate(s/c) \leftarrow totalVisitors(c), totalCustomers(s) \quad (8)$$

Some explanation of the most complex rules: rule (4) computes the number of clicks KPI as the average number of daily click during last week; rule (8) computes the customer conversion rate as the ratio between the number of customers and that of visitors during last week.

3.3 Modeling Users

The UL lets users express their emotions when considering data characterization made in the domain layer. It is also responsible for describing the emotions intensity. Importantly, this layer is user-dependent, which means that all thresholds are the user's own ones.

Example 4. With reference to Example 3, a possible user's emotional behavior could be the following:

$$joy() \leftarrow goodNumbClicks() \tag{9}$$
$$joy() \leftarrow goodPPVSO() \tag{10}$$
$$trust() \leftarrow LaunchAdvCamp() \tag{11}$$
$$trust() \leftarrow GetNewCustomers() \tag{12}$$

3.4 Modeling Actions

The AL models the actions to be taken when an emotion is triggered.

All emotions lead to one or another impulse to act (or not to act). According to Ronsenberg [17], the process involving emotions is: (i) recognize the emotion, (ii) identify the need beyond the emotion, (iii) trigger an adequate action. Taking into account the connections between emotions and actions discussed in [9], we propose some rules for emotion-based triggering of actions in the BI application scenario. Let k be the KPI which just triggered the emotion:

- *Joy.* The user has presumably satisfied the goal G associated to k, and her mental state prepares her to face new goals. According to Maslow's work, which postulates that basic needs should be fulfilled first, this should be done by climbing up the goal tree. Thus, the action triggered is to evaluate the KPI related to G', supergoal of G.
- *Anger.* The user has presumably found that the goal G associated to k is not satisfied. This emotion needs an immediate action. Again based on Maslow's work, the suggestion is to move towards more basic needs. Thus, the action triggered is to evaluate the KPI related to G', a subgoal of G.
- *Love.* The user is in a calm state which encourages further exploration. Thus, the action triggered is to evaluate the KPI related to a sibling goal of k.
- *Curiosity.* The user wants to know more, thus, the action triggered is a roll-up of the query associated to k.
- *Trust.* The user is confident with the data she just saw, thus, the action triggered is a drill-down of the query associated to k.

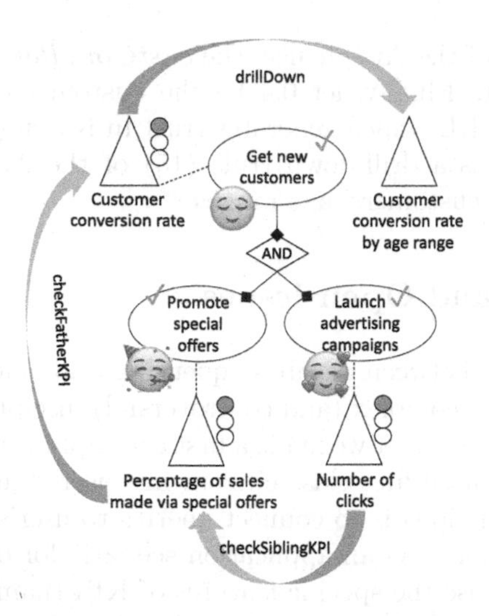

Fig. 4. A simple flow of emotions

- *Surprise.* The user has seen unexpected data and wants to get more information about the events; thus, the action triggered is a slice-and-dice of the query associated to k.
- *Sadness.* In this case, trying to satisfy other needs is not convenient. The action triggered is to move to a different goal tree or, if this is not available, stop the analytical session.

Example 5. Even the rules of the AL can be expressed in Datalog, for instance:

$$checkSiblingKPI() \leftarrow love() \tag{13}$$

$$checkFatherKPI() \leftarrow joy() \tag{14}$$

$$drillDown() \leftarrow trust() \tag{15}$$

3.5 The AUDE Model at Work

In this section we simulate how the different pieces fit together in a simple scenario. The reader can use Fig. 4 as a reference of KPIs, goals, emotions, and actions. Suppose Karen starts by checking the *numbClicks*(x) KPI, which evaluates to 2500 via rule (4) of the DL. This triggers rules (3) and (5), which makes both *LaunchAdvCamp*() and *goodNumbClicks*() true. In turn, this triggers rule (9) and (11) of the UL, so the expected emotions of Karen are joy and trust. These two emotions, together, give rise to love (via rule (1) of the EL). The suggested action for love is to evaluate a sibling KPI (rule (13) of the AL), hence, the *percSaleViaSpecOff*(x) KPI is evaluated. Now, let 0.35 be the value of the *percSaleViaSpecOff*(x) KPI; this triggers rules (6) of the DL and (10) of the UL, which raises joy in Karen. The suggested action for joy is to evaluate the father KPI (rule (14) of the AL), hence, the *custConvRate*(x) KPI is evaluated via rule (8) of the DL. Finally, let 0.4 be the customer conversion rate. This triggers rule (7) of the DL, which generates trust in Karen (rule (12) of the UL). The suggested action is a drill-down (rule (15) of the AL), so Karen will for instance drill-down to customers' age ranges.

4 Conclusions and Open Issues

Studying the interplay between database querying and emotions is challenging, even because it involves complex (and controversial) disciplines such as psychology and sociology. In this paper we made a first attempt in this direction by defining a layered model whose first-class citizens are users, queries, emotions, and actions. The underlying idea is to connect queries to user's emotions first, then user's emotions to actions. As an application scenario for our approach we proposed BI, mainly because the specific features of KPI (namely, their connection to goals, their tree-like structure, and their threshold-based definition) allowed us to establish an intuitive connection with emotions and actions. Remarkably, by using Datalog for expressing the model, we delivered a uniform formalization for all layers.

Clearly, the path to efficient and effective emotion-aware querying is still very long. There is a lot of questions that need an answer, among these:

- In the BI application scenario, the action triggered by an emotion may be an OLAP query instead of a KPI, in which case target values are not defined. Then how to connect values to emotions?
- The BI scenario provides a clear set of possible actions (either evaluate a KPI or apply an OLAP operator to formulate a query). In a more general setting, like the one described in Example 1, would actions still correspond to queries? If so, how to connect each emotion to a query?
- Different classifications of emotions have been proposed, some relying on a few primary emotions, some also including a large set of emotion nuances and combinations. While here we chose a very detailed classification, this may be too detailed to be used in practice —and not all emotions will be suited to any domain. How to pick a representative set of emotions to be used given a specific domain?
- Conflicts are part of the human nature, so contrasting emotions may rise in a user. The classification of emotions we adopted here deals with this by defining secondary emotions such as *confusion*, which is a mix of surprise and anticipation; however, not all classifications do the same. Adopting a classification that does not explicitly cope with contrasting emotions would require checking the emotional model for conflicts.
- While Datalog offers a uniform and elegant formalization, it may not be an efficient solution for implementation. Which architecture should be adopted to implement the approach?

References

1. Chomicki, J.: Preference formulas in relational queries. ACM Trans. Database Syst. **28**(4), 427–466 (2003)
2. Colombetti, G.: From affect programs to dynamical discrete emotions. Philos. Psychol. **22**, 407–425 (2009)
3. Ding, R., Han, S., Xu, Y., Zhang, H., Zhang, D.: QuickInsights: quick and automatic discovery of insights from multi-dimensional data. In: Proceedings SIGMOD, pp. 317–332 (2019)
4. Drushku, K., Aligon, J., Labroche, N., Marcel, P., Peralta, V.: Interest-based recommendations for business intelligence users. Inf. Syst. **86**, 79–93 (2019)
5. Ekman, P.: An argument for basic emotions. Cogn. Emot. **6**(3–4), 169–200 (1992)
6. Ekman, P.: Facial expressions of emotion: new findings, new questions. Psychol. Sci. **3**(1), 34–38 (1992)
7. El, O.B., Milo, T., Somech, A.: Automatically generating data exploration sessions using deep reinforcement learning. In: Proceedings SIGMOD, pp. 1527–1537 (2020)
8. Gkitsakis, D., Kaloudis, S., Mouselli, E., Peralta, V., Marcel, P., Vassiliadis, P.: Assessment methods for the interestingness of cube queries. In: Proceedings DOLAP, pp. 13–22 (2023)
9. Goleman, D.: Emotional intelligence. Bantam Books (2006)
10. Idreos, S., Papaemmanouil, O., Chaudhuri, S.: Overview of data exploration techniques. In: Proceedings SIGMOD, pp. 277–281 (2015)
11. Li, Y., O'Donnell, J., García-Castro, R., Vega-Sánchez, S.: Identifying stakeholders and key performance indicators for district and building energy performance analysis. Energy Build. **155**, 1–15 (2017)
12. Ma, P., Ding, R., Han, S., Zhang, D.: MetaInsight: automatic discovery of structured knowledge for exploratory data analysis. In: Proceedings SIGMOD, pp. 1262–1274 (2021)
13. Maté, A., Trujillo, J., Mylopoulos, J.: Specification and derivation of key performance indicators for business analytics: a semantic approach. Data Knowl. Eng. **108**, 30–49 (2017)
14. McLeod, S.: Maslow's hierarchy of needs. Simply Psychol. **1**, 1–18 (2007)
15. Negre, E., Ravat, F., Teste, O.: OLAP queries context-aware recommender system. In: Proceedings DEXA, pp. 127–137 (2018)
16. Plutchik, R.: The nature of emotions: human emotions have deep evolutionary roots, a fact that may explain their complexity and provide tools for clinical practice. Am. Sci. **89**(4), 344–350 (2001)
17. Rosenberg, M.B., Chopra, D.: Nonviolent communication: a language of life: life-changing tools for healthy relationships. PuddleDancer Press (2015)
18. Rubin, D.C., Talarico, J.M.: A comparison of dimensional models of emotion: evidence from emotions, prototypical events, autobiographical memories, and words. Memory **17**(8), 802–808 (2009)
19. Sarawagi, S.: Explaining differences in multidimensional aggregates. In: Proceedings VLDB, pp. 42–53 (1999)
20. Sarawagi, S.: User-adaptive exploration of multidimensional data. In: Proceedings VLDB, pp. 307–316 (2000)
21. Sarawagi, S., Agrawal, R., Megiddo, N.: Discovery-driven exploration of OLAP data cubes. In: Proceedings EDBT, pp. 168–182 (1998)
22. Sathe, G., Sarawagi, S.: Intelligent rollups in multidimensional OLAP data. In: Proceedings VLDB, pp. 531–540 (2001)
23. Tang, B., Han, S., Yiu, M.L., Ding, R., Zhang, D.: Extracting top-k insights from multi-dimensional data. In: Proceedings SIGMOD, pp. 1509–1524 (2017)

Fairness in Data Management

Exploring Biases for Privacy-Preserving Phonetic Matching

Alexandros Karakasidis(✉) and Georgia Koloniari

Department of Applied Informatics, University of Macedonia, Thessaloniki, Greece
{a.karakasidis,gkoloniari}@uom.edu.gr

Abstract. Soundex has been proposed as an alternative for Privacy-Preserving Record Linkage featuring significant performance in terms of result quality and efficiency, without, however, the corresponding attention to evaluate its performance with respect to fairness. In this paper, we focus on race and gender biases and examine the behavior of Soundex using a real world dataset and Apache Spark for processing. We compare these results with two other well known phonetic algorithms, namely NYSIIS and Metaphone. Our evaluation indicates that no biases are induced with respect to gender. On the other hand, regarding race, biases have been observed for all examined algorithms.

Keywords: Privacy · Fairness · Record Linkage

1 Introduction

In today's globalized interconnected reality, databases within organizations and enterprises hold data of diverse people irrespective of their characteristics such as age, gender or race. Often, for a variety of purposes, ranging from enterprise analytics to healthcare and security applications, such datasets have to be linked so that new insights are discovered. Nevertheless, this process is not trivial. Data may not be freely exchanged, as such an action would violate the privacy of each individual described by these data. To this end, Privacy-Preserving Record Linkage (PPRL) has arisen.

The goal of PPRL is to detect data pertaining to the same real world entities, i.e. individuals, across different databases while preserving their privacy through de-identification. As common unique identifiers are unavailable, combinations of attributes are used to form a composite key (e.g. name, surname, address), called quasi-identifiers. As these may exhibit low quality due to typos, approximate matching methods are required. Many approaches have been proposed offering a variety of qualitative and performance characteristics [2]. One of the most efficient approaches relies on the use of phonetic codes [4–6], managing to provide high matching quality and performance even for big data.

As phonetic codes have been designed to perform approximate matching by mapping each name to a code based on its pronunciation, they are language-oriented, with the most well known ones, such as Soundex [11], being based on

A. Abelló et al. (Eds.): ADBIS 2023, CCIS 1850, pp. 95–105, 2023.
https://doi.org/10.1007/978-3-031-42941-5_9

the English language. As a result, it is expected that they will exhibit bias when dealing with diverse data. Bias in this case refers to increased or decreased errors when approximately matching records of individuals belonging to a particular gender or race group. As in a privacy-preserving context plain text data are unavailable for a practitioner to assess the appropriateness of such algorithms, in this paper, we try to answer the following research questions:

RQ1: Does Soundex exhibit bias with names from individuals belonging to different races and genders in a privacy-preserving setup, and in which cases?

RQ2: How does bias exhibited by Soundex compare against other phonetic algorithms in a privacy-preserving setup, and would their use be preferable?

To answer these research questions while taking into account that, to the best of our knowledge, there has been no evaluation of the behavior of these algorithms with respect to gender and race in terms of biases, we provide an empirical survey of three of the most common phonetic encoding algorithms with respect to race and gender in a privacy-preserving setup. In particular, using a PPRL protocol designed for Soundex, we measure bias by comparing its performance with respect to gender and race. Then we explore the performance of alternative phonetic encoding methods, namely NYSIIS and Metaphone, in the same setup. In particular, we measure matching quality using precision and recall and study in detail the behavior of mismatches per gender and race, so as to design our future steps that will mitigate any biases when applying phonetic algorithms in a privacy-preserving big data setup.

The rest of this paper is organized as follows. Section 2 contains related works. Section 3 provides the required background and describes our methodology and Sect. 4 holds our experimental evaluation. We conclude in Sect. 5.

2 Related Work

In the last years, fairness and bias in machine learning [9] are in the spotlight [12,14]. In terms of record linkage, there is some work towards identifying and addressing bias. In [7], authors identify bias in string matching operators when applied to names, while authors in [10] focus on evaluating demographic bias in word embeddings for named entity recognition. Efthymiou et al. [3] propose the use of a greedy algorithm using fairness criteria based on equal matching decisions. More recently, Makri et al. [8] explore biases in svm-based record linkage and recommended the use of ensembles to improve performance.

Bias is also considered in the field of PPRL posing an additional constraint, as no additional data from the datasets to be linked should be revealed, not accounting for the resulting actually matching records. Vatsalan et al. [16] explore different forms of fairness-bias in PPRL when using logistic regression. Wu et al. [17] study new notions of differential privacy and their applications on Bloom filters, under fairness-bias constraints in order to achieve better matching performance. In this work, we investigate the existence of bias in phonetic-based PPRL so as to assess whether further measures need to be considered.

3 Methodology

In this section, we formally define the problem we address and lay out the methodology we use in our evaluation.

3.1 Problem Formulation

Without loss of generality, let us consider two data sources, Alice (A) and Bob (B), who respectively hold r^A and r^B records each. *PPRL* is defined as the problem of identifying (linking) all pairs of r^A and r^B records that refer to the same real world entity, so that no more information is disclosed to either A, B or any third party involved in the process besides the identifiers of the linked r^As and r^Bs, which are usually encrypted.

As no common identifiers exist, a set of m common fields, called *matching fields*, are selected to form a composite key that is used for matching. Moreover, as data is often noisy, exact matching is not sufficient and must rely on a similarity or distance function instead. Let $\mathcal{D}$ be the domain of each matching attribute, $sim_j : \mathcal{D} \times \mathcal{D} \rightarrow [0..1]$ a similarity function and $t_j > 0$ a user-defined threshold. Given records r_i^A and r_i^B with matching attributes $r_i.1, \ldots, r_i.m$, the matching function $M : \mathcal{D} \times \mathcal{D} \rightarrow \{0, 1\}$ is defined as:

$$M(r_i^A, r_i^B) = \begin{cases} 1, & \text{iff } sim_j(r_i^A.j, r_i^B.j) \geq t_j, \forall j \in [1, m] \\ 0, & \text{otherwise.} \end{cases} \tag{1}$$

If $M(r_i^A, r_i^B) = 1$, then the pair (r_i^A, r_i^B) is considered a match.

3.2 Phonetic Codes for String Matching

Soundex [11] encodes names into codes which maintain the name's original first letter followed by three numbers. Longer codes are stripped off, while shorter ones are padded with zeros. E.g., "Stephen" maps to S315, while "Hawking" maps to H252. NYSIIS [15], the second most popular phonetic encoding algorithm employed for data matching [1], only produces letters in its encoding which is limited to six letters. For the aforementioned names, NYSIIS respectively produces STAFAN and HACANG. Metaphone [13] has been designed to address issues of the previous two algorithms. Producing arbitrary length strings, applying Metaphone to the same names results to "STFN" and "HKNK".

Using a phonetic code may be formalized via Eq. 1 as follows. The similarity function sim_j, instead of the attribute values, will have as input their corresponding phonetic codes, returning 1 upon match (identical codes) or 0, otherwise. As such, the matching threshold t_j is set to 1, so that the two codes compared are considered as matches only when they are identical.

As all these phonetic codes provide approximate representations of the respective encoded string, they are tolerant against typos and misspellings. Furthermore, this generalization behavior that they exhibit, allowing more than one strings to map to the same encoding, make them appropriate for use for privacy-preserving matching, as generalization is one of the fundamental methods for enhancing privacy [2]. However, apart from their inherent privacy characteristics, this generalization behavior also increases the capacity of false positives being introduced in the matching process.

3.3 Protocol Operation

For our evaluation, we consider the PPRL protocol of [6] originally designed for big data environments using Soundex and Apache Spark [18] which eliminates information leak based on the primitives of Secure Multiparty Computation. Matching takes place at a third party called Carol. Both data holders, Alice and Bob, and Carol are Honest but Curious without Collusion, trying to learn as much information as possible without collaborating or deviating from the protocol, while none of them has access to all available information.

Initially, Alice and Bob agree on a set of common fields in their schemas to be used as matching fields. Each party separately prepares its own data. First, each of the matching fields is converted to its phonetic equivalent. Then, all of these codes are concatenated in the order agreed by Alice and Bob and a padding text is appended to form a single string encoded through a secure hash function e.g. SHA256. This is performed by a series of map operations in the Apache Spark environment used to process our data.

Afterwards, Alice and Bob inject random noise before transmitting their datasets to Carol, so as to create fuzziness as an additional privacy enhancing measure. Fake records are generated by exploiting a dictionary. For each fake field to be produced, a word is randomly selected from the dictionary and encoded using the phonetic algorithm. The resulting records are merged with the real ones, through Spark's Union operation. Then, Alice and Bob securely deliver the resulting datasets to Carol who joins them. Each party's records participating in the result are, then, securely delivered to their owner, who independently joins the received records with their actual records to remove noise and eventually securely exchange matching records with each other.

For our empirical study, we will alter this protocol, so that it allows the use of either Metaphone or NYSIIS as the string encoding method instead of Soundex.

4 Empirical Assessment

Our empirical assessment consists of two aspects. First, we study the matching performance of each approach. Next, we dive deeper, and explore the nature of mismatches with respect to gender and race so as to gain better insights that will help us design measures that mitigate any biases. We used Precision which is the fraction of the correctly linked record pairs among the linked record pairs and

Table 1. Records/Race.

Race	Count	Share
A: ASIAN	67455	1.24%
B: BLACK / AFRICAN AMERIC	1196045	22.00%
I: AMERIC. INDIAN or ALASKA NAT	39623	0.73%
M: TWO or MORE RACES	35678	0.66%
P: NAT. HAWAIIAN / PACIFIC ISLANDER	337	0.01%
W: WHITE	4096616	75.36%
Total	5435754	100.00%

Table 2. Records/Gender.

Gender	Count	Share
F	3014612	55.46%
M	2421142	44.54%
All	5435754	100.00%

Recall defined as the fraction of the correctly linked record pairs divided by the number of correctly linkable record pairs ($Precision = \frac{\text{TP}}{TP+FP}$ and $Recall = \frac{\text{TP}}{TP+FN}$.). To explore the methods' biases, we measure and compare Precision and Recall per race and per gender, as well as against average Precision and Recall using macro-averaging so as not to be affected by data skewness.

4.1 Experimental Setup

We ran our experiments on a cluster hosted in the cloud by the IaaS service of GRNET[1] featuring 11 virtual machines powered by Ubuntu 16.04 LTS with 16 GBs of RAM and 8-core Xeon CPUs at 2.3 GHz. We employed Apache PySpark 3.1.1, a fast and general engine for large-scale data processing [18], to process our data residing in Apache Hadoop 2 HDFS, due to its high throughput and the Jellyfish library for phonetic encoding. One of the virtual machines was the dedicated Master, while the rest operated as Workers. The driver's memory was 12 GBs, while we allocated 2 GBs of memory and one core to each executor, totaling to 70 executors. Finally, we fixed the number of data partitions to 100, so as to exploit the available executors. For all results, we report the average of 3 executions of each setup.

We used real world data from North Carolina's publicly available voter's database. We determined three matching fields, namely: 'last_name', 'first_name', 'midl_name', which we assume may convey race and gender information. We cleaned the database by removing records with missing or undefined gender or race. Then, we deduplicated the resulting records using the matching fields as a candidate key resulting in 5435754 rows comprising Alice's dataset. As our focus is on approximate matching so as to simulate real-world conditions with errors in data, Bob's dataset was derived from Alice's by corrupting his records by randomly choosing a field of each row and randomly performing an edit operation, so that join operations between Alice's and Bob's data yield zero matching records. For noise generation, we created fake records uniformly

[1] https://okeanos-knossos.grnet.gr/home/.

sampling words from a dataset[2] of 10000 English words and then encoding them according to the protocol.

The characteristics of our dataset with respect to gender and race are illustrated in Table 1 which maintains the number of rows per race and their corresponding percentage within the overall dataset. It is easy to see that the vast majority belongs to the White race and then the Black race follows. As such, it would be more than interesting to examine the behavior of all the underrepresented minorities. Next, Table 2 holds the number of records per gender. As we can see, there are more Females than Males within the dataset. We have maintained this group imbalance so as to simulate real world conditions with a practitioner not having knowledge of datasets data distributions.

4.2 Experimental Results

Let us now discuss the results of our experimental assessment, starting first with the matching quality of the three phonetic algorithms per gender and race, and then going into more detail to investigate the types of mismatches.

Matching Quality per Race. We first report Precision and Recall per race, as illustrated in Fig. 1a and Fig. 1b respectively. With regards to Recall, Soundex, (blue line), exhibits the best performance, Metaphone (green line) follows and NYSIIS (red line) exhibits the worst performance. Soundex is generally balanced for all races. Its average Recall is 0.61, while there is slightly higher performance for individuals belonging to two or more races (M) with 0.623. On the other hand, there is also a minor underperformance with respect to Asians (A) with 0.604 and Pacific Islanders (P) with 0.596. In general, however, we can say that Soundex is not particularly biased in favor, or against, a particular race group. Metaphone also exhibits more or less balanced behavior except for the case of Asians, again. While average Recall is 0.473, and for all other race groups Recall is between 0.47 and 0.484, for Asians it slightly drops to 0.447. Finally, NYSIIS has the lowest average performance with 0.224. In contrast to the previous two cases, members of the Asian group do not perform worse. On the contrary, members of the White (W) group have slightly lower performance, equal to 0.211.

Results get more interesting in terms of Precision (Fig. 1a). First of all, as expected, methods with lower Recall now exhibit higher Precision and vice-versa. Soundex exhibits the lowest Precision that may be due to the fact that the number of distinct encodings required for this dataset supersedes its capacity.

Examining biases, we observe that for all algorithms the biases introduced are more prominent in contrast to the case of Recall. For Soundex, with an average Precision of 0.688, there is evident bias in favor of the Asian group first, reaching 0.782 and the Pacific group, at 0.724, next. On the other hand, the most widely negatively affected group is the White race group with a performance of 0.603, with the Indian (I) group following with 0.645 and the Multiracial

[2] Available at: https://www.mit.edu/~ecprice/wordlist.10000.

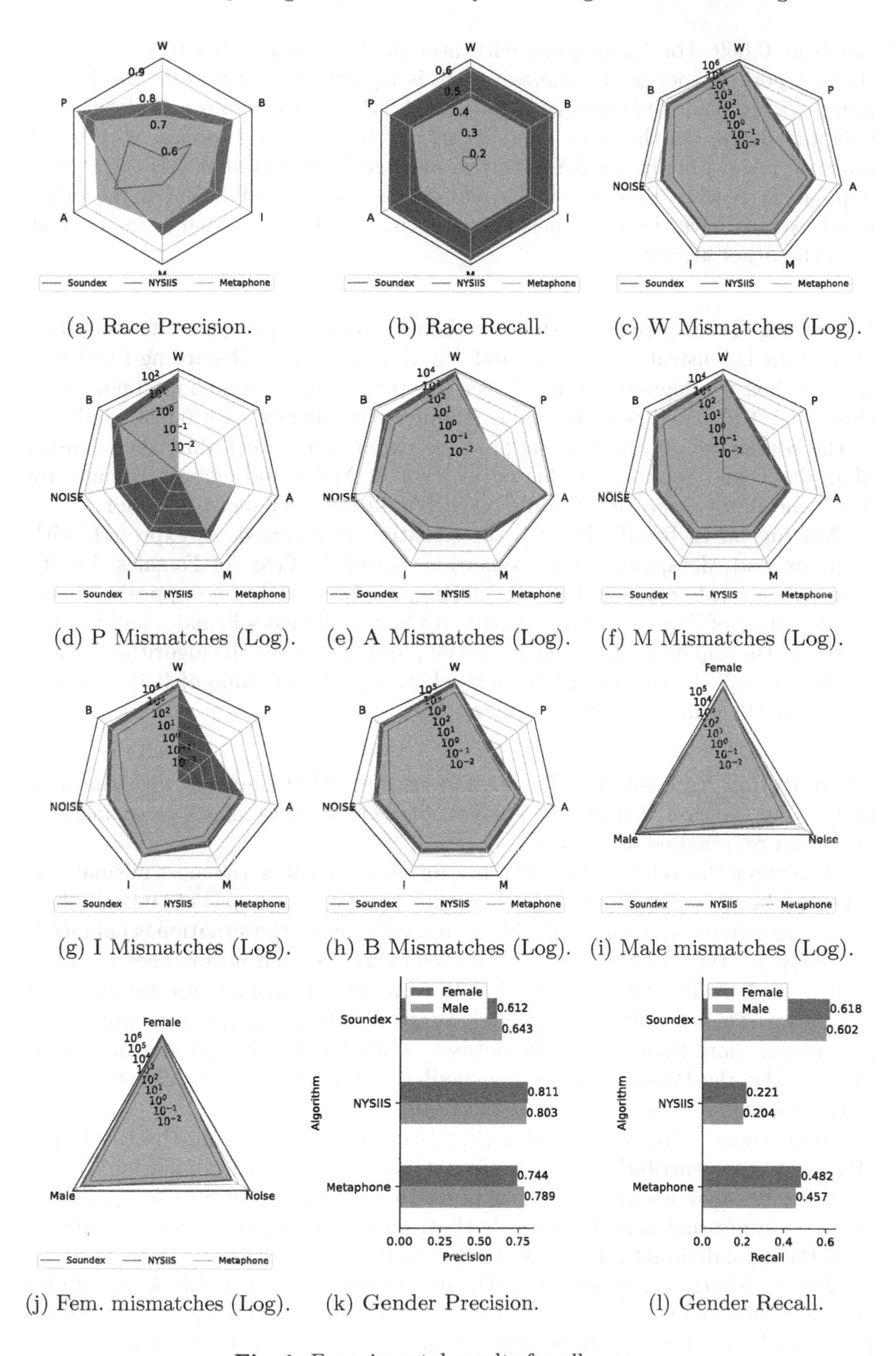

(a) Race Precision. (b) Race Recall. (c) W Mismatches (Log).

(d) P Mismatches (Log). (e) A Mismatches (Log). (f) M Mismatches (Log).

(g) I Mismatches (Log). (h) B Mismatches (Log). (i) Male mismatches (Log).

(j) Fem. mismatches (Log). (k) Gender Precision. (l) Gender Recall.

Fig. 1. Experimental results for all cases.

group at 0.669. For Metaphone, with average Precision at 0.810, we may see that its performance is also skewed; there is highly positive bias for the Pacific group with 0.862, while the Asian and Black (B) groups also overperform with 0.85 and 0.828. On the other hand, Whites with 0.744 and Indians with 0.773 underperform. Finally, for NYSIIS with average Precision at 0.837, the Pacific population reaches Precision 0.929 and the Black group 0.868. On the other hand, Asians have the worst performance with 0.766, a fact that is in contrast with the other algorithms.

Matching Quality per Gender. Precision and Recall per gender for the three algorithms is illustrated in Fig. 1k and Fig. 1l, respectively. Regarding Precision, NYSIIS has the highest average Precision, 0.807, with Metaphone being very close with 0.767 and Soundex being significantly inferior with 0.628, similarly to the results we observed in the previous experiment. Regarding their biases, Metaphone and Soundex are biased in favor of Males and against Females by 0.045 and 0.031 respectively, while NYSIIS exhibits a balanced behavior.

Moving on to Recall (Fig. 1l), the situation is reversed, as expected, with Soundex and Metaphone being somewhat biased in favor of Females, but to a smaller extent, compared to the difference observed in the opposite direction in terms of Precision. Interestingly, NYSIIS still favors Females and further increases the gap with Males. This may be partially due to the algorithm's average low Recall. In this case, Soundex achieves a Recall value of 0.61, NYSIIS reaches 0.213 while Metaphone achieves 0.47.

Mismatches Assessment. Let us now explore what happens with the mismatches in the case of Race. We analyze mismatches between all races and also fake data represented as Noise (N).

Analyzing the White group (Fig. 1c), for all algorithms, the most mismatches occur within the same group and then with the Black group. Less mismatches, for all algorithms are with the Pacific group. Otherwise, the situation is balanced. Similarly, for the Black group (Fig. 1h), again we see high mismatches with the Whites and within the group itself, while the fewest mismatches are with the (P) group. These results are expected, as the White group is the largest group comprising more than 75% of the dataset, while the Black group is the second in size. Also the Pacific group is the smallest group comprising only the 0.01% of the dataset. The behavior of all algorithms is consistent.

Next, there is the Indian group (Fig. 1g). In this case, again the Black and White groups contribute to most mismatches due to their population. Then, there are mismatches within the group, a situation consistent for all algorithms. What is interesting here, however, is that, only Soundex produces mismatches with Pacific individuals. This may be due to its shorter representation.

For the Multiracial group (Fig. 1f), the situation is similar. Black and White groups inhibit the mismatches due to their size. On the other hand, both Metaphone and Soundex seem to produce mismatches with the Pacific group, while

NYSIIS does not. Interestingly, mismatches within the same group are not higher, as it occurred with the previous cases.

For the Asian group (Fig. 1e), we can see that the number of mismatches occurred within the Asian group rivals the number of mismatches with the White group, although the latter is significantly larger. Also, there is increased false matching with the Black group. Less mismatches occur again with the Pacific group, while Noise, also has a significant contribution here.

Last but not least (Fig. 1d), there is the Pacific group. Here, the situation is quite complicated. First of all, for Soundex, most false matches, as expected occur with the Black and White groups then the Noise follows. Interestingly this also happens with Metaphone, but not in the case of NYSIIS. In this case, Blacks contribute to false matches more than Whites, while Noise does not affect the process. What is more interesting, however, is that there are no mismatches within the same group for any of the algorithms. Furthermore, there are false matches with the Indian group, but only for Soundex. Soundex and Metaphone also lead to false matches with both Asian and Multiracial groups. However, this does not hold for NYSIIS which does not incur any false matches.

Finally, let us examine what happens in terms of number of mismatches with respect to gender, for each algorithm. We can see that, for the case of Females, as illustrated in Fig. 1j, the most mismatches occur with Soundex, something expected, as it features the lowest Precision of the three algorithms. Here, the largest number of mismatches are with records of other Females, then with Males and eventually with the added Noise. The situation is similar for both Metaphone and NYSIIS, although absolute values are lower. For the case of Males (Fig. 1i), again Soundex exhibits the highest number of mismatches. Again most mismatches are with other Males. The same situation holds with the other two phonetic codes, with NYSIIS exhibiting a more balanced behavior. Overall, noise seems to have the smallest impact, while mismatches occur per gender.

Summary. Overall we have the following results. For RQ1, Soundex is not particularly biased with respect to gender. Regarding race, now, while Recall does not diverge, for Precision, there is evident negative bias mostly against names of the White and then of the Indian group, while there are positive biases for the Asian and the Pacific group.

For RQ2, regarding gender, Metaphone behaves as Soundex, while NYSIIS is even more balanced. For Race, Metaphone behaves again as Soundex, favoring the Pacific and Asian groups first, while disfavoring the White group, again in terms of Precision. On the other hand, the Asian group performs worse in terms of Recall, which is not evident with Soundex, but not the Pacific group. NYSIIS, without significant differences for Recall, for Precision, significantly favors the Pacific group and then the Black group, while the Asian group is disfavored.

Finally, mismatches seem to be mostly influenced by the population of the different classes, while the fact that in most cases most mismatches are within the same group is encouraging as it seems that despite their weakness, phonetic encodings partly manage to correctly map the diverse name pronunciations.

5 Conclusions

In this paper, we explored the behavior of phonetic codes for privacy-preserving record linkage in a big data setup. With respect to RQ1, we discerned cases of bias for Soundex and the other phonetic algorithms as well, regarding mostly race. With respect to RQ2, Metaphone exhibits superior precision but worst recall compared to Soundex and may be a promising alternative, though it also exhibits similar biases. Based on these observations, our future work will focus on developing methods for mitigating these biases.

References

1. Christen, P.: Data Matching - Concepts and Techniques for Record Linkage, Entity Resolution, and Duplicate Detection. Springer (2012). ISBN: 978-3-642-31163-5
2. Christen, P., Ranbaduge, T., Schnell, R.: Linking Sensitive Data - Methods and Techniques for Practical Privacy-Preserving Information Sharing. Springer (2020)
3. Efthymiou, V., Stefanidis, K., Pitoura, E., Christophides, V.: Fairer: entity resolution with fairness constraints. In: CIKM, pp. 3004–3008. ACM (2021)
4. Gkoulalas-Divanis, A., Vatsalan, D., Karapiperis, D., Kantarcioglu, M.: Modern privacy-preserving record linkage techniques: an overview. IEEE Trans. Inf. Forensics Secur. **16**, 4966–4987 (2021)
5. Karakasidis, A., Koloniari, G.: Efficient privacy preserving record linkage at scale using Apache Spark. In: 2022 IEEE International Conference on Big Data (Big Data), pp. 402–407. IEEE (2022)
6. Karakasidis, A., Koloniari, G.: More sparking soundex-based privacy-preserving record linkage. In: Foschini, L., Kontogiannis, S. (eds.) International Symposium on Algorithmic Aspects of Cloud Computing, pp. 73–93. Springer, Cham (2022). https://doi.org/10.1007/978-3-031-33437-5_5
7. Karakasidis, A., Pitoura, E.: Identifying bias in name matching tasks. In: EDBT, pp. 626–629 (2019)
8. Makri, C., Karakasidis, A., Pitoura, E.: Towards a more accurate and fair SVM-based record linkage. In: 2022 IEEE International Conference on Big Data (Big Data), pp. 4691–4699. IEEE (2022)
9. Mehrabi, N., Morstatter, F., Saxena, N., Lerman, K., Galstyan, A.: A survey on bias and fairness in machine learning. ACM Comput. Surv. (CSUR) **54**(6), 1–35 (2021)
10. Mishra, S., He, S., Belli, L.: Assessing demographic bias in named entity recognition. CoRR abs/2008.03415 (2020)
11. Odell, M., Russell, R.C.: The soundex coding system. US Patents 1261167 (1918)
12. Pessach, D., Shmueli, E.: A review on fairness in machine learning. ACM Comput. Surv. (CSUR) **55**(3), 1–44 (2022)
13. Philips, L.: Hanging on the metaphone. Comput. Lang. **7**(12), December 1990
14. Pitoura, E.: Social-minded measures of data quality: fairness, diversity, and lack of bias. J. Data Inf. Quality (JDIQ) **12**(3), 1–8 (2020)
15. Taft, R.: Name search techniques. Tech. rep, New York State Identification and Intelligence System, Albany, N.Y. (1970)

16. Vatsalan, D., Yu, J., Henecka, W., Thorne, B.: Fairness-aware privacy-preserving record linkage. In: Garcia-Alfaro, J., Navarro-Arribas, G., Herrera-Joancomarti, J. (eds.) DPM/CBT -2020. LNCS, vol. 12484, pp. 3–18. Springer, Cham (2020). https://doi.org/10.1007/978-3-030-66172-4_1
17. Wu, N., Vatsalan, D., Verma, S., Kâafar, M.A.: Fairness and cost constrained privacy-aware record linkage. IEEE Trans. Inf. Forensics Secur. **17**, 2644–2656 (2022)
18. Zaharia, M., et al.: Apache Spark: a unified engine for big data processing. Commun. ACM **59**(11), 56–65 (2016)

Multi-Objective Fairness in Team Assembly

Rodrigo Borges, Otto Sahlgrens, Sami Koivunen, Kostas Stefanidis(✉), Thomas Olsson, and Arto Laitinen

Tampere University, Tampere, Finland
{rodrigo.borges,otto.sahlgren,sami.koivunen,konstantinos.stefanidis, thomas.olsson,arto.laitinen}@tuni.fi

Abstract. Team assembly is a problem that demands trade-offs between multiple fairness criteria and computational optimization. We focus on four criteria: (i) fair distribution of workloads within the team, (ii) fair distribution of skills and expertise regarding project requirements, (iii) fair distribution of protected classes in the team, and (iv) fair distribution of the team cost among protected classes. For this problem, we propose a two-stage algorithmic solution. First, a multi-objective optimization procedure is executed and the Pareto candidates that satisfy the project requirements are selected. Second, N random groups are formed containing combinations of these candidates, and a second round of multi-objective optimization is executed, but this time for selecting the groups that optimize the team-assembly criteria. We also discuss the conflicts between those objectives when trying to understand the impact of fairness constraints in the utility associated with the formed team.

1 Introduction

Given a set of optimization criteria and constraints, team assembly targets at selecting, from a pool of candidates who each have a set of skills, a set of individuals that jointly fulfils the requirements of a predefined project. Decision-makers have to establish a clear understanding of project requirements and teams' envisioned tasks so that they can be translated into computationally tractable formal requirements, respectively, as well as choose between various ways of assigning candidates into teams [9]. Moreover, team assembly is often a socially, ethically and legally sensitive activity, especially when conducted in high-stakes domains, such as formal education or professional work contexts. A particularly salient set of concerns relates to unfair bias which can disadvantage members of protected groups (such as gender or ethnic groups) and marginalized communities. Whether technical or social in terms of its origin [7], the bias introduced in (or reproduced by) team assembly algorithms can result in unfair treatment of candidates in the team assembly process, even unlawful discrimination.

Existing work has developed methods for improving team-assembly algorithms in different respects, such as reducing the cost of team assembly [2], distributing the workload more equitably among candidates [1] and improving the representation of different demographic groups in the resulting teams [2].

A. Abelló et al. (Eds.): ADBIS 2023, CCIS 1850, pp. 106–116, 2023.
https://doi.org/10.1007/978-3-031-42941-5_10

Whereas a large body of work is devoted to developing methods for identifying and mitigating wrongful bias and unfairness in algorithms and software [13], research that addresses these issues in the context of computational team assembly remains scarce. Most existing approaches are designed for incremental solutions where teams are formed by selecting one candidate in sequence after the other, and optimize only a single distributive desideratum. Our work is motivated by the observation of two problems with this approach. On the one hand, decision-makers often have multiple objectives that need to be balanced or prioritized [11], and it is unlikely that a single fairness-objective can capture a holistic set of contextual values relevant to a given team-assembly process. On the other hand, an incremental approach to team-assembly can be undesirable in certain team-assembly contexts, such as when choosing one candidate at time t_1 closes off the possibility to choose another more suitable candidate later at time t_2.

To address these issues, we formulate team-assembly as a multi-objective optimization procedure motivated by the assumption that fairness-aware team-assembly should achieve several objectives constitutive a more holistic notion of fairness in team-assembly. We describe our framework and illustrate its benefits by employing four criteria for fairness-aware team-assembly. Ideally, a team assembly algorithm would compare every possible team-composition in light of these criteria and choose the one that minimizes a target objective. However, this approach can be expensive especially when the candidate pool is large. To address this issue, we propose a two-step team-assembly procedure: First, a multi-objective optimization procedure is executed and the Pareto candidates that satisfy the project requirements are selected. Second, N random groups are formed containing combinations of these candidates, and a second round of multi-objective optimization is executed, but this time for selecting the groups that optimize the team-assembly criteria. The choice between teams is determined according to a combination of all fairness criteria. This algorithm is not as cheap as selecting the best candidates incrementally, and it is not as expensive as testing all possible groups that can be formed. Instead, the proposed algorithm filters the best candidates among the ones that fulfill the project requirements, and it forms several random groups containing these candidates. When selecting a group that is already formed one can directly access the fairness metrics, and it is easier for the algorithm to minimize a given criterion or a set of criteria.

2 Related Work

Team-Assembly. Research on computational team-assembly is diverse, partly due to the variety of application areas and computational approaches. [10] proposes a team recommender that groups individuals within a social network based on pre-defined skill requirements. [14] presents an approach to form and recommend emergent teams based on how software artifacts are changed by developers, while [17] proposes building teams based on the personality of the team members using a classifier to predict the performance of the constructed teams. [16] frames team-assembly as a group recommendation, where it forms a team of users, each of whom has specific constraints, and recommends items to that team.

Bias and Fairness in Team-Assembly. Research on fair machine learning developed various ways for identifying and addressing unfair bias. In most works, fairness is framed as a local resource allocation problem where a given good should distribute efficiently without violating some pre-defined fairness constraint(s). Examples of metrics include Statistical Parity [6], which requires that the distribution of positive outcomes is statistically independent of so-called protected attributes (e.g., gender), and Equalized Odds [8], which requires parity in group-relative error rates. Different techniques can be applied throughout the system pipeline to mitigate bias in data, algorithms, or output distributions [15]. While research on fairness specifically in computational team-assembly contexts remains scarce, there are some notable exceptions. For example, [4] formulates the task of team-formation as an instance of fair allocation: a procedure for assigning students to projects should involve fair division, which is defined in terms of balanced workloads and tasks in the resulting teams. Another example is [2], which examines the fair team-formation problem in an online labour marketplace. To the best of our knowledge, [12] presents the most similar setting to our work, exploring a problem where teams have multidisciplinary requirements and the selection of members is based on the match of their skills and the requirements. For assembling multiple teams and allocating the best members in a fair way between the teams, it suggests a heuristic incremental method as a solution to create team recommendations for multidisciplinary projects.

3 Motivation

Team Assembly as One-Shot Subset Selection. Our approach is designed for subset selection cases where a team is formed by choosing an optimal set of individuals from a larger set of candidates, where project-to-team fit is evaluated by considering project requirements and candidates' skills. Our motivation is that, subset selection has received comparably less attention in research on fairness in algorithmic decision-making (see, however, [5]). Also, existing approaches to fairness-aware team assembly have largely focused on an *incremental* approach to selecting candidates, which can undesirable or suboptimal in certain cases since the overall composition of the team can be known only by selecting all candidates. Hence, we address a gap in the research literature by focusing on subset selection in an *one-shot team assembly* setting.

Multi-Objective Fairness in Team Assembly. We approach fairness-aware team assembly from the perspective of multi-objective optimization, observing the limitations of previous works that employ a single fairness metric. In particular, using a single measure does not allow the decision-maker to evaluate resulting team-compositions from a holistic evaluative perspective nor to identify trade-offs that may arise between their (un)desirable properties [11]. Our approach takes these notions into account, and recognizes that team assembly procedures can be multi-faceted in terms of the values they should promote and the goods and opportunities that are distributed therein. For example, in real-life contexts of team assembly, the decision-maker is not only distributing access to the team, but also allocating tasks and responsibilities between accepted team-members. Our notion of *multi-objective fairness* captures this idea, and we use it to denote the general sentiment that multiple goods and opportunities should be distributed fairly with due regard also for the overall utility generated.

For fairness-aware team assembly, we apply 4 objectives: (a) *Fair Representation*: The distribution of protected attributes within a team should be fair in terms of being as equitable as possible. (b) *Fair Workload Distribution*: The distribution of tasks within a team should be fair in terms of being as equal as possible. (c) *Fair Expertise Distribution*: The distribution of skills within a team should be fair in terms of being as equal as possible. (d) *Fair Cost Distribution*: The distribution of the cost within a team should be as fair as possible considering the protected attributes associated with candidates. Each objective equalize some benefit, or resource that many consider important in team assembly.

4 Problem Formulation

Let $S = \{s_1, \ldots, s_m\}$ be a set of skills, A be a binary sensitive attribute that can assume values A_0 or A_1, $U = \{u_1, \ldots, u_k\}$ be a set of individuals, i.e., the candidate pool, and P be a set of requirements for a project, i.e., a subset of the skill set ($P \subset S$). An individual $u \in U$ is represented as a combination of a cost profile (u^S) containing the hiring cost associated with their skills, and a value (u^A) associated with a sensitive attribute. The cost profile is obtained through a function θ that returns the cost of a certain skill, for example, $u^S = (\theta(s_1, u), \theta(s_2, u), \ldots, \theta(s_m, u))$ represents user u according to their cost in skills $\{s_1, s_2, \ldots, s_m\}$. We assume a user has a certain skill as long as the cost associated with that skill is greater than 0.

Any set of more than 2 and less than $|U|$ individuals is considered a team T. And the number of project requirements that are fulfilled by a team is referred to as *coverage*. The aim of the team-assembly method is to select among all teams that cover the project requirements, the team that minimizes five objectives: team cost, workload uneven distribution, expertise uneven distribution, representation parity, and cost difference. These objectives are described next.

Team Cost. The total cost of hiring a team for a project is defined as:

$$Cost(T, P) = \sum_{u \in T} \sum_{j=1}^{|S \cap P|} \theta(s_j, u). \tag{1}$$

It can be described as the summation of the cost associated with each team member's skills that match the project requirements. It is worth mentioning that one candidate can contribute to more than one task in the project, in the case when their skills coincide with more than one project requirement.

Workload Uneven Distribution is calculated as the standard deviation of the cost associated with each member of the team:

$$Workload(T, P) = \sqrt{\frac{1}{|T|} \sum\nolimits_{u \in T} \left(\sum\nolimits_{j=1}^{|S \cap P|} \theta(s_j, u) - \frac{Cost(T)}{|T|} \right)^2}. \quad (2)$$

A team in which the total cost is well distributed among members (low variance) is considered fair, whereas a team in which the cost is concentrated among a few members (high variance) is considered unfair.

Expertise Uneven Distribution. It is important to ensure that not just the costs are well distributed among candidates, but that the costs are also well distributed among project requirements. The unevenness of expertise distribution is calculated as:

$$Expertise(T, P) = \sqrt{\frac{1}{|P|} \sum\nolimits_{j=1}^{|S \cap P|} \left(\sum\nolimits_{u \in T} \theta(s_j, u) - \frac{Cost(T)}{|P|} \right)^2}. \quad (3)$$

In a similar fashion to the workload distribution, this objective measures the standard deviation of the cost associated with each project requirement. A fair team is expected to distribute their cost among requirements as even as possible, thus resulting in a low standard deviation value.

Representation Parity. It measures the difference between the occurrences of A_0 and A_1 within a team as potential values for a sensitive attribute A. The objective is calculated as:

$$Representation(T, A) = \frac{\sqrt{(|f(T, A_0)| - |f(T, A_1)|)^2}}{|T|}, \quad (4)$$

where function $f(T, A_0)$ returns a set containing the members of T associated with sensitive attributes A_0, as well as in the case of attribute A_1. A low Representation Parity indicates a fair distribution of attribute A whereas a high value indicates a majority of members associated with one of the classes, A_0 or A_1.

Cost Difference. It measures the difference between the cost allocated to two categories, A_0 and A_1, within a team. The total cost of team members associated with a certain sensitive attribute, named Cost Attribute (CA), is calculated as:

$$CA(T, P, A_0) = \sum\nolimits_{u \in f(T, A_0)} \sum\nolimits_{j=1}^{|S \cap P|} \theta(s_j, u), \quad (5)$$

in the case of attribute A_0. And the Cost Difference objective is calculated as:

$$CostDiff(T, A, P) = \frac{\sqrt{(CA(T, P, A_0) - CA(T, P, A_1))^2}}{Cost(T, P)}. \quad (6)$$

As mentioned before, our goal is to select a team T that fulfils the project P requirements and that minimizes the multi-objective condition:

$$\text{argmin}_T(Cost(T, P), Workload(T, P), Expertise(T, P), Representation(T, A), CostDiff(T, A)).$$

4.1 Multi-Objective Fairness in Team Assembly

In this section, we propose a method designed for assembling teams with multiple fairness constraints. The method assumes a pool of candidates from which the team will be selected, a project and a sensitive attribute associated with each of the candidates. The method formulates fairness-aware team-assembly as a multi-objective optimization problem that is performed in two stages: first, project requirements are considered as objectives, and the best candidates are selected for the next phase. Second, multiple teams are formed with these candidates and fairness constraints are calculated for each of the teams. This time the fairness constraints are assumed as objectives, and the team that minimizes these constraints while fulfilling the project requirements is selected as the fairer.

Given a candidate pool (U) and set of project requirements (P), the first action is to filter candidates with at least one skill required by the project. The filtering process removes all users for which $|u \cap P| = 0$, and the remaining candidates are referred to as U_p. The candidates in U_p are then submitted to a multi-objective optimization step with the aim of selecting the best candidates for this specific project according to the Pareto dominance concept. According to this concept, a candidate dominates another if they perform better in at least one of the project requirements. A candidate is considered non-dominated if they are not dominated by any other candidate in the population, and the set of all non-dominated candidates compose the *Pareto candidates* subset.

At this point, the $paretoCandidates$ subset contains the non-dominated candidates considered as the most suitable for the given project, but our notion of a fair team can only be assessed when having a formed team. The next step is to form a reasonable amount of teams containing a fixed number of Pareto candidates selected randomly. The number of random teams (N) as well as the size of these teams (M) are provided as parameters for the method. Once the teams are formed, it is possible to calculate their coverage, as well as their fairness objectives with Eqs. 1, 2, 3, 4 and 6. The teams that fulfil the project requirements are filtered out and considered in the following step. $Cost$, $Workload$, $Expertise$, $Representation$ and $CostDifference$ are then calculated, and each team end up being represented according to the values calculated for its objectives. A second round of multi-objective optimization is executed. This time, teams are being compared instead of candidates, with the same understanding as in from the first case. Given two arbitrary teams, two are the possibilities: (i) one team dominates the other if it has at least one objective with a lower value than the other team, or (ii) the two teams do not dominate each other. The non-dominated teams are referred to as *Pareto teams*. Finally, all objectives measured for Pareto teams are summed up as an indicator of an *overall unfairness*, and the method selects the team T with the lower unfairness value.

5 Experiments

We evaluated the proposed method in a dataset obtained from the *freelancer*[1] website, in which candidates register themselves to be hired as freelance workers. The dataset contains 1,211 candidates who self-declared their costs and their expertise in 175 skills [2]. In the information available in the dataset, users are associated with skills in a binary fashion, but no information is provided about the cost of each skill separately. We decided that the cost declared by the users is the same for every skill in which they have the expertise, meaning that if user u declared a cost c and they are hired for a project in which they will contribute with two skills, then the total cost associated with this user is $2 \times c$. [2] attributes a hypothetical binary sensitive attribute to each candidate, and generates several versions of the same dataset, associating candidates with this attribute in different proportions. We decided to use the dataset in which members are equally represented in the candidate pool (50/50), and the dataset in which members are more unevenly represented (10/90). The dataset contains also the requirements for 600 projects.

We applied two other team-assembly methods to the same task for the sake of comparison. The first method, named *Incremental*, selects the most suitable candidates incrementally until the project requirements are fulfilled. The second method, named *Fair Allocation*, operates in a similar fashion, but this time

Table 1. Cost, Workload, Expertise, Representation, Cost Difference, and number of formed teams for Incremental, Fair Allocation and Multi-Objective methods. Multi-Objective can optimize different criteria, and its results are presented according to seven objectives: Random, Top-Cost, Top-Workload, Top-Expertise, Top-Representation, Top-Cost Difference and Top-Sum. The best results for each objective are in boldface, and the second-best results are in underlining.

Classes Dist.		Algorithm	Cost	Workload	Expertise	Representation	Cost Difference	Teams
50/50		Incremental	**23.311 (15.548)**	0.035 (0.034)	0.025 (0.026)	0.480 (0.378)	0.610 (0.345)	486
		Fair Allocation	29.201 (20.453)	0.042 (0.040)	0.032 (0.032)	0.238 (0.268)	0.447 (0.286)	486
	Multi-Obj.	Random	40.862 (23.391)	0.045 (0.045)	0.035 (0.036)	0.343 (0.339)	0.419 (0.351)	506
		Top-Cost	26.099 (16.876)	0.035 (0.038)	0.026 (0.027)	0.450 (0.360)	0.595 (0.333)	506
		Top-Workload	42.294 (27.074)	**0.018** (0.029)	0.032 (0.036)	0.434 (0.367)	0.471 (0.357)	506
		Top-Expertise	37.943 (23.646)	0.039 (0.039)	**0.011 (0.020)**	0.445 (0.363)	0.525 (0.359)	506
		Top-Repres.	28.503 (17.531)	0.036 (0.038)	0.028 (0.027)	**0.143 (0.170)**	0.393 (0.251)	505
		Top-Cost Diff.	41.458 (25.980)	0.040 (0.040)	0.037 (0.036)	0.188 (0.197)	**0.091 (0.196)**	506
		Top-Sum.	39.147 (22.969)	0.032 (0.035)	0.028 (0.029)	0.144 (0.174)	0.107 (0.206)	506
10/90		Incremental	**23.423 (15.693)**	0.035 (0.034)	0.025 (0.026)	0.758 (0.341)	0.875 (0.239)	486
		Fair Allocation	31.212 (20.802)	0.046 (0.040)	0.034 (0.030)	**0.413 (0.370)**	0.600 (0.328)	484
	Multi-Obj.	Random	40.939 (25.097)	0.044 (0.047)	0.035 (0.038)	0.722 (0.354)	0.748 (0.350)	506
		Top-Cost	26.159 (17.047)	0.035 (0.038)	0.026 (0.027)	0.729 (0.347)	0.843 (0.253)	506
		Top-Workload	42.525 (27.463)	**0.018 (0.029)**	0.033 (0.036)	0.828 (0.297)	0.836 (0.285)	505
		Top-Expertise	38.140 (23.855)	0.039 (0.040)	**0.011 (0.020)**	0.795 (0.327)	0.842 (0.298)	506
		Top-Repres.	30.231 (19.608)	0.039 (0.041)	0.030 (0.032)	0.435 (0.390)	0.594 (0.353)	506
		Top-Cost Diff.	34.961 (22.256)	0.040 (0.039)	0.032 (0.031)	0.470 (0.377)	**0.459 (0.439)**	506
		Top-Sum.	37.661 (23.700)	0.035 (0.038)	0.027 (0.030)	0.432 (0.389)	0.460 (0.436)	505

[1] https://www.freelancer.com/.

considering users associated with a sensitive attribute and forcing as much as possible that the formed team has a fair distribution of this attribute among its members. For more details, see at [3].

Multi-Objective, Incremental and Fair Allocation were evaluated for forming a team for each of the 600 projects in the freelancer dataset, and the average value obtained for each of the objectives presented in Sect. 4 were calculated. The results are reported in Table 1, separately for the two datasets containing different proportions of the sensitive attributes, 50/50 and 10/90. The results of Multi-Objective are presented according to five different optimization objectives, named configurations: Top-Cost, Top-Workload, Top-Expertise, Top-Representation and Top-Cost Difference, along with a Random selection variation. On average, the first round of multi-objective optimization reduced the number of candidates by 77%, meaning that the candidates dominated others with compatible skills represent 23% of the total. In the second round of multi-objective optimization, the teams were reduced by 98% on average. The teams that dominate the others represent only 2%. This reflects how much the teams formed randomly can be internally equivalent or redundant.

In general, the Incremental method was the most efficient in minimizing the total cost and the size of the formed teams. The Multi-Objective method was able to assemble a slightly higher number of teams than other methods, probably because of forming teams in one-shot instead of incrementally. When configured to minimize a specific objective, the Multi-Objective method was efficient in selecting the teams, except when the objective was the Cost, and when the objective was the Representation and the dataset contained an uneven distribution of classes (10/90). In the former case, Incremental was the most efficient method, and in the latter case, the Fair Allocation performed better.

In the context of Multi-Objective fairness, it is preferable that a team presents a good balance of objectives instead of an extremely low value for one objective despite the others. E.g., when configured to optimize the Expertise objective (Top-Expertise) in the uneven dataset (10/90), Multi-Objective selected, on average, teams with a fairly high Representation (0.795) and Cost Difference (0.842) values. The Top-Sum configuration, on the other hand, selected teams with the second-best average values (highlighted with underline in Table 1) for three out of five objectives, for both datasets. The two objectives in which the configuration did not perform well were Cost and Expertise, which leads us to the next step of analyzing potential conflicts between objectives in the results.

Tension Between Objectives. Conflicts between objectives might emerge in situations when one objective is minimized, and another (or a set) goes up as a side effect. We noticed that minimizing the Workload objective increases the total cost of teams in both datasets (see Top-Workload configuration of Multi-Objective in Table 1). These were the configurations in which the average cost of selected teams presented the highest values, 42.294 in the case of the 50/50 dataset, and 42.525 in the case of the 10/90 dataset. These values are 181.4% and 181.5% higher than the lowest cost obtained in the results, respectively. Workload and Expertise, however, are fairness objectives that do not take into account the sensitive attribute associated with candidates, differently from Representation and Cost Difference, which are both calculated according to how many team members belong to each of the classes derived from this attribute. Top-Sum configuration of Multi-Objective performed especially well regarding those two objectives, obtaining relatively close values to the lowest ones obtained in the experiments. In general, ensuring an equal distribution of costs among protected groups (Cost Difference) had a bigger impact on the total cost than ensuring that both groups are equally represented in the teams (Representation).

Impact of Class Distribution. The trade-offs between objectives as well as their absolute values can vary depending on how the classes derived from the sensitive attribute are distributed within the candidate pool. Workload and Expertise objectives were not impacted by the difference in the datasets, but Representation and Cost Difference, on the other hand, presented substantially different values depending on the proportion in which users are distributed in classes. The lowest average value calculated for the Representation objective increased from 0.143 in the 50/50 dataset, to 0.413 in the 10/90 dataset, an increase of approximately 289%, and the Cost Difference objective increased even more, its lower value went from 0.091 to 0.459, an increase of more than 500%. When focusing on Top-Sum configuration of the Multi-Objective team-assembly method, a significant decrease of approximately 82.5% (from 0.610 to 0.107) was observed in the Cost Difference if compared to the Incremental method, and of approximately 76% if compared to the Fair Allocation method, in the case of the 50/50 dataset. In the case of the 10/90 dataset, however, the differences were slightly different, the Cost Difference was reduced by approximately 23.4% (from 0.6 to 0.46) when compared to the Fair Allocation method, but the Representation went higher, increasing some 4.6% (from 0.413 to 0.432). If compared to the Incremental method, the Top-Sum configuration performed better on those objectives: the Cost Difference was reduced by 47% (from 0.875 to 0.460), and the Representation was reduced by 40% (from 0.758 to 0.432).

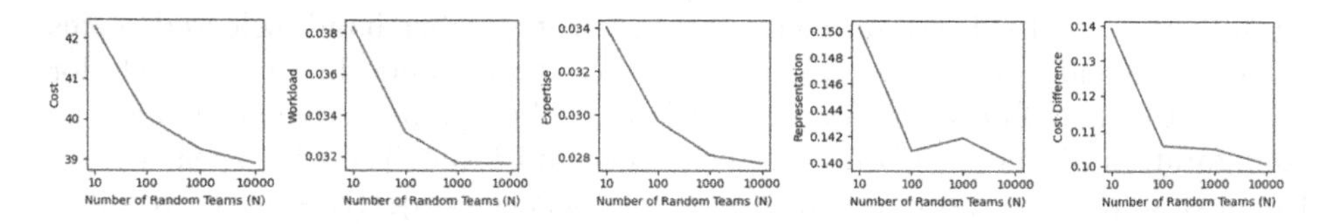

Fig. 1. The impact of N in cost, workload, expertise, representation and cost difference.

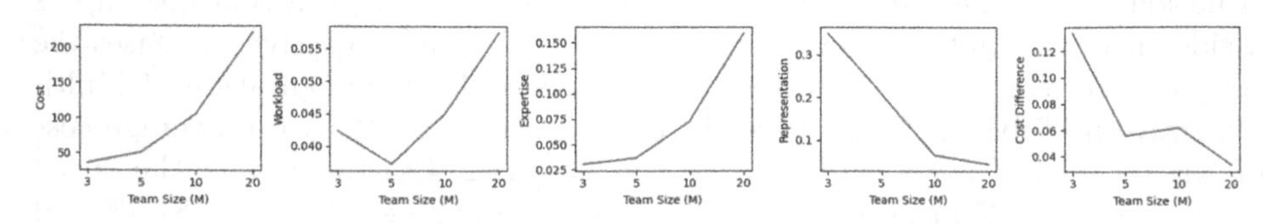

Fig. 2. The impact of M in cost, workload, expertise, representation and cost difference.

Impact of the Number of Random Teams. Multi-Objective receives N and M as parameters. First, N was set equal to 10, 100, 1,000 and 10,000, and the impact of these decisions on the team-assembly objectives can be seen in Fig. 1. One could expect that the number of random teams has a direct impact on the probability of the method forming a team that fulfils the team-assembly criteria, simply because when there are more teams there are more options to choose from. But this holds true to a certain limit, after which the improvement in the results is ordinary. We decided to configure the method to form 1,000 random teams once this is a point where all curves curve get more stable, as one can see in Fig. 1. M was then set equal to 3, 5, 10 and 20, and the impact of these decisions on the team-assembly objectives can be seen in Fig. 2. It is evident how the Cost, Workload and Expertise objectives increase as the teams get bigger, except for the workload distribution when teams are formed with 5 members. In this case, the workload gets slightly better distributed among team members than when teams are formed with 3 members. On the other hand, Representation and Cost Difference objectives present the opposite behaviour, they get lower as the teams get bigger, probably because it is more likely to get even distributions when there are more members to distribute among classes.

6 Conclusions

In this paper, we argued in favour of a wider notion of fairness in the context of team assembly by framing the task of forming teams as a multi-objective optimization procedure. We have also proposed an algorithm for assembling teams with multiple fairness constraints that assembled teams in a one-shot fashion, as opposed to incremental methods proposed previously in the literature. Our method is flexible enough that it can be applied to situations when one single objective needs to be minimized (or maximized), as well as in situations when all objectives need to be optimized jointly.

References

1. Anagnostopoulos, A., Becchetti, L., Castillo, C., Gionis, A., Leonardi, S.: Online team formation in social networks. In: WWW (2012)
2. Barnabò, G., Fazzone, A., Leonardi, S., Schwiegelshohn, C.: Algorithms for fair team formation in online labour marketplaces. In: Companion of WWW (2019)
3. Borges, R., Sahlgrens, O., Koivunen, S., Stefanidis, K., Olsson, T., Laitinen, A.: Computational team assembly with fairness constraints. CoRR abs/2306.07023 (2023)
4. Bulmer, J., Fritter, M., Gao, Y., Hui, B.: FASTT: team formation using fair division. In: Canadian AI (2020)
5. Cachel, K., Rundensteiner, E.: Fins auditing framework: group fairness for subset selections. In: AAAI/ACM Conference on AI, Ethics, and Society (2022)
6. Dwork, C., Hardt, M., Pitassi, T., Reingold, O., Zemel, R.: Fairness through awareness. In: Innovations in Theoretical Computer Science Conference (2012)

7. Friedler, S.A., Scheidegger, C., Venkatasubramanian, S.: The (im)possibility of fairness: different value systems require different mechanisms for fair decision making. Commun. ACM **64**(4), 136–143 (2021)
8. Hardt, M., Price, E., Srebro, N.: Equality of opportunity in supervised learning. In: NIPS (2016)
9. Harris, A.M., Gómez-Zará, D., DeChurch, L.A., Contractor, N.S.: Joining together online: the trajectory of CSCW scholarship on group formation. Proc. ACM Hum. Comput. Interact. **3**(CSCW), 1–27 (2019)
10. Lappas, T., Liu, K., Terzi, E.: Finding a team of experts in social networks. In: SIGKDD (2009)
11. Lee, M.S.A., Floridi, L.: Algorithmic fairness in mortgage lending: from absolute conditions to relational trade-offs. Mind. Mach. **31**(1), 165–191 (2021)
12. Machado, L., Stefanidis, K.: Fair team recommendations for multidisciplinary projects. In: WI (2019)
13. Mehrabi, N., Morstatter, F., Saxena, N., Lerman, K., Galstyan, A.: A survey on bias and fairness in machine learning. ACM Comput. Surv. **54**(6), 1–35 (2021)
14. Minto, S., Murphy, G.C.: Recommending emergent teams. In: MSR (2007)
15. Pitoura, E., Stefanidis, K., Koutrika, G.: Fairness in rankings and recommendations: an overview. VLDB J. **31**, 431–458 (2021). https://doi.org/10.1007/s00778-021-00697-y
16. Stefanidis, K., Pitoura, E.: Finding the right set of users: generalized constraints for group recommendations. In: PersDB (2012)
17. Yilmaz, M., Al-Taei, A., O'Connor, R.V.: A machine-based personality oriented team recommender for software development organizations. In: EuroSPI (2015)

Data Science

A Comparative Study of Assessment Metrics for Imbalanced Learning

Zakarya Farou[1(✉)], Mohamed Aharrat[1], and Tomáš Horváth[1,2]

[1] Department of Data Science and Engineering, Institute of Industry - Academia Innovation, ELTE Eötvös Loránd University, Pázmány Péter sétány 1/C, 1117 Budapest, Hungary
{zakaryafarou,ahr9oi,tomas.horvath}@inf.elte.hu

[2] Institute of Computer Science, Pavol Jozef Šafárik University, Jesenná 5, 04001 Košice, Slovakia

Abstract. There are several machine learning algorithms addressing class imbalance problem, requiring standardized metrics for adequete performance evaluation. This paper reviews several metrics for imbalanced learning in binary and multi-class problems. We emphasize considering class separability, imbalance ratio, and noise when choosing suitable metrics. Applications, advantages, and disadvantages of each metric are discussed, providing insights for different scenarios. By offering a comprehensive overview, this paper aids researchers in selecting appropriate evaluation metrics for real-world applications.

Keywords: Imbalanced learning · Assessment metrics · Classification

1 Introduction

Choosing suitable evaluation metrics for learning algorithms is crucial, especially in imbalanced learning problems with skewed datasets and unequal error costs. Traditional measures like classification accuracy can be unreliable in such scenarios, as they are biased towards the majority class [4]. For example, in a highly imbalanced dataset where the majority class represents 99.9% of instances, a default majority classifier would achieve high accuracy by predicting all instances as negative [11]. To address this, specific assessment metrics are needed. While there are binary and multi-class classification metrics, most multi-class metrics are extensions of the binary case and involve complex computations.

The rest of the paper is structured as follows: Sect. 2 describes the class imbalance problem from the definition to the causes, Sect. 3 introduces the evaluation measures, Sect. 4 presents the dataset, experiments, results, and discussion, and Sect. 5 provides the research's conclusion.

2 Problem Definition

The class distribution of a dataset determines its imbalance, where an imbalance occurs when one or more classes have significantly fewer samples compared

A. Abelló et al. (Eds.): ADBIS 2023, CCIS 1850, pp. 119–129, 2023.
https://doi.org/10.1007/978-3-031-42941-5_11

to others, ranging from small to extreme ratios like (1:100) or (1:1000). This imbalance can be referred to as rare event prediction, extreme event prediction, or severe class imbalance. For the rest of this paper, we continue to label this problem as class imbalance for clarity.

The imbalance can arise due to measurement errors or biased sampling, which can also be an inherent characteristic of the problem domain [4]. Various techniques [4,5,9,12,14] have been proposed to address the imbalance, with specific techniques needed for severe imbalances to prevent the under-representation of minority classes. Recently, there have been several contributions to imbalanced learning in predictive modeling; we continuously witness new promising algorithms and methods. Hence, it is essential to have a standardized set of performance measurement metrics to evaluate the effectiveness of such algorithms [9]. However, there has been more focus on binary than multi-class imbalance. This paper presents detailed descriptions of established and new metrics for binary and multi-class problems, categorized into threshold, ranking, and probabilistic metrics [9].

3 Assessment Metrics for Imbalance Learning

Evaluation metrics can be categorized into single-class and multi-class focus metrics. Single-class metrics, like precision, recall, and F_1 score, are sensitive to class importance and suitable for imbalanced learning scenarios [11]. Multi-class focus metrics, such as accuracy and error rate, consider overall classifier performance but may not account for varying class importance [6]. Also, the F_1 score can be used to evaluate multi-class classifiers by calculating it independently for each class and averaging the results. For class-imbalance problems, it is essential to consider class ratios when using multi-class metrics. This paper categorizes single and multi-class metrics into three main measurement types. It includes metrics from previous studies [11] and new ones introduced in recent years, providing their use cases, advantages, and disadvantages. Their purpose and functionality is explained, emphasizing their applicability to class imbalance problems. The metrics discussed in this paper are based on the confusion matrix (CM) [6], which summarizes the performance of a classification algorithm by cross-tabulating observed and predicted classes. The CM includes true positive (TP), true negative (TN), false positive (FP), and false negative (FN) values, representing correctly and incorrectly classified instances for each class.

3.1 Threshold Metrics

Various threshold metrics are employed in imbalanced learning scenarios to assess classifier performance, all of which are detailed in [10]. These metrics encompass single-class focus measures like sensitivity (SN), specificity (SP), precision (PR), as well as combined metrics such as geometric mean (Gm), F_1 score (F_1), macro-averaged accuracy (MAA), and adjusted geometric mean (AGm).

SN quantifies the accurate identification of positive samples, prioritizing detecting conditions or events over their absence [5]. Conversely, SP is crucial in situations where correctly identifying negative samples outweighs identifying positive ones [14]. PR focuses on accurately classifying positive samples, emphasizing minimizing false positives in applications like email spam detection [12]. GM calculates the geometric norm of SN and SP, allowing customization to highlight the positive class. The F_1 score combines PR and SN through the harmonic mean, delivering a comprehensive evaluation, especially in medical diagnosis [5]. MAA treats all classes equally, while AGm is estimated as GM$-$(100%$-$SP)$\times N_n$ according to [3], where N_n is the proportion of negative examples in the dataset. AGm considers class prevalence to enhance SN while minimizing the decrease in SP [2,6], and seems particularly beneficial in highly imbalance problems.

3.2 Ranking Metrics

The area under the ROC curve (AUC) and H-measure are ranking metrics commonly used in imbalance learning.

AUC is an analysis technique [7] that evaluates the classifier's ranking performance by measuring its ability to distinguish between positive and negative instances. It examines the classifier's performance at different thresholds, allowing for a better understanding of its behavior and the associated benefits and costs [9]. AUC which evaluates the classifier's performance based on the overall discrimination power can be affected by class imbalance, although it is considered relatively insensitive compared to some other metrics.

The H-measure [8], is a ranking metric considering TP, TN, FP, and FN rates. Since its introduction, researchers adopted the H-measure for successfully overcoming the problem of capturing performance across classifiers. It is also particularly advantageous in imbalanced datasets as it considers the overall agreement between predicted and actual class labels, accounting for class imbalance and potential biases in the data.

3.3 Probabilistic Metrics

Log and Brier losses [13] are two commonly used metrics for evaluating predicted probabilities in classification problems.

Log loss, also known as cross-entropy, measures the quality of predicted probabilities by taking the negative logarithm of the predicted probability for the true class. It can be informative in imbalanced datasets as it penalizes low probabilities for the TP class. However, it may be less informative in highly imbalanced scenarios where the majority class has a significant influence. On the other hand, the Brier loss or score is preferred for assessing imbalanced learning. It focuses on the minority class and calculates the mean squared error (MSE) [1] between predicted probabilities and target values. It provides a comprehensive evaluation, considering correct and incorrect predictions with an emphasize on minority class, which is often of greater interest in imbalanced datasets, and is well-suited for assessing classifier performance in imbalanced scenarios.

The covered metrics offer comprehensive assessments of model performance, addressing challenges in imbalanced datasets and accounting for the varying importance of different classes.

4 Experiments

In this section, we present the experiments conducted on synthetic imbalanced datasets generated from a normal distribution. These datasets were used to evaluate the performance of the mentioned metrics and compare their results. We considered three characteristics that significantly impact classification tasks: imbalance ratio, class separation, and noise in the data. The synthetic dataset consisted of ten thousand records with five features and two classes, each with different imbalance ratios. The experiments encompassed a range of imbalance ratios, ranging from (1:1) to severe imbalances such as (0.999:0.001). Throughout the experiments, we designated the minority class as positive and the majority class as negative. It's worth noting that the class separation factor multiplied the hypercube size, with smaller values indicating more challenging instances to classify accurately. Figure 1 illustrates the range of feature values.

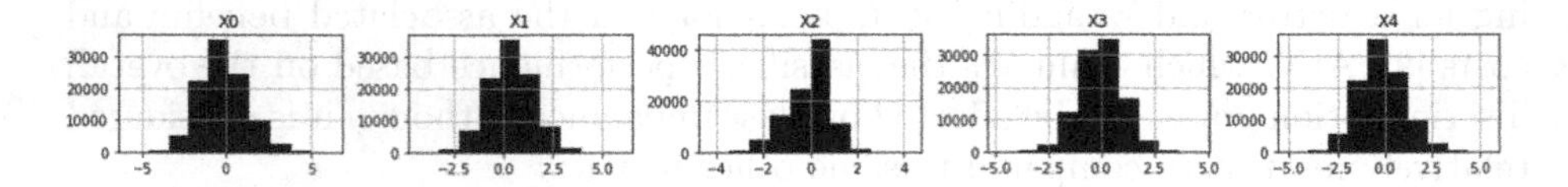

Fig. 1. Feature distribution of the first experiment.

The generated data is divided into train and test sets containing 8,000 and 2,000 samples. We train a support vector machine (SVM) to classify instances from the test set. SVM is implemented using the scikit-learn [15] API with default parameter settings.

4.1 Binary Classification

In the initial experiment, we trained SVM to classify datasets with increasing class separation factors (α) of 1, 5, and 10. Additionally, we varied the imbalance ratio (β) in favor of the majority class using different weight combinations, and recorded the results in Table 1. The accuracy measure showed consistent behavior across the models and class separation factors, reaching a maximum value of 99.55. Relying solely on accuracy may suggest efficient instance prediction. However, as discussed earlier, accuracy alone is inadequate for evaluating model performance in imbalanced class distributions. SN decreases as β increases, converging to 0 when $\alpha = 1$. For higher values of α, convergence occurs only at the highest β. The same trend applies to Gm, F_1, and MAA scores. This is because classification becomes easier with higher α values. AGm generally maintains a

Table 1. Results for binary classification with respect to α and β

α	β	ACC	SN	SP	PR	Gm	F_1	MAA	AGm	AUC	H
1	(1, 1)	94.85	90.15	99.50	99.44	94.71	94.57	94.82	99.49	95.15	0.84
	(0.7, 0.3)	83.30	49.24	97.85	90.76	69.42	63.85	73.55	97.81	74.0	0.36
	(0.8, 0.2)	87.45	37.37	99.50	94.77	60.98	53.60	68.43	99.40	68.55	0.30
	(0.9, 0.1)	93.25	25.82	100.0	100.0	50.81	41.04	62.91	99.73	63.45	0.22
	(0.99, 0.01)	98.85	4.16	100.0	100.0	20.41	8.00	52.08	96.81	52.33	0.03
	(0.999, 0.001)	99.55	0.00	100.0	NaN	0.0	NaN	50.00	90.00	50.0	0.00
5	(1, 1)	96.45	93.36	99.50	99.46	96.38	96.31	96.43	99.49	96.12	0.89
	(0.7, 0.3)	87.35	63.93	97.35	91.19	78.89	75.17	80.64	97.32	81.23	0.51
	(0.8, 0.2)	90.55	54.38	99.25	94.61	73.46	69.06	76.81	99.18	77.43	0.46
	(0.9, 0.1)	94.85	43.95	99.94	98.76	66.28	60.83	71.95	99.79	72.76	0.39
	(0.99, 0.01)	98.85	4.16	100.0	100.0	20.41	8.00	52.08	96.81	52.04	0.03
	(0.999, 0.001)	99.55	0.00	100.0	NaN	0.00	NaN	50.00	90.00	50.0	0.00
10	(1, 1)	99.20	98.89	99.50	99.49	99.19	99.19	99.19	99.50	99.22	0.97
	(0.7, 0.3)	94.95	90.15	97.00	92.78	93.51	91.44	93.57	96.99	94.34	0.82
	(0.8, 0.2)	95.65	83.76	98.51	93.12	90.83	88.19	91.13	98.49	91.54	0.77
	(0.9, 0.1)	97.25	70.87	99.89	98.47	84.14	82.42	85.38	99.80	85.62	0.67
	(0.99, 0.01)	99.10	25.00	100.0	100.0	50.00	40.00	62.50	98.00	62.09	0.22
	(0.999, 0.001)	99.55	0.00	100.0	NaN	0.00	NaN	50.00	90.00	50.0	0.00

similar value range for the first two α values but exhibits improved performance at higher α values, similar to SN except for small α values.

Regarding PR, the mean values for the models with separation factors $\alpha = 1$, 5, and 10 are 96.99, 96.80, and 96.77, respectively. These values indicate that the models prioritize classifying majority class instances when there is a significant overlap with the minority class. In terms of SP, the values are generally similar with minimal differences. In addition to the threshold metrics, we also computed the ranking metrics. The AUC scores show that α influences the classifier's performance. Higher values of α (5, 10) with low to moderate imbalance ratio result in high AUC scores. However, severe imbalance ratios exhibits lower AUC scores. The reason is that in highly imbalanced datasets, the classifier prioritizes the majority class, leading to a higher TNR and lower TPR, decreasing AUC scores. Class separability influences the shape and position of the ROC curve. Higher separability improves class distinction, resulting in increased TPR, lower FPR, and a higher AUC score. However, class separability has a minor impact compared to imbalance ratios. Regarding the H-measure, we observe that it is high when α is high, and the class distribution is balanced, and that the performance decreases as the imbalance ratio increases, indicating sensitivity to both α and β.

Table 2. Results for binary classification with respect to γ, β, and $\alpha = 5$.

γ	β	ACC	SN	SP	PR	Gm	F_1	MAA	AGm	AUC	H
0.1	(1, 1)	93.50	90.95	96.07	95.91	93.48	93.36	93.51	96.07	94.24	0.79
	(0.7, 0.3)	83.55	56.06	96.33	87.68	73.49	68.39	76.20	96.30	76.44	0.40
	(0.8, 0.2)	86.45	46.94	98.75	92.14	68.09	62.20	72.85	98.68	73.93	0.37
	(0.9, 0.1)	89.10	27.24	99.59	91.86	52.08	42.02	63.41	99.42	63.88	0.21
	(0.99, 0.01)	93.90	0.81	99.94	50.00	09.05	1.61	50.38	99.20	50.07	0.01
	(0.999, 0.001)	95.10	0.00	100.0	NaN	0.00	NaN	50.00	98.98	50.04	0.00
0.2	(1, 1)	87.45	90.51	84.36	85.32	87.38	87.84	87.44	84.37	87.98	0.61
	(0.7, 0.3)	81.45	29.56	98.14	83.72	53.87	43.70	63.85	98.05	71.23	0.30
	(0.8, 0.2)	79.55	47.41	95.30	83.20	67.22	60.40	63.85	98.05	64.11	0.20
	(0.9, 0.1)	84.55	14.73	99.51	86.66	38.28	25.18	57.12	99.34	57.42	0.10
	(0.99, 0.01)	89.10	1.36	99.94	75.00	11.67	2.67	50.65	99.54	51.75	0.01
	(0.999, 0.001)	89.65	0.00	100.0	NaN	0.00	NaN	50.00	99.51	50.0	0.00
0.3	(1, 1)	83.45	79.02	87.92	86.88	83.35	82.76	83.47	87.92	83.65	0.50
	(0.7, 0.3)	83.45	79.02	87.92	86.88	83.35	82.76	83.47	87.92	68.13	0.22
	(0.8, 0.2)	78.45	97.29	29.49	78.18	53.57	86.70	63.39	29.51	63.12	0.18
	(0.9, 0.1)	81.20	12.88	99.30	83.07	35.77	22.31	56.09	99.15	56.43	0.08
	(0.99, 0.01)	84.15	0.31	99.94	50.00	5.61	0.06	50.12	99.64	50.05	0.00
	(0.999, 0.001)	84.90	0.00	100.0	NaN	0.00	NaN	50.00	99.67	50.14	0.00

Furthermore, we conducted a similar analysis using β weights to account for γ. For that, we used the *flip_y* parameter of *make_classification()* provided by scikit-learn [15]. By setting γ, i.e., *flip_y* parameter to specific values (0.1, 0.2, and 0.3), we introduced random noise to the class labels of the generated dataset, creating 18 datasets with $\alpha = 5$ and allowing us to simulate more realistic and diverse scenarios for our experimentation. The results are presented in Table 2.

According to the computed metrics, the measurements generally decrease when introducing noise. As γ increases, certain metrics detect the models' poor performance. Accuracy and SP tend to approach one as the weight of the minority class decreases, which is expected. Other metrics behave as anticipated, but it is important to note that the values decrease more rapidly with higher γ, as expected since the model struggles to correctly classify instances in the presence of noise.

However, some metrics exhibit exceptional values. For instance, SP sharply decreases when $\beta = 0.2$ and $\gamma = 0.3$, which is not observed with smaller noise values. This may be attributed to the random nature of the introduced noise, which can impact the effectiveness of the metrics. Moreover, it can be observed from Table 2 that MAA exhibits lower sensitivity to noise compared to other metrics. MAA provides a more reliable reflection of the true performance of

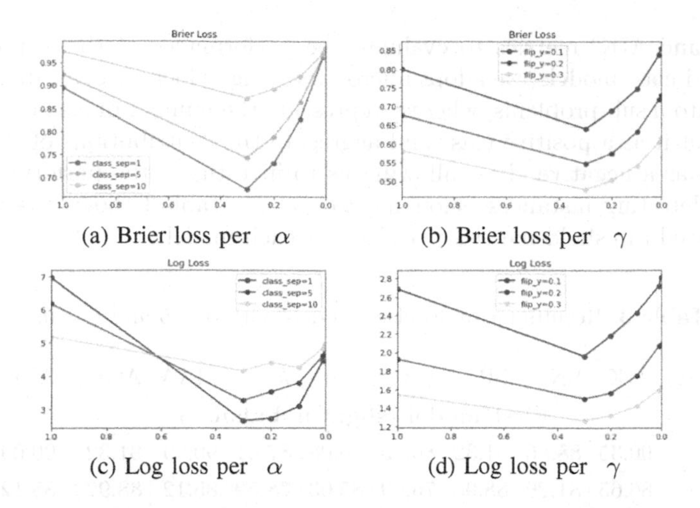

(a) Brier loss per α (b) Brier loss per γ

(c) Log loss per α (d) Log loss per γ

Fig. 2. Log and Brier losses for binary classification with different α and γ.

a classifier in the presence of different classification challenges, including class separation and noise in the data.

As γ increases, the model's performance in terms of AUC scores decreases. This is reflected in higher FPR and slightly lower TPR than the initial AUC scores without noise. The impact becomes more pronounced when considering the increasing imbalance ratio (β), indicating that the models are more susceptible to noise in the data. The comparison between H-Measure and AUC scores in Table 2 reveals that both metrics show decreasing values with varying noise levels. This effect is further exacerbated by an imbalance ratio (β) increase, indicating their sensitivity to noise in the data. Moving on to the probabilistic metrics, Log loss, and Brier score, we adjust the labels to a [0, 1] system, with 1 representing the positive label. The Brier score ranges from 0 to 1, where a score of 0 indicates a perfect model. The Log loss has an unlimited upper bound, approaching infinity for worse scores. Figure 2 illustrates the Brier score and Log loss for the models with varying α and γ. It can be observed that models with lower weights assigned to the minority class exhibit higher error values, with the lowest error values concentrated at the bottom of the curve-like shape. This indicates an increase in misclassification as the imbalance becomes more severe. Regarding the Log loss, a smaller α leads to more noticeable shifts in the loss values, resulting in more significant errors when γ is smaller.

4.2 Multi-class Classification

In the final experiment, we analyze models trained on datasets with four classes exhibiting varying class distributions, including high imbalance scenarios such as (0.4:0.3:0.2:0.1), (0.5:0.3:0.19:0.01), and (0.48:0.48:0.01:0.01). The class separation factor is set to 5, and the noise value is set to 0.1. For each dataset, a single model is trained, resulting in a total of three models. We then compute threshold and AUC metrics to evaluate the performance of these models. To assess multi-class models, we adopt micro averaging, which involves dividing the problem into n sub-problems, where n represents the number of classes. For each class, we set it as a positive class and aggregate the contributions of the rest of the classes as a negative class, allowing us to investigate the performance of a model in detecting instances belonging to a class. Table 3 shows the values of the calculated threshold metrics per class for each model.

Table 3. Results for multi-class models with $\alpha = 5$ and $\gamma = 0.1$.

Class	ACC	SN	SP	PR	Gm	F_1	MAA	AGm	AUC
First model (slight imbalance)									
1	90.35	88.86	91.32	86.99	90.08	87.91	90.09	91.32	90.09
2	86.65	81.26	88.98	76.09	85.03	78.59	85.12	88.97	85.12
3	88.5	70.25	93.06	71.68	80.86	70.96	81.66	93.03	81.66
4	96.30	70.05	99.33	92.36	83.41	79.67	84.69	99.25	84.69
Second model (moderate imbalance)									
1	90.5	92.2	88.80	89.17	90.48	90.66	90.5	88.8	90.50
2	88.4	83.9	90.28	78.32	87.03	81.01	87.09	90.28	87.09
3	90.3	67.46	95.62	78.22	80.32	72.44	81.54	95.58	81.54
4	98.8	25.0	100	100	50.0	40.0	62.5	98.048	62.50
Third model (high imbalance)									
1	91.15	90.18	92.06	91.4	91.11	90.79	91.12	92.06	91.12
2	89.15	92.87	85.79	85.59	89.25	89.08	89.32	85.77	89.32
3	98.1	0.0	100	NaN	0.0	NaN	50.0	97.44	50.00
4	98.4	26.19	99.95	91.67	51.16	40.74	63.07	98.81	63.07

In the first model, the overall performance is relatively good across all classes as illustrated in Fig. 3a, which is expected as the class distribution is relatively balanced. Thus a balanced class distribution allows the model to learn effectively and make accurate predictions for all classes. The performance in the second model (moderate imbalance) shows some variations across the classes. Class 4, which has the highest imbalance, exhibits lower TP (Fig. 3b), SN, Gm, and F_1. This suggests the model struggles to identify positive instances for the underrepresented class. The moderate imbalance may pose challenges, particularly for the underrepresented class, but the model still performs satisfactorily. The results of the third model (high imbalance) show significant variations and challenges. Class 3, which has the highest imbalance, shows a null SN and undefined PR, indicating the model's inability to identify positive instances for this class (Fig. 3c). This is likely due to the severe class imbalance, where the model focuses heavily on the majority classes and struggles to capture instances from the minority class. Gm and F_1 also suffer in this scenario. Class 4, which has the

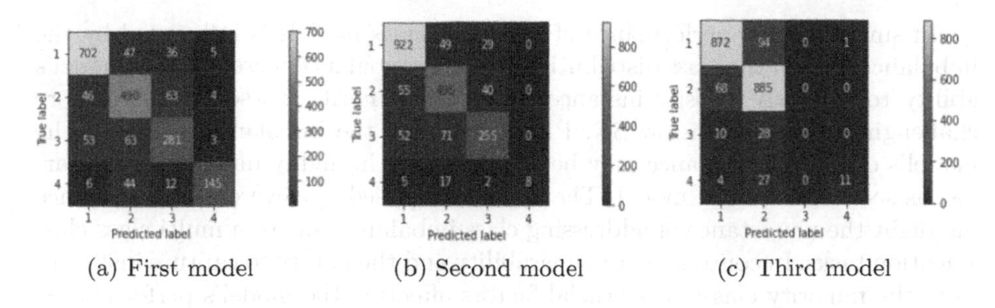

(a) First model (b) Second model (c) Third model

Fig. 3. Confusion matrices for multi-class models

same imbalance ratio as class 3, shows a moderate performance with relatively poor SN and high PR. Although the same number of instances represents them, the model captures instances from class 4 better than class 3. The model's overall performance is compromised, as reflected by lower ACC, MAA, and AGm values. The AUC values are also relatively low, indicating limited class separability.

Table 4. Comparison of imbalanced learning evaluation metrics.

Metric	Advantages	Disadvantages
ACC	Useful for mildly imbalanced learning problems with equal importance placed on both classes	Assumes equal distribution
SN	Critical for prioritizing accurate identification of positive samples over negative samples	Focuses on a single class
SP	Important for prioritizing accurate identification of negative samples	Focuses on a single class
PR	Useful when the cost of FPs is high	Focuses on a single class
Gm	Useful for highly imbalanced classes	Sensitive to changes in class distribution
F_1	Useful when both P and SN are important	Consider using F score for flexible P and SN weighting, as F_1 may not be suitable in all cases
AGm	Consider class prevalence, especially for highly imbalanced classes	Relative measure rather than absolute, making interpretation less intuitive
AUC	Provides a concise and interpretable summary of model performance for easy comparison with other models	Lacks error distribution details in some contexts
H-measure	Enables fair comparison by using an equal cost-weight function for all classifiers	Does not consider predicted probabilities, which may be relevant in certain contexts
LogLoss	Evaluate the model's performance in positive and negative classes, which is vital in imbalance learning	Strongly impacted by the majority class in datasets with high β values
BrierScore	Insensitive to variations in the class distribution, unlike certain metrics that exhibit a bias towards the majority class	Less informative for highly imbalanced problems and sensitive to outliers in predicted probabilities

In summary, the performance of the multi-class models is influenced by the imbalance level in the class distributions. As the imbalance increases, the model's ability to correctly classify instances from the minority classes becomes more challenging, resulting in low SN, PR, and F_1 for the imbalanced classes. The model's overall performance may be compromised in highly imbalanced scenarios, as seen in the third model. The challenges posed by severe class imbalance highlight the importance of addressing class imbalance issues in multi-class classification tasks. In contrast, class separability and the ability to capture instances from the minority classes are crucial factors affecting the model's performance.

4.3 Summary

We have compared various metrics for evaluating machine learning models on imbalanced datasets. The advantages and disadvantages of these metrics are summarized in Table 4.

5 Conclusion

This paper presents a comprehensive analysis of metrics used in imbalance learning for binary and multi-class classification problems. We emphasize the significance of considering class separability, imbalance ratio, and noise when choosing suitable metrics to evaluate classifier performance on imbalanced datasets. Our study reveals the relevance of different metrics in different scenarios, highlighting the importance of thoughtful metric selection for accurate assessments of classifier performance. Making informed decisions about classifier usage in real-world applications necessitates the careful consideration of appropriate metrics.

References

1. Allen, D.M.: Mean square error of prediction as a criterion for selecting variables. Technometrics **13**(3), 469–475 (1971)
2. Batuwita, R., Palade, V.: A new performance measure for class imbalance learning. Application to bioinformatics problems. In: 2009 International Conference on Machine Learning and Applications, pp. 545–550. IEEE (2009)
3. Batuwita, R., Palade, V.: Adjusted geometric-mean: a novel performance measure for imbalanced bioinformatics datasets learning. J. Bioinform. Comput. Biol. **10**(04), 1250003 (2012)
4. Farou, Z., Kopeikina, L., Horváth, T.: Solving multi-class imbalance problems using improved tabular GANs. In: Yin, H., Camacho, D., Tino, P. (eds.) Intelligent Data Engineering and Automated Learning – IDEAL 2022. IDEAL 2022. Lecture Notes in Computer Science, vol. 13756. Springer, Cham (2022). https://doi.org/10.1007/978-3-031-21753-1_51
5. Farou, Z., Mouhoub, N., Horváth, T.: Data generation using gene expression generator. In: Analide, C., Novais, P., Camacho, D., Yin, H. (eds.) IDEAL 2020. LNCS, vol. 12490, pp. 54–65. Springer, Cham (2020). https://doi.org/10.1007/978-3-030-62365-4_6

6. Fernández, A., García, S., Galar, M., Prati, R.C., Krawczyk, B., Herrera, F.: Learning from imbalanced data sets, vol. 11. Springer, Cham (2018). https://doi.org/10.1007/978-3-319-98074-4
7. Gonçalves, L., Subtil, A., Oliveira, M.R., de Zea Bermudez, P.: ROC curve estimation: an overview. REVSTAT-Statist. J. **12**(1), 1–20 (2014)
8. Hand, D.J.: Measuring classifier performance: a coherent alternative to the area under the ROC curve. Mach. Learn. **77**(1), 103–123 (2009)
9. He, H., Garcia, E.A.: Learning from imbalanced data. IEEE Trans. Knowl. Data Eng. **21**(9), 1263–1284 (2009)
10. Hossin, M., Sulaiman, M.N.: A review on evaluation metrics for data classification evaluations. Int. J. Data Min. Knowl. Manage. Process **5**(2), 1 (2015)
11. Japkowicz, N.: Assessment metrics for imbalanced learning. Imbalanced Learning: Foundations, Algorithms, and Applications, pp. 187–206 (2013)
12. Jouban, M.Q., Farou, Z.: Tams: Text augmentation using most similar synonyms for SMS spam filtering (2022)
13. Kull, M., Perello-Nieto, M., Kängsepp, M., Silva Filho, T., Song, H., Flach, P.: Beyond temperature scaling: Obtaining well-calibrated multi-class probabilities with Dirichlet calibration. Advances in Neural Information Processing Systems 32 (2019)
14. Morris, T., Chien, T., Goodman, E.: Convolutional neural networks for automatic threat detection in security X-ray images. In: 2018 17th IEEE International Conference on Machine Learning and Applications (ICMLA), pp. 285–292. IEEE (2018)
15. Pedregosa, F., et al.: Scikit-learn: machine learning in python. J. Mach. Learn. Res. **12**, 2825–2830 (2011)

A Hybrid GNN Approach for Predicting Node Data for 3D Meshes

Shwetha Salimath([✉]), Francesca Bugiotti, and Frédéric Magoules

CentraleSupélec, Gif-sur-Yvette, France
shwetha.salimath@student-cs.fr

Abstract. Metal forging is used to manufacture dies. We require the best set of input parameters for the process to be efficient. Currently, we predict the best parameters using the finite element method by generating simulations for the different initial conditions, which is a time-consuming process.

In this paper, introduce a hybrid approach that helps in processing and generating new data simulations using a surrogate graph neural network model based on graph convolutions, having a cheaper time cost. We also introduce a hybrid approach that helps in processing and generating new data simulations using the model. Given a dataset representing meshes, our focus is on the conversion of the available information into a graph or point cloud structure. This new representation enables deep learning. The predicted result is similar, with a low error when compared to that produced using the finite element method. The new models have outperformed existing PointNet and simple graph neural network models when applied to produce the simulations.

Keywords: Neural Network · Graph Neural Network · Deep Learning · Finite element methods · Metal forging simulations · 3D Mesh Data

1 Introduction

Metal forging is the process used to shape metals using compressive forces, and multiple parameters influence this process. Hot forging seals minor cracks and redistributes impurities leading it to be the most used in the industry. But this process has a high cost associated with the manufacturing of the forging die. This is due to the need of setting the production environment by tuning the important initial parameters by multiple iterations. In the beginning, tuning was done by producing samples resulting in lots of energy, time, and material wastage. This led to the use of simulation software to provide the best set of initial parameters for product manufacturing.

The Finite Element Method (FEM) has been used as a significant part of designing feasible metal forging processes [2]. The FEM calculates and gives us the simulation depending on the set of feasible conditions to select the best set

GNN: graph neural network.

A. Abelló et al. (Eds.): ADBIS 2023, CCIS 1850, pp. 130–139, 2023.
https://doi.org/10.1007/978-3-031-42941-5_12

of input parameters. The objective of FEM [6] is to solve partial differential equations resulting in a system of algebraic equations. Large meshes require a lot of computational time and resources for optimizing and running the process, attaching a very high time cost to the product.

Artificial neural networks [7] (ANN) are machine learning models used mostly for solving problems on conventional regression and statistical models [1]. The graph neural network (GNN) model [15] allows the processing of the data represented as a graph. They have two main purposes, (i) graph-focused, or (ii) node-focused.

In this paper, we propose a hybrid approach that uses FEM and a deep learning model together to create quicker simulations for finding the best set of input conditions. In this hybrid approach, we introduce a GNN model which is trained on a dataset of meshes. These meshes are generated by using FEM simulations for a subset of initial conditions. The trained model is then used to predict the simulations for the rest. Once trained, the model can generate one simulation in 300 ms, while the FEM would take about 45 min. The proposed model would thus be 99.9% faster than the FEM software. The simulations generated from the trained model, though not completely accurate, are good. We achieve an average mean absolute error of 10 N/meter at a mesh point, for which the actual wear ranges from 0–2000 N/m.

The paper is organized as follows. In Sect. 2 we discuss the literature review. This is followed by Sect. 3, which explains the whole process in detail. In Sect. 4 we compare our models with baseline models and end with a conclusion and future work in Sect. 5.

2 Related Work

There has been a rapid development in using 3D data for deep learning as it has numerous applications in different domains like robotics, autonomous driving, medical, and analyzing 3D objects in manufacturing industries. We can represent the 3D data as a point cloud, meshes, depth images, or grids.

Point clouds have been the most popular form of 3D representation. PointNets [12] are ANN used as a baseline for classification and segmentation tasks of 3D objects. The PointNet model upscales and then downscales the point cloud features using 1D convolution layers with activation function and max pooling.

Solving a FEM simulation is a difficult task. Attempts have been made to solve the partial differential equations using deep learning models known for their powerful function-fitting capabilities [4]. These are in the field of biomechanics to simulate phenomena in anatomical structures [11].

For applications related to automated analysis of the generated meshes, PointNet, and GNN like MeshNet and graph convolution networks are slowly being introduced, once trained, are efficient and time-saving. MeshNet [3] is used to learn features from the mesh data, which helps to solve the irregularity problems in representing the shape.

Graph convolution layers [9] use neighbor degree and node features and scale them linearly in terms of the number of graph edges to learn hidden layer representation. The graph features can be extracted without the need to perform extra transformations. The Edge convolution layer [20] uses the k-nearest neighbor of patch centers for constructing sequential embedding by extracting global features and pairwise operations for local neighborhood information. The SAGE convolution layer [5] uses sampling and aggregation of features from the neighborhood. Thus, with each iteration, more information is gained due to the aggregation, which could be mean, pooling, or graph convolution function. All of these methods were used for the classification of graphs or segmentation of graph nodes and not in generating simulations for meshes or graphs.

3 Methodology

The main objective is to create simulations as produced by the FEM for a new set of initial conditions in metal forging. The output parameter is wear at each node of the mesh, which tells us about the damage caused in the forging die during the process.

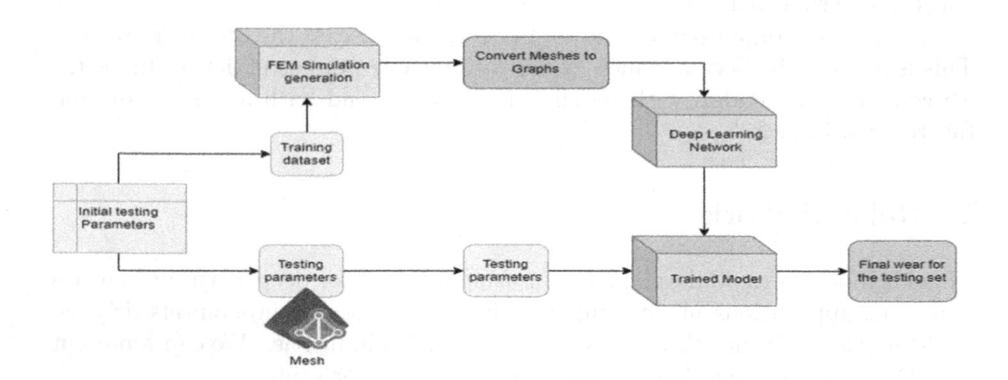

Fig. 1. Block diagram of the process

We start with the set of initial conditions as parameters. A very small subset is used to generate FEM simulations which will be used for training the GNN model. We need to extract mesh node data to create a graph or point cloud [14], used in deep learning models. Once the model is trained, we pass on the rest of the set to get the simulations. The process is represented in Fig. 1.

3.1 Dataset

The die designs of the Yoke metal forging process are provided as data by Transvalor. Mesh is composed of cells and points. We work with unstructured meshes, having sparse or arbitrary cell numbering within the mesh. We use

Transvalor packages to produce an unstructured mesh from the FEM simulations. These meshes are then analysed and converted into graphs or point clouds by extracting information using pyvista.

Each cell of the mesh has information such as temperature, pressure, displacement stress, etc. stored in them. We only require the "wear" which is our output feature to be predicted. Cells can be of two types, 2D or 3D. The meshes in our case are made of 3D tetrahedron cells. We take the x, y, and z coordinates of each mesh point instead of the cell as they do not have coordinates of their own. A point is a place of contact of cells with its neighbor. Thus we need to convert the cell data into point data. This is done by averaging the values of all cells attached at the point of contact.

The meshes were huge, about 40 Megabytes each, and are densely packed, having around twenty-seven thousand nodes. To bring our computation time and cost further down, we use just the nodes on the external surface as the "wear" of a particular area is an external feature.

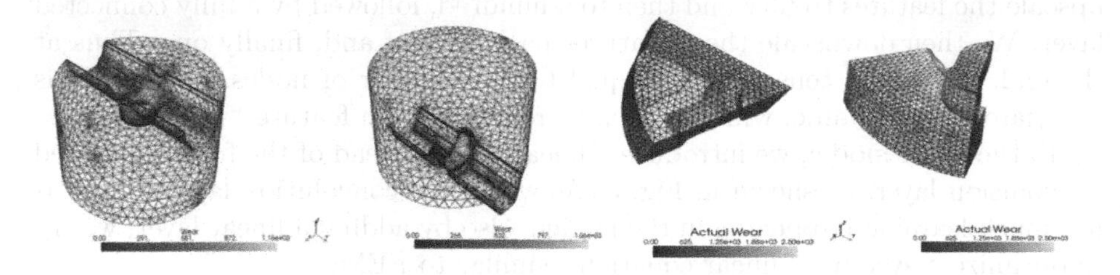

Fig. 2. External surface of the mesh

It can be seen in Fig. 2 that many nodes in the mesh have zero "wear". The yellow color represents maximum wear and the purple zero wear. Having a sparse output vector with a skewed dataset could affect our deep learning model. Thus, taking only the external surface and also selecting the training initial conditions a bit far from one another helps better fit the model. After considering just the surface points, the upper deformable die now consists of around seven thousand points, and the lower deformable die is about nine thousand points. We have a total of 40 meshes in our dataset.

$$G(V, E) \leftarrow V, Elist \tag{1}$$

We initialize the node vector V consisting of node features $(nf_1, nf_2, nf_3, nf_4, nf_5)$, such as x, y, and z coordinates, with initial parameters as temperature, and friction coefficient for which wear is needed to be calculated. The edges of the mesh are converted into an adjacency list using cells $[(n_1, n_3), (n_4, n_{10}), etc.]$ called *Elist*. We use them to create the graph to be used as input to the model.

Similarly, we later tested for a new dataset of meshes, to check the network's credibility on new data. The new mesh is comparatively smaller, about

3 Megabytes each. There are about two thousand nodes in the external surface of the mesh. There are a total of sixty-four meshes in the new dataset.

3.2 Deep Learning Models

The GNN model designs are used as a surrogate model to predict the final mesh with wear features. We have used PointNet and a GNN model with graph convolution layer [16] as a baseline model.

There are two main network architectures used. We first have a model consisting of five edge convolution layers, followed by Rectified Linear (ReLU) activation function to add non-linearity to the layers. Since "wear" can only be positive or zero, thus using a ReLU function helps us as it only allows values greater than or equal to zero to pass to the next layer by deactivating the neuron with negative output, thus training the model faster. The convolution is performed on the node features while also taking into account its neighboring node features.

The graph and node feature vector are both given to the model. We first upscale the features to fifty and then to a hundred, followed by a fully connected layer. We then downscale these features back to fifty and, finally one. Thus at the end, we have a tensor of size equal to the number of nodes. Each node is associated with a value, which is then stored as a mesh feature "wear".

In the other model, we introduce a linear layer instead of the fully connected convolution layer, as shown in Fig. 3. As with only convolution layers, there is not much learning happening in the model. Also by adding a linear layer, we try to optimize a system of linear equations, similar to FEM.

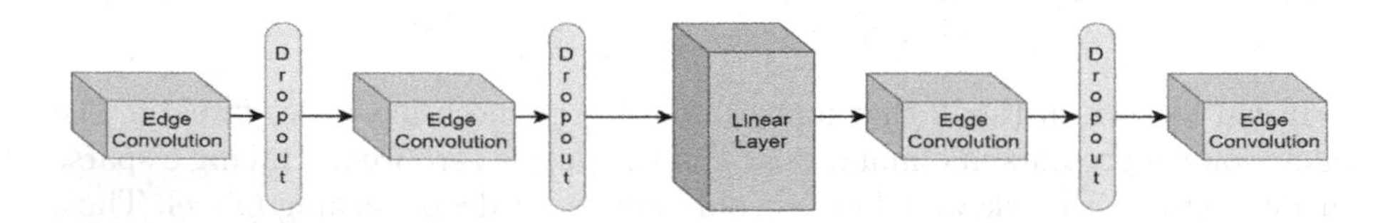

Fig. 3. GNN model using edge convolution and liner layers

On adding linear functions in between the convolutions with different positional combinations, the best position was to replace it in the between instead of the fully connected convolution layer. This not only increased the accuracy slightly but also did not require a lot of additional training time, since the total number of layers is still the same. The linear layer has input and output dimensions equal to nodes in the graph by taking a transpose of the output vector of the convolution layer. We now have a linear equation for each node, with all its convoluted features. An increase in the number of features in the liner layer did not lead to any more increase in accuracy.

Dropout layers [17] are used in between the convolution and linear layers to regularise the model, by preventing over-fitting. We are trying to create a generalized model, to make sure we do not over-fit the model when trained on a different set of data. By adding a dropout layer we randomly drop out or ignore some output node value, thus each layer is now different. It also helps to make the model more is robust by making the network adapt to correct mistakes from previous layers as each time there is a random dropout.

Different gradient descent algorithms are used to optimize the objective function consisting of model parameters by minimizing the error. The parameters are updated in the opposite direction of the gradient to reach a local or global minimum [13]. Adam optimization [8] gives better results than with just stochastic gradient [13] as it has a different learning rate associated with each parameter unlike in stochastic gradient. It is important to have a low learning rate and weight decay to not over-fit the model too early in the iteration, and to allow it to reach its correct minimum. For backpropagation, the Mean Absolute Error (MAE) [19] and the Mean Square Error (MSE) [19] are calculated over an iteration over a single node.

The other two models tested were of the same architecture as shown in Fig. 3, but we replace the edge convolution layer with the SAGE convolution layer. The model is built and trained using the deep graph library [18] with PyTorch [10] backend.

4 Results and Discussions

In this section, we discuss the criteria used to compare the models and analyze the results of our main models for both the upper and lower deformable die. In Table 1, the error percentage is represented to check the performance of the models. The error percentage is calculated as

$$Error\% = \frac{MAE}{Mean_{wear}} \quad (2)$$

Table 1. Error percentage on the old and new dataset

Model	Old dataset		New dataset	
M	lower DD	upper DD	lower DD	upper DD
Mean	90.82	48.34	305	265
Maximum	1100	857	4105	4162
Graph Convolution	39 %	65 %	21.3 %	11.3 %
PointNet	34 %	32.2 %	8.1 %	1.8 %
Edge Convolution(L)	9.3 %	13 %	6.5 %	1.8 %
SAGE convolution(L)	8.8 %	13 %	2.5 %	1.5 %

The error percentage may seem high, but it is because the mean is very low compared to higher points. This is due to the sparsity of the output vector nodes with zero wear. Table 1 also shows that even though the error was low for the upper deformable die compared to the lower one, the error percentage lets us know that is due to the overall values and the mean, in general being low. Both the SAGE and edge convolution model with the linear layer have performed very well.

For the new dataset, although both the edge and SAGE models have the best performance, PointNet has also performed quite well compared to the graph convolution. This could mean that the graph convolution network is less susceptible to changes, that is the network parameters need to be optimized again for the new dataset, which is a time-consuming process. The PointNet trains the fastest followed by GNN, SAGE, and EDGE models, which have a similar training time. Calculating the error percentage helps to better understand the results in terms of the value for the company.

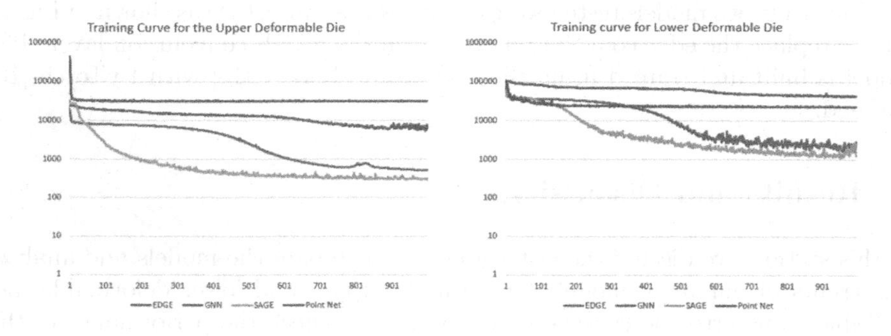

Fig. 4. Learning curve for our main models

The learning curve is plotted over the logarithm of average MSE error over all the training models. The learning rate for the upper deformable die is more smooth than that for the lower die shown in Fig. 4. The loss decreases quickly at the start except for the point cloud model and then after 700 epochs, the curve is still decreasing but at a very low pace. We could get a smoother curve with a smaller learning rate. The SAGE overall is smoother compared to the edge convolution model. We have run the model for about a thousand epochs to test for model stability. This also allows us to understand if there is a possibility of attaining a global minimum or local minimum.

The model created can be used for any similar kind of mesh data structure, without having to change the code. The number of neurons in the linear layer depends on the input dimension of nodes in the graph, which is automatically adjusted. The optimization and the model initialization parameters are the same for all four types of meshes studied. The only drawback is that sometimes the learning curve may not be smooth and very rarely it would get stuck at local minima.

4.1 Results for EDGE and SAGE Model Convolution with Linear Layer

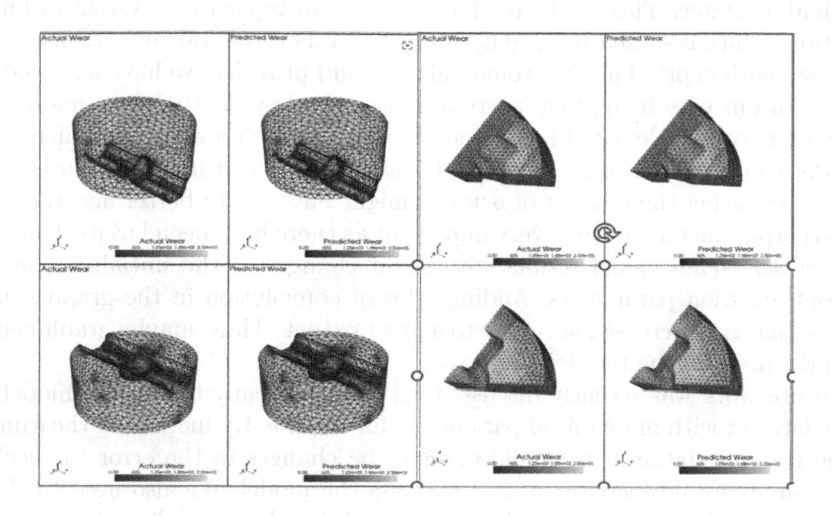

Fig. 5. Comparison of actual and predicted meshes for both old and new dataset

From the 3D figure representation of wear values at each node shown in Fig. 5, we see that the prediction is quite similar to the actual value. The pattern of the node values has been matched, with slight over-predictions at some points. We were mostly able to correctly detect the nodes with zero "wear" and the area of maximum wear. Though the values are not perfectly matched, the results are good and we are able to understand the "wear" distribution correctly over the die. Since the new die is a cylindrical sector instead of a cylinder, the mesh seen from different sides, is different, unlike the previous case when all the meshes were cylindrical. We can thus conclude the model is robust.

4.2 Time Cost Analysis for the Hybrid Process

Generating a thousand simulations with FEM would take around twenty days, can now be reduced to two to three days using the hybrid approach. We only need FEM simulations for around twenty to thirty initial conditions and the rest can be generated using the trained model. Thus, a cheaper time cost of about 85% could be achieved. The trained model takes only 300 ms to predict a new set of initial conditions. We could thus check for more initial conditions.

The time required by the neural network is proportional to the number of nodes in the mesh, as that would mean more features and parameters to be optimized. For training the upper die, the same network would take around three hours but for the lower die, it would be around five hours. It is similar for the completely new dataset as well. Since the new meshes are comparatively small, the network requires around one and a half hours to train.

5 Conclusion and Future Work

The main conclusion is that adding a linear layer to the model has increased the final accuracy. This might be due to trying to replicate a system of linear equations which is similar to the output of a FEM. For a neural network, accuracy increases with depth, but this would only be valid provided we have a large data source. In our case increasing layers increases the loss, as there are many new parameters to be calculated for the new layers, but with fewer data which leads to more complexity. If initially we had a large number of unique features then maybe increasing the number of neurons might have led to better accuracy.

Hyperparameter tuning is very important as there have been large differences between the same model results with slight changes in the initializing model and optimization parameters. Adding a lot of convolution in the graph neural network has not increased accuracy to a great extent. Thus, maybe graph neural networks need not be too deep.

Future work was recently discussed with the company to run the model on more data-set with more initial parameters for the die. By increasing the sample in the training data-set, we could observe the changes in the error to check if the accuracy would increase, or if it over-fits the model. We also need to check for more completely new mesh shapes, to check for the generality of the model.

A check on the MSE loss after every 100 epochs can be made. If the learning curve is still following a decreasing trend, then continue the process. If it is almost constant with very small fluctuations around the mean, stop the process, as it means we have reached our optimization minima. Also, a different combination of graph convolutions and graph attention layers can be used to create a new model. More structural features could be extracted from the graph. It is important to make sure that these features are not too correlated with each other, as in that case, it would decrease the accuracy of the model.

Acknowledgment. This research was conducted as part of an Internship at Mathematics and Informatics (MICS) lab, CentraleSupélec institutions under the guidance of Prof. Bugiotti and Prof. Magoules. We thank Dr. Jose Alves, Scientific Developer at Transvalor S.A for providing the datasets for conducting this research.

References

1. Abiodun, O.I., Jantan, A., Omolara, A.E., Dada, K.V., Mohamed, N.A., Arshad, H.: State-of-the-art in artificial neural network applications: A survey. Heliyon **4**(11), e00938 (2018)
2. Behrens, B.A.: Finite element analysis of die wear in hot forging processes. CIRP Ann. **57**(1), 305–308 (2008)
3. Feng, Y., Feng, Y., You, H., Zhao, X., Gao, Y.: MeshNet: mesh neural network for 3D shape representation. In: Proceedings of the AAAI Conference on Artificial Intelligence, vol. 33, pp. 8279–8286 (2019)
4. Guo, Y., Cao, X., Liu, B., Gao, M.: Solving partial differential equations using deep learning and physical constraints. Appl. Sci. **10**(17), 5917 (2020)

5. Hamilton, W., Ying, Z., Leskovec, J.: Inductive representation learning on large graphs. In: Advances in Neural Information Processing Systems, vol. 30 (2017)
6. Harari, I., Magoulès, F.: Numerical investigations of stabilized finite element computations for acoustics. Wave Motion **39**(4), 339–349 (2004)
7. Hassoun, M.H., et al.: Fundamentals of Artificial Neural Networks. MIT Press, Cambridge (1995)
8. Kingma, D.P., Ba, J.: Adam: a method for stochastic optimization. arXiv preprint arXiv:1412.6980 (2014)
9. Kipf, T.N., Welling, M.: Semi-supervised classification with graph convolutional networks. arXiv preprint arXiv:1609.02907 (2016)
10. Paszke, A., et al.: Pytorch: an imperative style, high-performance deep learning library. Advances in Neural Information Processing Systems, vol. 32 (2019)
11. Phellan, R., Hachem, B., Clin, J., Mac-Thiong, J.M., Duong, L.: Real-time biomechanics using the finite element method and machine learning: Review and perspective. Med. Phys. **48**(1), 7–18 (2021)
12. Qi, C.R., Su, H., Mo, K., Guibas, L.J.: PointNet: deep learning on point sets for 3D classification and segmentation. In: Proceedings of the IEEE Conference on Computer Vision and Pattern Recognition, pp. 652–660 (2017)
13. Ruder, S.: An overview of gradient descent optimization algorithms. arXiv preprint arXiv:1609.04747 (2016)
14. Rusu, R.B., Cousins, S.: 3D is here: point cloud library (PCL). In: 2011 IEEE International Conference on Robotics and Automation, pp. 1–4. IEEE (2011)
15. Scarselli, F., Gori, M., Tsoi, A.C., Hagenbuchner, M., Monfardini, G.: The graph neural network model. IEEE Trans. Neural Networks **20**(1), 61–80 (2009). https://doi.org/10.1109/TNN.2008.2005605
16. Shivaditya, M.V., Alves, J., Bugiotti, F., Magoules, F.: Graph neural network-based surrogate models for finite element analysis. arXiv preprint arXiv:2211.09373 (2022)
17. Srivastava, N., Hinton, G., Krizhevsky, A., Sutskever, I., Salakhutdinov, R.: Dropout: a simple way to prevent neural networks from overfitting. J. Mach. Learn. Res. **15**(1), 1929–1958 (2014)
18. Wang, M., et al.: Deep graph library: a graph-centric, highly-performant package for graph neural networks. arXiv preprint arXiv:1909.01315 (2019)
19. Wang, W., Lu, Y.: Analysis of the mean absolute error (MAE) and the root mean square error (RMSE) in assessing rounding model. In: IOP Conference Series: Materials Science and Engineering, vol. 324, p. 012049. IOP Publishing (2018)
20. Wang, Y., Sun, Y., Liu, Z., Sarma, S.E., Bronstein, M.M., Solomon, J.M.: Dynamic graph CNN for learning on point clouds. ACM Trans. Graph. (tog) **38**(5), 1–12 (2019)

Transformers and Attention Mechanism for Website Classification and Porn Detection

Lahcen Yamoun[1,2](✉), Zahia Guessoum[1], and Christophe Girard[2]

[1] CReSTIC EA 3804, University of Reims Champagne Ardenne, Reims, France
FL_Yamoun@esi.dz
[2] Efficient IP, La Garenne-Colombes, France

Abstract. Detecting pornographic content on the web is an important challenge in protecting users from inappropriate content. The heterogeneity of the web, the diversity of used languages, and the existence of implicit pornography using language that cannot be detected by keywords, make this task difficult. There are very few published works on text-based web classification. In this paper, we propose a novel approach that addresses these challenges. We tackle web porn detection based on multiple pages for the same website, by incorporating an attention mechanism to treat pages according to their respective importance, and the first to use transformers for web porn detection. Our method outperforms various other approaches that do not incorporate attention. With our multilingual solution, we achieved the accuracy of 91.59% on a hand-labeled test set for the task of porn detection.

Keywords: Porn detection · Web classification · Transformers · Attention mechanism · Multilingual

1 Introduction

World Wide Web has revolutionized communication by providing an unparalleled reach and continuous availability. However, the ease with which anyone can publish content on the web has resulted in the abusive exploitation of freedom of expression, leading to the dissemination of harmful content, such as violence, drugs, and more specifically: pornography. Unfortunately, this issue is particularly concerning for children and adolescents. Statistics show that a vast majority of them come across pornographic websites before they reach 18 years old. For instance, 90% of boys and 60% of girls come across pornographic websites before that age[1]. A recent study [3] found that, on average, 63% to 68% of teens have watched porn in their lifetime. It is therefore urgent to provide an effective filtering of web content to protect children and other unsuspecting users from the negative effects of such content. Porn detection is an important and useful solution. This work deals with the task of detecting porn on the web based on textual data.

[1] https://everaccountable.com/blog/how-pornography-affects-teenagers-and-children/.

A. Abelló et al. (Eds.): ADBIS 2023, CCIS 1850, pp. 140–149, 2023.
https://doi.org/10.1007/978-3-031-42941-5_13

The detection of pornographic textual content can be formulated as a binary text classification task with the labels: Porn and Non-Porn (white). However, several constraints exist, the classification of websites is a subject, unlike the generic classification of texts and images, not much studied and published, and the existing solutions remain private. Also, even if the detection of flagrant pornographic sites remains a task where perfect results can be achieved (see [19]), pornographic content can be found implicitly, usually in teenage blogs, and in sites designed in such a way as to avoid filtering systems. In these cases, it becomes difficult to detect porn material and generalize the trained models. It requires a good understanding of the semantics of the text by the NLP models. There are also constraints related directly to the web, such as the heterogeneity of the content and the lack of the logical structure usually found in textual content, the multitude of languages, the length of the texts exceeding the limits accepted by deep learning models.

In this work, we propose a new website classification approach. We consider 100 languages and include blogs in our dataset to generalize our model as much as possible and treat edge cases.

The paper is organized as follows: We start in Sect. 2 by describing and analysing the existing works dealing with web classification in general and the detection of porn on the web in particular. And then in Sect. 3, we detail our approach, explaining the data collection and preparation, and the classification methodology. In Sect. 4 we present the obtained results, and we end up with a conclusion and perspectives in Sect. 5.

2 Related Work

Classification of web content is an increasingly important area of study, with significant implications for a range of applications, from content moderation to targeted advertising. However, despite extensive research on the classification of generic text, the classification of web content remains relatively underexplored. In particular, there is a dearth of research on the detection of pornographic content specifically, with few works focusing on web-based approaches. Sahoo et al. [15] identified several key criteria for classifying websites, including content-based features such as text, images, styles, and scripts, as well as blacklist features indicating whether a website is present in known blacklists or not. Moreover, lexical features, host-based features and third-party features such as Alexa ranking were identified as important factors in website classification.

Several textual content-based methods for web classification were proposed. In [10], in order to build a database, the authors extracted texts from a list of websites, and then according to a list of keywords per class they automatically labelled their database via a voting system. To train the classification model, they considered BoW embeddings and a feature selection method, and concluded that a fully connected two layer network gives better results than SVM or Random Forrest models. The idea of using keywords for classification is also used in [13]. LSTMs were used in [4] with text and some meta-tags of websites: title, description, keywords. Five layers LSTM model was trained on a maximum length sequences of 100, with BoW embeddings to achieve almost 85% accuracy on a classification of 23 classes database. The work of [6] is, to the best of our knowledge and according to the authors themselves, the only one using transformers (BERT was used) for the website classification. The authors did not give details on which text they used from the websites. However, on the 5000best dataset [1], containing 5000 websites classified in 32 categories, they compared the results obtained from BERT, from an LSTM with pretrained GloVe embeddings, and from an LSTM with pretrained GloVe embeddings and char-level embeddings concatenated. BERT gave the best results: 67.81% in terms of accuracy.

Porn text detection relying on web textual content was rarely studied before. Authors of [9] have proposed handcrafted features to build a decision tree allowing the referral of a page to one of the categories: Continuous Text, Discrete Text and Image pages. According to the category, they use a suitable classifier based or not on the context and the image. The considered text classifier methods rely on Bayes' theorem and keywords, and results of image-based classification and discrete text classification are fused. Recently, authors of [16] worked on porn/normal text classification. They used two databases of articles in English and Chinese from stories of harassment. Their work is inspired by Multiple Instance Learning [2,8] and they considered the context of sentences and prior knowledge of words. They also used the principle of attention at both levels of words and sentences. For prior knowledge, they thought of an automatically generated pornographic lexicon with the help of BoW aiding in the intermediate predictions of the sentences, which once aggregated, give the final classification. Attention principle is also used in [17] with BiLSTM and multi-head self-attention [18] to deal with the constraint of the text lengths in the web pages.

To the best of our knowledge, our work is the first to use transformers for the task of pornographic web detection, and the first to tackle the multilingual constraint for this task by avoiding any language dependency caused by techniques such as BoW or the use of pre-defined keywords. Also it is the first to consider the classification of a domain based on several pages, while incorporating the task of paying attention to important pages over others into the classification model.

3 Page-Level Attention-Based Solution for Web Textual Porn Detection

This section presents our study. It first describes the collection and preparation of data, and it then explains the methodology proposed for the classification.

3.1 Data Collection and Preparation

Our research tackled the problem of identifying pornographic websites through a binary classification task. Our goal was to distinguish between pornographic and non-pornographic websites. To effectively train and test our model, we used real-world data that included websites from various categories, such as news and sports, to represent non-pornographic websites. We utilized several databases, including 5000best [1] and DMOZ [7], which are general website classification databases, and UT-Capitol [14], which specifically focuses on adult websites.

The focus of our research is website classification, where each website is treated as a set of pages. Due to the varying number of pages per website, we limit our analysis to a subset of pages. We determined a maximum of 30 subpages per website, selecting them in a FIFO order starting from the index page and direct links from the index. This approach allows us to efficiently classify websites based on their content, with a focus on showcase pages as the most indicative of pornographic content.

In our study, the first step in extracting textual content was to eliminate dead links using HTTP get requests and reacting to responses. Once we had identified the well-reachable websites, we used the playwright browser automation framework to scrape their textual content. We chose to use a browser automation approach rather than traditional scraping techniques that rely on HTTP requests alone because it allowed for the execution of JavaScript on the client side rendered (CSR) pages, resulting in a more accurate rendering of the websites' contents. The textual data taken from each website are the description and keyword metadata, the page title, the titles from h1 to h5, the texts forming a link, and then the texts under the tags: div, span, p. All these texts are concatenated and separated by spaces. The domain name generally inspires the category of the site, for this reason we added it in front of the extracted text, except if this domain name appears in a list that we have pre-established, the reason for this choice is detailed below. Also when the subdomain of the URL is not in the set {"www", "mobile", "mob", "m"}, generally it includes important information not to be discarded, this case often comes in blogs for example where we can find a URL like *http://pornographic_reference.blogname.com*, for this reason the subdomain is also concatenated in front of the domain in the textual content extracted from the page.

Actually, the UT-Capitole dataset [14] we used in our work for porn websites, contains disproportionately a lot of blogs from *Blogspot* and *Tumblr*, and sites hosted at free solutions, we illustrate this in Table 1, where the percentage of domains in the adult category of this dataset is highlighted, showing the monopolization of blogspot mainly, which explains that UT-Capitole was created based on blogs coming from this site. In addition, in the dataset we crawled, we found that 100% of the pages coming from *Blogspot* and *Tumblr* are in porn category. This creates a bias that should not exist. To eliminate this, a list of blogs and web hosting names to be filtered was established from statistics on the most frequent domain names in our training dataset. Note that the list is not necessarily exhaustive. It's sufficient to cover the blogs and hosting solutions present in our training database to eliminate any potential intrinsic bias.

We have thus built a multilingual dataset consisting of 6476 domains, encompassing a total of 185430 web pages. The dataset features a high degree of linguistic diversity, with over 80 languages represented. For example, English rep-

Table 1. Relative frequencies of main domains in the Adult category of the UT-Capitole dataset [14].

Domain	blogspot	tumblr	appspot	000webhostapp	All other domains
Frequency (%)	79.33%	3.89%	0.29%	0.18%	16.31%

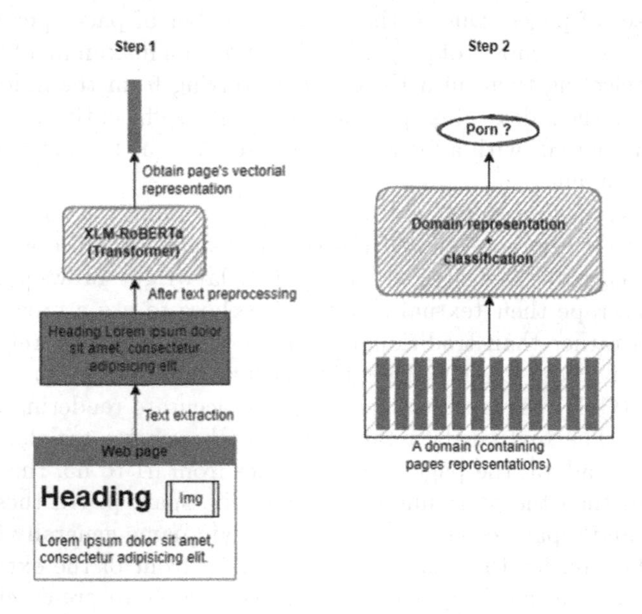

Fig. 1. The proposed website classification approach.

resents 65% of pages either exclusively in English or containing mainly English text. In terms of languages other than English: French represents 28%, Chinese represents 10.4%, Russian represents 8.2% and Japanese 5.4%.

3.2 Classification Methodology

To the best of our knowledge, this study is the first to utilize multiple pages of a website for its classification. The first step builds a vector representation of each individual page, while the second step aggregates the page vectors of the same domain to obtain the domain representation and classify it. The aggregation relies on an unsupervised attention approach, which treats pages differently according to the importance of each page. Figure 1 illustrates this approach.

Page Representation. A page, represented by the text extracted during the crawling, must be represented by a vector. Different text preprocessing techniques can be used, the proposed technique is described in Fig. 2. It uses as input a text that may be too long, and it provides a vector. It starts with a preprocessing step of the text that includes tokenization, truncation to a maximum size in terms of the number of tokens and eventually padding. The used

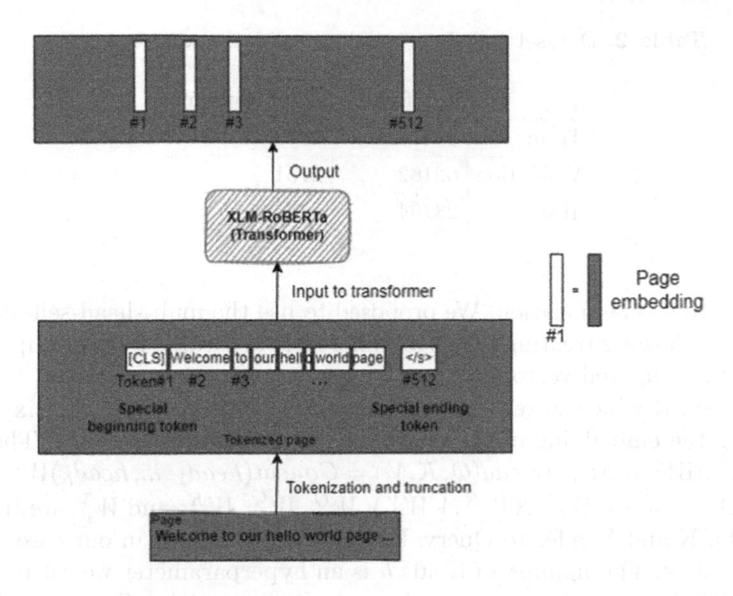

Fig. 2. Diagram illustrating page representation.

tokenization is a sub-word tokenization, the sentencepiece algorithm [11] is used to meet the needs of multilinguality. A special token "CLS" has been added at the beginning of the sequence of tokens to represent the whole text, and thus the page. Regarding the size, if needed a truncation of the first 512 tokens is done to deal with the size constraints of the XLM-RoBERTa transformer [5], and padding is done in the case of loading texts of different sizes in batch.

The XLM-RoBERTa transformer [5] was used to give a vector representation for the sequence given as input. It is based on earlier large multilingual Facebook's RoBERTa model [12] released in 2019. Using a multilingual transformer offers a number of advantages over a monolingual one. Firstly, it exploits a multitude of languages without heavy pre-processing, in instance translation. Secondly, in its pre-training logic, it can leverage the shared representations across languages, allowing it to transfer knowledge learned from one language to another, and thus improve the quality of text representation, even for low-resource languages where labeled data may be scarce. Thirdly, it can capture the nuances of language better than a monolingual model. Finally, it's less time-consuming and less error-prone than training several monolingual transformers, and picking the right one when a text arrives.

We used the pre-trained version of XLM-RoBERTa without any further fine-tuning. We input the tokenized sequence with the special token CLS at its beginning. The output of this model, for each input token, is a vector of dimension 768. We extract from the output the first vector corresponding to CLS, which represents the semantics of the whole sequence.

Domain Classification. At this step, each page is represented by a vector of size 786, we aim at using the embedding of the pages belonging to the domain

Table 2. Dataset split into training, validation and test sets.

	N. of pages	N. of domains
Train	99474	3473
Validation	62162	2151
Test	23794	852

to proceed to its classification. We proposed to use the multi-head-self-attention mechanism, allowing treating the pages differently according to their importance, by obtaining weighted vectors.

A domain d which is represented pages $p_{1..n}$, becomes the matrix D concatenating the embedding of the pages, i.e., $D = Concat(p_1, .., p_n)$. The equation of the MHSA: $MultiHead(Q, K, V) = Concat(head_1, ..., head_h)W^O$ Where $head_i = Attention(QW_i^Q, KW_i^K, VW_i^V)$; W^O, W_i^Q, W_i^K, and W_i^V are trainable weights. Q, K and V refer to Query, Key and Value, which in our case, are the same D matrix. The number of heads h is an hyperparameter we set to 12.

The MHSA provides a sequence of vectors (i.e., a matrix) $D_{attention}$. A residual connection with D is added before the classification. A vector V_d of size 786 representing the domain is defined as follows: $V_d = mean_{rows}(D + D_{attention})$

As shown in Fig. 3, we make the classification of V_d through a fully connected network. Its architecture did not have a significant impact on the results, but it is worth noting that we opted for two layers, each with 64 neurons and ReLU as nonlinearity. The output layer utilized two neurons with Softmax instead of one neuron with Sigmoid. Both MHA and FCN had a dropout rate of 0.15. The Binary cross entropy was used as loss function, and Adam for optimization with a starting learning rate of 0.001 and a rate decay of 0.1 every 10 epochs.

4 Experimental Results

In order to train and test our approach, we divided our dataset in three parts: training, validation and testing. We labelled the test part manually in order to be confident of the advanced result (see Table 2).

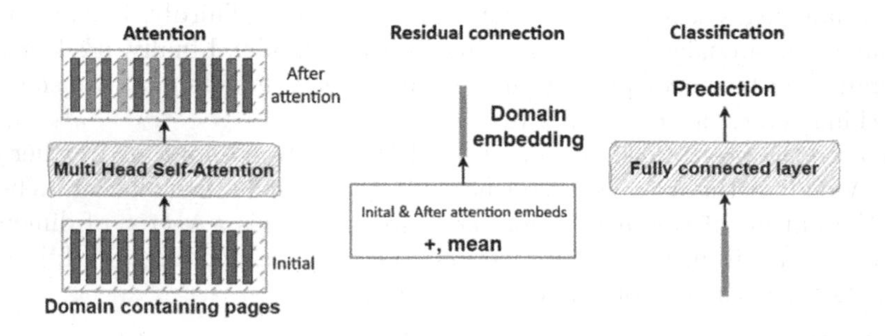

Fig. 3. Website classification.

Table 3. Comparison showing the relevance of our approach.

	F1-Score	Accuracy
Scenario 1 (proposed approach)	**90.15%**	**91.59%**
Scenario 2	89.01%	89.44%
Scenario 3	82.71%	83.80%
Scenario 4	81.88%	83.22%

Table 4. Results for the part of the dataset including only non-English sites.

	F1-Score	Accuracy
Original (No translation)	89.31%	90.3%
Translated to english	88.2%	89.12%

To measure the relevance of our approach, we compared the results of four scenarios: 1) The approach described in this paper, 2) An approach eliminating the attention + residual connection block, and simply averaging the embedding of the pages to obtain the domain embedding, 3) An approach classifying pages independently with a fully connected network, and after that averaging page predictions to get a domain-level prediction and 4) Consider only the index page for domain classification. The results are reported in Table 3.

Our approach gave the best results. The first scenario shows the benefit of using attention at the page level, since it brought an improvement of more than 1% in terms of f1-score and more than 2% in terms of accuracy compared to the sole average of the pages' embeddings. More data would further illustrate this difference. Also, the idea of making a unique prediction at a domain level using all the pages is better than treating them independently, which is illustrated by comparing the results of the first two scenarios with the third. And finally, using only the index page to rank a website gives worse results than exploring a set of pages for a given domain, the difference being more than 8% in terms of f1-score and accuracy in favor of our approach. Hence, the choice of using a set of pages to represent a domain, while incorporating an attention mechanism that weights the pages according to their importance, improves the results of website classification, particularly for the task of porn detection.

To evaluate the effectiveness of our solution for multilingual web content, we present in Table 4 the results obtained from websites in languages other than English. Specifically, we treat website pages in two different ways: (1) by analyzing the original pages in their respective languages, and (2) by translating the pages into English before analyzing them. This approach allows us to assess the performance of our solution in handling diverse linguistic contexts, which is a critical aspect of real-world web content analysis.

The results on the part of our test set of pages with languages other than English are very close to the results set out in Table 3, with a difference of $\sim 1\%$

only, showing that the approach does treat languages as a single entity, and does not act according to the language. This is confirmed in Table 4, where translation did not improve the results, and processing on the original languages directly gave even better results. The slight deterioration, when performing translation, can be explained by the heterogeneous nature of the web, where automatic translators have difficulty retaining the structure of the original language, since the extracted text from web pages as a whole does not have to be treated as a coherent text.

Note that we tried an end-to-end training: including fine-tuning the transformer, but the results were inconclusive. So we leave the improvement of this task for a future work, starting by obtaining more data.

5 Conclusion and Perspectives

In this study, we introduced an original approach for classifying websites, with a specific focus on detecting pornographic content. Our approach has broad applicability, and can be easily adapted for use in a range of filtering and categorization systems. We proposed the exploitation of several pages to represent a given domain, using an attention mechanism to weight the pages according to their importance in an unsupervised way. The results demonstrated a benefit on this choice. The proposed method is language-independent, since the representation of the pages is done through a multilingual transformer, and does not rely on keywords or language-dependent representations. To the best of our knowledge, our study is the first to propose this approach to website classification. Results demonstrate the efficacy of our method, highlighting the potential for future research in this area.

Moving forward, our results could be further improved by leveraging more reliable and less noisy databases. Obtaining such data would allow refining our approach by fine-tuning the transformer. This would likely lead to even more accurate and effective website representation. Also, in this work we proposed to use the MHA with a residual connection to obtain a sites' representations, it will be interesting to test other methods, such as RNN-Like architectures, or CNN.

References

1. Best websites. https://5000best.com/websites/. Accessed 25 Apr 2022
2. Angelidis, S., Lapata, M.: Multiple instance learning networks for fine-grained sentiment analysis. Trans. Assoc. Comput. Linguist. **6**, 17–31 (2018)
3. Bőthe, B., et al.: A longitudinal study of adolescents' pornography use frequency, motivations, and problematic use before and during the covid-19 pandemic. Arch. Sex. Behav. **51**(1), 139–156 (2022)
4. Buber, E., Diri, B.: Web page classification using RNN. Procedia Comput. Sci. **154**, 62–72 (2019). https://doi.org/10.1016/j.procs.2019.06.011. https://linkinghub.elsevier.com/retrieve/pii/S187705091930780X
5. Conneau, A., et al.: Unsupervised cross-lingual representation learning at scale. arXiv preprint arXiv:1911.02116 (2019)

6. Demirkıran, F., Çayır, A., Ünal, U., Dağ, H.: Website category classification using fine-tuned BERT language model. In: 2020 5th International Conference on Computer Science and Engineering (UBMK), pp. 333–336, September 2020. https://doi.org/10.1109/UBMK50275.2020.9219384
7. DMOZ: Dmoz open directory project. https://dmoz-odp.org/. Accessed 21 Jun 2023
8. Hellman, S., et al.: Multiple instance learning for content feedback localization without annotation. In: Proceedings of the Fifteenth Workshop on Innovative Use of NLP for Building Educational Applications, pp. 30–40 (2020)
9. Hu, W., Wu, O., Chen, Z., Fu, Z., Maybank, S.: Recognition of pornographic web pages by classifying texts and images. IEEE Trans. Pattern Anal. Mach. Intell. **29**(6), 1019–1034 (2007)
10. Karthikeyan, T., Sekaran, K., Ranjith, D., Vinoth, K.V., Balajee, J.M.: Personalized content extraction and text classification using effective web scraping techniques. Int. J. Web Portals **11**(2), 41–52 (2019). https://doi.org/10.4018/IJWP.2019070103. https://services.igi-global.com/resolvedoi/resolve.aspx?doi=10.4018/IJWP.2019070103
11. Kudo, T., Richardson, J.: Sentencepiece: a simple and language independent subword tokenizer and detokenizer for neural text processing. arXiv preprint arXiv:1808.06226 (2018)
12. Liu, Y., et al.: Roberta: a robustly optimized Bert pretraining approach. arXiv preprint arXiv:1907.11692 (2019)
13. Patel, A.D., Sharma, Y.K.: Web page classification on news feeds using hybrid technique for extraction. In: Satapathy, S.C., Joshi, A. (eds.) Information and Communication Technology for Intelligent Systems. SIST, vol. 107, pp. 399–405. Springer, Singapore (2019). https://doi.org/10.1007/978-981-13-1747-7_38
14. Prigent, F.: Blacklist université de toulouse 1 (ut-capitole). https://dsi.ut-capitole.fr/blacklists/. Accessed 21 Jun 2023
15. Sahoo, D., Liu, C., Hoi, S.C.H.: Malicious URL detection using machine learning: a survey. arXiv:1701.07179 [cs], August 2019
16. Song, K., Kang, Y., Gao, W., Gao, Z., Sun, C., Liu, X.: Evidence aware neural pornographic text identification for child protection. In: Proceedings of the AAAI Conference on Artificial Intelligence, vol. 35, no. 17, pp. 14939–14947 (2021). https://ojs.aaai.org/index.php/AAAI/article/view/17753
17. Sun, G., Zhang, Z., Cheng, Y., Chai, T.: Adaptive segmented webpage text based malicious website detection. Comput. Networks **216**, 109236 (2022). https://doi.org/10.1016/j.comnet.2022.109236. https://www.sciencedirect.com/science/article/pii/S1389128622003140
18. Vaswani, A., et al.: Attention is all you need. In: Advances in Neural Information Processing Systems, vol. 30 (2017)
19. Yamoun, L., Guessoum, Z., Girard, C.: Transformer RoBERTa vs. TF-IDF for websites content-based classification. In: Deep Learning meets Ontologies and Natural Language Processing, International Workshop in conjunction with ESWC, Hersonissos, Greece (2022). https://hal.archives-ouvertes.fr/hal-03725602

Analysing Robustness of Tiny Deep Neural Networks

Hamid Mousavi[1(✉)], Ali Zoljodi[1], and Masoud Daneshtalab[1,2]

[1] Mälardalen University, Universitetsplan 1, 722 20 Västerås, Sweden
{seyedhamidreza.mousavi,ali.zoljodi,masoud.daneshtalab}@mdu.se
[2] Tallinn University of Technology (Taltech), Akadeemia tee 15A, Tallinn, Estonia
masoud.daneshtalab@taltech.ee

Abstract. Real-world applications that are safety-critical and resource-constrained necessitate using compact and robust Deep Neural Networks (DNNs) against adversarial data perturbation. MobileNet-tiny has been introduced as a compact DNN to deploy on edge devices to reduce the size of networks. To make DNNs more robust against adversarial data, adversarial training methods have been proposed. However, recent research has investigated the robustness of large-scale DNNs (such as WideResNet), but the robustness of tiny DNNs has not been analysed. In this paper, we analyse how the width of the blocks in MobileNet-tiny affects the robustness of the network against adversarial data perturbation. Specifically, we evaluate natural accuracy, robust accuracy, and perturbation instability metrics on the MobileNet-tiny with various inverted bottleneck blocks with different configurations. We generate configurations for inverted bottleneck blocks using different width-multipliers and expand-ratio hyper-parameters. We discover that expanding the width of the blocks in MobileNet-tiny can improve the natural and robust accuracy but increases perturbation instability. In addition, after a certain threshold, increasing the width of the network does not have significant gains in robust accuracy and increases perturbation instability. We also analyse the relationship between the width-multipliers and expand-ratio hyper-parameters with the Lipchitz constant, both theoretically and empirically. It shows that wider inverted bottleneck blocks tend to have significant perturbation instability. These architectural insights can be useful in developing adversarially robust tiny DNNs for edge devices.

Keywords: Robustness analysis · Adversarial training · Adversarial data perturbation · Lipchitz constant

1 Introduction

Deep Neural Networks (DNNs) are increasingly employed in safety-critical applications [18]. Recent research indicates that DNNs are susceptible to adversarial data perturbations, which are small, imperceptible noises that add to the input data [6,19]. Adversarial data can be generated by different adversarial attack

A. Abelló et al. (Eds.): ADBIS 2023, CCIS 1850, pp. 150–159, 2023.
https://doi.org/10.1007/978-3-031-42941-5_14

methods to fool the DNNs [4,7,14]. In addition, the resource-constraint edge devices require to employ the tiny DNNs [1]. To this end, some tiny DNNs such as MobileNet-tiny [12] have been designed to deploy on edge devices with high accuracy on clean data. However, designing a robust and tiny DNN is a critical challenge in these applications [26]. To address the robustness of DNNs against adversarial data, adversarial training methods have been proposed as state-of-the-art defense methods [14,16,24]. Nevertheless, adversarial training methods have been analysed for large-scale DNNs such as WideResNet with high capacity (huge number of parameters) [10,22,23,25]. However, there are no comprehensive analyses of adversarial training for the tiny DNNs. In this paper, we present the first robustness analysis for tiny DNNs from the architectural perspective. Our analysis is based on MobileNet-tiny as the extensive architecture used for tiny applications. The baseline network architecture of MobileNet-tiny depicts in Fig. 1(a). It is composed of n inverted bottleneck block which is configured with two hyper-paramcters: width-multiplier and expand-ratio. We analyse the impact of increasing the width of MobileNet-tiny on the performance of the network by expanding the width-multiplier and expand-ratio hyper-parameters of inverted bottleneck blocks. These expanded blocks are illustrated in Fig. 1(b). The width-multiplier increases the width of the network by extending the number of output channels of the blocks, as shown in Fig. 1(b)-left. The expand-ratio hyper-parameter extends the number of middle channels inside the inverted bottleneck blocks, as shown in Fig. 1(b)-right. To comprehensively evaluate the baseline and expanded networks, we leverage different metrics, including natural accuracy, robust accuracy, and perturbation instability (the mathematical definition of the metrics is presented in the Sect. 3.1). Natural accuracy and robust accuracy measure the ratio of clean and adversarial data that can be correctly classified by trained DNNs. Perturbation instability shows the difference between the distribution of the predictions for natural and adversarial data without focusing on correct labels. To support our observations, we theoretically and empirically analyse the relationship between the width-multiplier and expand-ratio of inverted bottleneck blocks with the Lipchitz constant. The Lipchitz constant indicates the stability of the network output to data perturbations, and the larger Lipschitz constant value corresponds to the instability of the network. The following important insights have been discovered by our investigation:

1. Extending the inverted bottleneck blocks in MobileNet-tiny with both width-multiplier and expand-ratio hyper-parameters improves the natural and robust accuracy and increases the perturbation instability.
2. There is a threshold for expanding the width of the network that improves the natural and robust accuracy. Beyond the threshold, the improvement is negligible, and perturbation instability significantly increases.
3. The theoretical and experimental values for the Lipschitz constant upper bound show that increasing the width and depth of the MobileNet-tiny increase the perturbation instability

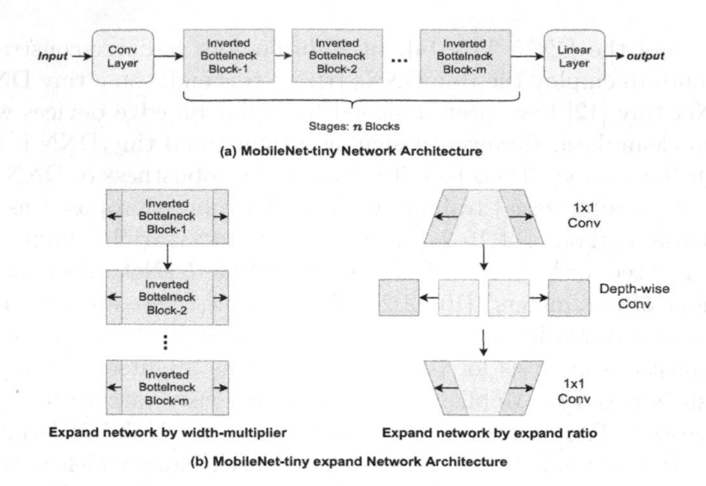

Fig. 1. (a) The baseline architecture of MobileNet-tiny with n inverted bottleneck blocks and (b) expanded networks by changing width-multiplier (left) and expand-ratio (right) hyper-parameters.

2 Related Works

2.1 Adversarial Data and Defences

The pretrained deep neural networks are vulnerable to adversarial data, which can be generated by the *Fast Gradient Sign Method (FGSM)* [7], *Projected Gradient Descent (PGD)* [14], and *Carlini and Wagner (CW)* [3] approaches. In addition, *Auto Attack (AA)* [4] is the ensemble of four attacking methods that generate powerful adversarial perturbed data. Adversarial training is the current state-of-the-art defense method against adversarial data. The first adversarial training method employed clean and adversarial data to train a robust deep neural network [7]. The robustness of the network can be increased by encouraging similar logits for clean and adversarial data [11]. To enhance robustness, adversarial training is reformulated as a min-max optimization problem, and the network is trained exclusively on adversarial data [14]. Theoretically, TRADES [24] regularizes the loss function for clean data by incorporating a robust loss term and making a trade-off between them. Improved variants of TRADES have been proposed to consider regularization terms and reduce the distance between the distribution of natural data and their adversarial counterpart [20]. We use TRADES as the default adversarial training method because it uses both natural and robust loss terms to improve robustness.

2.2 Robustness from Architectural Perspective

Researchers have investigated the relationship between robustness against adversarial data and the architecture of DNNs [10,22,23]. Xie et al. [23] have studied the impact of the depth of large-scale DNN networks on adversarial training. They have found that the number of layers in the WideResNet network has a much more substantial effect on robust accuracy than natural accuracy. Their experiments shed light on the intricate relationship between DNN architecture and robustness against adversarial data. Furthermore, Xie et al. [5] have also explored the role of batch normalization layers in the performance of adversarial training, particularly in large-scale datasets such as ImageNet-1K [5]. Their experiments demonstrate that proper batch normalization techniques markedly impact robust accuracy. In addition to the number of layers, the impact of the width of robust accuracy of the large-scale WideResnet network has been studied in [22]. B. Wu et al. [22] showed that the robustness against adversarial data is related to natural accuracy and perturbation stability parameters. Their studies illustrated that increasing the width of WideResNet improves natural accuracy but disprove the perturbation stability. They also elaborate that increasing the DNN width in large-scale networks reduces the overall robust accuracy. H. Huang et al. [10] obtain a study on the impact of DNN architecture on robustness against adversarial data. According to their results, increasing the network capacity (number of parameters) does not necessarily increase its robustness against adversarial data. They also indicate that reducing the capacity in the last blocks of the network may increase the robustness. H. Huang et al. [10] prove that with a constant number of parameters, we can find a DNN architecture with the optimum robustness. Although the research as mentioned earlier studies analysed the robustness of DNNs against adversarial data, but they used large-scale networks such as WideResNet with 127 million parameters. Due to the increase in the number of edge devices, it is necessary to analyse the robustness of tiny networks such as MobileNet-tiny. In this paper, we analyse the robustness of these tiny networks to find meaningful insights into designing adversarially robust tiny networks.

2.3 Tiny Deep Learning

By increasing the usage of tiny edge devices, the demand for DNNs with lower resource consumption and inference time is growing [17]. Recently, tiny deep learning networks [2,13,15] have been proposed to reduce DNNs computation cost and latency. Network pruning [8] and weight quantization [13] are two common approaches to compress existing networks without manipulating the number of layers and hyper-parameters. On the other hand, designing tiny DNN networks from scratch [2,15,17] is another widely used technique in tiny deep learning. Tiny DNN networks can be either designed manually [17] or using AutoML approaches [2]. We use MobileNet-tiny, which is designed based on neural architecture search method [12]. It has only $0.4M$ parameters and is significantly smaller than WideResNet networks. From an architectural perspective, there has not been any robustness exploration for these tiny networks. This paper analyses their robustness and finds some insights into designing tiny networks.

3 Exploring Robustness

To analyse robustness for configurations of MobileNet-tiny, we need to define the baseline architecture and metrics used for evaluation (Sect. 3.1). In Sects. 3.3 and 3.4, we demonstrate the results for expanding the network based on the width-multiplier and expand-ratio. In Sect. 3.5, we theoretically and empirically show the relation between these hyper-parameters with the Lipschitz constant.

3.1 Baseline Network and Evaluation Metrics

We take the MobileNet-tiny [12] network as the baseline architecture. Figure 1(a) show the overall architecture of this network. This architecture consists of 6 inverted bottleneck blocks that we expand them to generate a wider network. We denote the width-multiplier and expand-ratio for all inverted bottleneck blocks as W and E, respectively. For the baseline network, the width-multiplier and expand-ratio are set to 0.35 and 6, respectively. We explore the impact of W and E while other hyper-parameters are fixed. In terms of metrics, we consider different aspects of the performance of the tiny network as follows.

Natural Accuracy: the ratio of examples that are correctly classified as:

$$Acc_{Nat} = \frac{\#\{x : \forall x \in D, f(\theta; x) == y\}}{\#examples} \tag{1}$$

D, $f(\theta; .)$, and y indicate the test dataset, network, and correct labels.

Robust Accuracy: ratio of adversarial data that are correctly classified as:

$$Acc_{Rob} = \frac{\#\{x : \forall \hat{x} \in \mathcal{B}(x, \epsilon), f(\theta; \hat{x}) == y\}}{\#examples} \tag{2}$$

where $\mathcal{B}$ indicates the l_p norm ball around the natural example x.

Perturbation Instability: the difference between the prediction of the network for natural and adversarial examples. We use the KL-divergence statistical measure to compute the perturbation instability as:

$$Pert_{Inst} = \mathbb{E}_{x \sim D} KL(f(\theta; x), f(\theta; \hat{x})) \tag{3}$$

where $\mathbb{E}$ and $\hat{x}$ indicate the expectation function and adversarial example.

3.2 Experimental Setting

We train the explored networks using TRADES [24] on the CIFAR-10 training data. For adversarial training settings, we use l_∞ norm by setting the maximum perturbation size to $\epsilon = 8/255$, and use 10-steps PGD with step size $\alpha = 2/255$. For robustness evaluation, we use a 20-PGD attack to generate adversarial data with the same perturbation size ($\epsilon = 8/255$) on CIFAR-10 test data.

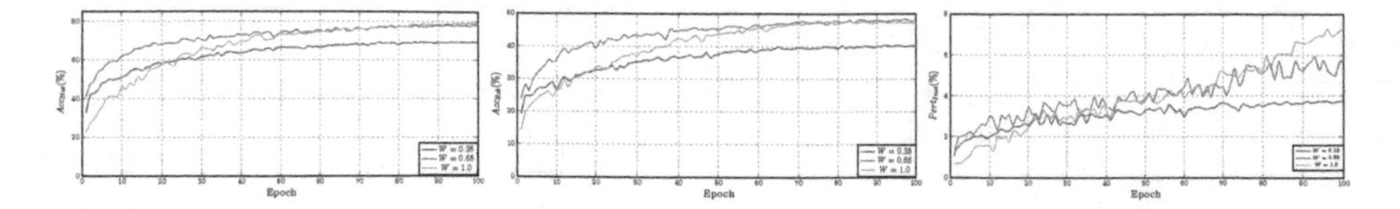

Fig. 2. The dynamics of natural accuracy, robust accuracy and perturbation instability with regard to training epochs and fixed expand-ratio ($E = 6$) on CIFAR-10 dataset.

3.3 Exploring Different Width-Multipliers

We first explore the impact of different width-multipliers on the baseline MoblieNet-tiny architecture. For each inverted bottleneck block with the same expand ratios ($E = 6$), we explore different width multipliers ($W = \{0.35, 0.65, 1.0\}$. Since we need to have a tiny network that is suitable for edge devices, we do not use a larger value than 1.0 for the width-multiplier. The dynamics of natural accuracy, robust accuracy, and perturbation instability measures with regard to the training epochs for these three adversarially trained networks are plotted in Fig. 2. By following the dynamics of the metrics in different epochs, we find that increasing the width-multiplier leads to improve natural and robust accuracy, but it also increases the perturbation instability. The other important finding is that the improvement of natural and robust accuracy is negligible (or sometimes reduced) after a threshold for width-multiplier. Table 1 shows the results for training MobilNet-tiny with different width-multiplier and fixed expand-ratio ($E = 6$). As shown in the table, increasing the width multiplier from 0.35 in the baseline network to 0.65 improves the natural and robust accuracy by 9.01% and 8.09% but moving from 0.65 to 1.0 have 0.91% improvement in natural and hurt robust accuracy by 0.89%. In addition by increasing the width-multiplier from 0.35 to 0.65 and 1.0 the perturbation instability increase by 20.13% and 26.46% It means that adversarial training increases the difference between the prediction of the network for natural and adversarial examples.

3.4 Exploring Different Expand-Ratios

We also explore the impact of the expand-ratio hyper-parameter on the robustness of MobileNet-tiny. The baseline MobileNet-tiny has a fixed width-multiplier of $W = 0.35$ for all inverted bottleneck blocks. We investigate different values for expand-ratio as $E = \{6, 10, 20, 29\}$. Like the width-multiplier, we do not significantly alter the expand-ratio to remain in the tiny regime. The dynamics of natural accuracy, robust accuracy, and perturbation instability metrics in the training epochs are plotted in Fig. 3. Furthermore, Table 2 indicates the best results for different expand-ratios. We find that the expand-ratio has a similar effect as the width-multiplier. However, increasing it until a threshold improves the natural and robust accuracy, but it compromises the perturbation stability. To compare the impact of the width-multiplier and expand-ratio, we set these hyper-parameters to have a similar number of parameters. To this end, we made

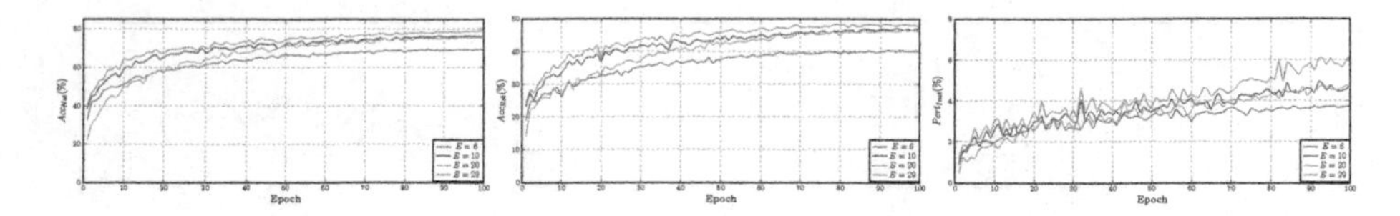

Fig. 3. The dynamics of natural accuracy, robust accuracy and perturbation instability with regard to training epochs and fixed width-multiplier ($W = 0.35$) on CIFAR-10 dataset.

a network by increasing the width-multiplier to 0.65 with the same expand ratio as the baseline and created another network by using 29 for expand-ratio and the same width multiplier as the baseline. Both networks have almost 1.08 million parameters. The expanded network with width-multiplier shows 0.17% better robust accuracy than the expanded network with expand-ratio. In terms of perturbation instability, expanding the network with width-multiplier, increases the instability by 2.59% compared to the expand-ratio.

3.5 Theoretical and Empirical Lipschitz Constant

Recent works [9,21] formally prove the relation between Lipschitzness and perturbation instability. They show that smaller Lipschitzness (small Lipschitz constant) leads to decreased perturbation instability and improved robustness. In this section, we first theoretically show the relation between width-multiplier and expand-ratio hyper-parameters in MobileNet-tiny with perturbation instability. Then we empirically analyse this relation to support theoretical findings. The Lipschitz constant L of MobileNet-tiny architecture measures the rate of change in the output of the network by changing the input as:

$$||f(\theta; x) - f(\theta; \hat{x})|| \leq L \cdot ||x - \hat{x}|| \tag{4}$$

The expected Lipschitz constant for MobileNet-tiny with n inverted bottleneck blocks with width h and m middle channels is upper bounded by:

$$L(f(\theta; x) \leq \left(\sqrt{W \cdot h} + \sqrt{E \cdot m}\right)^n \tag{5}$$

where W and E show the width-multiplier and expand-ratio hyper-parameters in MobileNet-tiny. This formulation is the conclusion of the theorem in [10] for

Table 1. The results of the expanded network by altering width-multiplier and fixed expand-ratio ($E = 6$) (Last-checkpoint)

Expand-Ratio	*Width-Multiplier*	*#MACs*	*#Params*	Acc_{Nat} (%)	Acc_{Rob}(%)	$Pert_{Inst}$	*Lipchitz L*
$E = 6$	$W = 0.35$	15.98	0.404	69.23	39.91	3.75	67.24
	$W = 0.65$	45.08	1.088	78.24	48.00	5.63	80.77
	$W = 0.1$	90.68	2.278	79.15	47.11	7.27	85.03

Table 2. The results of the expanded network by altering expand-ratio and fixed width-multiplier ($W = 0.35$) (Last-checkpoint).

Width-Multiplier	*Expand-Ratio*	*#MACs*	*#Params*	Acc_{Nat} (%)	Acc_{Rob}(%)	$Pert_{Inst}$	*Lipchitz L*
$W = 0.35$	$E = 6$	15.979	0.404	69.23	39.91	3.75	67.24
	$E = 10$	23.73	0.523	75.98	46.81	4.76	73.4
	$E = 20$	43.206	0.8218	75.54	46.3	4.51	71.26
	$E = 29$	58.63	1.086	78.72	47.83	6.28	82.88

WideResNet large-scale network. This establishes the connection between hyper-parameters in inverted bottleneck blocks and the Lipschitz constant and perturbation instability. This theoretical analysis shows that increasing the width-multiplier and expand-ratio increases the perturbation instability. Additionally, this formulation shows that adding more inverted blocks to the baseline network (more depth) exponentially increases the perturbation instability. Our empirical Lipschitz constant evaluation supports our theoretical findings:

$$L = \mathbb{E}_{x \sim D} \max_{\hat{x} \in \mathcal{X}} \frac{||f(\theta; x) - f(\theta; \hat{x})||}{||x - \hat{x}||} \tag{6}$$

where $\mathcal{X}$ is the ϵ-ball around the x and $\hat{x}$ is adversarial data generated by PGD. We compute this metric for different hyper-parameter configurations. The results are indicated in Fig. 4 and Tables 1, 2. We can observe that when the width of the network increase by using a larger width-multiplier and expand-ratio, the empirical Lipschitz constant also increases. Theoretical and empirical analysis of network perturbation instability agrees.

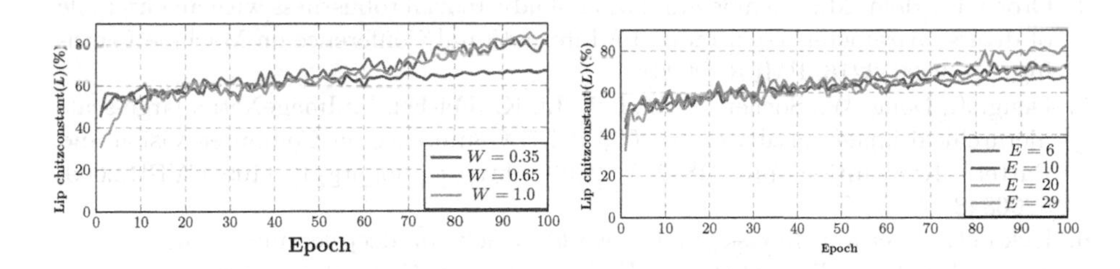

Fig. 4. The dynamics of Lipschitz constant with regard to training epochs and altering width-multipliers (left) and expand-ratios(right)

4 Conclusion

This paper analyses the robustness of tiny deep neural networks (MobileNet-tiny) from an architectural perspective. Specifically, we explore how the width of the inverted bottleneck blocks affects the robustness. To generate different architectures, we change the width-multiplier and expand-ratio hyper-parameters that increase the number of channels in inverted bottleneck blocks. Our findings are: 1) Although increasing the width of the blocks in MobileNet-tiny can improve the natural and robust accuracy, it also increases the perturbation instability. 2) After a threshold, expanding the width of the network cannot only improve the natural and robust accuracy but also increase the perturbation instability. We also find theoretically and empirically the relationship between width-multiplier and expand-ratio with the Lipchitz constant, which directly relates to perturbation instability. It shows that increasing the number of blocks and expanding the width of the network increase the Lipchitz constant. Our work provides valuable insights into designing robust tiny networks against adversarial data.

Acknowledgement. This work was supported in part by the European Union through European Social Fund in the frames of the "Information and Communication Technologies (ICT) program" and by the Swedish Innovation Agency VINNOVA project "AutoDeep" and "SafeDeep". The computations were enabled by the supercomputing resource Berzelius provided by National Supercomputer Centre at Linköping University and the Knut and Alice Wallenberg foundation.

References

1. Banbury, C., et al.: MicroNets: neural network architectures for deploying tinyml applications on commodity microcontrollers. In: Proceedings of Machine Learning and Systems, vol. 3, pp. 517–532 (2021)
2. Cai, H., Zhu, L., Han, S.: Proxylessnas: direct neural architecture search on target task and hardware. arXiv preprint arXiv:1812.00332 (2018)
3. Carlini, N., Wagner, D.: Towards evaluating the robustness of neural networks. In: 2017 IEEE Symposium on Security and Privacy (SP), pp. 39–57. IEEE (2017)
4. Croce, F., Hein, M.: Reliable evaluation of adversarial robustness with an ensemble of diverse parameter-free attacks. In: International Conference on Machine Learning, pp. 2206–2216. PMLR (2020)
5. Deng, J., Dong, W., Socher, R., Li, L.J., Li, K., Fei-Fei, L.: ImageNet: a large-scale hierarchical image database. In: 2009 IEEE Conference on Computer Vision and Pattern Recognition, pp. 248–255 (2009). https://doi.org/10.1109/CVPR.2009.5206848
6. Eykholt, K., et al.: Robust physical-world attacks on deep learning visual classification. In: Proceedings of the IEEE Conference on Computer Vision and Pattern Recognition, pp. 1625–1634 (2018)
7. Goodfellow, I.J., Shlens, J., Szegedy, C.: Explaining and harnessing adversarial examples. arXiv preprint arXiv:1412.6572 (2014)
8. Han, S., Mao, H., Dally, W.J.: Deep compression: compressing deep neural networks with pruning, trained quantization and huffman coding. arXiv preprint arXiv:1510.00149 (2015)
9. Hein, M., Andriushchenko, M.: Formal guarantees on the robustness of a classifier against adversarial manipulation. In: Advances in Neural Information Processing Systems, vol. 30 (2017)
10. Huang, H., Wang, Y., Erfani, S., Gu, Q., Bailey, J., Ma, X.: Exploring architectural ingredients of adversarially robust deep neural networks. Adv. Neural. Inf. Process. Syst. **34**, 5545–5559 (2021)

11. Kannan, H., Kurakin, A., Goodfellow, I.: Adversarial logit pairing. arXiv preprint arXiv:1803.06373 (2018)
12. Lin, J., Chen, W.M., Lin, Y., Gan, C., Han, S., et al.: Mcunet: tiny deep learning on IoT devices. Adv. Neural. Inf. Process. Syst. **33**, 11711–11722 (2020)
13. Loni, M., Mousavi, H., Riazati, M., Daneshtalab, M., Sjödin, M.: TAS: ternarized neural architecture search for resource-constrained edge devices. In: 2022 Design, Automation & Test in Europe Conference & Exhibition (DATE), pp. 1115–1118. IEEE (2022)
14. Madry, A., Makelov, A., Schmidt, L., Tsipras, D., Vladu, A.: Towards deep learning models resistant to adversarial attacks. arXiv preprint arXiv:1706.06083 (2017)
15. Mousavi, H., Loni, M., Alibeigi, M., Daneshtalab, M.: Pr-darts: pruning-based differentiable architecture search. arXiv preprint arXiv:2207.06968 (2022)
16. Rade, R., Moosavi-Dezfooli, S.M.: Reducing excessive margin to achieve a better accuracy vs. robustness trade-off. In: International Conference on Learning Representations (2022). https://openreview.net/forum?id=Azh9QBQ4tR7
17. Sandler, M., Howard, A., Zhu, M., Zhmoginov, A., Chen, L.C.: Mobilenetv 2: Inverted residuals and linear bottlenecks. In: Proceedings of the IEEE Conference on Computer Vision and Pattern Recognition, pp. 4510–4520 (2018)
18. Shafique, M., et al.: Robust machine learning systems: challenges, current trends, perspectives, and the road ahead. IEEE Design Test **37**(2), 30–57 (2020)
19. Szegedy, C., et al.: Intriguing properties of neural networks. arXiv preprint arXiv:1312.6199 (2013)
20. Wang, Y., Zou, D., Yi, J., Bailey, J., Ma, X., Gu, Q.: Improving adversarial robustness requires revisiting misclassified examples. In: International Conference on Learning Representations (2020)
21. Weng, T.W., et al.: Evaluating the robustness of neural networks: an extreme value theory approach. arXiv preprint arXiv:1801.10578 (2018)
22. Wu, B., Chen, J., Cai, D., He, X., Gu, Q.: Do wider neural networks really help adversarial robustness? Adv. Neural. Inf. Process. Syst. **34**, 7054–7067 (2021)
23. Xie, C., Yuille, A.: Intriguing properties of adversarial training at scale. arXiv preprint arXiv:1906.03787 (2019)
24. Zhang, H., Yu, Y., Jiao, J., Xing, E., El Ghaoui, L., Jordan, M.: Theoretically principled trade-off between robustness and accuracy. In: International Conference on Machine Learning, pp. 7472–7482. PMLR (2019)
25. Zhu, Z., Liu, F., Chrysos, G., Cevher, V.: Robustness in deep learning: the good (width), the bad (depth), and the ugly (initialization). Adv. Neural. Inf. Process. Syst. **35**, 36094–36107 (2022)
26. Zi, B., Zhao, S., Ma, X., Jiang, Y.G.: Revisiting adversarial robustness distillation: robust soft labels make student better. In: Proceedings of the IEEE/CVF International Conference on Computer Vision, pp. 16443–16452 (2021)

Temporal Graph Management

Semantic Centrality for Temporal Graphs

Landy Andriamampianina[1,2](✉), Franck Ravat[1], Jiefu Song[1], and Nathalie Vallès-Parlangeau[3]

[1] IRIT-CNRS (UMR 5505) - Université Toulouse Capitole, 2 Rue du Doyen Gabriel Marty, 31042 Toulouse Cedex 09, France
{landy.andriamampianina,franck.ravat,jiefu.song}@irit.fr

[2] Activus Group, 1 Chemin du Pigeonnier de la Cépière, 31100 Toulouse, France
landy.andriamampianina@activus-group.fr

[3] Université de Pau et des Pays de l'Adour, Pau, France
nathalie.valles-parlangeau@univ-pau.fr

Abstract. Centrality metrics in graphs, such as degree, help to understand the influence of entities in various applications. They are used to quantify the influence of entities based on their relationships. Time dimension has been integrated to take into account the evolution of relationships between entities in real-world phenomena. For instance, in the context of disease spreading, new contacts may appear and disappear over time between individuals. However, they do not take into account the semantics of entities and their relationships. For example, in the context of a disease spreading, some relationships (such as physical contacts) may be more important than others (such as virtual contacts). To overcome this drawback, we propose centrality metrics that integrate both temporal and semantics aspects. We carry out experimental assessments, with real-world datasets, to illustrate the efficiency of our solution.

Keywords: Centrality Metrics · Degree · Influential Entities · Temporal Graphs

1 Introduction

Understanding the influence of entities is a major issue in various real-world phenomena such as social, biological, epidemiological phenomena [8]. For example, in a social network, it is important to identify the individuals who have the greatest influence to understand the spread of information [4]. To do so, several centrality metrics from Graph Theory quantify the influence of vertices based on the graph structure, such as the degree, betweenness or closeness centrality [8]. For instance, in Fig. 1 a), the vertex A is the most influent because it has the highest degree, i.e., the highest number of direct relationships. Yet, real-world phenomena may be dynamic in nature. So the influence of entities may also evolve. For instance, in Fig. 1 b), the most influential vertex changes between the two timestamps t_1 and t_2 because of the addition and removal of certain relationships. Current research is therefore aimed at extending classic metrics

A. Abelló et al. (Eds.): ADBIS 2023, CCIS 1850, pp. 163–173, 2023.
https://doi.org/10.1007/978-3-031-42941-5_15

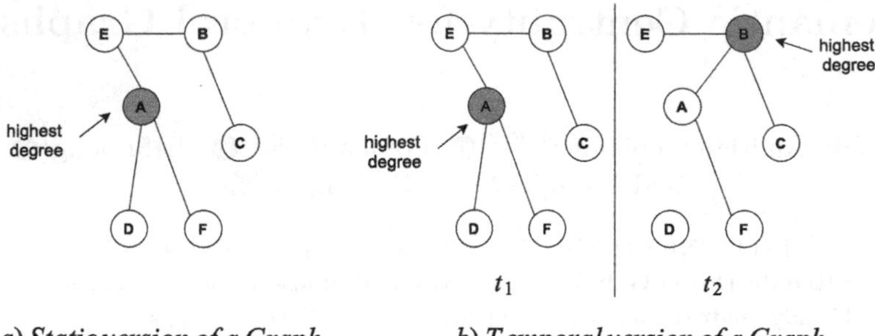

Fig. 1. Degree Centrality in static graphs VS temporal graphs.

by integrating temporal aspects (e.g. timestamps, time intervals, and temporal paths) [3]. However, these temporal metrics do not take into account the semantics of their application. They do not distinguish classes of entities and types of relationships. For instance, in a social network, some classes of entities (e.g., such as students, professionals) and some relationship types (e.g., family, friends, co-workers) may have different levels of influence. Therefore, our research question is the following: how can we improve the concept of centrality to consider both temporal and semantic aspects of an application?

The contribution of our work is the new concept of Semantic Temporal Centrality. It extends the basic concept of centrality to include both temporal and semantic aspects of an application. We use the degree metric as a simple baseline for understanding the importance of both aspects in centrality [8]. In this paper, we first review the calculation methods of degree in dynamic applications (Sect. 2). Then, we define our new concept using as an instantiation example, a graph model including both temporal evolution and semantics (Sect. 3). Finally, we evaluate the efficiency of our solution through experiments (Sect. 4).

2 Related Work

Existing calculation methods for centrality metrics in dynamic applications depend on the chosen metric and on how the time dimension is integrated in the graph. In the case of degree metric, the classic calculation method is the snapshot-based approach: the graph is modelled through a sequence of static graphs observed at different time points [7]. The centrality degree is calculated for each snapshot. The calculated degrees are generally exploited to produce a centrality ranking or distribution of vertices for each snapshot [5].

Another calculation method of degree metric is the time-aggregated approach. It consists in aggregating the time-series of degree values (calculated according to the snapshot-based approach) to obtain a single degree for each vertex. The aggregated degree value summarizes the overall influence of a vertex over time [4,5]. Generally, the average of degree values is used, but some research works propose also to use the minimum or maximum degree [6]. Therefore, instead of generating several rankings or distributions as in the snapshot-based approach, the aggregated-approach provides a single result.

The above-mentioned degree metrics have several limitations. On the one hand, they are only calculated at a single time point because graph snapshots are based on discrete-time. They do not integrate calculation over a time interval to consider their evolution over a time. On the other hand, current degree calculation methods do not take into account the semantics of vertices and edges because most graph models do not. Yet, vertex and edge semantics help distinguish the levels of influence of entity classes and relationship types from a qualitative perspective. We therefore extend the current methods for calculating degrees by introducing a time interval into the temporal parameter and the concept of vertex and edge label to represent the semantics in an application.

3 Proposition

3.1 Preliminaries

We propose to use the model of temporal graph presented in [1] to exemplify our concept of Semantic Temporal Degree. The temporal graph $TG = \langle E, R \rangle$ is composed of sets of entities E and relationships R. It represents the semantics of entities and relationships by a set of labels L_E describing entity classes and a set of labels L_R describing relationship types. Each entity has an explicit semantic described by a label $l_{E'}$ and, similarly, each relationship with a label $l_{R'}$.

The temporal graph captures the evolution of entities and relationships over time. An entity is then described by an identifier *id* and a set of states $\{s_1, ..., s_m\}$. A relationship r_i links a couple of entity states (s_k, s_j) and is composed by a set of states $\{s_1, ..., s_n\}$. Each state s of an entity or relationship is associated to a valid time interval $T^s = [t_b, t_f[$, which indicates the stability period of the state. The set of relationships incident to an entity e_i is denoted, $R(e_i) = \{r_1, ..., r_p\}$. The function $\Sigma : r_i \rightarrow \{s_1, ..., s_n\}$ returns for a relationship r_i all of its states.

Some operators allow querying the temporal graph to select the semantics and time window of interest in the degree calculation [2]. On the one hand, the operator $matching_{predicate} : TG_{input} \times w \rightarrow TG_{output}$ allows extracting a subgraph representing the entities and relationships valid during a user-defined time-window w. On the other hand, the operator $matching_{pattern} : TG_{input} \times (l_{E_i} \times l_{R_j} \times l_{E_k}) \rightarrow TG_{output}$ allows extracting a subgraph respecting user-defined conditions on the semantics of entities and relationships.

Example 1. We consider the temporal graph in Fig. 2 to analyse the number of interactions (degree) between individuals of a university in the context of disease spreading over a time period of 6 days. In the university, we distinguish students from teachers. Each entity is then represented by a grey vertex labelled with its semantics (*student* or *teacher*). Moreover, we distinguish two types of interactions between students and teachers: virtual contact (*call* relationship) and

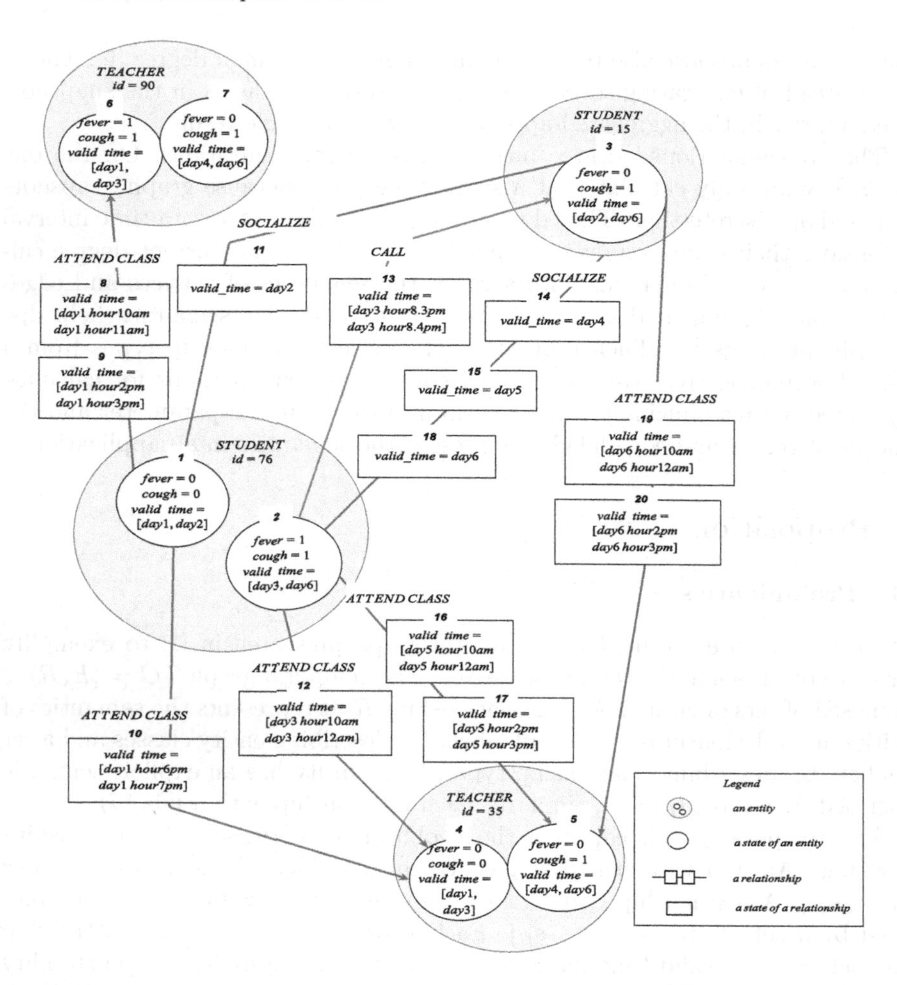

Fig. 2. An example of temporal graph

physical contact (*socialize* or *attend class* relationship). Taking into account the semantics during analyses can make a great difference. In this context, we can ignore *virtual contacts* and put more emphasis on *physical contacts* of teachers and students, who generally meet more people during classes. The health status of students and teachers can change over time due to the disease transmission. Each entity (grey vertex) is then composed of one or several states (white vertices). For instance, the student 76 has two different health status: the state 1 in which he has no fever and no cough from day 1 to day 2, and the state 2 in which he has fever and cough from day 3 to day 6. Moreover, new interactions between students and teachers may appear and disappear over time. Each interaction represents a state of a labelled relationship between two entity states and is illustrated by a white rectangle. For instance, when the student 15 is in the state 3 and the student 76 is in the state 2, they socialized three days in the period (rectangles numbered 14, 15 and 18).

3.2 Centrality Metrics

Existing research works only consider the degree metric at a specific time point and do not consider the semantics of their application (Sect. 2). We therefore propose three centrality metrics extending the current degree metric with the possibility of calculating its evolution within a time interval and by including the semantics of entities and relationships.

Definition 1 (semantic temporal degree). *The semantic temporal degree of an entity e_i in a temporal graph TG is calculated as a function of a time window w (a time point or a time interval), and as a function of a type $l_{R'}$ of the incident relationships to the entity. The semantic temporal degree of an entity at a time point $w = t_x$ is a single value representing the number of relationship states having the type $l_{R'}$, incident to e_i, and valid at the time point. The semantic temporal degree of an entity during a time interval $w = [t_{start}, t_{end}]$ is a series of values describing the changes in the number of relationship states having the type $l_{R'}$, incident to e_i, and valid during the time interval. If the entity does not exist during the time window w, the semantic temporal degree of the entity is not calculated. It is then denoted as follows:*

$$deg(e_i, w, l_{R'}) = \begin{cases} |\cup_{\forall r_j \in R(e_i)} \Sigma(r_j)|, \text{ where } \forall r_j, l_{r_j} = l_{R'}, \forall s \in \Sigma(r_j), T^s \circ w, \text{ if } w = t_x \\ \{y_1, ..., y_m\}, \text{ where } y = deg(e_i, t_x, l_{R'}), t_x \in w, \text{ if } w = [t_{start}, t_{end}] \\ \text{not defined, otherwise} \end{cases}$$

The calculation of semantic temporal degree is detailed in Table 1. It involves the extraction of a subgraph respecting a given semantic pattern using the $matching_{pattern}$ operator and the extraction of a subgraph which is valid in the given time window using the $matching_{predicate}$ operator.

Example 2. Consider the temporal graph in Fig. 2. We want to know how evolve the interactions of the student 76 over the whole time period (6 days). To do so, we can compute its temporal degree: $deg(76, [day1, day6], \emptyset) = \{deg(76, day1, \emptyset), deg(76, day2, \emptyset), deg(76, day3, \emptyset), deg(76, day4, \emptyset), deg(76, day5, \emptyset), deg(76, day6, \emptyset)\} = \{3, 1, 2, 1, 3, 1\}$. It is graphically illustrated in Fig. 3 a). The student 76 has peaks of interactions at the days 3 and 5. This could indicate that on those days the student is at a higher risk of contracting the disease or that they play an important role in spreading the disease to others. However, we do not know through what type of interactions. Therefore, we compute the semantic temporal degree of the student 76 according to the different types of interactions from *day*1 to *day*6. In Fig. 3 b), we observe that the student actively socialized and attended class over time, but called rarely. Moreover, the number of socializations of the student are almost the same over time. The

Table 1. Semantic Temporal Degree

Metric:	$deg(e_i, w, l_{R'})$
Input:	a temporal graph TG
	an entity e_i
	a time window w
	a relationship label $l_{R'}$
Output:	a single value or a set of values or an empty set
Actions:	1. $TG_1 \leftarrow matching_{pattern}(TG, l_{E'}, l_{R'}, \emptyset)$
	where $l_{E'}$ *is the label of the entity* e_i
	2. If $w = t_x$
	3. $TG_2 \leftarrow matching_{predicate}(TG_1, \{(l_{E'}.id = id^{e_i})$ $AND\,(T^{l_{R'}} \circ t_x)\,\})$
	4. Apply to TG_2: $deg(e_i, w, l_{R'}) = \mid \cup_{\forall r_j \in R(e_i)} \Sigma(r_j)\mid$
	5. Else if $w = [t_{start}, t_{end}]$
	6. $Y \leftarrow \emptyset$
	8. For each $t_x \in w$
	9. $TG_2 \leftarrow matching_{predicate}(TG_1, \{l_{E'}.id = id^{e_i})$ $AND\,(T^{l_{R'}} \circ t_x)\})$
	10. $Y \leftarrow Y \cup \{deg(e_i, t_x, l_{R'})\, in\, TG_2\}$
	11. $deg(e_i, w, l_{R'}) = Y$

spikes and drops of degree values observed in Fig. 3 a) are therefore not caused by calls and socializations. We should then focus on the other semantics of relationships: *attend class*. As we notice in Fig. 3 b), the student can attend multiple classes throughout the day. This could indicate that class interactions of the student play an important role in spreading the disease rather than other interaction types.

As discussed in Sect. 2, existing research work uses time-aggregated calculation methods to summarize the degree metric values for a time interval into a single value. We provide the same calculation possibility for the semantic temporal degree during a time interval by computing its average.

Definition 2 (average semantic temporal degree). *The average semantic temporal degree of an entity* e_i *in a temporal graph* TG *during the time interval* w *is the average where each value is the semantic temporal degree of the entity weighted by its duration* δ *in the time interval. It is denoted as follows:*

$$deg_{average}(e_i, w, l_{R'}) = \frac{\delta_1 \times deg_1(e_i, w_1, l_{R'}) + ... + \delta_n \times deg_n(e_i, w_n, l_{R'})}{\sum_{p \in [1,n]} \delta_p}$$

Example 3. Let us compute the average semantic temporal degree of the student 76 over the entire period for its interactions with the semantics *attend class*. We have: $deg_{average}(76, [day1, day6], ATTENDCLASS) =$

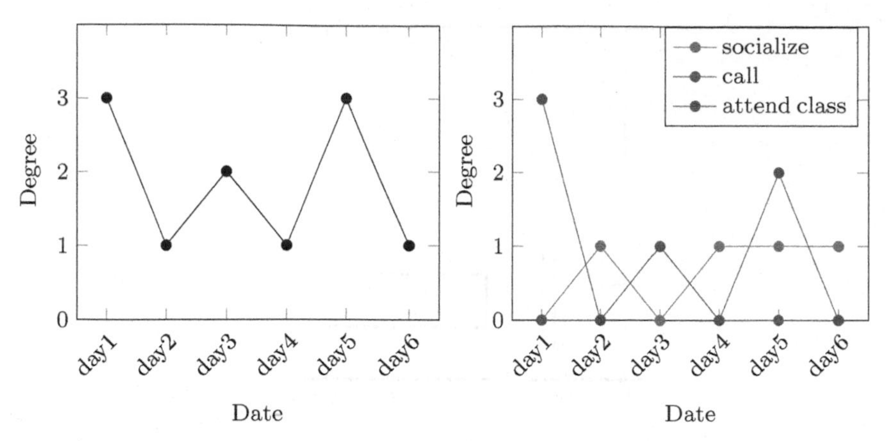

a) Temporal degree of student 76 b) Semantic temporal degree of student 76

Fig. 3. Centrality degree of student 76 in Fig. 2

$\frac{1day \times 3 + 1day \times 1 + 1day \times 2 + 3days \times 0}{6days} = 1$. In average, the student 76 attends one class in average in the week.

Existing research work generally focuses solely on the degree of a vertex. We enable the calculation of the distribution of a class of vertices through the *semantic temporal degree distribution* metric. The advantage of the latter is that it highlights the level of heterogeneity of the semantic temporal degree within a class of entities, which cannot be observed with the simple centrality ranking. Specifically, the semantic temporal degree of a chosen entity can be positioned in the distribution of its entity class to understand its relative importance.

Definition 3 (semantic temporal degree distribution). *The semantic temporal degree distribution of an entity class E' labelled $l_{E'}$ in a temporal graph TG is a set of five values, namely the minimum (min), the three quartiles (q_x) and the maximum (max) of the whole set of average semantic temporal degree values of entities with the semantics $l_{E'}$ during a time window w. The 1st quartile, denoted as q_1, represents the average semantic temporal degree at which 25% of entities falls below and 75% falls above. The 2nd quartile or median, denoted as q_2, represents the average semantic temporal degree at which 50% of entities falls below and 50% falls above. The 3rd quartile, denoted as q_3, represents the average semantic temporal degree at which 75% of entities falls below and 25% falls above. The semantic temporal degree distribution of an entity class E' is then denoted as follows:*

$$\begin{aligned} deg_{distribution}(l_{E'}, w, l_{R'}) = \{ & min(\cup_{e_i \in E'} deg_{average}(e_i, w, l_{R'})), q_1(\cup_{e_i \in E'} deg_{average}(e_i, w, l_{R'})), \\ & q_2(\cup_{e_i \in E'} deg_{average}(e_i, w, l_{R'})), q_3(\cup_{e_i \in E'} deg_{average}(e_i, w, l_{R'})), \\ & max(\cup_{e_i \in E'} deg_{average}(e_i, w, l_{R'}))\} \end{aligned}$$

It is graphically represented by a box plot (Fig. 4). The bottom and top of the box represent the first and third quartiles (q_1 and q_3), respectively. The line

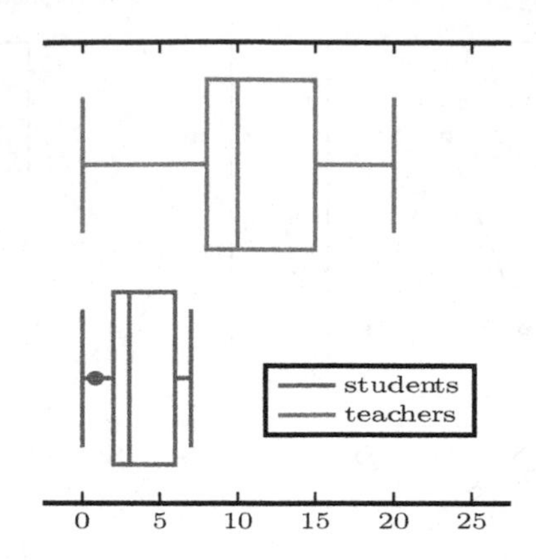

Fig. 4. Semantic Temporal Degree Distribution

inside the box represents the median (q_2). The minimum and maximum values are represented by the end points of the whiskers.

Example 4. The blue box plot in Fig. 4 presents the semantic temporal degree distribution of students during one week for their class interactions ($deg_{distribution}(STUDENT, week1, ATTENDCLASS)$). It shows that 25% of students have in average less than 2 class interactions in a week. 75% of students have in average less than 7 class interactions in a week. Half of students has in average more than 3 class interactions in a week and the other half has in average less than 3 class interactions in a week. We position the average semantic temporal degree of the student 76 (calculated in the Example 3) as a blue dot on the blue box plot. We observe that the average semantic temporal degree of the student is very low compared to the rest of students. He may therefore have no influence on the disease transmission. The red box plot in Fig. 4 presents the distribution of teachers during the same week as students. We notice that teachers have many more interactions than students and are therefore probably at greater risk of contracting or transmitting the disease.

4 Experimental Evaluation

In this section, we describe our experimental assessment which aims at evaluating the efficiency of our proposal. Details of our experimental assessment are available on the website https://gitlab.com/2573869/degreecentrality. A usability evaluation of the proposed centrality metrics in a real application is also available in the website.

4.1 Technical Environment

The experiments are conducted on a PowerEdge R630, 16 CPUs x Intel(R) Xeon(R) CPU E5-2630 v3 @ 2.40 Ghz, 63.91 GB. One virtual machine installed on this hardware, that has 32 GB in terms of RAM and 100 GB in terms of disk size, is used for our experiments. The Neo4j graph database is installed in the virtual machine to store and query temporal graph datasets. The programming language used to implement our metrics is Python 3.7.

4.2 Datasets

The first real-world dataset we use is the Social experiment dataset[1]. The dataset traces the changes in the students' symptoms and in their interactions over time to study epidemiological contagion at an university dormitory. Students may have different types of interactions: physical interactions (such as *proximity*) and virtual interactions (such as *calls*) which may occur several times each day and other interaction types (such as *socialize*). The number of interactions of students can be analysed to identify potential "spreaders" of the epidemic. The second real-world dataset we use is a dataset from an E-commerce website[2]. The dataset presents customers' actions on items (view, add to cart and buy) which are made frequently within an hour. It also traces the changes in items' characteristics which are more rare over time. The number of actions done on items can be analysed to identify the popular items that can be recommended to customers.

4.3 Experimental Methodology and Results

Methodology. First, we transform each dataset into the temporal graph model presented in Sect. 3.1. The Social Experiment dataset contains 33 934 vertices and 2 168 270 edges. The E-commerce dataset is composed of 4 315 375 vertices and 4 447 430 edges. All transformation details are available in [1]. Then, we evaluate the computation time of the basic application of centrality metrics: ranking entities of a dataset for a chosen time interval. In our case, we choose to compute the average semantic temporal degree metric for each entity of each dataset. For this metric, we choose to put the parameters that make the computation the longest: we integrate all relationship labels and the whole time span of each entity. We define 5 scale factors by varying the number of edges involved in the metric calculation. From 1 to 5, the scale factor is 20%, 40%, 60%, 80%, 100% of the number of edges of each dataset respectively. We run the metric calculation 10 times for each scale factor and make the average elapsed computation time over the 10 executions. To sum up, our experiments account for a total of 100 executions: 2 datasets $\times$ 5 scale factors $\times$ 10 executions.

Results. Figure 5 shows the computation time to rank the number of interactions of students in the Social Experiment dataset (in blue) and the computation time to rank the number of actions on items in the E-commerce dataset (in

[1] http://realitycommons.media.mit.edu/socialevolution.html.
[2] https://www.kaggle.com/retailrocket/ecommerce-dataset?select=item_properties_part2.csv).

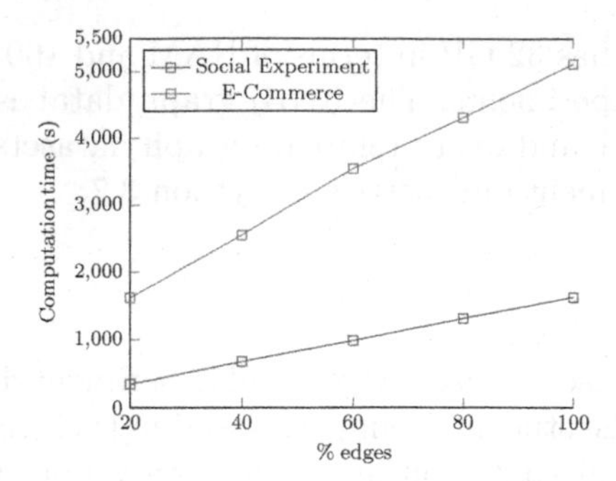

Fig. 5. Computation time of ranking the average semantic temporal degree values (Color figure online)

red) based on the average semantic temporal degree metric. We observe that the computation time of the analysis increases linearly with the number of edges for both datasets.

5 Conclusion and Future Work

In this paper, we proposed three degree-based metrics to understand the centrality of entities in a dynamic context. From the conceptual point of view, we complete the classic degree metrics by adding both temporal and semantic aspects. An instantiation of the proposed metrics is done through a temporal graph model. To validate their efficiency, we applied the proposed metrics on two real datasets. We observed that the computation time of the metrics increases linearly with the edge number. In our future work, we plan to make new applications of our metrics by using new datasets and to extend other graph metrics such as betweenness centrality or local clustering coefficient.

References

1. Andriamampianina, L., Ravat, F., Song, J., Vallès-Parlangeau, N.: Graph data temporal evolutions: from conceptual modelling to implementation. Data Knowl. Eng. **139**, 102017 (2022). https://doi.org/10.1016/j.datak.2022.102017
2. Andriamampianina, L., Ravat, F., Song, J., Vallès-Parlangeau, N.: Querying temporal property graphs. In: Franch, X., Poels, G., Gailly, F., Snoeck, M. (eds.) CAiSE 2022, pp. 355–370. Springer, Heidelberg (2022). https://doi.org/10.1007/978-3-031-07472-1_21
3. Ghanem, M., Magnien, C., Tarissan, F.: Centrality metrics in dynamic networks: a comparison study. IEEE Trans. Network Sci. Eng. **6**(4), 940–951 (2019). https://doi.org/10.1109/TNSE.2018.2880344
4. Ishfaq, U., Khan, H.U., Iqbal, S.: Identifying the influential nodes in complex social networks using centrality-based approach. J. King Saud Univ. Comput. Inf. Sci. **34**(10, Part B), 9376–9392 (2022). https://doi.org/10.1016/j.jksuci.2022.09.016

5. Meng, Y., Qi, Q., Liu, J., Zhou, W.: Dynamic evolution analysis of complex topology and node importance in Shenzhen metro network from 2004 to 2021. Sustainability **14**(12), 7234 (2022). https://doi.org/10.3390/su14127234
6. Rost, C., Gomez, K., Christen, P., Rahm, E.: Evolution of degree metrics in large temporal graphs (2023). https://doi.org/10.18420/BTW2023-23
7. Uddin, S., Piraveenan, M., Chung, K.S.K., Hossain, L.: Topological analysis of longitudinal networks. In: 2013 46th Hawaii International Conference on System Sciences, Wailea, HI, USA, January 2013, pp. 3931–3940. IEEE (2013). https://doi.org/10.1109/HICSS.2013.556
8. Wan, Z., Mahajan, Y., Kang, B.W., Moore, T.J., Cho, J.H.: A survey on centrality metrics and their network resilience analysis. IEEE Access **9**, 104773–104819 (2021). https://doi.org/10.1109/ACCESS.2021.3094196

A Knowledge Graph for Query-Induced Analyses of Hierarchically Structured Time Series Information

Alexander Graß[1,2](✉), Christian Beecks[2,3], Sisay Adugna Chala[1,2], Christoph Lange[1,2], and Stefan Decker[1,2]

[1] Fraunhofer Institute for Applied Information Technology FIT, Sankt Augustin, Germany
{alexander.grass,sisay.chala,christoph.lange, stefan.decker}@fit.fraunhofer.de
[2] RWTH Aachen University, Aachen, Germany
[3] University of Hagen, Hagen, Germany

Abstract. This paper introduces the concept of a knowledge graph for time series data, which allows for a structured management and propagation of characteristic time series information and the ability to support query-driven data analyses. We gradually link and enrich knowledge obtained by domain experts or previously performed analyses by representing globally and locally occurring time series insights as individual graph nodes. Supported by a utilization of techniques from automated knowledge discovery and machine learning, a recursive integration of analytical query results is exploited to generate a spectral representation of linked and successively condensed information. Besides a time series to graph mapping, we provide an ontology describing a classification of maintained knowledge and affiliated analysis methods for knowledge generation. After a discussion on gradual knowledge enrichment, we finally illustrate the concept of knowledge propagation based on an application of state-of-the-art methods for time series analysis.

Keywords: Knowledge Graph · Time Series · Knowledge Discovery · Exploratory Data Analysis · Machine Learning

1 Introduction

Time series represent the input of many analytical scenarios [6] including autonomous inspections of electrocardiograms to forecasts of econometric data or activity recognition in wearable devices. With an increasing size and complexity of stored time series data, the question of how to support users in generating semantic insights from queries, especially for different levels of expertise, has therefore become an important task. Although, as a response to this problem, many time series management systems [8] have witnessed an advancement regarding their integrated analytical capabilities, their core requirement remains an optimization of data manipulation and retrieval to scale to vast amounts

A. Abelló et al. (Eds.): ADBIS 2023, CCIS 1850, pp. 174–184, 2023.
https://doi.org/10.1007/978-3-031-42941-5_16

of data. One consequence is a conduction of data analyses by additional software solutions, where combined workflows commonly equal an independent and sequential processing of data. Neglected synergy effects, such as an inclusion of previously derived knowledge, might consequently lead to a redundant execution of common subroutines and a missed opportunity to exploit outcomes in future query-induced analyses. As an example consider a scenario involving two separate tasks, e.g., anomaly detection and regression, where identified anomalies can be used as valuable information to improve regression or vice versa. Although a variety of tools exist to manually [11] or automatically [5] model these types of layered analyses as pipelines, relationships as the ones from the example are not always apparent. In addition, required analysis tasks might be a result of independent use cases or can not be defined in a single workflow. While a preservation of revealed time series knowledge can thus help with temporal offsets, a classification of such knowledge and affiliated analysis methods supports a generation of nested or stacked analysis pipelines. In light of these difficulties, we propose to combine the mentioned concepts in a unified knowledge graph for time series data. Opposed to graph-based solutions that are designed for concrete data analysis scenarios [3,12], we pursue the approach of a more general concept. The goal is to enable a hierarchically structured knowledge repository, which is gradually built from queries expressing analysis tasks. Feedback is given based on existing information or dynamically derived insights, which further extend the collected amount of reusable information. The resultant graph should therefore reflect a chain of condensed information, starting with a reference of raw data, over extracted conclusions up to conclusions recursively derived from previous conclusions. We identified the following requirements to be essential or desirable for the envisioned solution:

1. **Ontology.** Creation of an ontology for different types of knowledge, as well as for affiliated data analysis methods, to enable pre-configured or automatically generated analysis workflows.
2. **Hierarchical knowledge representation and enrichment.** Successively generated and meaningful time series information should be incrementally integrated into the knowledge graph to enable a hierarchical representation of knowledge. This also includes a layer-wise maintenance of data provenance to link different granularities of revealed insights to the original input data.
3. **Integration of domain expertise.** Although many analysis methods are domain-agnostic and thus can be applied without depending on domain expertise, it is desirable to validate or optimize knowledge extraction and exploitation based on domain-specific facts [18].

The remaining paper is structured as follows: In Sect. 2, we introduce the concept of a time series to knowledge graph mapping for generic knowledge discovery and exploitation. It mainly corresponds to the basis for an elaboration of the former requirements and potential solutions. Besides a description of the graph construction phase and the associated ontology, the section focuses on a predefined and automatic utilization of classified information. In Sect. 3, we elucidate our approach by means of knowledge propagation using state-of-the-art time series analysis methods. We conclude in Sect. 4, where we discuss the pros and cons of our approach as well as future work.

2 Graph-Based Management of Time Series Knowledge

Characteristics of time series are often correlated, e.g. reflected by motifs or inter-dimensional dependencies and potentially associated with domain-specific events. Although a storage of such characteristics and domain information as a collection of descriptive features does allow for more than one database solution, we emphasize on the fact that time series characteristics can even be related hierarchically. One simple example for this statement could be a point outlier occurring within a repeating pattern. A natural representation is therefore to consider time series knowledge as linked information which temporally either occurs sequentially or in parallel. With regard to a storage solution, in our work we thus propose to use graph databases, since knowledge graphs have proven to be a popular choice for an efficient and dynamic management of linked information including associated data analysis [7]. We intended to use knowledge graphs as a query-based and generically structured system for time series knowledge discovery and exploitation, where task-specific queries lead to a utilization of existing or derivable knowledge. While repeated queries are answered based on previously generated information, different tasks automatically result in a potential enrichment of insights by subsequently integrating the outcome of analysis methods. This gradual processing of available facts ensures a continuous enlargement of reusable information on different hierarchy levels. To guarantee a correct implementation of this idea, the following subsections will discuss (i) the knowledge graph construction from time series data, structured by a compatible ontology, (ii) the processing of requests and associated execution of knowledge retrieval and (iii) the classification of stored knowledge and analysis methods to facilitate a generic enrichment of information.

2.1 Knowledge Graph Construction and Ontology

As time series characteristics such as anomalies, patterns and trends commonly occur locally in specific time intervals, one possible representation equals a time-based segmentation of a time series annotated with a meaningful description of each individual segment. A time series to graph mapping could therefore corresponds to a time series segmentation, where each indexed segment is reflected by one node of the graph. Formally, we represent a time series $T = \langle t_1, \dots, t_n \rangle$ of length n, divided into m not necessarily equally-sized segments, as an undirected graph $G = (V, E)$ with nodes $v \in V$ and edges $e \in E$. Each node $v_i \in V$ equals a segment $\langle t_k, \dots, t_l \rangle$ of the time series T, where $1 \leq k \leq l \leq n$ and $1 \leq i \leq m$. In addition, every node $v_i \in V$ is connected with a root node $v_0 \in V$, which represents the complete time series T, by defining an edge $e(v_0, v_i) \in E$ for each v_i. The defined mapping describes a high-level representation of segmented time series data and can thus be used for any segmentation strategy. A

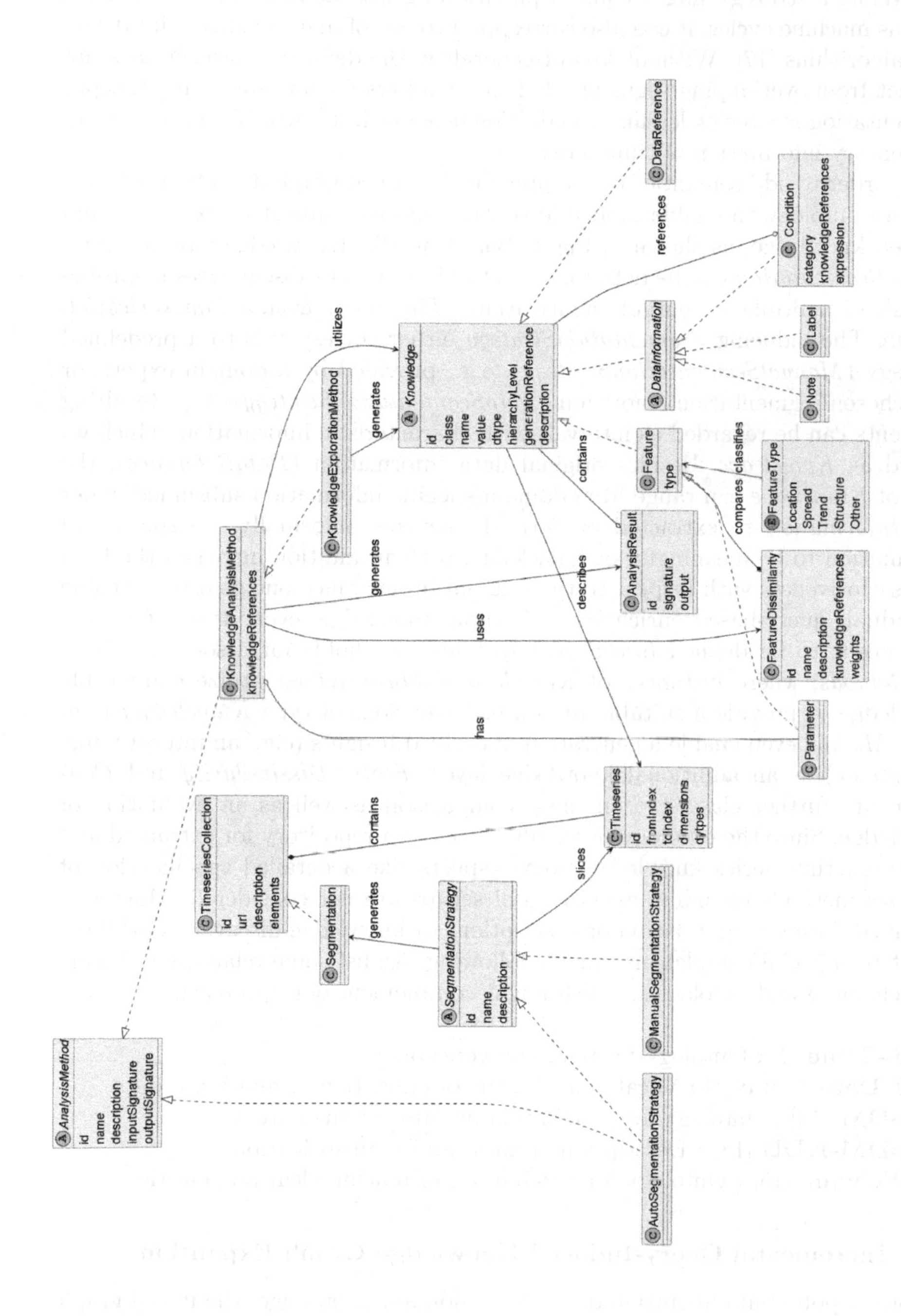

Fig. 1. Knowledge Graph Ontology (UML representation): Class defintions are: (A) for abstract class, (C) for class and (E) for enum.

segmentation strategy is defined as a concept to derive the formalized partitioning. While a strategy might equal a partitioning into domain-specific intervals such as machine cycles, it can also correspond to one of many changepoint detection algorithms [17]. Without loss of generality, the defined approach does not restrict from overlapping segments. It further allows for multiple, simultaneous segmentation strategies leading to additional nodes in G, which represent complementary information of time series T.

In order to add semantics to the previously defined graph definition, we propose an ontology to enable a generic storage and enrichment of extracted and derived knowledge, as shown in Fig. 1. Based on this Knowledge Graph Ontology, a *Segmentation* can be defined as a set of individual subsequences regarding a single or multiple, potentially multivariate, *Timeseries* from a *TimeseriesCollection.* The inducing *SegmentationStrategy* either corresponds to a predefined strategy (*ManualSegmentationStrategy*), e.g., provided by a domain expert, or to a chosen segmentation algorithm (*AutoSegmentationStrategy*) [17]. Resulting segments can be regarded as intervals with characteristic information, which we regard as *Knowledge.* Besides original data information (*DataReference*), the kind of *Knowledge* can range from domain-specific information subsumed under *DataInformation* to extracted or derived *Features.* Informally, we specify all information to be meaningful or knowledge, if their addition increases the level of descriptiveness with respect to the original input. Since one goal is to enable a gradual, query-based enrichment of graph knowledge, every type of knowledge needs to be defined (see Sect. 2.3). This also holds for associated *AnalysisMethods*, where instances of *KnowledgeAnalysisMethod* utilize compatible knowledge to provide a suitable response. Extensions of type *KnowledgeExplorationMethod* even enable a generation of derived insights (e.g., an interpretable *Parameter*) as an additional knowledge layer. *FeatureDissimilarity* and *Condition* are further classes to enable a comparison as well as an evaluation of knowledge. Since the ontology primarily serves as a repository for extracted and generated time series knowledge, some aspects like a detailed specification of analysis methods including experimental setups are not considered. Moreover, extended data set information or descriptions of knowledge discovery workflows might be a useful complement. In the following, we list some related vocabularies, schemata and ontologies, which might complement our approach:

OWL-Time [2] **:** Ontology for temporal concepts.
RDF Data Cube [1] **:** Vocabulary for the organization of multidim. data.
OntoDM [13] **:** Ontology for a definition of data mining entities.
OntoDM-KDD [14] **:** Description of data mining investigations.
ML-Schema [15] **:** Ontology for a definition of machine learning entities.

2.2 Incremental Query-Induced Knowledge Graph Expansion

Despite a potential integration of available domain knowledge, the initial graph construction only provides information about a reasonable partitioning defined by the intrinsic rules of the segmentation strategy. As illustrated in Sect. 3, this

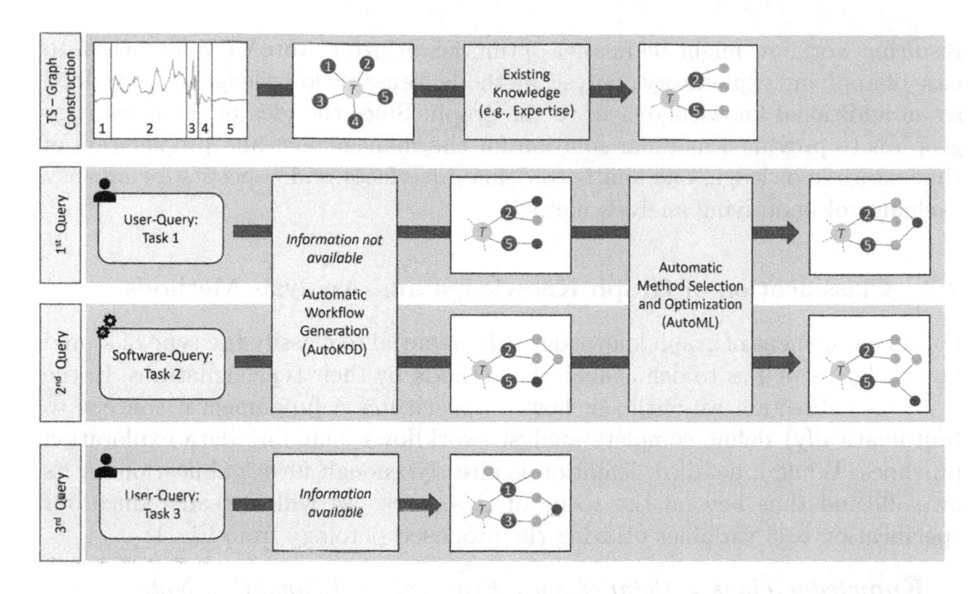

Fig. 2. Segmentation-based time series to graph mapping and three consecutive queries.

also allows to understand the peculiarity of certain segments depending on the used strategy. As the vision is however to successively capture all meaningful properties of a time series, we plan to make use of existing concepts from the field of Knowledge Discovery in Databases (KDD) [14]. To enable a generic generation of further knowledge based on task-specific queries, it is necessary to specify a suitable data analysis workflow. As this can be done manually for every potential task, automated approaches follow the idea of a self-organized workflow generation, given a clear definition of available methods, inputs and outputs [11]. An outline of such a definition is given in Sect. 2.3. The resultant workflow thus represents a task-specific implementation plan for analysis pipelines, which represent appropriate mappings from query to response. Furthermore, recent developments in automated machine learning (short AutoML) [10] allow for an additional level of optimization, by means of an autonomous selection and fine-tuning of suitable methods including corresponding hyperparameters. A conceptualization of automatic workflow generation in combination with the proposed knowledge graph is illustrated in Fig. 2. Starting with an initial version of the graph that is constructed following the steps from Sect. 2.1, we validate for each query, whether the answer can be provided by the existing graph knowledge (e.g. as it is the case for query duplicates). Otherwise potential answers are either given from a processing of referenced data, or by already stored knowledge. Regarding the illustration this enrichment is indicated by a propagation of knowledge via appropriate nodes in the graph (red nodes before and after the AutoML stage). The way how this is done depends on the predefined solution for workflow generation, where in addition each analysis method of the resulting workflow might be further optimized utilizing AutoML approaches. In case of applying exploratory analysis methods, reusable knowledge will be added as an additional knowledge layer to the graph. Since the idea of the knowledge graph is to provide a modular solution for the management and propagation of time series knowledge, one aim is to define interfaces and concepts for an easy exchange of underlying analysis methods.

2.3 Classification of Graph Knowledge and Analysis Methods

For an enrichment of graph knowledge it is essential to classify the type of stored knowledge as well as to define analysis methods by their type signatures. In the previous subsection we outlined, how it constitutes a fundamental concept to (automatically) define complete analysis workflows including data exploration pipelines. While a detailed definition is already enough for a publication on its own [9] and thus beyond the scope of this work, we will instead reduce our specification to 5 variables utilizing the proposed ontology from Fig. 1:

1. $\boldsymbol{Knowledge.class} = DataReference|Feature|Label|Condition|Note$
2. $\boldsymbol{Knowledge.dtype} = Primitive|String|Array(Primitive|String)$
3. $\boldsymbol{KnowledgeAnalysisMethod.inputSignature} =$
   $Array((Knowledge.class, Knowledge.dtype))$
4. $\boldsymbol{KnowledgeAnalysisMethod.outputSignature} =$
   $Array((Knowledge.class, Knowledge.dtype))$
5. $\boldsymbol{KnowledgeExplorationMethod.outputSignature} =$
   $Array((Knowledge.class, Knowledge.dtype))$

with *Array* being considered an n-dimensional list and *Primitive* a common primitive data type including $Integer$, $Float$ and *Double*. As therefore knowledge can be identified by its concrete class and associated data type, we can determine appropriate analysis methods based on their type signatures. In contrast to popular programming languages we extend the definition of type signatures by the instantiated knowledge class, in order to semantically interpret knowledge propagation and to match available information to suitable analysis methods in a meaningful way. For a simplified example, let us illustrate the concept based on a query requesting time series classification in absence of existing labels. In order to define labels we use a clustering algorithm, applied to multivariate floating-point data indicating features of each series. The resultant output is a definition of associated labels and can therefore be regarded as derived knowledge. To perform the actual task of classification we can reuse the derived knowledge as a labeled input for the training of the corresponding classification method.

Knowledge Exploration
$Method_1 = KnowledgeExplorationMethod(name = "ClusteringMethod")$
$Method_1.inputSignature = Array(Feature, Array(Float))$
$Method_1.outputSignature = Array(Label, Array(String))$

Knowledge Propagation and Exploitation:
$Method_2 = KnowledgeAnalysisMethod(name = "ClassificationMethod")$
$Method_2.inputSignature =$
$Method_1.inputSignature \cup Method_1.outputSignature$
$Method_2.outputSignature = Array(Label, Array(String))$

As this example shows a definition of signatures for analysis methods according to the presented schema, and thus can be used to apply predefined data analysis pipelines for automated knowledge generation, the following section illustrates a propagation of knowledge while using these methods.

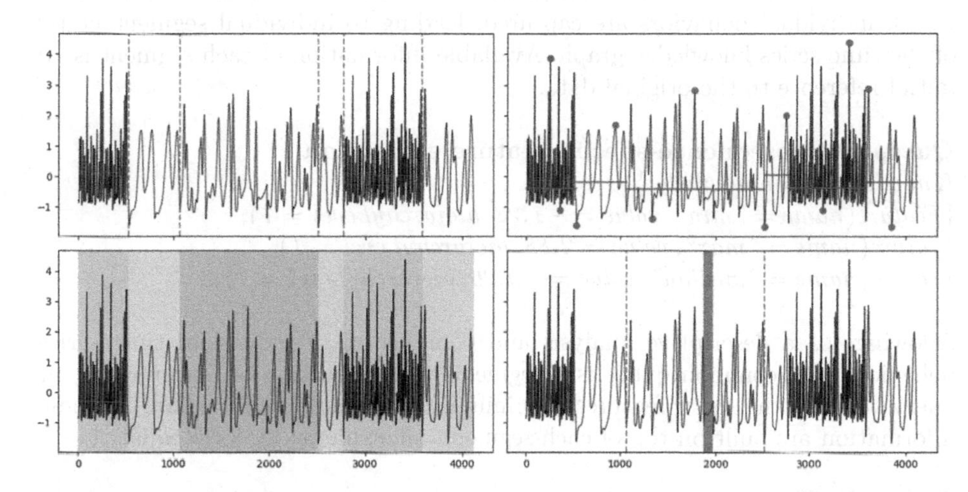

Fig. 3. Multiple tasks of exploratory time series analysis: (top left): segmentation based on changepoint detection [17], (top right): feature extraction within each segment including min, max and median, (bottom left): subsequence clustering based on extracted features [16], (bottom right): anomaly detection for a selected subsequence [19].

3 Knowledge Propagation Example

In this section, we will elucidate the concept of knowledge enrichment and propagation by exemplifying the processing of analysis requests for the WordSynonyms data set from the UCR Time Series Classification Archive [4]. To generate a potential representative time series T, we randomly sampled over multiple classes identifying signals corresponding to specific behaviors. The resulting time series, analyzed in the context of four different analysis tasks involving state-of-the-art approaches [16,17,19], is depicted in Fig. 3. While we decided to use univariate data for illustration purposes, there is no restriction towards multi-dimensional time series. The hereafter presented use cases in form of queries show an implicit representation of parts from the knowledge graph involving a segment-based propagation of recursively derived knowledge and traceable by the associated hierarchy level. Each query extends the set of available knowledge, where the according knowledge types match the ones from the ontology in Fig. 1. As a representative example, an incremental enrichment of generated knowledge is demonstrated based on segment 3, where $Knowledge_i^k$ equals the set of knowledge for segment node i at propagation level k.

***Initial graph construction*: Time series segmentation**
$Knowledge_3^0 = \{DataReference(hierarchyLevel = 0)\}$

The knowledge graph is constructed from a segmentation via PELT [17] using an RBF-kernel function for changepoint detection. As shown in the top left of Fig. 3, individual behaviors are captured, leading to individual segment nodes of the time series knowledge graph. Available information of each segment is an initial reference to the original data.

***Query 1*: Subsequence-specific feature extraction**
$Knowledge_3^1 = Knowledge_3^0 \cup$
$\{Feature(name = "min", value = -1.23, hierarchyLevel = 1),$
$Feature(name = "max", value = 2.88, hierarchyLevel = 1),$
$Feature(name = "median", value = -0.26, hierarchyLevel = 1)\}$

Allowing for a descriptive analysis and comparison of individual time series behaviors, one common query is a segment-wise extraction of features. In our query we limit the set of features to minimum, maximum and median. As these information are built on top of each segment, their hierarchy level equals 1.

***Query 2*: Clustering of subsequences**
$Knowledge_3^2 = Knowledge_3^1 \cup$
$\{Label(name = "cluster\_id", value = 2, hierarchyLevel = 2)\}$

Using the features from *Query 1*, we cluster the segments using DBSCAN [16]. Illustrated in the bottom left part of Fig. 3, every cluster is indicated by one specific color. As the output of a clustering method is a grouping via labels, the type of derived knowledge is *Label*. Since the features from *Query 1* where used as input to the clustering algorithm, the corresponding hierarchy level is 2 indicating the depth of propagation.

***Query 3*: Anomaly detection for a defined subsequence**
$Knowledge_3^3 = Knowledge_3^2 \cup$
$\{Label(name = "anomaly", value = [1869, 1968], hierarchyLevel = 1)\}$

As a final exemplary query, we identified anomalies in the third segment of the time series using [19]. Again the hierarchy level is 1, as anomalies are retrieved only using initial knowledge from the segmentation. Knowledge of a subsequent propagation step could be a *Note* from a domain expert, defining this anomaly to be the result of a specific event.

4 Conclusion and Future Work

In this paper we presented a graph-based solution for time series knowledge, which enables a structured management and query-driven consolidation of generated time series knowledge. In addition to a contribution of a top-level ontology, describing associated entities, we conceptualized the process of a recursive knowledge enrichment by combining existing ideas from automated data analysis with a generic graph structure for knowledge propagation. Finally, we clarified the utilization of our solution based on an example of knowledge propagation. Apart from a more in-depth evaluation of suitable KDD methods, as a next step we intend to further emphasize on the inclusion of domain information and its optimization potential regarding semantic knowledge interpretation.

References

1. The RDF Data Cube Vocabulary (2014). https://www.w3.org/TR/vocab-data-cube/
2. Time Ontology in OWL (2022). https://www.w3.org/TR/owl-time/
3. Cheng, Z., et al.: Time2graph: revisiting time series modeling with dynamic shapelets. In: Proceedings of the AAAI Conference on Artificial Intelligence, vol. 34, pp. 3617–3624 (2020)
4. Dau, H.A., Keogh, E., Kamgar, K., et al.: The ucr time series classification archive, October 2018. https://www.cs.ucr.edu/eamonn/time_series_data_2018/
5. Dwivedi, S., Kasliwal, P., Soni, S.: Comprehensive study of data analytics tools (rapidminer, weka, r tool, knime). In: 2016 Symposium on Colossal Data Analysis and Networking (CDAN), pp. 1–8. IEEE (2016)
6. Hamilton, J.D.: Time Series Analysis. Princeton University Press, Princeton (2020)
7. Hogan, A., Blomqvist, E., Cochez, M., et al.: Knowledge graphs. ACM Comput. Surv. **54**(4), 1–37 (2021). https://doi.org/10.1145/3447772, https://doi.org/10.1145%2F3447772
8. Jensen, S.K., Pedersen, T.B., Thomsen, C.: Time series management systems: a survey. IEEE Trans. Knowl. Data Eng. **29**(11), 2581–2600 (2017)
9. Johnson, T., Lakshmanan, L.V., Ng, R.T.: The 3w model and algebra for unified data mining. In: VLDB, pp. 21–32 (2000)
10. Karmaker, S.K., Hassan, M.M., Smith, M.J., et al.: Automl to date and beyond: challenges and opportunities. ACM Comput. Surv. **54**(8), 1–36 (2021)
11. Kietz, J.U., Serban, F., Bernstein, A., et al.: Designing KDD-workflows via HTN-planning for intelligent discovery assistance (2012)
12. Lu, Y., et al.: Grab: finding time series natural structures via a novel graph-based scheme. In: 2021 IEEE 37th International Conference on Data Engineering (ICDE), pp. 2267–2272. IEEE (2021)
13. Panov, P., Džeroski, S., Soldatova, L.: ONTODM: an ontology of data mining. In: 2008 IEEE International Conference on Data Mining Workshops, pp. 752–760. IEEE (2008)
14. Panov, P., Soldatova, L., Džeroski, S.: OntoDM-KDD: ontology for representing the knowledge discovery process. In: Fürnkranz, J., Hüllermeier, E., Higuchi, T. (eds.) DS 2013. LNCS (LNAI), vol. 8140, pp. 126–140. Springer, Heidelberg (2013). https://doi.org/10.1007/978-3-642-40897-7_9

15. Publio, G.C., Esteves, D., Ławrynowicz, A., et al.: Ml-schema: exposing the semantics of machine learning with schemas and ontologies (2018)
16. Schubert, E., Sander, J., Ester, M., Kriegel, H.P., et al.: DBSCAN revisited, revisited: why and how you should (still) use DBSCAN. ACM Trans. Database Syst. (TODS) **42**(3), 1–21 (2017)
17. Truong, C., Oudre, L., Vayatis, N.: Ruptures: change point detection in python. arXiv preprint arXiv:1801.00826 (2018)
18. Von Rueden, L., Mayer, S., Beckh, K., et al.: Informed machine learning-a taxonomy and survey of integrating prior knowledge into learning systems. IEEE Trans. Knowl. Data Eng. **35**(1), 614–633 (2021)
19. Yeh, C.C.M., Zhu, Y., Ulanova, L., et al.: Matrix profile I: all pairs similarity joins for time series: a unifying view that includes motifs, discords and shapelets. In: 2016 IEEE 16th International Conference on Data Mining, pp. 1317–1322 (2016)

Consistent Data Management

Eventually-Consistent Replicated Relations and Updatable Views

Joachim Thomassen and Weihai Yu(✉)

UIT - The Arctic University of Norway, Tromsø, Norway
weihai.yu@uit.no

Abstract. Distributed systems have to live with weak consistency, such as eventual consistency, if high availability is the primary goal and network partitioning is unexceptional. Local-first applications are examples of such systems. There is currently work on local-first databases where the data are asynchronously replicated on multiple devices and the replicas can be locally updated even when the devices are offline. Sometimes, a user may want to maintain locally a copy of a view instead of the entire database. For the view to be fully useful, the user should be able to both query and update the local copy of the view. We present an approach to maintaining updatable views where both the source database and the views are asynchronously replicated. The approach is based on CRDTs (Conflict-free Replicated Data Types) and guarantees eventual consistency.

Keywords: Data replication · eventual consistency · updatable views · lenses · CRDT

1 Introduction

Local-first software suggests a set of principles for software that enables both collaboration and ownership for users. Local-first ideals include the ability to both work offline and collaborate across multiple devices [11].

CRDT [13], or Conflict-free Replicated Data Type, has been a popular approach to constructing eventually-consistent local-first systems. With CRDT, a site updates its local replica without coordination with other sites. The states of replicas converge when they have applied the same set of updates. CRR [14], or Conflict-free Replicated Relation, is an application of CRDT to relational databases.

A user may want to maintain copies of views in her local devices, instead of the entire database. This may reduce both the amount of data stored on the device as well as the overhead of data communication and local data processing. For a view to be fully useful, the user should be able to both query and update the local copies of the views. The updates must then be translated and applied back to the original source database.

Supporting updatable views has been an active research topic for decades [2,2–4,8,10]. A view V is a sequence of query operations $\mathscr{Q}_v = [q_1, q_2, \ldots]$ over a source database s, i.e. $v = \mathscr{Q}_v(s)$. After an update u_v on view v, the new state of the view becomes $v' = u_v(v)$. A translation $T_\uparrow$ of u_v to source database s results in a sequence of updates in base relations $T_\uparrow(s, u_v) = [u_1, u_2, \ldots]$. According to [4], $T_\uparrow(s, u_v)$ *exactly*

A. Abelló et al. (Eds.): ADBIS 2023, CCIS 1850, pp. 187–196, 2023.
https://doi.org/10.1007/978-3-031-42941-5_17

translates $u_v(v)$ iff $\mathcal{Q}_v(T_\uparrow(s,u_v)(s)) = u_v(\mathcal{Q}_v(s))$ and the integrity constraints defined in the source database are preserved.

There is at present no existing work on supporting updatable views in a distributed setting where both the source database and the views are asynchronously replicated and local data can be updated even when the replicas are offline. One challenge is that the translation $T_\uparrow(s,u_v)$ is dependent on the state s of the source database. The same view update u_v may have different translations at replicas in different concurrent states.

We present an approach based on delta-state CRDT [1,5] (Sect. 2.1), applied to replication of relational databases [14] (Sect. 2.2). The states of the source database and the views form a join-semilattice [6,13]. The updates in the database and the views are represented as join-irreducible states in the join-semilattice. A view update is translated into a set of join-irreducible states in the source database and the translation is independent of the current state of the database. When the database (and equally view) replicas have applied the same set of updates (i.e. join-irreducible states), their states converge.

The paper is organized as the following. In Sect. 2, we briefly review the necessary background of CRDT and CRR. In Sect. 3, we give a high-level overview of our approach. In Sect. 4, we use examples to describe how we translate view updates to the source database. In Sect. 5, we discuss related work. Finally, in Sect. 6, we conclude.

2 Technical Background

In this section, we briefly review necessary technical background on CRDT and CRR.

2.1 CRDT

A CRDT [13] is a data abstraction specifically designed for data replicated at different sites. A site queries and updates its local replica without coordination with other sites. The data are always available for update, but the data states at different sites may diverge. From time to time, the sites send their updates asynchronously to other sites with an anti-entropy protocol. The sites also merge the received updates with their local data. A CRDT guarantees *strong eventual consistency* [13]: a site merges incoming remote updates without coordination with other sites; when all sites have applied the same set of updates, their states converge.

We adopt delta-state CRDTs [1,5]. The possible states must form a join-semilattice [6], which implies convergence. Briefly, the states form a *join-semilattice* if they are partially ordered with $\sqsubseteq$ and a join $\sqcup$[1] of any two states (that gives the least upper bound of the two states) always exists. State updates must be inflationary. That is, the new state supersedes the old one in $\sqsubseteq$. The merge of two states s_1 and s_2 is the result of $s_1 \sqcup s_2$. With delta-state CRDTs, it is sufficient to only send and merge join-irreducible states. Basically, *join-irreducible* states are elementary states: every state in the join-semilattice can be represented as a join of some join-irreducible state(s).

Since a relation instance is a set of tuples, the basic building block of CRR is a general-purpose delta-state set CRDT ("general-purpose" in the sense that it allows

[1] To avoid being confused with the join $\bowtie$ of relations, in the rest of the paper, we use the term *merge* for $\sqcup$.

$$\mathsf{CLSet}(E) \stackrel{\text{def}}{=} E \hookrightarrow \mathbb{N}$$

$$\mathsf{insert}^{\delta}(s,e) \stackrel{\text{def}}{=} \begin{cases} \{e \mapsto s(e)+1\} & \text{if } \neg\mathsf{in?}(s(e)) \\ \{\} & \text{otherwise} \end{cases}$$

$$\mathsf{delete}^{\delta}(s,e) \stackrel{\text{def}}{=} \begin{cases} \{e \mapsto s(e)+1\} & \text{if } \mathsf{in?}(s(e)) \\ \{\} & \text{otherwise} \end{cases}$$

$$(s \sqcup s')(e) \stackrel{\text{def}}{=} \mathsf{max}(s(e), s'(e))$$

$$\mathsf{in?}(s,e) \stackrel{\text{def}}{=} \mathsf{odd?}(s(e))$$

Fig. 1. CLSet CRDT [14]

both insertion and deletion of elements). We use CLSet (causal-length set, [14,15]), a general-purpose set CRDT, where each element is associated with a *causal length*. Intuitively, insertion and deletion are inverse operations of one another. They always occur in turn. When an element is first inserted into a set, its causal length is 1. When the element is deleted, its causal length becomes 2. Thereby the causal length of an element increments on each update that reverses the effect of a previous one.

As shown in Fig. 1, the states of a CLSet are a partial function $s\colon E \hookrightarrow \mathbb{N}$, meaning that when e is not in the domain of s, $s(e) = 0$. Using partial function conveniently simplifies the specification of insert, $\sqcup$ and in?. Without explicit initialization, the causal length of any unknown element is 0. insert^{δ} and delete^{δ} in Fig. 1 are delta-mutators that returns a join-irreducible state instead of the entire state.

An element e is regarded as being in the set when its causal length is an odd number. A local insertion has effect only when the element is not in the set. Similarly, a local deletion has effect only when the element is actually in the set. A local effective insertion or deletion simply increments the causal length of the element by one. For every element e in s and/or s', the new causal length of e, after merging s and s', is the maximum of the causal lengths of e in s and s'.

2.2 CRR

The relational database supporting CRR consists of two layers: an Application Relation (AR) layer and a Conflict-free Replicated Relation (CRR) layer (Fig. 2). The AR layer presents the same database schema and API as a conventional relational database. Application programs interact with the database at the AR layer. The CRR layer supports conflict-free replication of relations.

An AR-layer relation schema R has an augmented CRR-layer schema $\tilde{R}$. In Fig. 2, site A maintains both an instance r_A of R and an instance $\tilde{r}_A$ of $\tilde{R}$. A query q is performed on r_A without any involvement of $\tilde{r}_A$. An update operation u on r_A triggers an additional operation $\tilde{u}$ on $\tilde{r}_A$. The operation $\tilde{u}$ is later propagated to remote sites through an anti-entropy protocol. Merge with an incoming remote operation $\tilde{u}'(\tilde{r}_B)$ results in an operation $\tilde{u}'$ on $\tilde{r}_A$ as well as an operation u' on r_A.

CRR has the property that when both sites A and B have applied the same set of operations, the relation instances at the two sites are equivalent, i.e. $r_A = r_B$ and $\tilde{r}_A = \tilde{r}_B$.

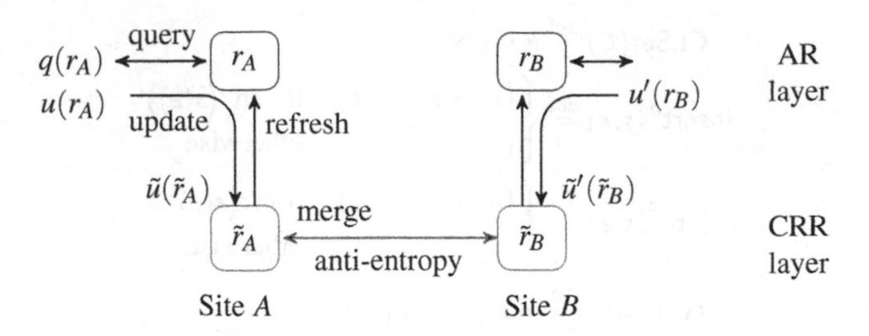

Fig. 2. A two-layer relational database system [14]

CRR adopts several CRDTs. Since a relation instance is a set of tuples, we use the CLSet CRDT (Fig. 1) for relation instances. We use the LWW (last-write wins) register CRDT [9,12] for individual attributes in tuples.

The join-irreducible states in a CRR relation $\tilde{r}$ are simply the tuples as the result of the insertions, deletions and updates. In the rest of the paper, we use the term *delta* for the tuple as the join-irreducible state of an operation. As we apply delta-state CRDTs, the tuples of the latest changes are sent to remote sites in the anti-entropy protocol.

For an AR-layer relation $R(K, A_1, A_2, \ldots)$, where K is the primary key, there is a CRR-layer relation $\tilde{R}(\tilde{K}, K, L, T_1, T_2, \ldots, A_1, A_2, \ldots)$. $\tilde{K}$ is the primary key of $\tilde{R}$ and its values are globally unique. L is the causal-lengths (Fig. 1) of the tuples in $\tilde{R}$. T_i is the timestamp of the last update on attribute A_i. In other words, the $(\tilde{K}, L)$ part represents the CLSet CRDT of tuples and the (A_i, T_i) parts represent the LWW register CRDT of the attributes.

When inserting a new tuple t into r, we insert a new tuple $\tilde{t}$ into $\tilde{r}$, with the initial $\tilde{t}(L) = 1$. When deleting t from r, we increment $\tilde{t}(L)$ with 1. Tuple t is in r, $t \in r$, if $\tilde{t}(L)$ is an odd number. That is,

$$\mathsf{in\_ar?}(\tilde{t}) \stackrel{\text{def}}{=} \mathsf{odd?}(\tilde{t}(L))$$

When updating $t(A_i)$ in r, we update $\tilde{t}(A_i)$ and $\tilde{t}(T_i)$ in $\tilde{r}$.

An update delta on an relation instance $\tilde{r}'$ at a remote site is actually a tuple $\tilde{t}'$. If a tuple $\tilde{t}$ in the local instance $\tilde{r}$ exists such that $\tilde{t}(\tilde{K}) = \tilde{t}'(\tilde{K})$, we update $\tilde{t}$ with $\tilde{t} \sqcup \tilde{t}'$ where the merge $\sqcup$ is the join operation of the join-semilattice (Sect. 2.1). Otherwise, we insert $\tilde{t}'$ into $\tilde{r}$. The merge $\tilde{t} \sqcup \tilde{t}'$ is defined as:

$$\tilde{t} \sqcup \tilde{t}' \stackrel{\text{def}}{=} \tilde{t}'', \text{ where } \tilde{t}''(L) = \mathsf{max}(\tilde{t}(L), \tilde{t}'(L)), \text{ and}$$

$$\tilde{t}''(A_i), \tilde{t}''(T_i) = \begin{cases} \tilde{t}'(A_i), \tilde{t}'(T_i) & \text{if } \tilde{t}'(T_i) > \tilde{t}(T_i) \\ \tilde{t}(A_i), \tilde{t}(T_i) & \text{otherwise} \end{cases}$$

After the update of $\tilde{r}$, we update r as the following. If $\mathsf{in\_ar?}(\tilde{t})$ evaluates to false, we delete t (where $t(K) = \tilde{t}(K)$) from r. Otherwise, we insert or update r with $\pi_{K, A_1, A_2, \ldots}(\tilde{t})$.

3 Approach Overview

We consider distributed database systems where data are replicated at multiple sites. For the purpose of, say, high availability, the sites may update the data without coordination with other sites. The system is said to be *eventually consistent*, or convergent, if, when all sites have applied the same set of updates, the sites have the same state. The system is said to be *strongly eventually consistent* [13], if the sites unilaterally resolve any possible conflict, i.e., without coordination with other sites. We focus on strongly eventually-consistent relational database systems.

We restrict on which views can be updated, similar to [3,8,10]. More specifically, a view can only project away non-primary-key attributes that are given default values or can remain unspecified with NULL when inserted without given value. Moreover, when joining two relations, the join attribute(s) must contain one of the primary keys.

For a source database schema S, we define a view V with $V = \mathcal{Q}_v(S)$. Suppose when the database state is initially s_0, the view state is $v_0 = \mathcal{Q}_v(s_0)$ (Fig. 3). Concurrently, the view applies updates with delta state $\Delta v'$ and the source database applies updates with delta state $\Delta s'$. The new states in the view and the database become $v_1 = v_0 \sqcup \Delta v'$ and $s_1 = s_0 \sqcup \Delta s'$ respectively. When the database receives $\Delta v'$, it applies the translated delta $T_\uparrow(\Delta v')$ to s_1 and the new state becomes $s_2 = s_1 \sqcup T_\uparrow(\Delta v')$. Similarly, when the view receives $\Delta s'$, it applies the translated delta $T_\downarrow(\Delta s')$ to v_1 and the new state become $v_2 = v_1 \sqcup T_\downarrow(\Delta s')$. One important property of the translations $T_\downarrow$ and $T_\uparrow$ is that they are independent of the target state in which the translation results are going to be applied.

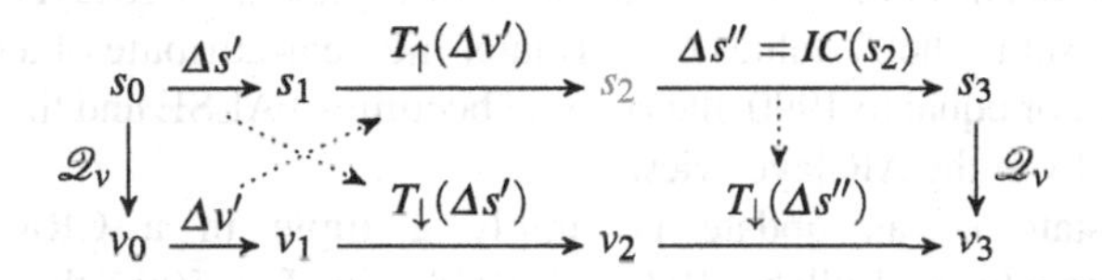

Fig. 3. Delta-states in source database and view

Unlike traditional work on updatable views, we do not restrict to side-effect-free view updates. However, we do respect integrity constraints, including the ones defined by application programs, for instance, functional dependencies enforced with triggers.

In Fig. 3, if the state s_2 violates an integrity constraint, s_2 is never visible to the application. Instead, the view immediately applies some additional delta (as side effect of $T_\uparrow(\Delta v')$), $\Delta s'' = IC(s_2)$ for integrity-constraint preservation, and the new state s_3 does not violate any integrity constraint. Finally, the view applies the translation of $\Delta s''$. Our approach guarantees that the updates in Fig. 3 commute. That is,

$$\mathcal{Q}_v(s_0 \sqcup \Delta s' \sqcup T_\uparrow(\Delta v') \sqcup \Delta s'') = \mathcal{Q}_v(s_0) \sqcup \Delta v' \sqcup T_\downarrow(\Delta s') \sqcup T_\downarrow(\Delta s'')$$

Since the merge operation $\sqcup$ is commutative, when the different replicas of the source database (or the view) have applied the same set of delta states, their final states converge.

We have implemented a prototype of CRR and updatable views with SQLite. We do not include the implementation and experiments in this paper due to space limit.

R_a album	quantity
Disintegration	6
Show	3
Galore	1
Paris	4
Wish	5

R_t track	year	instore
Lullaby	1989	TRUE
Lovesong	1989	TRUE
Trust	1992	FALSE

R_{ta} track	album
Lullaby	Galore
Lullaby	Show
Lovesong	Galore
Lovesong	Paris
Trust	Wish

Fig. 4. Example database

4 Translation of View-Update Delta States

The translation from source database to views, $T_{\downarrow}$, is traditionally know as incremental maintenance of materialized views. In this section, we focus on $T_{\uparrow}$, the translation of view-update delta states to the source database. We describe the translation through examples. The example database (Fig. 4) is adapted from [3,8].

We start with select and project views. In Fig. 5, the base relation R_t (top left) is first augmented to a CRR-layer relation $\tilde{R}_t$ (top right). $\tilde{R}_t$ has an attribute L for the causal lengths of the tuples. In addition, every non-primary-key attribute is associated with a timestamp attribute, indicating the last time at which the attribute value was set.

A project view has the same causal-length and timestamp attributes as the base relation, unless the attribute is projected away. A select view has two more attributes σ and T_σ that tell the last time the select predicate was evaluated. Initially, all σ values are TRUE and the timestamp value T_σ of a tuple is the maximum of the timestamp values of the attributes that occur in the select predicate. For tuples in CRR-layer $\tilde{v}_1$ in Fig. 5(a), the T_σ values are set to the T_y values of $\tilde{r}_t$. If later the year-attribute of a tuple is set to a value greater than or equal to 1990, the σ value becomes FALSE and the corresponding tuple disappears from the AR-layer view.

The delta state of an update is simply a tuple in a CRR-layer relation or view. For update $v_1\langle\text{Lullaby}, 1989 \nearrow 1988\rangle$ in Fig. 5(a), the delta state is $\tilde{v}_1'\langle\text{Lullaby}, 1988, 5.1, 1, \text{TRUE}, 5.1\rangle$. Here, $T_y = 5.1$ is the timestamp at which the new year-value is set. Since the year-attribute is used in the select predicate, T_σ is also set to 5.1.

For deletion $-v_1\langle\text{Lovesong}, 1989\rangle$, the L attribute of the delta state is incremented with 1. As it is an even number, the tuple is regarded as being deleted in the AR layer.

For insertion $+v_1\langle\text{Catch}, 1989\rangle$, the initial L value is 1 and all timestamps are set according to the current time. For all insertions in select views, the T_σ value must be TRUE.

Recall that a project view is updatable only if it keeps the primary key of the base relation. Moreover, CRR-layer base and view relations keep all tuples regardless of whether they have been deleted or not. Therefore, for every tuple in a CRR-layer select-and-project view, there is exactly one tuple in the CRR-layer base relation.

Delta states of a view can be translated almost directly to the base relation. The only exception is for the attributes that are projected away. The instore-attribute of the Catch-tuple, which is missing in view V_1, is set to its default value (suppose it is FALSE). Its timestamp value T_i is set to 0.0, the smallest possible timestamp value. This means that a default value (or NULL) cannot override any value that is explicitly given.

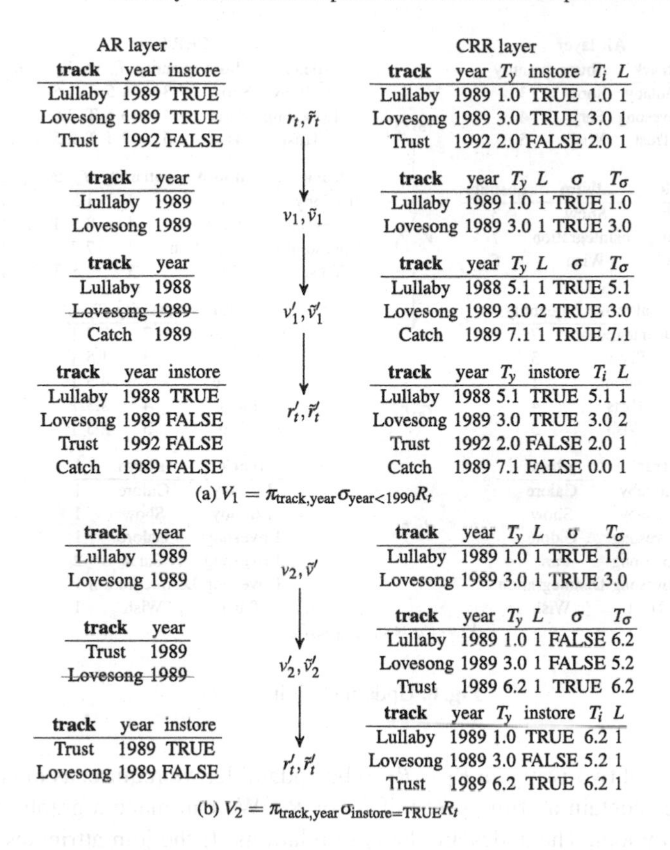

Fig. 5. Updating select and project views

Figure 5(b) shows two additional cases. The first case shows that a deletion in some views can be handled differently. Here, we have an opportunity to achieve a least-effect translation of deletions in a view, when the select predicate includes a boolean attribute, such as the instore-attribute in $\sigma_{\text{instore=TRUE}}$. Now, for the deletion $-v_2\langle\text{Lovesong}, 1989\rangle$, instead of deleting the Lovesong-tuple in the base relation (i.e. by incrementing the L value), we set the σ value to FALSE. When translating to the base relation, we set the boolean value of the attribute as the negation in the select predicate. That is,
$T_\downarrow(\tilde{v}_2\langle\text{Lovesong}, 1989, 3.0, 1, \text{FALSE}, 5.2\rangle) = [\tilde{r}_t\langle\text{Lovesong}, 1989, 3.0, \text{FALSE}, 5.2\rangle]$.

In this particular example, setting the Lovesong-track to be not-in-store is less destructive than deleting the track. When a select predicate uses multiple boolean attributes, we choose to update the truth value of the leftmost one in the view definition.

The next case that Fig. 5(b) shows actually applies generally to updates in both view and base relations. An update of (part of) a primary-key value is regarded as a deletion and an insertion. In the figure, the update $v_2\langle\text{Lullaby} \nearrow \text{Trust}, 1989\rangle$ is interpreted as $[-v_2\langle\text{Lullaby}, 1989\rangle, +v_2\langle\text{Trust}, 1989\rangle]$.

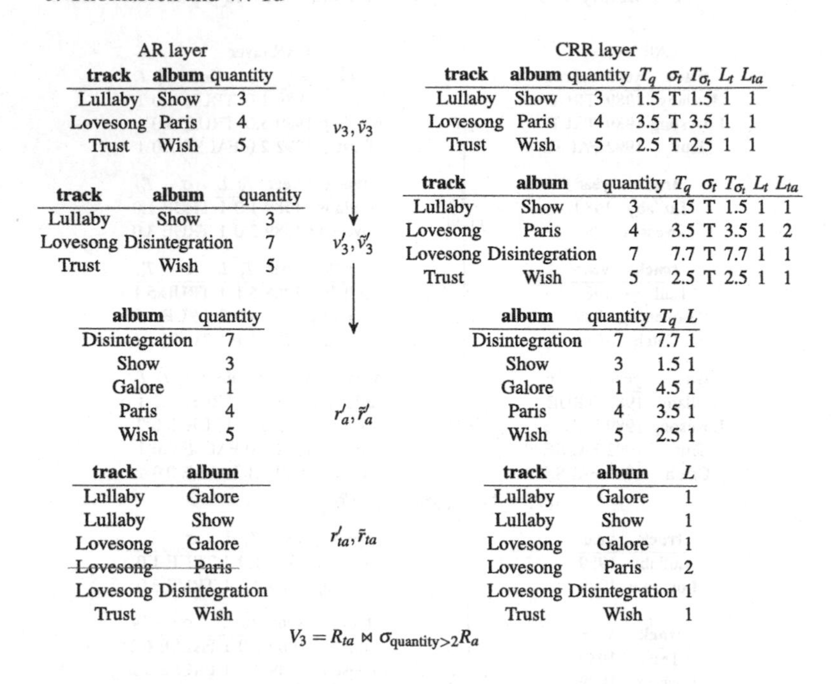

AR layer

track	**album**	quantity
Lullaby	Show	3
Lovesong	Paris	4
Trust	Wish	5

CRR layer

track	**album**	quantity	T_q	σ_t	T_{σ_t}	L_t	L_{ta}
Lullaby	Show	3	1.5	T	1.5	1	1
Lovesong	Paris	4	3.5	T	3.5	1	1
Trust	Wish	5	2.5	T	2.5	1	1

$v_3, \tilde{v}_3$

track	**album**	quantity
Lullaby	Show	3
Lovesong	Disintegration	7
Trust	Wish	5

track	**album**	quantity	T_q	σ_t	T_{σ_t}	L_t	L_{ta}
Lullaby	Show	3	1.5	T	1.5	1	1
Lovesong	Paris	4	3.5	T	3.5	1	2
Lovesong	Disintegration	7	7.7	T	7.7	1	1
Trust	Wish	5	2.5	T	2.5	1	1

$v_3', \tilde{v}_3'$

album	quantity
Disintegration	7
Show	3
Galore	1
Paris	4
Wish	5

$r_a', \tilde{r}_a'$

album	quantity	T_q	L
Disintegration	7	7.7	1
Show	3	1.5	1
Galore	1	4.5	1
Paris	4	3.5	1
Wish	5	2.5	1

track	**album**
Lullaby	Galore
Lullaby	Show
Lovesong	Galore
~~Lovesong~~	~~Paris~~
Lovesong	Disintegration
Trust	Wish

$r_{ta}', \tilde{r}_{ta}$

track	**album**	L
Lullaby	Galore	1
Lullaby	Show	1
Lovesong	Galore	1
Lovesong	Paris	2
Lovesong	Disintegration	1
Trust	Wish	1

$V_3 = R_{ta} \bowtie \sigma_{\text{quantity}>2} R_a$

Fig. 6. Updating a join view

For a view of two-way join $R_1 \bowtie R_2$ to be updatable, we require, as in [10], that the join attributes contain a primary key of R_1 or R_2. We can make a graph from a view of a multi-way join. The nodes are the base relations. If the join attributes of $R_i \bowtie R_j$ contains the primary key of R_j, there is a link from R_i to R_j in the graph. Currently, we require, also as [10], that the view graph is a tree. The primary key of the view is the primary key of the root relation of the tree. Since the primary keys of the base relations are not projected away, for a tuple t_v in the view, we can find the tuples in the base relations that contribute to t_v via their primary-key values.

For view $V = R_1 \bowtie R_2$, the set of attributes of the CRR layer $\tilde{V}$ is the union of the sets of attributes of the CRR-layer $\tilde{R}_1$ and $\tilde{R}_2$. For tuple $\tilde{t}_v$ in CRR-layer $\tilde{v}$, tuple t_v is in AR-layer v, if the L values of both $\tilde{r}_1$ and $\tilde{r}_2$ are odd and the σ values of both $\tilde{r}_1$ and $\tilde{r}_2$ are TRUE, i.e.,

$$\mathsf{in\_ar?}(\tilde{t}_v) \stackrel{\text{def}}{=} \mathsf{odd?}(\tilde{t}_v(L_{r_1})) \wedge \mathsf{odd?}(\tilde{t}_v(L_{r_2})) \wedge \tilde{t}_v(\sigma_{r_1}) \wedge \tilde{t}_v(\sigma_{r_2})$$

In Fig. 6, there is only one update in the view, $v_3\langle \text{Lovesong}, \text{Paris} \nearrow \text{Disintegration}, 4 \nearrow 7\rangle$. Since the album-attribute is part of the primary key of the view, the update is interpreted as a deletion $-v_3\langle \text{Lovesong}, \text{Paris}, 4\rangle$ and an insertion $+v_3\langle \text{Lovesong}, \text{Disintegration}, 7\rangle$.

For the deletion, we delete the corresponding tuple in the root base relation. Hence the tuple $\langle \text{Lovesong}, \text{Paris}\rangle$ is deleted from r_{ta}.

For the insertion, we first insert $\langle \text{Lovesong}, \text{Disintegration}\rangle$ into the root relation r_{ta}. Then, since there is already a Disintegration-tuple in r_a, we set the quantity-attribute to the new value 7.

5 Related Work

[2,7] study the consistency of updatable views via mapping of states between source databases and views, where a source database is modeled as the product of the view and a complementary. When a chosen complementary is kept constant (side-effect free) [2] or "shrinking" (under a partial order) [7], there is an unambiguous translation of a view update to the source database. [2] did not aim for computational algorithms that translate view updates to source databases.

To translate the updates from a view to a source database, [10] directly associates tuples and attributes in view relations with base relations in the source database. [4] makes the translation based on the tractability and functional dependency of attributes via view dependency graphs. [7] translates view programs (sequence of updates equipped with if-then-else statements) to base programs. [3,8] make bi-directional translation of every query operation (known as a lens) that defines the views. In most of the work on updatable views, translation of view updates is based on the attribute values. For example, since the view dependency graphs in [4] are defined on attributes, deletions are defined with predicates on attributes, for instance, "delete from V where $A = 7$". The source tuples can then be identified with queries on attributes with similar predicates. This may work well in a non-distributed system. In a distributed system where the source database and the view can be replicated, different replicas in different states may make different translations.

Our work is different from the previous work in that we use delta states (i.e. join-irreducible states in a join semilattice) to represent state updates. The translation is independent of the state to which the update is to be applied.

Regarding the restrictions on views that are updatable, [10] is the closest to our work, which are probably the most restrictive. There are at least two reasons for these restrictions. The first one is practical. Most related work assumes that all information about integrity constraints is available when a view is created, which is practically not true. In particular, the only functional dependencies that can be expressed in SQL is primary-key constraints. The second reason is that we are currently not able to express aggregate results (such as COUNT and MAX) as join-irreducible states.

In their seminal work [4], Dayal and Bernstein pointed out that a view update can be correctly (exactly) translated to the source relations if and only if there is a *clean source* of the update. It is possible to verify if a source is clean with the use of view dependency graphs. With the restrictions of the view that can be updated (Sect. 3), we guarantee that every update in a view has a clean source.

Unlike previous work, we allow translations of view updates to have side effects (Fig. 3). Avoiding side effect is probably more important in earlier work, which expects virtual (i.e. non-materialized) views. In fact, avoiding side effect is impossible without knowing all integrity constraints, such as the functional dependencies embedded in the view dependency graphs [4]. Notice that concurrent updates at different replicas may temporarily violate integrity constraints (like uniqueness and referential constraints) anyway [14]. We detect violations and repair constraints at the time of merge [14].

6 Conclusion

We presented an approach to asynchronously replicating both source databases and views. The local replicas of the database and the view can be updated even when they are offline. The approach guarantees eventual consistency. That is, the view updates are correctly translated to the source database, and when the replicas have applied the same set of updates, their states converge.

References

1. Almeida, P.S., Shoker, A., Baquero, C.: Delta state replicated data types. J. Parallel Distrib. Comput. **111**, 162–173 (2018)
2. Bancilhon, F., Spyratos, N.: Update semantics of relational views. ACM Trans. Database Syst. **6**(4), 557–575 (1981)
3. Bohannon, A., Pierce, B.C., Vaughan, J.A.: Relational lenses: a language for updatable views. In: Proceedings of the Twenty-Fifth ACM SIGACT-SIGMOD-SIGART Symposium on Principles of Database Systems (PODS), pp. 338–347. ACM (2006)
4. Dayal, U., Bernstein, P.A.: On the correct translation of update operations on relational views. ACM Trans. Database Syst. **7**(3), 381–416 (1982)
5. Enes, V., Almeida, P. S., Baquero, C., Leitão, J.: Efficient Synchronization of State-Based CRDTs. In: IEEE 35th International Conference on Data Engineering (ICDE), April 2019
6. Garg, V.K.: Introduction to Lattice Theory with Computer Science Applications. Wiley, Hoboken (2015)
7. Gottlob, G., Paolini, P., Zicari, R.V.: Properties and update semantics of consistent views. ACM Trans. Database Syst. **13**(4), 486–524 (1988)
8. Horn, R., Perera, R., and Cheney, J. Incremental relational lenses. Proc. ACM Program. Lang. **2**, 74:1–74:30. ICFP (2018)
9. Johnson, P., Thomas, R.: The maintamance of duplicated databases. Internet Request Comments RFC 677, January 1976
10. Keller, A.M.: Algorithms for translating view updates to database updates for views involving selections, projections, and joins. In: Proceedings of the Fourth ACM SIGACT-SIGMOD Symposium on Principles of Database Systems, 25–27 March 1985, Portland, Oregon, USA, pp. 154–163. ACM (1985)
11. Kleppmann, M., Wiggins, A., van Hardenberg, P., McGranaghan, M.: Local-first software: you own your data, in spite of the cloud. In: Proceedings of the 2019 ACM SIGPLAN International Symposium on New Ideas, New Paradigms, and Reflections on Programming and Software, (Onward! 2019), pp. 154–178 (2019)
12. Shapiro, M., Preguiça, N.M., Baquero, C., Zawirski, M.: A comprehensive study of convergent and commutative replicated data types. Rapport de recherche 7506, January 2011
13. Shapiro, M., Preguiça, N.M., Baquero, C., Zawirski, M.: Conflict-free replicated data types. In: 13th International Symposium on Stabilization, Safety, and Security of Distributed Systems, (SSS 2011), pp. 386–400 (2011)
14. Yu, W., Ignat, C.-L.: Conflict-free replicated relations for multi-synchronous database management at edge. In: IEEE International Conference on Smart Data Services (SMDS), pp. 113–121, October 2020
15. Yu, W., Rostad, S.: A low-cost set CRDT based on causal lengths. In: Proceedings of the 7th Workshop on the Principles and Practice of Consistency for Distributed Data (PaPoC), pp. 5:1–5:6 (2020)

Mining Totally Ordered Sequential Rules to Provide Timely Recommendations

Anna Dalla Vecchia, Niccolò Marastoni, Sara Migliorini, Barbara Oliboni, and Elisa Quintarelli(✉)

University of Verona, Strada Le Grazie, 15, Verona, Italy
anna.dallavecchia@studenti.univr.it,
{niccolo.marastoni,sara.migliorini,barbara.oliboni,
elisa.quintarelli}@univr.it

Abstract. In this paper we show the importance of mining totally ordered sequential rules, and in particular we propose an extension of sequential rules where not only the antecedent precedes the consequent, but their itemsets are labelled with an explicit representation of their relative order. This allows us to provide more precise timely recommendations. Our technique has been applied to a real-world scenario regarding the provision of tailored suggestions for supermarket shopping activities.

Keywords: Sequential Rules · Data mining · Recommendations

1 Introduction

In the era of increasing and rapid data generation, the ability to extract personalized, timely and often hidden knowledge has been a challenging task for years. Frequent pattern mining has received a lot of attention in many applications, such as Market Basket Analysis and Recommender Systems, where frequent correlations among data or events are relevant and useful in many decision-making processes [2].

In order to consider also the order between events in knowledge extraction processes, sequential pattern mining techniques have been proposed. In this context, discovering hidden sequential patterns allows the extraction of frequent sequential rules that are more appropriate for predictions and are usually formalized with implications of the form $X \rightarrow Y$, with support and confidence that satisfy the established minimum thresholds [6,10,11].

Many algorithms proposed in the literature consider partially ordered sequential rules, where the antecedent X is related to events that happened before the consequent Y and thus do not consider the relative order in which events appear in the antecedent [7]. However, in many real scenario, the relative order between events does matter and can be leveraged to provide more effective timely recommendations. For this reason we propose a mechanism to extract totally ordered sequential rules with explicit reference to the relative order between items.

The recommendation of the next-item in a sequence has attracted a lot of attention in recent years and is usually implemented by using collaborative filtering techniques [1,8]. For instance, to recommend additional items that a customer may be interested in, Amazon examines the sequences of items that similar

A. Abelló et al. (Eds.): ADBIS 2023, CCIS 1850, pp. 197–207, 2023.
https://doi.org/10.1007/978-3-031-42941-5_18

customers have purchased and rated, then it compares them with the preferences of the specific user and finds the best next suggestions. In order to provide personalized, timely and contextual recommendations, the order among the past activities and the past habits related to subsequent purchase activities should be considered.

To the best of our knowledge the only proposal that addresses the problem of adding explicit time references in totally ordered event rules is [3]. Since the authors focus on time-sensitive applications, they are interested in totally ordered episodes rules with a precise reference to the time interval required to avoid a negative event (e.g. the congestion in traffic management). For this reason, the antecedent of their rule is minimal, i.e. it does not contain sequences with a repeated sub-sequence. In our opinion, some real world scenario can leverage the knowledge of repeated sequences over time (e.g. repetitions in buying habits).

Motivating Example. Let us consider the sequence of items that three users purchased in a supermarket in different grocery runs which are graphically depicted in Fig. 1. In order to simplify the example we suppose each user buys a single item during each grocery run.

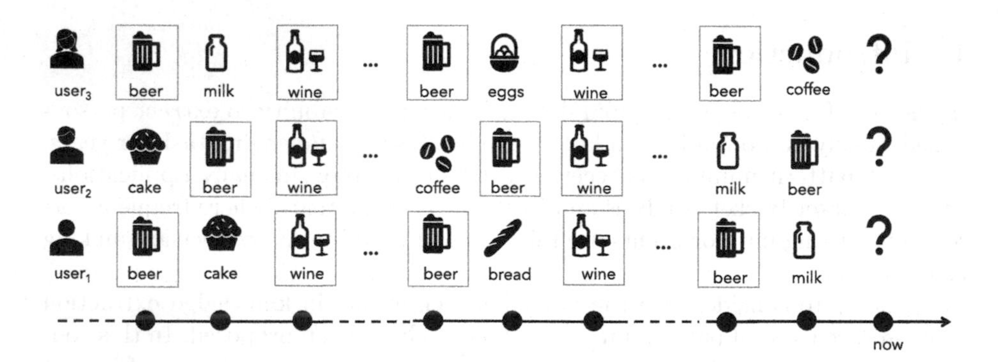

Fig. 1. Motivating example

Given these sequences, sequence mining algorithms would infer that the sequential pattern `(beer, wine)` is frequent and the sequential rule `beer` → `wine` could be mined too, stating that beer is often bought before wine. However, if we would recommend the best item to buy during the next purchase event (e.g. now) to each user, based on their history, the rule `beer` → `wine` would represent a correct timely recommendation only for $\texttt{user}_2$, because the others usually buy wine, but not as the purchase right after the beer. Indeed, the correct rules that we want to mine with our approach are the ones reported in Fig. 2.

In this case differently from $\texttt{user}_2$, $\texttt{user}_1$ and $\texttt{user}_3$ share a common rule pattern in which the antecedent is composed by the item `beer` followed by a not

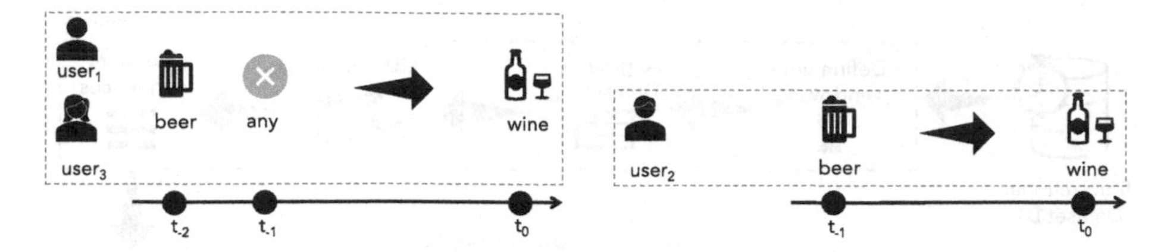

Fig. 2. Motivating example rules

specified item during the next purchase (the rule presents a hole, thus it is not complete), while the consequent is the item `wine`. Differently from [3] we do not need to know the time interval before the next purchase activity to recommend (it is not a time sensitive activity) and in the antecedent we could have a repeated sub-sequence. In addition, mined rules can contain frequent itemsets, instead of a single item, for each purchasing activity t_i.

The aim of this paper is to propose a framework to discover totally ordered sequential rules from historical logs that allow us to provide tailored suggestions to users. These rules are mined using an extension of the Apriori algorithm [2], called ALBA (Aged LookBack Apriori) which is able to deal with ordered sequential association rules that can also contain holes. The algorithm is further enriched with an aging mechanism to deal with the gradual obsolescence of past user behaviours.

The paper is organized as follows. Section 2 describes the proposed approach. Section 3 presents our Recommendation System for providing personalized recommendations to users. Section 4 describes our proposal in practice by using some examples, and Sect. 5 concludes the paper.

2 Aged LookBackApriori

This section describes our approach together with the ALBA (Aged LookBack-Apriori) algorithm, an extension of the classical Apriori algorithm [2] that we develop to infer totally ordered sequential rules where the items in the antecedent are enriched with an explicit relative order. These rules are then fed into a user-tailored recommendation system able to provide personalized suggestions to users. As shown in Fig. 3 the input is a transactional dataset D_0 (in our scenario, the log of transactions related to different purchases). The dataset D_0, where transactions are temporally ordered through a temporal marker i, is fed into LBA, proposed in [9], which considers a temporal window τ_w to construct an augmented data set D_{τ_w}. Given D_0, D_1 extends D_0 by looking back 1 temporal unit, i.e., each transaction $t'_i \in D_1$ is obtained as the concatenation of t_{i-1}, if any, and t_i in D_0, and so on (as shown in Fig. 4). This way, we iteratively build a dataset D_j, where each transaction t is related to a time window of length $j+1$.

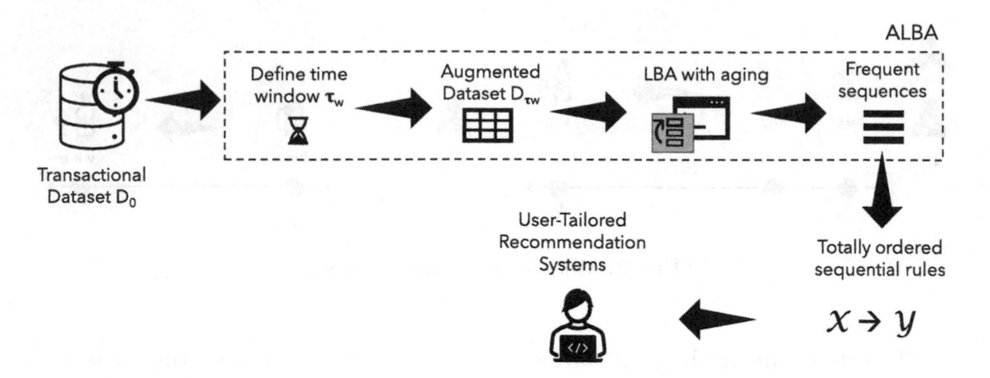

Fig. 3. Architecture of the proposed approach based on ALBA algorithm.

This augmented dataset is then fed into a novel algorithm, called ALBA, that extends LBA with an aging mechanism, in order to extract all the sequences whose support is equal or greater than a set threshold for the support (`minsupp`).

This process outputs a set of frequent sequences S, used to generate a set of totally ordered sequential rules R, formalized as implications $X \rightarrow Y$, where X and Y are two sets of ordered items, such that $X \cap Y = \emptyset$, according to specific thresholds for confidence and support [12].

D_0		D_1	
		t_{-1}	t_0
t0	A		A
t1	B	A	B
t2	A	B	A
t3	B	A	B
t4	C	B	C
t5	B	C	B
t6	A	B	A
t7	B	A	B
t8	C	B	C
t9	B	C	B
t10	C	B	C

W	A_0	A_{-1}	B_0	B_{-1}	C_0	C_{-1}
t0	1	0	0	0	0	0
t1	0	1	1	0	0	0
t2	1	0	0	1	0	0
t3	0	1	1	0	0	0
t4	0	0	0	1	1	0
t5	0	0	1	0	0	1
t6	1	0	0	1	0	0
t7	0	1	1	0	0	0
t8	0	0	0	1	1	0
t9	0	0	1	0	0	1
t10	0	0	0	1	1	0
confidence:	0.3	0.3	0.5	0.5	0.3	0.2

W_A	A_0	A_{-1}	B_0	B_{-1}	C_0	C_{-1}
t0	0.1	0	0	0	0	0
t1	0	0.2	0.2	0	0	0
t2	0.3	0	0	0.3	0	0
t3	0	0.4	0.4	0	0	0
t4	0	0	0	0.5	0.5	0
t5	0	0	0.6	0	0	0.6
t6	0.7	0	0	0.7	0	0
t7	0	0.8	0.8	0	0	0
t8	0	0	0	0.9	0.9	0
t9	0	0	1	0	0	1
t10	0	0	0	1	1	0
confidence:	0.1	0.1	0.3	0.3	0.2	0.1

W_{AC}	A_0	A_{-1}	B_0	B_{-1}	C_0	C_{-1}
t0	0.5	0	0	0	0	0
t1	0	0.6	0.6	0	0	0
t2	0.7	0	0	0.7	0	0
t3	0	0.8	0.8	0	0	0
t4	0	0	0	0.9	0.9	0
t5	0	0	1	0	0	1
t6	1.1	0	0	1.1	0	0
t7	0	1.2	1.2	0	0	0
t8	0	0	0	1.3	1.3	0
t9	0	0	1.4	0	0	1.4
t10	0	0	0	1.4	1.4	0
confidence:	0.2	0.3	0.4	0.5	0.3	0.2

Fig. 4. Example of augmented dataset construction and aging mechanism.

2.1 Aging Mechanism

With the aim to enhance the importance of recent frequent patterns, the LookBackApriori algorithm has been extended with an aging mechanism.

Normally, the support of an itemset is calculated at each iteration of the algorithm, to decide if a particular combination of items is frequent. This is done by counting the occurrences of these combinations at each temporal marker and then dividing it by the number of timestamps. Without an aging mechanism, the support of a sequence appearing frequently in recent times and another sequence that appeared the same amount of times a while ago would be the same. For this reason, we modify the computation of the support of each sequence to give more regard to the recent sequences. To this purpose, we build a matrix W of dimensions $n \times m$, where n is the number of transactions (i.e. temporal units) in the temporal dataset D and m is the number of different items in D.

We indicate with W_i the ith row and with W^l the lth column. The element at row i and column l is W_i^l and its value will be 1 if item l appears at time unit i, 0 otherwise. To compute the support of item l we can simply sum all elements of column W^l and divide the result by n. In order to penalize older sequences, we introduce the following variation: we multiply each row W_i, with $i < n - \tau_w$, by an aging factor $\alpha_i = \frac{i+1}{n-\tau_w+1}$, where τ_w is the temporal window. This guarantees that the items in the temporal window will still be represented with 1, while older items slowly decay but never quite reach 0. The 1 added in the denominator is simply to start the aging process at the first element outside of the temporal window.

In Fig. 4 we show an example of a dataset D_0 and the corresponding dataset D_1 built by considering a temporal window of length 2. In dataset D_1, the frequent sequences "AB" and "BC" appear 3 times each. However, the sequence "BC" appears more often in recent times, compared to "AB", which mainly appears at the beginning of the dataset. Using the normal matrix W to calculate the support of each item, the LookBackApriori algorithm will generate both "AB" and "BC" as frequent sequences without distinction. Conversely, by using the aged W_A matrix, the support for the item A_{-1} (i.e., A is present at time t_{-1}) is lower than the one of C_0 (i.e., C is present at time t_0), so sequences starting with "A" (such as "AB") will also have lower support, while sequences that end with "C" (i.e., "BC") are less penalized.

The biggest downside of this aged matrix is that the calculated support ceases to be a measure between 0 and 1, since any item that is present in every transaction will never be 1.0, due to the aging factor (unless the temporal window is as big as the whole dataset). This causes obvious problems when trying to compare the efficacy of our approach to the non-aged version of the algorithm. Thus, we build a *make-up* matrix to adjust the value of support to lie between 0 and 1 by first calculating how much compensation is needed in each column. In order to obtain 1 as support for an item l that appears n times in a dataset of length n, we need the sum of each element W_i^l to be n. The compensation needed can thus be calculated as: $c = n - \frac{n+\tau_w}{2}$, so each non-zero element W_i^l

of the aged matrix can be augmented by c/n. The resulting aged matrix with compensation is shown as W_{AC} in Fig. 4.

Looking at the resulting support values, it is easy to compare this final matrix with the original (un-aged) one and see that only two elements have been penalized, A_0 and B_0, as the sequences that end in A or B tend to be older than their originally equal counterparts (sequences that start with A or B).

In the literature, exponential decay has been used for many applications since it can be easily computed [4], but in our scenario it led to worse results.

3 Recommender System

Starting from the frequent sequences extracted by ALBA, we set a confidence threshold `minconf` and discover the totally ordered sequential rules r_i with confidence greater than or equal to `minconf`, and whose consequent contains items at temporal unit 0. Such rules will be of the form:

$$r_i : I_{-(\tau_w-1)} \wedge \cdots \wedge I_{-2} \wedge I_{-1} \rightarrow I_0 \; [s_i, c_i]$$

where s_i and c_i are the support and confidence of r_i, respectively. This shows, with support s_i and confidence c_i, the correlation between the itemset for the current temporal unit 0 (see the itemset I_0) and the past itemsets, considering at most τ_w temporal units, where τ_w is the chosen temporal window.

We remind that the sequence of itemsets in the antecedent do not need to be complete. For example, a mined rule stating that *"after buying hamburger and beer twice in consecutive grocery runs, the user buys salad during the successive one"* has the form:

$$\{hamburger, beer\}_{-2} \wedge \{hamburger, beer\}_{-1} \rightarrow \{salad\}_0 \; [s_r, c_r]$$

Conversely, another rule with an incomplete antecedent stating that *"after buying pizza, the user buys hamburger and beer but only after two grocery runs"* becomes:

$$\{pizza\}_{-2} \rightarrow \{hamburger, beer\}_0 \; [s_r, c_r]$$

Given the set of mined totally ordered sequential rules r_i, in order to discover the best rule $\overline{r}$ for predicting the answer to the query *"What do I need to buy today?"*, we need to find a match between these rules and a portion of the user log L containing past purchase activities limited to the considered temporal window:

$$L = \langle I'_{-(\tau_w-1)}, \ldots, I'_{-2}, I'_{-1} \rangle$$

This partial log will be hereafter called query for the purchase activity at temporal unit 0. For example, if during the previous 3 activities the user log/query is $L = \langle \{pizza, cola\}_{-3}, \{salad, wine\}_{-2}, \{hamburger, beer\}_{-1} \rangle$, then the answer to the query is the consequent of the best matching rule, which represents the most appropriate purchase for the user at the current moment, knowing what they have purchased the previous three times. Notice that if the user constantly declares their activities (e.g. by connecting with an app for online shopping), the query will contain information for each temporal unit; otherwise, some temporal units may be missing.

3.1 Rule Ordering

The set of mined rules is ordered using the following criteria: 1) confidence, 2) completeness, 3) support, and 4) size. Ordering by confidence and support is straightforward, as they are float values. On the other hand, ordering by completeness means that the rules with at least one item per time unit in the considered temporal window will be prioritized over those that lack information in other units. This choice reflects the fact that the itemsets appearing in a rule, covering a certain temporal window, represent frequent habits for the covered period that are relevant to determine the next suggestion. Conversely, a hole in the antecedent represents an itemset that is not relevant for the identification of the next suggestion. Let us define the subset of empty itemsets in a rule r as $I^r_\varnothing : \{I_i \in r \mid I_i = \varnothing\}$. We can define the completeness order between two rules r_1, r_2 as

$$r_1 >_c r_2 \rightarrow |I^{r_1}_\varnothing| < |I^{r_2}_\varnothing| \text{ and } r_1 =_c r_2 \rightarrow |I^{r_1}_\varnothing| = |I^{r_2}_\varnothing|$$

If two rules r_1, r_2 have the same support, confidence and $r_1 =_c r_2$, then the rule containing more items will be prioritized: $r_1 >_s r_2 : |r_1| > |r_2|$.

Given an ordered set of rules w.r.t. the above criteria, the algorithm needs to find the best rule that matches the query by means of the following criteria:

1. *EXACT_MATCH* $\rightarrow$ the query is exactly the antecedent of the rule:
   Query: $\langle\{pizza, cola\}_{-3}, \{salad, wine\}_{-2}, \{hamburger, beer\}_{-1}\rangle$
   Match: $\{pizza, cola\}_{-3} \wedge \{salad, wine\}_{-2} \wedge \{hamburger, beer\}_{-1}$
2. *MATCH* $\rightarrow$ all of the items in the matching rule appear in the right time slot in the query and the rule is complete:
   Query: $\langle\{pizza, cola\}_{-3}, \{salad, wine\}_{-2}, \{hamburger, beer\}_{-1}\rangle$
   Match: $\{pizza\}_{-3} \wedge \{salad, wine\}_{-2} \wedge \{hamburger\}_{-1}$
3. *PARTIAL_MATCH* $\rightarrow$ some time slots can be empty in the rule and/or some of the items in the query appear in the right time slot in the matching rule, while others have similar counterparts in the right time slot. For managing similarity, we have defined a simple list of weighted correspondences between similar items.
   Query: $\langle\{pizza, cola\}_{-3}, \{salad, wine\}_{-2}, \{hamburger, beer\}_{-1}\rangle$
   Match: $\{pizza, water\}_{-3} \wedge \{hamburger, wine\}_{-1}$

The search algorithm then iterates over the ordered rules and returns the best match (or no match at all) according to the criteria above. If an exact match is encountered, the algorithm immediately returns the corresponding rule $\overline{r}$. The iteration proceeds until the end and collects the other types of matching rules in their respective lists. At the end, the best rule $\overline{r}$ is the first rule in the first non-empty list in the order established above and is of the form:

$$\overline{r} : I_{-(\tau_w-1)} \wedge \cdots \wedge I_{-2} \wedge I_{-1} \rightarrow I_0$$

If all lists are empty, the algorithm returns NULL.

Note that the antecedent of a sequential rule is the explanation of the current suggestion, i.e., it is the recent behaviour of the user that leads to the suggestion. It is important to highlight that, when using sequential rules to provide recommendations, the sequential relationship between each itemset is important. Indeed, the user performs activities in particular temporal units and under precise contextual conditions, thus the order is important.

4 Running Example

We now present some examples to showcase the differences between our approach and others presented in the literature. Since the introduction of the aging mechanism in ALBA is innovative and not implemented in other proposals, we mine sequential rules with LBA (i.e., without applying the data aging) to compare running time and memory consumption of the two approaches. The aging mechanism does not affect the run time or the memory consumption as it only adds one simple linear transformation to the matrix at each iteration.

Although our approach has also been tested to suggest the fitness activities a user should do to sleep well [9], in this work, we use a toy dataset representing a market basket. We analyze the log of three users presented in Table 1, containing all the foodstuffs bought in 16 consecutive grocery shopping trips.

Table 1. Toy dataset used for the comparison.

T	User1	User2	User3
1	frozen food, eggs	frozen food, eggs	frozen food, eggs
2	beer, yogurt, milk	coca cola, yogurt, juice	wine, yogurt, soya milk
3	eggs	eggs	eggs
4	beer, yogurt, milk, pasta, flour	coca cola, yogurt, juice, rice, sugar	wine, yogurt, soya milk, rice, sugar
5	frozen food, eggs	frozen food, eggs	frozen food, eggs
6	beer, yogurt, milk	coca cola, yogurt, juice	wine, yogurt, soya milk
7	eggs	eggs	eggs
8	beer, yogurt, milk, pasta	coca cola, yogurt, juice, rice	wine, yogurt, soya milk, rice
9	frozen food, eggs, salt	frozen food, eggs, salt	frozen food, eggs, salt
10	beer, yogurt, milk	coca cola, yogurt, juice	wine, yogurt, soya milk
11	eggs	eggs	eggs
12	beer, yogurt, milk, pasta, flour	coca cola, yogurt, juice, rice, sugar	wine, yogurt, soya milk, rice, sugar
13	frozen food, eggs	frozen food, eggs	frozen food, eggs
14	beer, yogurt, milk	coca cola, yogurt, juice	wine, yogurt, soya milk
15	eggs	eggs	eggs
16	beer, yogurt, milk, pasta	coca cola, yogurt, juice, rice	wine, yogurt, soya milk, rice

A sequential pattern mining algorithm identifies frequent subsequences of items in the data, by applying two main steps: candidate generation and pattern pruning. In the first step the algorithm generates a set of candidate patterns by exploring the search space of possible sequences, while in the second one it removes any pattern that does not reach the minimum support threshold. We also follow this approach, but with a slight difference in the candidate generation step: after the augmented dataset step, the order of the itemsets is encoded in the itemsets themselves, which allows us to generate fewer candidates than usual sequence mining algorithms, resulting in less memory usage and computation time. This can be seen in Fig. 5, where we compare the benchmarks of LBA with those of GSP [11]. During the experiments the augmented dataset was provided as input to GSP in order to obtain the same results as LBA.

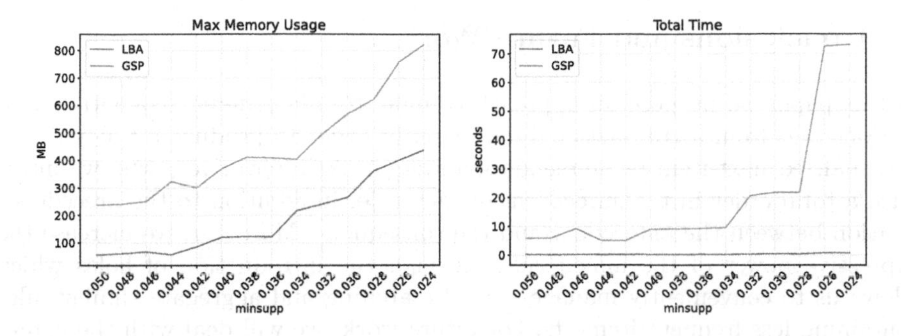

Fig. 5. Memory usage and total time in LBA and in GSP

Thanks to the relative order of mined information, we can extract more interesting rules. We generated a dataset of sequences by applying our augmented dataset algorithm to *User1* of Table 1 with $t_w = 5$ and then fed it to a generic sequential rule mining algorithm (ERMiner [5]) after removing the time units. ERMiner extracts the rule $eggs \rightarrow pasta$ with confidence 1, which intuitively works since eggs are always eventually followed by pasta in the set temporal window. At the same time, it is easy to observe that the user seldomly buys pasta on the subsequent grocery run after buying pasta. Conversely, LBA extracts the rule $\{frozen\ food\}_{-3} \rightarrow \{pasta\}_0$ with confidence 1. Thanks to this rule, we are able to say that *User1* will often buy *pasta* three time units after buying *frozen food*, thus providing a timely recommendation. This level of detail is impossible to achieve with conventional sequential rule mining algorithms.

Table 2. Rules for single users produced by our approach and rules obtained with collaborative filtering (CF).

	Rules
$user_1$	$\{EGG, FROZEN\}_{-3} \rightarrow \{BEER, MILK, PASTA, YOGURT\}_0$
$user_2$	$\{EGG, FROZEN\}_{-3} \rightarrow \{COLA, JUICE, RICE, YOGURT\}_0$
$user_3$	$\{EGG, FROZEN\}_{-3} \rightarrow \{SMILK, WINE, RICE, YOGURT\}_0$
CF	$\{EGG, FROZEN\}_{-3} \rightarrow \{YOGURT\}_0$

Collaborative Filtering (CF) is another technique often used to make recommendations when multiple users are involved. However, looking at Table 2, we can observe that the rule extracted with this technique is more general, as it takes into consideration the combined dataset of all the users. Notice that the three users have the same grocery shopping periodicity and a similar shopping basket each time. Despite that, the collaborative filtering recommendations cannot be as specific as ours; indeed, only the *yogurt* item is recommended, since it is the only common item in the consequent of the rules of all users.

5 Conclusions and Future Work

In this paper we propose an approach to mine totally ordered sequential association rules from a transactional database, in order to produce a tailored and personalized next-item recommendation system. As innovative parts, we introduce a total order in the antecedent of each rule, in addition to the precedence relation between the antecedent and the consequent. Moreover, we increase the expressive power of the mined rules through the introduction of holes which allows us to conveniently manage partial matching and aggregate similar rules containing less frequent itemsets. For future work, we will deal with the extension of the matching algorithm in presence of missing information in the input transactional dataset.

References

1. Adomavicius, G., Tuzhilin, A.: Toward the next generation of recommender systems: a survey of the state-of-the-art and possible extensions. IEEE Trans. Knowl. Data Eng. **17**(6), 734–749 (2005)
2. Agrawal, R., Srikant, R.: Fast algorithms for mining association rules in large databases. In: Bocca, J.B., Jarke, M., Zaniolo, C. (eds.) Proceedings of VLDB'94, pp. 487–499. Morgan Kaufmann (1994)
3. Ao, X., Luo, P., Wang, J., Zhuang, F., He, Q.: Mining precise-positioning episode rules from event sequences. IEEE Trans. Knowl. Data Eng. **30**(3), 530–543 (2018). https://doi.org/10.1109/TKDE.2017.2773493
4. Cormode, G., Shkapenyuk, V., Srivastava, D., Xu, B.: Forward decay: a practical time decay model for streaming systems. In: 2009 IEEE 25th International Conference on Data Engineering, pp. 138–149 (2009)
5. Fournier-Viger, P., Gueniche, T., Zida, S., Tseng, V.S.: ERMiner: sequential rule mining using equivalence classes. In: Blockeel, H., van Leeuwen, M., Vinciotti, V. (eds.) IDA 2014. LNCS, vol. 8819, pp. 108–119. Springer, Cham (2014). https://doi.org/10.1007/978-3-319-12571-8_10
6. Fournier-Viger, P., Lin, J.C.W., Kiran, R.U., Koh, Y.S.: A survey of sequential pattern mining. Data Sci. Pattern Recogn. **1**(1), 54–77 (2017)
7. Fournier-Viger, P., Wu, C., Tseng, V.S., Cao, L., Nkambou, R.: Mining partially-ordered sequential rules common to multiple sequences. IEEE Trans. Knowl. Data Eng. **27**(8), 2203–2216 (2015)
8. Linden, G., Smith, B., York, J.: Amazon.com recommendations: item-to-item collaborative filtering. IEEE Internet Comput. **7**(1), 76–80 (2003)

9. Marastoni, N., Oliboni, B., Quintarelli, E.: Explainable recommendations for wearable sensor data. In: Wrembel, R., Gamper, J., Kotsis, G., Tjoa, A.M., Khalil, I. (eds.) DaWaK 2022. LNCS, vol. 13428, pp. 241–246. Springer, Cham (2022). https://doi.org/10.1007/978-3-031-12670-3_21
10. Pei, J., et al.: Mining sequential patterns by pattern-growth: the prefixspan approach. IEEE Trans. Knowl. Data Eng. **16**(11), 1424–1440 (2004)
11. Srikant, R., Agrawal, R.: Mining sequential patterns: generalizations and performance improvements. In: Apers, P., Bouzeghoub, M., Gardarin, G. (eds.) EDBT 1996. LNCS, vol. 1057, pp. 1–17. Springer, Heidelberg (1996). https://doi.org/10.1007/BFb0014140
12. Zhang, C., Lyu, M., Gan, W., Yu, P.S.: Totally-ordered sequential rules for utility maximization. CoRR abs/2209.13501 (2022). https://doi.org/10.48550/arXiv.2209.13501

Data Integration

Streaming Approach to Schema Profiling

Chiara Forresi, Matteo Francia, Enrico Gallinucci(✉), and Matteo Golfarelli

University of Bologna, Cesena, Italy
{chiara.forresi,m.francia,enrico.gallinucci,matteo.golfarelli}@unibo.it

Abstract. Schema profiling consists in producing key insights about the schema of data in a high-variety context. In this paper, we present a streaming approach to schema profiling, where heterogeneous data is continuously ingested from multiple sources, as is typical in many IoT applications (e.g., with multiple devices or applications dynamically logging messages). The produced profile is a clustering of the schemas extracted from the data and it is computed and evolved in real-time under the overlapping sliding window paradigm. The approach is based on two-phase k-means clustering, which entails pre-aggregating the data into a coreset and incrementally updating the previous clustering results without recomputing it in every iteration. Differently from previous proposals, the approach works in a domain where dimensionality is variable and unknown apriori, it automatically selects the optimal number of clusters, and detects cluster evolution by minimizing the need to recompute the profile. The experimental evaluation demonstrated the effectiveness and efficiency of the approach against the naïve baseline and the state-of-the-art algorithms on stream clustering.

Keywords: Schema profiling · Stream clustering · K-means · Approximation algorithm

1 Introduction

Recent years have shown an increasing interest in the streaming paradigm, which gives users the possibility to monitor and analyze in near real-time a never-ending flow of data. This interest has been pushed by Internet-of-Things and Big Data applications, where a multitude of sources can generate huge quantities of data in a continuous manner. Depending on the application, the variety of data flowing in a stream can be significant. Real-world examples include applications collecting logs without fixed structures from multiple sources [7] or IoT systems collecting messages from a variety of devices, each with its own features and measurements [6]. In such scenarios, technicians are interested in characterizing the incoming traffic and knowing which kinds of error messages or measurements are currently flowing in the stream; this can be done by analyzing the high level of heterogeneity in the schemas of the messages, which conveys information about the nature of the data. The family of techniques that produce key insights about

A. Abelló et al. (Eds.): ADBIS 2023, CCIS 1850, pp. 211–220, 2023.
https://doi.org/10.1007/978-3-031-42941-5_19

the intensional aspect of data (i.e., the schema) is known as *schema profiling* and has been applied in the past to discern the variety within schemaless collections in NoSQL databases [7], but it has never been applied in streaming applications.

In this paper, we propose a streaming approach to schema profiling, with the goal to produce real-time insights about the schemas of data elements in a streaming application. The approach is based on the overlapping sliding window paradigm and the produced *profile* consists in clustering the schemas (extracted from the elements) in the window, which will help the user paint a high-level picture of the schema variety in the stream. The reference clustering algorithm is k-means, which must be adapted to the streaming application. Our proposal is a two-phase algorithm [16] to be run on every window. The first phase consists in building a *coreset* of pre-aggregated schemas by relying on a compact data structure (called *summaries*). In the second phase, the profile is built by clustering the summaries in the coreset or by updating the profile obtained in the previous iteration; here, we introduce rules to incrementally detect changes in the (number of) clusters and consider a reclustering policy to recompute clustering if the rules do not improve the result.

Several approaches have been proposed so far to carry out k-means clustering on a stream of data [16], which essentially introduce some degrees of approximation to compensate for the need of computing clustering in real-time; most importantly, approximation is introduced by pre-aggregating the original data elements and/or by incrementally updating the clustering result, without recomputing every time that the window slides. However, existing works are not applicable to our context, mostly because they either have limited capabilities to capture the evolution of clusters in time or they are designed for a Euclidean space with fixed dimensionality (as Euclidean distance loses significance in a variable high-dimensional space). Most algorithms work under the same two-phase method used in this paper. The most influential work about pre-aggregating data into a compact data structure is BIRCH [15], whose clustering features have been widely adopted and extended by related works. CSCS [14] is the one most similar to our. Differently from our approach, CSCS (i) requires a fixed predefined number of clusters k, (ii) uses a compaction technique that works only with data in a fixed domain (whereas the domain of attributes is not known apriori), and (iii) does not capture all kinds of cluster evolution. The most recent work on streaming k-means clustering supporting a variable number of clusters is FEAC-S [2]; however, FEAC-S does not involve a pre-aggregation of the data (it directly computes and updates clusters, which is more time-consuming) and, even though it formally supports all kinds of cluster evolution, it relies on pseudo-random criteria to decide when the evolution operations should be applied. Overall, no streaming k-means approaches support a variable and unknown dimensionality. A related field is the one of continuous profiling which generically consists in continuously monitoring the evolution of properties of some data properties [12]; however, existing approaches in this area focus on other aspects like functional dependencies [4] and data stream classification [13]. Real-time profiling has been proposed to dynamically extract patterns from unstructured logs [5], which have no notion of schemas. To the best of our knowledge, this is the first work to address the problem of profiling schemas in a streaming context.

The outline of the paper is as follows. Section 2 introduces the basic concepts used in the paper. Section 3 presents our approach, which is experimentally evaluated in Sect. 4. Conclusions are drawn in Sect. 5.

2 Basic Concepts

In this work, the first citizen is the schema of a message in the stream. Streaming applications typically exchange messages in the form of JSON documents, i.e., collections of key-value pairs, where keys are the schema elements (i.e., *attributes*). It is common to refer to nested attributes with *paths*, i.e., a dot-concatenation of the nested key with all the parent keys. The schema information that we are interested in is the union of all paths appearing in a document.

Definition 1 (Schema). *A schema is a set of attributes* $A = \{a_1, a_2, \ldots, a_n\}$*, where* a_i *is a string representing the attribute's path.*

The real-time analysis of the documents' schemas being ingested in a stream is carried out under the sliding window paradigm: the analytical algorithm is designed to run on a subset of stream elements (i.e., the window), which progressively moves in time to discard older elements and include newer ones. The *window* is typically defined by two parameters [1]: the *window length* (i.e., the amount of data to include in the window) and the *window period* (i.e., the frequency of the algorithm's execution). Both of these parameters can be expressed in terms of either time or the number of elements. For the sake of simplicity, we will use the number of elements. In our setting, we consider overlapping windowing with the window length (i.e., w_len) being set as a multiple of the window period (i.e., w_per), so that both the stream and the window can be defined in terms of *panes*, i.e., smaller subsets of stream elements of length w_per.

The clustering algorithm we take into account is k-means++ [3] (i.e., a more efficient version than traditional k-means with a non-random initial setting, which leads to a faster convergence). It is a type of unsupervised learning providing a straightforward method to categorize a given dataset into k clusters, where k is a user-defined parameter and each cluster is represented by a *centroid* (usually computed as the average of all the elements in a cluster).

Definition 2 (Clustering result). *Given a dataset of elements* U*, a clustering result is a set of clusters* $C = (c_1, c_2, \ldots, c_k)$ *where* $c_i = \langle E, \hat{e} \rangle$*;* $c_i.E \subset U$ *is the set of elements assigned to* c_i *and* $c_i.\hat{e}$ *is the centroid.*

The objective function of k-means is to find a solution C with minimum *cost*, meant as the sum of squared distances (according to a distance function d, usually Euclidean) between each element $e \in c_i.E$ and the centroid of its assigned cluster (i.e., $c_i.\hat{e}$). The algorithm starts by selecting k centroids from U

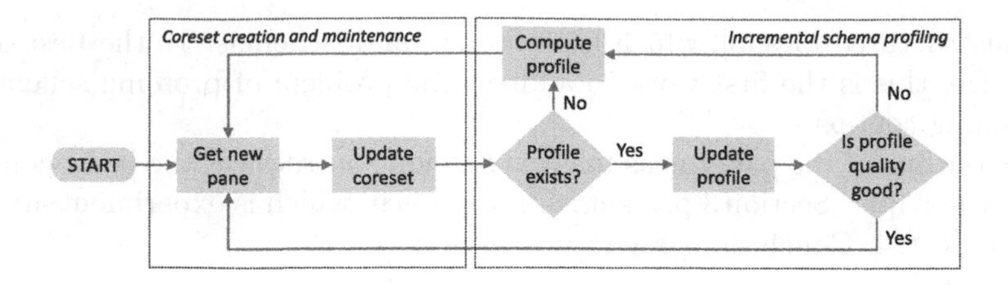

Fig. 1. Overview of the real-time schema profiling approach.

and assigns each element to the nearest centroid; then, the centroids are updated based on the assigned elements, and the procedure is iteratively repeated until convergence (which occurs when elements are not reassignments or when a maximum number of iterations is reached).

In this work, k is not known apriori; thus, we rely on the Ordered Multiple Runs of k-Means approach [11] (here called OMRk++), which consists in running k-means++ for all values of k between 2 and $\sqrt{|U|}$ and choosing the solution with the value of k that maximizes the *simplified silhouette* (SS). The SS of an element e assigned to cluster c_i (i.e., $e \in c_i.E$) is calculated as $SS(e) = \frac{y(e)-x(e)}{max(x(e),y(e))}$, where $x(e) = d(e, c_i)$ is the distance between the element and its centroid, and $y(e) = \arg\min_{c \in C \setminus c_i} d(e, c)$ is the minimum of the distances between the element and the other centroids. The SS of the clustering solution is the average of the values for every element, i.e., $SS(C) = avg_{e \in U} SS(e)$.

3 Real-Time Schema Profiling

Schema profiling [7] aims to provide insights into the structure of the schemas in a given dataset. In the context of this paper, the profile we aim to provide consists in grouping schemas into a non-predefined number of clusters with the OMRk++ technique (i.e., the clustering result in Definition 2), which allows to easily distinguish the main kinds of schema in the data. In real-time schema profiling, the challenge is to provide this result on a stream of continuously incoming data, shifting the focus on the latest data window and keeping the stream's pace. The naïve way to produce the intended result is to simply run OMRk++ on each data window. This is obviously inefficient and inapplicable in a streaming context, as (i) the clustering solution would be recomputed from scratch each time, without considering the previous result and analyzing each stream element multiple times (depending on the window settings), and (ii) the clustering algorithm would run on a (potentially very) large quantity of elements.

The approach proposed in this paper is called DSC (Dynamic Stream Clustering) and consists in returning an approximate solution to the problem, which uses a fixed amount of memory and runs with very high performances. The approximations enabling a faster execution are the following.

1. The cardinality of the data in input to the clustering algorithm is reduced to a *coreset*, i.e., a meaningful set of summaries (defined in Sect. 3.1). By

clustering fewer data, DSC is able to converge in shorter times. The coreset is maintained in time by removing expired summaries and adding new ones.
2. Instead of clustering the data in the coreset for every window, DSC incrementally updates the clustering result by reflecting the changes observed in the coreset. Unlike previous work, DSC captures multiple evolution phenomena of clusters including changing the number of clusters.

The approach is structured in two phases, sketched in Fig. 1. In the coreset creation and maintenance phase, a new pane of data is retrieved from the stream and the coreset is updated (i.e., summaries that are not anymore in the window are removed, and new summaries are added). In the incremental schema profiling phase, the profile is created by running OMRk++ on the coreset (if no profile existed yet) or updated. Remarkably, the update of the profile goes beyond the simple update of centroids after the removal of expired summaries and the assignment of new ones, but it includes the detection of changes in the number of clusters (i.e., if two clusters need to be merged, or if a cluster needs to be split). To prevent the updated profile from drifting too much from the optimal solution (i.e., the execution of OMRk++ on the coreset), we adopt heuristics to decide if the profile has to be recreated again from scratch.

3.1 Clustering Schemas with Summaries

Two-phase stream clustering algorithms rely on a compact data structure that summarizes a set of stream elements. The most notable is the *clustering feature*, introduced in BIRCH [15] and used/extended in several related works [16]. In this paper, we introduce the notion of *summary*.

Definition 3 (Summary). *A* summary *of a multiset of schemas $\mathcal{A}$ is a compact structure $s^{\mathcal{A}}$ giving a fuzzy summarization of the attributes found in $\mathcal{A}$. Given an attribute a, $s^{\mathcal{A}}.p(a) \in [0,1]$ is the percentage of schemas in $\mathcal{A}$ that contain a. The number of schemas summarized by $s^{\mathcal{A}}$ is $s^{\mathcal{A}}.card = |\mathcal{A}|$. The union of the attributes in $\mathcal{A}$ is a set indicated with $s^{\mathcal{A}}.A$.*

Like clustering features, summaries are additive, i.e., $s^{\mathcal{A}_i+\mathcal{A}_j} = s^{\mathcal{A}_i} + s^{\mathcal{A}_j}$. The proof is omitted for space reasons. To simplify the notation, the superscript referring to the multiset of schemas summarized by s is omitted when not needed.

The additivity of summaries allows defining the centroid of a cluster of summaries as the sum of all the summaries in the cluster.

Definition 4 (Centroid). *Let $c_i = \langle S, \hat{s} \rangle$ a cluster of a multiset of summaries $c_i.S$; then, the centroid is $c_i.\hat{s} = \sum_{s \in S} s$.*

In most k-means applications, the distance function typically used is the Euclidean one. In our context, clustering must be carried out on summaries of elements (i.e., schemas) whose dimensionality (i.e., the set of represented attributes) is not known apriori. As we recall, attributes are simply strings with no formal limitation and no order, i.e., it is not a Euclidean space. Thus, to measure the distance between two summaries we rely on the Jaccard distance [10], which enables the comparison of two elements independently of their dimensionality. Given two generic sets A and B, the Jaccard similarity is defined as $sim_J(A,B) = \frac{|A \cap B|}{|A \cup B|}$, and the distance is derived as $d_J(A,B) = 1 - sim_J(A,B)$. Since summaries are fuzzy, we adopt its weighted variant [8].

Definition 5 (Weighted Jaccard similarity and distance). *Given two summaries* s_i *and* s_j*, their weighted Jaccard similarity is:* $sim_{WJ}(s_i, s_j) = \frac{\sum_{a \in s_i.A \cup s_j.A} min(s_i.p(a), s_j.p(a))}{\sum_{a \in s_i.A \cup s_j.A} max(s_i.p(a), s_j.p(a))}$. *The weighted Jaccard distance is:* $d_{WJ}(s_i, s_j) = 1 - sim_{WJ}(s_i, s_j)$.

The purpose of using summaries is to group together similar schemas (based on sim_{WJ}) in order to build the coreset given in input to the clustering algorithm. Each summary of similar schemas is associated with an *id* and stored in a *bucket*. To support deletion/insertion of old/new summaries, buckets contain the list of summaries in different panes with the same *id*; then, the clustering algorithm run on the *representative* summary of each bucket.

Definition 6 (Coreset, bucket). *A* coreset B *is a set of at most* $2m$ *buckets. A* bucket $b \in B$ *is defined as* $b = \langle id, S, r \rangle$*, where* $b.id$ *is an identifier,* $b.S$ *is the list of summaries (one for each pane of the window) whose schemas are identified by* $b.id$*, and* $b.r$ *is the representative summary of* b*, corresponding to the sum of the summaries in* $b.S$ *(i.e.,* $b.r = \sum_{s \in b.S} s$*). The threshold* m *regulates the size of the coreset, reducing the number of buckets to* m *when it exceeds* $2m$*.*

3.2 Incremental Schema Profiling

In the incremental schema profiling phase of the algorithm, the coreset is used to produce the clustering profile of the current window of schemas. In the very first iteration, this is done by running OMRk++ on the summaries in the coreset, which initially consists of one pane alone. In the next iterations, the current profile is incrementally updated in accordance with the updates in the coreset. A reclustering policy is adopted to prevent the updated profile from drifting too much from the optimal solution (i.e., the execution OMRk++ on the coreset); when activated, the current profile is recreated from scratch by running again OMRk++ on the summaries in the coreset. When (re)creating the profile, the input data U to OMRk++ consists of the set of representatives of each bucket, i.e., $U = \{b.r : b \in B\}$; then, each bucket b is associated with a single cluster.

The challenge lies in updating a previous profile incrementally, aiming to approximate the profile that could be generated from scratch using the current window's data as closely as possible. This task is complex and subject to the concept-drift problem, which is usually dealt with by related works through a reclustering policy: it consists in evaluating the quality of the current solution and deciding whether it is still acceptable or must be recomputed. For instance, [14] measures the Kullback-Leibler divergence [9], which measures how the new solution represents the old data; when the loss in precision goes beyond a given threshold, the reclustering policy is activated. In the following, we enumerate the phenomena impacting a clustering solution in an evolutionary streaming context.

- *Sliding*: the elements of a cluster shift in the multidimensional space, causing the centroid to shift as well.
- *Fading out/in*: an existing cluster disappears or a new one appears.
- *Splitting*: the elements of a cluster grow in number/volume and become best captured by two or more clusters rather than one.
- *Merging*: the elements of two clusters shift closer to each other and become best captured by a single cluster.

Related works usually adopt basic update rules that can only partly capture the above-mentioned phenomenon. Applied to our context, they consist in i) associating newly created buckets with the cluster whose centroid is closest on d_{WJ} and recomputing the centroids; ii) deleting clusters that are left empty (because old buckets have expired and new buckets have been assigned with other clusters). These rules only capture the sliding and fading out of clusters.

In this work, these basic rules are extended with more advanced ones to capture the splitting, merging, and fading in phenomena. The challenge in defining these rules is to remain as faithful as possible to the behavior of the original k-means. It is known that k-means tends to create spherical and homogeneous clusters; in particular, the Simplified Silhouette (i.e., the objective function of OMRk++ in choosing the best k) aims at minimizing cluster scattering (i.e., having well-grouped elements) and maximizing cluster separation (i.e., clusters distant from each other). By controlling these properties, the following rules are implemented.

1. **Splitting rule.** The splitting of a cluster is determined by analyzing its *scattering*, i.e., the average distance between its summaries and its centroid. When a cluster exhibits unusually high scattering compared to others, it is split into two or more clusters by running OMRk++ on its elements.
2. **Merging rule.** The merging of two clusters is guided by the *separation* between them, i.e., the distance between their centroids. Two clusters are merged if (i) their separation is unusually high compared to other couples of clusters, and (ii) when they overlap, i.e., when the distance between the centroids is less than the sum of their radii (measured as the scattering of the clusters).

Notice that the splitting rule implicitly captures the fading in phenomenon as well, thus completing the covering of all mentioned phenomena. While these rules adhere to the principle of k-means (and iteratively applied, as more split/merge operations may be necessary), they do not guarantee an improvement of the profile. Thus, we validate the resulting profile by checking if the SS has improved compared to before applying the rule; otherwise, the reclustering policy is triggered and the profile recomputed from scratch.

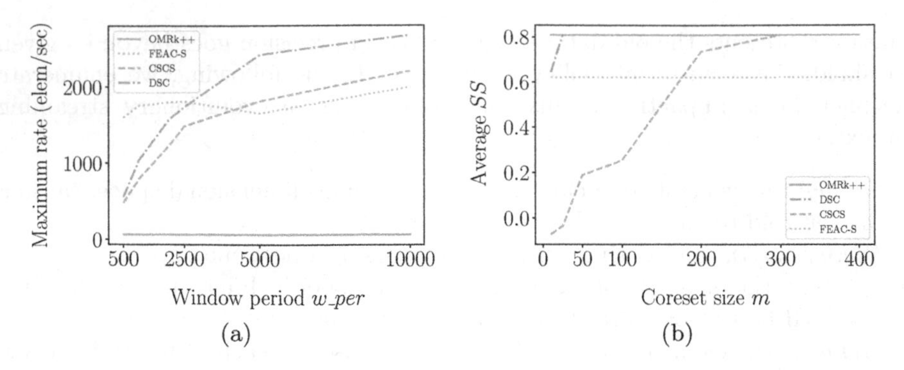

Fig. 2. (a) Average execution time by different window periods (i.e., number of elements in the pane) for $m = 100$. (b) Average SS by different values for m (coreset size).

4 Experiments

The experimental evaluation is done to assess the effectiveness and efficiency of DSC. Tests are executed on a single machine with a RAM of 16 GB and an i7-4790 CPU @3.60 GHz. The comparison is carried out against the reference baseline – which consists in running OMRk++ over the whole window of schemas (not summaries) at every iteration (i.e., for every new pane of data) – and the two closest state-of-the-art proposals for streaming k-means, i.e., CSCS [14] and FEAC-S [2]. To avoid excessive execution times, the maximum value of k for the baseline OMRk++ is set to 20. CSCS has been adapted to work in our context as follows: (i) to support variable k, the clustering of the coreset is done multiple times for different k as in OMRk++; (ii) since the variable domain of attributes breaks the compaction technique to define bucket identifiers, the schema elements in the window are retained and buckets redefined whenever the attribute domain changes. All algorithms have been implemented in Scala.

The evaluation is based on a generator simulating a stream of JSON schemas generated by multiple seeds (i.e., sources) with a given degree of variability: each seed generates schemas with around 10 attributes, chosen with probability 0.9 from a pool of exclusive or shared attributes, in order to create similar and slightly overlapped clusters. The generator works either in a static or evolutionary manner, where the latter implements mechanisms to simulate the phenomena discussed in Sect. 3.2; these mechanisms are gradually activated after a short period (approximately 5 to 10 panes), in order to test the incremental update procedure. The generated datasets are referred to as D_{static} (with 10 seeds), D_{fadein} (where the 10th seed is turned on only later), $D_{fadeout}$ (where a seed is turned off), D_{slide} (where a seed produces schemas with progressively different attributes), D_{split} (where a seed progressively generates two separate patterns of schemas), and D_{merge} (where two seeds progressively generate schemas converging towards the same attributes).

The default parameters for both the DSC and CSCS algorithms are: $m = 100$ (coreset size) and $l = 15$ (number of functions to compute identifiers).

Table 1. Algorithms' performances on the different datasets.

Dataset	Measure	OMRk++	DSC	CSCS1 [14]	CSCS3 [14]	FEAC-S [2]
D_{static}	Time (s)	35.90	0.42	0.48	0.36	**0.13**
	SS	**0.81**	**0.81**	0.24	**0.81**	0.62
D_{fadein}	Time (s)	33.76	0.41	0.40	0.36	**0.12**
	SS	**0.81**	0.80	0.39	0.75	0.57
$D_{fadeout}$	Time (s)	32.08	0.38	0.44	0.31	**0.12**
	SS	**0.81**	**0.81**	0.23	0.78	0.46
D_{slide}	Time (s)	35.08	0.42	0.58	0.47	**0.13**
	SS	**0.80**	**0.80**	0.42	**0.80**	0.56
D_{split}	Time (s)	28.22	0.43	0.52	0.49	**0.13**
	SS	**0.80**	0.79	0.32	0.78	0.46
D_{merge}	Time (s)	35.74	0.44	0.54	0.47	**0.13**
	SS	**0.78**	**0.78**	0.30	0.76	0.47

The efficiency of the algorithms is measured on D_{static} by varying the number of panes per window, whose size is fixed to $w\_len = 10000$. In Fig. 2a we show the maximum stream rate supported in terms of elements per second; it is calculated as the number of elements in the pane over the execution time in seconds, which intuitively indicates the maximum rate tolerated by the algorithm to keep up with the flow of data in the stream. The results in Fig. 2a show that (i) DSC runs constantly faster than CSCS, thus it is able to support a higher stream rate; (ii) since both DSC and CSCS pre-aggregate elements into a coreset, they are able to scale better and to increase the maximum rate as the pane size increases; differently, FEAC-S directly inserts every element into the closest cluster, thus its execution time is linear with respect to the number of elements in the pane and its maximum rate remains constant; (iii) OMRk++ lies at the bottom at only 65 elem/sec.

The effectiveness of the algorithms is first tested on D_{static}; here, we measure the simplified silhouette (SS) of the obtained solution by varying the size of the coreset m on D_{static} (which affects only DSC and CSCS, thus the SS for OMRk++ and FEAC-S is constant). The length and period of the window are $w\_len = 2000$ are $w\_per = 200$. Figure 2b shows the averages of the values obtained in every pane iteration. Interestingly, DSC is able to converge to the baseline reference values with a much smaller coreset size than CSCS, indicating that the way the coreset is created and maintained is more accurate – whereas the quality of the result by FEAC-S is quite lower than the baseline's and DSC's.

Table 1 compares the algorithms over all datasets. The length and period of the window are $w\_len = 2000$ and $w\_per = 200$. As CSCS struggles in providing good results with $m = 100$, it is tested also with $m = 100$ (CSCS1) and $m = 300$ (CSCS3). The results show that DSC is able to return results adhering to the OMRk++ baseline in all datasets (it is not shown for space constraints, but changes in the number of clusters are all captured) under much better performances. DSC also proves better than CSCS, with great improvements on the effectiveness side obtained with comparable (if not better) execution times – especially notable in the datasets containing a change in the number of clusters, which CSCS cannot capture. FEAC-S shows faster execution times, but the accuracy of its results is very distant from the baseline's and DSC's.

5 Conclusions

In this paper, a streaming approach to schema profiling has been presented. It proposes DSC; an approximate k-means algorithm with variable k to compute the profile, designed to pre-aggregate schemas and incrementally update the profile, thus minimizing the need to run the clustering algorithm. DSC extends state-of-the-art streaming algorithms by supporting the domain of schema attributes, which is variable and unknown, and fully capturing the evolution of clusters. Experiments demonstrate the effectiveness of DSC in providing an accurate profile and higher performances than the closest related works. Future work will be aimed to (i) further improve the efficiency by simplifying the coreset creation and maintenance phase, (ii) evaluate DSC on real-world datasets, to further validate its effectiveness and possibly extended the rules to capture cluster evolution, and (iii) adopt DSC in analytical applications to support users beyond the monitoring of the stream and actively aiding the formulation of queries.

References

1. Akidau, T., et al.: Streaming Systems: The What, Where, When, and How of Large-Scale Data Processing. O'Reilly Media, Inc., Sebastopol (2018)
2. de Andrade Silva, J., et al.: An evolutionary algorithm for clustering data streams with a variable number of clusters. Expert Syst. Appl. (2017)
3. Arthur, D., et al.: k-means++: the advantages of careful seeding. SIAM (2007)
4. Breve, B., et al.: Dependency visualization in data stream profiling. Big Data Res. (2021)
5. Du, M., et al.: Spell: streaming parsing of system event logs. IEEE Computer Society (2016)
6. Emmi, L.A., et al.: Digital representation of smart agricultural environments for robot navigation. In: CEUR Workshop Proceedings (2022)
7. Gallinucci, E., et al.: Schema profiling of document-oriented databases. Inf. Syst. (2018)
8. Grefenstette, G.: Explorations in automatic thesaurus discovery (1994)
9. Kullback, S., et al.: On information and sufficiency. Ann. Math. Stat. (1951)
10. Levandowsky, M., et al.: Distance between sets. Nature (1971)
11. Naldi, M.C., et al.: Comparison among methods for k estimation in k-means. IEEE Computer Society (2009)
12. Naumann, F.: Data profiling revisited. In: SIGMOD Rec. (2013)
13. Seyfi, M., et al.: H-DAC: discriminative associative classification in data streams. Soft. Comput. (2023)
14. Youn, J., et al.: Efficient data stream clustering with sliding windows based on locality-sensitive hashing. IEEE Access (2018)
15. Zhang, T., et al.: BIRCH: an efficient data clustering method for very large databases. ACM Press (1996)
16. Zubaroğlu, A., et al.: Data stream clustering: a review. Artif. Intell. Rev. (2021)

Using ChatGPT for Entity Matching

Ralph Peeters(✉) and Christian Bizer

Data and Web Science Group, University of Mannheim, Mannheim, Germany
{ralph.peeters,christian.bizer}@uni-mannheim.de

Abstract. Entity Matching is the task of deciding if two entity descriptions refer to the same real-world entity. State-of-the-art entity matching methods often rely on fine-tuning Transformer models such as BERT or RoBERTa. Two major drawbacks of using these models for entity matching are that (i) the models require significant amounts of fine-tuning data for reaching a good performance and (ii) the fine-tuned models are not robust concerning out-of-distribution entities. In this paper, we investigate using ChatGPT for entity matching as a more robust, training data-efficient alternative to traditional Transformer models. We perform experiments along three dimensions: (i) general prompt design, (ii) in-context learning, and (iii) provision of higher-level matching knowledge. We show that ChatGPT is competitive with a fine-tuned RoBERTa model, reaching a zero-shot performance of 82.35% F1 on a challenging matching task on which RoBERTa requires 2000 training examples for reaching a similar performance. Adding in-context demonstrations to the prompts further improves the F1 by up to 7.85% when using similarity-based example selection. Always using the same set of 10 handpicked demonstrations leads to an improvement of 4.92% over the zero-shot performance. Finally, we show that ChatGPT can also be guided by adding higher-level matching knowledge in the form of rules to the prompts. Providing matching rules leads to similar performance gains as providing in-context demonstrations.

Keywords: Entity Matching · Large Language Models · ChatGPT

1 Introduction

Entity matching is the task of discovering entity descriptions in different data sources that refer to the same real-world entity [4]. While early matching systems relied on manually defined matching rules, supervised machine learning methods have become the foundation of most entity matching systems [4] since the 2000s. This trend was reinforced by the success of neural networks [2] and today most state-of-the art matching systems rely on pre-trained language models (PLMs), such as BERT or RoBERTa [5,9,10].

The downsides of using PLMs for entity matching are that (i) PLMs need a lot of task-specific training examples for fine-tuning and (ii) they are not very robust concerning unseen entities that were not part of the training data [1,10].

Large autoregressive language models (LLMs) [13] such as GPT, ChatGPT, PaLM, or BLOOM have the potential to address both of these shortcomings.

A. Abelló et al. (Eds.): ADBIS 2023, CCIS 1850, pp. 221–230, 2023.
https://doi.org/10.1007/978-3-031-42941-5_20

Due to being pre-trained on huge amounts of text as well as due to emergent effects resulting from the model size [12], LLMs often have a better zero-shot performance compared to PLMs such as BERT and are also more robust concerning unseen examples [3]. Initial research on exploring the potential of LLMs for data wrangling tasks was conducted by Narayan et al. [8] using the GPT-3 LLM. This paper builds on the results of Narayan et al. and extends them with the following contributions:

1. We are the first to systematically evaluate the performance of ChatGPT (gpt3.5-turbo-0301) on the task of entity matching, while Narayan et al. applied the earlier GPT-3 model (text-davinci-002).
2. We systematically compare various prompt design options for entity matching while Narayan et al. tested only two designs.
3. We extend the results of Narayan et al. on in-context learning for entity matching by introducing a similarity-based method for selecting demonstrations from a pool of training examples. Furthermore, we analyze the impact of in-context learning on the costs (usage-fees) charged for running entity matching prompts against the OpenAI API.
4. We are the first to experiment with the addition of higher-level matching knowledge to prompts as an alternative to in-context demonstrations. We show that guiding the model by stating higher-level matching rules can lead to the same positive effect as providing in-context examples.

2 Experimental Setup

Narayan et al. [8] have measured the performance of GPT-3 using a range of well-known entity matching benchmark datasets [11]. All of these datasets are available on the Web for quite some time and are widely discussed in various papers and on various webpages. Thus, it is very likely that the training data of GPT-3 and ChatGPT contains information about these benchmarks which could give the language models an advantage. In order to eliminate this potential of leaking information about the test sets, we use the WDC Products [10] benchmark which has been published in December 2022 and is therefore newer than the training data of the tested models.

In order to understand the potential of LLMs for challenging entity matching use cases, we use a difficult variant of the WDC Products benchmark for the experiments which contains 80% corner-cases (hard positives and hard negatives). The types of products that are contained in our evaluation dataset range from computers and electronics over bike parts to general tools and thus examplify different product categories. The products are described by the attributes *brand, title, description* and *price*. In order to keep the costs of running benchmark experiments against the OpenAI API in an acceptable range, we down-sample the WDC Products benchmark to 50 products and retain the high ratio of corner-cases by using the original benchmark creation code. Table 1 shows statistics about the original and down-sampled versions of the benchmark.

Table 1. Statistics of the WDC Products benchmark datasets.

Dataset Type	Purpose	# Pairs	# Pos	# Neg
Original Validation	RoBERTa baseline	4,500	500	4,000
Original Training	RoBERTa baseline	19,835	8,741	11,364
Sampled Validation	Evaluation of prompts	433	50	383
Sampled Training	In-context sample selection	2025	898	1,127

API Calls and Costs: We use the down-sampled validation set to report the impact of the various prompt design decisions and the down-sampled training set as a source of in-context demonstrations for the corresponding experiments. Thus, one evaluation run results in 433 API calls to the OpenAI API. For all experiments we use the ChatGPT version *gpt3.5-turbo-0301* and set the temperature parameter to 0 to make experiments reproducible as stated in the OpenAI guidelines. We further track the cost associated with each pass of the validation set by using the Tiktokenizer[1] python package to calculate the cost associated with each prompt and corresponding ChatGPT answer.

Serialization: For the serialization of product offers into prompts, we follow related work [8] and serialize each offer as a string with pre-pended attribute names. Figure 1 shows examples of this serialization practice for a pair of product offers and the attribute *title*.

Evaluation: The responses gathered from the model are natural language text. In order to decide if a response refers to a positive matching decision regarding a pair of product offers, we apply simple pre-processing to the answer and subsequently parse for the word *yes*. In any other case we assume the model decides on not matching. This rather simple approach turns out to be surprisingly effective as the high recall values in Table 2 and manual inspection of the answers suggest. This approach has also been used by Narayan et al. [8].

Replicability: All data and code used in this paper are available at the project github[2] meaning that all experiments can be replicated. In addition, we contributed the down-sampled datasets and three selected prompts to the OpenAI evals[3] library.

3 General Prompt Design

Designing the prompt input to large language models to convey the task description, input, as well as additional information is one of the main challenges for achieving good results [7]. Careful prompt design is important as it can have a

[1] https://github.com/dqbd/tiktokenizer.
[2] https://github.com/wbsg-uni-mannheim/MatchGPT.
[3] https://github.com/openai/evals/blob/main/evals/registry/evals/product-matching.yaml.

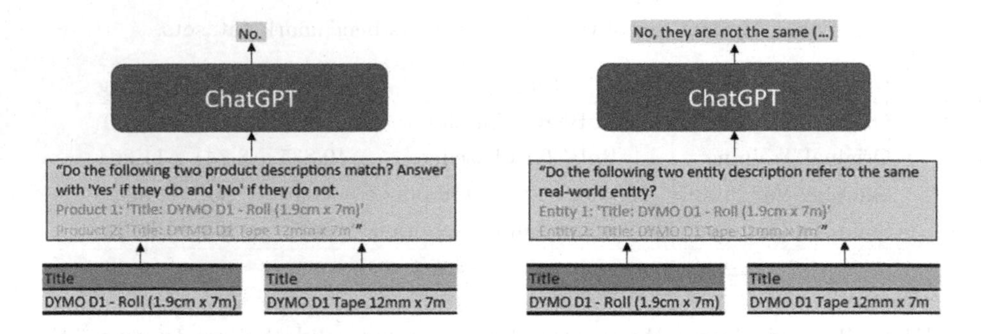

Fig. 1. Examples of prompt designs and product offer serializations.

large impact on the overall task performance [8,14]. In this Section, we experiment with various prompt designs for ChatGPT in a zero-shot setting: We describe the task to the model and ask for a matching decision for each of the examples in our validation set. The prompt designs that we use can be categorized as follows and are illustrated by the example prompts in Fig. 1:

- **General**: These prompts describe the task as the matching of entity descriptions to real-world entities. The product offers are referred to as *entities*. An example of a *general* prompt is the right prompt in Fig. 1.
- **Domain**: The domain-specific prompts describe the task as matching of product descriptions and refers to the examples as *product offers*. An example of this type of prompt is the left prompt in Fig. 1.
- **Complex**: Prompts in this category use more complex language, specifically they use the formulations "refer to the same real-world product" or "refer to the same real-world entity". An example is the right prompt in Fig. 1.
- **Simple**: This type of prompt uses less complex language and replaces the formulations from *Complex* with a simple "match". An example is the left prompt in Fig. 1.
- **Free**: This category reflects prompts that do not restrict the models answers in any way. An example is the right prompt in Fig. 1.
- **Forced**: In contrast to *Free*, these kinds of prompts explicitly tell the model to answer the stated question with "Yes" and "No". An example is the left prompt in Fig. 1
- **Attributes**: We vary using the three attributes *brand* (B), *title* (T) and *price* (P) in the combinations T, BT and BTP when serializing product offers into single strings.

Table 2 shows the results of the experiments with general prompt designs and associated average cost for querying a single pair with each design. The recall values for all prompts with ChatGPT are equal to or above 98% which suggests that, in combination with the lower precision values, the model is inclined to overestimating matching pairs in these cases. Interestingly, the prompt design of Narayan et al. [8] that we also evaluate using ChatGPT, conversely results in a

Table 2. Results of the general prompt design experiment with associated cost.

Prompt	P	R	F1	Δ F1	cost (¢) per pair
general-complex-free-T	49.50	**100.00**	66.23	–	0.11
general-simple-free-T	70.00	98.00	81.67	15.44	0.10
general-complex-forced-T	63.29	**100.00**	77.52	11.29	0.14
general-simple-forced-T	75.38	98.00	85.22	18.99	0.13
general-simple-forced-BT	79.66	94.00	**86.24**	20.01	0.13
general-simple-forced-BTP	71.43	70.00	70.70	4.47	0.13
domain-complex-free-T	71.01	98.00	82.35	16.12	0.11
domain-simple-free-T	61.25	98.00	75.38	9.15	0.10
domain-complex-forced-T	71.01	98.00	82.35	16.12	0.14
domain-simple-forced-T	74.24	98.00	84.48	18.25	0.13
domain-simple-forced-BT	76.19	96.00	84.96	18.73	0.13
domain-simple-forced-BTP	54.54	84.00	66.14	-0.09	0.13
Narayan-complex-T	85.42	82.00	83.67	17.44	0.10
Narayan-simple-T	**92.86**	78.00	84.78	18.55	0.10

more balanced precision and recall, the latter being significantly lower than the ones we observe for our prompts. The main difference between Narayan et al.'s and our prompts is that they provide the examples to be matched before the task description while we do it the other way around. Comparing the F1 values of our general and domain-specific prompts with ChatGPT, three patterns emerge: (i) Formulating the prompt with domain-specific wording leads to generally more stable results, (ii) Using simpler language works better than more complex wording in all but one case and (iii) forcing the model to answer with a short "Yes" or "No" leads to a significant increase in every scenario. While the addition of brand information increases F1 by up to 1% F1 percentage point, adding the price as well leads to a significant decrease in performance likely due to the format and currency of the prices not being normalized in the dataset.

Baselines: We compare the results of ChatGPT on our benchmark dataset to results of GPT-3 *gpt3.5-davinci-002* which has been used by Narayan et al. [8], as well as to results for RoBERTa-base fine-tuned with different amounts of training data. A fine-tuned RoBERTa-base corresponds to the state-of-the-art entity matching system Ditto [5] with all pre-processing and data augmentation options turned off. Results of the baseline methods are presented in Table 3. Using four of the prompt designs on the earlier gpt3.5-davinci-002 model shows that this model generally performs significantly worse compared to ChatGPT while having an about ten times higher cost per queried pair. The comparison with the fine-tuned RoBERTa-base baseline shows that ChatGPT in a zero-shot setting is able to reach a similar performance or even surpass RoBERTa fine-tuned with 2K training pairs. RoBERTa trained with 20K pairs is finally able to surpass most

Table 3. Results of the baseline experiments.

Model	Configuration	P	R	F1	Δ F1	cost (¢) per pair
gpt3.5-davinci-002	domain-complex-forced-T	59.70	80.00	68.38	2.15	1.36
	domain-simple-forced-T	72.34	68.00	70.10	3.87	1.29
	general-complex-forced-T	43.10	100.00	60.24	-5.99	1.40
	general-simple-forced-T	65.50	80.00	70.18	3.95	1.29
RoBERTa	fine-tuned on sampled training set (2K pairs)	85.99	80.00	82.72	16.49	–
RoBERTa	fine-tuned on original training set (20K pairs)	86.79	92.00	89.32	23.09	–

zero-shot prompts but its recall remains 6–8% lower. The training data for both RoBERTa models contains product offers for the same products that are also part of the validation set, i.e. these products are considered in-distribution. It has been shown [10] that such fine-tuned models experience a significant drop in performance when applied to pairs containing out-of-distribution products. The performance of ChatGPT in the zeroshot setup essentially corresponds to results on out-of-distribution data as no training happens, suggesting that ChatGPT is generally more robust concerning unseen products.

4 In-Context Learning

In the second set of experiments we analyse the impact of adding matching and non-matching product offer pairs as task demonstrations [6] to the prompts in order to help the model to understand and subsequently perform the task correctly. We experiment with three different heuristics for selecting task demonstrations:

- **Hand-picked**: Hand-picked demonstrations are a set of up to 10 matching and 10 non-matching product offer pairs which were hand-selected by a human domain expert from the pool of the training set.
- **Random**: Demonstrations are drawn randomly from the labeled training set while making sure that they do not contain any of the products that are part of the product offer pair to be matched.
- **Related**: Related demonstrations are selected from the training set by calculating the Jaccard similarity between the pair to be matched and all positive and negative pairs in the training set. The resulting similarity lists are sorted and the most similar examples are selected.

In addition to the three selection heuristics, we also vary the amount of demonstration (shots) from 6 over 10 to 20 with an equal amount of positive and negative examples in order to evaluate the impact on performance and API cost. Due to their length, we do not provide examples of in-context prompts in this paper but refer the reader to the project github which contains all prompts.

Table 4 shows the results of the in-context experiments. We compare the results to the zero-shot baseline of using domain-specific, complex language as well as forcing the model to answer with a simple "Yes" or "No" (see Sect. 3). For all three selection heuristics, providing 3 positive and 3 negative examples as task demonstrations leads to improvements over the zero-shot baselines of at least 2% F1. Random demonstrations have the smallest impact with a maximum increase of 3.89% (10 demonstrations) while the hand-picked demonstrations lead to an increase of up to 4.92% (10 demonstrations) over the zeroshot baseline. Providing 20 related examples as demonstrations has the largest impact and improves the F1 score by nearly 8% over the baseline. Across all in-context experiments, providing demonstrations consistently leads to an increase in precision while the recall decreases. This points to the model becoming more cautious when predicting positives. The more examples are provided the more pronounced this effect becomes. Providing task demonstrations is helpful in all cases as it provides the model with clear guidance on how the solutions to the task should look as well as patterns that correlate with the correct answer. The provision of related demonstrations increases this effect, as the model is steered towards patterns that are relevant for the decision at hand.

Table 4. Results of the in-context learning experiments and associated cost.

Selection heuristic	Shots	P	R	F1	Δ F1	Cost (¢) per pair	Cost increase	Cost increase per Δ F1
ChatGPT-zeroshot	0	71.01	**98.00**	82.35	–	0.14	–	–
ChatGPT-random	6	78.33	94.00	85.45	3.10	0.77	450%	145%
	10	79.66	94.00	86.24	3.89	1.13	707%	182%
	20	78.95	90.00	84.11	1.76	2.07	1379%	783%
ChatGPT-handpicked	6	76.19	96.00	84.86	2.51	0.72	414%	165%
	10	80.00	96.00	87.27	4.92	1.00	614%	125%
	20	79.66	94.00	86.24	3.89	2.03	1350%	347%
ChatGPT-related	6	80.36	90.00	84.91	2.56	0.68	386%	151%
	10	**89.58**	86.00	87.76	5.41	1.05	650%	120%
	20	88.46	92.00	**90.20**	7.85	1.97	1307%	167%
GPT3.5-handpicked	10	61.97	88.00	72.72	-9.63	10.54	7429%	771%
	20	61.43	86.00	71.67	-10.68	19.71	13979%	1309%
GPT3.5-related	10	67.69	88.00	76.52	-5.83	10.04	7071%	1213%
	20	61.43	86.00	71.67	-10.68	20.34	14429%	1351%

Cost Analysis: The performance gain resulting from task demonstrations comes with a sizable increase in API usage cost. While the zeroshot baseline prompt and model answer cost around 0.14¢ per matching decision, providing an additional 20 examples increases this cost by nearly 1400% to 2¢ per decision. Breaking the increase in cost down to the increase per percentage point of F1 (see rightmost column in Table 4) it becomes clear that providing related examples has the best price performance ratio, followed by hand-picked examples. This calculation does not factor in the cost of acquiring labeled pairs to be selected by the heuristics. As the handpicked demonstrations consists of only 20 labeled pairs, the labeling cost of this approach is significantly lower than the costs for the other two.

5 Providing Matching Knowledge

The last set of experiments focuses on providing explicit matching knowledge in the form of natural language rules and asking the model to use these rules for its decisions. Asking ChatGPT to explain matching decisions revealed that ChatGPT is able to identify product features and corresponding feature values. Following this finding, we formulate explicit rules for a set of common product features as well as a general rule capturing any possible additional features. The natural language formulation of the rule set that we add to the prompts is shown in Fig. 2. We experiment with using these rules in a zero-shot scenario as well as together with related demonstrations (see Sect. 4).

Task Desc.	Your task is to decide if two product descriptions match. The following rules regarding product features need to be observed:
Rules	1. The brand of matching products must be the same if available 2. Model names of matching products must be the same if available 3. Model numbers of matching products must be the same if available 4. Additional features of matching products must be the same if available
Task Desc.	Do the following two product descriptions match? Answer with 'Yes' if they do and 'No' if they do not.
Task Input	Product 1: 'Title: DYMO D1 – Roll (1.9cm x 7m)' Product 2: 'Title: DYMO D1 Tape 12mm x 7m'

Fig. 2. Example of a prompt containing matching rules.

Table 5 shows the results of adding matching rules to the prompts. Adding matching rules increases the zero-shot performance by 6% F1 to 88.29%. This performance is only 2% lower than the performance that was reached by providing 20 related task demonstrations. Interestingly, providing the rules in a zero-shot setting does not negatively impact the recall, which remains at 98%, but increases the precision of the model by nearly 10%. Combining matching rules and 20 related demonstrations in a single prompt slightly further improves the precision but leads to a 10% drop in recall and an overall lower F1. ChatGPT seems to be able to interpret matching rules and successfully applies them to improve matching results. Ultimately, task demonstrations and matching rules both serve the same purpose of guiding the model on how to match entities. Matching rules are more generic, while related demonstrations are rather product pair specific. This specificness might be the reason for the slightly higher performance. On the other hand, defining matching rules requires significantly less human effort compared to labeling a pool of examples for selecting related demonstrations. Adding explicit matching rules to prompts might thus be a promising approach for many real-world use cases.

Table 5. Results of providing explicit matching knowledge to ChatGPT.

Prompt	Shots	P	R	F1	Δ F1	Cost (¢) per pair	Cost increase	Cost increase per Δ F1
ChatGPT-zeroshot	0	71.01	**98.00**	82.35	–	0.14	–	–
ChatGPT-zeroshot with rules	0	80.33	**98.00**	88.29	5.94	0.28	100%	17%
ChatGPT-related	6	80.36	90.00	84.91	2.56	0.68	386%	151%
	10	89.58	86.00	87.76	5.41	1.05	650%	120%
	20	88.46	92.00	**90.20**	7.85	1.97	1307%	167%
ChatGPT-related with rules	6	90.70	78.00	83.87	1.52	0.79	464%	305%
	10	90.91	80.00	85.11	2.76	1.17	736%	267%
	20	**91.11**	82.00	86.32	3.97	2.09	1393%	351%

6 Conclusion

We have demonstrated the impact of various prompt designs on the performance of ChatGPT on a challenging entity matching task. We have shown that the model can achieve competitive performance in a zero-shot setting compared to PLMs like RoBERTa which require to be fine-tuned using thousands of labeled examples. Due to the relative shortness of the prompts and the associated low API fees, using ChatGPT for entity matching can be considered a promising alternative to fine-tuned PLMs which require the costly collection and maintenance of large in-domain training sets. ChatGPT can further be considered more robust as it demonstrates competitive performance even in zero-shot settings while fine-tuned PLMs struggle with out-of-distribution entities which where not seen during training [10]. The provision of task demonstrations further increases the performance, especially if the selected demonstrations are textually similar to the pair of entities to be matched. If closely related demonstrations are not available, providing randomly selected demonstrations also has a significant positive effect. The manual selection of just 20 demonstration pairs by a domain expert resulted in a strong positive effect and requires a significantly lower effort than having the domain expert label thousands of pairs for PLM-based matchers. Finally, providing the model with a set of explicit matching rules has a similar effect as providing textually related demonstrations. The provision of explicit, higher-level matching knowledge to LLMs via prompts is a promising direction for future research as it has the potential to significantly reduce the labeling effort required for achieving state-of-the-art entity matching results.

References

1. Akbarian Rastaghi, M., Kamalloo, E., Rafiei, D.: Probing the robustness of pre-trained language models for entity matching. In: Proceedings of the 31st ACM International Conference on Information and Knowledge Management, pp. 3786–3790 (2022)
2. Barlaug, N., Gulla, J.A.: Neural networks for entity matching: a survey. ACM Trans. Knowl. Discov. Data **15**(3), 52:1–52:37 (2021)
3. Brown, T., Mann, B., Ryder, N., Subbiah, M., Kaplan, J.D., et al.: Language models are few-shot learners. Adv. Neural. Inf. Process. Syst. **33**, 1877–1901 (2020)

4. Christophides, V., Efthymiou, V., Palpanas, T., Papadakis, G., Stefanidis, K.: An overview of end-to-end entity resolution for big data. ACM Comput. Surv. **53**(6), 127:1–127:42 (2020)
5. Li, Y., Li, J., Suhara, Y., Doan, A., Tan, W.C.: Deep entity matching with pre-trained language models. Proce. VLDB Endow. **14**(1), 50–60 (2020)
6. Liu, J., Shen, D., Zhang, Y., Dolan, B., Carin, L., et al.: What makes good in-context examples for GPT-3? In: Proceedings of Deep Learning Inside Out: The 3rd Workshop on Knowledge Extraction and Integration for Deep Learning Architectures, pp. 100–114. Association for Computational Linguistics (2022)
7. Liu, P., Yuan, W., Fu, J., Jiang, Z., Hayashi, H., et al.: Pre-train, prompt, and predict: a systematic survey of prompting methods in natural language processing. ACM Comput. Surv. **55**(9) (2023)
8. Narayan, A., Chami, I., Orr, L., Ré, C.: Can foundation models wrangle your data? Proc. VLDB Endow. **16**(4), 738–746 (2022)
9. Peeters, R., Bizer, C.: Supervised contrastive learning for product matching. In: Companion Proceedings of the Web Conference 2022, pp. 248–251 (2022)
10. Peeters, R., Der, R.C., Bizer, C.: WDC products: a multi-dimensional entity matching benchmark. arXiv preprint arXiv:2301.09521 (2023)
11. Primpeli, A., Bizer, C.: Profiling entity matching benchmark tasks. In: Proceedings of the 29th ACM International Conference on Information and Knowledge Management, pp. 3101–3108 (2020)
12. Wei, J., Tay, Y., Bommasani, R., Raffel, C., Zoph, B., et al.: Emergent abilities of large language models. Trans. Mach. Learn. Res. (2022)
13. Zhao, W.X., Zhou, K., Li, J., Tang, T., Wang, X., et al.: A survey of large language models. arXiv preprint arXiv:2303.18223 (2023)
14. Zhao, Z., Wallace, E., Feng, S., Klein, D., Singh, S.: Calibrate before use: improving few-shot performance of language models. In: Proceedings of the 38th International Conference on Machine Learning, pp. 12697–12706 (2021)

CONSchema: Schema Matching with Semantics and Constraints

Kevin Wu, Jing Zhang, and Joyce C. Ho(✉)

Emory University, Atlanta, GA 30322, USA
{kevin.wu2,jing.zhang2,joyce.c.ho}@emory.edu

Abstract. Schema matching aims to establish the correspondence between the attributes of database schemas. It has been regarded as the most difficult and crucial stage in the development of many contemporary database and web semantic systems. Manual mapping is a lengthy and laborious process, yet a low-quality algorithmic matcher may cause more trouble. Moreover, the issue of data privacy in certain domains, such as healthcare, poses further challenges, as the use of instance-level data should be avoided to prevent the leakage of sensitive information. To address this issue, we propose CONSchema, a model that combines both the textual attribute description and constraints of the schemas to learn a better matcher. We also propose a new experimental setting to assess the practical performance of schema matching models. Our results on 6 benchmark datasets across various domains including healthcare and movies demonstrate the robustness of CONSchema.

Keywords: schema matching · constraint matching · semantic matching

1 Introduction

Schema matching in relational databases can be viewed as one of the most essential elements of data integration. The purpose is to identify correspondences among concepts across heterogeneous and potentially distributed data sources. For example, a wide variety of database systems collect similar data and each system has been customized for the company. This results in similar collections of data being stored in different formats, terminologies, and even logically arranged ways. As such, data exchange and integration can be hindered by these customized databases. Thus, schema matching becomes necessary across various domains including sharing health records [17] and merging documents with different formats [1]. Although schema matching is well-studied [1], the existing methods entail significant manual labor or fail to generalize across domains [21].

Given the rising focus on privacy across various sectors such as healthcare, there is a need to focus on schema-level rather than instance- or hybrid levels (i.e., no exchange of information related to instance-level records). Under the schema-level paradigm, only table and attribute information such as the name, description, meta-data, and summary statistics are shared. Meta-data and summary

A. Abelló et al. (Eds.): ADBIS 2023, CCIS 1850, pp. 231–241, 2023.
https://doi.org/10.1007/978-3-031-42941-5_21

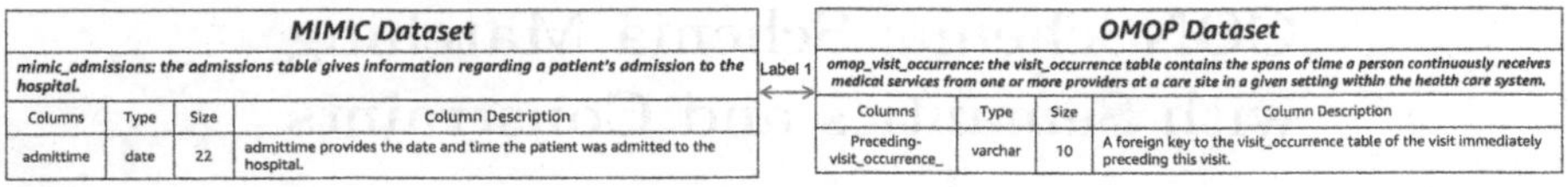

MIMIC Dataset			
mimic_admissions: the admissions table gives information regarding a patient's admission to the hospital.			
Columns	Type	Size	Column Description
admittime	date	22	admittime provides the date and time the patient was admitted to the hospital.

Label 1 ⟷

OMOP Dataset			
omop_visit_occurrence: the visit_occurrence table contains the spans of time a person continuously receives medical services from one or more providers at a care site in a given setting within the health care system.			
Columns	Type	Size	Column Description
Preceding-visit_occurrence_	varchar	10	A foreign key to the visit_occurrence table of the visit immediately preceding this visit.

Fig. 1. Example of an identified schema match between the mimic_admission table and admittime source field in MIMIC-III and the omop_visit table and preceding_visit_time field in OMOP using both semantics and constraints.

statistics pose fewer privacy risks and are often shared for federated databases and privacy-preserving learning [4]. Several approaches have been proposed to automate schema-level matching, including constraint-based approaches [3,9,16] and linguistic-based approaches [18,21]. Unfortunately, both approaches entail background knowledge to manually define the mapping between the two relations, assume the content of the elements will be the same across the two schemas, or fail to adequately capture the similarities between the field descriptions. This can yield suboptimal performance for new applications.

Deep learning (DL) has been proposed as a new paradigm for data integration [19] given its success in other applications such as computer vision and natural language processing. DITTO, a state-of-the-art (SOTA) entity matching model, utilizes a pre-trained Transformer-based language model that can solve classification problems with entity matching [13]. However, DITTO may not perform well across different domains. SMAT, another DL model, generates a schema-level embedding using the element names and descriptions to identify the matching relations [21]. These models demonstrate the potential of DL to encode the textual information present in the attribute names and descriptions, yet ignore constraints such as data types, ranges, and key constraints.

We propose CONSchema to fuse the constraint information such as the data type, range, and key constraints with the textual information (Fig. 1 shows an example) to improve DL-based matching. The central insight is a classification model can then learn the interaction between the attribute similarity and the constraint relatedness, without requiring manual mapping. Furthermore, existing strategies do not assess the generalizability of the matching model to unseen elements within the schema. Often, training samples include either the source or target schema elements, thereby offering an optimistic assessment of the predictive performance. We introduce a *unseen partition* experimental setting and our experiments on 6 datasets demonstrate the robustness of CONSchema. The CONSchema is publicly available in the GitHub repository.[1]

2 Related Work

We briefly summarize the existing schema-level matching work focused on relational databases. Instance-level and hybrid-level models require privacy-preserving mechanisms for sensitive domains like healthcare and are beyond the

[1] https://github.com/kwu78/CONSchema.

scope of this work. Other related schema matching can be found in the survey [1]. Thus, we focus on existing schema-level matching methods [1].

Linguistic-level approaches calculate similarity based on the name of the attributes and/or the description of the attributes. Previous heuristic methods [3,7,15] provided decent solutions for schema matching with combinations of matchers. However, the numerical representations of the schema with distance metrics would not handle the semantic heterogeneity. Recent DL models have been introduced to perform linguistic matching. ADnEV proposed to post-process the matching results from other matchers [18]. Unfortunately, the quality of the matchers impacts the ADnEV performance. SMAT [21] utilized attention-over-attention to pretrain a language model for the attributes, and obtained SOTA performance on several schema-level matching benchmark datasets.

The constraint-based approach relies on the meta-data of the attributes such as the data types and value ranges, uniqueness, optionality, relationship types, and cardinalities. A measure of similarity can be determined by data types and domains, key characteristics (e.g., unique, primary, foreign), and relationships [9,16]. However, precise matching requires rich constraint information. The hybrid approach combining constraint-based and instance-based approaches [2,6] has been popularized to achieve flexible and robust matchers. Unfortunately, instance-based approaches can result in privacy leakage.

3 CONSchema

3.1 Problem Statement

Given two table descriptions S_{TS} and S_{TT}, two attributes' names N_{F1} and N_{F2}, their descriptions S_{F1} and S_{F2}, and their constraints C_{F1} and C_{F2} (i.e., data type, value ranges, primary key, and foreign key) from the source and target schema respectively, we construct two sets of sequences: (1) the source sequence set $S_S = <N_{F1}>, <S_{TS} + S_{F1}>, <C_{F1}>$, and (2) the target sequence set $S_T = <N_{F2}>, <S_{TT} + S_{F2}>, <C_{F2}>$. For Fig. 1, the source example can be constructed as the sequence set "the admissions table gives information regarding a patient's admission to the hospital", "admittime", "admittime provides the date and time the patient was admitted to the hospital", "date" and size 13. For training, there is an annotated label $L(S_S, S_T)$ where 0 and 1 denotes two fields are not related (i.e., not mapped to each other) and related (i.e., corresponding attribute-to-attribute matching), respectively. The task objective is then classifying the relatedness between the two attributes.

3.2 Model

Textual Similarity Embedding. The textual embedding captures the relatedness between the two attributes' names and descriptions. The idea is that the semantic similarity between the two attributes serves as the proxy for relatedness. For example, SMAT constructs two sentence pairs where a sentence consists

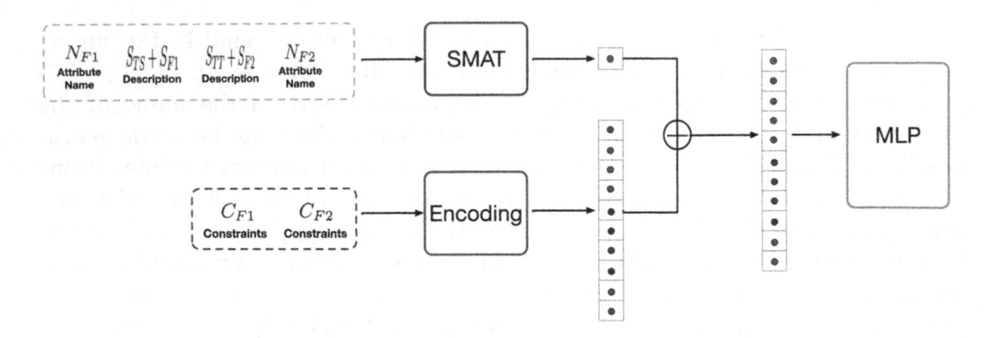

Fig. 2. Illustration of CONSchema's structure.

of the attribute name and description (e.g., <$N_{F1}, S_{TS}+S_{F1}$>). The model then learns the textual similarity between the two sentence pairs and is trained using the labels without encoding domain knowledge explicitly. CONSchema uses the last SMAT layer to serve as the attribute embedding (a 2-dimensional vector) that captures the semantic similarity between the two attributes. SMAT is chosen due to its superior performance for schema matching [21].

Constraint Encoding. The key idea behind CONSchema is to fuse the schema constraints (i.e., C_{F1} and C_{F2}) to the textual embedding. This is done by encoding the constraints into a numerical vector format where each column represents a different constraint such that a downstream classifier can then learn the importance without requiring previous knowledge. For our experiments, we focus on the data types (e.g., varchar, datetime, int, numeric), the data size for the contents (2 versus 128 characters), and key constraints. To represent the data type, we use a one-hot encoding where the value is 1 for the corresponding feature and 0 elsewhere. For example, if the attribute type is a String, then the isString feature will be set to 1. Similarly, key constraints are encapsulated using the one-hot encoding mechanism (e.g., isPrimaryKey, isForeignKey). The raw data size serves as a numeric element.

For the Fig. 1 scenario, the "admittime" attribute is a date type with a size of 22, thus the isDate feature is set to be 1 while the other data types remain 0 (i.e., isVarchar, isInt2, isInt4). In addition, the size is set to 22 (i.e., size = 22). For the OMOP attribute, since it is a varchar of size 10, the vector representation is all 0 except isVarchar = 1 and size = 10. This representation avoids the need to create ad-hoc rules for each domain. Other constraints such as uniqueness, optionality, and functional dependencies can be captured in a similar fashion, but such information is not available in the 6 datasets used for our experiments.

Final Classification. The textual similarity embedding and the constraint encoding representations are concatenated together to create the final vector representation. This fused vector encapsulates the semantic relation and the constraints of the two attributes and is then used with the annotated labels to train a

Table 1. Summary statistics of the 6 datasets used in our experiments. The columns under # capture the conversion statistics for table, attribute, and related, respectively. The next 5 columns under % represent the data type distribution where B/I2/I4 denotes boolean, int2, and int4; Fl/Arr/Oth represents float, array, and other; and PK/SK denotes primary and secondary keys. The last 3 columns provide the character length of the textual descriptions.

	#			%					Length		
	Tab.	Attr.	Rel.	String	Date	B/I2/I4	Fl/Arr/Oth	PK/SK	Min	Avg	Max
MIMIC [11]	25	240	129	47	19	-/9/15	-/-/10	-/-	64	255	688
Synthea [20]	12	11	105	77	14	-/4/1	-/-/4	-/-	45	219	688
CMS [5]	5	96	196	53	14	-/12/21	-/-/-	-/-	54	232	688
Real Estate [8]	3	76	66	46	-	14/-/29	11/-/-	-/-	4	12	20
IMDB [12]	23	129	45	33	19	3/-/33	5/7/-	16/20	63	132	306
Thalia [10]	21	167	52	70	14	-/-/16	-/-/-	-/-	14	22	35

multi-layer perceptron (MLP).[2] MLP can capture non-linear interactions (especially with an increasing number of hidden layers) and is relatively lightweight compared to an end-to-end DL model that will incur significant training and inference overhead. Moreover, the MLP seamlessly integrates with existing DL frameworks, offering adaptability and unlocking the potential of the models. In this work, the MLP consists of two linear layers and a ReLU layer. A softmax is then applied to the outputs of MLP in order to obtain a prediction of schema matching. Figure 2 shows the entire architecture of CONSchema.

4 Experiments

Our experiments are designed to evaluate the accuracy and robustness of the model to *unseen* attributes. Existing evaluation strategies involves randomly partitioning the attribute pairs into training, validation, and test datasets. Thus, the source attribute likely occurs in at least 1 pair sample in the training dataset and provides an optimistic assessment of the model performance as partial information on the test pairs has been seen by the model. Instead, we introduce an unseen partition evaluation strategy. We randomly partition the source attributes and pair them with all the target attributes, ensuring source attributes in test never appear in training or validation, named as {dataset}_S. The same strategy is applied on the target attributes and denoted as {dataset}_T.

Our experiments evaluate the schema-matching models under our unseen partition using a ratio of 80-10-10 for train, validation, and test, respectively. The hidden units and learning rate sweep are done over the ranges [24, 36, 48, 64] and [1e−5, 5e−5, 1e−4], respectively. Since the DL-based models are sensitive to the initialization of the parameters, we train 5 versions of the model using different initial weights and report the mean value across the 5 initializations.

[2] We explored other models such as random forest and logistic regression and the results follow similar trends with MLP providing the largest performance boost.

Table 2. Comparison of precision (P), recall (R), and F1 (F) on the 6 datasets under the unseen partition evaluation strategy. The best performance is **bolded** and the second best is underlined.

Datasets	DITTO			SMAT			Con-MLP			CONSchema		
	P	R	F1	P	R	F1	P	R	F1	P	R	F1
MIMIC_S	0.002	0.323	0.004	**0.261**	0.467	0.284	0.041	**1.000**	0.079	0.247	0.550	**0.298**
MIMIC_T	0.001	0.285	0.002	**0.137**	0.650	**0.226**	0.008	**1.000**	0.016	0.122	0.750	0.209
Synthea_S	0.004	0.282	0.008	0.409	**0.720**	0.457	0.077	0.300	0.122	**0.430**	0.680	**0.510**
Synthea_T	0.003	0.314	0.006	0.259	1.000	0.411	0.256	1.000	0.408	**0.345**	**1.000**	**0.513**
CMS_S	0.156	0.321	0.210	0.289	0.821	0.426	0.214	0.333	0.261	**0.430**	**0.933**	**0.575**
CMS_T	0.008	0.763	0.015	0.097	0.200	0.089	0.009	0.316	0.017	**0.140**	**0.316**	**0.194**
IMDB_S	0.149	0.203	**0.172**	0.107	0.125	0.110	0.079	**0.375**	0.130	**0.224**	0.150	0.162
IMDB_T	0.056	**0.867**	**0.355**	0.092	0.422	0.150	0.018	0.333	0.058	**0.143**	0.444	0.216
Real Estate_S	0.138	0.109	0.122	**0.900**	0.167	0.279	0.214	**1.000**	**0.353**	0.470	0.400	0.352
Real Estate_T	**0.613**	0.533	**0.553**	0.084	0.500	0.111	0.065	0.500	0.115	0.082	**0.667**	0.146
Thalia_S	0.117	0.269	0.163	0.120	0.400	0.181	0.167	**1.000**	0.286	**0.252**	0.760	**0.374**
Thalia_T	0.052	0.880	0.095	0.109	0.431	0.164	0.116	**1.000**	0.208	**0.176**	0.600	**0.273**

Datasets. We assess the models on the OMAP benchmark, a schema-level matching healthcare dataset [21] mapping 3 databases to the Observational Medical Outcomes Partnership (OMOP) Common Data Model standard to facilitate evidence-gathering and informed decision-making [17], and 3 popular schema matching benchmark datasets, IMDB, Real Estate, and Thalia, used for several existing studies [12]. For each dataset, the element table name with its descriptions, attribute column name with its descriptions, attribute data type, and attribute key constraints are used to construct the sequence. The label annotation is based on the final ETL design, where a 1 denotes the table-column in the source schema was mapped to a table-column in the target schema. The summary statistics for the 6 datasets are summarized in Table 1.

Baseline Methods. CONSchema is compared against 3 matching models: (1) **DITTO** [13], a SOTA entity matching model based on the pre-trained Transformer model that matches using a sequence-pair classification problem; (2) **SMAT** [21], a SOTA schema matching model generating embeddings from the attribute name and description and then feeding the embedding to a MLP to conduct the classification task; and (3) **CON-MLP**, an MLP model using only the constraint encoding as an input. The optimal hyperparameters are determined using grid search and evaluation on the validation dataset.

5 Results

5.1 Unseen Partition Evaluation

Table 2 summarizes the results under our unseen evaluation strategy, where source ({dataset}_S)/ target ({dataset}_T) attributes in the test dataset are

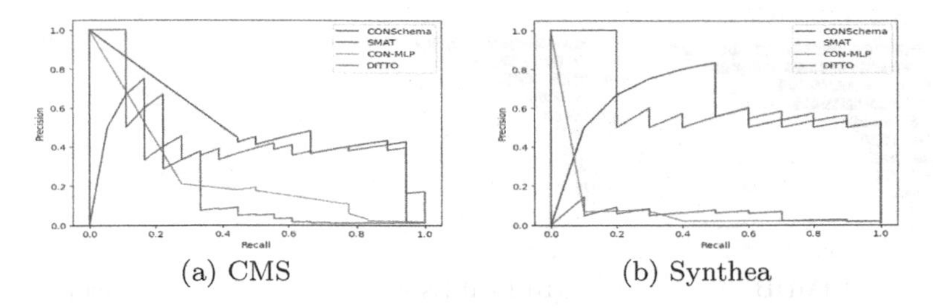

(a) CMS (b) Synthea

Fig. 3. Precision-Recall Curves for two of the datasets.

guaranteed not to be seen during training. In comparison with the random partition evaluation for MIMIC, Synthea, and CMS in [21], we observe that the recall for both DITTO (i.e., 0.462, 0.40, 0.636) and SMAT (i.e., 0.846, 0.950, and 0.909 respectively) is lower. This suggests the performance under random splits tends to overestimate the recall performance as having seen some of the pairings with the attributes can help the model generalize better on the test set.

We observe that CONSchema achieves the highest precision across 4 of the 6 datasets and second best for MIMIC and Real Estate on two different partitions. It also yields the best F1 score for MIMIC, Synthea, CMS, and Thalia, and the second-best F1 score for Real Estate and IMDB. Furthermore, the F1 scores for CONSchema are all better than SMAT for the 6 datasets, except **MIMIC_T**. This provides evidence that the constraints offer further information to more accurately identify the correspondences. Moreover, there are no huge differences on CONSchema between target and source partitions demonstrating its robustness.

Semantic embeddings can work even without long and well-formed textual descriptions. As shown in Table 1, the real estate database attributes are short (average of 12 characters). Examples of the textual descriptions include "water", "agent name", "type", and "firm city" which correspond to the attributes "water", "agent_name", "type", and "firm_city". From Table 2, we observe that DITTO and SMAT, which rely only on textual descriptions, can achieve reasonable performance compared to the longer counterparts such as CMS and Synthea.

The Con-MLP results illustrate the importance of our constraint vector representation. Without any textual similarity information, the model achieves better F1 scores across all but IMDB datasets when compared with DITTO. The F1 score is also better than SMAT for IMDB, Real Estate, and Thalia. To better understand the benefits of the constraint representation, Table 1 summarizes the mean, or frequency, of each encoded column for its corresponding dataset. We observe that Synthea has the highest proportion of the *Varchar* datatype when compared to the other datasets. Thus, CON-MLP is unable to achieve high recall or F1 as the constraint representation offers little information. In contrast, Real Estate, IMDB, MIMIC, and Thalia have a diversity of data types, thereby yielding improved recall scores compared to SMAT. The constraint statistics also illustrate the importance of appropriately specifying the data type and data

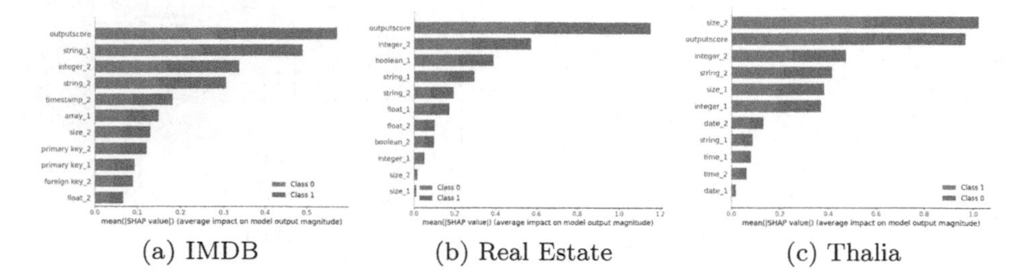

(a) IMDB (b) Real Estate (c) Thalia

Fig. 4. Illustration of the SHAP values to explain the impact of the features in CONSchema.

range in the database schema. Ambiguous information is likely to hurt CONSchema more than helping it to achieve better results.

To better understand the trade-off in precision and recall, Fig. 3 plots the precision-recall curve for two datasets. For CMS, the precision of CONSchema is consistently higher than SMAT until the higher recall rates. DITTO and CON-MLP have comparable precision at the lower recall and DITTO drops below CON-MLP for recall > 0.3. For Synthea, we observe slightly different dynamics where at the lower recall (<0.2), SMAT outperforms CONSchema in terms of precision. However, for recall between 0.2 and 0.5, CONSchema outperforms SMAT significantly in terms of precision. For recall > 0.5, the two methods yield similar precision. The plots suggest for midrange recall, the constraints are particularly helpful to differentiate the positive matches. The plots also suggest that solely using constraints can generate comparable precision at lower recall rates.

5.2 Explaining CONSchema Matching Decisions

To better understand the predictions of CONSchema, we investigate the importance of the features and how they differ across the three datasets. Our analysis is based on the SHapley Additive exPlanations (SHAP) framework [14] to better understand the impact with respect to the label. SHAP is a popular explainable artificial intelligence framework that is model-agnostic. It is an additive feature attribution method and explains the change in the expected model prediction when conditioning on that feature. The SHAP analysis is performed on the best-performing model from the 5 different versions.

Figure 4 provides the summary plots for 3 of the 6 datasets where the features are sorted in descending order of their overall impact on the model output. The Y-axis labels with 1 (i.e., size_1) represent the constraints of the source dataset, whereas the 2 represent the constraints of the target dataset. The plots illustrate that the SMAT output score is one of the most important features across all datasets, which is not surprising given the results from Table 2. However, the importance of the constraints on the precision is illustrated both for IMDB and Thalia. For IMDB (Fig. 4a), which has the richest constraint diversity, we observe that the data type (string versus integer) is almost equivalent in importance to SMAT. Similarly, for Thalia (Fig. 4c), the size of the target data type is more important than SMAT. On Real Estate (Fig. 4b), we observe boolean, integer, and string all have top SHAP values. This further illustrates the importance of constraint diversity towards improving performance.

5.3 CMS Case Study

We also performed a qualitative study on the CMS dataset by assessing three different scenarios. The first positively maps the CMS icd9_dgns_cd attribute from table inpatientclaims (varchar type of size 100 with a description of "claim diagnosis code 1 - claim diagnosis code 10") to the OMOP cause_source_concept_id element from table death (int4 type of size 10 with a description of "a foreign key to the concept that refers to the code used in the source. note this variable name is abbreviated to ensure it will be allowable across database platforms."). CONSchema correctly identifies the match over SMAT even though the descriptions are dissimilar as the constraints indicatethey might be potentially related.

The second is a negative pair where the CMS clm_from_dt attribute from the inpatientclaims table (date type of size 13 with the description "claims start date") does not map to the OMOP condition_start_datetime element from the condition occurrence table (date type of size 296 with the description "the date and time when the instance of the condition is recorded"). CONSchema incorrectly identifies a match whereas SMAT does not. We observe both the text (the start date of a claim and the start time of a medical condition) and the constraints are similar, thus leading to an incorrect conclusion by CONSchema.

The last scenario is a positive pair that matches the CMS sp_cncr attribute from the beneficiary summary table (int2 type of size 5 with the description "chronic condition: cancer") and the OMOP cohort_definition_id element from the cohort table (int4 type of size 10 with description "a foreign key to a record in the cohort definition table containing relevant cohort definition information"). Both SMAT and CONSchema incorrectly classify this sample as the textual description of OMOP is too broad (no text related to the chronic condition cancer), and the constraint type encoding does not convey enough information.

6 Conclusion

This paper proposes CONSchema, a model that incorporates schema constraints and textual descriptions to achieve better schema-level matching. As it does not utilize instance-level information and avoids directly encoding domain knowledge regarding the source and target systems, CONSchema can be used for privacy-sensitive applications. Moreover, our constraint encoding can encompass categorical-style features (type of data, type of constraint) and numeric representations (size of data) common across a variety of relational database schemas. We also propose an evaluation strategy to better understand the generalizability of existing models and demonstrate the robustness on 6 datasets.

There are several limitations of our work. First, the F1 scores are too low to be used in practice. Yet, the improvement in precision can facilitate less manual matching by prioritizing the predicted positive cases. Next, utilizing one-hot encoding to represent the constraints can yield sparse inputs for large number of constraints. This can be addressed by utilizing an auto-encoder to reduce variations or inconsistencies in the constraints originating from diverse data sources. Another limitation is the need for sufficient labels. We posit that contrastive learning techniques and data augmentation approaches may reduce the need for annotations and improve predictive performance. We also note a stronger evaluation strategy is to use the zero-shot learning framework where the model is not trained on any of the source or target attributes, and leave this for future work. Finally, CONSchema has only been demonstrated for relational schemas and should be extended to encompass a variety of data (e.g., nested data models and unstructured data) and data discovery tasks.

Acknowledgements. This work was supported by the National Science Foundation award IIS-2145411.

References

1. Alwan, A.A., Nordin, A., Alzeber, M., Abualkishik, A.Z.: A survey of schema matching research using database schemas and instances. Int. J. Adv. Comput. Sci. Appl. **8**(10), 2017 (2017)
2. Atzeni, P., Bellomarini, L., Papotti, P., Torlone, R.: Meta-mappings for schema mapping reuse. Proc. VLDB Endow. **12**(5), 557–569 (2019)
3. Aumueller, D., Do, H.H., Massmann, S., Rahm, E.: Schema and ontology matching with coma++. In: Proceedings of the 2005 ACM SIGMOD International Conference on Management of Data, pp. 906–908 (2005)
4. Azevedo, L.G., de Souza Soares, E.F., Souza, R., Moreno, M.F.: Modern federated database systems: an overview. ICEIS **1**, 276–283 (2020)
5. Centers for Medicare & Medicaid Services: CMS 2008–2010 data entrepreneurs' synthetic public use file (de-synpuf) (2011)
6. Chen, C., Golshan, B., Halevy, A.Y., Tan, W.C., Doan, A.: Biggorilla: an open-source ecosystem for data preparation and integration. IEEE Data Eng. Bull. **41**(2), 10–22 (2018)
7. Do, H.H., Rahm, E.: Coma-a system for flexible combination of schema matching approaches. Proc. VLDB, 610–621 (2002)
8. Doan, A.: Learning to map between structured representations of data (2002)
9. Fagin, R., Kolaitis, P.G., Popa, L., Tan, W.C.: Schema mapping evolution through composition and inversion. In: Schema Matching and Mapping, pp. 191–222 (2011)
10. Hammer, J., Stonebraker, M., Topsakal, O.: Thalia: test harness for the assessment of legacy information integration approaches. In: Proceedings of ICDE, pp. 485–486 (2005)
11. Johnson, A.E., et al.: Mimic-iii, a freely accessible critical care database. Sci. Data **3**, 160035 (2016)
12. Leis, V., Gubichev, A., Mirchev, A., Boncz, P., Kemper, A., Neumann, T.: How good are query optimizers, really? Proc. VLDB Endow. **9**(3), 204–215 (2015)

13. Li, Y., Li, J., Suhara, Y., Doan, A., Tan, W.C.: Deep entity matching with pre-trained language models. arXiv preprint abs/2004.00584 (2020)
14. Lundberg, S.M., Lee, S.: A unified approach to interpreting model predictions. In: Proceedings of NeurIPS, pp. 4765–4774 (2017)
15. Madhavan, J., Bernstein, P.A., Rahm, E.: Generic schema matching with cupid. In: vldb. vol. 1, pp. 49–58 (2001)
16. Mecca, G., Papotti, P., Santoro, D.: Schema mappings: from data translation to data cleaning. In: A Comprehensive Guide Through the Italian Database Research Over the Last 25 Years, pp. 203–217 (2018)
17. Observational Health Data Sciences and Informatics: The book of OHDSI (2019)
18. Shraga, R., Gal, A., Roitman, H.: Adnev: cross-domain schema matching using deep similarity matrix adjustment and evaluation. Proc. VLDB **13**(9), 1401–1415 (2020)
19. Thirumuruganathan, S., Tang, N., Ouzzani, M., Doan, A.: Data curation with deep learning. In: EDBT, pp. 277–286 (2020)
20. Walonoski, J., et al.: Synthea: an approach, method, and software mechanism for generating synthetic patients and the synthetic electronic health care record. JAMIA **25**(3), 230–238 (2017)
21. Zhang, J., Shin, B., Choi, J.D., Ho, J.C.: Smat: an attention-based deep learning solution to the automation of schema matching. In: Proceedings of ADBIS, pp. 260–274 (2021)

Data Quality

Context-Aware Data Quality Management Methodology

Flavia Serra[1,2(✉)], Veronika Peralta[2], Adriana Marotta[1], and Patrick Marcel[2]

[1] Universidad de la República, Montevideo, Uruguay
{fserra,amarotta}@fing.edu.uy
[2] Université de Tours, Blois, France
{veronika.peralta,Patrick.Marcel}@univ-tours.fr

Abstract. Data quality management (DQM) is a complex task involving activities for data quality (DQ) assessment and improvement. Many DQ methodologies address DQM (sometimes partially), and are made up of several stages, where many DQM activities are carried out. According to the literature, most of these activities are influenced by the context of data. However, very few state-of-the-art DQ methodologies consider the context of data, and when they do, context is addressed only at few stages. In this work, we propose a context-aware data quality management (CaDQM) methodology, that clarifies the influence of context in most DQM activities. In particular, context components are identified at early stages and are used at all stages of the CaDQM.

Keywords: Context · Data Quality · Data Quality Management · Data Quality Methodology

1 Introduction

Data quality management (DQM) is a complex task [4] that involves many activities for assessing and improving data quality (DQ). As early as 1995, Gassman et al. [12] claimed that DQ must be monitored continually, and many methodologies have been proposed for this purpose (e.g., [6,11,18,22]). A DQ methodology provides a set of guidelines and techniques that, from input information that describes a given application context, defines a rational process to assess and improve the quality of data [4]. It is made up of several stages, organized into phases [4], and each stage involves a set of activities. Many other more recent works (e.g., [5,9,13,14,21]) point out that most activities (as DQ dimensions selection, DQ measurement and assessment, selection of strategies for DQ improvement) addressed in DQ methodologies are influenced by the context. However, very few state-of-the-art methodologies consider the context of data, limiting its use to initial stages (e.g., [5,9]) or not specifying how to use it (e.g. [13]). In addition, such methodologies deal with a simple form of context (e.g. it is reduced to the task at hand in [22]) or a static context (set at the initial

A. Abelló et al. (Eds.): ADBIS 2023, CCIS 1850, pp. 245–255, 2023.
https://doi.org/10.1007/978-3-031-42941-5_22

stage and no longer updated, as in [6]). We claim that context should be identified, used and maintained throughout all the stages of a DQ methodology. This allows not only objective and subjective DQ assessment, but also control and improvement activities adapted to the context of data.

In this paper we propose CaDQM, a Context-aware Data Quality Management methodology, that clarifies the influence of the context on DQM activities. The context is defined by a set of components, namely: the application domain, user characteristics, tasks at hand, data filtering needs, DQ and systems requirements, business rules, general metadata, DQ metadata, and other data (the latter are data related to the evaluated data, and therefore complementing them). Context components were identified and justified in previous works [19,20]. In CaDQM, context components are identified at early stages, and can be updated at later ones, and as this happens, they influence all stages of the methodology. The possibility of updating the context whenever necessary, makes our methodology adaptable to any application domain and data usage. What's more, when DQM is guided by the context, the decisions made are explicit, which limits biases, i.e. hidden causes for certain decisions, during the execution of DQM activities. In addition, all management decisions can be justified and documented, allowing the possibility of reproducing each of the DQM activities.

The reminder of this paper is organized as follows: Sect. 2 analyzes DQ methodologies and their use of context. Section 3 presents CaDQM and describes the use of context at all stages. In Sect. 4, we present a case study where CaDQM is applied. Finally, Sect. 5 concludes and presents future work.

2 Related Work

This section presents an overview of state-of-the-art DQ methodologies and their use of context. Some of them are dedicated to specific domains, such as health [16], linked data [10,14], e-government [21], decisions making [13], data governance [3], and service oriented architectures [17]. In addition, DQ framework proposals generally include a DQ methodology [3,10,16,21]. Some of them use decision models [15,21,22]. For example, Total Data Quality Management (TDQM) [22] adapts the Plan-Do-Check-Adjust (PDCA) cycle [1], instantiating its stages as follows: plan (defining), do (measuring), check (analyzing), and act (improving) DQ. In the ISO 8000-61:2016 reference model for DQM, the PDCA cycle is instantiated as follows: plan (according to data requirements), do (implementing data processing), check (monitoring and measuring DQ and process performance, and reporting the results), and act (taking actions to continually improve process performance). In [21], the Observe-Plan-Do-Check-Adjust (OPDCA) cycle [1], is instantiated as follows: observe (the existing situation and the need for improvement), plan (the target DQ maturity level and the DQ improvement plan), do (implement the DQ improvement plan), check (whether the overall DQ values have been achieved), and adjust (set new DQ state and improve the DQM process if necessary). These 3 works implicitly use context at initial stages, limited to DQ requirements [15,21,22] and business rules [21] identification.

Two detailed surveys [4,9] compare respectively 13 and 12 methodologies (6 in common), which mainly address two phases, for DQ assessment and improvement. Only 3 of the reviewed methodologies propose a prior phase where the context of data can be identified and/or defined, namely, Comprehensive Data Quality (CDQ) [4,6], Heterogeneous Data Quality Methodology (HDQM) [7] and Hybrid Information Quality Management (HIQM) [8]. Although they do not explicitly mention the context of data, they collect some contextual information at initial stages. Such information is not updated at later stages and is reduced to DQ requirements [6–8] and business rules [6].

To the best of our knowledge, there is no methodology that make use of context throughout all its stages. In addition, the few ones that consider contextual information at initial stages, address few context components (DQ requirements, business rules or task at hand), which are not updated at later stages.

3 A Context-Aware DQM Methodology

We propose a context-aware data quality management (CaDQM) methodology, which is inspired by CDQ methodology [4,6]. According to the authors, CDQ is a complete methodology that has 3 phases: state reconstruction, assessment, and improvement. In the first phase, the relationships among organizational units, processes, services, and data are reconstructed. The second phase sets new targets about DQ levels, and evaluates corresponding costs and benefits, while in the third phase the improvement process is selected by performing a cost-benefit analysis [4]. CaDQM also has 3 phases, namely DQ planning, DQ assessment, and DQ improvement (described in Subsects. 3.1–3.3), and explicitly uses the context of data at each of its phases. While in the DQ planning phase most of the context components are identified and defined, in the DQ assessment phase the context can also be updated. At the same time, the context of data influences the stages present throughout the whole methodology.

The most important differences between CaDQM and CDQ are the following: i) CDQ collects contextual information only in the state reconstruction phase, while CaDQM identifies it at both DQ planning and DQ assessment phases, ii) the influence and use of context on the activities provided by CaDQM are explicitly described, while in CDQ it is not clear how contextual information is used, iii) in CaDQM the 3 phases define a cycle, while in CDQ, the initial phase (state reconstruction), can only be executed once. In addition, in the CDQ methodology, stages and DQM activities are only listed, while CaDQM presents a well-defined process, making explicit the transitions among the stages and the relationships between stages and context. Figure 1 shows this process. Stages (numbered from ST1 to ST9) are represented as colored boxes and their flows with plain arrows. Encompassing dotted boxes delimit the three phases. Dotted arrows represent the relationships with context, and labels distinguish when context is defined or updated and when the context influences a stage. This figure also lists, at the right side, the activities, inputs and outputs of each stage.

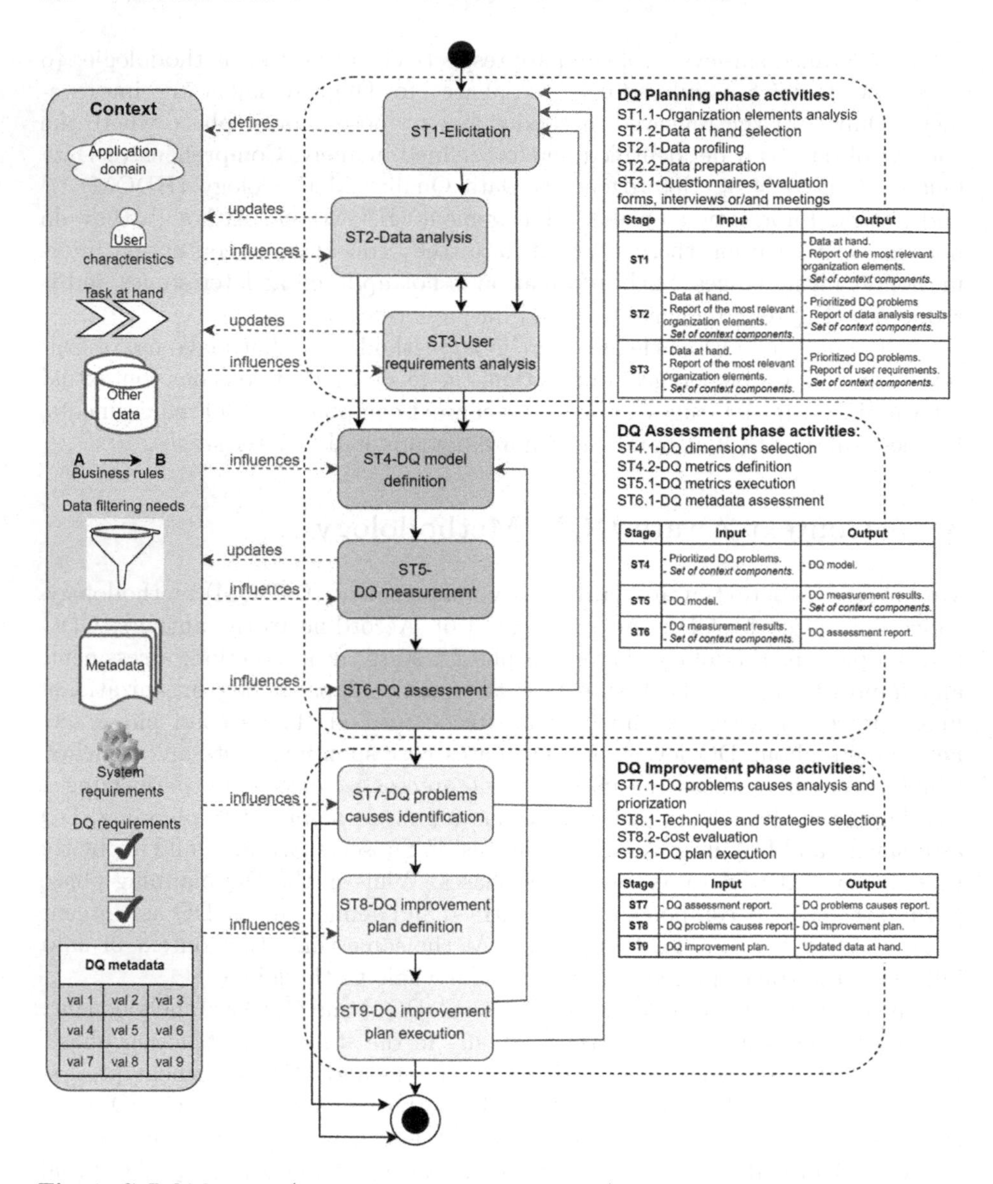

Stage	Input	Output
ST1		- Data at hand. - Report of the most relevant organization elements. - *Set of context components.*
ST2	- Data at hand. - Report of the most relevant organization elements. - *Set of context components.*	- Prioritized DQ problems - Report of data analysis results - *Set of context components.*
ST3	- Data at hand. - Report of the most relevant organization elements. - *Set of context components.*	- Prioritized DQ problems. - Report of user requirements. - *Set of context components.*

Stage	Input	Output
ST4	- Prioritized DQ problems. - *Set of context components.*	- DQ model.
ST5	- DQ model.	- DQ measurement results. - *Set of context components.*
ST6	- DQ measurement results. - *Set of context components.*	- DQ assessment report.

Stage	Input	Output
ST7	- DQ assessment report.	- DQ problems causes report.
ST8	- DQ problems causes report	- DQ improvement plan.
ST9	- DQ improvement plan.	- Updated data at hand.

Fig. 1. CaDQM stages (activities, inputs and outputs), and relationships with context.

Next subsections present phases, describing stages and activities and specifying the influence of the context. Context components are highlighted in italics.

3.1 DQ Planning Phase

This phase is composed by 3 stages: Elicitation, Data Analysis, and User Requirements Analysis. The two latter can be executed in any order, even in parallel, since context components that arise at one stage can enrich the analysis carried out at the other. One of them can be even omitted when it is necessary to opt for a data-based or user-based approach.

ST1-Elicitation Stage: At this stage, all organizational information, such as its application domain, business processes (including all participating tasks), services, data sources, DQ issues (as reported DQ problems or DQ requirements), is analyzed and collected. Furthermore, characteristics (e.g., profiles, permissions, etc.) of users and stakeholders of each of the aforementioned elements are registered. In particular, different data usages are identified, as well as users who use data are distinguished from those who modify it, as they have different DQ requirements. At the end of this stage, data whose quality will be evaluated (called *data at hand*) are identified. In addition, a report is generated, documenting the most relevant elements of the organization for DQM.

Context Definition: Once data at hand are identified, the components that define their context are selected of the organizational information initially identified. Firstly, *other data* that describe the data at hand and *metadata* arise from data sources. *Application domain* in which data at hand are used, the tasks that involve them (called *tasks at hand*), and users (this includes all *users characteristics*) who execute such tasks are context components. In addition, all *data filtering needs*, arising from both users and their tasks, are specified.

Based on the identified application domain, *business rules* that arise from standards, norms and constraints, and restrict the data usage, are selected. Furthermore, systems in which the task at hand is executed are also analyzed, in particular its requirements, since the possibility of executing certain DQ metrics strongly depends on the *system requirements.* On some occasions, it is also possible to carry out an initial identification of DQ problems and *DQ requirements* reported by the users involved in data usage. Finally, all the contextual information collected determines the set of components that defines the context of data at hand.

ST2-Data Analysis Stage: In this stage data preparation tasks are performed. In particular, data at hand characteristics and their relationships to other data are discovered by applying data profiling tasks. In these tasks it is possible to identify DQ problems (e.g., amount of null values, data duplication, data freshness, etc.). At the end of this stage, DQ problems are prioritized and data analysis results are reported.

Context Influence and Update: The context defined so far guides the activities developed at this stage. That is, the particular characteristics of the application domain, the task at hand and DQ requirements help selecting the most appropriate data analysis activities (e.g. data preparation tasks for the Financial domain are not the same as that performed for the Health domain). New *metadata* may emerge from data profiling tasks, and the identified DQ problems may determine new *DQ requirements* and preliminary DQ values (also called

DQ metadata). The latter allow giving an idea of the current state of the organization, regarding the quality of the data at hand. Furthermore, *other data* that describes the data at hand, and *business rules* can arise from new relationships that emerge from data analysis. Finally, this new set of components allow updating the context definition.

ST3-User Requirements Analysis Stage: Activities carried out at this stage can be questionnaires, evaluation forms, interviews, regular meetings, etc. The information exchange with data users and stakeholders allows identifying more DQ problems and expectations about DQ levels. At the end of this stage, DQ problems are prioritized and user requirements analysis results are reported.

Context Influence and Update: Users characteristics are the most influential context component at this stage, because it allows determining which are the most appropriate activities and guides their execution. For example, for some type of user, it could be more suitable to set up questionnaires, while for others it might be necessary to define regular meetings. Besides, the rest of the context components can support the definition of the interviews, questionnaires, etc.

Interaction with data users and stakeholders also allows determining new *metadata* (e.g., data owners, publication managers or creation date), establishing *data filtering needs*, and associating *other data* (possibly from other organization sectors or even from other organizations). In addition, new *business rules* that arise from the day-to-day data usage can also be communicated by users, as well as DQ problems that may imply new *DQ requirements*. Finally, the context definition is updated, adding these new context components.

3.2 DQ Assessment Phase

This phase is composed by 3 stages: DQ Model Definition, DQ measurement, and DQ assessment.

ST4-DQ Model Definition Stage: At this stage a set of DQ dimensions is selected, which represent data characteristics that will be assessed. Any set of DQ dimensions can be considered among the ones proposed in the literature (e.g., [2,22]). DQ metrics are defined based on the selected DQ dimensions. Aggregated DQ metrics may also be specified, typically defined as weighted averages of other metrics of more granularity (e.g. combining DQ of several columns to report the DQ of a dataset). At the end of this stage a DQ model, defined by DQ dimensions and DQ metrics, is reported.

Context Influence: DQ dimensions selection is a very important activity, since different DQ dimensions will lead to different interpretations of the current state of the quality of the data at hand. Therefore, the context of data is decisive for the DQ model definition, since DQ dimensions selection is strongly influenced by the context components. Furthermore, the prioritized DQ problems, input of this stage, also have an important role, since DQ problems also influence on the DQ dimensions selection. Note that DQ problems were also prioritized according to the context.

DQ requirements and business rules are the most influential components in this activity, leading the selection of DQ dimensions. For example, a DQ requirement about data precision leads to the accuracy dimension, while a business rule stating a functional dependency, leads to the consistency dimension. Furthermore, context components can be used for defining DQ metrics, i.e. the algorithms that implement DQ metrics sometimes make use of elements of context components such as metadata and thresholds.

ST5-DQ Measurement Stage: The implementation and execution of measurement methods, that implement the DQ metrics defined in the DQ model, are the only activities of this stage. At the end of this stage, a report is generated with measurement results.

Context Update and Influence: The DQ values resulting from DQ measurement constitutes DQ metadata, which is a context component (since DQ measurements of certain data may be used as context for other data). Such DQ metadata obtained at this step also give context (i.e. influence) to the calculation of the aggregated DQ metrics.

ST6-DQ Assessment Stage: At this stage, DQ metadata obtained at the measurement stage are assessed. The term "assessment" is used when measurements are compared to reference values, in order to enable a diagnosis of quality [4]. At the end of this stage, results of the assessment determines the quality level of the data at hand, and a report is generated with assessment results.

Context Influence: DQ assessment is performed based on the context of the data at hand. For example, DQ metadata are compared to thresholds (i.e. reference values), that arise from DQ requirements, as well as from system requirements, business rules, and data filtering needs, among others. This is the most context-dependent stage, since DQ assessment is the most subjective DQM activity.

3.3 DQ Improvement Phase

This phase is composed by 3 stages: DQ Problems Causes Identification, DQ Improvement Plan Definition, and DQ Improvement Plan Execution.

ST7-DQ Problems Causes Identification Stage: Once DQ assessment is carried out, it is possible to identify which and where are DQ problems. Therefore, the causes of such DQ problems are analyzed at this stage, and once they are identified, they can be prioritized. In addition, the cause of a DQ problem with low priority could have high priority. For example, outdated data issues caused by lack of user training. In this case, it is important to solve the cause avoiding new DQ problems. At the end of this stage, results of each of these activities are reported.

Context Influence: The causes of DQ problems are objective, however their identification and prioritization must be done based on the context of data. In fact, some context components may be more relevant than others. For instance, DQ requirements that make it possible to identify DQ problems also guide the identification and prioritization of their causes.

ST8-DQ Improvement Plan Definition Stage: At this stage the DQ improvement plan is defined, taking into account DQ problems causes prioritization carried out at the previous stage. Techniques and strategies that ensure actions on DQ problems are selected and incorporated into the plan. In addition, the costs of their execution are evaluated, also taking into account the costs generated by poor DQ. At the end of this stage, a DQ improvement plan is reported.

Context Influence: The influence of the data context on the activities of this stage, as at the previous stage, is also associated with the relevance of context components. For example, techniques and strategies selection can be conditioned by users characteristics, since low-skilled users will restrict the improvement techniques to be executed. In addition, DQ metadata assessed as bad or very bad can influence the improvement plan definition, having to select techniques or strategies that allow eliminating the causes that lead to poor DQ.

ST9-DQ Improvement Plan Execution Stage: The only activity of this stage is the execution of the improvement plan. It is totally context independent, because all activities or contextual decisions are performed in the previous stage. As result of this stage, the data at hand are updated, correcting reported DQ problems according to their priorities and the priorities of their causes. In addition, depending on the improvement plan, other activities can also be carried out, such as improvements in information systems or data users training, among others.

Finally, as evidenced in Fig. 1, the proposed process is cyclic. Indeed, after the improvement phase, it is typical to return to the assessment phase for updating the DQ model and performing further measurement. When it is not necessary to update the DQ model, only the measurement and assessment stages are executed one more time to verify the results of the improvement plan. In addition, it is always possible to return to the planning phase, identify more organization elements and set more context components.

4 Case Study

We experimented our approach within a regional research project in collaboration with the Regional University Hospital Center of the city of Tours, in France. Our goal is to test CaDQM within a real small-sized project.

The project concerns the preparation and analysis of medical data for the stratification of Amyotrophic Lateral Sclerosis patients. As baseline, we use preliminary data profiling and transformation activities (within an ETL project), conducted by a data scientist without following a particular DQ methodology.

The application of the first 2 phases of CaDQM, DQ Planning and DQ Assessment, with an explicit use of context allowed a more robust analysis of data, leading to the identification of more DQ problems, compared to the preliminary work of the data analyst. In particular, having the data context helped in the execution of data profiling tasks, since it allowed the identification of relevant data and the establishment of priorities among them. Furthermore, the use of CaDQM allowed the definition of a DQ model funded on context, which perfectly documents DQ assessment goals and is a valuable input for the setting and implementation of an improvement plan. The complete case study can be consulted in https://github.com/fserrasosa/caseStudy-healthcare/wiki

5 Conclusions

This paper proposes a context-aware data quality management (CaDQM) methodology, which bridges the gap among the recognized importance of data context for DQM, and the lack of context usage in DQ methodologies.

CaDQM is composed of 3 phases, DQ planning, DQ assessment, and DQ improvement, which explicitly use the context of the data. It defines a DQM process through a set of stages and transitions among them, and interactions with context. CaDQM allows DQ experts to carry out DQM tasks in an ordered manner, taking advantage of all relevant input information (embedded in context) and keeping trace of DQ values (also embedded in context) and taken decisions. This also ensures the possibility of reproducing such tasks. We are currently working on the definition of the reports obtained as output of the different stages of CaDQM.

A small-sized real project, in the health domain, allowed us to obtain preliminary results about the applicability of CaDQM regarding the use of the context and encouraging feedback from users.

Our long-term outlook is to validate CaDQM via various user studies. Firstly, we are working on a theoretical case study that will be applied in a DQ course, whose students (professionals of data science and other areas), will solve a problem with and without following CaDQM, allowing us to compare their results. In addition, we plan to test CaDQM in a large-scale project in the domain of digital government, involving various public services in Uruguay. They currently apply a framework[1] for DQM in Digital Government, which does not consider data context. Our challenge is to include CaDQM in their framework, obtaining direct user feedback and comparing to previous results. We emphasize that DQM in digital government is extremely important, since it has a strong impact on the quality of public services. At the same time, this application domain serves as a rich validation platform, where CaDQM capabilities can be largely

[1] Research project developed by the e-Government and Information and Knowledge Society Agency (AGESIC) of Uruguay: https://www.gub.uy/agencia-gobierno-electronico-sociedad-informacion-conocimiento/.

exploited, since it includes many information systems, data collections, users and stakeholders, among others.

References

1. Foresight university: Shewhart-deming's learning and quality cycle. https://foresightguide.com/shewhart-and-deming/. Accessed Mar 2023
2. Iso/iec 25012 standard. https://iso25000.com/index.php/en/iso-25000-standards/iso-25012. Access June 2023
3. Al-Salim, W., et al.: Analysing data quality frameworks and evaluating the statistical output of united nations sustainable development goals' reports. Renew. Energy Environ. Sustain. 7 (2022)
4. Batini, C., et al.: Methodologies for data quality assessment and improvement. CSUR **41**(3), 1–52 (2009)
5. Batini, C., Scannapieco, M.: Methodologies for information quality assessment and improvement. In: Data and Information Quality, pp. 353–402. Springer (2016)
6. Batini, C., et al.: A comprehensive data quality methodology for web and structured data. In: ICDIM, pp. 448–456 (2007)
7. Batini, C., et al.: A data quality methodology for heterogeneous data. IJDMS **3**, 60–79 (2011)
8. Cappiello, C., et al.: Hiqm: a methodology for information quality monitoring, measurement, and improvement, pp. 339–351 (2006)
9. Cichy, C., Rass, S.: An overview of data quality frameworks. IEEE Access **7**, 24634–24648 (2019)
10. Debattista, J., et al.: Luzzu-a methodology and framework for linked data quality assessment. JDIQ **8**(1), 1–32 (2016)
11. English, L.P.: Improving Data Warehouse and Business Information Quality: Methods for Reducing Costs and Increasing Profits. Wiley, USA (1999)
12. Gassman, J.J., et al.: Data quality assurance, monitoring, and reporting. Control. Clin. Trials **16**(2), 104–136 (1995)
13. Günther, L.C., et al.: Data quality assessment for improved decision-making: a methodology for small and medium-sized enterprises. Procedia Manuf. **29**, 583–591 (2019)
14. Gürdür, D., et al.: Methodology for linked enterprise data quality assessment through information visualizations. JIII **15**, 191–200 (2019)
15. Standard ISO 8000–61:2016. Data quality - part 61: Data quality management: Process reference model. Technical report (2022)
16. Kerr, K., Norris, T.: The development of a healthcare data quality framework and strategy. In: ICIQ, pp. 218–233 (2004)
17. Petkov, P., Helfert, M.: A methodology for analyzing and measuring semantic data quality in service oriented architectures. In: 14th International Conference on Computer Systems and Technologies, pp. 201–208 (2013)
18. Pipino, L.L., et al.: Data quality assessment. ACM **45**(4), 211–218 (2002)
19. Serra, F., Peralta, V., Marotta, A., Marcel, P.: Modeling context for data quality management. In: ER 2022. p. 325–335 (2022)
20. Serra, F., Peralta, V., Marotta, A., Marcel, P.: Use of context in data quality management: a systematic literature review (2022). https://arxiv.org/abs/2204.10655

21. Tepandi, J., et al.: The data quality framework for the estonian public sector and its evaluation. In: TLDKS, vol. 10680, pp. 1–26. Springer (2017)
22. Wang, R.Y.: A product perspective on total data quality management. ACM **41**(2), 58–65 (1998)

Thesaurus-Based Transformation: A Classification Method for Real Dirty Data

Maxime Perrot[1,2](✉), Mickaël Baron[2], Brice Chardin[2], and Stéphane Jean[3]

[1] Bimedia, Roche-sur-Yon, France
[2] LIAS, ISAE-ENSMA, Chasseneuil, France
maxime.perrot@ensma.fr
[3] LIAS, Université de Poitiers, Poitiers, France
Stephane.Jean@ensma.fr

Abstract. In this paper, we consider a retail store classification problem supported by a real word dataset. It includes yearly sales from several thousand stores with dirty categorical features on product labels and product hierarchies. Despite the fact that classification is a well-known machine learning problem, current baseline methods are inefficient due to the dirty nature of the data. As a consequence, we propose a practical thesaurus-based transformation. It uses an intermediary global approximate classification of products, based on local products hierarchies. Activities for a subset of stores are human-labeled to serve as ground truth for validation, and enable semi-supervision. Experiments show the effectiveness of our approach compared to baseline methods. These experiments are based on datasets and solutions made available for reproducibility purposes.

1 Introduction

Bimedia is a French company which markets hardware (cash registers) and software for convenience stores such as grocery stores, tobacco stores, bakeries, florists, etc. It supplies more than 6,000 stores with a wide range of activities, archiving approximately 60 million transactions monthly. Given the freedom granted to each customer to manage their own product catalog, Bimedia's data collection is heterogeneous in terms of quality and quantity depending on stores size, activities and customs.

Store activities (e.g., grocery or tobacco) are filled in by Bimedia's sales staff when contracts are set up. However, due to the large number of stores, this information is difficult to manage manually, while stores tend to diversify their activities over the years. As a result, the level of confidence in the activities entered in the database is low. Yet, the activities of client stores are valuable, for example to report accurate revenue breakdowns.

In this paper, we address the problem of identifying store activities from their sales. From a scientific point of view, this is a retail store classification problem supported by a real word dataset. Despite the fact that classification is a well-known problem in machine learning, we show in this paper that current

A. Abelló et al. (Eds.): ADBIS 2023, CCIS 1850, pp. 256–265, 2023.
https://doi.org/10.1007/978-3-031-42941-5_23

baseline solutions are inefficient to accurately classify stores in this case study. This is mainly due to the dirty nature of the considered real dataset. From our experience, this is also the case for numerous other datasets in practice.

To overcome the limitations of state-of-the-art solutions, we propose a preprocessing step named ThesaurusBT. This method is based on a thesaurus built using business knowledge. This thesaurus enables an intermediary global approximate classification of products, based on multiple examples of hierarchical structures used to group products. Even if the proposed thesaurus is specific to our domain of applications, the same methodology with a different thesaurus could be used for different use cases. Our main contributions in this paper are the following.

1. We propose a thesaurus-based transformation method that uses business knowledge to overcome the dirtiness of data.
2. We evaluate experimentally the impact of our method compared to state-of-the-art solutions.
3. For the purpose of facilitating reproducibility, we offer access to both real datasets and implemented solutions used in our experiments.

This paper is organized as follows. Section 2 details the problem addressed in this paper and the considered use case. Section 3 analyses contributions reporting similar problems and solutions. Section 4 presents our thesaurus-based transformation method. Sections 5 and 6 describe our implementation, experiments and results obtained. We conclude in Sect. 7 and introduce future work.

2 The Store Classification Problem

In Bimedia's operation, store owners manage their product catalogs with some flexibility on the hierarchical structure used to group products into *families*. There exists two types of families.

- *Global families* are defined by Bimedia. They cover products which are either: 1) subject to special legislation that restrict potential providers, such as tobacco, vape and press, or 2) dematerialized, such as money transfer, prepaid phone cards or gift cards.
- *Local families* are defined by store owners. Any name can be used to label these families. Moreover, products of these families are frequently defined without a global product ID such as a normalized barcode.

The objective is to classify stores according to their activities. Over a third of the stores offer products and services related to activities such as restaurants, bakeries, grocery stores, florists, bars, and more. Products related to these activities belong to local families. These are difficult to handle because of the differences in the naming of the products (synonyms, acronyms or spelling mistakes), differences in the codification systems (normalized scanned barcodes, hand-typed barcodes with potential typos, or even unspecified) and the arrangements into families (by brand, by product types with different granularity, etc.). Therefore, it is almost impossible to compare products between stores and identify corresponding activities without complex data transformations.

Table 1. Dataset sample[4] of products names and families

	Store ID	Product name	Product family
1	ccb...2d6	Tray of Fry	On-site catering
2	3e1...cf7	large fry	Catering to take away 10 - 707140
3	609...ba4	TRAY FRY	ALIM5.5
4	379...949	FRY PLATE ONLY	BAR
5	379...949	MID. TRAY FRY TAKEOUT	Catering to take away
6	aa1...590	Fries Tray	Snack
7	bc3...3d3	TRAY OF FRY	CATERING 10
8	ab1...8ef	NOODLE FRY	Tabletry
9	8ab...c19	PIK FRY	Confectionery 20

We illustrate the considered problem with a simple example from real data of the provided dataset in Table 1. The first seven of the nine selected rows refer to the same product: a portion of fries, but these are expressed by different strings and arranged in various families depending on the stores or even on the way of consumption (as in rows 4 and 5 of the table, corresponding to the same store). Moreover, it is not possible to rely on a product ID (e.g., a barcode) to map products between stores, as this identifier is generated locally for this type of product. A data transformation step is required. The original dataset has more than 2 millions unique values for product names, and more than 6,000 stores with variable revenues and sales quantities. As the illustrated problem is recurring, classifying store activities becomes complex[1].

The store classification problem can be broken down into three sub-problems: 1) dirty categorical variables encoding: as illustrated in Table 1, sales data is structured under categories with a two-level (products and families) local hierarchy and dirty textual values, 2) variable input shapes: raw input sales data is, for each store, of variable length and non-sequential, its transformation is therefore not trivial, 3) multi-labeling classification: one or more business activities can be assigned to stores, depending on their sales.

Table 2 illustrates some of the raw data extracted from transactions for three stores that we wish to classify, along with the expected output. In this sample, we can observe that stores sell a various number of products, resulting into a different number of rows per store. The hardest part is not to assign labels to stores, but to identify relationships between products from different stores. In this paper, we consider the problem as a whole, including the multi-labeling task, because we are not able to produce a consistent validation dataset for products (2 millions products), while it is feasible for stores activities (6,000 stores). Consequently, the multi-labeling task will mainly be included in this study as a way to evaluate the quality of the data transformation step.

[1] Irrelevant columns are not displayed, irrelevant information are cropped and textual values are translated, as literally as possible, from French to English in all Figures.

Table 2. Dataset sample of sales with targeted activities predictions

Store ID	Barcode	Product	Family	Quantity	Activities
db...5c2	31...01	Corn Bread	Bread	4	Bakery
	31...02	Bread	Bread	2338	
	31...04	Baguette	Bread	13377	
	31...51	Cookie	Pastry	2378	
e9...f43	00...03	Larks Pie	Bakery	6315	Bakery
	00...19	Traditional	Bakery	135445	
	31...07	Almond cream galette	Patisserie	3	
bb...49f	04...94	ZIPPO GASOLINE	Other smoking items	40	Tobacco store Coffee shop
	11...10	Clearomiser Q16 PRO	Misc 20	3	
	31...45	COFFEE	HOT	83253	
	_A...36	PHILIP MORRIS 20	Cigarettes	26125	
	_A...65	NEWS RED 20	Cigarettes	6126	

3 Related Work

Many methods exist to encode text-based categorical variables with synonyms and morphological variability, such as Bert [11], Gamma-Poisson [4], Similarity Encoding [6], MinHash [3], PairClass [12], and others [5,13]. Based on the experiments conducted by Cerda and Varoquaux [5], the MinHash technique provided the best overall performance for this task, better than large language models.

Processing methods for arbitrary length inputs can be distinguished into two groups: input processing and problem transformation [2]. Due to the difficulty to implement and adapt problem transformation techniques to our context, those are not covered. Input processing techniques include truncation [8], padding [9], aggregation, PCA [1] and models with feature selection, such as XGBOOST [7] and CatBoost [10]. Since our dataset is non-sequential, many other solutions cannot be considered. The basic approach to deal with arbitrary length inputs in our case is the aggregation data pivoting transformation. Truncation, padding and others do not make sense given the shape of our dataset. Experiments conducted by Borisov *et al.* [2] highlight CatBoost as an efficient solution for feature selection.

4 Thesaurus-Based Transformation

The basis of our proposition is a thesaurus-based transformation, abbreviated as ThesaurusBT, that groups products into categories before labeling store activities. Product categorization is performed for two reasons. First, being able to categorize products sold by stores, independently of the store's catalog, serves multiple purposes, including generating dashboards and targeted advertising. This capability makes our method superior to non-explanatory encoding solutions for companies seeking understandable product labeling. Second, as previously mentioned, the primary challenge posed by the multi-labeling of store activities does not lie in the labeling process itself, but in the data transformation required to encode product sales. Specifically, it involves encoding dirty textual categorical variables such as products or families labels, which often have a high cardinality. Classifying products represents a solution for this task.

To categorize products, our approach is based on a thesaurus that links recurring terms occurring in family labels to coarse-grained categories defined by the company. In our use case, we have defined a total of 60 product categories, plus an *unknown* category for unidentified product types. Examples of these categories are: *takeaway catering*, *clothes*, or *tobacco*. Our assumption is that, on average over the whole dataset, family labels can be correctly mapped to predefined categories. We consider this assumption to be reasonable when the company, as is the case with Bimedia, possesses significant insight about the kind of products sold by their stores. Thus, ThesaurusBT requires some business knowledge to be applicable, namely:

1. a list of categories of products that the company wants to identify,
2. a dictionary mapping common keywords—i.e. n-grams that are commonly used by store owner—to product categories,
3. (optional) a list of ambiguous keywords, that are commonly associated with several categories, along with differentiation rules. For example, the word *drink* could be used to describe both alcoholic and non-alcoholic drinks. As the VAT (Value Added Tax) applied to alcoholic and non-alcoholic drinks differs under French law, it can be used to resolve ambiguous keywords.

The goal of ThesaurusBT is to automatically create a mapping dictionary between products sold by stores and categories provided by the company. Our method is based on the following principle. While store owners retain the freedom to assign labels to their products and to group them into families, due to the vast number of stores, it is likely that a significant proportion of them will feature some keywords identified by business experts.

The first processing step is a partial cleaning of textual values of dirty categorical variables to encode. This cleaning step includes conversion to lower case, removal of extra spaces, accents, special characters, stop words—including domain-specific stop words.

The second step identifies products that can be categorized using simple rules. In Bimedia's case, this is performed on global families, which are normalized and supervised by the company. In the provided dataset, the identification and the labeling of global families products is based on the family identifier field and a mapping between family identifiers and categories. The remaining processing steps only apply to product from families with dirty labels—local families in Bimedia's case.

The third step creates unique identifiers for product instances. This identification does not have to be exact: some instances can be incorrectly merged or kept separate without significantly impacting the whole process. Best results are still obtained when erroneous identification are minimized. In our dataset, this step starts with the identification of store-internal product identifiers and generic product identifiers, such as normalized barcodes. Internal identifiers generated by Bimedia software are local to a store, and cannot be used to map a product between one store and another. The creation of a unique product identifier in this case is therefore performed using the product name.

The fourth step lists family labels for each product identifier. It then computes the number of occurrences of each predefined keyword (n-gram) within this list. The most frequent keyword is used to identify the category.

The result of these steps is a mapping from products to categories. In our use case, ThesaurusBT categorizes 94% of products appearing in the dataset. This process does not guarantee a correct categorization of products, and its accuracy is difficult to assess due to the lack of an annotated dataset. We consider this categorization usable in practice if, when included as a transformation step, it improves the accuracy of another classification workflow. This validation is described in our experiments, where we compared ThesaurusBT with existing methods.

5 Experimental Design

The originality of our approach concerns the data transformation step that we use for the multi-labeling task (finding the activities of store). As a consequence, the aim of the experiments is to compare the efficiency of baseline machine learning approaches with our method on the data transformation step. We assume that the performance of the data transformation step has a direct impact on the accuracy of the multi-labeling results for store activities. Datasets and implementations are available online[2].

We provide two sets of real data supplied by Bimedia. The sales dataset is composed of anonymized data over a one-year period for 2325 selected stores. This dataset variables are: store identifier, product barcode, product label, family identifier, family label, total amount sold during the year and VAT rate. The dataset contains 11,637,397 rows, 817,385 unique product names and 8,427 unique family labels.

The labeled store activity dataset is composed of 400 stores with their activities annotated by experts. These stores were randomly selected from the previous sales dataset and their activities were annotated based on sales aggregated by product family and information extracted from external sources. This dataset contains two variables: a store identifier and activity tags. There are 9 possibles activity tags and 29 unique combinations of those in this dataset.

We use accuracy and macro F1 scores to evaluate multi-labeling performance. Considering that store activities are very unbalanced, we need a model that is efficient in identifying both rare activities, such as hotels, and common ones, such as tobacco shops. The macro F1 score is a reliable metric for identifying models that meet these criteria. For our experiments, 300 annotated stores are randomly selected as the training dataset and the remaining 100 are the test dataset.

We divide our experiments into three workflows: a *basic approach* **(1)**, a *literature approach* **(2)** and a *business specific approach* **(3)**. These are illustrated

[2] https://forge.lias-lab.fr/thesaurusbt.

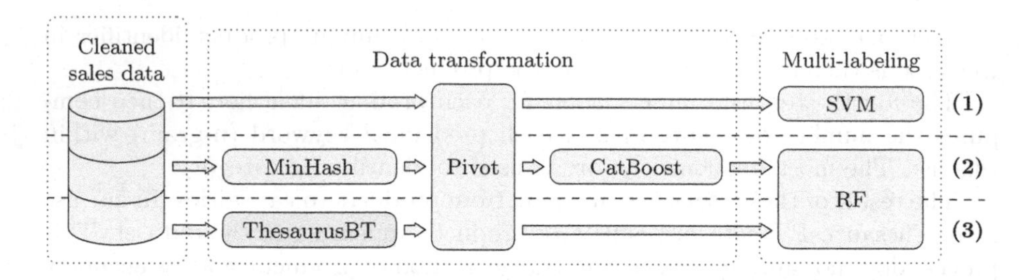

Fig. 1. Processing steps of considered workflows

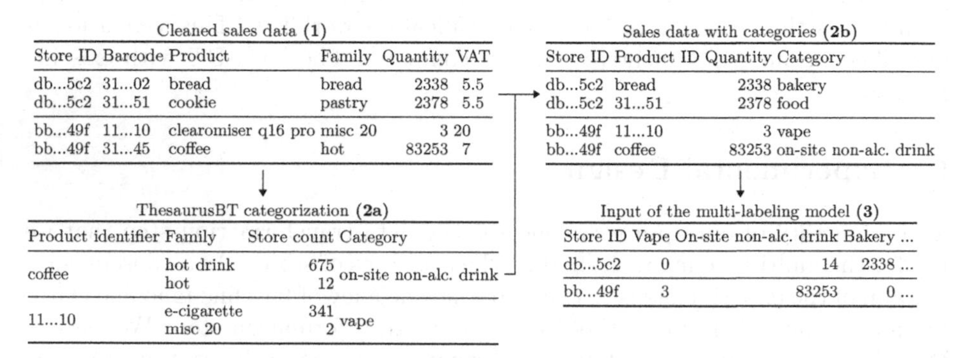

Cleaned sales data **(1)**

Store ID	Barcode	Product	Family	Quantity	VAT
db...5c2	31...02	bread	bread	2338	5.5
db...5c2	31...51	cookie	pastry	2378	5.5
bb...49f	11...10	clearomiser q16 pro	misc 20	3	20
bb...49f	31...45	coffee	hot	83253	7

Sales data with categories **(2b)**

Store ID	Product ID	Quantity	Category
db...5c2	bread	2338	bakery
db...5c2	31...51	2378	food
bb...49f	11...10	3	vape
bb...49f	coffee	83253	on-site non-alc. drink

ThesaurusBT categorization **(2a)**

Product identifier	Family	Store count	Category
coffee	hot drink	675	on-site non-alc. drink
	hot	12	
11...10	e-cigarette	341	vape
	misc 20	2	

Input of the multi-labeling model **(3)**

Store ID	Vape	On-site non-alc. drink	Bakery ...
db...5c2	0	14	2338 ...
bb...49f	3	83253	0 ...

Fig. 2. Transformation steps of the business workflow

in Fig. 1 and further described in this section. Additional workflows are also considered and evaluated, mixing the processing steps from the three aforementioned workflows, excluding some non-relevant combinations. Hyperparameters for the CatBoost feature selection, Random Forest and SVM are optimized for each workflow using grid search with cross-validation.

Basic Approach. The basic approach is a three-step workflow using baseline methods: 1) data cleaning (conversion to lower case, removal of extra spaces, accents, special characters and stop words, lemmatisation), 2) pivoting with the store identifier as the key, product family labels as columns, and the sum of product sold as values, and 3) multi-labeling with a Binary Relevance model with SVM as its base classifier.

Literature Approach. The literature approach consists of five steps. It uses the best methods reported in the literature: 1) data cleaning (same as for the basic approach), 2) MinHash encoding on the product family labels, 3) pivoting with the store identifier as the key and product family signatures generated by MinHash as columns, 4) feature selection using CatBoost, and 5) multi-labeling with a Binary Relevance model with Random Forest as its base classifier.

Business-Specific Approach. The business-specific approach consists of four steps: 1) data cleaning (same as for the basic approach), 2) application of

Table 3. Multi-labeling classification experimental results

Data transformation	Classifier	Classifier input dimensions	Accuracy	Macro F1	Execution time (min)
Pivot	RF	6574	0.64	0.482	0.41
	SVM	6574	0.77	0.611	0.23
Pivot–CatBoost	RF	56	0.79	0.609	5.4
	SVM	56	0.77	0.643	5.23
MinHash–Pivot	RF	5104	0.65	0.483	52.18
	SVM	5104	0.77	0.613	52.16
MinHash–Pivot–CatBoost	RF	50	0.79	0.619	57.18
	SVM	50	0.80	0.549	57.02
ThesaurusBT–Pivot	RF	61	0.85	0.663	9.16
	SVM	61	0.80	**0.732**	9.01
ThesaurusBT–Pivot–CatBoost	RF	34	**0.87**	0.667	14.33
	SVM	34	0.80	0.719	14.16

the ThesaurusBT method, resulting in a mapping from products to categories, 3) pivoting with the store identifier as the key and categories (identified by ThesaurusBT) as columns, and 4) multi-labeling with a Binary Relevance model with Random Forest as its base classifier.

The sequence of transformations for the business-specific approach is shown in Fig. 2. ThesaurusBT creates a new product identifier (subfigure Fig. 2-2a) using either its barcode or its label, depending on the rules defined to detect generated local product identifiers or global IDs. For each product, family labels are listed along with the corresponding number of store. Each product is then processed by ThesauruBT to search for the most common keyword within family labels, weighted by store count. When a match occurs, the product is tagged with the corresponding category, or *unknown* if there is no match. ThesaurusBT product categories are merged with the sales dataset (subfigure Fig. 2-2b) before pivoting (subfigure Fig. 2-3). This pivot table is built with the store identifier as the key, categories identified by ThesaurusBT as columns, and the sum of quantities sold as values. This result is then used as input for the multi-labeling model.

6 Results and Discussion

Table 3 reports total execution times for each workflow. These workflows, implemented in Python, were executed on an i7-11800H 2.30 GHz CPU and 16 GB of RAM. Execution times include both training and scoring. The execution time of the literature workflow is dominated by the hash calculation step of the MinHash method (52 min). Other steps with execution times higher than a minute are ThesaurusBT (9 min) and CatBoost (5 min). Data preparation based on ThesaurusBT is therefore significantly slower than a simple pivoting method (3 s).

The accuracy and macro F1 score of the multi-labeling classification task are presented in Table 3. Without CatBoost, the MinHash method does not significantly improve classification performance but reduces the complexity of the dataset from 6574 to 5104 variables. It does not succeed in improving the performances by deduplicating family labels with multiple morphological values (typos, misspellings, etc.) and increases the execution time significantly. This insignificant impact on performance can be explained by the characteristics of family labels, which include synonyms and variable strategies of arrangement of products (by type, by brands, etc.) with variable granularity. These cannot be captured by MinHash. Moreover, increasing the MinHash sensibility leads to the grouping of unrelated families.

CatBoost can significantly improve the performance of workflows with RF-based classifiers when it is used in conjunction with Pivot (+23% accuracy and +27% macro F1) or MinHash (+22% accuracy and +29% macro F1). These are positive results considering that this method also reduces the complexity of the dataset from 6574 (resp. 5104) to 56 (resp. 50) variables. The benefits of CatBoost are lower when used with SVM as the base classifier. When CatBoost is used in conjunction with ThesaurusBT, its feature selection has close to no impact on classification performance (between −1% and +2%).

A conclusion drawn from these results is that ThesaurusBT outperforms all other transformations. The improvement is especially visible with the macro F1 score. Based on these performances, we can assume that this method has successfully labeled a significant part of the products, partially solving the data transformation problem. If we focus on the macro F1 score, the results are heterogeneous due to the unbalanced nature of the dataset (for instance, there are many tobacco shops, but not many hotels or restaurants). In fact, the models are overfitted to recognize tobacco shops or newspapers, but underfitted to recognize hotels or restaurants. This directly affects the macro F1 score for all models that were not able to detect rare cases. The best performing models in that regard are those that include ThesaurusBT in their workflow. More generally, a significant improvement is achieved by incorporating ThesaurusBT, as shown in Table 3, since the business-specific solution is always the most efficient.

7 Conclusion

In this paper, we have considered the problem of assigning activity tags to stores according to their sales. Even if we have considered a specific use case provided by the Bimedia company, this is a general problem that affects software providers of different industries. Compared to the use cases found in the literature, the one proposed in this paper raises the challenges of dirty data as stores can use any label to name their products and families of products. As a baseline, we have proposed and implemented two approaches: one with baseline methods and one with the most efficient methods known in the literature. All the datasets and implementations are available online for reproducible purpose. These approaches are compared with the one we proposed named ThesaurusBT, a thesaurus-based transformation. This transformation is based on some business knowledge that a company such as Bimedia can provide and maintain over time. As we have shown in our experiments, this approach outperforms the baseline approaches for the task of labeling store activities using dirty data. As a drawback, the implementation of this solution incurs a significant business-specific development cost. By making this dataset available, Bimedia wishes to raise interest for this kind of problem, as having an efficient machine learning solution, with limited human involvement, usable by a medium-scale company would bring a significant improvement to this field.

As a future work, we plan to consider deep learning approaches. One difficulty is to have enough training data. Thus, automating the production of such data is a perspective of our work. Another challenge is to extend the business knowledge used by our approach. Currently, we only use lexical resources but we are convinced that more complex models such as ontologies could be useful to improve our method.

References

1. Abdi, H., Williams, L.J.: Principal component analysis. Wiley Interdisciplinary Rev. Comput. Stat., **2**(4), 433–459 (2010)
2. Borisov, V., Leemann, T., Seßler, K., Haug, J., Pawelczyk, M., Kasneci, G.: Deep neural networks and tabular data: a survey. IEEE Trans. Neural Networks Learn. Syst., 1–21 (2022)
3. Broder, A.Z.: On the resemblance and containment of documents. In: Proceedings. Compression and Complexity of SEQUENCES 1997, pp. 21–29. IEEE (1997)
4. Canny, J.: Gap: a factor model for discrete data. In: Proceedings of the 27th Annual International ACM SIGIR Conference on Research and Development in Information Retrieval, pp. 122–129 (2004)
5. Cerda, P., Varoquaux, G.: Encoding high-cardinality string categorical variables. IEEE Trans. Knowl. Data Eng. (2020)
6. Cerda, P., Varoquaux, G., Kégl, B.: Similarity encoding for learning with dirty categorical variables. Mach. Learn., 1477–1494 (2018). https://doi.org/10.1007/s10994-018-5724-2
7. Chen, T., Guestrin, C.: Xgboost: a scalable tree boosting system. In: Proceedings of the 22nd ACM SIGKDD International Conference on Knowledge Discovery and Data Mining, ppp. 785–794 (2016)
8. Crow, J.F., Kimura, M.: Efficiency of truncation selection. Proc. Natl. Acad. Sci. **76**(1), 396–399 (1979)
9. Dwarampudi, M., Reddy, N.V.: Effects of padding on lstms and cnns. arXiv preprint arXiv:1903.07288 (2019)
10. Prokhorenkova, L., Gusev, G., Vorobev, A., Dorogush, A.V., Gulin, A.: Catboost: unbiased boosting with categorical features. Advances in neural information processing systems, 31 (2018)
11. Tenney, I., Das, D., Pavlick, E.: BERT rediscovers the classical NLP pipeline. In: Proceedings of the 57th Annual Meeting of the Association for Computational Linguistics, pp. 4593–4601, July 2019
12. Turney, P.D.: A uniform approach to analogies, synonyms, antonyms, and associations. In: 22nd International Conference on Computational Linguistics (COLING-08) (2008)
13. Sun, W., Manber, U.: Fast text searching: allowing errors. Commun. ACM **35**(10), 83–91 (1992)

[illegible] of labeling [illegible] data. As a drawback, the implementation of this [illegible] the development [illegible] this [illegible] wishes to raise interest for this kind of [illegible] machine learning solution, with limited [illegible] would bring a significant improvement [illegible]

[illegible] deep learning approaches. One difficulty [illegible] the collection of such data [illegible] only [illegible] could [illegible]

References

1. [illegible] analysis. Wiley Interdisciplinary [illegible] (2010)
2. [illegible] Networks [illegible]
3. [illegible] Proceedings [illegible] Computing [illegible] pp. 21–29. IEEE (1987)
4. [illegible] of the 27th Annual [illegible] In Information [illegible]
5. [illegible]
6. [illegible] doi.org/10.1007/ [illegible]
7. [illegible] system, in: Proceedings [illegible] Knowledge Discovery and [illegible]
8. [illegible] Proc. Natl. Acad. Sci. [illegible]
9. [illegible]
10. [illegible] information [illegible]
11. [illegible] the classical NLP pipeline. In: [illegible] Association for Computational Linguistics [illegible]
12. [illegible] Computational Linguistics (COLING [illegible]
13. [illegible] Commun. ACM [illegible]

Metadata Management

Mobile Feature-Oriented Knowledge Base Generation Using Knowledge Graphs

Quim Motger[1(✉)], Xavier Franch[1], and Jordi Marco[2]

[1] Department of Service and Information System Engineering, Universitat Politècnica de Catalunya, Barcelona, Spain
{joaquim.motger,xavier.franch}@upc.edu

[2] Department of Computer Science, Universitat Politècnica de Catalunya, Barcelona, Spain
jordi.marco@upc.edu

Abstract. Knowledge bases are centralized repositories used for developing knowledge-oriented information systems. They are essential for adaptive, specialized knowledge in dialogue systems, supporting up-to-date domain-specific discussions with users. However, designing large-scale knowledge bases presents multiple challenges in data collection and knowledge exploitation. Knowledge graphs provide various research opportunities by integrating decentralized data and generating advanced knowledge. Our contribution presents a knowledge base in the form of a knowledge graph for extended knowledge generation for mobile apps and features extracted from related natural language documents. Our work encompasses the knowledge graph completion and deductive and inductive knowledge requests. We evaluated the effectiveness and performance of these knowledge strategies, which can be used as on-demand knowledge requests used by third-party software systems.

Keywords: knowledge base · knowledge graph · mobile app

1 Introduction

Knowledge-based chatbots are dialogue systems embedding real-time, centralized access to domain-specific information systems [16]. These knowledge bases (KB) assist in resolving user intent and entity recognition tasks for a particular knowledge domain. With the emergence of disruptive large language models (LLMs) like GPT-4 [18], these KBs offer great potential in terms of accuracy, scalability and performance efficiency for highly adaptive knowledge modelling [29]. While LLMs excel in various tasks they are pre-trained or fine-tuned for, re-training for dynamic knowledge adaptation requires significant time, energy consumption and economic expenses [19].

In this context, one of the main challenges is extending these KBs with advanced knowledge generation techniques. To this end, the use of knowledge graphs (KG) as the underlying infrastructure of a KB is becoming a research trend [8]. The combined use of KGs with machine/deep learning techniques

A. Abelló et al. (Eds.): ADBIS 2023, CCIS 1850, pp. 269–279, 2023.
https://doi.org/10.1007/978-3-031-42941-5_24

reveals the potential of leveraging graph-structured data towards effective and scalable extended knowledge generation [24,25]. Nevertheless, constructing and exploiting a KG for a given domain is still a challenging task in terms of KG completion and the design of derived knowledge strategies [8]. Mobile applications and app stores exemplify highly-adaptive knowledge domains [20]. Google Play, the leading global app store [3], releases an average of 75K new apps monthly [4]. Among their top 1000 downloaded apps, 62% are updated at least once a month [1]. Given this context, feature-based knowledge generation, which leverages the knowledge exposed by documented app functionalities, emerges as a relevant research area, benefitting from the abundance of large natural language corpora (e.g., app descriptions, user reviews) in these repositories [10,13].

In this paper, we present the design and development of a KB in the form of a KG supporting feature-oriented extended knowledge generation in the field of mobile app catalogues. Our main contributions are: (1) a distributable knowledge graph in the field of mobile app repositories to support document-based and feature-oriented data management tasks; (2) a state-of-the-art method for feature-oriented extended knowledge generation in the context of mobile apps; and (3) an effective, scalable approach for designing and developing a semantic web-based KB in the context of LLM-based chatbots for real-time consumption.

2 Background

Based on the notation from Zhao et al. [31], we denote a KG instance as $G = (V, E)$, where $V = \{v_1, v_2, ..., v_n\}$ denotes the nodes of the graph and $E = \{e_1, e_2, ..., e_m\}$ denotes the edges in a labelled directed graph [8]. Each e is defined as (v_i, p, v_j), where v_i is the source node, v_j is the target node and p is the relation type from v_i to v_j. A particular type of labelled directed graphs are *Resource Description Framework* [22] or RDF graphs. They are built as semantic web networks based on subject-predicate-object (v_i,p,v_j) triples. The specification of V and E entities is built through a data-modelling vocabulary known as the RDF Schema, for which there are public repositories defining shared schemas for modelling structured data on the Internet. Knowledge representation and extraction in dynamic, large-scale KGs involve deductive and inductive techniques [8]. Deductive techniques entail precise, logic-based transformations $G \rightarrow G'$ based on particular data observations, while inductive techniques involve pattern generalizations within a given G which, although potentially imprecise, can be used to generate novel and complex predictions [8].

A particular instance G is a context-specific representation of real world entities and relations. This research focuses on the context of mobile software ecosystems [7], which involves *actors* (e.g., users, developers, platform vendors) and *entities* (e.g., apps, app stores, mobile OS platforms). From the user perspective, these ecosystems provide access to a *catalogue* or a set of *mobile apps*, which users access to build their own application *portfolio* [17]. A potential descriptor for mobile apps is the set of *features* they expose, representing functionalities from the user's perspective , implemented by one or multiple mobile apps in their portfolio [17]. Extended knowledge about app features plays a crucial role in various tasks, such as optimization ranking algorithms [26] and version modelling for app recommendation [15]. Research in this area primarily focuses on using available *documents* through natural language processing (NLP) techniques for tasks like feature extraction and topic modelling [13,15,26].

3 Research Method

3.1 Project Vision

Using the *Goal Question Metric* (GQM) template, we defined the following goal:

Analyse extended feature-based knowledge generation supported by KGs
for the purpose of adaptive KB generation for LLM-based dialogue systems
with respect to the correctness and efficiency of extended knowledge requests
from the point of view of dialogue system developers
in the context of mobile software repositories.

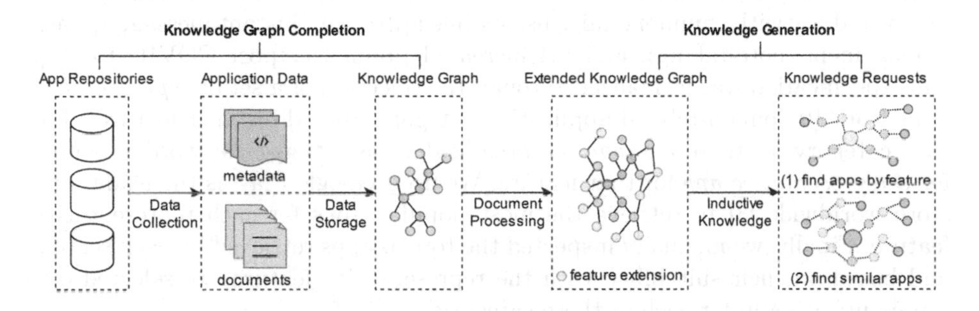

Fig. 1. Knowledge base: project vision

We define two types of feature-based knowledge generation requests:

1. Given a feature f, which mobile apps from the catalogue expose a set of features equivalent or similar to f?
2. Given a mobile app m from the catalogue, which mobile apps are more similar to m based on the features they expose?

Figure 1 summarizes the KB generation process. KG completion involves decentralized data collection from web-based app repositories, followed by modeling and storage in the KG. Knowledge generation extends the KG using deductive techniques based on NLP for feature extraction from available documents. The extended KG exposes knowledge requests for third-party software systems. Complementary materials and resources are available in the replication package[1]

[1] Available at https://doi.org/10.5281/zenodo.8038301.

3.2 Research Plan

To guide the design of this KB, we define the research question (RQ) as follows:

> **RQ.** How does KB development in mobile app repositories benefit from KGs and extended knowledge requests? Specifically, how does this approach contribute to correctness, efficiency, and adaptability?

To respond to this RQ, we conducted an evaluative sample study [27], structured through the following stages: (1) **sample study definition**, including scope, specifications and limitations of the sample data set; (2) **KG completion**, including data source selection and data collection techniques; (3) **knowledge generation**, including feature-oriented, NLP-based inductive and deductive knowledge strategies; and (4) **evaluation** of the aforementioned strategies.

3.3 Sample Study: Multicategory Android Mobile Apps

We limited the research scope to publicly accessible Android mobile apps in a repository, focusing on **trailing and sport activity** related apps due to: (1) feature heterogeneity (e.g., geolocation, biomonitoring, social networking); (2) integrated use with common, daily-use mobile apps (e.g., instant messaging, task management, note-taking); and (3) increased popularity post COVID-19 [14]. This specification was formalized through the selection of a set of *representative apps* from 10 domain-related application categories based on their features. For each category, we used one *representative feature* as a trigger keyword to search for a *representative app* for that feature. We used Google Play as the leader app store worldwide [3] to retrieve the most popular apps for each representative feature. Finally, we manually inspected the top 10 apps retrieved by each search, and based on their suitability with the representative feature, we selected one representative app for each of these categories.

4 Knowledge Graph Completion

Data collection → To support the KG completion and data source identification, we conducted a literature review focused on grey literature [11], including technical reports and specialized, mobile-oriented publishers. As a result, representatives of three types of mobile app data sources were covered, including: **app store programs** (i.e., Google Play); **sideloading repositories** (i.e., F-Droid[2]); and **app search engines** (i.e., AlternativeTo[3]). For each type, we have developed two data collection operations. The **query** operation uses a set of domain-related keywords as input to search for related apps based on keyword search algorithms from data sources. The resulted set is used as the input for the **scan** operation, which collects all available data items (i.e., *metadata* and *documents*)

[2] https://f-droid.org/es/.
[3] https://alternativeto.net/.

for each app. We considered 4 document types categorized into 2 categories: *(i)* *Proprietary Documents* (i.e., *summaries*, *descriptions* and *changelogs*), defined by developers and platform vendors; *(ii)* *User Documents* (i.e., *reviews*), defined by users. We used web scraping, an essential tool for web-based knowledge design [12], to collect data from static and dynamic web sources like F-Droid and AlternativeTo, where alternative access methods are unavailable. Whenever possible, we utilized API consumption mechanisms, including non-official REST APIs that wrap the web scraping process, enabling direct HTTP-based access.

Data normalization → We followed the semantic schema definition process for knowledge graph design as described in [8]. We decided to use an RDF semantic schema [2], being a directed edge-labelled graph schema focused on the standardization and interoperability of web-based bodies of knowledges [8], which are some of the main purposes of the resulted knowledge base. For the data schema definition, we used existing schemas from Schema.org[4] as the most popular public schema repository for RDF graphs.

Data and schema integration → For schema integration [28], we combined key matching techniques and manual inspection of app repository schemas, ensuring conformance with Schema.org's RDF data schemas. For data integration, entity recognition relied on key matching using app *name* and *package* properties. Strategies for data inconsistencies prioritized completeness and correctness, including app repository prioritization based on data quality and document merging for increased natural language content availability.

5 Knowledge Generation

5.1 KG Definitions

We define a KG as $G = (V, E) = (M \cup D \cup F, E_{md} \cup E_{mf})$. Nodes include mobile apps (M), documents (D), and user annotated features (F), where $D = PD \cup UD$, including proprietary documents (PD) and user documents (UD). Edges are represented by E_{md}, linking apps to their related documents, and E_{mf}, linking apps to user annotated features. The extended KG, $G' = (V \cup F', E \cup E_{df} \cup E_{ff})$, includes the set of extracted features F' from documents, the links E_{df} between documents and their respective extracted features, and explicit semantic similarity between features (f_i, f_j) represented by E_{ff}. Deductive techniques process D using NLP to (1) extend the set of features (F') and (2) link mobile apps to their features through transitivity based on the documents these features were extracted from (E_{df}). These transformations are **batch processes**, as they are expected to be executed only once for a given G. Inductive techniques use the extended G' to resolve the knowledge requests depicted in Sect. 3.1. Hence, these are **on-demand processes** given that, for a given G', they are expected to be executed once for each user request. Figure 2 illustrates a partial representation of G' focused on the Strava app and a subset of its annotated features, extracted features and related apps linked to Strava through direct or indirect connections via each feature f_i.

[4] https://schema.org/.

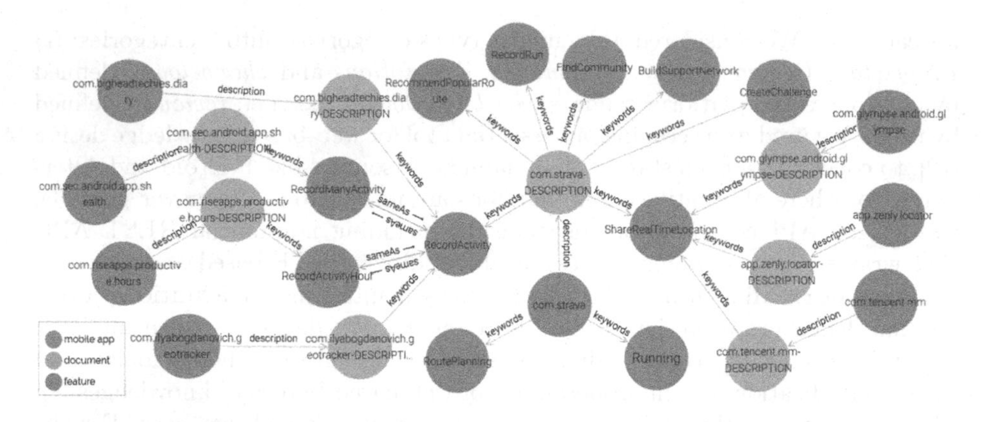

Fig. 2. Partial representation of G' focused on Strava app

5.2 Deductive Knowledge: Extracting Features

ExtractFeatures $\rightarrow$ Each document $d_i \in D$ is processed through a NLP-based feature extraction process in the form of a pre-trained RoBERTa-based transformer pipeline built for this purpose [6]. This pipeline utilizes linguistic annotations and PoS-based expression patterns to identify natural language features in the documents. As a result, for each $d_i \in D$ we extend G through $V \leftarrow V \cup F'(d_1) \cup F'(d_2) ... \cup F'(d_n)$ and $E \leftarrow E \cup E_{df}(d_1) \cup E_{df}(d_2) ... \cup E_{df}(d_n)$, where $F'(d_i) \subset F'$ includes all features extracted from d_i (e.g., *RecordActivity*) and $E_{df}(d_i) \subset E_{df}$ includes all links between those features and d_i.

LinkSimilarFeatures $\rightarrow$ Each feature $f_i \in F \cup F'$ is processed through a semantic similarity pipeline to compute a similarity score $sim(f_i, f_j)$ for all $(f_i, f_j) \in F \cup F'$ using a TF-IDF-based search index built upon the database management system for scalability and applicability in large repositories [21]. For all $sim(f_i, f_j) >= thr$, where thr is a domain-dependent similarity score threshold , we extend G through $E \leftarrow E \cup E_{ff}(f_1) \cup E_{ff}(f_2) ... \cup E_{ff}(f_n)$. $E_{ff}(f_i)$ models the link between all (f_i, f_j) for which $sim(f_i, f_j) >= thr$ (e.g., *RecordActivity* is semantically linked to *RecordManyActivities* and *RecordActivityHours*).

5.3 Inductive Knowledge: Finding Apps

FindAppsByFeature $\rightarrow$ Two feature-to-app similarity methods were designed and compared. The **index-based** method uses a term-to-document similarity index for each PD type (i.e., *summary, description, changelog*). For a given f_i (e.g., *ShareRealTimeLocation)*, the index returns a ranked list of $m_j \in M$ (e.g., *app.zenly.locator*) sorted by $sim(f_i, m_j)$ in descending order. This approach is parameterized to allow similarity evaluations filtered by a given document type (e.g., *descriptions*), or to compute an aggregated similarity score based on all available $E_{md}(m_j)$ from $d_k \in PD$. The **graph-based** method relies on a Python adaptation [23] of the SimRank* algorithm [30], designed for scalable similarity search, which has proven to be an effective approach for its adoption in large-scale graphs. For a given f_i, the graph-based method computes a similarity score based on its shared paths with respect to $E_{df}(d_i) \cup E_{ff}(f_j)$ for all $d_i \in D$. This method is parameterized to allow similarity based on $d_i \in PD$ (i.e., without reviews), or based on $d_j \in D$ (i.e., with reviews).

FindSimilarApps → Given a mobile app $m \in M$ (e.g., *com.strava*), we have applied the same algorithm implementation approach as in *FindAppsByFeature*. The **index-based** method relies on a document-to-document index for each PD type. For a given m_i, the index returns a ranked list of m_j sorted by $sim(m_i, m_j)$ in descending order. The index-based approach is parameterized for a given document type or for aggregated results as in *FindAppsByFeature*. On the other hand, the **graph-based** (adapted from SimRank* algorithm) computes for a given m_i a similarity score based on its shared paths with respect to $E_{md}(m_j) \cup E_{df}(d_k) \cup E_{ff}(f_l)$ for all $m_j \in M$, parameterized by document type.

6 Evaluation

We focus the evaluation on the *functional correctness* and *performance efficiency* [9] of the inductive knowledge requests. The evaluation plan is defined as follows:

1. Run *ExtractNLFeatures* for all $d_i \in PD$.
2. Create/update the graph database indexes with the extended G′.
3. Run *LinkFeatures* using all thresholds (thr) within [0.1, 0.2, ..., 0.9].
4. For each thr, run *FindAppsByFeature* using the *representative features*.
5. For each thr, run *FindSimilarApps* using the *representative apps*.
6. Compute the precision[5] rate@k ($k <= 20$) for both inductive algorithms.
7. Run *ExtractNLFeatures* for all $d_i \in UD$ and repeat 2→6.

Functional correctness → Fig. 3 reports precision rate@k for both inductive algorithms obtained with the optimal value for the similarity threshold score in Step 3 (i.e., $thr = 0.7$). At each k value, the plots show how many recommended apps found by each algorithm matched the category of the input item, which we considered as ground-truth for evaluation (i.e., *representative feature* for *FindAppsByFeature*, and *representative app* for *FindSimilarApps*). Both figures report the aggregated results for all categories. For *FindAppsByFeature*, index-based algorithms show similar precision using description, summary, or aggregated document types, while the graph-based version performs poorly. However, for *FindSimilarApps*, the graph-based version (without reviews) is the best algorithm. Including reviews in the feature extraction algorithm worsens the quality of the prediction. For the best algorithm version of each inductive technique (i.e., index-based with description for *FindAppsByFeature*, graph-based without reviews for *FindSimilarApps*), both algorithms report a precision rate $>= 80\%$ at the top positions of the returned ranked list (i.e., $k <= 3$). At $k = 5$

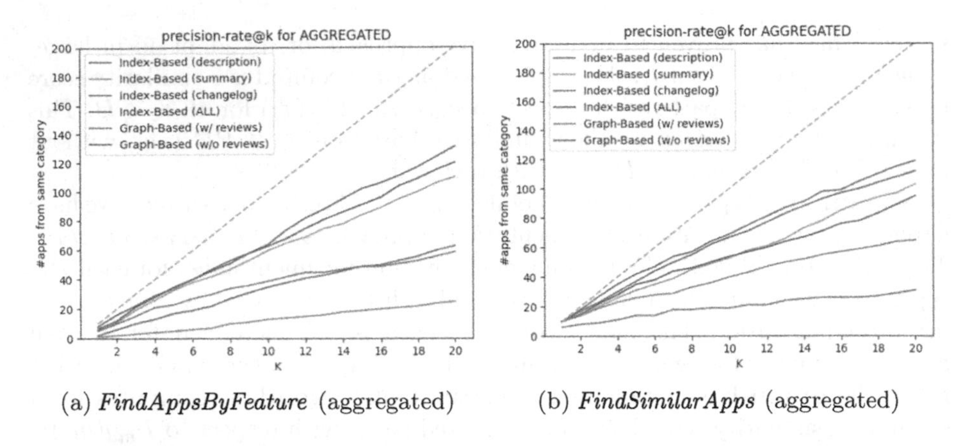

(a) ***FindAppsByFeature*** (aggregated) (b) ***FindSimilarApps*** (aggregated)

Fig. 3. Precision rate@k (k = 20) for inductive knowledge algorithms

FindSimilarApps still reports a precision rate of 80%, while *FindAppsByFeature* reports 72% of matched suggestions.

Performance efficiency → Table 1 reports the average execution time for both deductive (i.e., *ExtractNLFeatures*, *LinkFeatures*) and inductive (i.e., *FindAppsByFeature*, *FindSimilarApps*) algorithms. Batch processes (i.e., deductive algorithms) for the whole sample data set are particularly time-consuming, especially when processing user reviews and their extracted features (up to 44 h including *ExtractNLFeatures* and *LinkFeatures*). However, updates on the KG instance (i.e., adding a new app) only require a few seconds for *ExtractNLFeatures* and less than 1 s for *LinkFeatures* on average. On the other hand, on-demand processes (i.e., inductive algorithms) take on average less than 1 s, with the exception of *FindSimilarApps* when processing all reviews.

Table 1. Average execution time for inductive and deductive algorithms

T(s)		w/o reviews		w/ reviews	
		/item	total	/item	total
ExtractNLFeatures		2.48 s	1570 s s (˜0.5 h)	105.9 s	67,690 s (˜19 h)
LinkFeatures		0.29 s	3362 s s (˜1 h)	0.68 s	88,753 s (˜25 h)
FindAppsByFeature	Index-Based	0,11 s	1.06 s	-	-
	Graph-Based	0.39 s	3.91 s	0.69 s	6.86 s
FindSimilarApps	Index-Based	0.09 s	0.87 s	-	-
	Graph-Based	0.41 s	4.08 s	1.21 s	12.24 s

[5] We focus on precision to maximize recommendation quality at the top positions.

7 Discussion

Evaluation results show that the designed knowledge requests are effective for on-demand knowledge retrieval, with high prediction precision at the top positions. In the context of integrating the designed KB to support LLM-based dialogue systems, prediction quality for top k values (i.e., $k >= 5$) is especially relevant for reporting simplified snapshots from each knowledge request response [5]. The graph-based similarity measure performs well for comparing nodes with shared semantics (i.e., app-app in *FindSimilarApps*). While the term-to-document similarity measure is better for comparing different node types (i.e., app-feature in *FindAppsByFeature*). Finally, the use of reviews did not improve the quality of the results, while it also added a great overhead in batch process execution.

Regarding validity threats, we primarily focus on construct and internal validity. Construct validity concerns the KG construction, where potential biases may arise from the selection of representative features and app repositories. Nevertheless, we argue that the criteria for selecting these representatives and sources (i.e., popularity, availability of data, quality of data) minimizes the risk of biased coverage and low completeness. Concerning knowledge generation, the extended knowledge process relies on the quality of the feature extraction process, whose design and development was recently published as a NLP-based feature extraction tool [6]. Finally, we considered as ground-truth for inductive techniques the application categories based on the original app query data collection method, relying on the accuracy of the app repositories search algorithms, for which we argue that the criteria for selecting those repositories (i.e., popularity and relevance) reduces the impact of this potential threat.

8 Conclusions

This research aims at providing a holistic perspective on the end-to-end process of building a KB of mobile apps in the form of a KG. The technical infrastructure (i.e., software components), the resulting resources (i.e., data artefacts) and the findings from this research are also intended to lay the groundwork for the integration of additional knowledge requests, either beyond the object of analysis (i.e., NL documents), the descriptor to which similarity is based on (i.e., features), or the knowledge requests formalization (i.e., feature vs. app and app vs. app similarity). Moreover, evaluation results prove that on-demand inductive knowledge can be effectively and efficiently integrated for real-time consumption use cases. We envisage focusing future research efforts on (1) formalization of the concept of feature, and (2) conducting a user-study evaluation.

Acknowledgment. With the support from the Secretariat for Universities and Research of the Ministry of Business and Knowledge of the Government of Catalonia and the European Social Fund. This paper has been funded by the Spanish Ministerio de Ciencia e Innovación under project/funding scheme PID2020-117191RB-I00/AEI/10.13039/501100011033.

References

1. 42matters AG: Google Play Store App Update Frequency Statistics (2023). https://42matters.com/google-play-aso-with-app-update-frequency-statistics
2. Brickley, D., Guha, R.V.: (2014). https://www.w3.org/TR/2014/REC-rdf-schema-20140225/
3. Ceci, L.: Biggest app stores in the world 2022, August 2022. https://www.statista.com/statistics/276623/. Accessed 22 Nov 2022
4. Ceci, L.: Number of monthly android app releases worldwide 2023, March 2023. https://www.statista.com/statistics/1020956/android-app-releases-worldwide/
5. Chen, N., et al.: Mobile app tagging. In: Proceedings of the 9th WSDM (2016)
6. Gallego, A., et al.: TransFeatEx: a NLP pipeline for feature extraction. In: REFSQ 2023, CEUR Workshop Proceedings (2023)
7. Grua, E.M., et al.: Self-adaptation in mobile apps: a systematic literature study. In: 2019 IEEE/ACM 14th SEAMS (2019)
8. Hogan, A., et al.: Knowledge graphs. ACM Comput. Surv. (2021)
9. ISO/IEC: System and software quality models ISO/IEC 25010 (2011)
10. Johann, T., et al.: SAFE: a simple approach for feature extraction from app descriptions and app reviews. In: 25th International RE Conference (2017)
11. Kamei, F., et al.: Grey literature in software engineering: A critical review. Information and Software Technology (2021)
12. Khvatova, T., Dushina, S.: Scientific online communication: the strategic landscape of researchgate users. IJTHI (2021)
13. Kumari, S., Memon, Z.A.: Extracting feature requests from online reviews of travel industry. Acta Scientiarum - Technology 44 (2022)
14. Kwon, J.Y., et al.: Analysis of strategies to increase user retention of fitness mobile apps during and after the covid-19 pandemic. IJERPH (2022)
15. Lin, J., et al.: New and improved: Modeling versions to improve app recommendation. In: Proceedings of the 37th International ACM SIGIR (2014)
16. Motger, Q., Franch, X., Marco, J.: Software-Based Dialogue Systems: Survey. Taxonomy and Challenges. ACM Comput. Surv. (2022)
17. Motger, Q., et al.: Integrating adaptive mechanisms into mobile applications exploiting user feedback. In: Research Challenges in Information Science (2021)
18. OpenAI: Gpt-4 technical report (2023)
19. Patterson, D., et al.: Carbon emissions and large neural network training (2021)
20. Petrik, D., Schönhofen, F., Herzwurm, G.: Understanding the design of app stores in the iiot. In: IEEE/ACM IWSiB, pp. 43–50 (2022)
21. Raatikainen, M., et al.: Improved management of issue dependencies in issue trackers of large collaborative projects. IEEE TSE (2022)
22. RDF Working Group: Resource Description Framework (RDF). https://www.w3.org/RDF/. Accessed 22 Nov 2022
23. Reyhani, M., et al.: Effectiveness and efficiency of embedding methods in task of similarity computation of nodes in graphs. Applied Sciences (2021)
24. Rožanec, J.M., et al.: XAI-KG: Knowledge Graph to Support XAI and Decision-Making in Manufacturing. Lecture Notes in Business Information Processing (2021)
25. Schlichtkrull, M., Kipf, T.N., Bloem, P., van den Berg, R., Titov, I., Welling, M.: Modeling relational data with graph convolutional networks. In: Gangemi, A., Navigli, R., Vidal, M.-E., Hitzler, P., Troncy, R., Hollink, L., Tordai, A., Alam, M. (eds.) ESWC 2018. LNCS, vol. 10843, pp. 593–607. Springer, Cham (2018). https://doi.org/10.1007/978-3-319-93417-4_38

26. Shen, S., et al.: Towards release strategy optimization for apps in google play. In: Proceedings of the 9th Asia-Pacific Symposium on Internetware (2017)
27. Stol, K.J., Fitzgerald, B.: The abc of software engineering research. ACM Trans. Softw. Eng, Methodol. (2018)
28. Wang, L.: Heterogeneous Data and Big Data Analytics. Automatic Control Inf. Sci. **3**(1), 8–15 (2017)
29. Xu, P., et al.: MEGATRON-CNTRL: controllable story generation with external knowledge using large-scale language models. In: EMNLP 2020 (2020)
30. Yu, W., et al.: Simrank*: effective and scalable pairwise similarity search based on graph topology. VLDB J. (2019)
31. Zhao, Y., et al.: A supervised learning community detection method based on attachment graph model. In: Franch, X., Poels, G., Gailly, F., Snoeck, M. (eds.) Advanced Information Systems Engineering. CAiSE 2022. LNCS, vol 13295. Springer, Cham (2022). https://doi.org/10.1007/978-3-031-07472-1_22

NoSQL Schema Extraction from Temporal Conceptual Model: A Case for Cassandra

Maryam Mozaffari(✉), Anton Dignös, Hind Hamrouni, and Johann Gamper

Free University of Bozen-Bolzano, Bozen-Bolzano, Italy
{Maryam.Mozaffari,Anton.Dignos,Hind.Hamrouni,
Johann.Gamper}@unibz.it

Abstract. NoSQL data stores have been proposed to handle the different breed of scale and challenges caused by Big Data. While a suitable schema design is of vital importance in NoSQL databases, in contrast to relational databases no standard schema design procedure exists yet. Instead, manual schema design is applied by using often vague and generic rules of thumb, which must be adapted to each application. Additionally, many applications require the management and processing of temporal data, for which NoSQL databases lack explicit support. To overcome such limitations, in this paper we propose an MDA-approach for mapping an existing conceptual UML class extension with temporal features into a NoSQL wide-column store schema. Then, we evaluate the schemas generated by our approach with the Cassandra wide-column store.

1 Introduction

NoSQL systems are designed for large scale applications and offer a high level of flexibility in the schema design. While relational systems provide known principles and standard procedures for translating a conceptual data model to a normalized logical data model, NoSQL systems do not provide a fully standard schema design procedure. On the other hand, they do not enforce a fixed and explicit schema upfront.

NoSQL databases can be categorized into four main groups [5]: key-value, document, graph, and wide-column stores. In this paper, we focus on schema design for wide-column stores, which store data in tables of records. Each record has a uniquely identifying key and may contain an arbitrary set of columns. While NoSQL schema design is already a challenging task for static data, the support for temporal data adds additional complexity since several time-dependent aspects need to be considered.

This paper presents a systematic schema design framework for temporal data, which has two desirable features: (1) it allows to systematically extract the NoSQL physical schema from a temporal conceptual model and (2) offers support for modeling and storing temporal data, which is ubiquitous but poorly supported in today's NoSQL systems. To this end, we propose a set of transformation rules to extract a physical NoSQL schema from a conceptual UML model extension enriched with temporal features.

Supported by the Autonomous Province of Bozen-Bolzano with research call "Research Südtirol/Alto Adige 2019" (project ISTeP).

A. Abelló et al. (Eds.): ADBIS 2023, CCIS 1850, pp. 280–290, 2023.
https://doi.org/10.1007/978-3-031-42941-5_25

We adopt OMG's Model Driven Architecture (MDA) [12], which provides a framework for mapping an input model into an output model. This mapping defines a set of transformation rules from the source to the target model, which are divided into three levels of abstraction: Computation Independent Model (CIM), Platform Independent Model (PIM), and Platform Specific Model (PSM). Since the input of our process is a temporal UML class diagram (PIM) and the output is a physical NoSQL model (PSM), we focus on PIM and PSM.

To summarize, the technical contributions of this paper are as follows: (1) a systematic MDA-based approach to extract a NoSQL wide-column store model from a temporal conceptual UML model, (2) Two mapping solutions, termed *referencing* and *denormalization* that guide model transformation, (3) Evaluating the schemas generated by our approach using the Cassandra wide-column store.

2 Related Work

Despite intensive research on NoSQL schema design for static data, research on temporal NoSQL schema design is still in an early stage and has received scant attention. On the other hand, traditional temporal models, techniques and algorithms are inappropriate to deal with novel characteristics of big data, chiefly due to scalability, performance, heterogeneity and volume issues. Hence, with a few exceptions there has been little research in this direction for NoSQL systems.

The work by Eshtay et al. [8] proposes solutions for embedding temporal properties in key-value and column stores. For key-value stores, the solution is to add attributes for valid start time, valid end time, and transaction time in each key-value pairs. In column stores, they propose adding time intervals to express valid time and timestamps for transaction time. While this work focuses on adding temporal aspects to key-value and column stores, it does not deal with schema extraction from a conceptual model.

To manage and process temporal data in wide-column stores, Hu et al. [11] propose to transform each original table into two table representations: one explicit history representation (EHR), which directly associates each data version with an explicit temporal interval, and one tuple timestamping representation (TTR) in which each tuple has an explicit time interval. Moreover, temporal operators for processing queries in wide-column stores are proposed. This work targets the HBase wide-column store and it does not address the problem of schema extraction from a conceptual models.

Several works, such as [2,9], investigate how to capture the evolving history of a temporal JSON schema and how to properly support co-existing schema versions, such that users can navigate and restore old versions if necessary. Furthermore, the authors of [7] propose a model transformation from a temporal object relational database (TORDB) into a document-oriented MongoDB database. Finally, some research contributions have been made on the management of temporal graph data. In [1] the authors present a new conceptual model for temporal graphs and define a set of translation rules to convert the temporal conceptual model into a logical property graph. Some works [4,6] propose a data model and query language for temporal graph databases.

To the best of our knowledge, so far no study focuses on mapping temporal conceptual UML models into NoSQL wide-column stores.

3 Model Transformation

We now present an MDA-based approach which applies a model transformation, where a PIM (temporal UML class diagram) is transformed through a set of mappings into a PSM (physical Cassandra model with temporal features).

3.1 Source Metamodel: Temporal UML Class Diagram

An UML class diagram (http://www.omg.org/spec/UML/2.5/) is defined as a set of classes linked with relationships. Each class consists of structural features (attributes) and behavioral features (operations). Since operations typically do not directly relate to database elements, we consider only structural features. There are four types of relationships between classes: association, dependency, generalization, and realization, while aggregation and composition are two specific kinds of associations. Dependency and realization do not directly relate to database elements. Linked classes in composition are highly dependent on each other, but independent of each other in aggregation. So we consider aggregation as a ordinary association relation (with the same mapping rules). Therefore, we focus on the relation types association ($assoc$), generalization ($gener$), and composition ($compo$).

We use the temporal conceptual UML model as proposed in [3], which uses a stereotype to indicate temporal features. Temporal entities, relations, and attributes are marked with the stereotype ≪Temporal≫$\{Dur, Freq\}$, where the tag *Dur* (*durability*) refers to the persistence of instances, which can be either *instantaneous* (i.e., associated with a time point) or *durable* (i.e., associated with a time interval). The tag *Freq* (*frequency*) of an entity or relationship refers to the cardinality of instances, which can be either *single* (i.e., instances are unique independently of time) or *intermittent* (i.e., instances are unique at any given time, but may not be unique at different times).

In temporal relations, the cardinalities specify the snapshot cardinality, i.e., the cardinality for each snapshot of the data. To express cardinalities over the lifespan of classes in a temporal relation, we use the *lifespan cardinality* [10]. We use the ≪LifespanCR≫$\{LC, RC\}$ stereotype in a temporal relation, where the two tags LC (left cardinality) and RC (right cardinality) allow to specify the lifetime multiplicity of the left and right class, respectively. The lifespan cardinality must be larger or equal than the snapshot cardinality, and the lifespan cardinality can only be "one-to-many" or "many-to-many", since a cardinality of "one-to-one" would not allow to evolve over time. Next, we formally define the temporal conceptual UML class diagram.

Definition 3.1 (Class). *A class c in a class diagram (CD) is an n-tuple $c = (N, A, PK, \ll Temporal \gg \{Dur, Freq\})$, where N is the name of the class, A is a set of attributes, PK is a special attribute of the class that uniquely identifies an object, and the optional ≪ Temporal ≫ specifies the temporal dimension of the class using Dur and Freq. Similarly, an attribute $a \in A$ is as an n-tuple $a = (N, T, \ll Temporal \gg \{Dur, Freq\})$, where N is the attribute name, T is the attribute type, and the optional ≪Temporal≫ is the temporal stereotype similar to classes.*

Definition 3.2 (Relationship). *A relationship r between two classes c_1 and c_2 can be represented as an n-tuple $r = (N, K, CR, \ll Temporal \gg \{Dur, Freq\},$*

≪LifespanCR≫$\{LC, RC\}$)*, where* N *is the name of the relationship,* $K \in \{assoc, compo, gener\}$ *is the type of the relationship,* $CR = \{(c_1, cr_1), (c_2, cr_2)\}$ *is the cardinality of the relationship between the classes. Note that for the kind generalization,* cr *is null since no cardinality for generalization exists. The optional ≪ Temporal ≫ stereotype specifies the temporal dimension of the relationship, and the optional ≪ LifespanCR ≫ specifies the participation constraint of the classes* c_1 *and* c_2 *in the relationship over their lifespan using the tags* LC *and* RC.

Example 3.1 (Source Data Model). Figure 1 shows an example of a temporal UML class diagram for a company database. *Department* is a non-temporal class, has an attribute *DID* of type Integer that is the PK and an attribute *DName* of type String. *Employee* is a temporal class that is time-varying using time intervals ($Dur = durable$). Over disjoint intervals the same *DID* can exist multiple times ($Freq = intermittent$), i.e., employees are recorded over their contract period, and when they leave the company and return the same *DID* is used. Class *Employee* has five attributes, among which *ESalary* is temporal and recorded using time intervals. Temporal class *Project* is durable and single, since once a project is finished it is never started again. Relationship *worksIn* is temporal, and its history is recorded using time intervals with cardinality *many-to-one*, i.e., an employee at any given time can only work for one project, but many employees can work for the same project at the same time. The lifespan cardinality is *many-to-many*, i.e., employees may switch to different projects during their contract period.

3.2 Target Metamodel: Temporal Cassandra Model

Apache Cassandra is a NoSQL wide-column store designed for the management of large volumes of data that automatically distributes data across the nodes in a cluster or "ring". Its query language, called CQL (Cassandra Query Language), is very similar to traditional SQL, but does not support joins, and thus to perform joins either *data denormalization* or *application layer join* (on the client side) have to be applied.

A database schema in Cassandra is represented by a top-level container, called *keyspace*. Within a keyspace, CQL tables are defined to store and query data for a particular application. A CQL table is divided into three sets of attributes and can be represented with the following triple notation: $TableName : [PartKey][CKey][NonKeyCols]$

A partition key (*PartKey*) defines partitions over the table, where a partition is entirely stored on one node in the cluster. Each row in a partition may optionally (required in case of not unique *PartKey*) have a clustering key (*CKey*) that uniquely identifies a row within a partition. Rows within a partition are ordered by *CKey*. The combination of *PartKey* and *CKey* is the primary key that uniquely identifies a row in a table. The third element are non-key columns (*NonKeyCols*) in the table. Each column of a table is assigned a data type that can be simple (primitive), such as `TEXT` or `INT`, complex, such as collections (`set`, `list`, `map`), or a user-defined type (UDT). Data retrieval is completely dependent on the *PartKey* of tables and can optionally also contain (ordered) attributes from the *CKey*.

The temporal UML class diagram we use supports timestamping by time point as well as time intervals and, since Cassandra does not support an interval or period datatype, to model interval we use two separate timestamps for start ("$StartTime$") and end ("$EndTime$"). These timestamps can be used to manage transaction or valid time, but to simplify presentation, we only assume valid time in this work.

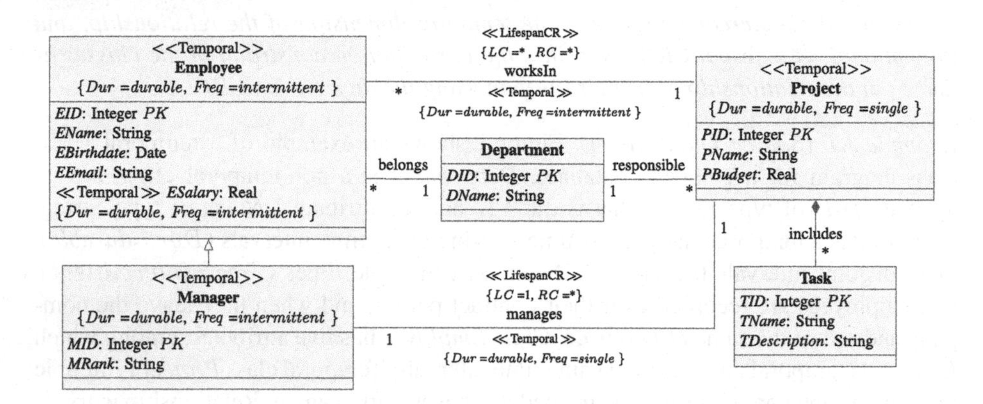

Fig. 1. Example temporal UML class diagram

3.3 Mapping Rules from Temporal UML Diagrams to Cassandra

We now define the mapping rules between source and target metamodels. Since Cassandra does not support join operations, it is important that the proposed mapping model ensures data access across relationships. We provide two alternative solution to express relationships between tables, where developers can choose the one that better fits their requirements based on performance and application queries. The first solution, denoted as *referencing*, uses reference columns that are single-valued or multi-valued (of type SET) columns in a table whose values match the values in the referencing table. The second solution, denoted as *denormalization*, creates new tables that include redundant copies of the data related to tables participating in the relationship. In the mappings we use c_1 and t_1 to denote the left class and its mapped table respectively, and c_2 and t_2 for the right class and corresponding table.

Mapping 1 (UML Classes). This rule is concerned with the mapping of classes, attributes, and the ≪Temporal≫ stereotype.

1. Each class c is transformed into a table t where $t.N = c.N$ and $t.PartKey = c.PK$.
2. Each non-temporal attribute a in class c is mapped into a column col, where $col.N = a.N$, $col.Type = a.Type$; col is added to $NonKeyCols$ of t.
3. Each temporal attribute a in class c is mapped into a complex multi-valued column that can store multiple fields in a single column and, depending on the Dur, contains the following:
   - if $Dur = instantaneous$, it includes attribute a and an attribute $TimePoint$;
   - if $Dur = durable$, it includes attribute a and two attributes $StartTime$ and $EndTime$.

4. If class c includes the ≪ Temporal ≫ stereotype, we transform c according to its *Dur* and *Freq*:
   - if Dur = $instantaneous$ and $Freq$ = $single$, we add one additional column $TimePoint$ to the $NonKeyCols$ of t;
   - if Dur = $instantaneous$ and $Freq$ = $intermittent$, we add one additional column $TimePoint$ and use it as $CKey$ of t;
   - if Dur = $durable$ and $Freq$ = $single$, we add two additional columns $StartTime$ and $EndTime$ to the $NonKeyCols$ of t;
   - if Dur = $durable$ and $Freq$ = $intermittent$, we add one additional column $StartTime$ and use it as $CKey$, and one additional column $EndTime$ is added to the $NonKeyCols$ of t.

Mapping 2 (Non-temporal Association Relationships). Depending on the cardinalities of the association we distinguish between different mapping rules (MR).

MR 2.1 (One-to-One). For $r = (N, assoc, \{(c_1, 1), (c_2, 1)\})$, a new column col referencing *PartKey* of t_1 (the table representing c_1) is added to $NonKeyCols$ of t_2 (the table representing c_2) and vice versa.

MR 2.2 (One-to-Many). For $r = (N, assoc, \{(c_1, 1), (c_2, *)\})$, we have the following solutions for referencing and denormalization.

Referencing. We apply the following steps: (a) a new column referencing $PartKey$ of t_1 is added to $NonKeyCols$ of t_2 and (b) a new multi-value column (of type SET) referencing *PartKey* of t_2 is added to $NonKeyCols$ of t_1.

Denormalization. We apply the following steps: (a) a new column referencing $PartKey$ of t_1 is added to the $NonKeyCols$ of t_2, and (b) a new table t with the following schema is created: $t : [t_1.PartKey][t_2.PartKey][t_2.NonKeyCols, t_2.CKey]$

MR 2.3 (Many-to-Many). For $r = (N, assoc, \{(c_1, *), (c_2, *)\})$, we have the following solutions for referencing and denormalization.

Referencing. A new multi-value column (of type SET) referencing $PartKey$ of t_2 is added to $NonKeyCols$ of t_1, and vice versa.

Denormalization. Two new tables t and t' are created:
$t : [t_1.PartKey][t_2.PartKey][t_2.NonKeyCols, t_2.CKey]$
$t' : [t_2.PartKey][t_1.PartKey][t_1.NonKeyCols, t_1.CKey]$

Mapping 3 (Temporal Association Relationships). Depending on the lifespan cardinalities of the temporal association relation we distinguish between two mapping rules. Recall that there is no *one-to-one* cardinality for the lifespan (cf. Section 3.1).

MR 3.1 (One-to-Many). For $r = (N, assoc, CR, \ll\text{Temporal}\gg\{Dur, Freq\}, \ll\text{LifespanCR}\gg\{1, *\})$, we have the following solutions.

Referencing. We apply the following two steps: (a) a new column referencing $PartKey$ of t_1 is added to the $NonKeyCols$ of t_2; (b) a new multi-value column, that can attach multi data fields to a single column, is added to $NonKeyCols$ of t_1, where:

- if $Dur = instantaneous$, it includes $t_2.PartKey$ and $TimePoint$.
- if $Dur = durable$, it includes $t_2.PartKey$, $StartTime$ and $EndTime$.

Denormalization. (a) a new column referencing $PartKey$ of t_1 is added to the $NonKeyCols$ of t_2; (b) a new table t is created, where:

- if $Dur = instantaneous$ and $Freq = single$ then
  $t : [t_1.PartKey][t_2.PartKey][TimePoint, t_2.NonKeyCols, t_2.CKey]$
- if $Dur = instantaneous$ and $Freq = intermittent$ then
  $t : [t_1.PartKey][t_2.PartKey, TimePoint][t_2.NonKeyCols, t_2.CKey$
- if $Dur = durable$ and $Freq = single$ then
  $t : [t_1.PartKey][t_2.PartKey][StartTime, EndTime, t_2.NonKeyCols, t_2.CKey]$
- if $Dur = durable$ and $Freq = intermittent$ then
  $t : [t_1.PartKey][t_2.PartKey, StartTime][EndTime, t_2.NonKeyCols, t_2.CKey]$

MR 3.2 (Many-to-Many).
Relation $r = (N, assoc, CR, \ll\text{Temporal}\gg\{Dur, Freq\}, \ll\text{LifespanCR}\gg\{*, *\})$ are mapped as follows.

Referencing. (a) a new multi-value column col is added to $NonKeyCols$ of t_2, where:

- if $Dur = instantaneous$, it includes $t_1.PartKey$ and $TimePoint$;
- if $Dur = durable$, it includes $t_1.PartKey$, $StartTime$ and $EndTime$.

(b) we act vice versa of step *(a)* for t_1 and t_2.

Denormalization. Two new tables t and t' are created:

- if $Dur = instantaneous$ and $Freq = single$ then
  $t : [t_1.PartKey][t_2.PartKey][TimePoint, t_2.NonKeyCols, t_2.CKey]$
  $t' : [t_2.PartKey][t_1.PartKey][TimePoint, t_1.NonKeyCols, t_1.CKey]$
- if $Dur = instantaneous$ and $Freq = intermittent$ then
  $t : [t_1.PartKey][t_2.PartKey, TimePoint][t_2.NonKeyCols, t_2.CKey]$
  $t' : [t_2.PartKey][t_1.PartKey, TimePoint][t_1.NonKeyCols, t_1.CKey]$
- if $Dur = durable$ and $Freq = single$ then
  $t : [t_1.PartKey][t_2.PartKey][StartTime, EndTime, t_2.NonKeyCols, t_2.CKey]$
  $t' : [t_2.PartKey][t_1.PartKey][StartTime, EndTime, t_1.NonKeyCols, t_1.CKey]$
- if $Dur = durable$ and $Freq = intermittent$ then
  $t : [t_1.PartKey][t_2.PartKey, StartTime][EndTime, t_2.NonKeyCols, t_2.CKey]$
  $t' : [t_2.PartKey][t_1.PartKey, StartTime][EndTime, t_1.NonKeyCols, t_1.CKey]$

Mapping 4 (Composition Relationships). Composition is a specific type of a one-to-many association, where one composite class is composed of component classes. The component class cannot exist without its composite class, and thus the primary key of a component must include the primary key of the composite. Depending on whether we have a non-temporal or temporal composition, we have the following two mappings.

MR 4.1 (Non-temporal). For $r = (N, compo, \{(c_1, 1), (c_2, *)\})$, we apply MR 2.2 with partial changes to steps (a) and (b).

Referencing. We apply MR 2.2 with the following changes. (a) a new column referencing the $PartKey$ of t_1 is added to the $CKey$ of t_2; (b) same as MR 2.2.

Denormalization. We apply MR 2.2 with the following changes. (a) a new column col referencing $PartKey$ of t_1 is added to the $CKey$ of t_2; (b) same as MR 2.2.

MR 4.2 (Temporal). For $r = (N, compo, CR, \ll \text{Temporal} \gg \{Dur, Freq\}, \ll \text{LifespanCR} \gg \{LC, RC\})$ we apply MR 3.1 with partial changes.

Referencing. We apply MR 3.1 with the following changes: (a) a new column referencing $PartKey$ of t_1 is added to the $CKey$ of t_2; (b) the same as MR 3.1.

Denormalization. We apply MR 3.1 with the following changes: (a) a new column referencing $PartKey$ of t_1 is added to the $CKey$ of t_2; (b) the same as MR 3.1.

Mapping 5 (Generalization Relationships). Generalization represents inheritance, i.e., a sub-class inherits all attributes and operations of its super-class. In practice, it is common that the primary key of the super-class is the same as the primary key of the sub-class, while this is not strictly necessary, we will adopt this common practice in our mapping. Note that we do not define a mapping for temporal generalization, since it is not supported by the temporal UML model that we use as a source model.

MR 5.1. For $r = (N, gener, \{(c_1, null), (c_2, null)\})$, where c_1 is the super-class and c_2 is sub-class, we have the following transformations.

Referencing. The $PartKey$ of t_2 (table for the sub-class) references the $PartKey$ of t_1 (table for the super-class).

Denormalization. All columns of table t_1 except $PartKey$ are added to the $NonKeyCols$ of table t_2, and the $PartKey$ of t_1 references the $PartKey$ t_2.

Example 3.2 (Mapping). Consider the temporal UML class diagram from Fig. 1. By applying our mapping rules, we obtain the Cassandra physical model shown in Fig. 2. For illustrative purposes, we applied *referencing* in some and *denormalization* in other cases.

```
Department: [DID] [ ] [EIDs (set), PIDs (set), DName]
UDT Salary: (Amount, StartTime, EndTime)
Employee: [EID] [EStartTime]
  [EEndTime, DID, EName, EBirthdate, EEmail, ESalaries (set)]
Project: [PID] [ ] [DID, MID, TIDs (set), PName, PBudget, PStartTime, PEndTime]
Manager: [MID] [MStartTime] [MEndTime, MRank]
Task: [TID] [PID] [TName, TDescription]
Project_by_Employee: [EID] [PID, WStartTime]
  [WEndTime, PName, PBudget, PStartTime, PEndTime]
Employee_by_Project: [PID] [EID, WStartTime] [WEndTime,
  EName, EBirthdate, ESalaries (set), EStartTime, EEndTime]
Project_by_Manager: [MID] [PID]
  [MStartTime, MEndTime, PName, PBudget, PStartTime, PEndTime]
```

Fig. 2. Example temporal physical Cassandra model

The classes *Department*, *Employee*, *Task*, *Project*, and *Manager* from the UML class diagram are mapped to CQL tables (Mapping 1) including their attributes. For instance, class *Department* is non-temporal and thus table *Department* has no timestamps. Class *Employee*, on the other hand, is temporal with *Dur* = *durable* and therefore contains a *StartTime* and *EndTime*, and *Freq* = *intermittent* and thus *StartTime* is the *CKey*. Attribute *Salary* in class *Employee* is temporal (with *Dur* = *durable*) and thus we define and use a salary UDT with multi data fields. For the non-temporal association relationships *belongs* and *responsible* we apply *referencing* in MR 2.2, and thus both tables *Employees* and *Project* contain the *DID* from *Department*. Similarly, for the non-temporal composition relationship *includes* we apply *referencing* from MR 4.1 and thus, *Project* contains a set of *TID* and *TID* is the *CKey* of *Task*. We use *referencing* (from MR 5.1) for the generalization relationship between *Manager* and *Employee*, and thus *MID* is a reference to *EID*. For the temporal association relationship *works* we use *demoralization* and MR 3.2 due to its *many-to-many* lifespan cardinality, resulting in the creation of two new tables *Projects_by_Employee* and *Employees_by_Project*. These denormalized table allow efficient access to project data based on an employee (*EID*) and vice versa (*PID*). Similarly, we apply *demoralization* and MR 3.1 for the relationship *manages*, resulting in *MID* being added to the *NonKeyCols* of table *Project* and the creation of table *Projects_by_Manager*.

4 Experimental Evaluation

Setup and Dataset. We used an Intel(R) Xeon(R) CPU E5-2667 v3 @ 3.20GHz machine with 94GB main memory, Ubuntu 64-bit (20.04.5 LTS), and the official docker release for Apache Cassandra version 4.1.1. We run a three node cluster on the same machine. To access Cassandra we use Python 3.11.3. As a dataset, we use the *MySQL* Employee database (https://github.com/datacharmer/test_db) with six tables and a total of 4M records. We first, created a temporal conceptual UML diagram (from the existing ER diagram) and then, applied our two mapping approaches *referencing* and *denormalization*. We compare the different solutions based on space usage and query time for read and update queries with different characteristics as a workload.

Results. In terms of space usage, the mapping based on *referencing* requires 232 MB, while *denormalization* requires 364 MB, thus the latter requires 56% more space.

Queries	Q1	Q2	Q3	Q4	Q5	Q6
Denorm.	0.02	0.71	0.02	0.24	0.22	392.62
Referencing	0.02	375.32	0.31	0.09	0.14	0.08

Fig. 3. Runtime in seconds for the workload queries

The results for runtime are shown in Fig. 3. Query Q1 is a simple data access query by key. As expected, both solutions perform the same, despite records for *referencing* are larger, since they need to store references to other tables. Query Q2 retrieves employees given a department (join). *referencing* is much slower as compared to *denormalization*, since *denormalization* can retrieve the relevant employees from a denormalized table using the department key, and *referencing* needs an application side join, i.e., first, retrieve the employee-key set from the department table using the department key and then, perform a query for each employee key to retrieve the employee data. Since there are only a few departments, but many employees working for the same department, it results in many queries that need to be performed, which explains the very large difference (520x) in runtime. Query Q3 retrieves the managers given a department. This query also requires an application side join for *referencing*, but we see a much smaller difference in runtime between the two solutions (15x). This is because in contrast to Q2, for *referencing* this query require less queries in the client side join, since there are less managers than employees for a given department. For update queries, as expected we see the opposite picture, where generally denormalization requires more time to reflect the changes in the denormalized tables. Q4 adds a new salary to an employee and Q5 sets an employee as a new manager of a department. For *referencing*, these queries require to add a new salary to the salary set and update manager references respectively, resulting in an update of one or two tables. For *denormalization* it requires to update the demoralized tables as well, explaining the higher runtime. Q6 changes the name of a given department, for *referencing* this implies updating a single table, whereas for *denormalization* it requires a change to many denormalized tables, since department has many relationships, and its name appears and needs to be changed in many denormalized tables.

In summary, the *denormalization* solution of our mapping requires more space as compared to *referencing*, but has an improved query time over relationships, particularly when the number of queries that need to be performed in the application side join for *referencing* is large. On the other hand, for update queries *referencing* is faster and *denormalization* suffers when entities with many relationships need to be changed.

5 Conclusion

In this paper we provide a systematic mapping from a temporal conceptual model in the form of a temporal UML diagram to a wide-column store schema in the form of a physical Cassandra schema. Thereby, we provide two solutions based on *referencing* and *denormalization*, where the former features a smaller data size but higher runtime for joins and latter avoids joins by using denormalized data for relationships.

Future work points in several directions. First, we want to consider workloads in the mapping and provide recommendations for the best solution to use and second, we want to implement an automatic mapping tool from the conceptual temporal UML diagram to the final physical Cassandra schema

References

1. Andriamampianina, L., Ravat, F., Song, J., Vallès-Parlangeau, N.: Graph data temporal evolutions: from conceptual modelling to implementation. Data Knowl. Eng. **139** (2022)
2. Brahmia, S., Brahmia, Z., Grandi, F., Bouaziz, R.: τjschema: A framework for managing temporal json-based nosql databases. In: DEXA (2). LNCS, vol. 9828, pp. 167–181 (2016)
3. Cabot, J., Olivé, A., Teniente, E.: Representing temporal information in UML. In: Stevens, P., Whittle, J., Booch, G. (eds.) UML 2003. LNCS, vol. 2863, pp. 44–59. Springer, Heidelberg (2003). https://doi.org/10.1007/978-3-540-45221-8_5
4. Campos, A., Mozzino, J., Vaisman, A.A.: Towards temporal graph databases. CoRR abs/1604.08568 (2016)
5. Davoudian, A., Chen, L., Liu, M.: A survey on nosql stores. ACM Comput. Surv. (CSUR) **51**(2), 1–43 (2018)
6. Debrouvier, A., Parodi, E., Perazzo, M., Soliani, V., Vaisman, A.: A model and query language for temporal graph databases. VLDB J. **30**(5), 825–858 (2021). https://doi.org/10.1007/s00778-021-00675-4
7. El Hayat, S.A., Bahaj, M.: Modeling and transformation from temporal object relational database into mongodb: rules. Adv. Sci. Technol. Eng. Syst. J **5**, 618–625 (2020)
8. Eshtay, M., Azzam, S., Aldwairi, M.: Implementing bi-temporal properties into various nosql database categories. Int. J. Comput. (2019)
9. Goyal, A., Dyreson, C.E.: Temporal JSON. In: CIC, pp. 135–144. IEEE (2019)
10. Gregersen, H., Jensen, C.S.: Conceptual modeling of time-varying information. Technical report, Technical Report TimeCenter TR-35, Aalborg University, Denmark (1998)
11. Hu, Y., Dessloch, S.: Defining temporal operators for column oriented NoSQL databases. In: Manolopoulos, Y., Trajcevski, G., Kon-Popovska, M. (eds.) ADBIS 2014. LNCS, vol. 8716, pp. 39–55. Springer, Cham (2014). https://doi.org/10.1007/978-3-319-10933-6_4
12. Soley, R., et al.: Model driven architecture. OMG white paper **308**(308), 5 (2000)

Contributions from ADBIS 2023 Workshops and Doctoral Consortium

Databases and Information Systems: Contributions from ADBIS 2023 Workshops and Doctoral Consortium

Adam Przybyłek[1(✉)], Aleksandra Karpus[1], Allel Hadjali[2], Anton Dignös[3], Carmem S. Hara[4], Danae Pla Karidi[5], Ester Zumpano[6], Fabio Persia[7], Genoveva Vargas-Solar[8], George Papastefanatos[5], Giancarlo Sperlì[9], Giorgos Giannopoulos[5], Ivan Luković[10], Julien Aligon[11], Manolis Terrovitis[5], Marek Grzegorowski[12], Mariella Bonomo[13], Mirian Halfeld Ferrari[14], Nicolas Labroche[15], Paul Monsarrat[16], Richard Chbeir[17], Sana Sellami[18], Seshu Tirupathi[19], Simona E. Rombo[13], Slavica Kordić[20], Sonja Ristić[20], Tommaso Di Noia[21], Torben Bach Pedersen[22], and Vincenzo Moscato[9]

[1] Faculty of Electronics, Telecommunications and Informatics, Gdańsk University of Technology, Gdańsk, Poland
adam.przybylek@gmail.com, alekarpu@pg.edu.pl

[2] LIAS - ISAE-ENSMA/University of Poitiers, Chasseneuil, France
allel.hadjali@ensma.fr

[3] Free University of Bozen-Bolzano, Bozen-Bolzano, Italy
anton.dignoes@unibz.it

[4] Universidade Federal do Paraná, Curitiba, Brazil
carmemhara@ufpr.br

[5] Athena Research Center, Curitiba, Greece
{danae,gpapas,giann,mter}@athenarc.gr

[6] DIMES, University of Calabria, Arcavacata di Rende (CS), Italy
e.zumpano@dimes.unical.it

[7] University of L'Aquila, L'Aquila, Italy
fabio.persia@univaq.it

[8] CNRS, Univ Lyon, INSA Lyon UCBL, LIRIS UMR5205, Lyon, France
genoveva.vargas-solar@cnrs.fr

[9] University of Naples "Federico II", Napoli, Italy
{giancarlo.sperli,vmoscato}@unina.it

[10] Faculty of Organizational Sciences, University of Belgrade, Beograd, Serbia
ivan.lukovic@fon.bg.ac.rs

[11] Université de Toulouse-Capitole, IRIT, (CNRS/UMR 5505), Toulouse, France
julien.aligon@irit.fr

[12] University of Warsaw, Warszawa, Poland
m.grzegorowski@mimuw.edu.pl

[13] University of Palermo, Palermo, Italy
mariella.bonomo@community.unipa.it, simona.rombo@unipa.it

[14] Université d'Orléans, LIFO, INSA CVL, Orléans, France
mirian@univ-orleans.fr

[15] University of Tours, Tours, France
nicolas.labroche@univ-tours.fr

A. Abelló et al. (Eds.): ADBIS 2023, CCIS 1850, pp. 293–311, 2023.
https://doi.org/10.1007/978-3-031-42941-5_26

[16] RESTORE, Paris, France
paul.monsarrat@univ-tlse3.fr
[17] Univ Pau & Pays Adour, E2S-UPPA, LIUPPA, EA3000, Anglet, France
richard.chbeir@univ-pau.fr
[18] Aix Marseille Univ, Université de Toulon, CNRS, LIS, Marseille, France
sana.sellmi@univ-amu.fr
[19] IBM Research Europe, Dublin, Ireland
SESHUTIR@ie.ibm.com
[20] Faculty of Technical Sciences, University of Novi Sad, Novi Sad, Serbia
{slavica,sdristic}@uns.ac.rs
[21] Polytechnic University of Bari, Bari, Italy
tommaso.dinoia@poliba.it
[22] Aalborg University, Aalborg, Denmark
tbp@cs.aau.dk

1 Introduction

The 27th European Conference on Advances in Databases and Information Systems (ADBIS) aims at providing a forum where researchers and practitioners in the fields of databases and information systems can interact, exchange ideas and disseminate their accomplishments and visions. ADBIS originally included communities from Central and Eastern Europe, however, throughout its lifetime it has spread and grown to include participants from many other countries throughout the world. The ADBIS conferences provide an international platform for the presentation of research on database theory, development of advanced DBMS technologies, and their advanced applications. The ADBIS series of conferences aims at providing a forum for the presentation and dissemination of research on database theory, development of advanced DBMS technologies, and their advanced applications. ADBIS 2023 in Barcelona continues after Turin (2022), Tartu (2021), Lyon (2020), Bled (2019), Budapest (2018), Nicosia (2017), Prague (2016), Poitiers (2015), Ohrid (2014), Genoa (2013), Poznan (2012), Vienna (2011), Novi Sad (2010), Riga (2009), Pori (2008), Varna (2007), Thessaloniki (2006), Tallinn (2005), Budapest (2004), Dresden (2003), Bratislava (2002), Vilnius (2001), Prague (2000), Maribor (1999), Poznan (1998), and St. Petersburg (1997).

This year, ADBIS, attracted five workshops and a doctoral consortium.

- *1st Workshop on Advanced AI Techniques for Data Management and Analytics (AIDMA)*, organized by Anton Dignös (Free University of Bozen-Bolzano, Italy), Vincenzo Moscato (University of Naples "Federico II", Italy), Giancarlo Sperlì (University of Naples "Federico II", Italy), Fabio Persia (University of L'Aquila, Italy), Nicolas Labroche (University of Tours, France), Julien Aligon (Université de Toulouse-Capitole, IRIT, France), Danae Pla Karidi (Athena Research Center, Greece), Giorgos Giannopoulos (Athena Research Center, Greece), George Papastefanatos (Athena Research Center,

Greece), Manolis Terrovitis (Athena Research Center, Greece), Paul Monsarrat (RESTORE, France), Seshu Tirupathi (IBM Research Europe, Ireland), Torben Bach Pedersen (Aalborg University, Denmark), Sana Sellami (Aix Marseille University, France), Richard Chbeir (University Pau & Pays Adour, France) and Allel Hadjali (Engineer School ENSMA, France)
- *4th Workshop on Intelligent Data - from data to knowledge (DOING)*, organized by Mirian Halfeld Ferrari (Université d'Orléans, INSA CVL, LIFO EA, France) and Carmem S. Hara (Universidade Federal do Paraná, Curitiba, Brazil).
- *2nd Workshop on Knowledge Graphs Analysis on a Large Scale (K-Gals)*, organized by Mariella Bonomo (University of Palermo, Italy) and Simona E. Rombo (University of Palermo, Italy).
- *5th Workshop on Modern Approaches in Data Engineering and Information System Design (MADEISD)*, organized by Ivan Luković (University of Belgrade, Faculty of Organizational Sciences, Serbia), Slavica Kordić (University of Novi Sad, Faculty of Technical Sciences, Serbia), and Sonja Ristić (University of Novi Sad, Faculty of Technical Sciences, Serbia).
- *2nd Workshop on Personalization and Recommender Systems (PeRS)*, organized by Marek Grzegorowski (University of Warsaw, Poland), Aleksandra Karpus (Gdańsk University of Technology, Poland), Tommaso Di Noia (Politechnic University of Bari, Italy), and Adam Przybyłek (Gdańsk University of Technology, Poland).
- The *ADBIS Doctoral Consortium*, organized by Genoveva Vargas Solar (LIRIS, French Council of Scientific Research, France) and Ester Zumpano (DIMES, Universitá della Calabria, Italy).

The ADBIS satellite events had its own international program committee, whose members served as the reviewers of papers included in this volume. This volume contains papers on the contributions of all workshops and the doctoral consortium of ADBIS 2023. In the following, for each event, we present its main motivations and topics of interest and we briefly outline the papers selected for presentations. The selected papers will then be included in the remainder of this volume. Some acknowledgements from the organizers are finally provided

2 AIDMA 2023: 1st Workshop on Advanced AI Techniques for Data Management and Analytics

2.1 Description

Artificial Intelligence (AI) methods are now well-established and have been fully integrated by the data management and analytics (DMA) community as an innovative way to address some of its challenges. For example, and without being exhaustive, one may cite the personalization of queries on databases, the management of time series big data streams, the recommendation of dashboards in business intelligence, the indexing of large amounts of data possibly in distributed systems based on ML-based optimization methods, the guided exploration of novel data or the explanation of the provenance of data in queries.

The AIDMA workshop fully embraces this new trend in data management and analytics. It aims at gathering researchers from both AI, data management and analytics to address the new challenges of the domain. These challenges are now of prime importance for several reasons.

Data are increasingly complex and, most of the time, are not simple tabular data. For example, one may consider multivariate temporal relationships between observations as time series, possibly with no synchronicity between observations. Others may view more general relations as graphs, such as in social networks, that carry multi-modal information such as text, sound, videos and user-related information with different links between them. Indeed, multimedia data allow fast and effective communication and sharing of information about peoples' lives, their behaviors, work, and interests, but they are also the digital testimony of facts, objects, and locations. In such a context, Social Media Networks (SMNs) actually represent a natural environment where users can create and share multimedia content such as text, images, video, audio, and so on. Within these "interest-based" networks, each user interacts with the others through multimedia content (text, image, video, audio) and such interactions create "social links" that well characterize the behaviors of the users. Here, in addition to social information (e.g., tags, opinions, insights, evaluations, perspectives, ratings, and user profiles) multimedia data can play a "key-role" especially from the perspective of Social Network Analysis (SNA): representing and understanding multimedia characteristics and user-to-item interaction mechanisms can be useful to predict user behavior, model the evolution of multimedia content and social graphs, design human-centric multimedia applications and services, just to make a few examples. AIDMA workshop aims to address these critical challenges by bringing together scientists, engineers, practitioners, and students in an interdisciplinary forum where they can share experiences and ideas, discuss the results of their research on models, methodologies and algorithms, and forge new collaborations that will lead to the development of new applications and tools to transform information extracted from SMNs into actionable intelligence.

The remarkable achievements of AI and DMA in the past and present have spurred a growing demand from high-stake domains such as Renewable Energy Systems for comprehensive data management and analysis solutions. To address these issues, AIDMA invites paper submissions in any field connected to AI tools and applications. Regarding the RES sector, the workshop focuses on complex and developing areas in data management, machine learning, and artificial intelligence techniques and applications, providing a thorough grasp of the various machine-learning methods and equipment that may be applied to the design, upkeep, and improvement of renewable energy systems, such as wind and solar parks. One of the workshop's main focuses will be the most recent developments in machine learning and how they apply to renewable energy systems, giving attendees a chance to learn about the various models and their efficacy in improving the performance of renewable energy systems. Such domains need efficient approaches that can be trusted by field experts, as the outcome of the analysis may directly impact the quality of life of millions of people worldwide.

Another direction for AIDMA workshop concerns medicine and healthcare management systems that now heavily rely on the recent progress of machine and deep learning to support complex decision tasks. Complex data acquisition, preprocessing and then analysis pipelines are set up to answer questions related to patients' medical situation or in a prospective way to learn more about diseases, healthcare paths, treatments, and physiological, chemical or biological processes involved in medicine and healthcare. As such, these decision systems delegate to end-to-end data-driven systems with no or little regulation, decisions that generally have a crucial impact on the life of patients.

These new approaches must consider the user as a first-class citizen in the data management and analysis process. This relates to the ability of the user to inflect the process and to understand the rationale of a data analysis, which is addressed in the explainability domain. Finally, interacting with users implies being able to manage efficiently the uncertainty attached to data as well as in any artificial intelligence process. Uncertainty needs to be managed at various levels of the data management process: data collection, data querying, machine learning and data analytics. For instance, the presence of uncertainty can be the source of semantics errors during query evaluation. Moreover, traditional machine learning and deep learning models do not consider uncertainty in data and predictions, while they are noise-prone. Then, quantifying uncertainty is a critical challenge for most machine learning techniques.

2.2 Keynote Speaker

The AIDMA workshop welcomes Michele Linardi (PhD in Computer Science), who is an assistant professor at Cergy Paris University (CYU - ETIS laboratory, France). His research interests span the areas of time series analytics and databases, with a great interest in machine learning for temporal data.

Multivariate time series analysis (MTS), such as classification, forecasting, and anomaly detection, are omnipresent problems in many scientific domains. In this context, several state-of-the-art solutions rely on deep learning (DL) architectures such as CNN (Convolutional Neural Network), LSTM (Long Short-Term Memory Network), and attention-based architecture like Transformer.

Despite their effectiveness and usage, DL models remain uninterpretable black boxes, where the user feeds an input and obtains an output without understanding the motivations behind that decision. In countless real-world domains, from legislation and law enforcement to healthcare and precision agriculture, diagnosing what aspects of a model's input drive its output is essential to ensure that decisions get driven by appropriate insights in the context of its use. In this sense, explainable machine learning techniques (xAI) aim to provide a solid descriptive approach to DL models, and it is at the cusp of becoming a compulsory requirement in all use cases.

In this talk, we introduce and present the main state-of-the-art xAI methods adopted in the MTS DL models. Among several technical and fundamental aspects, we will show how xAI solutions become effective when they can leverage the causal relationships (between target and predictors) occurring over MTS variables. We will also present how xAI solutions can be effective in DL Domain Adaptation which is an omnipresent problem in various scientific fields, including Healthcare.

2.3 Selected Papers

The paper *"Data exploration based on local attribution explanation: a medical use case"* [11] presents a new approach that considers the explanation description space as a first-class citizen for data exploration. This approach is of particular interest as it combines the determination of a model in the data space, and then benefits from the explanations of the complex relationships extracted by this model as a basis for unsupervised study of the data. The underlying research question relates to the informative and discriminating power of the influence explanation space, which generally embodies only the most distinctive features for a specific class and their importance attached to a specific value of the features.

The paper *"Explainability based on feature importance for better comprehension of machine learning in healthcare"* [10] presents a framework for analysis of features' importance (FI) in the context of healthcare and specifically for Systemic Inflammatory Response Syndrome. The originality of the paper lies in the use of the FI based on the decision tree selection procedure as well as Shap, which means using both ante-hoc/post-hoc methods. Other problems addressed in the paper relate to the impact of the ML pipeline on explanation such as the classifier or the sampling to limit unbalancing of data.

The paper *"An empirical study on the robustness of active learning for biomedical image classification under model transfer scenarios* [19] proposes a study of the effects of transferring an actively sampled training data set from an acquisition model to different successor models for biomedical image classification tasks. The research shows that training a successor model with an actively-acquired data set is most promising if the acquisition and successor models are of similar architecture.

In the paper *"Holistic Analytics of Sensor Data From Renewable Energy Sources: A Vision Paper"* [20] the authors present their vision for a next-generation time series management system that efficiently can manage vast amounts of time series across edge, cloud, and client.

The paper *"Evaluating the Robustness of ML Models To Out-of-Distribution Data Through Similarity Analysis"* [26] addresses the perception accuracy problem from a data out-of-distribution (OOD) point-of-view and proposes a method for analyzing datasets from a use-case scenario perspective, detecting and quantifying OOD data on the dataset level.

3 DOING 2023: 4th Workshop on Intelligent Data – From Data to Knowledge

3.1 Description

The DOING Workshop focuses on transforming data into information and then into knowledge. It gathers researchers from natural language processing, databases, and artificial intelligence. DOING 2023 received 9 submissions, out of which 3 were accepted as full papers and 2 as short papers, resulting in an acceptance rate of 50%. Each paper received three reviews from members of the program committee. The final program featured works in the areas of information extraction from textual data, data classification, and quantum technology applied to natural language processing and classification.

This workshop is an event sponsored by the French network MADICS[1]. More specifically, it is an event of the action DOING[2] in MADICS and of the DOING working group in the regional French network DIAMS[3].

The workshop is the result of the collective effort of a large community, which we gratefully acknowledge. We thank the ADBIS conference chairs, who worked hard to support the workshop organization. We are also grateful to the members of the program committee, who did an outstanding job, providing timely and thoughtful reviews. Finally, we are grateful to the authors who submitted their work to DOING 2023.

3.2 Selected Papers

Three full papers were presented at DOING 2023. The first one, entitled *"Labeling Portuguese Man-of-War Posts Collected from Instagram"* [33], details the process of choosing labels for classifying posts from Instagram as legitimate occurrences of the cnidarian Portuguese Man-of-War on the Brazilian coast. It is shown that labels for manual labeling may not always be adequate for training a machine-learning classification model. The article reports on the need for analyzing the labeling process in order to adapt it and generate an adequate dataset for training a classification model. Experimental analysis are presented to show the effect of this process on the quality of the resulting model.

The second paper, entitled *"Semantic Business Trajectories Modeling and Analysis"* [2] presents a method for constructing business trajectories based on online news data. Methods for topic modelling and named entity recognition are used to extract information from the news and semantically enrich the trajectory data. The goal of the study is to provide spatial, temporal and contextual views of businesses over time, which can also be used to understand their impacts on the environment, such as noise pollution and carbon emissions.

The third paper, *"A Text Mining Pipeline for Mining the Quantum Cascade Laser Properties"* [21], tackles the problem of extracting information on quantum

[1] https://www.madics.fr/.
[2] https://www.madics.fr/actions/doing/.
[3] https://www.univ-orleans.fr/lifo/evenements/RTR-DIAMS/.

cascade laser properties from scientific literature, such as working temperature, optical power, lasing frequency, material design and barrier thickness. The proposed method is based on an extension of the ChemDataExtractor pipeline with parsing rules to manage special characters and formats that are specific to the application. Experimental analysis shows that the proposed rules improve the performance of the extracted information.

Two short papers completed the program of DOING 2023. Both are in an exciting area of research of quantum computing. The paper *"Ensemble Learning based Quantum Text Classifiers"* [6], relates quantum computing with natural language processing. It describes a comparative evaluation of text classification using different models in the Lambeq Quantum NLP toolkit. Finally, "Exploring the Capabilities and Limitations of VQC and QSVC for Sentiment Analysis on Real-World and Synthetic Datasets" [4], evaluates the performance of two quantum classifiers for sentiment analysis applications. The evaluation is based on implementing a quantum version of a Support Vector Classifier and a Variational Quantum Classifier for running experiments over a synthetic dataset and IMDB.

4 K-Gals 2023: 2nd Workshop on Knowledge Graphs Analysis on a Large Scale

4.1 Description

Knowledge Graphs are powerful models to represent networks of real-world entities, such as objects, events, situations, and concepts, by illustrating the relationships between them. Information encoded by knowledge graphs is usually stored in graph databases, and visualized as graph structures. Although these models have been introduced in the Semantic Web context, they have recently found successful applications also in other contexts, e.g., the analysis of financial, social, geospatial and biomedical data.

Knowledge Graphs often integrate datasets from various sources, which frequently differ in their structure. This, together with the increasing volumes of structured and unstructured data stored in a distributed manner, bring to light new problems related to data/knowledge representation and integration, data querying, business analysis and knowledge discovery.

The ultimate goal of K-GALS is to provide participants with the opportunity to introduce and discuss new methods, theoretical approaches, algorithms, and software tools that are relevant to the Knowledge Graphs-based research, especially when it is focused on a large scale. In this regard, interesting open issues include how Knowledge Graphs may be used to represent knowledge, how systems managing Knowledge Graphs work, and which applications may be provided on top of a Knowledge Graph, in the distributed.

4.2 Selected Papers

The 2023 edition of K-GALS includes five papers, all accepted as research full papers after a very selective peer review process, which are published in this volume. Research reported in the accepted papers spans from the most theoretical foundations [37], to notions related to software development [34], and applications in contexts such as sustainability [16,18] and biomedicine [5].

In the manuscript *"P2KG: Declarative Construction and Quality Evaluation of Knowledge Graph from Polystores"* [37] the authors provide a systematic way to build and evaluate Knowledge Graphs from diverse data sources. The approach proposes a new generic language to specify the mapping rules to guide users to translate data from relational and graph sources into a meaningful Knowledge Graph. The mapping language specifies how elements of the Knowledge Graph are defined as views over one or more data sources. The authors define three types of constraints, derived from source data, user specifications, and common rules of a good Knowledge Graph, and translate them into unified expressions in the form of graph functional dependencies and extended graph dependencies to evaluate the quality, correctness, and consistency, given the mapping rules applied to construct the Knowledge Graph.

The manuscript *"An ArchiMate-Based Thematic Knowledge Graph for Low-Code Software Development Domain"* [34] illustrates how ArchiMate modelling language can be applied at the ontology level of a Knowledge Graph for representing knowledge that concerns an enterprise. In particular, the ArchiMate-based Knowledge Graph is proposed for representing knowledge about low-code software development, with the purpose to make this knowledge available for enterprise management. Navigating the Knowledge Graph enables stakeholders to exploit knowledge about different aspects of lowcode development and promote the use of the low-code development approach in their enterprises.

The authors of the paper *"Using Knowledge Graphs to Model Green Investment Opportunities"* [18] discuss how Knowledge Graphs can be used to represent investment opportunities for the green transition, and they refer to a case study investment for the green transition targeting African regions. The described project aims to develop a system that automatically detects and analyzes green investment opportunities. In the Knowledge Graph, each funding initiative has been described as a subgraph itself, with nodes and relationships used to encode relevant facts, such as the type of investment or the provided budget.

The authors of the paper *"Knowledge Graphs Embeddings for Link Prediction in the Context of Sustainability"* [16] propose a framework for building a collaborative platform that can support the System Dynamics experts in identifying new partnerships and innovative solutions. The approach uses a Knowledge Graph that can track the links among stakeholders and store their metadata. It is in charge of leveraging machine learning methodologies and models for data analysis, as well as the prediction of connections among stakeholders and potential work tables to facilitate the achievement of the Sustainability Development Goals (SDGs) in specific contexts.

Finally, the manuscript *"A Knowledge Graph to Analyze Clinical Patient Data"* [5] describes how Knowledge Graphs can be used to represent patient data coming from clinical folders information. The main aim is to provide suitable classifications of patients, in order to allow a deeper understanding of possible (side) effects that the same treatment may cause on different patients. The Knowledge Graph building and related software services for its analysis are implemented upon the Neo4J NoSQL database.

5 MADEISD 2023: 5th Workshop on Modern Approaches in Data Engineering and Information System Design

5.1 Description

For decades, there has been an open issue of how to support the information management process to produce useful knowledge and tangible business values from the data being collected. One of the hot questions in practice is still how to effectively transform large amounts of daily collected operational data into useful knowledge from the perspective of declared company goals, and how to set up the information design process aimed at the production of effective software services. Nowadays, we have great theoretical potential for applying new and more effective approaches in data engineering and information system design. However, it is more likely that the actual deployment of such approaches in industry practice is far behind their theoretical potential.

The main goal of the Modern Approaches in Data Engineering and Information System Design (MADEISD) workshop is to address open questions and real potentials for various applications of modern approaches and technologies in data engineering and information system design to develop and implement effective software services in a support of information management in various organization systems. The intention was to address the interdisciplinary character of a set of theories, methodologies, processes, architectures, and technologies in disciplines such as Data Engineering, Information System Design, Big Data, NoSQL Systems, Data Streams, Internet of Things, Cloud Systems, and Model Driven Approaches in development of effective software services. In this issue, from 14 submissions, after a rigorous selection process, we accepted 7 papers for publication at ADBIS 2023.

5.2 Selected Papers

This edition of MADEISD workshop includes the following 7 papers, where the first 3 of them have been accepted as full papers and the rest 4 as short papers.

In the paper *"Towards Automatic Conceptual Database Design based on Heterogeneous Source Artifacts"* [3], the authors present an early prototype of the tool named DBomnia. It is the first online web-based tool enabling the automatic derivation of conceptual database models from heterogeneous source artifacts, including business process models and textual specifications. DBomnia employs other pre-existing tools to derive conceptual models from sources of the same type and then integrates those models, whereby the main challenge is related to the integration of conceptual models that are automatically generated and can be considered unreliable. The authors come to the initial results showing that the implemented tool derives models that are more precise and complete compared to the models derived from sources of the same type.

The authors of the paper *"Lightweight Aspect-Oriented Software Product Lines with Automated Product Derivation"* [32] advocate that applications of the aspect-oriented software product lines are facing two obstacles: establishing software product lines is challenging and aspect-oriented programming is not that widely accepted. To address the obstacles, the authors propose an approach to establishing lightweight aspect-oriented software product lines with automated product derivation. The authors evaluate the approach as simple and accessible since developers decide about variation points directly in the code without any assumption on the development process and applied management. Also, the proposed approach allows for variability management by making the code readable, configurable, and adaptable mainly to scripts and code fragments in a modular and concise way, while the use of annotations helps in preserving feature models in code. The authors present the practical usability of the approach on a battleship game and the data preprocessing pipeline product lines.

The authors of the paper *"Comparison of selected neural network models used for automatic liver tumor segmentation"* [24] consider the problem of automatic and accurate segmentation of liver tumors, as crucial for diagnosing and treating hepatocellular carcinoma or metastases. The authors present a study focusing on tumor segmentation as a more critical aspect from a medical perspective, compared to liver parenchyma segmentation, which is the focus of most authors in publications. The authors trained four state-of-the-art models and used them to compare with UNet in terms of accuracy. Based on polar coordinates and Visual Image Transformer (ViT), two of them have been adopted for the specified task. Experiments on a selected public dataset have demonstrated that the proposed ViT-based network can accurately segment liver tumors from CT images end-to-end, outperforming many existing methods. The authors state that the obtained results seem to be clinically relevant.

Index tuning in databases is a critical task that significantly impacts database performance. However, manually configuring indexes is often time-consuming and, as a rule, inefficient. In the paper *"Intelligent index tuning using reinforcement learning"* [27], the authors consider using the reinforcement learning approach to create database indexes. The authors developed an agent that can learn to make optimal decisions for configuring indexes in a chosen database. They also proposed an evaluation method to measure database performance regarding the loading, querying, and processing power of multiple query streams simultaneously. It is to be used in the proposed reinforcement learning algorithm.

The following paper also addresses the problem of index recommendation in databases. The authors of the paper *"Automatic Indexing for MongoDB"* [12] present a new method for automated index suggestions for MongoDB, based solely on the queries, called aggregation pipelines, without requiring data or usage information. The solution handles complex aggregations and is suitable for cloud and standalone databases. The authors validated the algorithm and showed that all suggested indexes were used. The authors state that this is MongoDB's first query-based solution for automated indexing.

The paper *"From High-Level Language to Abstract Machine Code: An Interactive Compiler and Emulation Tool for Teaching Structural Operational Semantics"* [35] is devoted to the development of an abstract machine compiler specifically for the Structural Operational Semantics (SOS) framework. This compiler serves as a formal framework for programming language specification and analysis. The authors put focus on the development process, such as translating high-level language constructs into low-level machine instructions and implementing memory models. Their ultimate objective is to provide valuable insights into the design and implementation of an abstract machine compiler and emulator with potential applications in language design and verification. The tools developed can also be applied in educational settings to aid the teaching process.

The author of the paper *"Decentralised Solutions for Preserving Privacy in Group Recommender Systems"* [31] states that privacy in Group Recommender Systems (GRS) has always been a core issue since most recommendation algorithms rely on user behavior signals and contextual information that may contain sensitive information. Existing works in this domain mostly distribute data processing tasks without addressing privacy, and the solutions that address privacy for GRS, e.g., k-anonymization and local differential privacy, remain centralized. In the paper, the author identifies and analyzes privacy concerns in GRS and provides guidelines on how decentralized techniques can address privacy problems.

6 PeRS 2023: 2nd Workshop on Personalization and Recommender Systems

6.1 Description

Recommender systems play an essential role in our daily lives, whether we read news, use social media, or make online purchases. As a result, this field has garnered increasing attention from academics and industry professionals. User Modelling is intricately intertwined with Recommender Systems as it enables personalization, a crucial aspect of developing innovative recommendation techniques. However, User Modelling encompasses a broader scope, encompassing areas such as user representation, personalized search, adaptive educational systems, and intelligent user interfaces.

In 2022, Aleksandra Karpus and Adam Przybyłek established the workshop on Personalization and Recommender Systems (PeRS) as part of the FedCSIS multiconference, aiming to provide a platform for academics and industry practitioners to connect, collaborate, and learn from one another. In 2023, PeRS (https://pers.lasd.pl) joined ADBIS, further expanding its reach and opportunities for advancing the field of User Modelling and Recommender Systems.

6.2 Selected Papers

PeRS'23 attracted 14 submissions. After a rigorous review, two full and five short papers were accepted.

The paper titled *"Overcoming the Cold-Start Problem in Recommendation Systems with Ontologies and Knowledge Graphs"* [23] presents a solution to the cold-start problem in recommendation systems by leveraging Ontologies and Knowledge Graphs. The authors propose incorporating a semantic layer into text-based methods. The implicit and explicit characteristics of item text attributes are captured by generating a knowledge graph using ontologies, enriching the item profile with semantically related keywords. The proposed approach is evaluated against state-of-the-art text feature extraction techniques, demonstrating improved recommendation performance in scenarios with limited user data and interactions. Overall, this research enhances recommendation systems by providing a novel method combining ontologies and knowledge graphs with text-based approaches.

Frikha et al. [14] propose a personalized service recommendation framework in the paper titled *"A Recommendation System for Personalized Daily Life Services to Promote Frailty Prevention"*. The framework aims to assist senior citizens in selecting appropriate services to prevent frailty and improve autonomy. It uses a multidimensional evaluation of the elderly person, considering both user's status information and service-related data, and defines a four-step recommendation process using knowledge-based and rule-based approaches to generate personalized recommendations that align with the individual's needs. The proposal's effectiveness is evaluated using different representative scenarios, and an example is described in detail to demonstrate the developed recommendation process, algorithms, and relevant rules.

The paper entitled *"Design-focused Development of a Course Recommender System for Digital Study Planning"* [30] identifies important factors to consider when creating a course recommender system (CRS) for higher education. The researchers analyzed and summarized students' selection criteria and processes to establish the requirements for the system. They then developed a design prototype and evaluated it through think-aloud user tests. The paper's main contribution is identifying six guidelines for designing an effective CRS.

In the paper titled *"Systematic Literature Review on Click Through Rate Prediction"* [25], the authors present the findings of a comprehensive literature review that will assist researchers in getting a head start on developing new solutions. The most prevalent models were variants of the state-of-the-art DeepFM model.

The paper titled *"Neural Graph Collaborative Filtering: analysis of possibilities on diverse datasets"* by Kobiela et al. [22] focuses on assessing the possibilities of the NGCF (Neural Graph Collaborative Filtering) technique on diverse datasets. The main contribution of this work involves replicating the research conducted by Wang et al. [36] and expanding it to encompass numerous new and diverse datasets. Additionally, the paper prepares the groundwork for achieving reproducible results by providing Python scripts available in the GitHub repository.

Galka et al. [15] in their paper *"Performance and reproducibility of BERT4Rec"* aim to asses the BERT4Rec's reproducibility on MovieLens1M and measure its performance on Netflix Prize dataset in comparison to other recommender systems. Overall findings suggest that while using a proper implementation, BERT4Rec can still be called a state-of-the-art solution, however, additional work is needed to increase reproducibility of the original model.

The paper entitled *"Recommender Chatbot as a Tool for Collaborative Business Intelligence in Tourism Domain"* [8] introduces a framework for a virtual assistant that incorporates the following crucial elements: a conversational unit, a task identification, data exploration component, and a recommendation model. Tourists often search for information from different sources and compare found facts, leading to spending significant amounts of time comparing descriptions and photos across different platforms before deciding. To overcome this challenge, virtual assistants have become increasingly common.

6.3 Acknowledgments

We express our gratitude to Piotr Szczuko for enabling the participation of a workshop co-chair and 11 students from the AI-TECH project at ADBIS'23.

This work has been partially supported by Statutory Funds of Electronics, Telecommunications and Informatics Faculty, Gdansk University of Technology. This work was partly supported through the European Regional Development Fund as part of the Project entitled: Academy of Innovative Applications of Digital Technologies under Grant The Operational Programme "Digital Poland 2014-2020" number POPC.03.02.00-00-0001/20-00.

7 DC 2023: ADBIS Doctoral Consortium

7.1 Description

The ADBIS 2023 Doctoral Consortium (DC) is a forum for PhD students to present their research projects to the scientific community. DC papers describe the status of PhD student's research, a comparison with relevant related work, their results, and plans on how to experiment, validate and consolidate their contribution. The PhD Consortium allows students to establish international collaborations with members and participants of the ADBIS community.

The ADBIS 2023 DC received 22 submitted papers evaluated through a selective peer-review process by an international program committee. After a thorough evaluation, ten papers were selected for presentation, seven of which are included in this volume, giving an overall acceptance rate of 45%.

The ADBIS 2023 DC program also included a panel titled "Being productive and developing relevant research vs achieving well-being and a well-balanced life. How to have everything and survive without leaving your skin on the line?". Accomplished professionals in academia, industry, and entrepreneurship did the panel discussion. Finally, recognizing the importance of Diversity and Inclusion

(D&I) according to the initiative promoted by the conference, the DC organized a data-driven activity to create inclusion awareness in research together with the D&I conference program.

7.2 Selected Papers

The program of this DC edition included eigth oral presentations with sessions where PhD students received feedback from a selected group of mentors about the opportunities and perspectives they perceive and suggest to pursue their work. The following papers composed the DC program:

The article *"Towards a researcher-in-the-loop driven curation approach for quantitative and qualitative research methods"* by Alejandro Adorjan [1], describes challenges and initial results concerning the modelling researcher in-the-loop curation for qualitative research methodologies.

The article *"Tackling the RecSys Side Effects via Deep Learning Approaches"* by Erica Coppolillo [9] presents research activities aimed at analyzing and mitigating the impact of Recommender Systems in changing users' preferences and affecting niche items.

In the article *"Deep learning techniques for television broadcast recognition"* [7] Federico Candela proposes several deep-learning techniques and the advancement of research in this area for TV program classification on long-form video.

The article *"Intelligent Technologies for Urban Progress: Exploring the Role of AI and Advanced Telecommunications in Smart City Evolution"*, by Enea Vincenzo Napolitano [28], contributes to the understanding of how AI and advanced telecommunications can drive urban progress and foster the sustainable evolution of smart cities.

In the article *"Process Mining solutions for public administration"* [13], Simona Fioretto, shows the current AI application for public administration (PA), and presents process discovery techniques that could practically improve PA effectiveness.

In the article *"Automatic Discovery of Zones of Interests with Maritime Trajectory Mining"* [17], Omar Ghannou describes a method and an algorithm for trajectory mining to identify specific trajectory points called zones of interest (ZOI).

In the article *"Towards reliable machine learning"* [29], Simona Nisticò focuses on the Explainable Artificial Intelligence topic giving an overview of some of the open problems and describing some methodologies proposed to face them.

Acknowledgments. We would like to thank all the contributing speakers, and the members of our Program Committee for timely providing their reviews, and, last but not least, the ADBIS-2023 General chairs, Oscar Romero and Robert Wrembel, the Program Chairs, Alberto Abelló and Panos Vassiliadis, and the Workshop Chairs, Johann Gamper and Francesca Bugiotti for their trust and support. We, Genoveva Vargas Solar and Ester Zumpano, doctoral consortium co-chairs, hope readers will enjoy reading the DC papers and follow their evolution in other publications!

References

1. Adorjan, A.: Toward a researcher-in-the loop driver curation approach for quantitative and qualitative research methods. In: New Trends in Database and Information Systems - ADBIS 2023 Short Papers, Doctoral Consortium and Workshops: AIDMA, DOING, K-Gals, MADEISD, PeRS, Barcelona, Spain, 4–7 September, 2023, Proceedings (2023)
2. Arslan, M., Cruz, C.: Semantic business trajectories modeling and analysis. In: New Trends in Database and Information Systems - ADBIS 2023 Short Papers, Doctoral Consortium and Workshops: AIDMA, DOING, K-Gals, MADEISD, PeRS, Barcelona, Spain, 4–7 September, 2023, Proceedings (2023)
3. Banjac, G., Brdjanin, D., Banjac, D.: Towards automatic conceptual database design based on heterogeneous source artifacts. In: New Trends in Database and Information Systems - ADBIS 2023 Short Papers, Doctoral Consortium and Workshops: AIDMA, DOING, K-Gals, MADEISD, PeRS, Barcelona, Spain, 4–7 September, 2023, Proceedings (2023)
4. Belhadef, H., Benchiheb, H., Lebdjiri, L.: Exploring the capabilities and limitations of vqc and qsvc for sentiment analysis on real-world and synthetic datasets. In: New Trends in Database and Information Systems - ADBIS 2023 Short Papers, Doctoral Consortium and Workshops: AIDMA, DOING, K-Gals, MADEISD, PeRS, Barcelona, Spain, September 4–7, 2023, Proceedings (2023)
5. Bonomo, M., Ippolito, F., Morfea, S.: A knowledge graph to analyze clinical patient data. In: New Trends in Database and Information Systems - ADBIS 2023 Short Papers, Doctoral Consortium and Workshops: AIDMA, DOING, K-Gals, MADEISD, PeRS, Barcelona, Spain, 4–7 September, 2023, Proceedings (2023)
6. Bouakba, Y., Belhadef, H.: Ensemble learning based quantum text classifiers. In: New Trends in Database and Information Systems - ADBIS 2023 Short Papers, Doctoral Consortium and Workshops: AIDMA, DOING, K-Gals, MADEISD, PeRS, Barcelona, Spain, 4–7 September, 2023, Proceedings (2023)
7. Candela, F.: Deep learning techniques for television broadcast recognition. In: New Trends in Database and Information Systems - ADBIS 2023 Short Papers, Doctoral Consortium and Workshops: AIDMA, DOING, K-Gals, MADEISD, PeRS, Barcelona, Spain, 4–7 September, 2023, Proceedings (2023)
8. Cherednichenko, O., Muhammad, F.: Recommender chatbot as a tool for collaborative business intelligence in tourism domain. In: New Trends in Database and Information Systems - ADBIS 2023 Short Papers, Doctoral Consortium and Workshops: AIDMA, DOING, K-Gals, MADEISD, PeRS, Barcelona, Spain, 4–7 September, 2023, Proceedings (2023)
9. Coppolillo, E.: Tackling the recsys side effects via deep learning approaches. In: New Trends in Database and Information Systems - ADBIS 2023 Short Papers, Doctoral Consortium and Workshops: AIDMA, DOING, K-Gals, MADEISD, PeRS, Barcelona, Spain, 4–7September, 2023, Proceedings (2023)

10. Das, P.P., Wiese, L.: Explainability based on feature importance for better comprehension of machine learning in healthcare. In: New Trends in Database and Information Systems - ADBIS 2023 Short Papers, Doctoral Consortium and Workshops: AIDMA, DOING, K-Gals, MADEISD, PeRS, Barcelona, Spain, 4–7 September, 2023, Proceedings (2023)
11. Escriva, E., Doumard, E., Excoffier, J.B., Aligon, J., Monsarrat, P., Soulé-Dupuy, C.: Data exploration based on local attribution explanation: a medical use case. In: New Trends in Database and Information Systems - ADBIS 2023 Short Papers, Doctoral Consortium and Workshops: AIDMA, DOING, K-Gals, MADEISD, PeRS, Barcelona, Spain, 4–7 September, 2023, Proceedings (2023)
12. de Espona Pernas, L., Vichalkovski, A., Steingartner, W., Pustulka, E.: Automatic indexing for mongodb. In: New Trends in Database and Information Systems - ADBIS 2023 Short Papers, Doctoral Consortium and Workshops: AIDMA, DOING, K-Gals, MADEISD, PeRS, Barcelona, Spain, 4–7 September, 2023, Proceedings (2023)
13. Fioretto, S.: Process mining solutions for public administration. In: New Trends in Database and Information Systems - ADBIS 2023 Short Papers, Doctoral Consortium and Workshops: AIDMA, DOING, K-Gals, MADEISD, PeRS, Barcelona, Spain, 4–7 September, 2023, Proceedings (2023)
14. Frikha, G., Lorca, X., Pingaud, H., Taweel, A., Bortolaso, C., Borgiel, K., Lamine, E.: A recommendation system for personalized daily life services to promote frailty prevention. In: New Trends in Database and Information Systems - ADBIS 2023 Short Papers, Doctoral Consortium and Workshops: AIDMA, DOING, K-Gals, MADEISD, PeRS, Barcelona, Spain, 4–7 September, 2023, Proceedings (2023)
15. Galka, A., Grubba, J., Walentukiewicz, K.: Performance and reproducibility of bert4rec. In: New Trends in Database and Information Systems - ADBIS 2023 Short Papers, Doctoral Consortium and Workshops: AIDMA, DOING, K-Gals, MADEISD, PeRS, Barcelona, Spain, 4–7 September, 2023, Proceedings (2023)
16. Galluzzo, Y., Gennusa, F.: Knowledge graphs embeddings for link prediction in the context of sustainability. In: New Trends in Database and Information Systems - ADBIS 2023 Short Papers, Doctoral Consortium and Workshops: AIDMA, DOING, K-Gals, MADEISD, PeRS, Barcelona, Spain, 4–7 September, 2023, Proceedings (2023)
17. Ghannou, O.: Automatic discovery of zones of interests with maritime trajectory mining. In: New Trends in Database and Information Systems - ADBIS 2023 Short Papers, Doctoral Consortium and Workshops: AIDMA, DOING, K-Gals, MADEISD, PeRS, Barcelona, Spain, 4–7 September, 2023, Proceedings (2023)
18. Grani, G., Di Rocco, L., Ferraro Petrillo, U.: Using knowledge graphs to model green investment opportunities. In: New Trends in Database and Information Systems - ADBIS 2023 Short Papers, Doctoral Consortium and Workshops: AIDMA, DOING, K-Gals, MADEISD, PeRS, Barcelona, Spain, 4–7 September, 2023, Proceedings (2023)
19. Janusko, T., Gonsior, J., Thiele, M.: An empirical study on the robustness of active learning for biomedical image classification under model transfer scenarios. In: New Trends in Database and Information Systems - ADBIS 2023 Short Papers, Doctoral Consortium and Workshops: AIDMA, DOING, K-Gals, MADEISD, PeRS, Barcelona, Spain, 4–7 September, 2023, Proceedings (2023)
20. Jensen, S.K., Thomsen, C.: Holistic analytics of sensor data from renewable energy sources: a vision paper. In: New Trends in Database and Information Systems - ADBIS 2023 Short Papers, Doctoral Consortium and Workshops: AIDMA, DOING, K-Gals, MADEISD, PeRS, Barcelona, Spain, 4–7 September, 2023, Proceedings (2023)

21. Kerre, D., Laurent, A., Maussang, K., Owuor, D.: A text mining pipeline for mining the quantum cascade laser properties. In: New Trends in Database and Information Systems - ADBIS 2023 Short Papers, Doctoral Consortium and Workshops: AIDMA, DOING, K-Gals, MADEISD, PeRS, Barcelona, Spain, September 4–7, 2023, Proceedings. Barcelona, Spain (2023)
22. Kobiela, D., Groth, J., Sieczczyński, M., Wolniak, R., Pastuszak, K.: Neural graph collaborative filtering: analysis of possibilities on diverse datasets. In: New Trends in Database and Information Systems - ADBIS 2023 Short Papers, Doctoral Consortium and Workshops: AIDMA, DOING, K-Gals, MADEISD, PeRS, Barcelona, Spain, 4–7 September 2023, Proceedings (2023)
23. Kuznetsov, S., Kordík, P.: Overcoming the cold-start problem in recommendation systems with ontologies and knowledge graphs. In: New Trends in Database and Information Systems - ADBIS 2023 Short Papers, Doctoral Consortium and Workshops: AIDMA, DOING, K-Gals, MADEISD, PeRS, Barcelona, Spain, 4–7 September, 2023, Proceedings (2023)
24. Kwiatkowski, D., Dziubich, T.: Comparison of selected neural network models used for automatic liver tumor segmentation. In: New Trends in Database and Information Systems - ADBIS 2023 Short Papers, Doctoral Consortium and Workshops: AIDMA, DOING, K-Gals, MADEISD, PeRS, Barcelona, Spain, 4–7 September, 2023, Proceedings (2023)
25. Leszczełowska, P., Bollin, M., Grabski, M.: Systematic literature review on click through rate prediction. In: New Trends in Database and Information Systems - ADBIS 2023 Short Papers, Doctoral Consortium and Workshops: AIDMA, DOING, K-Gals, MADEISD, PeRS, Barcelona, Spain, 4–7 September, 2023, Proceedings (2023)
26. Lindén, J., Forsberg, H., Daneshtalab, M., Söderquist, I.: Evaluating the robustness of ml models to out-of-distribution data through similarity analysis. In: New Trends in Database and Information Systems - ADBIS 2023 Short Papers, Doctoral Consortium and Workshops: AIDMA, DOING, K-Gals, MADEISD, PeRS, Barcelona, Spain, 4–7 September, 2023, Proceedings (2023)
27. Matczak, M., Czochański, T.: Intelligent index tuning using reinforcement learning. In: New Trends in Database and Information Systems - ADBIS 2023 Short Papers, Doctoral Consortium and Workshops: AIDMA, DOING, K-Gals, MADEISD, PeRS, Barcelona, Spain, 4–7 September, 2023, Proceedings (2023)
28. Napolitano, E.V.: Intelligent technologies for urban progress: exploring the role of ai and advanced telecommunications in smart city evolution. In: New Trends in Database and Information Systems - ADBIS 2023 Short Papers, Doctoral Consortium and Workshops: AIDMA, DOING, K-Gals, MADEISD, PeRS, Barcelona, Spain, 4–7 September, 2023, Proceedings (2023)
29. Nistico', S.: Towards reliable machine learning. In: New Trends in Database and Information Systems - ADBIS 2023 Short Papers, Doctoral Consortium and Workshops: AIDMA, DOING, K-Gals, MADEISD, PeRS, Barcelona, Spain, 4–7 September, 2023, Proceedings (2023)
30. Ochs, M., Hirmer, T., Past, K., Henrich, A.: Design-focused development of a course recommender system for digital study planning. In: New Trends in Database and Information Systems - ADBIS 2023 Short Papers, Doctoral Consortium and Workshops: AIDMA, DOING, K-Gals, MADEISD, PeRS, Barcelona, Spain, 4–7 September, 2023, Proceedings (2023)

31. Paldauf, M.: Decentralised solutions for preserving privacy in group recommender systems. In: New Trends in Database and Information Systems - ADBIS 2023 Short Papers, Doctoral Consortium and Workshops: AIDMA, DOING, K-Gals, MADEISD, PeRS, Barcelona, Spain, 4–7 September, 2023, Proceedings (2023)
32. Perdek, J., Vranić, V.: Lightweight aspect-oriented software product lines with automated product derivation. In: New Trends in Database and Information Systems - ADBIS 2023 Short Papers, Doctoral Consortium and Workshops: AIDMA, DOING, K-Gals, MADEISD, PeRS, Barcelona, Spain, 4–7 September, 2023, Proceedings (2023)
33. Rocha, H.F., Nascimento, L.S., Camargo, L., Noernberg, M., Hara, C.S.: Labeling portuguese man-of-war posts collected from instagram. In: New Trends in Database and Information Systems - ADBIS 2023 Short Papers, Doctoral Consortium and Workshops: AIDMA, DOING, K-Gals, MADEISD, PeRS, Barcelona, Spain, 4–7 September, 2023, Proceedings (2023)
34. Rokis, K., Kirikova, M.: An archimate-based thematic knowledge graph for low-code software development domain. In: New Trends in Database and Information Systems - ADBIS 2023 Short Papers, Doctoral Consortium and Workshops: AIDMA, DOING, K-Gals, MADEISD, PeRS, Barcelona, Spain, 4–7 September, 2023, Proceedings (2023)
35. Steingartner, W., Sivý, I.: From high-level language to abstract machine code: An interactive compiler and emulation tool for teaching structural operational semantics. In: New Trends in Database and Information Systems - ADBIS 2023 Short Papers, Doctoral Consortium and Workshops: AIDMA, DOING, K-Gals, MADEISD, PeRS, Barcelona, Spain, 4–7 September, 2023, Proceedings (2023)
36. Wang, X., He, X., Wang, M., Feng, F., Chua, T.S.: Neural graph collaborative filtering. In: Proceedings of the 42nd International ACM SIGIR Conference on Research and Development in Information Retrieval, SIGIR'19, pp. 165–174. Association for Computing Machinery, New York (2019). https://doi.org/10.1145/3331184.3331267
37. Zheng, X., Dasgupta, S., Gupta, A.: P2kg: declarative construction and quality evaluation of knowledge graph from polystores. In: New Trends in Database and Information Systems - ADBIS 2023 Short Papers, Doctoral Consortium and Workshops: AIDMA, DOING, K-Gals, MADEISD, PeRS, Barcelona, Spain, September 4–7, 2023, Proceedings (2023)

AIDMA: 1st Workshop on Advanced AI Techniques for Data Management, Analytics

Data Exploration Based on Local Attribution Explanation: A Medical Use Case

Elodie Escriva[1,2(✉)], Emmanuel Doumard[3,4], Jean-Baptiste Excoffier[1], Julien Aligon[2], Paul Monsarrat[3,5,6], and Chantal Soulé-Dupuy[2]

[1] Kaduceo, Toulouse, France
elodie.escriva@kaduceo.com
[2] Université de Toulouse-Capitole, IRIT, (CNRS/UMR 5505), Toulouse, France
[3] RESTORE Research Center, Toulouse, France
[4] Université de Toulouse-Paul Sabatier, IRIT, (CNRS/UMR 5505), Toulouse, France
[5] Artificial and Natural Intelligence Toulouse Institute ANITI, Toulouse, France
[6] Oral Medicine Department, Toulouse, France

Abstract. Exploratory data analysis allows to discover knowledge and patterns and to test hypotheses. Modelling predictive tools associated with explainability made it possible to explore more and more complex relationships between attributes. This study presents a method to use local explanations as a new data space to retrieve precise and pertinent information. We aim to apply this method to a medical dataset and underline the benefit of using explanations to gain knowledge. In particular, we show that clusters based on local explanations, combined with decision rules, allow to better characterise patient subgroups.

Keywords: Medical Data Exploration · Explainable IA · Machine Learning

1 Introduction and Related Work

As data availability increased in the last decades, exploratory data analysis techniques have arisen to investigate data and discover patterns, make and test hypotheses with the help of statistics, graphical representation, clustering or predictive tools. In particular, Bottom-Up approaches consists in finding patterns and gaining insight by analysing data without making *a-priori* hypotheses [15,18]. Among the tools for exploratory data analysis, predictive approaches, primarily through machine learning, have made it possible to capture more complex statistical phenomena in the data that classical statistical techniques cannot understand. However, due to the lack of explanation of the predictions, the Machine Learning (ML) black box effect is a limitation for sensitive areas, such as those involving human lives. In the medical field, patients may legally ask for

A. Abelló et al. (Eds.): ADBIS 2023, CCIS 1850, pp. 315–323, 2023.
https://doi.org/10.1007/978-3-031-42941-5_27

the reasons behind a decision, which may be problematic when ML modelling is used in the decision-making process [9].

A way to better understand machine learning modelling and the prediction they produce lies in the Explainability domain (XAI). In particular, local explanations allow investigation of the reasons behind the model prediction for each instance. Local attribution methods like LIME [16], SHAP [13] or Coalitional-based methods [7] explain the prediction by computing the impact of each attribute for each instance. All these methods produce explanations called "influences", each with different strengths and weaknesses as detailed in [4].

Research has focused on the applicability, evaluation and uses of explanations, especially in the medical field. Influences can be used for multiples purposes: select attributes [11,17], find attributes relationships [2,10], determine subgroups and recommend instances based on influences [5,6], extract knowledge in data from influences [14]. Each paper shows that using influences is of great interest in the modelling pipeline and uses influences as a new data space to explore. However, not all papers strictly compare the benefit of using explanations to gain knowledge, compared to a classical analysis of raw data. And as explanations provide information on the modelling and complex interactions of the dataset, the contribution of influences must be assessed against raw data.

Then, our objective is to apply a bottom-up exploratory data analysis approach on a medical dataset, on both explanations and raw data, to highlight and compare the knowledge retrieved in both data spaces. We show that explanations can allow a deeper dataset investigation. This study can also show the usefulness of seeing explanations not only as an outcome but also as a tool.

The paper is structured as follows: we introduce our method and the dataset used in Sect. 2, demonstrate the usefulness of explanation-based analysis in Sect. 3 and discuss results and perspectives in Sect. 4.

2 Methods

2.1 Dataset

To enable reproducible results, we use an open-source dataset: Acute Inflammation dataset[1]. The Acute Inflammation dataset was created to develop an expert system for urinary disease. It consists of 120 patients, described by six attributes: Temperature *(35° C-42° C)*, Occurrence of nausea *(yes-no)*, Lumbar pain *(yes-no)*, Urine pushing (continuous need for urination, *yes-no*), Micturition pain *(yes-no)* and Burning of urethra, itch, swelling of urethra outlet (abbreviated as Urethra burning, *yes-no*). Each patient can have two different diseases of the urinary system: acute inflammation of urinary bladder (AIUB) and acute nephritis of renal pelvis origin. Patients may suffer from both diseases simultaneously, so this dataset is a multi-output problem. We only focus on the AIUB disease to have a binary classification problem. Medical staff defined AIUB as *"a sudden occurrence of pains in the abdomen region and the urination in form of*

[1] Dataset: https://archive.ics.uci.edu/ml/datasets/Acute+Inflammations.

constant urine pushing, micturition pains and sometimes lack of urine keeping. Temperature of the body is rising, most often not above 38C. The excreted urine is turbid and sometimes bloody" [3].

2.2 Modelling

The proposed method aims to analyse and explore datasets through modelling and influences. Based on a dataset of interest consisting of patients' medical records and their disease diagnosis, this method allows an understanding of interactions between patients' characteristics and the disease. It is divided into three parts, inspired by [5]:

(**1**) The first one consists of *ML predictive modelling*, to evaluate the risk of AIUB disease for each patient based on the understanding of the complex statistical relationship of the dataset. An XGBoost model, a boosted tree ensemble technique [1], is used for its efficiency. We use a nested cross-validation (CV) procedure to provide unbiased modelling (hyperparameters optimization with an inner 5-fold CV) and to evaluate performances and compute local explanations (through an outer 5-fold CV).

(**2**) Second step is the *explanation of the modelling* to provide individual explanations of the prediction for each patient, corresponding to individual risk and protective factors. TreeSHAP [12], a local attribution XAI method for tree-based predictive models, is used to compute influence explanations.

(**3**) Last step consists of *identifying subgroups of similar patients* to discover local patterns in the data and explain the subgroups characteristics. K-Medoids algorithm is used for the clustering task to ensure robustness against outliers, while the optimal number of groups was chosen with the Silhouette score. K-medoids algorithm is used on the influence explanations from step (2), with the advantages of taking into account the non-linear interactions discovered by the model while having all features at the same unit. Decisions rules for all clusters are computed with Skope-Rules algorithm [8]. Rules are computed to ensure perfect precision and recall of all rules: all instances of the cluster respect the rule, and all instances respecting the rule belong to the cluster.

3 Results

3.1 Raw Data Analysis

Populations and Statistical Tests. Table 1 shows the main characteristics of the dataset using raw data only, with results from statistical tests performed on AIUB and Non-AIUB patients: Student tests for quantitative attributes and Chi-squared test for qualitative attributes. Three attributes are defined as statistically significant to detect AIUB: Lumbar pain, Urine pushing and Micturition pain. Patients with lumbar pain seem to have less AIUB while having urine pushing and micturition pain correlate with an AIUB diagnosis.

Clustering and Rule-Based Analysis. To create homogeneous groups of patients, one method consists of performing clustering. The optimal number of clusters was 11, based on the silhouette scores in Table 2. Table 3 shows the rules defined by Skope Rules to describe each cluster. Rules have a median of 2.5 attributes per rule. All rules have perfect precision and recall with a maximum of three attributes, which is a small enough number of attributes to facilitate the interpretation of each rule. The most used attributes are urethra burning and temperature with six distinct occurrences, both previously defined as not significantly discriminating for AIUB diagnosis in Table 1. Only one cluster, Cluster 2, uses only significantly discriminating attributes. Also, having eleven clusters makes it challenging to easily understand the rules and clusters.

Table 1. Population characteristics. Mean and standard deviation are presented for quantitative attributes, and numbers and proportions for binary qualitative attributes. P-values were adjusted using Bonferroni correction to control family-wise error rate.

		Total	Non-AIUB	AIUB	p-value
	Nb patients	120	61 (50.8)	59 (49.2)	
Quanti.	Temperature	38.72 (±1.8)	39.15 (±1.9)	38.29 (±1.7)	0.0552
Quali.	Nausea	29 (24.2)	10 (16.4)	19 (32.2)	0.4224
	Lumbar pain	70 (58.3)	51 (83.6)	19 (32.2)	<0.01
	Urine pushing	80 (66.7)	21 (34.4)	59 (100.0)	<0.01
	Micturition pain	59 (49.2)	10 (16.4)	49 (83.1)	<0.01
	Urethra Burning	50 (41.7)	21 (34.4)	29 (49.2)	0.8814

Table 2. Silhouette Score for multiple numbers of clusters for Raw data.

K	2	3	4	5	6	7	8	9	10	11	12	13	14	15
Raw	0.56	0.44	0.37	0.42	0.46	0.51	0.54	0.54	0.56	**0.57**	0.56	0.56	0.56	0.56

3.2 XAI Analysis

Local Post-Hoc Explanations. An XGBoost model was also trained and explained through SHAP method [13]. The model had an accuracy of 98.33%, a sensitivity of 96.72%, a specificity of 100% and an AUC ROC Score of 99.06%. Figure 1 shows the SHAP mean absolute influences and the distributions of influences based on the attribute value. The three most important attributes were Micturition pain, Urine pushing and Temperature. Micturition pain and Urine pushing increases the risk of having AIUB. On the contrary, a higher temperature decreases the probability of having AIUB. In particular, having urine pushing also seems to have less impact on the prediction than not having urine pushing. In contrast, Nausea and Urethra burning have little to no impact on the predictions. For nausea, SHAP describes that having them increases the risk of AIUB for some patients and a subgroup of patients is identified.

Table 3. Decision Rules for clusters based on raw data, with the number of patient per cluster and the mean percentage of AIUB-risk.

	Rules	Nb	Mean %
1	Nausea = 1 & Urine pushing = 0	10	45.6
2	Lumbar pain = 0 & Urine pushing = 0	10	10.7
3	Nausea = 1 & Urethra burning = 1	9	72.2
4	Temperature < 39.85 & Micturition pain = 0 & Urethra burning = 1	10	13.0
5	Lumbar pain = 0 & Urethra burning = 1	20	97.1
6	Temperature < 38.95 & Temperature > 36.65 & Urine pushing = 0	13	11.0
7	Temperature < 38.95 & Lumbar pain = 0 & Micturition pain = 0	10	59.9
8	Nausea = 1 & Urine pushing = 1 & Urethra burning = 0	10	73.6
9	Temperature > 39.85 & Nausea = 0 & Urethra burning = 1	11	11.2
10	Lumbar pain = 0 & Micturition pain = 1 & Urethra burning = 0	10	97.1
11	Temperature < 36.65 & Urethra burning = 0	7	11.2

Figure 2 shows the distribution of influences only for patients having Nausea. Looking in details at these patients, they all suffer from lumbar pain, micturition pain and temperature above 40°C (which is higher than the dataset mean). There seems to be a subgroup of patients with a strong relationship between these four attributes. Moreover, for this subgroup of patients, there is a strong correlation between the attribute Urine Pushing and the presence of AIUB: when a patient has urine pushing, they have an AIUB; when they do not have urine pushing, there is no AIUB. This subgroup is probably best to study, as the nausea attribute may create a real-world bias due to its strong association with other attributes in the dataset.

Clustering and Rule-Based Analysis. As one subgroup is already discovered, clustering can help to find other subgroups of interest. For clustering on SHAP influences, the optimal number of clustering is set as 7, based on the silhouette score in Table 4. Table 5 shows rules defined by SkopeRules for clusters based on influences. These rules have a median of two attributes per rule and focus mainly on statistically relevant attributes. Only one rule consists of three attributes, and the most used attribute is Urine Pushing, with five occurrences. As shown before for the "Nausea subgroups", this attribute is the most important for patients with Nausea (clusters 4 and 6) and also for patients with lumbar pain (clusters 3 and 5). Urine Pushing does not appear in rules only for clusters 2 and 7, the two biggest clusters, where AIUB-risk is respectively very low and very high. These clusters may be interesting to study from a medical point of view to understand patients characteristics and why the Urine Pushing variable is not the most relevant variable to distinguish them from other clusters. Also, although Micturition pain is the most influential attribute for SHAP, it is not very present in the rules, mainly because this attribute seems replaced by the attribute Nausea in the clusters since there is a strong link between having Nausea and Micturition pain.

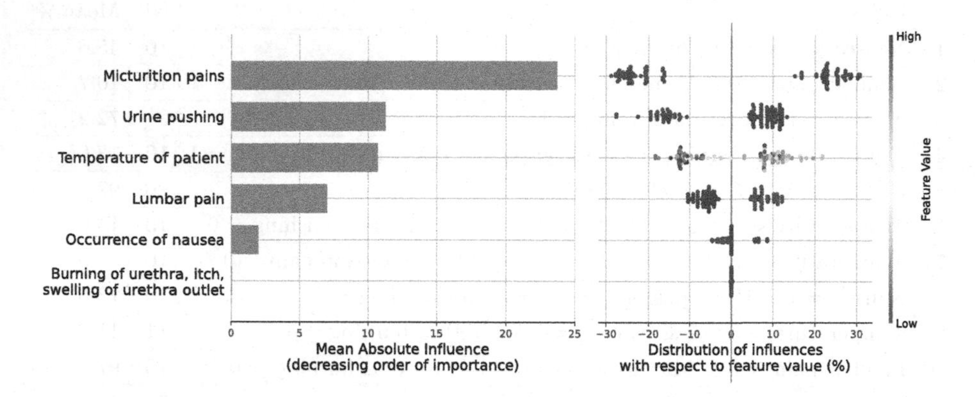

Fig. 1. SHAP mean absolute influences and Distribution of influences for the trained modelling.

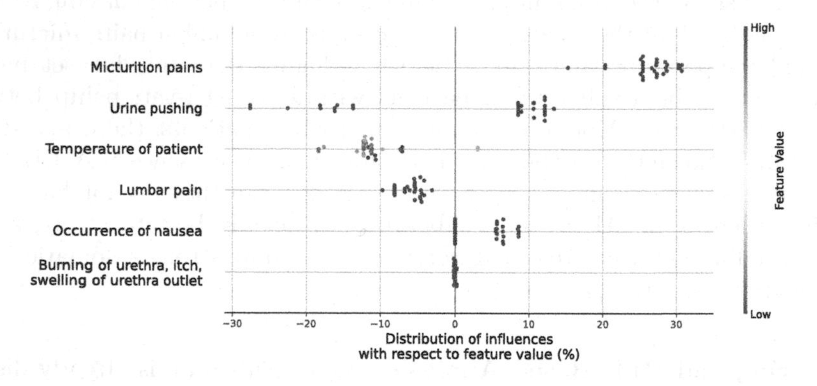

Fig. 2. Distribution of SHAP influences for patients with nausea.

Table 4. Silhouette Score for multiple numbers of clusters for XAI data.

K	2	3	4	5	6	7	8	9	10	11	12	13	14	15
XAI	0.59	0.59	0.52	0.62	0.69	**0.76**	0.74	0.69	0.63	0.61	0.61	0.61	0.62	0.67

Table 5. Decision Rules for clusters based on influences, with the number of patient per cluster and the mean percentage of AIUB-risk.

	Rules	Nb	Mean %
1	Temperature <= 38.89 & Urine pushing = 0	20	11.0
2	Micturition pain = 0 & Urethra burning = 1	21	12.0
3	Lumbar pain = 0 & Urine pushing = 0	10	10.7
4	Nausea = 1 & Urine pushing = 0	10	45.6
5	Lumbar pain = 0 & Urine pushing = 1 & Micturition pain = 0	10	59.9
6	Nausea = 1 & Urine pushing = 1	19	73.0
7	Lumbar pain = 0 & Micturition pain = 1	30	97.2

4 Discussion and Perspectives

In this study, both raw data and explainability methods detect patterns in the data, subgroups of patients and information about the relationship between the AIUB disease and patients' symptoms. In addition to the information known in the literature [3] and found in the raw data analysis, the explanation-based data analysis allowed risk and protective factors to be identified more concisely. Rules are mainly based on statistically significant attributes, adding interactions between attributes, and with the target class, compared to raw data analysis. The smaller number of clusters and attributes in each rule also simplifies the understanding of patient subgroups and the relationship of each attribute to the AIUB risk. With raw data, multiple clusters have similar mean percentages of AIUB risk and almost identical patients. The differences between these clusters are often based on attributes not important for detecting AIUB. This behaviour can be beneficial to study the dataset in-depth, less for discovering the attributes that truly impact the diagnosis of the disease and for capturing concise knowledge. The conciseness provided by influences also makes it easier to assign a new patient to a subgroup of patients to study their disease and risk factors. This advantage comes from the ability of ML modelling to capture more complex relationships than traditional statistical methods. Finally, the explanation data allowed the discovery of relevant subgroups of patients, including those with nausea. This subgroup has strong relationships between several attributes, and the presence of AIUB is based solely on the attribute Urine Pushing, making its study interesting for understanding the mechanisms of the disease in some patients. Finding this type of subgroup can help to investigate biases in the dataset, especially around the attribute Nausea.

However, the proposed method should be applied and tested in more complex medical contexts, with datasets having different characteristics, such as more observations, more attributes, more variability leading to lower model performances. Therefore, this is the principal axis of future work to identify the main improvement points so that the proposed method can be tested with practitioners in the loop and fully and reliably adopted by them.

Acknowledgement. We gratefully acknowledge the French ANRT and Kaduceo company for providing us with PhD grants (no. 2020/0964), and the Programme d'Investissements d'Avenir and the French ANR for the support of the national infrastructure ECELLFrance (PIA-ANR-11-INBS-005) and the PhD grant EUR CARe N°ANR-18-EURE-0003.

References

1. Chen, T., Guestrin, C.: XGBoost: a scalable tree boosting system. In: Proceedings of the 22nd ACM SIGKDD International Conference on Knowledge Discovery and Data Mining, pp. 785–794 (2016). https://doi.org/10.1145/2939672.2939785
2. Cooper, A., Doyle, O., Bourke, A.: Supervised clustering for subgroup discovery: an application to COVID-19 symptomatology. In: ECML-PKDD Proceedings (2021)
3. Czerniak, J., Zarzycki, H.: Application of rough sets in the presumptive diagnosis of urinary system diseases. In: AI and Security in Computing Systems (2003)
4. Doumard, E., et al.: A quantitative approach for the comparison of additive local explanation methods. Inf. Syst. **114**, 102162 (2023). https://doi.org/10.1016/j.is.2022.102162
5. Excoffier, J.B., Escriva, E., Aligon, J., Ortala, M.: Local explanation-based method for healthcare risk stratification. In: Medical Informatics Europe 2022. Studies in Health Technology and Informatics (2022)
6. Excoffier, J.B., Salaün-Penquer, N., Ortala, M., Raphaël-Rousseau, M., Chouaid, C., Jung, C.: Analysis of COVID-19 in patients in France during first lockdown of 2020 using explainability methods. Med. Biol. Eng. Compu. **60**, 1647–1658 (2022). https://doi.org/10.1007/s11517-022-02540-0
7. Ferrettini, G., Escriva, E., Aligon, J., Excoffier, J.B., Soulé-Dupuy, C.: Coalitional strategies for efficient individual prediction explanation. Inf. Syst. Front. (2021). https://doi.org/10.1007/s10796-021-10141-9
8. Gardin, F., Gautiern, R., Goix, N., Ndiaye, B., Schertzer, J.M.: Skope-rules (2019). https://github.com/scikit-learn-contrib/skope-rules
9. Hoofnagle, C.J., van der Sloot, B., Borgesius, F.Z.: The European union general data protection regulation: what it is and what it means. Inf. Commun. Technol. Law **28**, 65–98 (2019)
10. Lee, K., Ayyasamy, M.V., Ji, Y., Balachandran, P.V.: A comparison of explainable artificial intelligence methods in the phase classification of multi-principal element alloys. Sci. Rep. **12**, 11591 (2022)
11. Liu, Y., Liu, Z., Luo, X., Zhao, H.: Diagnosis of Parkinson's disease based on SHAP value feature selection. Biocybernetics Biomed. Eng. **42**, 856–869 (2022)
12. Lundberg, S.M., Erion, G.G., Lee, S.I.: Consistent individualized feature attribution for tree ensembles. arXiv preprint arXiv:1802.03888 (2018)
13. Lundberg, S.M., Lee, S.I.: A unified approach to interpreting model predictions. In: NeurIPS Proceedings (2017)
14. Monsarrat, P., et al.: Systemic periodontal risk score using an innovative machine learning strategy: an observational study. J. Personalized Med. **12**, 217 (2022). https://doi.org/10.3390/jpm12020217
15. Morgenthaler, S.: Exploratory data analysis. WIREs Comp Stats 1 (2009)

16. Ribeiro, M., Singh, S., Guestrin, C.: "why should i trust you?": explaining the predictions of any classifier. In: KDD Proceedings (2016)
17. Wang, H., Doumard, E., Soulé-Dupuy, C., Kémoun, P., Aligon, J., Monsarrat, P.: Explanations as a new metric for feature selection: a systematic approach. IEEE J. Biomed. Health Inform. **27**(8), 4131–4142 (2023)
18. Wirsch, A.: Analysis of a top-down bottom-up data analysis framework and software architecture design, Ph.D. thesis, MIT (USA) (2014)

Explainability Based on Feature Importance for Better Comprehension of Machine Learning in Healthcare

Pronaya Prosun Das[1(✉)], Lena Wiese[1,2], and ELISE STUDY GROUP

[1] Fraunhofer Institute for Toxicology and Experimental Medicine, Hannover, Germany
pronaya.prosun.das@item.fraunhofer.de
[2] Institute of Computer Science, Goethe University Frankfurt, Frankfurt a. M., Germany

Abstract. The use of Artificial Intelligence (AI) in healthcare is getting more prevalent, encompassing responsibilities like intelligent medical diagnoses and operative robots. The accuracy and performance of AI systems are prioritized by Machine Learning (ML) engineers while medical professionals are more interested in their applicability and usefulness in clinical settings. Unfortunately, medical practitioners often lack the necessary skills to interpret AI-based systems, limiting the usage of the tools that enhance healthcare solutions, automating routine analysis tasks and limiting expertise available for validation. Explainable Artificial Intelligence(XAI) is a field that focuses on methods to help understand and interpret ML models. However, most XAI research has been from a viewpoint of Computer Science (CS), with little focus on supporting other domains like healthcare. In this work, a straightforward solution is presented to increase the explainability of ML models to professionals from non-CS domains like healthcare experts. The suggested method integrates feature importance that assesses the influence of distinct features on AI-based system outcomes into standard ML workflows. This could permit medical experts to better understand AI-based systems, improving their ability to comprehend the usefulness and applicability of ML models.

ELISE STUDY GROUP: Louisa Bode [a]; Marcel Mast [a]; Antje Wulff [a, d]; Michael Marschollek [a]; Sven Schamer [b]; Henning Rathert [b]; Thomas Jack [b]; Philipp Beerbaum [b]; Nicole Rübsamen [c]; Julia Böhnke [c]; André Karch [c]; Pronaya Prosun Das [e]; Lena Wiese [e]; Christian Groszweski-Anders [f]; Andreas Haller [f]; Torsten Frank [f]
[a]Peter L. Reichertz Institute for Medical Informatics of TU Braunschweig and Hannover Medical School, Hannover, Germany
[b]Department of Pediatric Cardiology and Intensive Care Medicine, Hannover Medical School, Hannover, Germany
[c]Institute of Epidemiology and Social Medicine, University of Muenster, Muenster, Germany
[d]Big Data in Medicine, Department of Health Services Research, School of Medicine and Health Sciences, Carl von Ossietzky University Oldenburg, Oldenburg, Germany
[e]Research Group Bioinformatics, Fraunhofer Institute for Toxicology and Experimental Medicine, Hannover, Germany
[f]medisite GmbH, Hannover, Germany

A. Abelló et al. (Eds.): ADBIS 2023, CCIS 1850, pp. 324–335, 2023.
https://doi.org/10.1007/978-3-031-42941-5_28

Keywords: Explainable Artificial Intelligence (XAI) · Decision Tree · Feature Importance · SHAP · Healthcare

1 Introduction

The concept of interpretability refers to the level of transparency within a system's decision-making process [10]. If the operation of a machine learning model is easily comprehended, it is considered interpretable. On the other hand, explainability enables the inner characteristics of an intelligent platform or system more understandable to non-technical professionals. XAI focuses on developing methods to make an AI system explainable and interpretable. The interpretation and explanation of AI systems are viewed as domain-specific tasks, as different professions have different perspectives on what constitutes an "explainable" system. From the perspective of a machine learning engineer, interpretability is associated with understanding the inner workings of a system to adjust the technical parameters and enhance overall performance. From a viewpoint of a medical professional, interpretability involves comprehending the intrinsic characteristics of an AI-based system that mainly serves medical functions. On the other hand, explainability for a machine learning engineer involves presenting technical data in a comprehensible format to enable effective system evaluation, while explainability for healthcare or medical experts might involve understanding why a patient is provided with a specific treatment course. To consider a healthcare setting explainable, AI-based systems must account for the perspective of a physician, which poses several concerns including designing domain-agnostic XAI. Explanations must be context-specific to a specific domain, such as engineering, medicine, or healthcare. Additionally, certain explanations might be beneficial for one viewpoint although trivial for others.

Interpretability and explainability are important in AI-based solutions, encompassing the entire training workflow. ML engineers gain technical knowledge from the workflow, including pre-processing steps, used models, and evaluation criteria. Medical practitioners benefit from understanding performance metrics, comprehension of the data by ML models, and relevant medical data. XAI in a medical context introduces a multitude of complex problems, including establishing connections between the knowledge of a trained model and medical features, providing explanations for medical datasets [9], and understanding the impact of specific medical features on model efficiency and interpretation.

The challenges mentioned encompass a range of complex issues, such as the intricacy of incorporating XAI methodologies into established ML workflows [5], the absence of explainability in essential data and feature engineering procedures, the inadequacy of understanding the operation of a model in diverse medical settings, and the deficiency of comprehensive ML model explainability [16]. In order to seamlessly incorporate explainability and interpretability within the conventional ML workflow, Feature Importance (FI) based approaches are utilized in XAI strategies. These approaches facilitate the assignment of weights to individual features, thereby implying their respective significance in making accurate predictions or classifications by the ML model [6]. By leveraging FI techniques during data pre-processing, the selection of relevant features can be greatly enhanced.

Although it may be impractical to tackle all the challenges associated with XAI in a single paper, addressing the aforementioned problems can constitute a noteworthy advancement toward developing a more all-inclusive solution. This paper employs FI approaches as a means to facilitate XAI, offering a straightforward yet impactful approach for seamlessly integrating domain-agnostic explainability into the conventional ML workflow. This approach embraces a multifaceted perspective, allowing for comprehensive explainability at multiple levels. By seamlessly integrating the ML model's functioning with diverse intrinsic datasets across various settings, it facilitates a thorough understanding of the system's operation from multiple angles. The medical acceptance of AI-based systems can be strengthened using the resulting explainability, as elaborated in the forthcoming sections.

The paper is structured as follows: In Sect. 2, previous research on XAI, FI, and their application in practical ML scenarios is discussed. Section 3 presents the proposed methodology for integrating XAI into a standard ML workflow. The experimental results are outlined in Sect. 4. Section 5 offers a conclusion of the research.

2 Related Work

Significant endeavours have been undertaken to enhance the explainability and interpretability of ML models deployed in medical fields, as seen in previous studies [1,5,15]. ML practitioners can gain a deeper comprehension and assessment of crucial model components, including hyperparameters (e.g., input size, layer count) and parameters (such as weights), along with the resultant outputs (e.g., predictions, classifications) by enhancing the model's explainability. Moreover, this advancement facilitates medical professionals in effectively comprehending and validating the outcomes generated by ML models in line with their profound medical knowledge.

Numerous XAI techniques can be utilized in the medical field. Ante-hoc XAI methods are capable of achieving interpretability without any additional steps, making them more convenient for integration into existing ML workflows. These techniques encompass Generalized Additive Models (GAMs) [1], which are commonly employed to attain interpretability, even if it comes at the cost of slightly lower performance in comparison to more intricate ML models and Decision trees [2], Random Forests [2] which are characteristically interpretable models. Nevertheless, the comprehensive examination of their applicability in non-CS domains has been relatively limited in existing literature [5]. In a recent study [11], an XAI-based framework was proposed to integrate medical knowledge into

AI-based systems, with a focus on utilizing feature importance (FI) to facilitate the incorporation of medical knowledge in developing AI solutions. Another study [2] achieved interpretability by employing FI scores derived from Decision trees to assess feature importance in the classification of cervical cancer. However, the utilization of FI scores to enhance explainability from the viewpoint of medical experts was not explored, and the interpretability concerning the related dataset was not resolved.

Post-hoc XAI approaches are built for the sole purpose of achieving explainability and are utilized after training a machine learning (ML) model. Although they can be difficult to adopt, they offer the advantage of supporting multiple noninterpretable but high-performance classifiers [15]. As a post-hoc XAI method, Local Interpretable Model-agnostic Explanations (LIME) calculates feature importance (FI) scores to understand predictions produced by any ML classifier. However, it is essential to acknowledge that LIME relies on certain assumptions that may not hold true for different classifier types [14]. SHapley Additive exPlanation (SHAP) presents another post-hoc XAI technique that builds upon the concept of Shapley values, originally introduced in game theory. By quantifying the average estimated borderline contribution of a player in obtaining a payout across all possible player combinations, Shapley values offer valuable insights into the relative significance of distinctive features [8]. This method allows for a comprehensive analysis of each feature's contribution, enhancing our understanding of their influence on the model's predictions and overall performance. A comparative study [3] investigated the FI scores of SHAP and LIME, revealing that SHAP's FI scores were found to be more consistent than LIME's and were evaluated based on objective criteria such as separability, similarity, and identity, which hold significant importance when delivering explanations within a medical setting.

3 Methodology

The overall workflow of this work is illustrated in Fig. 1. This approach follows a conventional ML workflow and incorporates an additional stage focused on FI, which aids in achieving post-hoc explainability. In this work, the calculation of FI scores is seamlessly integrated into the existing workflow without modifying any preceding stages. These scores play a crucial role in enhancing the interpretability and explainability of both the ML model and the related dataset.

Comparative feature ranking: To ensure the relevance of features identified by ML models in the healthcare domain, careful validation is required [1]. The approach in this study involves generating FI scores through two different techniques. The primary approach incorporates FI scores from Decision trees, taking into account a feature's role in the classification process. The secondary approach utilizes Shapley values to assess the influence of a feature upon the model's final outcome. The resulting FI scores are then ranked in descending order to provide an overview of how the ML model ranks several features when

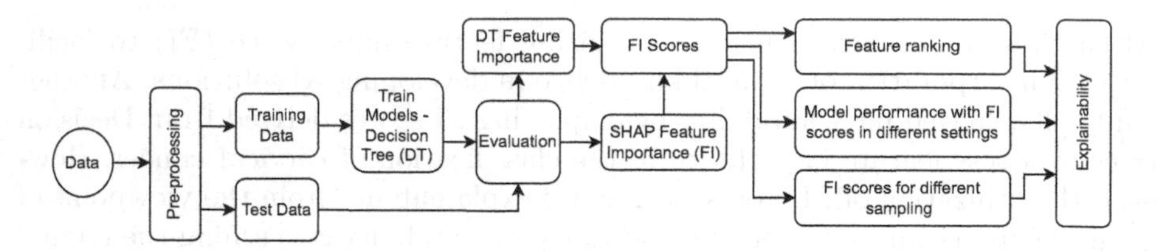

Fig. 1. The inclusion of a Feature Importance stage in a typical Machine Learning workflow to facilitate explainability.

generating an outcome. By utilizing the proposed approach, a meaningful comparison can be made between the features identified as significant by the classification model (as identified using the first method) and the features that have a notable impact on the outcome of the model (as identified using the second method). Mainly, explainability is achieved through the provision of features' relative ranking as perceived by the ML model, along with an assessment of their influence on the final outcome. Medical practitioners can use this information to understand which features are crucial in creating accurate medical prognoses. By emphasizing the features with significant values that hold substantial influence over the model's output, medical practitioners can prioritize their attention and decision-making process.

Feature Importance in a Variety of Medical Setups: Medical information is not equally available in all medical setups, and this can lead to variations in the methods adopted by medical professionals across various setups. Consequently, the effectiveness of AI-based solutions can be reduced. To ensure a robust solution, it is crucial to develop a gold-standard method that incorporates all relevant medical data. However, considering the varying degrees of data accessibility among different healthcare facilities, it is essential to create multiple versions of the solution. This approach improves the usefulness of the solution by considering the inclusion/exclusion of features in various setups. By evaluating changes in performance metrics and FI scores with different feature sets, a comprehensive understanding of an ML model can be achieved, highlighting its suitability across diverse medical setups. For instance, considering variations in patient data access between a general practitioner and an emergency room doctor, training the model on various feature subsets enables explainability and insights into feature impacts on model performance.

Understanding the Underlying Data: The way in which data is processed and used to train a model can have important implications in the medical field [9]. This paper explores the impact of data processing and augmentation methods on model performance in the medical field. By analyzing FI scores and performance metrics, insights are gained into the relationship between data variations and model outcomes. Medical experts can utilize this approach to validate feature rankings and understand their correlation with performance metrics across different augmentation methods. The paper provides discussions on the dataset, feature extraction, and modelling technique in Sects. 3.1, 3.2, and 3.3, respectively. Section 3.4 delves into the two FI methods employed: Decision trees-based FI and Shapley values-based FI.

3.1 Dataset

In this study, data from the pediatric intensive care unit at Hannover Medical School is used [17], which has been anonymized to protect patient privacy. The dataset contains information on 168 pediatric patients, including vital signs and laboratory test results, as well as information from medical devices. Each patient is identified by a unique study number, and each measurement has a corresponding timestamp. The age of the patients is also recorded, and blood pressure values have been added to the existing parameters. The dataset includes a gold standard for the presence of Systemic Inflammatory Response Syndrome (SIRS), which was established by two experienced pediatric intensive care physicians based on the SIRS diagnostic rules defined by the International Pediatric Sepsis Consensus Conference (IPSCC) [4]. The dataset covers 1,998 days of hospital stay for 168 patients, out of which 460 days were marked with a SIRS label according to the day-wise gold standard.

3.2 Feature Extraction

Our analysis involves examining a set of six health indicators that are considered relevant to the IPSCC criteria. These indicators include temperature, respiration rate, pulse rate, systolic and diastolic pressures, and leukocyte count. In addition, we incorporate information about the patient's birthdate, gender, and disease diagnosis to further explore the data. Birthdates are used to determine age groups. We extract these features from their respective datasets and divide them into hourly observations. Subsequently, we calculate the maximum, minimum, median, and mean values from the observations of each hour, resulting in a new dataset. Consequently, the processed dataset comprises 24 features, encompassing the maximum, minimum, median, and mean values for temperature, pulse, respiration, systolic and diastolic pressures, and leukocytes. Furthermore, the dataset includes four additional dimensions: study number, timestamp, age groups, and gender.

After aggregating all the data, it becomes evident that each patient has multiple observations within a single day, resulting in an imbalance in the number of observations for certain attributes. To address this issue of class imbalance, Imbalanced-learn is employed, with various data sampling approaches offered to correct the discrepancy between minority and majority classes [7]. The record counts for SIRS and NO SIRS diagnoses are presented in Table 1 after applying a sampling technique.

Table 1. A list of sampling techniques.

Approach	Sample Count	Ratio
No Sampling	14920	9240:5680
Random Over Sampling (ROS)	18480	9240:9240
Random Under Sampling (RUS)	11360	5680:5680
Adaptive Synthetic Over Sampling (ASOS)	18597	9357:9240
SMOTEtomek Combination Sampling (STOM)	18472	9236:9236

3.3 ML Model

A Decision Tree is a type of graph that represents sets of data samples using nodes and edges. Each node represents a set of samples and offers a corresponding impurity factor indicating class diversity. If all the samples in a node are part of the same label/class, the node is considered pure. During our experiment, we expanded nodes until either all the leaves became pure or until the leaves contained a number of samples below the minimum threshold required for splitting an internal node, which was set at 2. Decision trees are used in classification problems to decrease impurity by applying conditions on the edges. The interpretability of decision trees makes them a favourable choice compared to complex models like Artificial Neural Networks or Support Vector Machines, while still providing good performance scores for datasets [5,13,15].

3.4 Feature Importance (FI)

The process of FI involves identifying the crucial features that are considered by a ML model for making predictions or classifications. In this study, the FI scores were calculated using two methods: Decision Trees and SHAP. A feature's FI score in decision trees is measured by calculating how much it contributes towards reducing impurity during training [12]. The higher the FI score obtained, the more important that feature is considered by the decision tree model. On the other hand, Shapley values are calculated by determining the contribution of a feature to the model's output by including and excluding it and then weighting it based on the subsets of other features. By summing the values for all subsets of features, the weighted and permuted FI score is derived. We used TreeSHAP in this work. These FI scores, obtained through both approaches, play a crucial role in identifying the features that exert the greatest impact on the outcome of ML model.

4 Results

The main challenge in achieving explainability using FI was to effectively demonstrate the FI scores obtained from model training with various feature sets and data augmentation methods in a cohesive fashion. The objective was to establish their association with performance metrics like accuracy and F-scores, along with the features' relative ranking, in a way that can be applied across different domains. By integrating this information, it becomes possible to explain both the machine learning model itself and the underlying data, thereby expanding the range of explainability. At first, a number of models were trained using data sampled with various strategies as outlined in Table 1. The accuracies and F-scores for the different models are presented in Table 2.

Table 2. Accuracies and F-scores from different sampling strategies

Approach	Accuracy	F-scores
No Sampling	0.974	0.966
ROS	0.984	0.984
RUS	0.967	0.966
ASOS	0.982	0.982
STOM	0.980	0.980

From Table 2, we can see that Random Over Sampling gives us higher accuracy than other strategies. Therefore, ROS was chosen for the first two experiments. We implemented the three approaches for explainability and interpretability described in Sect. 3. The results are presented here. This contribution is significant as it establishes a foundation for a comprehensive workflow that facilitates the integration of Explainable Artificial Intelligence (XAI) in practical contexts, particularly in healthcare. This long-term workflow aims to streamline the process and make XAI more accessible and beneficial in various applied settings. Figure 2 displays the visualization of the FI scores allocated to various features when the ML model was trained with data that underwent ROS.

The results indicate that the age feature holds the highest rank, followed by leukocytes, according to both FI approaches. There is a resemblance in the obtained FI scores between the two methods, implying that features regarded as significant by the model (as determined by Decision Trees-based FI) will inherently exert a greater influence on the model's outcome (as indicated by Shapley values-based FI). The significance of these FI scores lies in the fact that they allow medical practitioners to comprehend and verify the feature ranking. This facilitates the integration of clinical expertise to enhance feature engineering methods and develop more refined models for future applications.

The ML model was trained using randomly oversampled data, with each iteration excluding the highest-ranked feature (based of SHAP FI Scores) from the previous iteration. The FI scores assigned to each feature after each iteration provide insights into the influence of excluding the top-ranked feature on the ML model. These results are visualized in Fig. 3.

The omission of the highest-ranked feature leads to a change in the relative ranking of other features. For example, in the case where all features are present

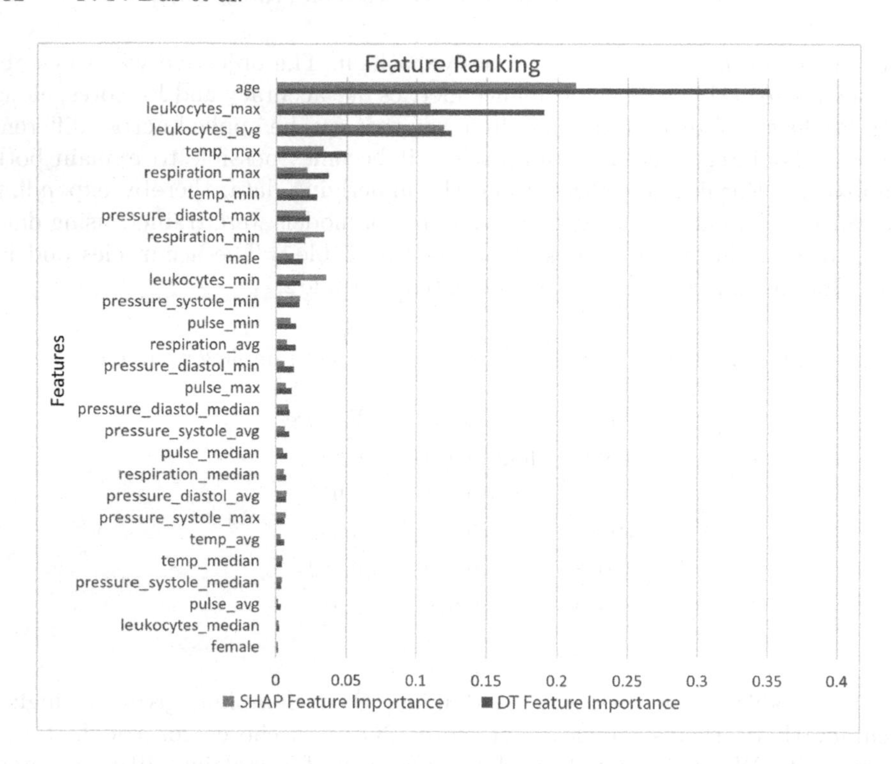

Fig. 2. Comparison of feature rankings based on Decision Tree FI and SHAP FI.

(labelled as "All"), the highest importance is assigned to age, accompanied by Leukocytes_max. However, when age is excluded (second bar), Leukocytes_max becomes the most important feature instead of age. This indicates that a feature's exclusion or inclusion may not follow a predetermined order based on a standardized method which usually incorporates all features. As the top-ranked features are progressively excluded from left to right, the total sum of FI scores in each model significantly decreases. Generating multiple instances involving a single model as well as validating the findings with medical expertise is necessary to understand the interplay between features. Healthcare professionals can choose an ML model based on performance metrics, evaluate available features in their medical setup, and assign appropriate importance scores to achieve desired outcomes.

The model underwent training using various sampling techniques, as explained in Sect. 3.2. FI scores were then plotted for each sampled version, as shown in Fig. 4. Each type of sampled data is represented by a specific colour in the FI score plot. The legend provides information about the performance metrics associated with each sampling technique.

The proposed method allows for the interpretation of the dataset by comparing the FI scores obtained using different data augmentation techniques. For instance, Table 2 shows that the FI scores obtained using the under-sampling

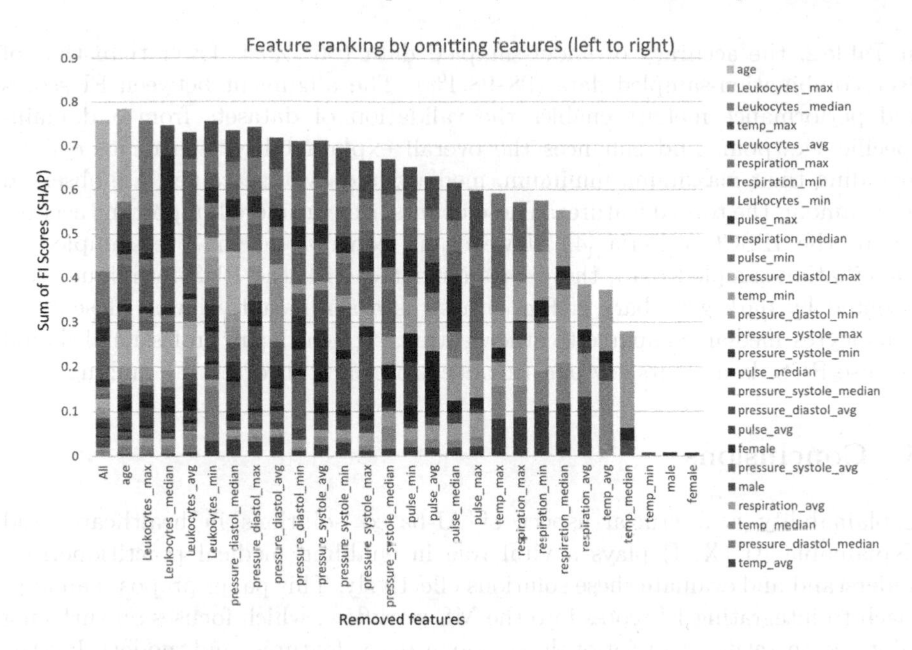

Fig. 3. The effect of removing the top-ranked features on FI scores.

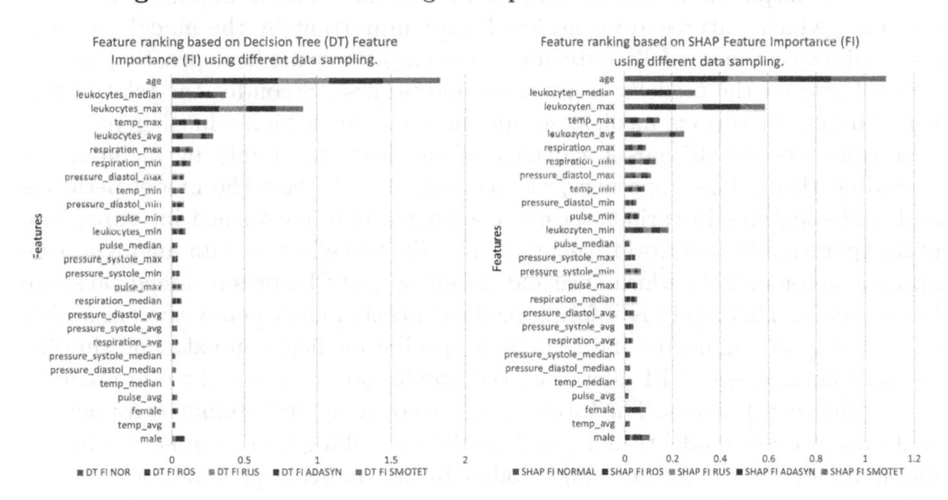

Fig. 4. Feature ranking for all sampling strategies. On the left, feature ranking based on Decision Tree FI and on the right, feature ranking based on SHAP FI.

(RUS) technique are not similar to those obtained using over-sampling or combination sampling techniques (such as SMOTE and ROS). This difference in scores can be attributed to the fact that under-sampling results in a smaller volume of data and consequently less diverse values for a feature. The sorted ordering of FI scores in under-sampling methods compared to combination sampling or over-sampling methods provides evidence of this observation. As depicted in Table 2, the accuracy of under-sampled data (96.7%) is lower than that of over/combination-sampled data (98–98.4%). The alignment between FI scores and performance metrics enables the validation of datasets from a domain-specific viewpoint and enhances the overall explainability. Leukocyte counts, including their maximum, minimum, median, and average values, are observed to be among the top 10 features. These counts are considered important according to the IPSCC criteria [4]. However, in contrast to the over-sampled or combination-sampled data, the data augmented with the RUS technique (represented by dark grey bars in Fig. 4) assigns a significantly lower FI score to Leukocytes_median compared to other features. A healthcare professional should not use RUS data for its low performance scores and incorrect FI ranking.

5 Conclusions

Explainability is a crucial aspect of AI-based solutions in healthcare, and eXplainable AI (XAI) plays a vital role in enabling medical practitioners to understand and evaluate these solutions effectively. This paper proposes an approach to integrating FI scores into the ML workflow, which focuses on surfacing information related to relationships among data, features, and models. Firstly, the model/output-based perspective focuses on the relative ranking of features, revealing which features are considered most important by the model and how they influence the model's outcome. This empowers medical practitioners to better interpret the model's decision-making process. Secondly, the relative feature ranking in various medical setups incorporates a hierarchical perspective that considers the diagnostic capacity of the features simply by including or excluding them. This approach provides insights into how the model performs and ranks features in various setups, allowing a better-informed interpretation of its operation in real-world scenarios. Thirdly, the effect of data augmentation approaches on model performance and the efficacy of FI scores in a medical setup is examined. This analysis informs medical practitioners about the suitability and validity of augmented data for their specific medical context. The simplicity yet effectiveness of FI allows for the application of these three approaches across different domains. The future goal is to develop an automated framework that encompasses model training and validation, taking into account the desired hierarchy level, to facilitate explainability from a multi-level standpoint.

Acknowledgement. The ELISE project is partially funded by the Federal Ministry of Health; Grant No. 2520DAT66A. This work was also partially supported by the Fraunhofer Internal Programs under Grant No. Attract 042-601000. Ethics approval for use of routine data was given by the Ethics Committee of Hannover Medical School (approval number 9819_BO_S_2021). We would like to thank our colleagues from the MHH Information Technology (MIT) from the Hannover Medical School for their support.

References

1. Caruana, R., Lou, Y., Gehrke, J., Koch, P., Sturm, M., Elhadad, N.: Intelligible models for healthcare: predicting pneumonia risk and hospital 30-day readmission. In: Proceedings of the 21th ACM SIGKDD International Conference on Knowledge Discovery and Data Mining, pp. 1721–1730 (2015)
2. Deng, X., Luo, Y., Wang, C.: Analysis of risk factors for cervical cancer based on machine learning methods. In: 2018 5th IEEE International Conference on Cloud Computing and Intelligence Systems (CCIS), pp. 631–635. IEEE (2018)
3. ElShawi, R., Sherif, Y., Al-Mallah, M., Sakr, S.: Interpretability in healthcare: a comparative study of local machine learning interpretability techniques. Comput. Intell. **37**(4), 1633–1650 (2021)
4. Goldstein, B., Giroir, B., Randolph, A., et al.: International pediatric sepsis consensus conference: definitions for sepsis and organ dysfunction in pediatrics. Pediatr. Crit. Care Med. **6**(1), 2–8 (2005)
5. Holzinger, A., Biemann, C., Pattichis, C.S., Kell, D.B.: What do we need to build explainable AI systems for the medical domain? arXiv preprint arXiv:1712.09923 (2017)
6. Hoyle, B., Rau, M.M., Zitlau, R., Seitz, S., Weller, J.: Feature importance for machine learning redshifts applied to SDSS galaxies. Mon. Not. R. Astron. Soc. **449**(2), 1275–1283 (2015)
7. Lemaître, G., Nogueira, F., Aridas, C.K.: Imbalanced-learn: a python toolbox to tackle the curse of imbalanced datasets in machine learning. J. Mach. Learn. Res. **18**(1), 559–563 (2017)
8. Lundberg, S.M., Lee, S.I.: A unified approach to interpreting model predictions. In: Advances in Neural Information Processing Systems, vol. 30 (2017)
9. Miller, D.D.: The medical AI insurgency: what physicians must know about data to practice with intelligent machines. NPJ Digit. Med. **2**(1), 62 (2019)
10. Miller, T.: Explanation in artificial intelligence: insights from the social sciences. Artif. Intell. **267**, 1–38 (2019)
11. Pawar, U., O'Shea, D., Rea, S., O'Reilly, R.: Explainable AI in healthcare. In: 2020 International Conference on Cyber Situational Awareness, Data Analytics and Assessment (CyberSA), pp. 1–2 (2020). https://doi.org/10.1109/CyberSA49311.2020.9139655
12. Quinlan, J.R.: Induction of decision trees. Mach. Learn. **1**, 81–106 (1986)
13. Quinlan, S., Afli, H., O'Reilly, R.: A comparative analysis of classification techniques for cervical cancer utilising at risk factors and screening test results. In: AICS, pp. 400–411 (2019)
14. Ribeiro, M.T., Singh, S., Guestrin, C.: "why should i trust you?" explaining the predictions of any classifier. In: Proceedings of the 22nd ACM SIGKDD International Conference on Knowledge Discovery and Data Mining, pp. 1135–1144 (2016)
15. Tjoa, E., Guan, C.: A survey on explainable artificial intelligence (XAI): Toward medical XAI. IEEE Trans. Neural Netw. Learn. Syst. **32**(11), 4793–4813 (2020)
16. Vaswani, N., Chi, Y., Bouwmans, T.: Rethinking PCA for modern data sets: theory, algorithms, and applications [scanning the issue]. Proc. IEEE **106**(8), 1274–1276 (2018)
17. Wulff, A., et al.: Clinical evaluation of an interoperable clinical decision-support system for the detection of systemic inflammatory response syndrome in critically ill children. BMC Med. Inform. Decis. Mak. **21**(1), 1–9 (2021)

An Empirical Study on the Robustness of Active Learning for Biomedical Image Classification Under Model Transfer Scenarios

Tamás Janusko[1](✉), Julius Gonsior[2], and Maik Thiele[1]

[1] Hochschule für Technik und Wirtschaft Dresden, Dresden, Germany
{tamas.janusko,maik.thiele}@htw-dresden.de
[2] Technische Universität Dresden, Dresden, Germany
julius.gonsior@tu-dresden.de

Abstract. Active learning (AL) is a popular training strategy that involves iteratively selecting training examples for annotation, typically those for which the current model is most uncertain. This allows to learn more effectively with fewer labeled examples beyond what could be achieved with random samples. However, AL research literature often over-simplifies the evaluation by assuming that the model to guide training data acquisition and the successor model used in the final deployment are identical. In real-world scenarios that is almost never the case since 1) due to performance reasons acquisition models often have less complexity compared to the successor models and 2) successor models are frequently replaced by potentially better performing models in productive environments. In this paper, we systematically study the effects of transferring an actively sampled training data set from an acquisition model to different successor models for biomedical image classification tasks. Our research shows that training a successor model with an actively-acquired data set is most promising, if acquisition and successor model are of similar architecture.

Keywords: Active Learning · Machine Learning · Image Classification

1 Introduction

High-quality training data is crucial for developing machine learning (ML) models. The process of data annotation, however, is extremely time-consuming and cost-intensive, especially in the domain of biomedical image classification, which heavily relies on domain expert annotators. To minimize the cost of data annotation projects, several methods have been developed, of which Active Learning (AL) is the most important representative.

The fundamental assumption of active learning is that each data sample has a different impact on the model to be trained. The goal of AL is to identify

Technical paper.

A. Abelló et al. (Eds.): ADBIS 2023, CCIS 1850, pp. 336–347, 2023.
https://doi.org/10.1007/978-3-031-42941-5_29

the best set of candidate samples that should be annotated to maximize the performance of a ML model.

AL involves an iterative process of selecting and annotating unlabeled data. Therefore, an initial ML model (*acquisition model*) is trained on a subset of the already labeled data. The acquisition model (AM) is used subsequently to identify the most informative samples that are then labeled by a human annotator. By doing so the labeled data set is continuously extended and the model is retrained on the updated data set.

As a result the AM and the training data derived by this model are strongly coupled. This approach is valid if the same AM is used in production as well. However, a data set acquired through AL may be utilized over a much longer time span than the (acquisition) model used to obtain the data. This raises the question whether the training data generated by AL can be transferred to successor model (SM), i.e. a model that may differ from the AM in size or type. A robust AL data set is therefore one that achieves a similarly good result for all SMs. In the case of biomedical image classification, there is another important argument why AM and SM are often not the same. Since AL is performed in an iterative manner with humans-in-the-loop, it is crucial that the AM, which has to be trained in each AL cycle, is fast enough to utilize the human annotators to full capacity and avoid waiting times. Thus, complex CNN architectures such as ResNet [7] or MobileNet [8] are ruled out as AMs. The distinction between generic acquisition and more specialized successor models is also useful as one can independently fine-tune successor model-specific hyperparameters on the acquired data set. From an XAI perspective AL's benefits are 1) the use of uncertainty as a self-declaration of the models view on the data, and 2) the fact, that an annotator is kept in the loop while each query gives insight into the AM's preferences. This way a human expert is able to determine whether the model is prone to selecting training instances with undesirable bias.

Contributions. In this paper we explore the transferability of training data generated by AL for biomedical image classification by investigating whether the labeled data derived from an AM trained through AL can be effectively used for training SMs. We also investigate the possibility of using simpler and faster AMs that can keep human annotators busy during the labeling task without sacrificing the final SM's performance. We identify where and under which circumstances AL is preferable to a random sampling baseline. Our findings contribute to a better understanding of the usefulness of AL for biomedical image classification, particularly the implications of introducing new SMs to a production system.

Outline. The remainder of this paper is organized as follows: In Sect. 2 we introduce the AL cycle, it's different query types and query strategies. The experimental setup to compare acquisition and successor model performance is presented in Sect. 3, followed by an experimental evaluation in Sect. 4 and a discussion in Sect. 5. Finally, we present related work in Sect. 6 and conclude in Sect. 7.

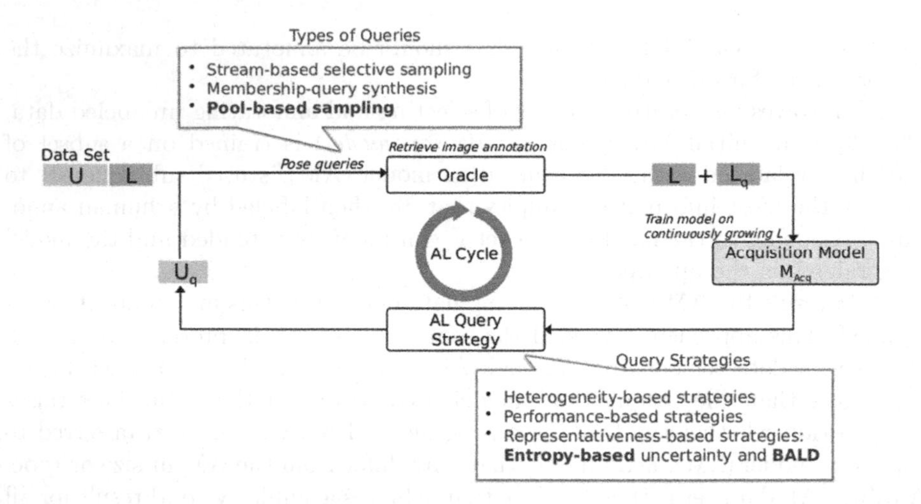

Fig. 1. Active Learning cycle using an acquisition model to label a set of samples L

2 Active Learning Foundations

The problem of active learning has been extensively studied in traditional machine learning literature and has more recently been explored in the context of deep learning. In this section we briefly summarized the core concepts of AL (Sect. 2.1) and especially emphasize different query types (Sect. 2.2) and query strategies (Sect. 2.3).

2.1 Active Learning Cycle

Figure 1 sketches the AL cycle or *acquisition phase* and its main steps. The AL process starts with two sets of data: The unlabeled sample set U and the already labeled data set L, with $|L| << |U|$. Initially, the AM is trained on the small labeled data set L. The main task of the AL cycle is to iteratively increase the set of labeled data L by identifying the most promising samples in U. The AL cycle is performed by two main actors: The AM and an *oracle*. The AM is continuously retrained on the old and newly labeled data L_q. The oracle maintains the label-providing entity, typically a human annotator.

A query is a selection of instances from the data set. Based on the utilized *query strategy* (see Sect. 2.3) an unlabeled sample U_q or a batch of unlabeled samples is chosen from the unlabeled *pool* U. The purpose of this task is to identify such samples that contribute most to the model to be trained. All other samples in U remain unlabeled, potentially saving a lot of human effort. The samples proposed by the query strategy U_q are then given to the oracle resulting in L_q. The newly labeled data L_q is added to L and the process starts again by retraining the AM on the extended data set. The AL cycle proceeds until a *stopping criterion* is met or until U is out of queries, meaning the whole data set is labelled. Stopping criteria can be performance thresholds of various metrics, such as classification accuracy, or simply the exhaustion of the labeling budget. For further references regarding details of AL see [1,15].

2.2 Types of Queries

Queries can be directed to an oracle in different forms. In the case of *stream-based selective sampling* unlabeled samples are drawn from the underlying data distribution individually and examined by the AM. It decides whether the sample should be forwarded to the oracle for annotation. *Membership-query synthesis* denotes the generation of new instances that resemble samples from the data set, typically used with generative models. These synthetic samples can be passed to the oracle instead of original data samples, effectively enlarging the training data set. *Pool-based sampling* is the most commonly used type of query, where a pool of samples is readily available. This is especially helpful when querying multiple samples at once according to a query strategy (see Sect. 2.3). For our experiments, described in Sect. 3, we will make use of the pool-based sampling approach.

2.3 Query Strategies

The presentation of queried samples to the oracle is preceded by the actual selection of samples. For the purpose of selection, several query strategies have been devised which can be categorized by whether they focus on the quality of query candidates, or their implications for model performance.

Heterogeneity-Based Strategies. To avoid quasi-redundant annotations, it is desirable to select those samples that bring most novel information into a model. *Uncertainty sampling* identifies those instances the AM is most uncertain about, which is the most heterogeneous region in the data space. In the case of a simple linear regression, these are the data points closest to the separating regression line. This approach focuses on the *exploitation* of a decision boundary. Prediction uncertainty in deep learning can be derived from ensembles of deep neural networks [11]. *Query-by-committee* is a disagreement-based approach, where multiple (usually smaller) models trained on the currently annotated data predict the labels of a set of unlabeled samples. The samples with the smallest prediction overlap are then selected for annotation [16]. Here heterogeneity refers to that of the predictions. *Expected model change* is approximated by selecting samples, that are most dissimilar to the previously annotated samples. However, this strategy works only for models with gradient-based training. Selecting diverse data points is often called *exploration* and has to be weighed against the exploitation strategy.

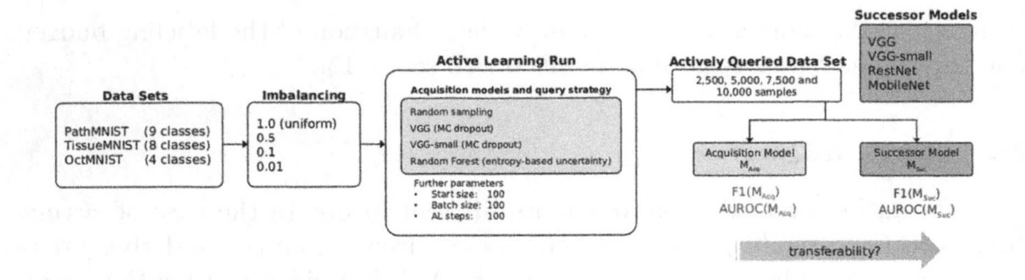

Fig. 2. Experimental Setup

Performance-Based Strategies. As heterogeneity-based strategies focus solely on the quality of the query candidates, performance based strategies take into account all unlabeled samples. *Expected error reduction* assumes, that the increase in model certainty can be linked to overall lower error rates. Given a set of query candidates, one can select samples that minimize the error rate on the remaining unlabeled instances. *Expected variance reduction* is also founded in the link between certainty and error rate but additionally takes into account the observation, that variance typically decreases with the error rate. This is used to addresses the computational expense of expected error reduction, as variance can be observed without retraining the model for each new query candidate.

Representativeness-Based Strategies. Representativeness-based query strategies combine the expected error/variance reduction and heterogeneity-based approach by assigning weights to each sample. The weights model the underlying distribution of the data, so representative samples are preferred, although the entire unlabeled data set is considered for the query.

3 Experimental Setup

To explore the transferability of training data generated by AL from an AM to a SM we use the experimental setup shown in Fig. 2 and which is described in detail in the following sections. In Sect. 3.1 we present our AL setup regarding the type and extent of AL queries, as well as utilized methods of uncertainty estimation for different AMs and implementation tools. In order to study the impact of AM choice, we consider three lightweight ML models along with random sampling as a baseline. Actively queried data sets from each AM are then used to train four different SMs. Details on model choice and implementation are presented in Sect. 3.2. Acquisition and successor training are carried out on three different base data sets, where for each data set four variants with different class distributions are derived. This is motivated by the fact that real-world data is also often skewed. Section 3.3 gives further information about the source, content and properties of the data.

3.1 AL Setup

In our experimental setup we use pool-based uncertainty sampling for our experiments with a randomly sampled initial pool of size 100 and 100 AL steps with query size of 100, which yields an AL set of size 10, 000. The entire unlabeled fraction of the data sets is made available as the pool to sample from. AL sets are persisted after 2, 500, 5, 000, 7, 500 and 10, 000 queried samples. This setup is used on each data set with every imbalance configuration for all AMs. It should be noted, that our approach is purely exploitative and relies only on the uncertainties retrieved from the AMs, with no exploratory elements whatsoever. For the CNN models we use the BaaL framework [4], which implements an elegant solution to derive an ensemble from a single network (and mathematically sound approximation of Bayesian inference for uncertainty estimation). MC Dropout [6] leverages dropout layers within a CNN. By enabling them during inference, a portion of nodes in the network is randomly deactivated. This turns the otherwise deterministic network into a slightly altered version of itself. Repeating this process yields an ensemble of CNNs, from whose predictions the uncertainty can be calculated given a heuristic. We use a dropout rate of 0.5 and entropy as the heuristic. Uncertainties for each AL step are computed from 15 MC sampling iterations, i.e. an ensemble of 15 CNNs. AL for random forest (RF) classifiers is implemented in the Scikit ActiveML framework [10]. Again, we use the entropy based uncertainty score.

3.2 Acquisition and Successor Models

Since it is essential for an AM to have a short runtime, we implement two lightweight VGG-type CNNs [17] with three stacks of convolution layers, each followed by a max-pooling layer, whereas the small CNN has two convolutional layers in each stack, 16 filters in the first stack, 32 in the second and 64 in the third. The larger CNN's stacks consist of two convolutional layers with 32 filters, two with 64 and three with 128. Kernel sizes are always 3×3 for convolutional layers, 2×2 for max-pooling layers. Batch size is set to 512 and we train each net for 15 epochs with a learning rate of 0.001.

Additionally we employ a vanilla RF classifier with default parameters as defined in Scikit-Learn [14]. Random sampling from the underlying data set is used as a baseline against which we benchmark the other AMs.

We examine the transferability of the queried data points among our acquisition models and additionally train larger networks on the AL data sets, namely ResNet50 [7] and MobileNetV1 [8]. SMs are trained from scratch, i.e. no pre-trained models are used. Therefore, results for the larger models are unlikely to be up with state-of-the-art reports, given the maximum training set size of 10, 000. For each configuration five separate models of a type are trained on the AL sets and then evaluated on the original held out sets. The scores are averaged to obtain more robust F1 and AUROC scores.

3.3 Data Sets

We use publicly available data from the medical imaging domain concerned with colon pathology, kidney cortex microscopy and retinal OCT. Thus experiments are conducted on images from the MedMNIST benchmark, namely PathMNIST, TissueMNIST and OctMNIST [18]. All data sets contain images scaled to 32×32 pixels by us, with 3 RGB, a single greyscale and 7 greyscale channels, respectively, and 9, 8 and 4 distinct classes.

While the majority of academic image data sets are uniformly distributed, many real-world data sets follow a long tail distribution. In order to also study the impact of AL on imbalanced data, data sets with desired imbalance are derived by sampling randomly from the original data sets to create class imbalances with factors 0.5, 0.1 and 0.01. E.g. the smallest class has 0.5 times the number of data points as the largest class with exponentially increasing sizes between. Additionally, sets with equal class distribution are sampled. These constructed data sets form the basis for the AL process.

4 Results

We present our observations regarding the impact of AMs in Figs. 3, 4, 5 and 6 - one Figure per imbalance specification. Each row corresponds to one specific MedMNIST data set and each column to one SM evaluated in the graphs. Each subplot contains eight graphs in total, where colors denote the AM used to derive the data the SMs were trained on, and solid/dashed lines distinguish between AUROC and F1-Scores. *VGG-type* denotes both VGG and VGG-small models. Generally, both F1-Score and AUROC paint a similar picture in terms of AM ranking. The surprisingly poor performance of the MobileNet successor model has to be noted, as it does not improve even with increasing training data. Also the fact, that the more potent ResNet and MobileNet appear less susceptible to different AMs than VGG-type successors. Random sampling (orange line) is the most favorable acquisition method for VGG-type successors on all 0.5-imbalanced data sets (Fig. 4) and mostly so for 0.1-imbalances (Fig. 5).

We observe a big lead of MobileNet on F1-Scores on the balanced OctMNIST, which is less pronounced on the other balanced data sets (Fig. 3), but also present on 0.5- and 0.1-imbalanced Path- and TissueMNIST (Figs. 4 and 5). ResNet SMs perform worst on balanced PathMNIST and OctMNIST, when trained with randomly sampled data. This can also be observed on 0.01-imbalanced OctMNIST (Fig. 6).

There is no constellation where the VGG AM consistently outperforms other models. Leads are punctual and only by small margins, like for the MobileNet successor on balanced TissueMNIST on the full data set (Fig. 3) and with MobileNet on 0.01-imbalanced Tissue- and PathMNIST on larger set sizes. It generally is middle of the pack, with the notable exception of VGG-type SMs on all OctMNIST and 0.5-imbalanced PathMNIST sets, where it benefits the least from increasing data set sizes. Especially VGG-type successors perform by far the worst with VGG acquisition on balanced OctMNIST.

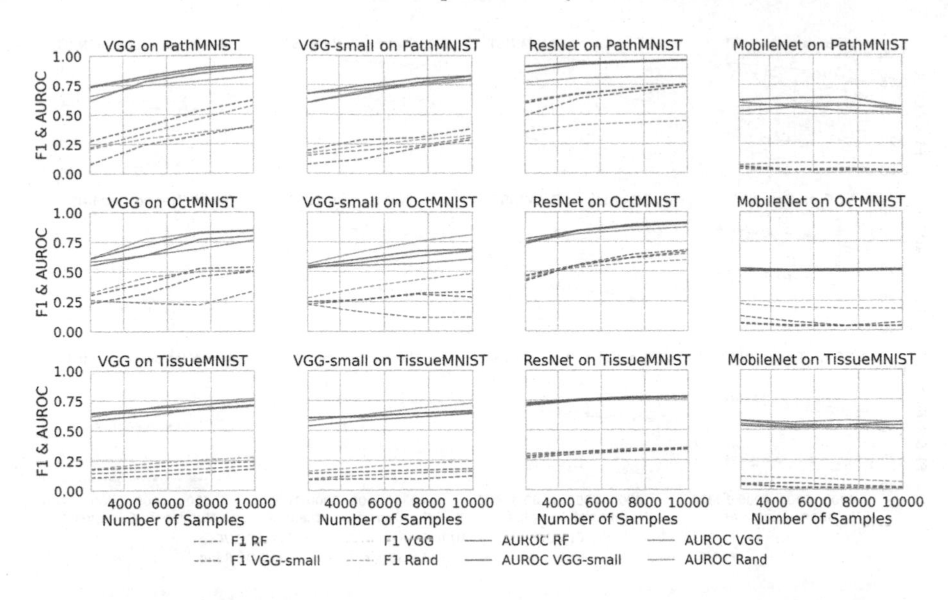

Fig. 3. AUROC and F1-Scores of Successor Models for Class Imbalance 1.0

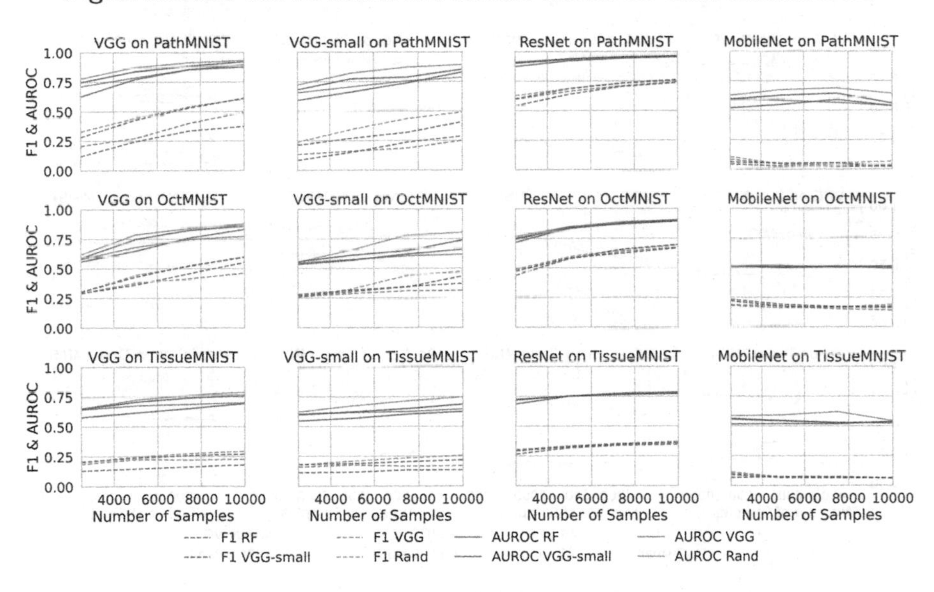

Fig. 4. AUROC and F1-Scores of Successor Models for Class Imbalance 0.5

The most surprising observation is that VGG-small consistently outperforms the larger VGG AM, often by a sizable margin, for example on 0.5-imbalanced PathMNIST (Fig. 4). It is also the second best scoring AM after random sampling in most cases and often the leading model, such as on balanced PathMNIST, or on all 0.01-imbalanced data sets for VGG-successors (Fig. 6).

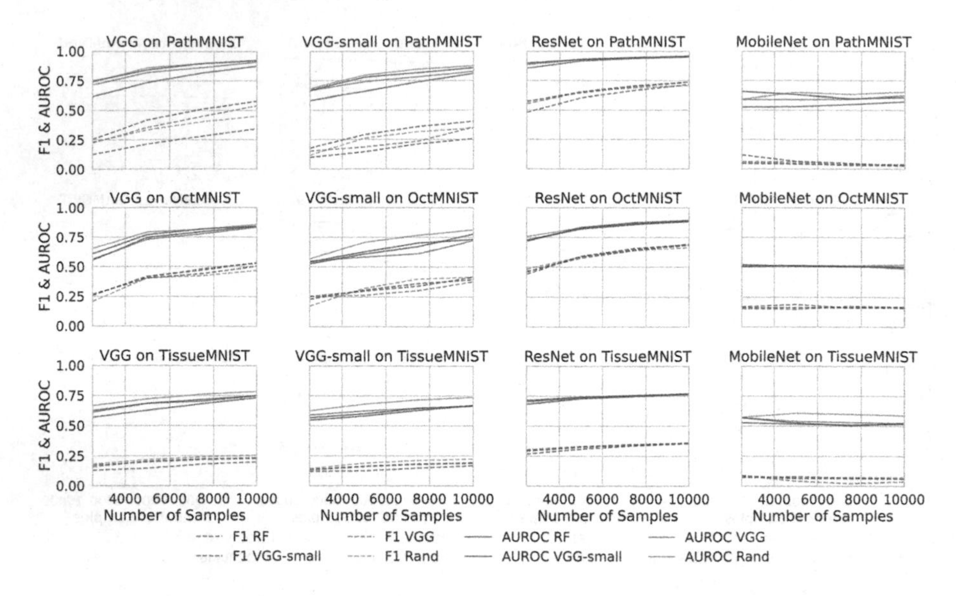

Fig. 5. AUROC and F1-Scores of Successor Models for Class Imbalance 0.1

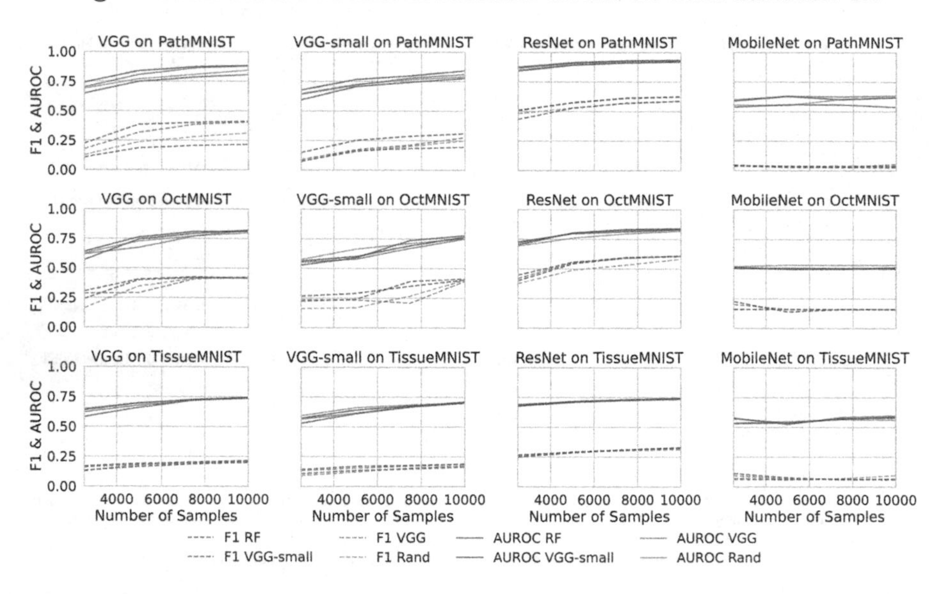

Fig. 6. AUROC and F1-Scores of Successor Models for Class Imbalance 0.01

Random forest is generally the worst performing AM, with the exception of a VGG-small SM on the heavily imbalanced OctMNIST (Fig. 6) and ResNet on the balanced OctMNIST (Fig. 3), where it takes the lead on larger data sets. However, on OctMNIST random forest consistently manages to perform averagely.

5 Discussion

We observed that all aspects of our investigation impact the transferability of AL data sets, whether it is the SM, the underlying data or its class imbalance. This is highlighted by the similarity of results from VGG-type SMs. However, AMs show relatively consistent behaviour across data sets, compared to their impact across SMs on a single data set.

Generally, AL appears to have a tendency to steer class distribution towards an equilibrium. Consolidation of scores with increasing class imbalance and number of samples is due to the fact, that the stock of informative samples from small classes is simply exhausted and therefor all AMs have acquired the same samples for those classes. This is also the reason for the graphs flattening out, as the information from underrepresented small classes cannot be compensated by samples from the remaining classes.

The fact that VGG-type SMs are the most receptive for VGG-type AMs poses the question where this behaviour stems from. And if there is a coupling of similar acquisition and successor models, can other model architectures benefit from having a slimmed-down version of themselves as AMs, as was observable with VGG-small acquisition and VGG SM combinations?

Regarding the divide of VGG SMs and more capable ResNet and MobileNet: The latter generally do not display the same susceptibility to different AMs than VGG successors. This can be explained by them either requiring more input to function at all or their higher capacity to deal with noisy data which would be deemed uninformative by simple models.

When comparing results from the different data sets it is apparent that the choice of AMs must be in accordance with data properties. The poor performance of the random forest model illustrates the need of an AMs capacity to match those properties. When used as a SM, random forest displayed no performance increase with additional training data. We interpret this as the model having hit its limit.

Although random sampling remains a strong baseline we observe cases where the SM does not fare well with random acquisition at all. Further, we have lightweight AMs performing very well across all settings and often outperforming random sampling. We interpret this as interaction of data and model characteristics that needs to be addressed and believe that the dichotomy of lightweight AMs and more potent SMs of the same type can be leveraged to build strong AL setups in a problem-oriented manner. Runtimes for all considered acquisition models are well within the bounds of what is justifiable in AL deployment, considering that 100 samples have to be labeled each step. This makes its application feasible even in settings, where no specialized hardware is available.

6 Related Work

Lowell et al. [13] evaluated the performance of AL methods for text classification and named entity recognition (NER) under varying acquisition and successor model combinations. The findings for text classification indicate that AL performs unreliably, with gains often marginal and inconsistent or worse performance for SMs compared to the AM. For NER, the results seem more favorable. However, the performance of individual AL acquisition models varies considerably over data sets and domains. Overall, Lowell et al. stated that the actively acquired data sets may be disadvantageous for training subsequent models and raise serious concerns regarding the efficacy of AL in practice.

Jelenić et al. [9] addressed the concerns regarding AL data set transferability. For NLP tasks with transformer-based language models they also found transferability to be highly dependent on the underlying data set. Additionally they devised a metric, *acquisition sequence mismatch*, to measure the similarity of acquired data point sequences which has proven to be indicative of successful transferability. Although our results also exhibit some inconsistencies regarding the benefits of acquired data sets, we derive a more positive view on AL (see Sect. 5).

In the research literature, AL is mostly studied on uniformly distributed data sets. However, there are some efforts to extend AL in order to cope with imbalanced data sets: Aggarwal et al. [2] modified the acquisition functions to find representative and diversified samples. They also introduced a sample balancing step, which reduces the propagation of imbalance of the unlabeled to the labeled data set. Similarly, Bengar et al. [5] introduced Class-Balanced Active Learning (CBAL), an optimization framework that takes class-balancing into account and which can be combined with most existing active learning algorithms.

Closely tied to AL is the notion of model uncertainty which we derived via MCDropout. While this is a reliable method, its computational costs limit the size of AMs in practice. Approaches such as SNGP [12] and DUQ [3] promise to achieve state-of-the-art results while only requiring a single inference run, thus drastically reducing computational costs.

7 Conclusion

In this work we investigated the transferability of AL data sets acquired by lightweight ML models and random sampling to other, more capable successor models. Experiments were conducted on three data sets from the medical imaging domain and with varying class imbalances. We find, that random sampling remains a strong baseline, often outperforming our best acquisition models. Data set imbalances are addressed by AL, but its use is limited in a sparse-data context, namely the low total number of instances in underrepresented classes.

Acquisition and successor models of similar architecture appear to be coupled. Future work is required to investigate the nature of this coupling and discern, whether slimmed-down versions of a successor model have advantages over non-related acquisition models.

References

1. Aggarwal, C.C.: Data Classification: Algorithms and Applications. Chapman & Hall/CRC, 1st edn. (2014)
2. Aggarwal, U., Popescu, A., Hudelot, C.: Minority class oriented active learning for imbalanced datasets. In: 2020 25th International Conference on Pattern Recognition (ICPR). IEEE (2021)
3. van Amersfoort, J., Smith, L., Teh, Y.W., Gal, Y.: Uncertainty estimation using a single deep deterministic neural network (2020)
4. Atighehchian, P., Branchaud-Charron, F., Freyberg, J., Pardinas, R., Schell, L., Pearse, G.: Baal, a Bayesian active learning library (2022). https://github.com/baal-org/baal/
5. Bengar, J.Z., van de Weijer, J., Lopez-Fuentes, L., Raducanu, B.: Class-balanced active learning for image classification. CoRR abs/2110.04543 (2021)
6. Gal, Y., Ghahramani, Z.: Dropout as a Bayesian approximation: representing model uncertainty in deep learning (2015)
7. He, K., Zhang, X., Ren, S., Sun, J.: Deep residual learning for image recognition (2015)
8. Howard, A.G., et al.: MobileNets: efficient convolutional neural networks for mobile vision applications (2017)
9. Jelenić, F., Jukić, J., Drobac, N., Šnajder, J.: On dataset transferability in active learning for transformers (2023)
10. Kottke, D., et al.: scikit-activeml: a library and toolbox for active learning algorithms. Preprints (2021)
11. Lakshminarayanan, B., Pritzel, A., Blundell, C.: Simple and scalable predictive uncertainty estimation using deep ensembles (2017)
12. Liu, J.Z., Lin, Z., Padhy, S., Tran, D., Bedrax-Weiss, T., Lakshminarayanan, B.: Simple and principled uncertainty estimation with deterministic deep learning via distance awareness (2020)
13. Lowell, D., Lipton, Z.C., Wallace, B.C.: Practical obstacles to deploying active learning. In: Proceedings of the 2019 Conference on Empirical Methods in Natural Language Processing and the 9th International Joint Conference on Natural Language Processing (EMNLP-IJCNLP), pp. 21–30. Association for Computational Linguistics, Hong Kong (2019)
14. Pedregosa, F., et al.: Scikit-learn: machine learning in Python. J. Mach. Learn. Res. **12**, 2825–2830 (2011)
15. Settles, B.: Active learning literature survey. Computer Sciences Technical Report 1648, University of Wisconsin-Madison (2009)
16. Seung, H.S., Opper, M., Sompolinsky, H.: Query by committee. In: COLT 1992, pp. 287–294. ACM, New York (1992)
17. Simonyan, K., Zisserman, A.: Very deep convolutional networks for large-scale image recognition (2014)
18. Yang, J., et al.: MedMNIST v2 - a large-scale lightweight benchmark for 2D and 3D0 biomedical image classification. Sci. Data **10**(1), 41 (2023)

Evaluating the Robustness of ML Models to Out-of-Distribution Data Through Similarity Analysis

Joakim Lindén[1,3](✉), Håkan Forsberg[1], Masoud Daneshtalab[1], and Ingemar Söderquist[2,3]

[1] Mälardalen University, Västerås, Sweden
joakim.linden@mdu.se
[2] Royal Institute of Technology, Stockholm, Sweden
[3] Saab AB, Linköping, Sweden

Abstract. In Machine Learning systems, several factors impact the performance of a trained model. The most important ones include model architecture, the amount of training time, the dataset size and diversity. We present a method for analyzing datasets from a use-case scenario perspective, detecting and quantifying out-of-distribution (OOD) data on dataset level.

Our main contribution is the novel use of similarity metrics for the evaluation of the robustness of a model by introducing relative Fréchet Inception Distance (FID) and relative Kernel Inception Distance (KID) measures. These relative measures are relative to a baseline in-distribution dataset and are used to estimate how the model will perform on OOD data (i.e. estimate the model accuracy drop). We find a correlation between our proposed relative FID/relative KID measure and the drop in Average Precision (AP) accuracy on unseen data.

Keywords: datasets · neural networks · similarity metrics · accuracy estimation

1 Introduction

Properly trained models for perception tasks such as object classification, detection and semantic segmentation, show great performance in today's state-of-the-art works [17,19]. It is however not often the case that extensive care has been put into designing the dataset used to train said models. Especially in dependable systems, the use of data driven perception functions like machine learning vision models, requires specific dataset management procedures to ensure the relevance and sufficiency of captured and/or generated data for the task [2]. To be more specific, one need to assure that the data used for training a model sufficiently spans the operating design domain (ODD) for the intended use of

This work was partially funded by Sweden's Innovation Agency and the Swedish Foundation for Strategic Research.

A. Abelló et al. (Eds.): ADBIS 2023, CCIS 1850, pp. 348–359, 2023.
https://doi.org/10.1007/978-3-031-42941-5_30

the function when operating in its real-world environment. This includes scenario diversity where different scene parameters having a visual impact on the rendered scene are varied; parameters like daylight conditions, weather, location etc. are examples of such scene-altering parameters.

The intention of this paper is to address the perception accuracy problem from a data OOD point-of-view. To have maximum control in our experiments, we exclude real-world captured data from our scope, and focus purely on the simulated environment. It is however a long-term objective to extend this work to incorporate captured data into the data curating process, creating a hybrid data approach that addresses also the inherent domain shift from synthetic to real-world captured data. The notion of dataset distance measures, along with a definition of a baseline in-distribution dataset, allows us to quantify a dataset's distance from the baseline dataset, and relate this to the internal variation of the baseline dataset. This makes it possible to explore different dimensions of the image space for the ODD and construct a well-balanced training set.

In this paper we direct our focus to an aviation use-case - visually detecting runways during approach - which could serve as a natural extension of the pilots' perception helping to reduce some of the workload present during critical stages of approach. The performance of our detection model is quantified by the MS COCO [8] evaluation metrics commonly used for object detection. With the ability to estimate the performance of our model in a certain part of our ODD it is possible to design our dataset to be (more) complete from the start and hence shorten accumulated model training time due to dataset updates. This will likely lead to a more controlled and efficient data management and model development phase. It is however not the primary focus of this paper to point to how performance drop estimates translate into requirements on data sampling, nor do we try to assess total completeness of datasets.

This paper is organized as follows: In Sect. 2 we present relevant related work. In Sect. 3 we explain our method for creating the baseline dataset and variations thereof. We also present the relative distance measures and how they're used in this context. In Sect. 4 we present the results from our experiments, including the sampling of new positional coordinates at different locations, the effects of parameter variations to dataset distances and further data visualizations for context. We also present the correlation between accuracy scores and similarity measures. In Sect. 5 we discuss the results and how to interpret the findings. Finally, Sect. 6 concludes this paper.

2 Related Work

Sun et al. [12] show that the amount of data trained upon increases the accuracy of the model on a logarithmic scale. Gaidon et al. [4] successfully train models on the virtual KITTI dataset, suggesting that the use of synthetically produced data indeed can deliver performance in the real-world scenario, given that the domain gap is sufficiently small. By training on synthetic data followed by real-world data fine-tuning, they find good performance in their automotive experiments.

Freemont et al. [3] propose the Scenic probabilistic programming language and describe the use of this to find corner cases in synthetically generated scenarios in general, and the automotive domain in particular. Scenic allows the user to programmatically construct a parameterized scenario where position and orientation of objects and their inter-relations can be controlled and views of the scenario can be sampled with these variations included in the process.

Yang et al. [18] discuss out-of-distribution detection methods in general, of which some are categorized as distance-based. An example is that of Masana et al. [10] where they propose to use metric learning for anomaly and novelty detection, or Techapanurak et al. [15] where they use scaled Cosine Similarity for a hyper parameter-free ood detection. Sun et al. [13] also explore non-parametric nearest-neighbor distance for OOD detection. Zilly et al. [20] investigate the correlation between Fréchet Distance (FD) and model accuracy for two different classification tasks, finding the performance to correlate to the distance between training and test sets. Guillory et al. [5] claim FD and Maximum Mean Discrepancy (MMD) distance measures do not reliably predict performance drop due to distribution shift in natural image content. Instead, they advocate the use of Average Confidence (AC) for this purpose.

Theis et al. [16] discuss different aspects of several different measures of similarity (in their context for evaluation of Generative Adversarial Networks (GANs)) including FID and KID, both of which are used in our study.

3 Method

The first step of our method regards creating a baseline scenario. We consider this to be our unperturbed scenario and we use it to define our in-distribution dataset. If it is possible having a real-world data source to guide the parameter choices of the baseline scenario, it greatly helps this step. In our use-case we use the OpenSky [11] Automatic Dependent Surveillance-Broadcast (ADS-B) data source to establish a funnel of coordinates for a normal approach to an airport. This part is detailed in Sect. 3.1. When the baseline scenario is defined we need to create the corresponding dataset by sampling visual representations of the scene from the simulator, in our case we use Xplane 11.

The second step in this method is to impose variations in the baseline scenario and create additional datasets for these augmented scenarios. The details of this step are layed out in Sect. 3.2.

The third step is to measure a distance between the baseline dataset and the augmented ones, thus quantifying the degree of out-of-distribution. The details of this part of the procedure are shown in Sect. 3.3.

Having a notion of distance between our baseline and augmented datasets, this metric may be used to estimate the expected drop in accuracy of our baseline model (i.e. trained on un-augmented baseline dataset) when exposed to the augmentations. We hypothesize that this accuracy drop estimation can translate into requirements on the amount of data augmentation needed to diminish the accuracy gap for the out-of-distribution test.

3.1 Sampling Methods

Data Filtering. The OpenSky historical database hosts lots of ADS-B information on air traffic movement, which we utilize to help define our in-distribution dataset. The ADS-B data is filtered to remove unwanted aircraft (operating in other modes) and we end up with a usable set of ADS-B points representing our baseline use-case in terms of aircraft positions.

In previous research [9] we go into more details on how we analyze this ADS-B dataset, using three different methods to draw new samples of aircraft position. In this paper however, we focus on one of the methods which is described below. This method works well to draw in-distribution samples and facilitates 'teleportation' of our scenario to a different geographic location without the need for ADS-B data for that particular site.

We divide the airspace into 8 bins, one nautical mile (NM) sized based on distance to runway. For each bin k, we calculate the average lateral position and standard deviation (lateral in this context is relative to runway extended centerline) ${\mu_k}^{Lat}, {\sigma_k}^{Lat}$ along with the relative altitude and standard deviation ${\mu_k}^{Alt}, {\sigma_k}^{Alt}$.

When we draw new samples we randomize distance to runway to be uniform within bin k. We find our appropriate altitude and lateral displacement by sampling the normal distributions $N({\mu_k}^{Lat}, {\sigma_k}^{Lat})$ and $N({\mu_k}^{Alt}, {\sigma_k}^{Alt})$ respectively.

3.2 Environment Parameter Variations

In the previous section we established a way of parameterizing our relative position to the object of interest, the runway. In this section we describe how to control other variations of our scene rendering which are not directly tied to aircraft position.

There are some limitations to the ADS-B data source. Specifically it does not include aircraft attitude information (i.e. roll, pitch and heading angles) so these still need to be estimated. Position and attitude is used to place a virtual camera with a cockpit-like view of the approaching runway - the object we are trying to detect. In our work we assert uniform variations of the attitude angles within limits of normal aircraft operations: roll angle limited to $\pm 10°$, pitch angle $\pm 3°$ (except the closest set of images, where the camera was tilted down an extra $15°$) and heading was within $\pm 3°$ of runway heading. The variations imposed by positional and attitude variations are all included in our baseline case, i.e. they are considered in-distribution. We use Scenic to help randomize aircraft attitude in the sampling process.

We are now ready for introducing variations into our environment. For this study we have considered variations in weather (clear vs cloudy) and daylight conditions (mid day vs dusk or evening). The reasons for these choices are that we expect variations due to the effect of clouds to be quite small, whereas the variations of daylight conditions are expected to show a greater dissimilarity to the baseline case. Finally, one more experiment with dusk conditions was included to cover a moderate variation case. It is desired to have variations spanning a wider range here to get a more general understanding of how these variations impact accuracy performance in later experiments.

We sample data from our simulated environment for our different cases (baseline and augmentations). For each case we sample 1512 images, split into 1000 for training and 512 for evaluation (see Fig. 1):

A_k Baseline scenario, clear weather, daylight conditions.
B_k Augmented scenario, cloudy weather, daylight conditions (approach lights sometimes on).
C_k Augmented scenario, clear weather, dark night conditions (approach lights always on).
D_k Augmented scenario, cloudy weather, dark night conditions (approach lights always on).

Finally we repeat this experiment for k different runway sites, such that we define a baseline case for each runway site and perform the same augmentation experiments. We include 4 different sites in the study.

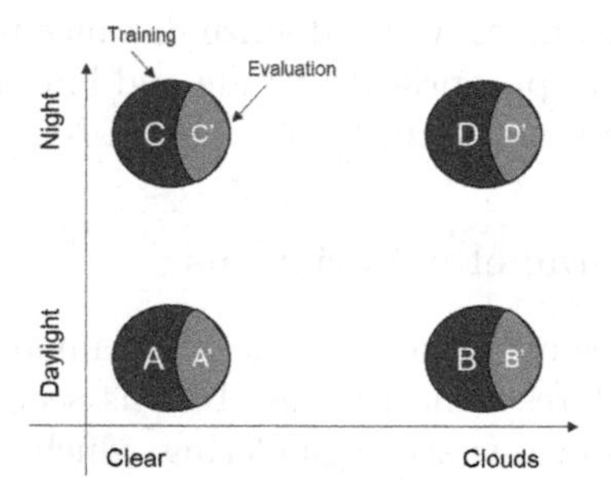

Fig. 1. Augmented and baseline datasets naming and organization. A–D are the named training datasets (represented by the blue parts of the diagram), whereas the sets A′–D′ (represented by the red parts of the diagram) are used for evaluation and distance measurements. Note that the evaluation data is never used for training. The training data in set A are sampled from the same scenario and distribution as the data in A′, so we can guarantee that A′ is in-distribution of A; the same applies to B and B′ etc. (Color figure online)

3.3 Similarity Measures

Different measures of distance exist and it is important to understand how and when to use a specific distance or similarity measure. Kullback-Leibler (KL) divergence [7] is a measure of distribution similarity defined as

$$KL(P||Q) = \sum_x P(x) \log(\frac{P(x)}{Q(x)}). \tag{1}$$

By evaluating this measure over distributions P and Q related to two different datasets D_P and D_Q we may quantify how well they align, and in a sense, whether one can be used in place of the other. t-distributed stochastic neighbor embedding (t-SNE) is a dimensionality reduction method which minimizes the KL divergence between two distributions of data point embeddings (one in high-dimensional space and the other in the reduced dimensional space). Examples of this are shown in Sect. 4.

A more common way in the field of machine learning to measure distances between image datasets is the FrÃĺchet Inception Distance (FID) [6] which is based on a metric distance function (the Wasserstein distance). This is based on the assumption that the feature vector representation of the different datasets is normally distributed, i.e. for two multivariate Gaussian distributed variables $X_1 \sim N(\mu_1, \Sigma_1)$ and $X_2 \sim N(\mu_2, \Sigma_2)$ the squared distance is calculated as

$$d^2 = ||\mu_1 - \mu_2||^2 + Tr(\Sigma_1 + \Sigma_2 - 2 * \sqrt{\Sigma_1 * \Sigma_2}), \tag{2}$$

where d is the distance, μ_k is the mean vector and Σ_k is the covariance matrix of the multivariate variable X_k. The vectors X_k are taken as the output of an intermediate layer of the Inception-v3 network [14], a 2048-dimensional vector.

Kernel Inception Distance [1] (KID) is another way to measure dataset similarity, which is based on the Maximum Mean Discrepancy (MMD). MMD is a distance on the space of probability measures. The distance is defined based on the notion of embedding probabilities in a Reproducing Kernel Hilbert Space (RKHS). The Hilbert space properties conveniently lends us a way of measuring distance, e.g. by the norm induced from the inner product. Let P be a probability measure on X_1 and Q be the same for X_2. Then

$$MMD^2(P, Q) = E_P[k(X_1, X_1)] - 2E_{P,Q}[k(X_1, X_2)] + E_Q[k(X_2, X_2)], \tag{3}$$

using a kernel function k. In our experiments we use a polynomial kernel function $k(x, y) = (\gamma x^\top y + c_0)^d$, with $d = 3$, $\gamma = 1/2048$ and $c_0 = 1$. Similar to FID, Inception-v3 intermediate layer outputs are used for the probability measure, however the Gaussian distribution assumption of X_k can be relaxed here.

Relative FID and KID. We now have the tools for quantifying similarity. In this work we will use the FID and KID distances for our measurements. We define the relative FID (RFID) measure by the following reasoning: We measure the FID distance between our sets A and A', which should be small since we have drawn them from the same scenario and distribution. We let $K_{FID} = 1/FID(A, A')$ be a normalizing constant and then define

$$RFID(A, B) = K_{FID} * FID(A, B), \tag{4}$$

i.e. we normalize the FID score based on what is the expected distance for in-distribution data. Note that by definition $RFID(A, A') = 1$. Analogously we define $K_{KID} = 1/KID(A, A')$ and

$$RKID(A, B) = K_{KID} * KID(A, B). \tag{5}$$

4 Results

In this section we show our experimental results. The results of the following sub-sections will be analyzed in the discussion section following the results.

4.1 Sampling Method

The proposed sampling method is used to create new coordinates for previously unseen aircraft positions. Figure 2 shows how our samples are distributed laterally. We also show the possibility of this sampling method to sample at new locations, where ADS-B data might be limited. This image is published in our previous work [9].

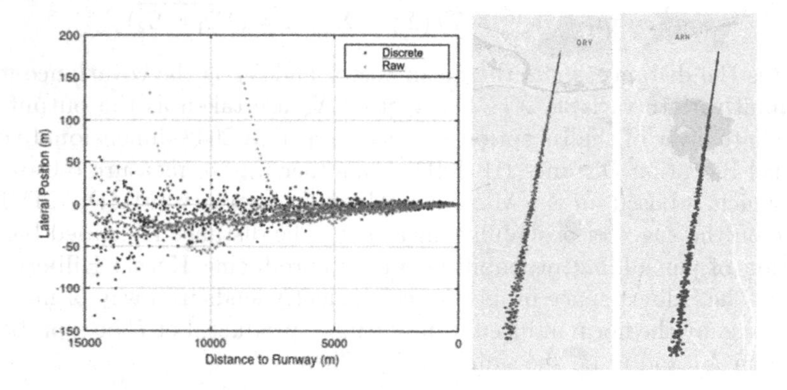

Fig. 2. Left: Data points from filtered ADS-B points (red) and generated samples (blue). Center: Generated samples at Paris-Orly. Right: Generated samples at Stockholm Arlanda. These sample points are used to set the viewing point for an aircraft approaching a runway. A visual image is generated at each view-point, rendered in Xplane flight simulator. Best viewed in colour. (Color figure online)

4.2 Environment Parameter Variations

In Fig. 3 we show Average Precision (AP) results from different training scenarios like training only on clear daylight data (A) and evaluating on different datasets (in- and out-of-distribution). We also show the recovery in accuracy when including some of the B, C and D data into training, which is expected since B', C' and D' are then no longer out-of-distribution.

4.3 Similarity Measures

For each airport site k we have trained a model (Faster R-CNN with ResNet-50-FPN-3x backbone pretrained on ImageNet) on the baseline dataset (A_k) for that

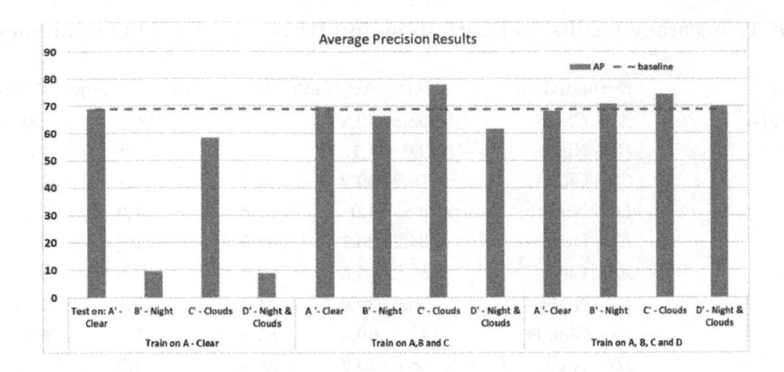

Fig. 3. Dashed line shows the baseline scenario performance. Blue bars indicate AP score on test sets (A'–D'). Accuracy drop is evident when tested on B', C' and D' OOD data. When our model is trained on data from sets A, B and C we see a clear recovery, which increase further if we add also data from set D. (Color figure online)

site. This model was then evaluated on A'_k, B'_k, C'_k and D'_k datasets. We can thus evaluate the absolute and relative performance drop of the model. We also calculate the $RFID(A_k, B'_k)$, $RFID(A_k, C'_k)$ and $RFID(A_k, D'_k)$ and similarly the same combinations for $RKID$ measures, as shown in Table 1. AP small, medium and large refers to the MS COCO [8] evaluation metrics commonly used for object detection. Here AP small only includes accuracy for objects smaller than 32^2 pixels, AP medium includes object sizes from 32^2 to 96^2 pixels and AP large includes all those objects larger than 96^2 pixels.

In Fig. 4 we show the correlation between our relative dataset similarity measures (RFID, RKID) and the drop in AP accuracy for large objects. The linear

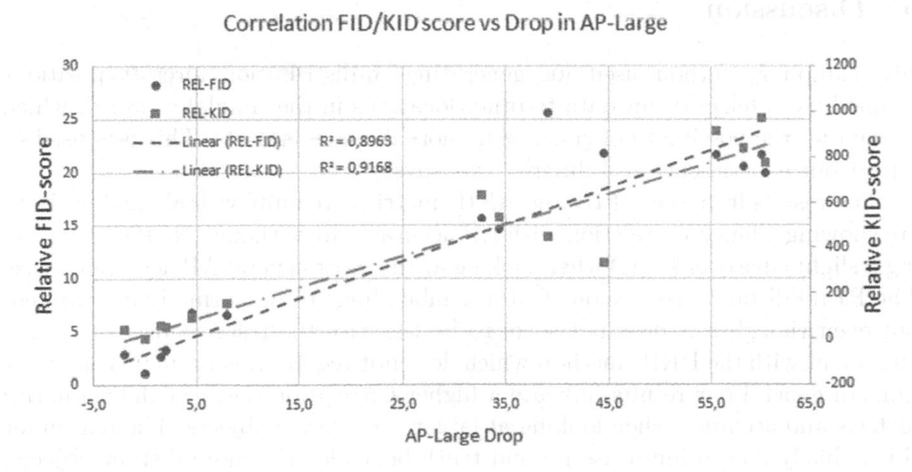

Fig. 4. Correlation between drop in AP for large objects and Relative FID/KID measures. Pearson correlation with AP-Large Drop is 0.93 for RFID and 0.94 for RKID.

Table 1. Accuracy results and corresponding RFID and KFID measurements.

Trained on	Evaluated on	AP	AP small	AP medium	AP Large	RFID	RKID
A_1 - Arlanda	A_1' - Clear	68,8	59,2	86,0	90,2	1,0	1,0
	B_1' - Night	9,6	4,1	10,2	55,8	14,6	539,1
	C_1' - Clouds	58,2	49,2	75,1	92,1	2,9	39,5
	D_1' - Night + Clouds	8,8	4,0	12,8	57,6	15,7	635,0
	E_1' - Dusk	34,2	26,7	42,5	82,3	6,6	158,7
A_2 - Doha	A_2' - Clear	83,2	75,5	89,4	90,4	1,0	1,0
	B_2' - Night	36,6	29,0	41,2	45,8	21,7	333,0
	C_2' - Clouds	75,1	60,4	84,8	85,8	6,7	92,3
	D_2' - Night + Clouds	36,6	29,3	38,3	51,3	25,5	448,5
A_3 - Paris-Orly (rwy 07)	A_3' - Clear	79,5	70,4	93,0	94,9	1,0	1,0
	B_3' - Night	29,4	28,4	31,8	34,3	19,9	775,5
	C_3' - Clouds	64,5	54,3	75,1	92,8	3,3	56,0
	D_3' - Night + Clouds	28,0	26,7	29,2	39,4	21,6	913,8
A_4 - Paris-Orly (rwy 25)	A_4' - Clear	77,5	69,2	90,6	89,8	1,0	1,0
	B_4' - Night	34,5	32,2	48,2	31,6	20,6	841,6
	C_4' - Clouds	74,6	67,9	84,6	88,3	2,6	60,3
	D_4' - Night + Clouds	33,7	29,8	50,6	29,6	21,7	971,8

regression lines included show the general correlation here. In Fig. 5 we show the corresponding results for AP score across all object sizes. The correlation is less pronounced in this case. In Fig. 6 we show the t-distributed stochastic neighbor embedding of our datasets. t-SNE is a way of visualizing high-dimensional data by embedding it in a lower dimensional space. In our case we use this statistical method to visually relate all the images in all our datasets.

5 Discussion

The sampling method used for generating in-distribution aircraft positions enabled us to teleport our data to other locations in the simulated world, which opened up the possibility of generating more diverse datasets, This was used to repeat our experiments at 4 different runways.

The results from our RFID and RKID metrics are quite well aligned, as both are showing a linear correlation with AP accuracy drop, though RFID was showing a slight edge over RKID when looking at the most general AP accuracy score. The RFID distance assume our feature embeddings to be normally distributed, but even though this does not seem to be the case the result shows surprising alignment with the RKID method which does not require this normality assumption. In general our results indicate a higher correlation between the similarity metrics and accuracy when looking at larger (i.e. closer) objects. The reason for this is likely due to imprecise ground truth-boxes for the more distant objects, where a few pixels offset can give a large reduction in accuracy.

We note that the drop in accuracy was recovered when the training dataset was expanded to include the cloudy and night time sets (B, C and D).

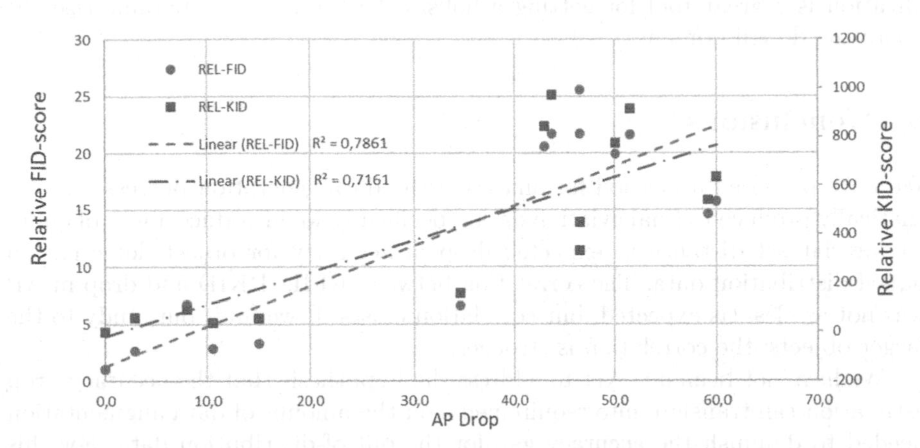

Fig. 5. Correlation between drop in AP and Relative FID/KID measures. Pearson correlation with AP Drop is 0.81 for RFID and 0.77 for RKID.

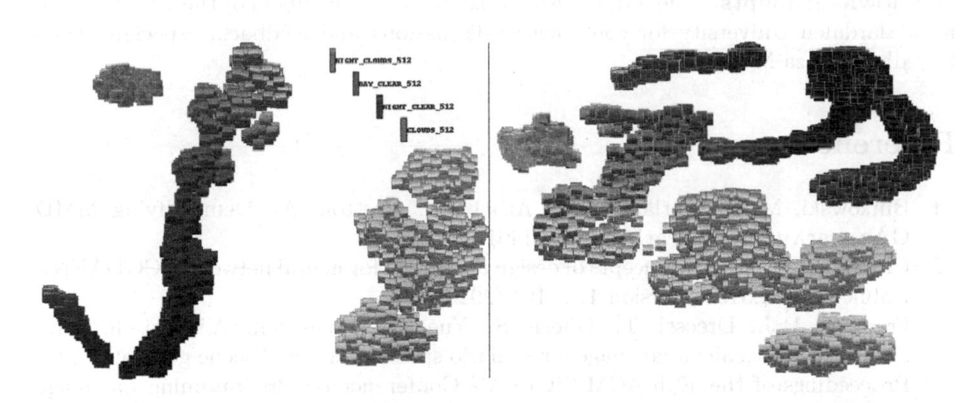

Fig. 6. T-SNE embeddings of images in datasets A'_1-D'_1 (left plot) and all datasets A'_1-D'_4 (right plot). Each colour of the image frame indicates a unique dataset. We can see some large clustering where clear and cloudy datasets gets lumped together. In the right plot we see a high degree of mixing between different sets, e.g. the day clear datasets for the different sites overlap substantially, but the night and day sets are quite distinct (as can be seen by regarding the brightness of the images). Best viewed in colour. (Color figure online)

The t-SNE embedding is also in general agreement with the RFID and RKID results. For instance we can see visually in Fig. 6 that night (B) and night with clouds (D) are not disjoint enough to be separate categories. The t-SNE visualization is a great tool for getting a holistic view on how to parametrize the operating design domain.

6 Conclusions

We have proposed a method for analyzing the data generation process for synthetically produced visual aviation data. Specifically we introduce a way of quantifying dataset distance to expected drop in accuracy for object detection on out-of-distribution data. The correlation between RFID/RKID and drop in AP was not as clear as expected, but correlation exists. If we limit our study to the larger objects, the correlation is stronger.

We have not been able yet to address the hypothesis that the accuracy drop estimation can translate into requirements on the amount of data augmentation needed to diminish the accuracy gap for the out-of-distribution data. For this more work is needed. It should also be noted that this method should be validated in future work with other models and use-cases to see the extent to which these results generalize. Finally, since we have not looked at natural image content, we cannot make certain statements on the validity of these results in a natural image context.

Acknowledgements. The authors would like to thank members of the HERO group at Mälardalen University for constructive discussions and feedback, especially from Seyedhamidreza Mousavi.

References

1. Bińkowski, M., Sutherland, D.J., Arbel, M., Gretton, A.: Demystifying MMD GANs. arXiv preprint arXiv:1801.01401 (2018)
2. Cluzeau, J.M., et al.: Concepts of design assurance for neural networks (CODANN). Public Rep. Extract Version **1**, 1–104 (2020)
3. Fremont, D.J., Dreossi, T., Ghosh, S., Yue, X., Sangiovanni-Vincentelli, A.L., Seshia, S.A.: Scenic: a language for scenario specification and scene generation. In: Proceedings of the 40th ACM SIGPLAN Conference on Programming Language Design and Implementation, pp. 63–78 (2019)
4. Gaidon, A., Wang, Q., Cabon, Y., Vig, E.: Virtual worlds as proxy for multi-object tracking analysis. In: Proceedings of the IEEE Conference on Computer Vision and Pattern Recognition, pp. 4340–4349 (2016)
5. Guillory, D., Shankar, V., Ebrahimi, S., Darrell, T., Schmidt, L.: Predicting with confidence on unseen distributions. In: Proceedings of the IEEE/CVF International Conference on Computer Vision, pp. 1134–1144 (2021)
6. Heusel, M., Ramsauer, H., Unterthiner, T., Nessler, B., Hochreiter, S.: GANs trained by a two time-scale update rule converge to a local nash equilibrium. In: Advances in Neural Information Processing Systems, vol. 30 (2017)

7. Kullback, S., Leibler, R.A.: On information and sufficiency. Ann. Math. Stat. **22**(1), 79–86 (1951)
8. Lin, T., et al.: Microsoft COCO: common objects in context. CoRR abs/1405.0312 (2014). http://arxiv.org/abs/1405.0312
9. Lindén, J., et al.: Curating datasets for visual runway detection. In: 2021 IEEE/AIAA 40th Digital Avionics Systems Conference (DASC), pp. 1–9. IEEE (2021)
10. Masana, M., Ruiz, I., Serrat, J., van de Weijer, J., Lopez, A.M.: Metric learning for novelty and anomaly detection. arXiv preprint arXiv:1808.05492 (2018)
11. Schäfer, M., Strohmeier, M., Lenders, V., Martinovic, I., Wilhelm, M.: Bringing up OpenSky: a large-scale ADS-B sensor network for research. In: IPSN-14 Proceedings of the 13th International Symposium on Information Processing in Sensor Networks, pp. 83–94. IEEE (2014)
12. Sun, C., Shrivastava, A., Singh, S., Gupta, A.: Revisiting unreasonable effectiveness of data in deep learning era. In: Proceedings of the IEEE International Conference on Computer Vision, pp. 843–852 (2017)
13. Sun, Y., Ming, Y., Zhu, X., Li, Y.: Out-of-distribution detection with deep nearest neighbors. In: International Conference on Machine Learning, pp. 20827–20840. PMLR (2022)
14. Szegedy, C., Vanhoucke, V., Ioffe, S., Shlens, J., Wojna, Z.: Rethinking the inception architecture for computer vision. In: Proceedings of the IEEE Conference on Computer Vision and Pattern Recognition, pp. 2818–2826 (2016)
15. Techapanurak, E., Suganuma, M., Okatani, T.: Hyperparameter-free out-of-distribution detection using softmax of scaled cosine similarity. arXiv preprint arXiv:1905.10628 (2019)
16. Theis, L., Oord, A.v.d., Bethge, M.: A note on the evaluation of generative models. arXiv preprint arXiv:1511.01844 (2015)
17. Wang, C.Y., Bochkovskiy, A., Liao, H.Y.M.: YOLOv7: Trainable bag-of-freebies sets new state-of-the-art for real-time object detectors. arXiv preprint arXiv:2207.02696 (2022)
18. Yang, J., Zhou, K., Li, Y., Liu, Z.: Generalized out-of-distribution detectionasurvey. arXiv preprint arXiv:2110.11334 (2021)
19. Zaidi, S.S.A., Ansari, M.S., Aslam, A., Kanwal, N., Asghar, M., Lee, B.: A survey of modern deep learning based object detection models. Digit. Signal Process. **126**, 103514 (2022)
20. Zilly, J., Zilly, H., Richter, O., Wattenhofer, R., Censi, A., Frazzoli, E.: The Frechet distance of training and test distribution predicts the generalization gap (2019)

Holistic Analytics of Sensor Data from Renewable Energy Sources: A Vision Paper

Søren Kejser Jensen and Christian Thomsen(✉)

Department of Computer Science, Aalborg University, Aalborg, Denmark
{skj,chr}@cs.aau.dk

Abstract. Modern Renewable Energy System (RES) installations, e.g., wind turbines, produce petabytes of high-frequency time series. State-of-the-art systems cannot cope with such amounts of data. Thus, practitioners generally store simple aggregates, e.g., 10-min averages. Based on discussions with practitioners, we present requirements and our vision for a next-generation time series management system that can efficiently manage vast amounts of time series across edge, cloud, and client.

1 Introduction

Modern Renewable Energy System (RES) installations, such as wind turbines, are monitored by up to hundreds of high-quality sensors sampled at high frequencies, e.g., 10 Hz, 50 Hz, or 100 Hz. Each of these sensors can perform several measurements and output several time series. Thus, a wind turbine produces thousands of high-frequency time series. A time series is a sequence of data points $\langle(t_1, v_1), (t_2, v_2), \ldots\rangle$ where $t_i < t_{i+1}$ and t_i represents the time when the value $v_i \in \mathbf{R}$ was measured by the sensor. Assuming a wind turbine can generate 2500 time series sampled at 100 Hz and that both time stamps and values require 8 bytes, one wind turbine generates more than 321 GiB of data each day. Thus, a park of 100 wind turbines generates more than 11 PiB of data every year.

Ingesting, managing, and analyzing such vast amounts of time series is infeasible with traditional Relational Database Management Systems (RDBMSs) [11]. Besides, RES installations are often located in remote areas with such limited connectivity that it is impossible to transfer the raw time series to the cloud. Thus, in addition to using lossless compression, manufacturers and owners of RES installations typically downsample the time series, e.g., by computing 10-min averages. However, this removes valuable fluctuations and outliers.

In addition, data scientists often use advanced, tailor-made, RES-specific scripts executed on their local PCs ("clients") to analyze the performance of and detect problems in RES installations. Thus, the data scientists have to download the relevant time series from the edge and/or cloud before they can do their analysis. This is not only cumbersome but also highly inefficient and thus very costly due to the large amount of bandwidth, storage, and computation required.

A. Abelló et al. (Eds.): ADBIS 2023, CCIS 1850, pp. 360–366, 2023.
https://doi.org/10.1007/978-3-031-42941-5_31

As an attempt to remedy these problems, several systems for ingesting, managing, and analyzing time series have been proposed [1,5,9,10]. However, these Time Series Management Systems (TSMSs) are generally designed to be deployed on the edge or in the cloud and do not take the client's computational power into account when executing queries. Thus, they cannot optimize across edge, cloud, and client. Instead users must connect the systems using additional software and determine both where to execute queries and when to transfer data.

Based on discussions with manufacturers and owners of RES installations, we present our vision for a next-generation TSMS that can efficiently ingest, manage, and analyze vast amounts of RES time series across edge, cloud, and client. To allow data scientists to focus on analytics and creating value instead of data management, the envisioned TSMS must provide the following functionality:

(R1) Error-Bounded Compression on the Edge: Time series must be compressed on the edge to reduce the required bandwidth and storage. To support accurate analytics, timestamps must be represented without any error while each value must be represented within a user-defined error bound (possibly 0%).

(R2) Continuous Transfer of Data Points from Edge to Cloud: The compressed time series must be continuously transferred to the cloud in a resource-aware manner, i.e., important data must be prioritized. The TSMS must also support persisting compressed time series on the edge if the connection fails.

(R3) Locality-Aware Execution of Queries and Scripts: To avoid transferring vast amounts of data from the edge and cloud to the client, the TSMS must efficiently execute queries and scripts across the edge, cloud, and client.

(R4) Integration with Data Analytics Tools: The TSMS must integrate with existing infrastructure and tools such that data scientists can continue using their advanced, tailor-made, RES-specific scripts. This makes it feasible for the TSMS to actually be adopted by manufacturers and owners of RES installations.

The rest of this paper is structured as follows. In Sect. 2, we describe the requirements in detail and our vision for a next-generation TSMS. In Sect. 3, we describe related work. Finally, in Sect. 4, we present our conclusion.

2 The Envisioned System

Data from modern RES installations is managed and analyzed by edge nodes, cloud nodes, and clients. Based on discussions with manufacturers and owners of RES installations, we have defined the following high-level requirements that a next-generation TSMS must meet to efficiently ingest, manage, and analyze high-frequency time series at the vast scale required for modern RES installations.

(R1) Error-Bounded Compression on the Edge: The vast amounts of raw time series data produced by sampling the many high-quality sensors in modern RES installations make it infeasible to store the raw time series on the edge nodes, transfer them to the cloud nodes, and even store them on the cloud nodes. Thus, state-of-the-art time series compression methods are required and the compression must be performed on the edge nodes. The compression to use depends on how the time series data will be used, e.g., lossy compression

is known to be significantly more effective than lossless compression [8] and can improve the precision of analytics [19], but more research is required for users to trust lossy compression [3]. Thus, a next-generation TSMS for RES must automatically use the best compression possible while taking the available resources and requirements of the analytics that will be performed into account.
(R2) Continuous Transfer of Data Points from Edge to Cloud: Due to the very limited hardware on the edge nodes, and to enable analytics across multiple RES installations, the data points ingested on the edge nodes must be transferred to the cloud nodes. From discussions with owners of RES installations, we know that their edge nodes are low-end commodity PCs, e.g., 4 CPU Cores, 4 GiB RAM, and an HDD. However, it is often impossible to transfer the raw time series due to the limited amount of bandwidth available as it can be as low as 500 Kbits/s to 5 Mbits/s. In addition, the connection may be down for an unknown period of time. Thus, a next-generation TSMS for RES must continuously transfer the highly compressed representations of the high-frequency time series to the cloud, decide which data must be transferred if the available bandwidth is limited, and also persist the data in case the connection is down.
(R3) Locality-Aware Execution of Queries and Scripts: Data scientists generally perform analytics using advanced, tailor-made, RES-specific scripts on the client, however, this can require downloading vast amounts of time series to the client before the analytics can be performed. Instead, it would be much more efficient to execute the queries or scripts where the data is located, i.e., the edge nodes and cloud nodes. For example, if a simple aggregate such as `MIN` or `MAX` is computed, the input may be terabytes of data while the result is only an 8 byte value. However, the client should also participate in query processing when it is more efficient to do so as even low-end commodity PCs today have multiple CPU cores and gigabytes of memory. For example, if multiple types of analytics have to be performed and the result of each is similar in size to the input data, it may be faster to download the data to the client and perform the analytics there. Thus, a next-generation TSMS for RESs must support executing queries and scripts that perform advanced analytics on the edge nodes, cloud nodes, and client and it must include a dynamic optimizer that automatically determines where to perform the analytics to use the least amount of resources.
(R4) Integration with Data Analytics Tools: As stated, data scientists employed by RES manufacturers and owners use tailor-made, specialized analytics tools to detect and understand problems in RES installations. These tools use a significant amount of custom code that has been refined for years. In addition, the computations performed by these tools are not easily expressable in SQL and are typically implemented using Python and its packages for scientific computing such as NumPy and pandas. Thus, to make it feasible for manufacturers and owners of RES installations to adopt the next-generation TSMS, it must provide effective and efficient integration with the current infrastructure and tools used by the data scientists so they can continue to use their existing tools.

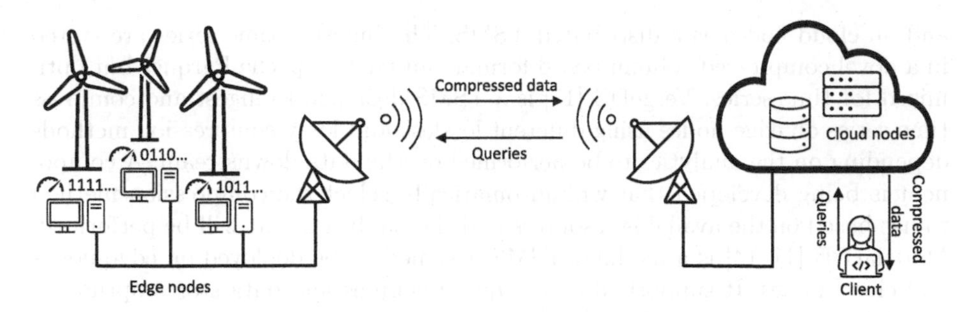

Fig. 1. Illustration of envisioned data collection and analysis in a RES domain

To meet these requirements, we envision a system as outlined in Fig. 1. A single-node version of the TSMS is deployed on each of the edge nodes. It collects sensor data and compresses it within an error bound (possibly 0%) based on the requirements of the analytics **(R1)**. The compressed data is continuously transferred from the edge nodes to the cloud with important data transferred first **(R2)**. A distributed version of the TSMS is deployed on the cloud nodes for scalable analytics. Data scientists perform analytics using their own client PCs and use the TSMS as a library that integrates with their existing tools **(R4)**. The library sends queries and scripts to the cloud, and when cloud nodes execute them, they may send queries, scripts, or data requests to the edge nodes **(R3)**. As the TSMS is deployed on the edge, cloud, and client, it can continue to operate on local data if a network connection is temporarily unavailable. For example, an edge node can simply write data to disk, and engineers working on the wind turbine can query its local data using their existing tools **(R3, R4)**.

3 Related Work

Many TSMSs have been proposed to manage the vast amounts of time series data being produced [1,5,9,10]. However, while some of them can be deployed across edge and cloud, none of them take a holistic approach to time series analytics.

Some TSMSs support ingesting time series on the edge nodes and then transferring them to the cloud. Respawn [2] is a TSMS that ingests sensor data on edge nodes, continuously computes aggregates from the ingested sensor data to improve query response time, and continuously transfers data to cloud nodes based on two strategies: transfer low-resolution aggregates to the cloud nodes and transfer important data based on its standard deviation. Queries are routed to the relevant edge nodes and cloud nodes. Storacle [4,6] is a TSMS designed to be deployed on edge nodes throughout a smart grid. Storacle uses a three-tiered storage model consisting of local memory, local storage, and cloud storage. Data is continuously transferred from the edge nodes to the cloud nodes. The latest data points are not immediately deleted when they are transferred to the next tier so they can be used to answer queries with low latency. Apache IoTDB [20] is designed to be deployed on edge nodes as an embedded or standalone TSMS and on cloud nodes as a distributed TSMS. The ingested time series are stored in a novel compressed column-based format similar to Apache Parquet but optimized for time series. VergeDB [15] is a TSMS designed to ingest and compress time series on edge nodes using different lossless and lossy compression methods depending on the analytics to be performed on the data downstream. A component is being developed that will automatically select the compression method to use based on the available resources and the analytics that will be performed. ModelarDB [11–14] is a modular TSMS designed to be deployed on edge nodes and cloud nodes. It supports different query engines and data stores optimized for different use cases and it is simple to extend the system with support for additional query engines and data stores due to its modularity. It uses multiple different types of models to efficiently compress high-frequency time series within a user-defined error bound (possibly 0%). Thus, both lossless and lossy compression are supported. User-defined model types can optionally be added to ModelarDB without recompiling the system. The models and accompanying metadata are continuously transferred from the edge nodes to the cloud nodes.

Some TSMSs support performing complex user-defined analytics directly in the TSMS while others utilize the computational capabilities of both the cloud nodes and the client when executing queries. NilmDB [16] stores time series in a novel data store named BulkData and metadata in SQLite [7]. Users can retrieve data points using queries and then analyze them on the client. Alternatively, the users can run the analysis on the server by submitting a Python script. This reduces the amount of data to be transferred. LittleTable [18] is a TSMS designed like an RDBMS but specialized for time series. For example, it lacks support for updates and `NULL` values. The client software for LittleTable is implemented using SQLite's [7] Virtual Table Interface. When LittleTable is queried, it guarantees that the data is transmitted to SQLite in ascending or descending order by primary key. The SQLite client can utilize this knowledge to, e.g., efficiently compute aggregates that `GROUP BY` time and metadata such as a device id. DuckDB [17] is not a TSMS but an embeddable RDBMS designed for analytics. Thus, it can be used like SQLite [7] but is optimized for Online Analytical Processing (OLAP) instead of Online Transaction Processing (OLTP). To do so, DuckDB combines a C/C++/SQL-API, a cost-based optimizer, Multiversion Concurrency control (MVCC), and a vectorized query engine. Thus, it allows OLAP queries to be efficiently performed on the client without requiring installation, configuration, and management of a complex standalone RDBMS.

Existing TSMSs thus only have simple integration between the edge and the cloud [2,4,6,11–15] **(R1, R2, R3)** or simple integration between the cloud and the client [16,18] **(R3, R4)**. We envision a TSMS that provides both to enable scalable, holistic analytics of RES sensor data across edge, cloud, and client.

4 Conclusion

Modern Renewable Energy System (RES) installations produce petabytes of high-frequency time series. State-of-the-art systems cannot cope with such vast amounts of data. In this paper, we presented our vision for a next-generation TSMS that can efficiently manage vast amounts of time series data. Based on discussions with practitioners, we presented requirements for such a system and an outline of how it should operate distributedly across edge, cloud, and client.

Acknowledgements. This research was supported by the MORE project funded by Horizon 2020 grant number 957345. In addition, we thank our industry partners for providing a large amount of detailed information about their domain.

References

1. Bader, A., Kopp, O., Michael, F.: Survey and comparison of open source time series databases. In: Proceedings of the BTW - Workshopband, pp. 249–268. GI (2017)
2. Buevich, M., Wright, A., Sargent, R., Rowe, A.: Respawn: a distributed multi-resolution time-series datastore. In: Proceedings of the RTSS, pp. 288–297. IEEE (2013)
3. Cappello, F., Di, S., Gok, A.M.: Fulfilling the promises of lossy compression for scientific applications. In: Nichols, J., Verastegui, B., Maccabe, A.B., Hernandez, O., Parete-Koon, S., Ahearn, T. (eds.) SMC 2020. CCIS, vol. 1315, pp. 99–116. Springer, Cham (2020). https://doi.org/10.1007/978-3-030-63393-6_7
4. Cejka, S., Mosshammer, R., Einfalt, A.: Java embedded storage for time series and meta data in Smart Grids. In: Proceedings of the SmartGridComm, pp. 434–439. IEEE (2015)
5. DB-Engines Ranking of Time Series DBMS (2023). https://db-engines.com/en/ranking/time+series+dbms
6. Faschang, M., et al.: Provisioning, deployment, and operation of smart grid applications on substation level. CSRD **32**(1–2), 117–130 (2017)
7. Gaffney, K.P., Prammer, M., Brasfield, L.C., Hipp, D.R., Kennedy, D.R., Patel, J.M.: SQLite: past, present, and future. PVLDB **15**(12), 3535–3547 (2022)
8. Hung, N.Q.V., Jeung, H., Aberer, K.: An evaluation of model-based approaches to sensor data compression. TKDE **25**(11), 2434–2447 (2013)
9. Jensen, S.K., Pedersen, T.B., Thomsen, C.: Time series management systems: a 2022 survey. In: Palpanas, T., Zoumpatianos, K. (eds.) Data Series Management and Analytics. ACM (Forthcoming) survey. In: Palpanas, T., Zoumpatianos, K. (eds.) Data Series Management and Analytics. ACM (Forthcoming)
10. Jensen, S.K., Pedersen, T.B., Thomsen, C.: Time series management systems: a survey. TKDE **29**(11), 2581–2600 (2017)
11. Jensen, S.K., Pedersen, T.B., Thomsen, C.: ModelarDB: modular model-based time series management with spark and Cassandra. PVLDB **11**(11), 1688–1701 (2018)
12. Jensen, S.K., Pedersen, T.B., Thomsen, C.: Demonstration of ModelarDB: model-based management of dimensional time series. In: Proceedings of the SIGMOD, pp. 1933–1936. ACM (2019)
13. Jensen, S.K., Pedersen, T.B., Thomsen, C.: Scalable model-based management of correlated dimensional time series in ModelarDB$_{+}$. In: Proceedings of the ICDE, pp. 1380–1391. IEEE (2021)
14. Jensen, S.K., Thomsen, C., Pedersen, T.B.: ModelarDB: integrated model-based management of time series from edge to cloud. TLDKS **53**, 1–33 (2023)

15. Paparrizos, J., et al.: VergeDB: a database for IoT analytics on edge devices. In: Proceedings of the CIDR (2021)
16. Paris, J., Donnal, J.S., Leeb, S.B.: NilmDB: the non-intrusive load monitor database. TSG **5**(5), 2459–2467 (2014)
17. Raasveldt, M., Mühleisen, H.: DuckDB: an embeddable analytical database. In: Proceedings of the SIGMOD. ACM (2019)
18. Rhea, S., Wang, E., Wong, E., Atkins, E., Storer, N.: LittleTable: a time-series database and its uses. In: Proceedings of the SIGMOD, pp. 125–138. ACM (2017)
19. Tirupathi, S., et al.: Machine learning platform for extreme scale computing on compressed IoT data. In: Proceedings of the BigData, pp. 3179–3185. IEEE (2022)
20. Wang, C., et al.: Apache IoTDB: time-series database for internet of things. PVLDB **13**(12), 2901–2904 (2020)

DOING: 4th Workshop on Intelligent Data - From Data to Knowledge

Labeling Portuguese Man-of-War Posts Collected from Instagram

Heloisa Fernanda Rocha[1(✉)], Lorena Silva Nascimento[2], Leonardo Camargo[1], Mauricio Noernberg[2], and Carmem S. Hara[1]

[1] Departamento de Informática, Universidade Federal do Paraná, Curitiba, PR, Brazil
{heloisarocha,camargo.s.leonardo,carmemhara}@ufpr.br

[2] Centro de Estudos do Mar, Universidade Federal do Paraná, Pontal do Paraná, PR, Brazil
{lorena.sn,m.noernberg}@ufpr.br

Abstract. The need for knowledge of biodiversity is constant, while research resources are often scarce. The Portuguese man-of-war (*Physalia physalis*) is a beautiful animal, but it poses a risk to the population, as it may cause severe burns if touched. Data on their occurrences on the coast are not always available in traditional sources. On the other hand, previous studies show that social media can be an effective source of information for conservation science. This paper reports the process of collecting and labeling Instagram posts, based on hashtag searches. Labels were given manually in order to distinguish posts that in fact refer to the animal from others, such as those that refer to ships or tattoos. We highlight the importance of choosing appropriate labels when the dataset is used for training machine learning models for automatically classifying new posts. An experimental study is presented to show the effect of unquestioning adoption of labels given by a specialist, in comparison with labels adapted for machine learning training.

Keywords: Physalia physalis · dataset construction · labeling · machine learning

1 Introduction

The Portuguese man-of-war (cnidarian *Physalia physalis*) is a multicellular organism whose tentacles have stinging cells that can release harmful toxins. This species occurs all over the Brazilian coast and accidents with these animals have been frequently reported [4].

Despite being of compulsory notification, data from SINAN (the Brazilian system for reporting injuries) do not present a specific indication for Portuguese man-of-war poisoning. Furthermore, there is evidence of underreporting of cases [4]. The limited information about these animals makes it difficult to use official public data in the work of researchers interested in the species.

This Work Was Partially Supported by CNPq (Process 407644/2021-0), and by CAPES/PrInt-UFPR.

A. Abelló et al. (Eds.): ADBIS 2023, CCIS 1850, pp. 369–381, 2023.
https://doi.org/10.1007/978-3-031-42941-5_32

An alternative is to use involuntary data from social media. This approach, called passive citizen science, consists of using data generated by non-professionals, collected and shared on the Internet, mainly on social media [6].

Among the advantages resulting from the use of social media data are [6]: abundant and almost always public content; less labor intensive, less time consuming, and with lower cost, especially if automated, compared to traditional methods; data available almost in real time and continuously; and data acquisition with wide geographic scope [14]. In addition, the use of information technology tools, such as machine learning (ML), makes it possible to explore the potential of social media data and turn it into useful information.

In the Caravela project [15], one of the specific goals is to compile data extracted from social media about occurrences of Portuguese man-of-war on the Brazilian coast. This process has been done manually [16] and the results showed that the posts collected from Instagram represented 60% of the entire dataset, which included other sources such as newspapers, iNaturalist[1], and reports in the literature. The goal of the study was to determine the effectiveness of Instagram as a data source for spatial and temporal information of their sightings, as well as obtain the number, size and types of interactions with other species and human beings.

In this paper, the goal is to report the annotation process for showing the importance of properly designing the labels that will be associated with each entry in order to prepare the dataset for training machine learning models. The labeling process is required to distinguish between posts that are legitimate sightings of the species from others, such as tattoos and Portuguese boats.

It has been a rich real world application experience. It contrasts with other machine learning works that consider benchmark datasets for which labels were already given. The design of the resulting labels involved a close interaction between a computer scientist and an oceanographer, with discussions on the requirements for the end user and for the training process to automate the classification task. Such a classifier can later be inserted into a search, classification and data storage system, which can continuously monitor the species.

The rest of the paper is organized as follows. Section 2 presents some related works that apply machine learning on data extracted from social media. The Portuguese man-of-war dataset construction is described in Sect. 3, followed by details on the annotation process in Sect. 4. An analysis of the dataset labeled by the oceanographer and the computer scientist, along with an experimental study is presented in Sect. 5. Section 6 concludes the paper presenting some future works.

2 Related Work

Several works highlight the advantages and possibilities of using involuntary data, such as those available on social media, for tasks related to conservation science [6,14]. There are also a number of studies that apply machine learning to detect aspects of the environment, such as: estimating the number of animals of a given species [8] and classifying land cover [7]. One can also find studies that used Instagram to monitor other marine species [12,19], but without the use of automated tools.

[1] https://www.inaturalist.org/.

However, as already noted in other studies [5,9], there are few works found that applied machine learning to involuntary data for wildlife classification tasks. In particular, when searching for works that applied natural language processing, it was not possible to find works using texts in Portuguese extracted from Instagram to identify wildlife. Here, we report on 4 related works that applied machine learning on involuntary data.

Mazars-Simon [13] developed a system that searches, classifies, identifies and stores data on sea turtles. He used a dataset with 22,500 images that were shared by conservationists for the project. To build his system, the author trained a convolutional neural network (CNN) to classify images into six categories. With the model trained, the system fetched images from Flickr and rated them. If positive for a sea turtle, he performed individual recognition by comparing the image extracted from Flickr with the images in his database. As a result of training the model for image classification, the author obtained an accuracy of 0.95.

In [11], the authors used texts from Google News and Twitter to train a classifier to identify news about endangered species. They used the MurmurHash3[2] as a feature extraction technique and a neural network as a classifier. For training the model, a dataset with 5,464 examples was used, unevenly distributed between the positive and negative classes. As a result, they obtained an F1 of 0.96.

Another work [5] experimented with different combinations of representation techniques and machine learning algorithms aiming at identifying wildlife observations in Twitter texts. A dataset with 2,798 tweets was used, equally distributed between positive and negative classes. The authors obtained the best performance with BERT with F1 of 0.96.

A trained neural network to identify Portuguese man-of-war posts in images extracted from Instagram was reported in [3]. A dataset with 12,300 images was used, equally distributed between the positive and negative classes. Part of the training images was obtained from Instagram, part was downloaded through search engines and specialized websites. The authors experimented with different architectures of CNNs pre-trained with Imagenet [18] and without pre-training. The pre-trained networks were refined with the data. As a result, they obtained an F1 of 0.95 with the pre-trained ResNet50 [10].

Although the works listed in this section bear some resemblance to the study problem of this paper, only one of the models obtained by the authors can be used in the classification problem presented here [3]. The others have characteristics and objectives different from the problem addressed in this paper. In [13] the goal is to classify images of turtles. In [11] the objective is to classify news as being or not about endangered species, which does not include the Portuguese man-of-war. In addition to the text used in the training being in English, the characteristics of the text itself, in this case news, are different from the characteristics of social media texts. In [5] the goal is to classify Twitter texts as being or not about wildlife observations. Furthermore, the authors used only texts in

[2] https://github.com/aappleby/smhasher/blob/master/src/MurmurHash3.cpp.

English to train the models and there were no texts on our species of interest in their training database. Moreover, none of these works present a detailed description of the annotation process or focus on the impact of choosing appropriate labels for training a machine learning model. In the case of [3], the goal is to classify images of the species. In fact, we used the best architecture reported by the authors in our experimental study (Sect. 5).

3 Dataset Construction

This section describes how the data used for training and validation of the machine learning models proposed in this work was collected and labeled. It is important to recall that our research interest is in occurrences of *Physalia physalis* on the Brazilian coast. Therefore, the dataset construction followed this premise.

In this work we used data extracted from Instagram. Instagram is a social media developed by Meta. Among its features is the sharing of media (i.e., photos and videos). Users can upload up to 10 media in a single post. They can also write a caption for the post and add its location. Moreover, users can add hashtags in the caption. A hashtag can be written with text, emojis and numbers. Spaces and special characters will not work. Figure 1 shows an example of an Instagram post about a Portuguese man-of-war.

Fig. 1. Example of an Instagram post about a Portuguese man-of-war. Source: Instagram.

Fig. 2. Screenshot of the #caravela-portugesa page. Source: Instagram.

Instagram posts have some characteristics that should be considered:

- A post can have one or more images or videos. It is common to find posts with different media, for example: a picture of a *Physalia physalis*, a video showing the waves and a picture of trash, all in the same post;
- As Instagram is an image-focused platform, posts do not always have caption;

- Although it is possible for the user to add location metadata to the post, it is not mandatory and sometimes the location of the occurrence is informed in the caption or comments of the post;
- There are linguistic variations in the caption. While some posts show correct use of grammar rules, social media text is informal in nature, with many grammatical errors, slang, abbreviations, neologisms, and language mixture. There are also posts that are formed almost exclusively by hashtags and emojis. Furthermore, the use of compound hashtags (e.g., #caravelaportuguesa, #goprobrasil) generate words that do not exist in the lexicon.

When a user has a public account and adds hashtags to a post, the post will be visible on the corresponding Hashtag Page. A Hashtag Page shows a *Top* section where the most popular posts tagged with the hashtag appear. The page also shows a *Recent* section, where the most recent posts tagged with the hashtag appear. In this section the posts appear in the order in which they were posted. Figure 2 shows a screenshot of the #caravelaportugesa page. It is important to know that each hashtag is unique: #cnidario is different from #cnidarios, #aguaviva is different from #águaviva, and so on. This means that one should not expect to find posts that have the hashtag #cnidarios in the #cnidario page, for example.

The first option considered for obtaining Instagram data was the use of Instagram's API. However, this API does not return location metadata. So we chose to use the Instaloader library[3]. This library uses scraping to get data from Hashtag Pages and returns data in .json format. More details on the data extraction process are published in the [2] work. The data of interest for our research are: the caption of the post, media, posting date (timestamp) and location (latitude, longitude, city, location name and country). We chose to use the data obtained through the following hashtags: #aguaviva, #caravelaportuguesa, #cnidarios, #cnidários, #cnidario, #cnidário and #physaliaphysalis.

The #aguaviva was chosen because it is a generic term sometimes used in posts about *Physalia physalis*. The search result for this hashtag returned about 151,000 posts, which include, in addition to *Physalia physalis*, tattoos, people, clothes, other cnidarians, among other things. The #caravelaportuguesa was chosen because it is the popular name of *Physalia physalis* in Portuguese language and thus more assertive than #aguaviva. The result of the search for this hashtag returned around 3,300 posts, which included, in addition to *Physalia physalis*, boats, handicrafts, tattoos, clothes, among other things. The #physaliaphysalis, being the scientific name of the species, was chosen because it has a high potential for obtaining positive posts about *Physalia physalis*. A search for this hashtag returned around 2,500 posts. Cnidaria is the name of the phylum to which *Physalia physalis* belongs. Thus the hashtag #cnidarios, and some variations like #cnidário, #cnidario and #cnidários, were chosen because they are common names of this phylum in the Portuguese language and for being terms with medium potential to return positive posts about *Physalia physalis*. A search

[3] https://github.com/instaloader/instaloader.

for #cnidarios returned around 3,400 posts which include *Physalia physalis* as well as other cnidarians such as *Velella velella* and medusa. All searches were executed on August 27, 2022. In all, we collected 6,204 posts dated between 2012 and 2021.

After downloading, the data were saved in spreadsheets to facilitate the annotation process. For posts that had geolocation (i.e., latitude and longitude) the BigDataCloud API[4] was used to obtain additional location data (i.e., city, location name, state, country). This data was used as a complement to the existing metadata location.

The data were also enriched by identifying the caption language using the Langdetect library[5]. This library has a function that returns a vector indicating which languages were detected in the text along with the probability of the text being written in the language in question. The Portuguese language was assigned to the post whenever it appeared with some probability, even if small. As previously mentioned, among the characteristics of the data extracted from Instagram are the existence of posts using words from more than one language, in addition to the use of compound hashtags that generate words that do not exist in the lexicon. These characteristics make the task of assigning the language for the caption harder. For this reason, the language assigned by the library was reviewed by the computer scientist, with the goal of identifying the presence of terms used in Portuguese that could characterize the caption as written by a speaker of that language. After the review, the number of posts considered in Portuguese increased from 3,440 (according to the library) to 3,849 (after the review).

Table 1. Instances of caption that the library did not detected the Portuguese language

Caption	Languages and Probabilities
Sou imortal!	Catalan 99%
#aguaviva Por USER #nemdói TELEFONE para orçamentos #tattoo #inked #inkwork #inkworld #inklife #inkedgirls #tatuagens #art #blackwork #sketch #brunotattoo #omegainkstudio	English 42%, Swedish 28% and Afrikaans 28%

Table 1 shows 2 instances of caption that the library did not detect the Portuguese language. Figure 3 shows a screenshot of the spreadsheet created with the downloaded and enriched data.

[4] https://www.bigdatacloud.com/.
[5] https://github.com/Mimino666/langdetect.

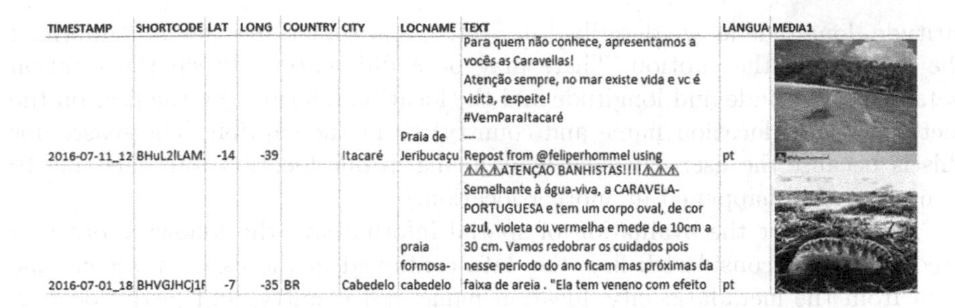

TIMESTAMP	SHORTCODE	LAT	LONG	COUNTRY	CITY	LOCNAME	TEXT	LANGUA	MEDIA1
2016-07-11_12	BHuL2lLAM:	-14	-39		Itacaré	Praia de Jeribucaçu	Para quem não conhece, apresentamos a vocês as Caravellas! Atenção sempre, no mar existe vida e vc é visita, respeite! #VemParaItacaré --- Repost from @feliperhommel using	pt	
2016-07-01_18	BHVGJHCj1F	-7	-35	BR	Cabedelo	praia formosa-cabedelo	⚠⚠⚠ATENÇAO BANHISTAS!!!!⚠⚠⚠ Semelhante à água-viva, a CARAVELA-PORTUGUESA e tem um corpo oval, de cor azul, violeta ou vermelha e mede de 10cm a 30 cm. Vamos redobrar os cuidados pois nesse período do ano ficam mas próximas da faixa de areia . "Ela tem veneno com efeito	pt	

Fig. 3. Screenshot of a spreadsheet created with the downloaded and enriched data.

4 Data Annotation

The data collected were manually annotated by a computer scientist and an oceanographer, who is a specialist in the species. However, the purpose of the annotation differed for each person. The oceanographer was interested in determining the effectiveness of Instagram for providing temporal and location information on sightings in the Brazilian coast. On the other hand, the computer scientist was interested in generating an annotated dataset for training a machine learning model in order to automatically classify new Instagram posts, and thus continually monitor the species.

This distinction resulted in a large number of posts classified as legitimate occurrences of the species by one annotator and negative by the other. The oceanographer used the following criteria to accept a post as a legitimate occurrence of *Physalia physalis*. It had to satisfy three conditions:

- Taxonomic identification: a media that clearly shows a *Physalia physalis*;
- Spatial information: a location in the Brazilian coast;
- Temporal information: include a timestamp associated with the post.

For the computer scientist, on the other hand, the lack of spatial and/or temporal information may not be important for a model that classifies posts based on their text and/or media. Thus, the dataset annotated by the specialist can include noise for training machine learning models.

Thus, before assigning a binary label (accepted or rejected) to the posts, they were annotated with their taxonomic and spatial information. This approach helped the annotators reach a consensus on the labels, resolve doubts and ambiguities, increase transparency and promote repeatability of the process. In addition, it made it possible to generate better classification models, as reported in Sect. 5.

Spatial Criterion. For the spatial criterion, we evaluated whether the post contained spatial information, and in particular if the location was on the Brazilian coast. Note that a post may have several metadata about location: latitude, longitude, city, location name and country. Besides, we enriched the metadata using latitude, longitude as we described in Sect. 3, and sometimes the user informed the location in the caption. There may be a difference between the location obtained by latitude and longitude and the location informed by the user on the metadata (city, location name and country) or in the caption. The reason for this is because the user can post while being in one location, but referring to something that happened in another location.

Therefore, for the evaluation of spatial information, the following order of precedence was considered: first the data contained in the caption, second the data from the metadata: city, location name and country, and lastly location obtained by latitude and longitude. It should be noted that the word or hashtag PORTUGAL alone was not used as location information, as it often appears as a reference to Portuguese culture, such as a post which talks about a visit to the Museu Paranaense[6]. It says: *"Navegar é preciso #caravelaportuguesa #portugal #brasil #museuparanaense"*[7]. Also, just the state (e.g., #bahia), #beach or names of very common beaches (e.g., *Praia Grande*) were not used alone to determine the location.

The spatial information was classified as follows:

- COAST-TEXT: The caption contains some information that allows identifying the location as being on the Brazilian coast (even if there is metadata indicating otherwise).
- COAST-GEO: The post contains metadata that allows identifying the location as being on the Brazilian coast.
- COUNTRYSIDE-TEXT: The caption contains some information that indicates that the post did not take place on the Brazilian coast, but still the location is in Brazil.
- COUNTRYSIDE-GEO: The post contains metadata indicating that the post did not take place on the Brazilian coast, but still the location is in Brazil.
- FOREIGNER: The post contains information in the caption or metadata that indicates that the post did not take place in Brazil.
- NOTHING: There is no information that allows the post location to be identified.

In some cases the location has been identified as being in the coastal city, but it is clear that it is not on the beach, as in the examples: *"Aquário Marinho do Rio de Janeiro"* and *"Centro Acadêmico de Vitória - CAV - UFPE"*[8]. These posts are classified as COUNTRYSIDE-TEXT when the information is part of the caption, and classified as COUNTRYSIDE-GEO when the information is part of the metadata.

[6] Museu Paranaense is a museum located in the Brazilian city of Curitiba.

[7] Translated as: "Sailing is necessary #caravelaportuguesa #portugal #brasil #museuparanaense".

[8] Translated as: "Marine Aquarium of Rio de Janeiro" and "Vitoria Academic Center - CAV - Federal University of Pernambuco".

Taxonomic Criterion. For the taxonomic criterion, we evaluated the posts media, especially if the images are pictures of *Physalia physalis*. The media of the posts were classified as follows. It is important to recall that a single post can have up to 10 media, so each media may have a different classification.

- REALISTIC: The media is a realistic picture of *Physalia physalis*, either on the beach sand or in the sea. In addition, when a post contains several images, they appear to be from the same occurrence.
- CLOSE: The media is a realistic picture of *Physalia physalis*, but it is a close-up image, showing only parts of *Physalia physalis*.
- DISPLACED: The media is a realistic picture of *Physalia physalis*, but displaced from its habitat. For example, it is in an aquarium or inside a plastic bag.
- EDITED: The media is a realistic picture of *Physalia physalis*, but with minor edits such as: inclusion of authorship, text or frame, but that do not disturb the visualization of *Physalia physalis*.
- COLLECTION: The images in the post are a collection of images from *Physalia physalis*, but it is possible to notice that they are not from the same occurrence and are possibly images taken from the internet.
- ART: The image is a representation of *Physalia physalis*, such as: compositions, drawings, photos with many edits and realistic tattoos.
- ACCIDENT: The media is a picture of a body part "burned" by a *Physalia physalis*.
- CNIDARIA: The media is a picture of another cnidarian, for example: *Velella velella*.
- VIDEO: The media is a video and therefore was not evaluated for this criterion.
- NOTHING: The media cannot be sorted in any other way described above.

Binary Label. Based on the location and taxonomic annotations, the posts were labeled as ACCEPTED or REJECTED as a legitimate occurrence of *Physalia physalis* on the Brazilian coast. For Instagram posts, the temporal criterion is always satisfied, because every post has an associated timestamp. The given label is also associated with a justification as described next.

To be *accepted* as a legitimate occurrence of *Physalia physalis* on the Brazilian coast, the post must meet taxonomic and spatial criteria. It means that: the post must have at least one image classified as: REALISTIC or COLLECTION or EDITED or DISPLACED or CLOSE (even if the post has other media that do not fit these five classes) and have the spatial information classified as: COAST-GEO or COAST-TEXT.

One post that does not meet one or more criteria is considered *rejected* as legitimate occurrence of *Physalia physalis* on the Brazilian coast. The rejection can be justified as:

- BECAUSE OF THE MEDIA: The post has the spatial information classified as: COAST-GEO or COAST-TEXT, but no media were classified as: REALISTIC or COLLECTION or EDITED or DISPLACED or CLOSE.

- BECAUSE OF THE LOCATION: The post has at least one media classified as: REALISTIC or COLLECTION or EDITED or DISPLACED or CLOSE, but the spatial information is classified as: COUNTRYSIDE-TEXT or COUNTRYSIDE-GEO or FOREIGNER or NOTHING.
- BECAUSE OF THE MEDIA AND LOCATION: The post does not meet both criteria.

5 Exploratory and Experimental Study

In this section we make a descriptive analysis of the data, with the goal of finding out the general aspects of the studied problem. To simplify the analysis, we joined all the data from the hashtags: #cnidario, #cnidarios, #cnidario and #cnidarios. So from now on, when we write #cnidario, we are considering all posts with these hashtags.

Of the 6,204 posts collected, 151 are repeated one or more times. That is, they appear in the search results for more than one of the searched hashtags with the same identifier. As discussed earlier, posts rejected only for lack of location information may introduce noise for training machine learning models. Thus, in this section we consider two annotated datasets: **original**, the one annotated with the oceanographer criteria, and **adapted**, which considers as positive, posts rejected only for lack of spatial information, that is, when spatial information is classified as NOTHING and at least one image meets the taxonomic criterion.

5.1 Distribution by Label

Table 2 shows the number of accepted and rejected posts per hashtag. In general, the number of positive posts for the occurrence of *Physalia physalis* on the Brazilian coast was small, mainly for the hashtags: #aguaviva, #cnidarios and #physaliaphysalis. We expected to find more positive posts in the search for #physaliaphysalis, but only 8% of the collected posts were accepted as legitimate occurrences of *Physalia physalis* for this hashtag. Despite the small number of positive occurrences found with #aguaviva (1%), the period surveyed was also the shortest (175 days), which makes room to invest more effort in the search for species in posts that have this hashtag. The search for occurrences using the hashtag #caravelaportuguesa had the best result, with 422 accepted posts by the specialist.

Table 2. Number of posts by Hashtag and Label

Hashtag	Label	Original Label		Adapted Label	
		No. Posts	%	No. Posts	%
#aguaviva	ACCEPTED	23	1%	36	2%
	REJECTED	1,657	99%	1,644	98%
#caravelaportuguesa	ACCEPTED	422	24%	600	36%
	REJECTED	1,364	76%	1,186	67%
#cnidario	ACCEPTED	77	3%	136	5%
	REJECTED	2,472	97%	2,413	95%
#physaliaphysalis	ACCEPTED	15	8%	54	29%
	REJECTED	174	92%	135	71%
TOTAL	ACCEPTED	537	9%	826	13%
	REJECTED	5,667	91%	5,378	87%

The table also shows that the number of rejected posts is much larger than the accepted ones. Comparing the two datasets, the adapted one contains 289 more accepted posts. Although this is a small number, it already has an effect on the quality of the machine learning model obtained, as we report next.

5.2 Experimental Study

To find out if there is any impact of considering as accepted posts rejected only for lack of spatial information, we performed some experiments.

For the training, validation and testing of the machine learning models we first filtered the dataset as follows. We deleted posts that have the following characteristics: empty text (12 posts); repeated posts (164 posts); with video only (643 posts); and location outside Brazil (1,991 posts). Moreover, we kept only posts identified as Portuguese language or emoji only in the caption because emojis can be translated to Portuguese. After applying the filters, 2,610 posts remained in the dataset, which were used in the two experiments described next.

Textual Analysis. We trained two models with the posts caption, one using original labels assigned by the specialist and the other with adapted labels. The experiments were performed using Python and libraries for preprocessing and text classification, such as: NLTK [1] and Scikit-learn [17].

To vectorize the text we used the TfidfVectorizer method from the Scikit-learn [17] library, keeping the default values of the method. Models were trained with Logistic Regression. As the dataset is unbalanced we set the parameter `class_weight` to `'balanced'`. This parameter automatically adjusts weights inversely proportional to the class frequencies in the input data.

We divided the dataset into 70% (train) and 30% (test), then the train dataset was divided into 70% (train) and 30% (validation) and applied cross-validation with 5 folds. The results obtained for classifying posts based on their captions are presented in the Table 3.

Table 3. Classification of Posts based on their captions and images. Standard deviation is displayed along with text training results.

Label	TEXT			IMAGE		
	Precision	Recall	F1	Precision	Recall	F1
original	0.637 (0.007)	0.875 (0.009)	0.737 (0.006)	0.913	0.913	0.913
adapted	0.764 (0.008)	0.897 (0.011)	0.825 (0.005)	0.941	0.932	0.936

Image Analysis. We trained two CNNs with the posts image, one using original labels assigned by the specialist and the other with adapted labels. As the posts can have more than one image, we selected one image per post to perform the training. In particular, for the positive posts, we selected one image that met the taxonomic criterion. The experiments were performed using Python and Keras[9].

We trained the CNNs by transferring the learning from a ResNet50 [10] pre-trained with Imagenet [18] and adapted the last layer to our problem. We chose the same architecture described as the best model in [3].

As the data was unbalanced, we applied the undersampling technique, which resulted in a dataset with 696 samples. This dataset was divided into 70% (train) and 30% (test), then the train dataset was divided into 70% (train) and 30% (validation).

The results for classifying posts based on their images are presented in Table 3. The experiments for both text and image show that the adapted dataset produced better classification models than the original dataset.

6 Conclusion

This paper presented an experience of constructing a dataset on the cnidarian *Physalia physalis* with posts extracted from Instagram. In particular, we showed that even for a binary classification problem, to distinguish between legitimate and false sightings of the species, the design of appropriate labels can affect the quality of the machine learning model trained with the dataset.

As future work, we intend to experiment with different preprocessing and neural networks such as: BERT and LSTM for text training. We also plan to work on multimodal machine learning in order to produce a classifier that takes into consideration both text and images in the same model. In addition, we are planning to make additional experiments with datasets with different training and test sizes.

References

1. Bird, S., Klein, E., Loper, E.: Natural Language Processing with Python. O'Reilly Media Inc. (2009). https://www.nltk.org/
2. Camargo, L., Rocha, H., Nascimento, L., Hara, C.: Coleta de dados do instagram sobre ocorrências de caravelas-portuguesas na costa brasileira. In: Anais da XVIII Escola Regional de Banco de Dados, pp. 51–59. SBC, Porto Alegre (2023). https://doi.org/10.5753/erbd.2023.229499

[9] https://keras.io.

3. Carneiro, A., do Nascimento, L., Noernberg, M., Hara, C., Pozo, A.: Portuguese man-of-war image classification with convolutional neural networks. arXiv preprint arXiv:2207.01171 (2022)
4. Cavalcante, M.M.E., Rodrigues, Z.M.R., Hauser-Davis, R.A., Siciliano, S., Haddad Júnior, V., Nunes, J.L.S.: Health-risk assessment of portuguese man-of-war (physalia physalis) envenomations on urban beaches in são luís city, in the state of maranhão, brazil. Revista da Sociedade Brasileira de Medicina Tropical 53 (2020)
5. Edwards, T., Jones, C.B., Corcoran, P.: Identifying wildlife observations on twitter. Eco. Inform. **67**, 101500 (2022)
6. Edwards, T., Jones, C.B., Perkins, S.E., Corcoran, P.: Passive citizen science: the role of social media in wildlife observations. PLOS ONE **16**(8), e0255416 (2021)
7. ElQadi, M.M., Lesiv, M., Dyer, A.G., Dorin, A.: Computer vision-enhanced selection of geo-tagged photos on social network sites for land cover classification. Environ. Model. Softw. **128**, 104696 (2020)
8. Foglio, M.: Animal wildlife population estimation using social media images collections. Master's thesis, University of Illinois, Chicago, Illinois, USA (2019)
9. Ghermandi, A., Sinclair, M.: Passive crowdsourcing of social media in environmental research: a systematic map. Glob. Environ. Chang. **55**, 36–47 (2019)
10. He, K., Zhang, X., Ren, S., Sun, J.: Deep residual learning for image recognition. In: Proceedings of the IEEE Conference on Computer Vision and Pattern Recognition (CVPR), pp. 770–778, June 2016
11. Kulkarni, R., Di Minin, E.: Automated retrieval of information on threatened species from online sources using machine learning. Methods Ecol. Evol. **12**(7), 1226–1239 (2021)
12. Leitão, A.T.T.S., de O Alves, M.D., dos Santos, J.C.P., Bezerra, B.: Instagram as a data source for sea turtle surveys in shipwrecks in Brazil. Anim. Conserv. **25**(6), 736–747 (2022)
13. Mazars-Simon, A.E.: The Wild in Live Project: A Human/Algorithm learning network to help citizen science in wildlife conservation. Master's thesis, Universidade de Coimbra (2019)
14. Morais, P., Afonso, L., Dias, E.: Harnessing the power of social media to obtain biodiversity data about cetaceans in a poorly monitored area. Front. Mar. Sci. **8** (2021). https://doi.org/10.3389/fmars.2021.765228
15. do Nascimento, L.S.: Monitoring jellyfish population by social media. Technical report, Universidade Federal do Paraná (2020), technical report, Pós-Graduação em Sistemas Costeiros e Oceânicos
16. do Nascimento, L.S., Hara, C.S., Jr., M.N., Noernberg, M.: Instagram como fonte de dados alternativa no monitoramento da #caravelaportuguesa (physalia phisalis, cnidaria). In: Livro de Memórias do IV SUSTENTARE e VII WIPIS: Workshop internancional de Sustentabilidade, Indicadores e Gestão de Recursos Hídricos (2022)
17. Pedregosa, F., et al.: Scikit-learn: machine learning in Python. J. Mach. Learn. Res. **12**, 2825–2830 (2011)
18. Russakovsky, O., et al.: ImageNet large scale visual recognition challenge. Int. J. Comput. Vision **115**, 211–252 (2015)
19. Sullivan, M., Robinson, S., Littnan, C.: Social media as a data resource for #monkseal conservation. PLoS ONE **14**(10) (2019)

Semantic Business Trajectories Modeling and Analysis

Muhammad Arslan(✉) and Christophe Cruz

Laboratoire Interdisciplinaire Carnot de Bourgogne (ICB), Dijon, France
{Muhammad.Arslan,Christophe.Cruz}@u-bourgogne.fr

Abstract. With the increasing availability of online news data through social media, there has been a growing focus on its potential as a source of business insights. However, interest in these insights has shifted towards more application-oriented analysis methods tailored to specific business purposes, leading to the emergence of semantically rich business trajectories. This article provides a definition of business trajectories and presents a method for constructing them using online news data. It also explores how trajectories can be enriched with semantic information to enable desired business interpretations and insights. Finally, the article discusses the potential of semantic business trajectories for environmental analysts in carrying out studies for various purposes.

Keywords: Business trajectories · Environment · Knowledge management · NLP

1 Introduction

Online news articles serve as a valuable resource for businesses seeking to identify opportunities across various industries [1]. These opportunities can range from a company advertising for labor to the sale of construction machinery, providing individuals or companies with the chance to capitalize on them [1]. These opportunities can be identified by grouping together multiple related key concepts, such as company, premises, project, and build, to form a business topic like "construction of premises."

By analyzing news articles using business topics, companies can classify business opportunities and gain valuable insights into the latest trends in their fields [2]. But business topics do not remain static over time. Instead, they follow a path of evolution or "trajectory" that can be influenced by a variety of external forces, such as changes in market trends, consumer behavior, technological advancements, or government regulations [2].

To obtain valuable insights from these trajectories, it is crucial to enhance them with supplementary information such as location and company details [2]. A semantic business trajectory (see Fig. 1) is a type of trajectory that has been enriched with annotations, which can provide additional contextual information. By representing this type of trajectory as a tuple (as mentioned below), businesses and researchers can gain a better understanding of the path of a particular business topic and how it evolves over time.

A. Abelló et al. (Eds.): ADBIS 2023, CCIS 1850, pp. 382–392, 2023.
https://doi.org/10.1007/978-3-031-42941-5_33

$(\boldsymbol{trajectoryIdentification}, \boldsymbol{businessTopic}, \boldsymbol{trajectoryAnnotations})$
where:

trajectoryIdentification is the trajectory's unique identifier,
businessTopic defines the set of business concepts,
trajectoryAnnotations is the set of annotations (such as location, company name, time, etc.) associated to the trajectory.

Annotating business trajectories with location data can help identify geographic territories where business opportunities may be concentrated or emerging, while company information can reveal the source of the information and the specific organizations involved.

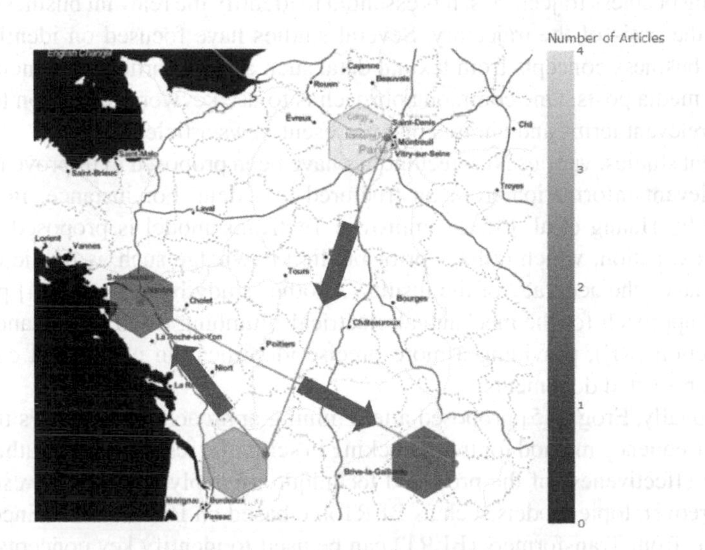

Fig. 1. A semantic business trajectory.

Natural Language Processing (NLP) techniques, such as topic modeling, can be used to identify key concepts from news articles, group them together to form business topics, and classify the articles under different topics. This process can be helpful in constructing business trajectories over time. Additionally, Named-Entity Recognition (NER) can be used to highlight important entities like locations and company names that are relevant to the topic, thereby providing semantic enrichment to the trajectories.

Semantic business trajectories have a wide range of applications beyond the business domain. They can also provide valuable insights into other areas, such as environmental science. By studying the impact of businesses across different geographical locations, we can gain a better understanding of their environmental impact. For example, we can analyze the carbon emissions resulting from various construction activities and identify potential ways to reduce them. This type of analysis is crucial for addressing climate change and other environmental challenges. Therefore, the use of semantic business trajectories can be an important tool for environmental scientists and researchers in a variety of fields.

The paper is structured as follows: Sect. 2 discusses the background of the research. Section 3 presents the proposed method for constructing semantic business trajectories. Section 4 discusses the presented method. The paper concludes Sect. 5 by discussing the potential directions for future research.

2 Background

2.1 From Business Concepts to Business Trajectories

Business trajectories are important in the field of Business Intelligence (BI) as they provide a way to track the evolution of business topics over time [1, 2]. However, before constructing business trajectories, it is essential to identify the relevant business concepts that form the basis of the trajectory. Several studies have focused on identifying and extracting business concepts from textual data, such as news articles, financial reports, and social media posts. One common approach is to use keyword extraction techniques to extract relevant terms and phrases that represent news articles.

In recent studies, various NLP techniques have been proposed to improve the extraction of relevant information from unstructured text data. For instance, in the study conducted by Huang et al. [3], an improved TextRank model is proposed for patent keywords extraction, which utilizes prior public knowledge such as public dictionary data to enhance the accuracy of the results. Another study by Wu et al. [4] proposes a rule-based approach for the mechanical-electrical-plumbing (MEP) NER and relationship extraction (RE), providing a more precise identification of relevant concepts in construction-related documents.

Additionally, Eroglu [5] proposed a text mining approach that employs the inverse document frequency method for trend tracking in scientific research. The author demonstrated the effectiveness of the proposed technique by applying it to a forest fire case study. Moreover, topic models such as BERTopic based on Bidirectional Encoder Representations from Transformers (BERT) can be used to identify key concepts as topics for text classification [6]. Traditional techniques, such as Latent Dirichlet Allocation (LDA), are limited in their ability to capture the evolution of topics over time. However, BERT-based dynamic topic modeling techniques overcome this limitation by modeling the evolution of topics and their representations over time. These studies illustrate the potential of NLP techniques in enhancing the extraction of valuable information from unstructured text data.

2.2 From Business Trajectories to Semantic Business Trajectories

The transformation of raw business trajectories into semantically rich trajectories is a challenge in this field. Several approaches and techniques [7–9] have been proposed that can help us to address this challenge and enrich business trajectories with semantic information. One of the primary techniques used to transform raw business trajectories into semantic trajectories is NER [10]. NER is a popular technique that identifies, and extracts named entities such as locations, organizations, and people from unstructured text data [9]. This technique has been used in various studies [10] to enrich business trajectories with location and company information.

Other techniques used to enrich business trajectories with semantic information include semantic web technologies such as Resource Description Framework (RDF) and Web Ontology Language (OWL), and knowledge graphs [11]. These techniques use ontologies and data models to represent the relationships between entities, providing a structured and semantically rich view of business trajectories. Several studies [12] have used these techniques to enrich business trajectories with additional information, such as spatio-temporal data and company details, which can provide valuable insights for businesses. By transforming raw business trajectories into semantically rich trajectories, businesses can gain a better understanding of their operations and how they are influenced by external factors, ultimately leading to more informed decision-making.

3 Semantic Business Trajectories

The literature review played a crucial role in identifying the competency questions (see Table 1) to model semantic business trajectories. Reviewing previous studies allowed us to identify the potential applications and use cases of semantic business trajectories, which formed the basis of the competency questions.

Table 1. Semantic business trajectories addressing competency questions.

No	Competency Questions	Required Information
1	What is the trend of business topics over time?	Business topics
2	What are the most common business topics overall, and how does their frequency change over time and across regions?	Business topics
3	How does the popularity of different business topics vary across regions?	Business topics and Locations
4	Which companies are most active in publishing news articles related to specific business topics in different regions?	Business topics, Locations, and Companies
5	What are the overall most important companies in terms of the frequency of news articles published about a business topic?	Business topics and Companies
6	Which companies are most important with respect to each region in terms of the frequency of news articles published about a business topic?	Business topics, Locations, and Companies

To address the competency questions using semantic business trajectories, the method is proposed. It mainly involves two stages:

1. Preprocessing the news data, and executing a topic model to identify relevant thematic concepts related to business trajectories using historical news article datasets, and

2. Semantically enriching the identified thematic topics using location and company information through NER.

To define the context of the study, the first step involves constructing a collection of relevant business concepts of interest. This is done by using a dataset of news articles that have been manually filtered to only include relevant articles. The news articles (8,300 in total), which were crawled from 95 different news sources such as lamontagne.fr, fusacq.com, businessimmo.com, sudouest.fr, actu.fr, ladepeche.fr, lavoixdunord.fr, etc., cover a time span of three months from Jan. 2023 to Mar. 2023 (see Fig. 2). To facilitate analysis, the news articles were translated from French into English using the Deep Translator Python library (https://pypi.org/project/deep-translator/) and underwent basic pre-processing steps such as stop-word removal, and filtering of special characters, weblinks, email addresses, and punctuations using the Natural Language Toolkit (NLTK) (https://www.nltk.org/).

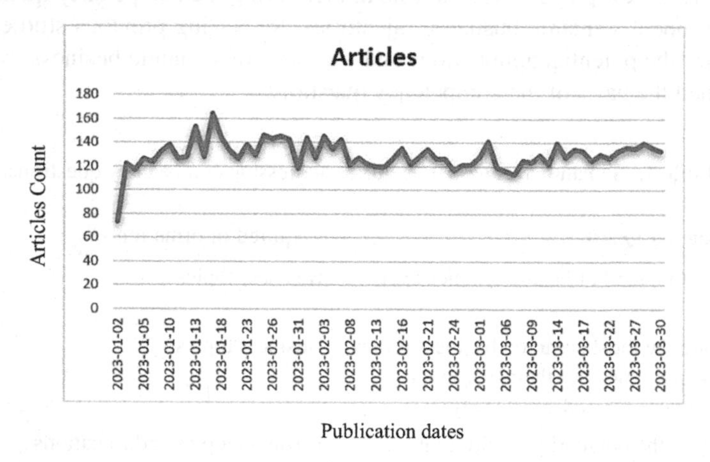

Fig. 2. Count of published news articles per date.

To identify the relevant thematic concepts from the text corpus, a topic model is applied. While there are many topic models available, BERTopic [6] is selected due to its efficiency and popularity in identifying underlying semantics in text and creating dense clusters for easily interpretable topics using transformers and c-TF-IDF (Term Frequency - Inverse Document Frequency) technique. The state-of-the-art BERTopic model [6] is trained on the news article corpus to extract topics (see Fig. 3). Each topic contains a group of related concepts. These concepts are represented in bar charts constructed from the c-TF-IDF scores for each topic representation. This allows for easy comparison of topics representations and their evolution over time (see Fig. 4). Once the topic model is trained, it can be used to predict the topics (as a set of related concepts) of new news articles. More details on BERTopic can be found in [6].

Figure 4 presents visualizations of business trajectories covering various business topics. It is evident from Fig. 4 that business topic 0 has the highest frequency and is the most dominant topic. For demonstration purposes, this topic's trajectory data has been used for further enrichment to gain more insights. For semantic enrichment, we have used

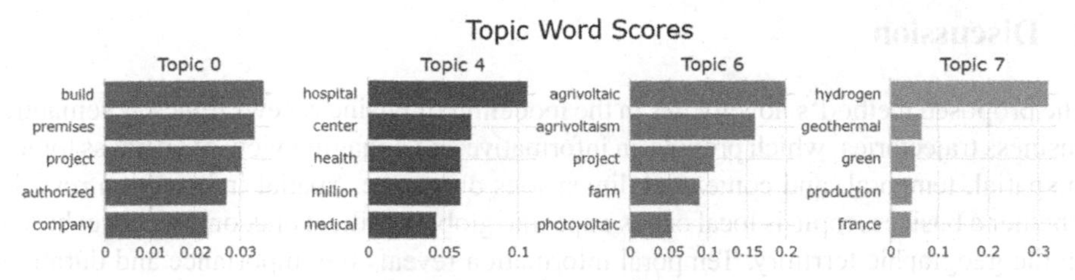

Fig. 3. Topic modeling.

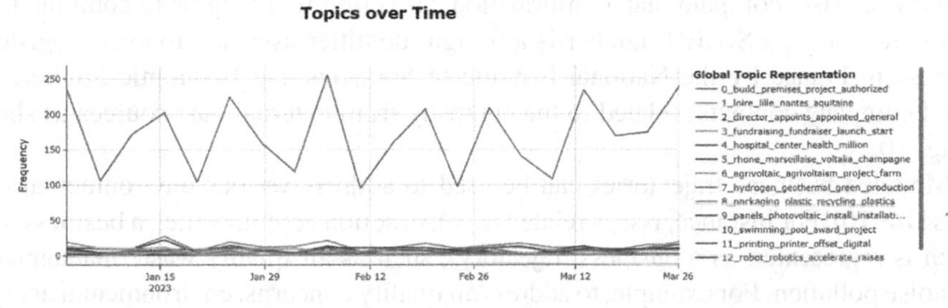

Fig. 4. Evolution of business trajectories over 3 months.

NER to identify and extract entities such as organizations and geographical locations from unstructured news data. There are various NER libraries available including SpaCy, Apache OpenNLP, and TensorFlow. Shelar et al. [13] found that SpaCy (https://spacy.io/) gives more accurate results as compared to Apache OpenNLP and TensorFlow to identify named entities in the text. Henceforth, SpaCy (https://demos.explosion.ai/) is used here (see Fig. 5).

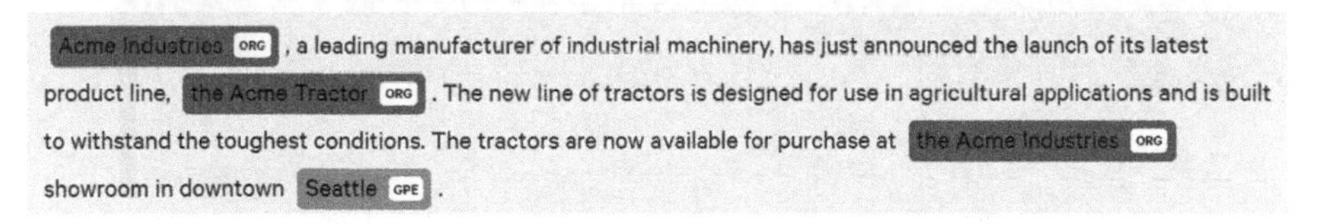

Fig. 5. NER.

By utilizing the location information extracted through NER, we can investigate the development of business topic 0 across various geographic regions (see Fig. 6). In Figs. 7 - 9, we present an analysis of the company information related to different geographic regions. Using NER, we extract company names from the news articles and associate them with their respective geographic locations. This allows us to track the presence and evolution of companies in different regions over time. By visualizing the figures, we can address the competency questions listed in Table 1. These visualizations allow us to gain insights into how the business topic of interest is evolving over time in different geographic regions and among different companies.

4 Discussion

The proposed method's novelty lies in the modeling of business news topics as semantic business trajectories, which provide an informative and engaging view of business topics in spatial, temporal, and contextual dimensions over time. Spatial information reveals whether a business topic is local or has a specific global political or economic facet based on the geographic territory. Temporal information reveals the importance and duration of the business topic. Contextual information of the trajectory shows how different geographic territories come up with the same business opportunities as competitors in the market. Also, company name information extraction will help us to combine more information (e.g., a SIREN number is a 9-digit identifier assigned to every registered business in France by the National Institute of Statistics and Economic Studies, and linked company contacts) related to the company from external data sources as shown in Fig. 10.

Modeled business trajectories can be used to address various environmental concerns [14–16]. For instance, issues related to construction activities (i.e., a business topic which is represented as a business trajectory), such as air quality waste management, and noise pollution. For example, to address air quality concerns, environmental analysts can use trajectory analysis to identify the geographic locations that are most affected by construction-related air pollution. This analysis can help environmental analysts develop targeted measures to reduce air pollution, such as endorsing cleaner construction materials, air filtration systems, or recommending construction activities during times of the day when air pollution levels are lower (Fig. 8).

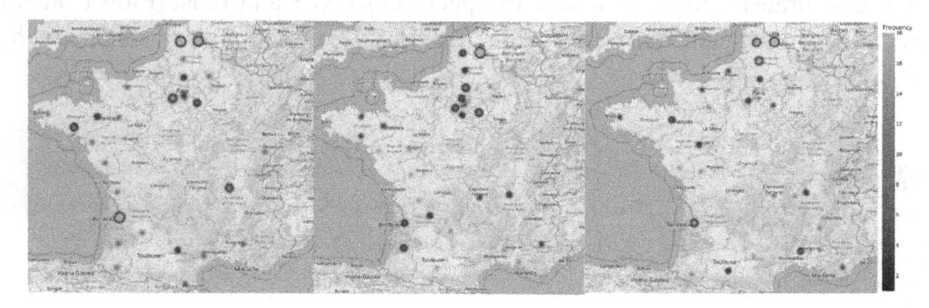

Fig. 6. Business topic evolution by publication count of news articles by regions.

Similarly, trajectory analysis can also be used to address waste management concerns. By analyzing the spatial and temporal distribution of construction activities identified through the published news articles (see Fig. 6), environmental analysts can identify areas where waste management is most needed and develop targeted strategies to reduce waste generation and ensure proper disposal of waste materials. Noise pollution can also be addressed using trajectory analysis. Environmental analysts can identify areas where noise levels are most significant due to increased construction activities (see Fig. 6) and recommend targeted strategies to reduce noise generation.

Environmental analysts could explore the use of semantic trajectory data from news articles to gain insights into how businesses are evolving in response to environmental

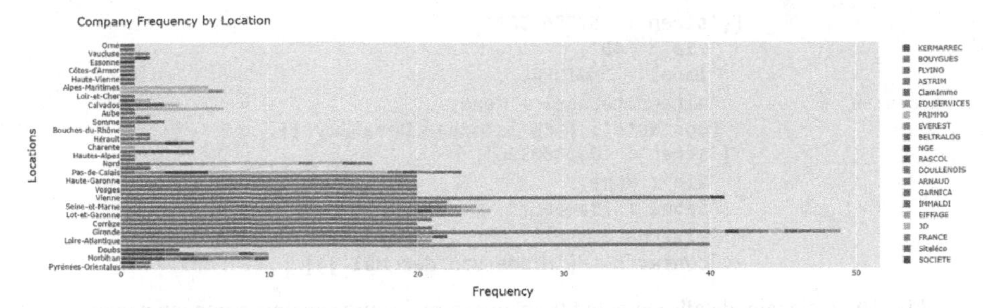

Fig. 7. Company and location information related to a business topic (Jan. 2023)

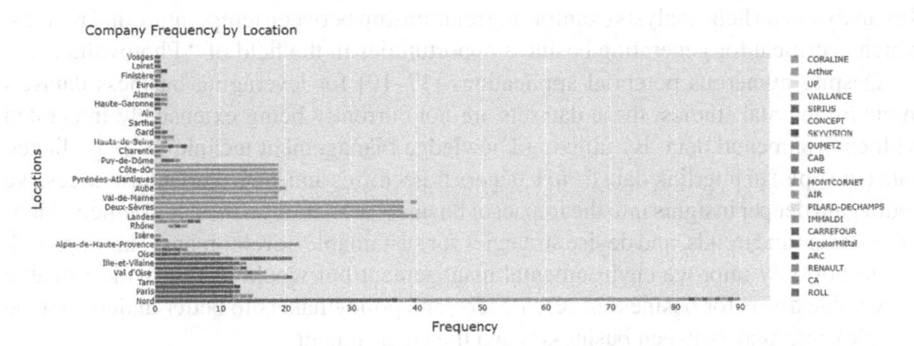

Fig. 8. Company and location information related to a business topic (Feb. 2023)

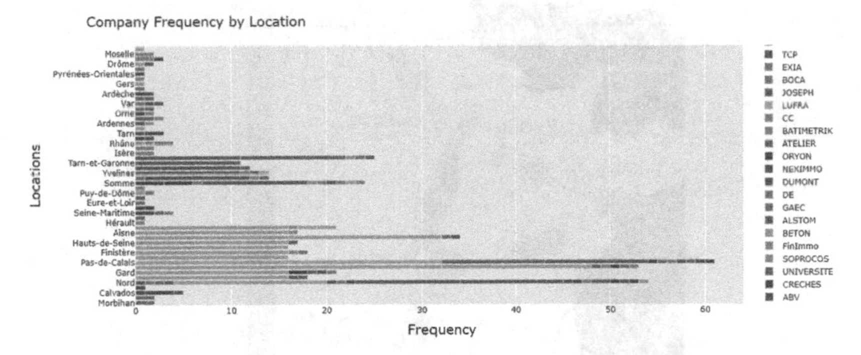

Fig. 9. Company and location information related to a business topic (Mar. 2023)

factors, such as temperature, in different geographical regions. This approach would provide a comprehensive understanding of the interplay between environmental factors and businesses over time. For instance, we visualize news articles related to topic 6 ("Photovoltaic") from Fig. 3, along with temperature values of various geographical regions during a 1-month period (see Fig. 11). By doing so, we can identify which regions are publishing business opportunities related to " Photovoltaic" over time, and how these regions are triggering opportunities in other regions (see Fig. 11). Furthermore,

```
[{'siren': '527561583',
  'zip': '49',
  'label': 'NATURAL',
  'alternateLabel': None,
  'contacts': ['Christopher Desailly']},
 {'siren': '000000526',
  'zip': None,
  'label': 'Barentz',
  'alternateLabel': None,
  'contacts': ['Hidde van der Wal']}]
```


Fig. 10. Company details extracted from an external source against company names.

this analysis can help analysts examine the relationship between temperature and regions, which is critical for generating business opportunities in the field of " Photovoltaic".

Despite numerous potential applications [17–19] for leveraging business datasets in environmental studies, these datasets are not currently being extensively integrated with environmental data. By utilizing knowledge management techniques (e.g., linked data concepts) to interlink data from business trajectories and environmental sources, we could gain deeper insights into the impact of business activities on the environment, identify patterns and trends, and devise strategies for sustainable development. This approach would not only improve environmental management but would also provide valuable knowledge assets for businesses, researchers, and policymakers to better understand the complex interplay between businesses and the environment.

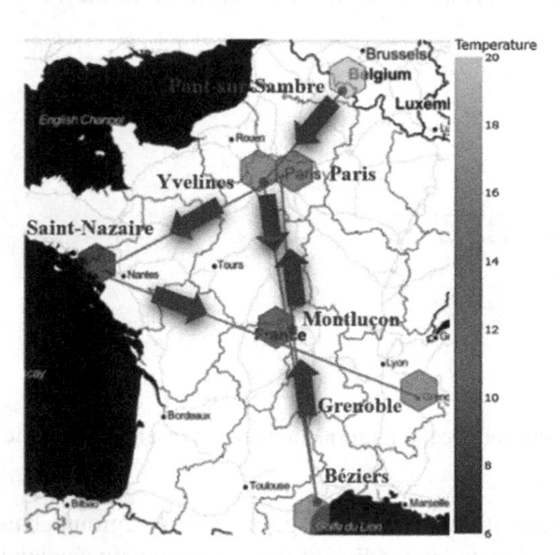

Fig. 11. Semantic business trajectory mapped with environmental data (i.e., temperature)

5 Conclusion

Online news articles offer a wealth of information about emerging business opportunities across various industries. By grouping related concepts into business topics, companies can gain valuable insights into the latest trends and changes in their fields. However, these business topics are not static and follow the path of "trajectory". To extract meaningful insights from these trajectories, it is essential to enrich them with additional information like location and company details. NLP techniques, including topic modeling and NER, are used for understanding business trajectories over time. Furthermore, semantic business trajectories have applications beyond the business domain and can offer valuable insights into areas like environmental science. By analyzing the impact of businesses across different geographical locations, researchers can gain a better understanding of their environmental impact and find ways to reduce carbon emissions and other environmental challenges. Therefore, the use of semantic business trajectories can be a crucial tool for researchers and scientists in various fields, including environmental science.

Acknowledgements. The authors thank the French government for the plan France Relance funding.

References

1. Arslan, M., Cruz, C.: Extracting business insights through dynamic topic modeling and NER. In: Proceedings of the 14th International Joint Conference on Knowledge Discovery (KEOD), pg. 215–222 (2022). https://doi.org/10.5220/0011552900003335
2. Arslan, M., Cruz, C.: Semantic taxonomy enrichment to improve business text classification for dynamic environments. In: INISTA, pp. 1–6. IEEE, Biarritz (2022)
3. Huang, Z., Xie, Z.: A patent keywords extraction method using TextRank model with prior public knowledge. Complex Intell. Syst. **8**(1), 1–12 (2022)
4. Wu, L.T., Lin, J.R., Leng, S., Li, J.L., Hu, Z.Z.: Rule-based information extraction for mechanical-electrical-plumbing-specific semantic web. Autom. Constr. **135**, 104108 (2022)
5. Eroglu, Y.: Text mining approach for trend tracking in scientific research: a case study on forest fire. Fire **6**(1), 33 (2023)
6. Grootendorst, M.: BERTopic: Neural topic modeling with a class-based TF-IDF procedure. arXiv preprint arXiv:2203.05794 (2022)
7. Solomon, Z., Ginzburg, K., Ohry, A., Mikulincer, M.: Overwhelmed by the news: a longitudinal study of prior trauma, posttraumatic stress disorder trajectories, and news watching during the COVID-19 pandemic. Soc. Sci. Med. **278**, 113956 (2021)
8. Zhang, Y., Zhang, H: FinBERT–MRC: financial named entity recognition using BERT under the machine reading comprehension paradigm. Neural Process. Lett. 1–21. (2023)
9. Novo, A.S., Gedikli, F.: Explaining BERT model decisions for near-duplicate news article detection based on named entity recognition. In: 17th International Conference on Semantic Computing (ICSC), pp. 278–281. IEEE (2023)
10. Arslan, M., Cruz, C.: Business insights using knowledge graphs by text analytics in dynamic environments. In: Proceedings of the 14th International Conference on Management of Digital EcoSystems, pp. 32–39 (2022). https://doi.org/10.1145/3508397.3564833
11. Memduhoglu, A., Basaraner, M.: An approach for multi-scale urban building data integration and enrichment through geometric matching and semantic web. Cartogr. Geogr. Inf. Sci. **49**(1), 1–17 (2022)

12. Arslan, M., Cruz, C.: Modeling virtual knowledge graphs using relevant news data by NLP methods for business analysis. In: 17th ICET, pp. 172–177. IEEE (2022). https://doi.org/10.1109/ICET56601.2022.10004674
13. Shelar, H., Kaur, G., Heda, N., Agrawal, P.: Named entity recognition approaches and their comparison for custom ner model. Sci. Technol. Libr. **39**(3), 324–337 (2020)
14. Lamba, P., Kaur, D.P., Raj, S., Sorout, J.: Recycling/reuse of plastic waste as construction material for sustainable development: a review. Environ. Sci. Pollut. Res. **29**(57), 86156–86179 (2022)
15. Atmaca, A., Atmaca, N.: Carbon footprint assessment of residential buildings, a review and a case study in Turkey. J. Clean. Product. 130691 (2022)
16. Szigeti, C., Major, Z., Szabó, D.R., Szennay, Á.: The ecological footprint of construction materials—a standardized approach from hungary. Resources **12**(1), 15 (2023)
17. Prakash, L.N.C.K., Suryanarayana, G., Jabbar, M.A.: Exploiting trajectory data to improve smart city services. Smart Urban Comput. Appl. **55** (2023)
18. Yu, D., Chen, Y.: The knowledge dissemination trajectory research of the carbon footprint domain: a main path analysis. Environ. Sci. Pollut. Res 1–18 (2022)
19. Arslan, M., Cruz, C.: Modeling semantic business trajectories of territories for multidisciplinary studies through controlled vocabularies. In: IEEE 39th International Conference on Data Engineering Workshops (ICDEW), pp. 170–177. IEEE (2023). https://doi.org/10.1109/ICDEW58674.2023.00032

A Text Mining Pipeline for Mining the Quantum Cascade Laser Properties

Deperias Kerre[1,2](✉), Anne Laurent[2], Kenneth Maussang[3], and Dickson Owuor[1]

[1] SCES, Strathmore University, Nairobi, Kenya
{dkerre,dowuor}@strathmore.edu
[2] LIRMM, Univ Montpellier, CNRS, Montpellier, France
anne.laurent@umontpellier.fr
[3] IES, Univ Montpellier, CNRS, Montpellier, France
Kenneth.Maussang@umontpellier.fr

Abstract. The development of the Terahertz laser technology in quantum cascade lasers (qcl) has brought about great potential for industrial applications. These lasers are based on the Terahertz electromagnetic waves, in the frequency range from about 100 GHz to 10 THz. There is need to understand the structure of the laser and its influence on the performance in order to optimize the design process. One way of collating this information is by having ontologies and knowledge bases capturing the various qcl designs and their performance characteristics. Majority of the laser design data is usually contained in scientific literature. The main drawback of such textual data sources is their unstructured nature. The complex nature of the laser design and the varying author language styles poses some level of difficulty in retrieving this information. Owing to this, the existing methods needs improvement in order retrieve the laser information at a high precision (with minimal number of incorrect records extracted) and minimized number of correct records not extracted. In this paper, we tackle this initial challenge by proposing a text mining pipeline for mining the qcl properties by extending the grammar rules of a conditional random field (CRF) based model using a rule-based approach. The properties of interest include: hetero-structure (laser stacking properties), working temperature, lasing frequency, laser thickness and the optical power. We evaluate the pipeline on sample open access journal papers from AIP, OPTICA and IOP Publishers.

Keywords: CRF model · Information Extraction · Knowledge Bases · Ontologies · Property Models · Quantum Cascade Lasers · Text Mining

1 Introduction

There exists a lot of information in scientific literature published daily on quantum cascade laser technologies. The literature documents the various laser designs and their performance properties [1]. The terahertz quantum cascade lasers have varying industrial application potential ranging from the biomedical

A. Abelló et al. (Eds.): ADBIS 2023, CCIS 1850, pp. 393–406, 2023.
https://doi.org/10.1007/978-3-031-42941-5_34

field, where the radiation can be used in detection of abnormal tissues, including cancers [2] and in the pharmaceutical field, where the lasers have been used in detecting organic compounds in drugs and identification of two or three dimensional distributions of molecules [3]. In electronics, the lasers can be used to pre-configure high speed telecommunication networks [4].

Quantum cascade lasers are complex hetero-structures. Most of the properties of the laser are defined by its growth sheet, i.e. the description of the different stacked layers: their thickness, the nature of the material, the order etc. The hetero-structural design of the laser constitutes the stacking properties i.e the different materials stacked together to form the laser while the opto-electronic characteristics of the laser entails the laser performance behaviour such as working temperature, power, frequency which is as a result of current injection in the laser.

Information regarding the description of the quantum cascade laser structures and performance is highly desired to give crucial insights for several purposes such as optimization of scientific design processes/implementations. The quantum cascade laser properties of interest to our study include: Working temperature, Optical Power, Lasing Frequency, Material design (Hetero-structure) and the Barrier thickness. These properties are crucial in evaluating the performance of the laser on various tasks. The quantum cascade laser working temperature, power, and frequency properties consists of value and a unit. For instance, "the quantum cascade laser lases at 9.7 THz, at a working temperature of 186 K with a maximum output power of 9 mW". The material design consist of a combination of chemical material names, in some cases with digits together with forward slashes. A sample statement containing this information is "We present two different terahertz quantum cascade laser designs based on GaAs/Al0.3Ga0.7As heterostructures". On the other hand, the quantum cascade laser barrier thickness consists of the thickness of the barriers in the hetero-structure. The property definition consist of the value (which in most cases consist of a sequence of numbers and forward slashes/commas) and a unit. A sample expression of this unit may be as follows: "The improved structure has layer sequence 31/93/14/73.4/23/155.4/11/110.2/14/84.7/20/155.4/17/110.1 Å".

One of the ways to capture the quantum cascade laser design and performance information from scientific literature is by designing ontologies and knowledge bases from the unstructured textual data. The main limitation of such textual data is their unstructured nature owing to the domain specific terminologies and different language styles by the authors. Some of the quantum cascade laser properties such as the barrier thickness poses difficulty in extraction due to the presence of special characters such as the forward slash (/) and the comma (,). In some cases, the properties are expressed in different ways such that there is need for contextualized rules to identify the property. The initial step of achieving structured ontologies and knowledge bases of quantum cascade laser design and performance properties is therefore implementing a text mining pipeline for extracting the quantum cascade laser properties from scientific literature.

There has been advances in the field of Information Extraction to structure the unstructured textual data in order to extract meaningful information from them. This has been accelerated by the adoption of the TDM Exception, a policy framework that advocates for the use of published resources for text and data mining purposes [5]. Examples include the use of machine learning algorithms in accelerated materials discovery [6]. With this breakthrough, there is enormous potential for applicability to other domains.

In this paper, we propose a text mining pipeline for mining the quantum cascade laser properties based on an extension of the ChemDataExtractor pipeline, a chemistry aware toolkit based on the CRF model [7]. We propose this as the first step in developing ontologies and knowledge bases for the quantum cascade laser domain. Our main contribution in this paper constitutes proposed efficient qcl property mining rules with improved precision and minimized number of correct records that are not extracted. This is achieved by defining new property parsing rules in form of property parsers using the rule based grammar approach [8]. We also extend the extraction capabilities by defining new property models for the qcl properties to be used along with the defined rules.

The rest of the paper is organized as follows: we first review the related works in Sect. 2, then we propose the workflow in the methodology in Sect. 3, we present the experimental and evaluation results in Sect. 4, and finally we conclude in Sect. 5.

2 Literature Review

Several works have been reported in the field of information retrieval in the materials science domain. The methodologies used can be broadly categorized into machine learning approaches and those that use a combination of machine learning and natural language processing principles. One of the crucial tasks under IR in materials science is the Chemical Name Entity Recognition (CNER).

CNER usually involves the identification of chemical and materials terms in the text. It can also be used to extract properties, physical characteristics, and synthesis actions. Early works on CNER focused on the on extraction of drugs and biochemical information [9,10]. Recently, CNER has gained alot of interest in the extraction of chemical and materials terms. The methods used in the CNER vary from traditional rule-based and dictionary look-up approaches to modern methodology built based on advanced machine learning (ML) and NLP techniques [11,12].

Examples of publicly available toolkits for extracting material terms include: those using rules and dictionaries-based approaches e.g LeadMine [13], ChemicalTagger [8], statistical models e.g OSCAR4 [14] and predominantly, the CRF model e.g ChemDataExtractor [7], ChemSpot [15], tmChem [16]. ChemDataExtractor has been extended/modified to extract several material terms and properties: semi-conductor bandgaps [17], thermo-electric materials [18], battery materials [19], refractive indices and dielectric constants [20], transition temperatures of magnetic materials [21] and an auto-populated ontology of material sciences [22]. Machine Learning techniques have also been utilized in CNER to identify chemical materials and their roles based on context information. Examples include bidirectional LSTM models [23,24] and a combination of deep convolutional and recurrent neural networks [25]. Others studies have also proposed mined datasets of inorganic materials synthesis recipes [26] and gold nanoparticle synthesis procedures, morphologies, and size entities [27]. Pre-trained BERT models have also been utilized in the extraction of battery materials [28] and for optical materials research [29].

Material science information has also been extracted from tables and figures. There has been attempts to parse tables from the scientific literature using heuristics and machine learning approaches [30]. Attempts have been reported on parsing article images, for instance ImageDataExtractor tool that uses a combination of OCR and CNN to extract the size and shape of the particles from microscopy images [31] and the Livermore SEM Image Tools for electron microscopy images using Google Inception-V3 network [32].

As noted from the literature review, several works have been reported on the applications of machine learning and NLP to materials discovery. In this study, the interest is more on "wafer fabrication" or hetero-structure properties, which is a critical step in the quantum cascade lasers development. Despite the great advancements reported in the literature, there is still a great potential for research in the materials science domain in order to achieve structured information regarding the quantum cascade lasers. The existing methodologies cannot be readily applied to mining these structures and the corresponding performance without modification/extension. Most of the natural language toolkits perform well in chemical terms, but when generalized from chemistry to the wider materials science, the grammar-based parsing rules used become less efficient. The BERT based models also need a lot of training data which involves manual annotation of the various properties by an expert. This may be cumbersome for large collections of articles. There is therefore need to extend the parsing capabilities of these techniques in order to adapt to the problem of mining the quantum cascade laser properties.

3 Methodology

In this section, we provide a detailed description of the workflow of the text mining pipeline for mining the qcl properties. The pipeline is based on an extension of the ChemDataExtractor, a chemical aware software toolkit [7]. We define new rules and make targeted extensions in order to fit to our domain of interest. The steps are as follows:

3.1 Document Retrieval and Processing

The first step in the text mining workflow is to acquire the scientific articles documenting the design of quantum cascade lasers. The study targets open access journals published by AIP, OPTICA and IOP publishers. The papers are retrieved using the keyword "quantum cascade lasers" and manually downloaded in the HTML format for further processing. The downloaded documents are then fed into ChemDataExtractor which uses the bespoke to process their information one document at a time. The downloaded HTML documents have a hierarchical structure with semantic markup tags. An example of such tags is the <head> tag which contains the metadata about the document such as title of the paper, the doi, authors etc. These tags are utilized by ChemDataExtractor to identify the key information about the papers such as the abstract, paragraphs, sentences etc. These files are then converted into plain text using the "reader" package in ChemDataExtractor which is then stored in the Document object of the toolkit for further processing.

3.2 Natural Language Processing

In this step, state-of-the-art Natural Language Processing techniques are applied to the document text. These capabilities are provided by the ChemDataExtractor toolkit. The techniques, which are tailored to the materials science domain include Sentence splitting, Tokenization, Part-of-Speech Tagging and Chemical-Named Entity Recognition (CNER). Sentence splitting, Tokenization and Part-of-Speech-Tagging were adopted from ChemDataExtractor without modification. The CNER rules are extended and adapted to the quantum cascade laser domain as described in the information extraction section.

3.3 Information Extraction

The ChemDataExtractor toolkit provides three ways of extracting information from text. These include: (i) Rule-based approach-which involves explicit crafting of statements that utilize regular expressions patterns and POS tags, (ii) automatic parsing and (iii) the modified snowball algorithm that can be trained in a semi-supervised manner on documents dataset and probabilistically used to extract information. In this paper, we adopt an extended rule-based approach by defining new property models and grammar logic for the qcl properties of interest. The property models are defined based on the user model concept [21].

The user model concept in ChemDataExtractor consists of a collection of defined property models for extracting different information. In general, a property model specifies the information to be extracted and the extraction rules to be used in retrieving the information. The information can be in form of physical quantities or chemical names. The user model consists of three models i.e the quantity models, general base model and the compound model.

The quantity model defines physical quantities such as time, electric charge, volume and the compound model defines chemical names together with the corresponding chemical name labels and roles. The general base model on the other hand contains user-defined fields, such as words, regular expressions, or other models. Every quantity model has the respective fields that will be populated upon data extraction from the document. The fields include the value, units, error, the standardized value and the specifier used to extract the data. For our text mining pipeline, we define five new property model to capture each of the properties of interest.

The property models for the working temperature, lasing frequency, power and laser thickness constitute quantity models while the hetero-structure model constitutes the compound model as the hetero-structure consist of material names. The Working Temperature property model inherits/nests the existing Temperature model in ChemDataExtractor. This handles the unit standardization process for the extracted temperatures. This is also the case for the OpticalPower model which inherits from the Power Model and the Heterostructure model which inherits from the Base model. In the quantum cascade laser literature, power readings are expressed in milliwatts (mW). We include this as an additional unit in the Power model. For the Lasing Frequency property model and the Barrier thickness, we define the Frequency model and the Barrier thickness model to handle the units. The fields to be populated from are also defined from scratch.

For all the property models, one of the important attributes is the parser. A property model can have one or more parsers. The parsers includes the defined grammar rules (logic) for relationship extraction. More information on parsers is given in relationship extraction section.

Phrase Parsing and Relationship Extraction: This is a key step that entails the extraction of suitable relationships. The relationship can be in the form of (i) a specifier expression/keyword and a chemical name or (ii) a specifier expression, a value and a unit. These relations are the ones that populate the specific records of the various qcl properties. ChemDataExtractor makes use of a hybrid approach to Chemical Named Entity Recognition (CNER); machine-learnt, dictionary-based and rule-based methods are all used.

The default parser of ChemDataExtractor, AutoSentenceParser, uses multiple specialized grammar rules that have been designed to extract more specific types of chemical information. In order to use the autosentence parser, a specifier expression is defined to capture the property relationship extraction rule. The rules are formed by combining the different keywords (Table 1) and parser elements (Table 2) in form of tokens.

Table 1. Quantum Cascade Laser Properties and the Keywords

Target Property	Keyword/Sentence	Unit
Working Temperature	heat-sink, Tmax, Maximum Temperature, Working Temperature	K
Optical Power	Optical power, Output Power, Peak Power	W
Hetero-structure (Material)	Growth, Grown in, Wafer, MBE, Laser-structure	N/A
Frequency	Laser Frequency, Lasing at, output Frequency	THz
Barrier Thickness	Layer thicknesses	Å

Table 2. The Parser Elements

Elements	Description	Elements	Description
R (Regex)	Match text with regular expression	T (Tag)	Match tags
W (Word)	Match case-sensitive token text	I (IWord)	Match case-insensitive token text
Any	Match any single token	H (Hide)	Ignore the matched tokens
Not	Match only if not followed by some text	FollowedBy	Match only if followed by some text
ZeroOrMore	Match zero or more of the expressions	OneOrMore	Match one or more of the expressions
Optional	Match if it exists	SkipTo	Skips to the next occurrence of text

The keywords adopted for each of the properties shown in Table 1 were settled upon based on consultation and advise by experts in the quantum cascade laser domain.

The default Autosentence parser however fails and under performs on some properties due to the high level of ambiguity and implicit knowledge carried within natural language. This has an implication on the precision of the extraction process and also leads to various correct records not being extracted. For instance, where several temperatures are mentioned in an article, there is need to define more precise rules for extracting the temperature of interest.

The Autosentence parser also requires a chemical compound in order to merge a complete record for extraction. This causes it to fail in cases where properties are mentioned without an associated chemical compound as this is the case with most of the qcl properties. The requirement to display a compound also leads to many false positives as the many records with characters in form of compounds are extracted.

Some of the properties such as the barrier/layer sequence have special characters such as the forward slash (/). A sample property of this is as follows: "42/67.8/23/96/34/73/40/206.2 nm". In some cases, the unit is also is put in brackets immediately after the value. Experimental analysis indicates failure by the AutoSentence parser in extracting these properties due to the unique combination of the special characters. In cases where the unit is mentioned after the property, the parser only extracts the last digits close to the unit and the unit (i.e. 206.2 nm in the example property given).

In order to extract the material design (hetero-structure), working temperature, the lasing frequency, barrier thickness and the optical power, we define five efficient grammar parsing rules for the respective defined property models for these properties in form of property parsers. The parsers capture the phrase extraction rules expressed in form of regular expressions. These expressions are based on the selected keywords describing the various qcl parameters. The grammar rules of the parsers are defined based on a set of parser elements indicated in Table 2 as defined in Chemdataextractor.

A parser typically consist of a prefix, value and the unit for the properties capturing physical quantities. The prefix contains the combined tokens of the various keywords used in identifying a property, the value contains the rules (in form of regular expressions) for matching the property value and the unit captures the units for the property.

For the qcl material parser, we define the 'heterostructure' as the main field to capture the value of the material design. We also only have the prefix (key phrases contextualizing the qcl material property) and the material attributes for the material design parser. The material consists of a series of regular expressions to match the qcl material names. The prefix and the material properties are combined to populate a complete heterostructure record.

Figure 1 shows a sample material design (hetero-structure) tree structure (upper) and extracted record output (lower) for a sample sentence from a journal paper: "We present two different terahertz quantum cascade laser designs based on GaAs/Al0.3Ga0.7As heterostructures that feature a depopulation mechanism of two longitudinal-optical phonon scattering events.".

```
b'<material>GaAs / Al0.15Ga0.85As</material>'

[{'QclMaterialDesign': {'heterostructure': 'GaAs / Al0.3Ga0.7As'}}]
```

Fig. 1. Sample Extracted Material Design (Heterostructure) Record.

For the barrier/layer thickness property parser, the rules are defined in such a way that the records are extracted with the special characters matched. The defined parsers interpret the manually defined grammar rules into an xpath parse tree from which the data model is constructed. The different parser elements, are combined with the "+" or "—" operators making the grammar rule flexible for update. The nested grammatical rules constitutes the specifier expression. The defined property rules are run for each document containing the qcl properties of interest. Algorithm 1 shows the workflow of the pipeline for extracting the qcl properties.

Algorithm 1: Mining the QCL properties

Input: D-Union{Ei}, input document object.
Output: : R-Union{Ri}
1 $S \leftarrow Union\{Si\}$`/* Prefix for the various keywords describing qcl properties */`
2 $M \leftarrow PropertyModel$`/* specifies the fields to be captured for a particular property. */`
3 $P \leftarrow Parser$ `/* grammar logic. */`
4 Set $D.model \leftarrow M$ and $parser \leftarrow P$ `/* Defining the parser and property model. */`
5 **for** *each document element Ei in D* **do**
6 | scan(Ts) **if** $Ti \leftarrow Si \subseteq S$ **then**
7 | | match Ri.
8 | **else**
9 | | skip to the next Ei and repeat step 6 and 7.
10 | **end**
11 **end**
12 **repeat** line 5-11 until all Ris are merged.
13 **return** *record R with the matched property relationship(s).*

We consider a document object D, capturing the various qcl properties of interest. D contains different document elements E1...En such as the title, paragraphs, sentences etc. A defined prefix S captures the possible expressions Si for a particular property. The expressions consists of keywords/a combination of the keywords used in context of a particular qcl property of interest. R consists of set of the property records which may consist of several of the individual record elements Ri. Ri captures the contextual property name, value and units. M and P consists of the defined property models and the parsers respectively. Before scanning though the document tokens, the property models and parsers have to be specified as shown in step 3 of the algorithm. The subsequent steps now involves searching for the matching token expressions for the prefix, property values, units and names which are merged into complete records.

4 Results and Discussions

4.1 Evaluation Metrics

In order to evaluate the performance of the proposed pipeline, we use the precision and recall as the evaluation metrics. In this context, the precision is the fraction of correct (relevant) records among all extracted records and the recall is the fraction of successfully extracted records among all correct (relevant) records in the articles. The word "correct" implies that the relationship of that record can be identified by a human when reading the corresponding sentence. In contrast, an "incorrect" (false) record suggests that a human expert cannot deduce the relationship of that record from the corresponding sentence. The metrics are determined as follows:

$$Precision = \frac{TP}{TP + FP} \tag{1}$$

$$Recall = \frac{TP}{TP + FN} \tag{2}$$

where TP is the true positive count (the number of correct records extracted), FP is the false positive count (the number incorrect records extracted), and FN is the false negative count (the number of correct records that are not extracted). The metrics are used to assess the chances of the pipeline leaving correct records unextracted and the chances of getting only the correct records in a given number of records.

4.2 Discussions

We use a sample of 43 open access articles as the evaluation dataset. The articles are randomly sampled from AIP, OPTICA and IOP publishers using the keyword "terahertz quantum cascade lasers". We restrict the sample to the articles describing proposed qcl designs and the corresponding performance characteristics. The articles consist of a total of 192 records manually extracted. The distribution of records is as follows: Optical power (33), Working Temperature (32), Lasing Frequency (44), Hetero-structure (66) and Barrier thickness 15 records. The records are manually extracted by an expert in quantum cascade lasers. The records are compared with those extracted by the pipeline in order to come up with the evaluation metrics. We compare the performance of our defined parsers and the default autosentence parser in chemdataextractor except for the sequence layer thickness property. The performance evaluation of the pipeline on the sequence layer thickness is done separately as the autosentence is not used in extracting this property. This is owed to the inability of the autosentence parser in extracting these records. A correct extracted record is one that can be identified to correspond to the one manually extracted by an expert. Table 3 shows the evaluation metrics for the default autosentence parser and Table 4 shows the performance evaluation metrics for our defined parsers.

Table 3. Performance Evaluation Metrics for the Default Autosentence Parser.

Property	Records Extracted	TP	FP	FN	Precision (%)	Recall (%)
Optical Power	16	11	5	22	68.75	33.33
Working Temperature	128	31	97	1	24.22	91.17
Lasing Frequency	16	10	6	34	62.50	22.73
Hetero-structure	140	50	90	16	35.71	75.76
Total	300	102	198	73	**47.80**	**55.75**

Table 4. Performance Evaluation Metrics for the Defined Parsers.

Property	Records Extracted	TP	FP	FN	Precision (%)	Recall (%)
Optical Power	25	24	3	9	96.00	72.73
Working Temperature	32	25	7	7	78.13	78.13
Lasing Frequency	37	29	8	15	78.34	65.91
Hetero-structure	60	59	1	7	98.33	89.39
Total	154	137	19	38	**87.70**	**76.55**

The autosentence parser achieves a precision of 68.75% and recall of 33.33% for the optical power property. This implies a higher number of correct records that are not extracted. Most the unextracted records are expressed in different contexts with varying keywords hence posing a chellenge to the default parser. The defined parser achieves a higher precision of 96.00% and has a higher recall of 72.73% indicating a lesser number of correct records that are not extracted. This is as indicated by the false negative (FN) values in Tables 3 and 4. The defined rules take into consideration the various contexts in which the power values are expressed.

For the working temperature, the autosentence parser has a higher false positive rate hence resulting to a lower precision of 24.22%. One the other hand, the working temperature parser achieves minimal incorrect and unextracted records hence attaining a higher precision and recall of 78.13% and 78.13% respectively. This clearly indicates that as more temperatures are mentioned in literature, there is increased difficulty in retrieving the temperature of interest. The autosentence parser extracts most of the temperatures mentioned including the working temperature but has increased number of incorrect records extracted.

The default parser results to a recall of 22.73% for the lasing frequency. The defined parser on the other hand has a recall of 65.91%. The defined logic has therefore higher chances of extracting the correct records due to the specialized grammar rules. This is also pointed out by the higher precision of 78.34% for the defined parser. The lasing frequency property is however expressed in many forms. This needs a wider definition of the rules hence the lower recall for both the autosentence parser and the defined parser. For the hetero-structure/material property, the default autosentence parser exhibits a higher false positive rate hence resulting to a precision of 35.71% and a recall of 75.76%. The higher false positive rate is attributed to the records having compound like names but are not necessarily qcl material names. The defined parser on the other hand achieves a precision of 98.33% and recall of 89.39%. The higher precision in attributed to the more specialized rule combination for the material design.

Overall, the default autosentence parser achieves a precision of 47.80% and a recall of 55.75%. The defined parsers on the other hand achieves a precision of 87.70% and a recall of 76.55%. This indicates a better performance of the defined parsers on the power, frequency, working temperature and the hetero-structure properties as shown in Table 3. For the barrier thickness grammar logic, 2 records are left unextracted. The unextracted records consists of a combination of values with units in different positions and not after the reading. This results to a precision of 72.22% and a recall of 86.67%. The performance of the defined parsing rules in extracting the qcl barrier thickness indicates a great potential of their applicability on this property as they constitute the initial attempt to extract such properties with special characters.

5 Conclusion

In this paper, we propose a text mining pipeline for mining the qcl hetero-structure and the opto-electronic properties based on efficient rule based grammar logic. This is achieved by defining new parsing rules for the properties of interest in order to minimize the number of incorrect records extracted and the number of correct records not extracted. The rules are also able to match readings with special characters such as the qcl barrier thickness. Experimental analysis of comparative performance indicates better performance by the proposed rules. The work is however limited on open access articles for the specified publishers and more articles will be needed in future for extensive experimentation. The grammar logic is also limited to descriptions where the unit immediately follows the readings. We aim to extend this in future work to capture situations where the barrier thickness values are separated by commas and no unit mentioned after the value. We also aim to explore the integration of the named entity recognition with ontology population techniques in order to generate ontologies for the extracted properties.

Acknowledgement. This work was supported by the CNRS (French Centre National de la Recherche Scientifique) through the founding of a project within the Programme "Dispositif de Soutien aux Collaborations avec l'Afrique sub-saharienne". The authors would also like to thank the Strathmore University, School of Computing and Engineering Sciences and the Strathmore University Doctoral Academy for their involvement in creating the opportunity for this work to be produced and lastly, Qingyang Dong (University of Cambridge, Cavendish laboratory-molecular engineering group) for the insightful discussions.

Availability of Materials. The source code and the materials used for the production of this work are publicly available at our GitHub repository: https://github.com/DeperiasKerre/qclProperties.

References

1. Kumar, S., Hu, Q., Reno, J.L.: 186 K operation of terahertz quantum-cascade lasers based on a diagonal design. Appl. Phys. Lett. **94**(13), 131105 (2009). https://doi.org/10.1063/1.3114418
2. Vafapour, Z., Keshavarz, A., Ghahraloud, H.: The potential of terahertz sensing for cancer diagnosis. Heliyon **6**(12), e05623 (2020). https://doi.org/10.1016/j.heliyon.2020.e05623
3. Shur, M., Liu, X.: Biomedical applications of terahertz technology. In: Advances in Terahertz Biomedical Imaging and Spectroscopy, vol. 11975, p. 1197502. SPIE, March 2022. https://doi.org/10.1117/12.2604800
4. Kanno, A., et al.: High-speed coherent transmission using advanced photonics in terahertz bands. IEICE Trans. Electron. **98**(12), 1071–1080 (2015). https://doi.org/10.1103/PhysRevMaterials.4.123802
5. Rosati, E.: The exception for text and data mining (TDM) in the proposed Directive on copyright in the Digital Single Market-technical aspects. Briefing Requested by the Juri Committee, European Parliament (2018). https://doi.org/10.1093/jiplp/jpy063
6. Liang, H., Stanev, V., Kusne, A.G., Takeuchi, I.: CRYSPNet: crystal structure predictions via neural networks. Phys. Rev. Mater. **4**(12), 123802 (2020). https://doi.org/10.1103/PhysRevMaterials.4.123802
7. Swain, M.C., Cole, J.M.: ChemDataExtractor: a toolkit for automated extraction of chemical information from the scientific literature. J. Chem. Inf. Model. **56**(10), 1894–1904 (2016). https://doi.org/10.1021/acs.jcim.6b00207
8. Hawizy, L., Jessop, D.M., Adams, N., Murray-Rust, P.: ChemicalTagger: a tool for semantic text-mining in chemistry. J. Cheminform. **3**, 1–13 (2011). https://doi.org/10.1186/1758-2946-3-17

9. Corbett, P., Copestake, A.: Cascaded classifiers for confidence-based chemical named entity recognition. BMC Bioinform. **9**(11), 1–10 (2008). https://doi.org/10.1186/1471-2105-9-S11-S4
10. García-Remesal, M., García-Ruiz, A., Prez-Rey, D., De La Iglesia, D., Maojo, V.: Using nanoinformatics methods for automatically identifying relevant nanotoxicology entities from the literature. BioMed Res. Int. **2013** (2013). https://doi.org/10.1155/2013/410294
11. Lafferty, J., McCallum, A., Pereira, F.C.: Conditional random fields: probabilistic models for segmenting and labeling sequence data (2001)
12. Hochreiter, S., Schmidhuber, J.: Long short-term memory. Neural Comput. **9**(8), 1735–1780 (1997). https://doi.org/10.1162/neco.1997.9.8.1735
13. Lowe, D.M., Sayle, R.A.: LeadMine: a grammar and dictionary driven approach to entity recognition. J. Cheminform. **7**(1), 1–9 (2015). https://doi.org/10.1186/1758-2946-7-S1-S5
14. Jessop, D.M., Adams, S.E., Willighagen, E.L., Hawizy, L., Murray-Rust, P.: OSCAR4: a flexible architecture for chemical text-mining. J. Cheminform. **3**(1), 1–12 (2011). https://doi.org/10.1186/1758-2946-3-41
15. Rocktäschel, T., Weidlich, M., Leser, U.: ChemSpot: a hybrid system for chemical named entity recognition. Bioinformatics **28**(12), 1633–1640 (2012). https://doi.org/10.1093/bioinformatics/bts183
16. Leaman, R., Wei, C.H., Lu, Z.: tmChem: a high performance approach for chemical named entity recognition and normalization. J. Cheminform. **7**(1), 1–10 (2015). https://doi.org/10.1186/1758-2946-7-S1-S3
17. Dong, Q., Cole, J.M.: Auto-generated database of semiconductor band gaps using chemdataextractor. Sci. Data **9**(1), 193 (2022). https://doi.org/10.1038/s41597-022-01294-6
18. Sierepeklis, O., Cole, J.M.: A thermoelectric materials database auto-generated from the scientific literature using ChemDataExtractor. Sci. Data **9**(1), 648 (2022). https://doi.org/10.1038/s41597-022-01752-1
19. Huang, S., Cole, J.M.: A database of battery materials auto-generated using ChemDataExtractor. Sci. Data **7**(1), 260 (2020). https://doi.org/10.1038/s41597-020-00602-2
20. Zhao, J., Cole, J.M.: A database of refractive indices and dielectric constants auto-generated using chemdataextractor. Sci. Data **9**(1), 192 (2022). https://doi.org/10.1038/s41597-022-01295-5
21. Court, C.J., Cole, J.M.: Auto-generated materials database of Curie and Néel temperatures via semi-supervised relationship extraction. Sci. Data **5**(1), 1–12 (2018). https://doi.org/10.1038/sdata.2018.111
22. Mavracic, J., Court, C.J., Isazawa, T., Elliott, S.R., Cole, J.M.: ChemDataExtractor 2.0: autopopulated ontologies for materials science. J. Chem. Inf. Model. **61**(9), 4280–4289 (2021). https://doi.org/10.1021/acs.jcim.1c00446
23. He, T., et al.: Similarity of precursors in solid-state synthesis as text-mined from scientific literature. Chem. Mater. **32**(18), 7861–7873 (2020). https://doi.org/10.1021/acs.chemmater.0c02553
24. Weston, L., et al.: Named entity recognition and normalization applied to large-scale information extraction from the materials science literature. J. Chem. Inf. Model. **59**(9), 3692–3702 (2019). https://doi.org/10.1021/acs.jcim.9b00470
25. Korvigo, I., Holmatov, M., Zaikovskii, A., Skoblov, M.: Putting hands to rest: efficient deep CNN-RNN architecture for chemical named entity recognition with no hand-crafted rules. J. Chem. **10**(1), 1–10 (2018). https://doi.org/10.1186/s13321-018-0280-0

26. Kononova, O., et al.: Text-mined dataset of inorganic materials synthesis recipes. Sci. Data **6**(1), 203 (2019). https://doi.org/10.1038/s41597-019-0224-1
27. Cruse, K., et al.: Text-mined dataset of gold nanoparticle synthesis procedures, morphologies, and size entities. Sci. Data **9**(1), 234 (2022). https://doi.org/10.1038/s41597-022-01321-6
28. Huang, S., Cole, J.M.: BatteryBERT: a pretrained language model for battery database enhancement. J. Chem. Inf. Model. **62**(24), 6365–6377 (2022). https://doi.org/10.1021/acs.jcim.2c00035
29. Zhao, J., Huang, S., Cole, J.M.: OpticalBERT and OpticalTable-SQA: text-and table-based language models for the optical-materials domain. J. Chem. Inf. Model. (2023). https://doi.org/10.1021/acs.jcim.2c01259
30. Milosevic, N., Gregson, C., Hernandez, R., Nenadic, G.: A framework for information extraction from tables in biomedical literature. Int. J. Doc. Anal. Recognit. (IJDAR) **22**, 55–78 (2019). https://doi.org/10.1007/s10032-019-00317-0
31. Mukaddem, K.T., Beard, E.J., Yildirim, B., Cole, J.M.: ImageDataExtractor: a tool to extract and quantify data from microscopy images. J. Chem. Inf. Model. **60**(5), 2492–2509 (2019). https://doi.org/10.1021/acs.jcim.9b00734
32. Kim, H., Han, J., Han, T.Y.J.: Machine vision-driven automatic recognition of particle size and morphology in SEM images. Nanoscale **12**(37), 19461–19469 (2020). https://doi.org/10.1039/D0NR04140H

Ensemble Learning Based Quantum Text Classifiers

Yousra Bouakba(✉) and Hacene Belhadef

LISIA Laboratory, University of Abdelhamid MEHRI, Constantine, Algeria
{yousra.bouakba,hacene.belhadef}@univ-constantine2.dz

Abstract. Quantum Natural Language Processing (QNLP) is a very young area of research, aimed at the design and implementation of NLP models that exploit certain quantum phenomena such as superposition, entanglement, and interference to perform language-related tasks on quantum hardware. To explore the structural relationships between quantum theory and natural languages Lambeq toolkit was created. This first high-level open-source Python toolkit for quantum natural language processing offers a fully automated quantum machine learning pipeline. Lambeq currently includes five compositional models that use varying degrees of syntactic information. While the previous studies focus on Discocat model, the aim of this study is to conducte an extensive evaluation of various classifiers based on different compositional models for text classification applications, focusing on the potential advantages of ensemble learning. The classifiers, including Spider, Cups, Stairs, Tree, and DisCoCat, were examined on two datasets: MC and RP. Performance evaluation was carried out using accuracy, precision, recall, and F1-score as the metrics. The results revealed notable variations in the performance of the classifiers across the datasets. Spider emerged as the top-performing classifier on the MC dataset, achieving remarkable scores of 100% in accuracy, precision, recall, and F1-score. However, on the RP dataset, Stairs outperformed the other classifiers with an accuracy of 68%, precision of 71%, recall of 68%, and F1-score of 69%. Furthermore, ensemble models using hard voting and soft voting techniques were constructed and evaluated. The ensemble models showcased improved performance compared to individual classifiers, with the soft voting ensemble achieving a score of 97% for all metrics on the MC dataset, and a precision of 72% and F1-score of 68% on the RP dataset. These findings highlight the potential benefits of ensemble learning in enhancing the overall performance of text classification tasks on both the MC and RP datasets using soft voting method, allowing for more accurate and reliable predictions.

Keywords: Ensemble Learning · Quantum machine learning · Quantum natural language processing · Quantum Text Classifier

A. Abelló et al. (Eds.): ADBIS 2023, CCIS 1850, pp. 407–414, 2023.
https://doi.org/10.1007/978-3-031-42941-5_35

1 Introduction

Machine translation, question answering, text classification, sentiment analysis [10] and many other NLP applications have been considerably enhanced by recent developments in neural language models based on Transformers architecture. As an example, the BERT Large uses 24 layers of transformers block with a hidden size of 1024 and number of self-attention heads as 16 and has around 340M trainable parameter. Because more parameters and data are needed to train this type of models efficiently, there is a cost in terms of time, resources, and computing power. This results an increase in model complexity. Therefore, it is necessary to search for new ways to process natural language with less complexity.

However, quantum computing has attracted a lot of attention recently. The fundamental idea is to use quantum mechanics to solve computational problems, in order to following different paths of computation at the same time, quantum computers can make use of a qubit's superposition of two quantum states $\mid 0\rangle and \mid 1\rangle$. A significant advance in quantum computing was made by quantum algorithms such as Short perform calculations with lower complexity than conventional methods. Naturally, the field of machine learning benefits from the ideas that quantum computing has to offer which led to the emergence of a new sub-field called Quantum Machine Learning (QML). The need to expand the NLP field in new directions coincided with the development of quantum machine learning, which resulted in the emergence of so-called Quantum Natural Language Processing (QNLP), which represents today one of the best applications of quantum computing.

The following is how this paper is structured: Sect. 2 presents the main related work for QNLP. In Sect. 3 describes the data used in this paper and a comparative evaluation of baseline classifiers. In Sect. 4 the proposed ensemble schema is presented and the results are discussed. Finally in Sect. 5, we draw the main conclusions and discuss possible future works.

2 Related Work

The concept behind the implementation of natural language on quantum hardware is to create a *link between language meaning and grammatical structure and quantum states.* This connection is made using the Categorical Distributional Compositional (DisCoCat) model [3], where the meanings of words are vectors in vector spaces and words grammatical roles are types in a Pregroups then a tensor product of vector spaces paired with the Pregroup composition is used for the composition of (meaning, type) pairs. Authors in [11] demonstrate that a quantum algorithm for calculating phrase similarity yields a quadratic speedup over conventional approaches under specific conditions. However, it requires quantum random access memory (QRAM), which is costly and has yet to be implemented. As a result, [2] and [6] articles presents quantum algorithms that can potentially be implemented in current NISQ devices (Noisy intermediate-scale quantum).

These last papers assume that NLP is quantum native, in the sense that the exponentially massive vector space necessary to represent sentences can only be realized naturally and practically in quantum computers. Following that, various investigations in the literature have shown that quantum speedup is more likely to impact NLP tasks like question answering [7], machine translation [8] and classification [5]. In order to close the gap between theory and practice, eventually contributing to actual real-world QNLP implementations, Cambridge Quantum Computing present Lambeq, 'an open-source, modular, extensible high-level Python library, which provides the necessary tools for implementing a pipeline for experimental QNLP. At a high level, the library allows the conversion of any sentence to a quantum circuit, based on a given compositional model and certain parameterisation and choices of ansatze' [4]. This first high-level open-source Python toolkit for quantum natural language processing is still in progress to offer a fully automated quantum machine learning pipeline.

3 Methodology

The aim of this study is to investigate the potential of employing ensemble learning to improve prediction of quantum text classification tasks. As a result, the study was divided into two distinct phases, the first in which a comparative evaluation of five classifiers on two well-Known QNLP datasets. Then, an ensemble learning was designed and implemented using two different voting techniques: Hard and Soft Voting.

3.1 Dataset

Table 1. Description of MC and RelPron datasets.

Dataset	Number of sentences	Vocabulary	Classes
Meaning classification (MC)	130 sentences	17 words	IT or Food
Relative Pronoun (RP)	105 sentences	115 word	Subject or Object

This research uses the MC and RP datasets (See Table 1), a specific datasets for QNLP tasks proposed by [5]. MC dataset is plain-syntax generated from a 17 words vocabulary using a simple CFG that can refer to one of two possible topics, food or IT. RP dataset extracted from the original RelPron dataset [9] for predicting whether a noun phrase contains a subject-based or an object-based relative clause, with 115 words vocabulary. The motivation for using RP dataset, is that it requires some syntactic understanding from the model, making it a suitable option for evaluating lambeq's models. Furthermore, the large amount of the vocabulary and, as therefore, the sparsity of words make this task a significantly more challenging benchmark than the MC dataset.

3.2 Baseline Classifiers

In order to establish a foundation for comparison, we employed several baseline classifiers based on Lambeq's compositional models, which generate semantic representations of sentences by combining the semantic representations of the words within them. Currently, the Lambeq toolkit provide five compositional models that use different levels of syntactic information. As described in the official documentation of Lambeq library there are three levels of syntactic information [1]:

1. No syntactic-information: include Spider reader, a bag-of-word model which renders simple monoidal structures without any form of tagging.
2. Word-sequence information: where words are formed sequentially from left right. Cups and Stairs are the two models available.
3. Full syntax-information: derived from grammatical derivations BobCatParser and TreeReader are two examples of syntax-based lambeq models. The Distributional Compositional Categorical model introduced by [3] is implemented by BobCatParser. Tree reader, on the other hand, provides a CCG derivation that follows a biclosed form, which may be directly understood as a series of compositions without any explicit translation into a pregroup form.

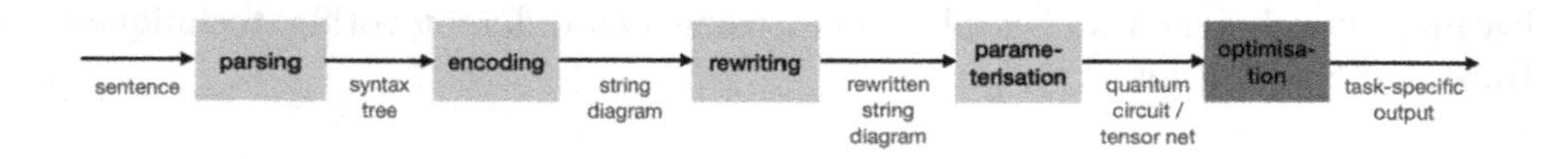

Fig. 1. Full Quantum Pipeline

Figure 1 describes a process involving the conversion of a sentence into a quantum circuit in lambeq. Initially, a syntax tree is obtained for a sentence using the parser, and then it is converted into a string diagram, representing the relationships between words based on a compositional model of choice. The string diagram can be simplified or transformed using rewriting rules, making it suitable for implementation on a quantum processing unit. The resulting diagram is converted into a quantum circuit or a tensor network, depending on the experiment type. The output of this pipeline, whether a quantum circuit or tensor network, can be used for training.

In our study, we investigate the usage of quantum classifiers with different semantic representations; as a consequence, we nominate our quantum classifiers with the same name as the compositional model used to identify them, such as the Stairs, Spiders, Cups, Tree, and DisCoCat classifiers. These classifiers served as reference models against which we evaluated the performance of our Ensemble model. The baseline classifiers were selected based on their varying degrees of syntactic information, simplicity in deployment, and the absence of literature on their effectiveness (reserve for the DisCoCat model).

3.3 Comparative Evaluation

To ensure fair and consistent comparisons, all baseline classifiers were trained and evaluated using the same experimental setup and evaluation metrics. The Removecups function was used to simplify string diagrams, and the IQP ansatz was utilized to generate the corresponding quantum circuit. The baseline classifiers hyperparameters were optimized using Simultaneous Perturbation Stochastic Approximation (SPSA) optimizer and ran on a Qiskit Aer similator, which is a high-performance simulator environment for exploring quantum computing methods and applications in the noisy intermediate scale quantum (NISQ) realm.

Table 2. Performance results of quantum text classifiers

Model		Accuracy	Precision	Recall	F1-score
Spider	MC data	100%	100%	100%	100%
	RP data	37%	43%	37%	36%
Stairs	MC data	60%	60%	60%	60%
	RP data	68%	71%	68%	69%
Cups	MC data	57%	59%	57%	54%
	RP data	32%	32%	32%	32%
Tree	MC data	87%	87%	87%	87%
	RP data	63%	68%	63%	64%
DisCoCat	MC data	70%	81%	70%	67%
	RP data	58%	70%	58%	57%

According to the Table 2 the classifiers revealed varying performance across the metrics. Spider achieved excellent performance with all metrics on the MC data. However, it performance significantly dropped on the RP data. Stairs and Tree showed consistent and reasonable performance on both datasets, indicating their ability to handle diverse data while maintaining a good balance between metrics. Cups exhibited moderate performance on the MC data but struggled on the RP data. DisCoCat showed relatively high accuracy and precision but lower recall and F1-score on both datasets, indicating a tendency to be more conservative in predictions. These findings highlight the strengths and weaknesses of each classifier. In summary, all classifiers perform worse on the RP dataset over on the MC dataset, as expected, because the RP dataset is more challenging. However, DisCoCat, Stairs and Tree models perform better which explains their robustness in comparison to other models.

4 Ensemble Learning

Ensemble learning is a powerful approach to machine learning that has been successfully applied in a wide range of domains. It involves combining several base learners (individual learning algorithms) to improve the generalization ability of the learning system. An ensemble is usually more accurate and robust than any individual base classifier because it can handle different types of data and errors, as well as capture multiple aspects of the data distribution. As it has been demonstrated that ensemble learning improves model performance, the model suggested in this work is a voting ensemble model formed by five baseline quantum text classifiers: Spider, Cups, Stairs, Tree and DisCoCat. At the top level, we have used a two different voting methods in order to combine the predictions produced by the bottom level. Figure 2 illustrates the applied scheme graphically.

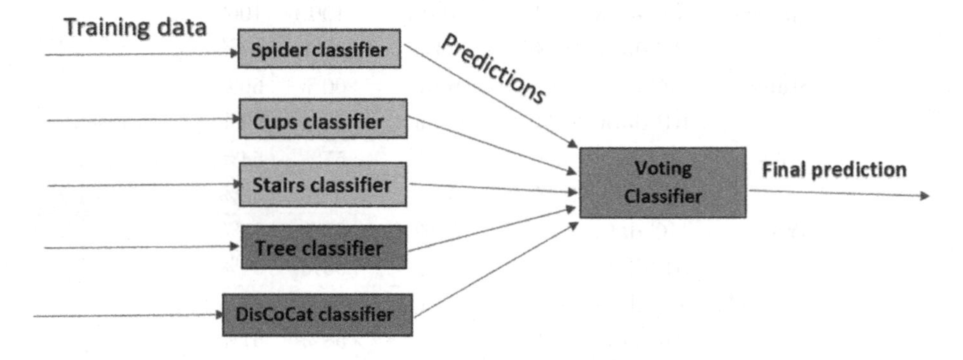

Fig. 2. A graphic illustration of the ensemble scheme utilized in the present work.

After splitting the dataset into training, validation and test data, each classifier was trained independently on the training and validation data. Then, the voting classifier feeds the test instance to each individual model and obtains its predicted class label as a distribution of probabilities. Finally, The voting classifier estimates the right label using either Hard or Soft voting methods.

In Hard Voting, the final prediction is the one with the highest frequency in the probability distributions predicted from the individual models. For example, if three models predict probability distributions of [0.1, 0.9], [0.8, 0.2] and [0.6, 0.4] then 0.9 is the highest frequency so model 1 is chosen and the result of the vote would be [0, 1] which means class two. In Soft Voting, for example, if three models predict probability distributions of [0.2, 0.8], [0.8, 0.2], and [0.6, 0.4], respectively, so the result of the soft vote would be $[(0.2+ 0.8 + 0.6)/3, (0.8 + 0.2 + 0.4)/3] = [0.53, 0.46]$, the result of the vote would be [1, 0] which means class one.

4.1 Results and Discussion

The primary findings are presented in this section based on the results obtained on the datasets given previously running on a Qiskit Aer simulator. Using the quantum pipeline (see Fig. 1) in all the experiments, the quantum text classifiers performance are presented in Table 3.

Table 3. Performance results of Ensemble quantum text classifiers

Dataset		Accuracy	Precision	Recall	F1-score
MC dataset	Hard Voting	93%	93%	93%	93%
	Soft Voting	97%	97%	97%	97%
RP dataset	Hard Voting	70%	65%	70%	66%
	Soft Voting	68%	72%	68%	68%

According to the Table 3, in the MC dataset, both hard voting and soft voting achieved high accuracy, precision, recall, and F1-scores. Soft voting performed slightly better, with higher values of 97% across all metrics. However, on the RP dataset, the performance of both voting methods was comparatively lower. Hard voting achieved higher accuracy and recall of 70% but had lower precision and F1-score. Soft voting, on the other hand, had higher precision of 72% but lower accuracy, recall, and F1-score. Overall, the ensemble model using soft voting consistently outperformed the hard voting approach on both datasets, showcasing its effectiveness in combining the predictions of the individual classifiers.

Comparing the results of the individual classifiers with the ensemble models on both the MC and RP datasets, we observe the following trends. On the MC dataset, the ensemble models using hard voting and soft voting consistently outperformed the individual classifiers (except Spider), achieving higher accuracy, precision, recall, and F1-scores of 97%. However, on the RP dataset, the ensemble model outperform the individual classifiers, achieving higher accuracy of 70% and precision of 72%. This suggests that the ensemble approach might face challenges in effectively combining predictions or addressing the complexities specific to the RP dataset.

5 Conclusion

Lambeq was established to meet the aims of QNLP by linking the theoretical and practical aspects. Consequently, the objective of this study was to conduct a comparative evaluation of different classifiers for text classification and explored the benefits of ensemble learning. The proposed schema involves training each individual classifier using different semantic representation of sentences, on two specific QNLP datasets, MC and RP. Afterwards, a Voting classifier was formed to make predictions on test data, using two different voting methods;

hard and soft. The key findings indicate that the ensemble models, utilizing both hard voting and soft voting, consistently outperformed the individual classifiers. Accordingly, Soft voting can yield a more accurate and dependable prediction by taking the confidence of each model's prediction into consideration with. Furthermore, The ensemble approach demonstrated improved accuracy, precision, recall, and F1-scores, highlighting the effectiveness of combining the predictions from multiple classifiers. Although this study presents beneficial insights, there is opportunity for further study and improvement. Further studies could look into the usage of advanced ensemble approaches to improve classification performance. Furthermore, examining the influence of various voting schemes might contribute to improve the proposed methodology.

References

1. Bouakba, Y., Belhadef, H.: Quantum natural language processing: a new and promising way to solve NLP problems. In: Salem, M., Merelo, J.J., Siarry, P., Bachir Bouiadjra, R., Debakla, M., Debbat, F. (eds.) ICAITA 2022. CCIS, vol. 1769, pp. 215–227. Springer, Cham (2023). https://doi.org/10.1007/978-3-031-28540-0_17
2. Coecke, B., de Felice, G., Meichanetzidis, K., Toumi, A.: Foundations for near-term quantum natural language processing. arXiv preprint arXiv:2012.03755 (2020)
3. Coecke, B., Sadrzadeh, M., Clark, S.: Mathematical foundations for a compositional distributional model of meaning. arXiv preprint arXiv:1003.4394 (2010)
4. Kartsaklis, D., et al.: lambeq: an efficient high-level python library for quantum NLP. arXiv preprint arXiv:2110.04236 (2021)
5. Lorenz, R., Pearson, A., Meichanetzidis, K., Kartsaklis, D., Coecke, B.: QNLP in practice: running compositional models of meaning on a quantum computer. J. Artif. Intell. Res. **76**, 1305–1342 (2023)
6. Meichanetzidis, K., Gogioso, S., De Felice, G., Chiappori, N., Toumi, A., Coecke, B.: Quantum natural language processing on near-term quantum computers. arXiv preprint arXiv:2005.04147 (2020)
7. Meichanetzidis, K., Toumi, A., de Felice, G., Coecke, B.: Grammar-aware question-answering on quantum computers. arXiv preprint arXiv:2012.03756 (2020)
8. Nieto, V.: I. Towards machine translation with quantum computers. Ph.D. thesis, Master's Thesis, University of Stockholm, Stockholm, Sweden (2021)
9. Rimell, L., Maillard, J., Polajnar, T., Clark, S.: RELPRON: a relative clause evaluation data set for compositional distributional semantics. Comput. Linguist. **42**(4), 661–701 (2016)
10. Sodré, A., Magalhães, D., Floriano, L., Pozo, A., Hara, C., Machado, S.: COVID-19 portal: machine learning techniques applied to the analysis of judicial processes related to the pandemic. In: Bellatreche, L., et al. (eds.) ADBIS 2021. CCIS, vol. 1450, pp. 109–120. Springer, Cham (2021). https://doi.org/10.1007/978-3-030-85082-1_10
11. Zeng, W., Coecke, B.: Quantum algorithms for compositional natural language processing. arXiv preprint arXiv:1608.01406 (2016)

Exploring the Capabilities and Limitations of VQC and QSVC for Sentiment Analysis on Real-World and Synthetic Datasets

Hacene Belhadef(✉), Hala Benchiheb, and Lynda Lebdjiri

Faculty of New Technologies of Information and Communication, University of Constantine 2, Constantine, Algeria
{hacene.belhadef,hala.benchiheb,linda.lebdjiri}@univ-constantine2.dz

Abstract. This study conducts a comparative analysis of two quantum classifiers, namely the Quantum Support Vector Classifier (QSVC) and the Variational Quantum Classifier (VQC), within the framework of sentiment analysis on both real-world and synthetic datasets. The primary aim is to assess their performance and delineate the current limitations and challenges associated with the application of these classifiers to complex tasks. The IMDB movie review dataset serves as a real-world example, while a generated dataset is employed for a simplified benchmark. The findings indicate that, although both classifiers exhibit potential in sentiment analysis, their performance on real-world datasets is impeded by factors such as limited qubit numbers, noise, and error rates in contemporary quantum hardware. This study underscores the necessity for advancements in quantum hardware and algorithms to enhance performance in sentiment analysis tasks and offers insights into potential avenues for future research.

Keywords: Quantum Computing · Quantum Algorithms · Quantum machine learning · Sentiment Analysis

1 Introduction

Sentiment analysis, an interdisciplinary field that combines linguistics and computer science, aims to identify and categorize sentiment in textual data. As a subfield of natural language processing, it has a wide range of applications, such as monitoring social media and evaluating customer satisfaction [1]. Machine learning has played a crucial role in sentiment analysis, enabling the discovery of valuable insights and patterns from vast amounts of unstructured text data. With the advent of quantum computing, new opportunities for machine learning have emerged, giving rise to the growing field of quantum machine learning (QML) [2]. QML combines the principles of quantum computing, quantum physics, and machine learning to harness the unique properties of quantum

A. Abelló et al. (Eds.): ADBIS 2023, CCIS 1850, pp. 415–424, 2023.
https://doi.org/10.1007/978-3-031-42941-5_36

systems and improve the performance, efficiency, and capabilities of traditional machine learning algorithms. This exciting development has led researchers to explore the potential benefits of using quantum computing techniques in Natural Language Processing (NLP) tasks such as sentiment analysis. However, QML is still a young field with many challenges and unanswered questions that must be addressed before its full potential can be realized.

2 Background on Quantum Machine Learning

Quantum machine learning is an emerging field that combines quantum physics, quantum computing, and machine learning to enhance or accelerate classical machine learning tasks like optimization, generative modeling, inference, classification, and regression. This field can be divided into two main types: quantum-enhanced machine learning and quantum machine learning. Quantum-enhanced machine learning utilizes quantum devices to perform classical machine learning tasks more quickly and effectively. On the other hand, quantum machine learning employs classical devices to learn from quantum data or simulate quantum systems [5]. Since traditional quantum computing algorithms rely on fault-tolerant quantum processors, which can handle a large number of qubits and quantum gates, this is not practical with the current quantum computers that can only implement a few tens of qubits with noisy and imperfect quantum gates. Therefore, an alternative design paradigm called quantum machine learning is emerging, specifically tailored to work with the current noisy intermediate-scale quantum (NISQ) computers. The approach involves a two-step methodology similar to classical machine learning, where a parametrized architecture for the quantum gates is first defined, and classical optimization is then used to adjust the gate parameters [3].

One of the potential applications of quantum machine learning is quantum classification, which is the task of assigning labels to data points based on their features. Quantum classification can be done using different quantum algorithms, such as Quantum Support Vector Classifier (QSVC) and Variational Quantum Classifier (VQC).

QSVC is an advanced version of the classical Support Vector Classifier (SVC) that incorporates quantum kernels in the classification process. By leveraging the unique properties of quantum computing, QSVC aims to potentially enhance the performance of classification tasks in machine learning. QSVC seamlessly integrates with traditional machine learning workflows while offering the benefits of quantum kernel functionality. Users can provide a custom quantum kernel or rely on a default option when working with QSVC for classification tasks. Quantum kernels play a crucial role in enabling QSVC to utilize the advantages of quantum computing. These kernels can be seen as a quantum-enhanced version of classical kernels, which are used to project data into higher-dimensional feature spaces, making it easier to separate and classify them [7]. In a pioneering study by Havlicek et al. (2019), the authors demonstrated the use of a quantum feature map and a quantum kernel for classification tasks in a quantum-enhanced feature space [7]. The quantum kernel leverages the power of quantum computing to efficiently compute similarity measures between quantum states, which represent the data in the quantum-enhanced feature space. This approach allows QSVC to potentially outperform classical methods in certain scenarios, as quantum kernels can capture complex patterns and correlations in the data that may be difficult to discern using classical techniques.

VQC are an important class of quantum algorithms that use parametrized quantum circuits to encode data and a classical optimizer to update the circuit parameters. They are similar to classical neural networks, acting as function approximators using a unitary evolution to map the quantum state containing problem features to the state containing the labels. VQCs are promising candidates for quantum neural networks in the NISQ era because of the low number of qubits needed to encode information and the small number of parameters required to adequately approximate the function of interest. It has been demonstrated that any unitary operator acting on multiple qubits can be expressed through ROTs and CNOTs, justifying the interest in using VQCs and exploring their capabilities to solve machine learning tasks. VQCs are currently being used in several machine learning tasks, where they are set in a classical neural network pipeline, and their weights are trained to minimize a selected loss function once a fixed architecture has been chosen [6].

Quantum machine learning is a promising field that could have significant impacts on various domains such as quantum materials, biochemistry, and high-energy physics. However, the field also faces many challenges and open questions, such as the scalability and reliability of quantum hardware, the design and analysis of quantum algorithms, the integration of quantum and classical software, and the validation and verification of quantum machine learning results [4].

3 Motivation and Scope of the Study

This study aims to evaluate the performance of two quantum classifiers, Quantum Support Vector Classifier (QSVC) and Variational Quantum Classifier (VQC), in the context of sentiment analysis. Sentiment analysis is a challenging problem that requires understanding the nuances and emotions of human language. Exploring the capabilities and limitations of VQC and QSVC for sentiment analysis on real-world and synthetic datasets is an important motivation for evaluating the potential of quantum machine learning for natural language processing tasks. To achieve this, we conduct a comparative analysis of QSVC and VQC using both real-world and synthetic datasets. For our real-world dataset, we use the IMDB movie review dataset, while a generated dataset serves as a simplified benchmark. Through this study, we aim to identify the current limitations and challenges when applying these classifiers to complex tasks. By comparing the performance of VQC and QSVC classifiers on different types of datasets, one can gain insights into the strengths and weaknesses of these quantum algorithms, as well as the effects of data quality and quantity on the results. Additionally, we investigate the impact of encoding methods such as angle and amplitude encoding on the accuracy and efficiency of the quantum models, which can provide new methods of data preprocessing and feature engineering for quantum machine learning applications. We hope that the results of this study will provide insights into potential future research directions and highlight the need for advancements in quantum hardware and algorithms to achieve better performance in sentiment analysis. This study aims to contribute to the development of quantum machine learning techniques for natural language processing tasks and shed light on the potential benefits and limitations of quantum computing in this field.

4 Datasets and Preprocessing

4.1 Real-World Dataset: IMDB Movie Reviews

For our real-world dataset, we chose the IMDB movie reviews dataset, which is widely used as a benchmark for sentiment analysis tasks. It consists of 50,000 movie reviews that are labeled as positive or negative based on the sentiment expressed by the reviewer. We randomly sampled 200 sentences from this dataset. We used this real-world dataset to evaluate our quantum classifiers and measure their performance in realistic sentiment analysis scenarios.

4.2 Synthetic Dataset: Generated Sentiment Analysis Data

In addition to the real-world IMDB dataset, we also use a synthetic dataset that we generate for sentiment analysis tasks. This dataset has 200 short sentences, each with up to 7 words, that are labeled as positive or negative. The sentences are constructed using a vocabulary of 60 unique words that are repeated in different combinations. The reason for using this synthetic dataset is to evaluate our quantum classifiers in a controlled setting and compare their performance with the real-world IMDB dataset.

4.3 Data Preprocessing and Encoding for Quantum Classifiers

For both datasets, we apply several preprocessing steps to ensure compatibility with the quantum computing framework. The preprocessing techniques include converting text to lower case, removing stop words, removing symbols and digits, and filtering out words shorter than three characters. For the real-world IMDB dataset, we further apply Part-of-Speech (POS) tagging to reduce sentence length and complexity. We retain specific POS tags that are more likely to carry sentiment information, such as nouns and adjectives. After preprocessing, the tokenized sequences are truncated or padded to guarantee a consistent length, and normalization techniques are applied to scale the data to a suitable range for quantum processing.

After preprocessing the data, we encode the classical information into quantum states using quantum feature maps. These feature maps transform classical data points into high-dimensional quantum states, which can then be processed by our quantum classifiers. By carefully selecting and designing quantum feature maps, we ensure that our variational quantum classifier and quantum support vector classifier can effectively capture the patterns and features present in the sentiment analysis datasets. The choice of encoding method can greatly impact the performance of the algorithm. In this study, we explored angle encoding and amplitude encoding methods.

Amplitude encoding is a technique that allows us to represent classical data as quantum states by using the amplitudes of the wavefunction. The basic idea of amplitude encoding is to map a vector x of length N to an n-qubit quantum state, where n is the smallest integer such that $N \leq 2^n$. The amplitudes of the quantum state are given by the components of x, normalized to have unit length.

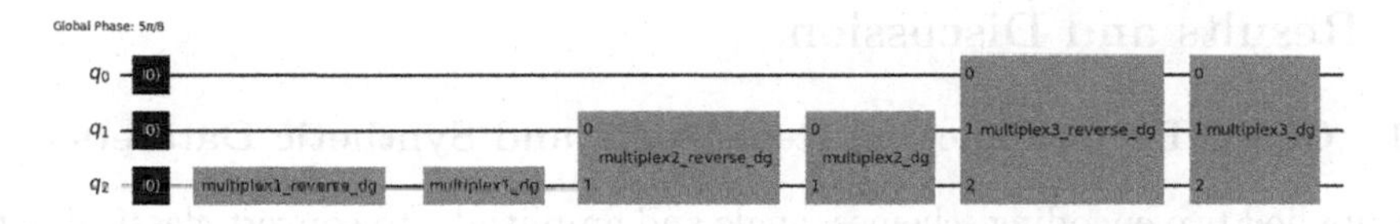

Fig. 1. Amplitude encoding circuit for a 3-qubit system.

Angle encoding is a method of transforming classical data into quantum states by applying rotation gates to qubits. The rotation angles are determined by the classical features, and the rotation gates can be chosen from Rx, Ry, or Rz. This way, each qubit encodes one feature of the classical data.

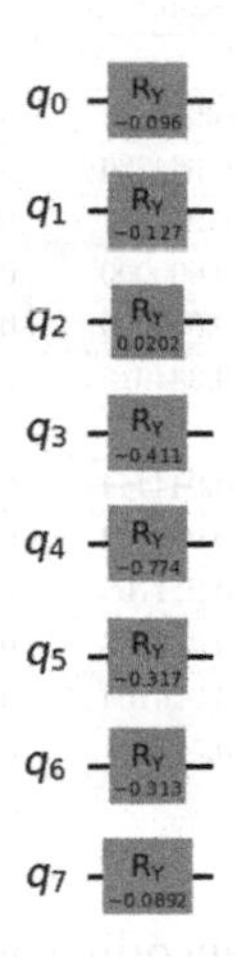

Fig. 2. Angle encoding circuit for an 8-qubit system.

5 Experimental Setup

In our experiment, we used the `statevector_simulator` from the Qiskit Aer Simulator, which gives the exact quantum state of the system without any noise or errors. We compared our quantum algorithms with different numbers of qubits, from 2 to 8. We expected that more qubits would improve the accuracy and F1 score of our quantum algorithms, but also increase the training time and the complexity of the quantum circuits. We aimed to find the optimal number of qubits that would balance the trade-off between performance and efficiency for our quantum algorithms.

6 Results and Discussion

6.1 QSVC Performance on Real-World and Synthetic Datasets

We applied two encoding schemes, angle and amplitude, to convert classical data into quantum states for the QSVC classifier. We experimented with two datasets: a synthetic dataset of simple sentences and a subset of 200 random sentences from the IMDB dataset (Table 1).

Table 1. QSVC Classifier Results With Amplitude encoding and Angle Encoding On Real and Artificial Dataset.

Metrics/Classifier	Number of Qubits	Accuracy Score	Precision Score	Recall Score	F1 Score	Training Time	Dataset Type
Amplitude Encoding							
QSVC	2	0.909091	0.967742	0.857143	0.909091	492.027178	
	4	0.863636	0.964286	0.771429	0.857143	510.556221	Artificial
	8	0.893939	0.966667	0.828571	0.892308	541.410265	
QSVC	2	0.60	0.000000	0.00	0.000000	207.233523	
	4	0.60	0.000000	0.00	0.000000	208.776821	Reel
	8	0.54	0.384615	0.25	0.303030	226.553183	
Angle Encoding							
QSVC	2	0.954545	0.944444	0.971429	0.957746	92.949024	
	4	0.909091	0.967742	0.857143	0.909091	105.963407	Artificial
	8	0.893939	0.911765	0.885714	0.898551	372.902979	
QSVC	2	0.469697	0.333333	0.296296	0.313725	131.386217	
	4	0.469697	0.346154	0.333333	0.339623	144.477326	Reel
	8	0.606061	0.517241	0.555556	0.535714	498.767965	

The results showed that angle encoding with 2 qubits achieved the highest accuracy (0.954545) and F1 score (0.957746) on the synthetic dataset, while angle encoding with 8 qubits achieved the highest accuracy (0.606061) and F1 score (0.535714) on the IMDB dataset. Amplitude encoding with 2 qubits also performed well on the synthetic dataset (accuracy: 0.909091, F1: 0.909091), but poorly on the IMDB dataset (accuracy: 0.60, F1: 0). Furthermore, angle encoding had shorter training times than amplitude encoding on both datasets. Angle encoding also outperformed amplitude encoding in terms of overall performance. The synthetic dataset of simple sentences was easier to classify than the IMDB dataset of random sentences for both encoding schemes.

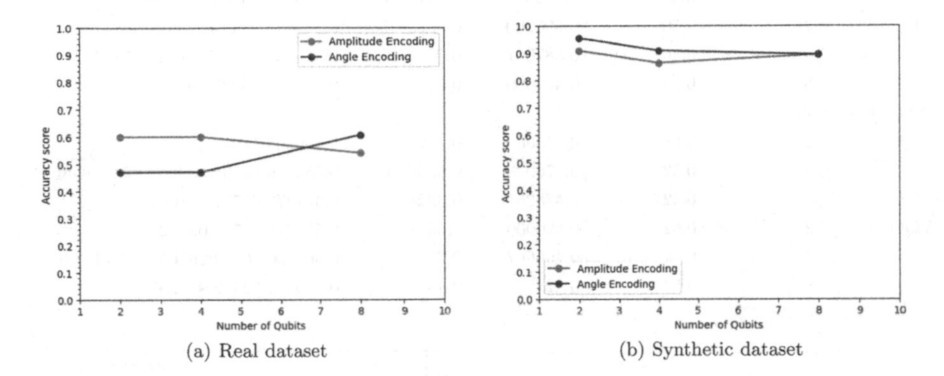

(a) Real dataset (b) Synthetic dataset

Fig. 3. Accuracy results of the QSVC classifier on real and synthetic datasets.

6.2 VQC Performance on Real-World and Synthetic Datasets

We also applied the same encoding schemes to the VQC classifier and experimented with the same datasets. The results showed that the best performance on the synthetic dataset was achieved by amplitude encoding with 2 qubits (accuracy: 0.7, F1: 0.882353), while the worst performance was achieved by angle encoding with 8 qubits (accuracy: 0.42, F1: 0.478261) (Table 2).

The results also showed that the best performance on the real dataset was achieved by angle encoding with 2 qubits (accuracy: 0.62, F1: 0.52), while the worst performance was achieved by amplitude encoding with 4 qubits (accuracy: 0.46, F1: 0.387097). Furthermore, the training time increased with the number of qubits for both encoding methods and datasets. In comparison, angle encoding performed better on the real dataset, while amplitude encoding performed better on the synthetic dataset.

Table 2. VQC Classifier Results With Amplitude encoding and Angle Encoding On Real and Artificial Dataset.

Metrics/ Classifier	Number of Qubits	Accuracy Score	Precision Score	Recall Score	F1 Score	Training Time	Dataset Type
Amplitude Encoding							
VQC	2	0.70	0.882353	0.535714	0.666667	400.945274	
	4	0.60	0.653846	0.607143	0.629630	898.741048	Artificial
	8	0.46	0.526316	0.357143	0.425532	3343.685296	
VQC	2	0.60	0.500000	0.6	0.545455	345.725651	
	4	0.46	0.387097	0.6	0.470588	851.156744	Reel
	8	0.58	0.480000	0.6	0.533333	3228.463683	
Angle Encoding							
VQC	2	0.68	0.875000	0.500000	0.636364	191.826026	
	4	0.52	0.576923	0.535714	0.555556	475.168384	Artificial
	8	0.42	0.478261	0.392857	0.431373	1772.641002	
VQC	2	0.62	0.520000	0.65	0.577778	273.808872	
	4	0.56	0.466667	0.70	0.560000	671.928565	Reel
	8	0.54	0.428571	0.45	0.439024	2442.915293	

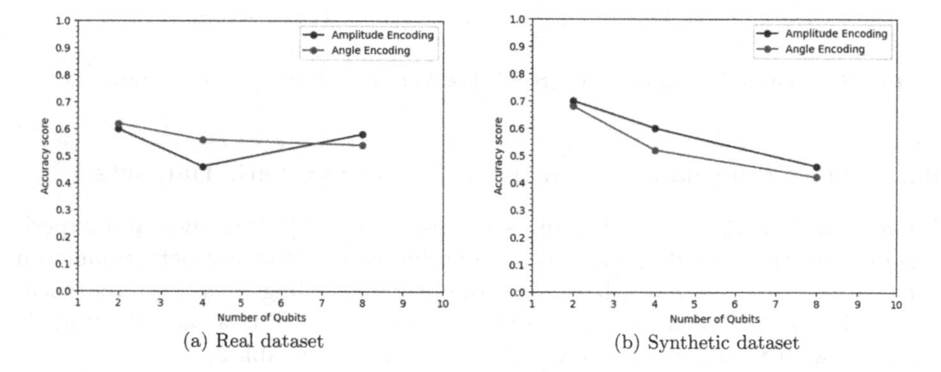

(a) Real dataset (b) Synthetic dataset

Fig. 4. Accuracy results of the VQC classifier on real and synthetic datasets.

7 Comparison Between QSVC's and VQC's Results

This study's findings indicate that the performance of QSVC and VQC models in text classification tasks is influenced by several factors, including the encoding method, the number of qubits, and the dataset used. Generally, QSVC outperformed VQC in terms of speed and accuracy in most cases, with the exception of the real dataset with amplitude encoding where VQC achieved a higher accuracy score when using 8 qubits. The efficiency of QSVC in binary classification can be attributed to its direct approach, as opposed to VQC which requires an optimization process to determine the optimal parameters for the variational circuit. With kernel-based training, the parameters are directly optimized and the best model is searched for within a subspace defined by the training data.

This method is guaranteed to identify the best measurement among all measurements. In contrast, variational training employs a general ansatz to parameterize the measurement and attempts to locate the optimal measurement within the subspace explored by the ansatz. However, this subspace may not contain the globally optimal measurement and optimization is typically non-convex [10]. Consequently, kernel-based training is guaranteed to identify minima that are either better or equivalent to those found by variational training [10]. Furthermore, QSVC can utilize the kernel trick to project input data into a higher-dimensional feature space where they are more readily separable [11], whereas VQC depends on the expressiveness and adaptability of the variational circuit to achieve a similar outcome.

8 Conclusion

This study explores the possibilities and obstacles of using quantum machine learning for text classification. Two quantum classifiers, QSVC and VQC, were evaluated using various encoding techniques, qubit numbers, and datasets to better comprehend the potential of quantum machine learning. Despite encouraging outcomes, several issues arose that require further investigation. For instance, processing only 200 rows of data took an excessively long time and increased with sentence complexity and vocabulary size. Additionally, the results for the real dataset were not as good as those for the synthetic dataset. Nonetheless, there are several potential developments in the field of quantum algorithms for sentiment analysis that could enhance their performance and address some of the current limitations of quantum hardware. One such development is the use of quantum cognition to develop new decision fusion strategies for predicting sentiment judgments. For example, a recent study suggested a fusion strategy based on quantum cognition that uses quantum superposition states and positive-operator valued measures to handle the incompatibility of sentiment judgments from different modalities [8]. Another potential advance is the creation of new quantum natural language processing (QNLP) techniques that can offer a quantum advantage for sentiment analysis tasks. For example, a recent study demonstrated the first application of QNLP for sentiment analysis using the lambeq QNLP toolkit and achieved perfect test set accuracy for three different simulations and decent accuracy for experiments conducted on a noisy quantum device [9]. These developments indicate that there is significant potential to enhance the efficacy of quantum algorithms for sentiment analysis by exploiting the distinctive characteristics of quantum mechanics and developing innovative techniques and approaches. As research in this field continues to progress, it is likely that we will see further advancements and breakthroughs that will enable even more powerful and accurate sentiment analysis using quantum computing.

References

1. Zhang, L., Liu, B.: Sentiment analysis and opinion mining. In: Encyclopedia of Machine Learning and Data Mining (2012)
2. Biamonte, J., Wittek, P., Pancotti, N., Rebentrost, P., Wiebe, N., Lloyd, S.: Quantum machine learning. Nature **549**(7671), 195–202 (2017). https://doi.org/10.1038/nature23474
3. Simeone, O.: An Introduction to Quantum Machine Learning for Engineers (2022). https://arxiv.org/abs/2205.09510
4. Cerezo, M., Verdon, G., Huang, H.Y., Cincio, L., Coles, P.J.: Challenges and opportunities in quantum machine learning. Nat. Comput. Sci. **2**(9), 567–576 (2022). https://doi.org/10.1038/s43588-022-00311-3
5. Dunjko, V., Taylor, J.M., Briegel, H.J.: Quantum-enhanced machine learning. Phys. Rev. Lett. **117**(13) (2016). https://doi.org/10.1103/physrevlett.117.130501
6. Giovagnoli, A., Ma, Y., Tresp, V.: QNEAT: natural evolution of variational quantum circuit architecture (2023). https://arxiv.org/abs/2304.06981
7. Havlíček, V., et al.: Supervised learning with quantum-enhanced feature spaces. Nature **567**(7747), 209–212 (2019). https://doi.org/10.1038/s41586-019-0980-2
8. Gkoumas, D., Li, Q., Dehdashti, S., Melucci, M., Yu, Y., Song, D.: Quantum cognitively motivated decision fusion for video sentiment analysis, arXiv preprint arXiv:2101.04406 (2021). [cs.CL]
9. Ganguly, S., Morapakula, S.N., Coronado, L.M.P.: Quantum natural language processing based sentiment analysis using lambeq toolkit, arXiv preprint arXiv:2305.19383 (2023). [quant-ph]
10. Schuld, M.: Supervised quantum machine learning models are kernel methods, arXiv preprint arXiv:2101.11020 (2021). [quant-ph]
11. QiskitMLKernal: Quantum Kernel Estimation. https://qiskit.org/ecosystem/machine-learning/tutorials/03_quantum_kernel.html. Accessed 04 June 2023

K-Gals: 2nd Workshop on Knowledge Graphs Analysis on a Large Scale

P2KG: Declarative Construction and Quality Evaluation of Knowledge Graph from Polystores

Xiuwen Zheng(✉), Subhasis Dasgupta, and Amarnath Gupta

University of California San Diego, La Jolla, USA
{xiz675,sudasgupta,a1gupta}@ucsd.edu

Abstract. Constructing knowledge graphs from heterogeneous data sources and evaluating their quality and consistency are important research questions in the field of knowledge graph. We propose mapping rules to guide users to translate data from relational and graph sources into a meaningful knowledge graph, and design a user-friendly language to specify the mapping rules. Given the mapping rules and constraints on source data, equivalent constraints on the target graph can be inferred, which is referred as data source constraints. Besides this type of constraints, we design other two types: user-specified constraints and general rules that a high-quality knowledge graph should adhere to. We translate the three types of constraints into uniform expressions in the form of graph functional dependencies and extended graph dependencies, which can be used for consistency checking. Our approach provides a systematic way to build and evaluate knowledge graphs from diverse data sources.

Keywords: Knowledge graph construction · Knowledge graph evaluation · Graph functional dependency

1 Introduction

Knowledge graphs (KGs) are increasingly used as complex data products derived by integrating information from multiple sources [3]. Informally, an entity-centric knowledge graph [5,10] is a graph whose nodes represent real-world entities together with their properties, and edges (predicates) represent relationships between pairs of these entities. The edges may also have their own properties.

The broad topic of this paper is the issue of *quality* in knowledge graphs that are constructed from more than one data source. We specifically focus on a situation where a KG is constructed from independent, heterogeneous sources like relational databases and ontological graphs. An important metric of knowledge graph quality is ***consistency***, the property that asserts that the KG does not have contradictions. In other words, if we assume that the data sources from which the KG is constructed are already consistent (and accurate), the construction process of the KG should not introduce any inconsistency in the KG. In addition to consistency, we would like to ensure that the construction algorithm maintains a set of structural properties of the target KG. For instance, the KG should have no isolated nodes.

A. Abelló et al. (Eds.): ADBIS 2023, CCIS 1850, pp. 427–439, 2023.
https://doi.org/10.1007/978-3-031-42941-5_37

To achieve this goal, we adapt schema mapping techniques [1,2,9] that have been extensively used in the information integration literature. However, we note that a fundamental difference between information integration and KG-construction, is that in the former, the source and the target schema both exist and the role of the schema mapping is to establish the correspondence across the source and target schemas, whereas in our case, an a priori target schema does not exist. We show that with a slight abuse of intent, graph functional dependencies (GFDs) can be effectively used to express KG construction constraints even when the data sources have heterogeneous data models.

This paper makes the following contributions toward knowledge graph construction and quality evaluation,

- It defines a new generic language to specify mapping rules between data sources and the target knowledge graph. It is easily adaptable to different data sources.
- We adapt the theory of graph functional dependencies (GFD) [7] to generate equivalent GFDs on target KG from original constraints on data sources given the mapping rules users applied to construct the KG.
- We extend GFDs to graph dependency (GD) expressions to specify user-defined constraints and general-rule constraints.
- We translate these different constraints to uniform Graph Functional Dependencies and our extended graph dependencies for ease of evaluation.

2 Preliminaries

Definition 1 (Knowledge Graph). *A knowledge graph G is a tuple $(V, E, L, R, A_V, A_E, U_V, U_E, \lambda, \mu)$ where*

- *V is a set of vertices such that designate entites;*
- *E is the set of relationships between entity pairs;*
- *L is the set of entity categories such that $\lambda : L \mapsto V$ is a labeling function that assigns a category to every entity $v \in V$;*
- *R is the set of relationship categories such that $\mu : R \mapsto E$ is a labeling function that assigns a category to every entity $e \in E$;*
- *A_V (resp. A_E) is the set of node attributes (resp. edge attributes) such $u_a^i \in U_A$ (resp. U_E) is the domain of attribute $a^i \in A_V$, U_E is similarly defined.*

Definition 2 (Ontological Knowledge Graph). *An ontological KG is a KG such that the entity categories L map to the concepts $\mathcal{C}$ and the relationship categories R belong to the relationships $\mathcal{R}$ of an existing ontology $\mathcal{O}$.*

Thus, if v is a vertex in the knowledge graph and l is its label, the l must correspond to some concept c in ontology $\mathcal{O}$. Similarly, if $r(v_1, v_2)$ is an edge in the knowledge graph then its predicate name r will correspond to the object properties defined in $\mathcal{O}$. In reality however, it is not always feasible that all KG concepts and relationships can be mapped to the ontology. Hence we consider that the mappings of $L \mapsto \mathcal{C}$ and $R \mapsto \mathcal{R}$ to the ontology are **partial**.

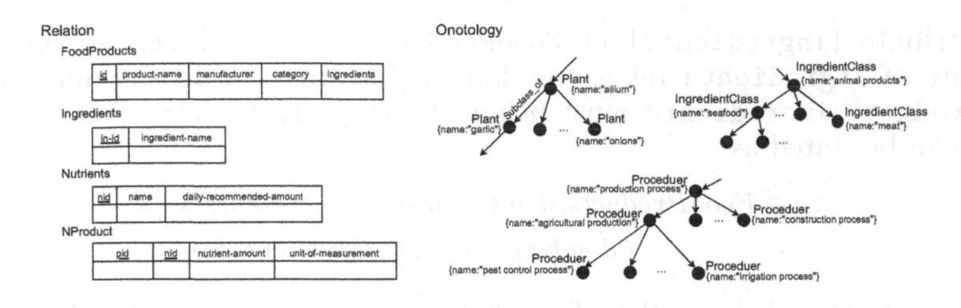

Fig. 1. Source data for running example.

Ontology. One of the data sources used in our knowledge graph is an ontology. For this paper, we assume that the ontology is expressed as a DL-Lite (specifically, $DL-Lite_{\mathcal{A},id}$) [4] that corresponds to OWL 2 QL, a tractable profile of the OWL ontology. DL-Lite is less expressive than other ontologies like OWL-Full (e.g., it cannot express subproperty relationships). However, as [4] elegantly elaborates, ontological expressions in DL-Lite cleanly translate to relational queries. In our setting, we translate a DL-Lite compatible OWL ontology into a faithfully encoded property graph. The translation process is out of scope for this paper, but see [2] for a comparable approach. However, we will present examples throughout this paper to illustrate the graph representation of the $DL-Lite$ ontology.

Graph Functional Dependency. Graph functional dependency (GFD) [7,8] is a concept in database theory that extends the idea of functional dependency from traditional relational databases to graph databases. A GFD is a constraint that describes a relationship between properties of nodes. GFDs are used to ensure data integrity in graph databases and to help ensure that queries can be executed efficiently.

A GFD is formally defined as a pair $(Q[\bar{x}], X \to Y)$ where Q defines a graph pattern; X and Y are two sets of literals of $\bar{x}$. The literal can be a constant literal, e.g., $x.A = c$, or variable literal, e.g., $x.A = y.A$ where $x, y \in \bar{x}$.

3 Running Example

We use a food knowledge graph as our running example (Fig. 1). We consider a relational data source having tables modeled after the 2022 USDA Database on Branded Food Products[1], and an ontology graph by extending the Food Ontology (FOODON) [6]. Our target KG incorporates information regarding the products, their ingredients, and their nutrient content from the relational source, as well as relevant ontology information concerning the product ingredients and procedures for producing the products.

The relational source (Fig. 1) has four tables. The `FoodProducts` table keeps the information for food products where `id` is the primary key and `[ingredients]` is a list of ingredient names. The `Ingredients` table has ingredients information with their ids and names. There is a domain constraint for

[1] https://fdc.nal.usda.gov/.

the attribute `[ingredients]` of `FoodProducts` table and `ingredient-name` attribute of `Ingredients` table: the domain for any item in the union list of `FoodProducts.ingredients` must belong to `Ingredients.ingredient-name`, which can be stated as:

$$\forall p \in FoodProducts, dom(unnest(p.ingredients)) \subset dom(ingredients.ingredient\text{-}name) \quad (1)$$

The `Nutrients` table keeps the information of nutrients where `nid` is the primary key. The `NProduct` table keeps the mapping information between products and nutrients which satisfies the PK-FK constraints.

```
NProduct.pid = FoodProducts.id
NProduct.nid = Nutrients.nid.
```

The ontology graph models food production procedures, ingredients class and plants. There are three node labels `Procedure`, `IngredientClass` and `Plant` and a node property `name` and transitive edge `Subclass_of`. It models information like "agricultural production" is a subclass of "production process". Since the ontology graph may contain information not relevant to food products(e.g., "pest control process"), only a subset of the ontology dataset should be preserved in the target knowledge graph.

4 The P2KG Mapping Language and Mapping Rules

The P2KG mapping language specifies how elements of the knowledge graph are defined as views over one or more data sources. A mapping statement has the following structure.

```
for row/node/edge/path <vars-1> in <source-query-1>
    for row/node/edge/path <vars-2> in <source-query-2> ...
      <knowledge-graph-construction-statement>
```

The mapping statement we employ uses multiple `for` constructs that may be nested, followed by a <knowledge-graph-construction-statement>. The syntax of the <knowledge-graph-construction-statement> is inspired by Cypher, which is a standard language used for querying graph databases. To construct nodes or edges, we use a subset of Cypher language that includes the Create, Merge, Where, and Set clauses, among others.

In the following section, we present a series of examples that demonstrate how P2KG mapping rules.

4.1 Mapping from Relations

There are several ways to map relations to a knowledge graph, and the choice of mapping rules depends on the intended purpose of the knowledge graph. In this section, we present different mapping rules, and then demonstrate how they can be used to create different knowledge graphs using the running dataset.

Rule 1: Mapping Table Columns to KG Nodes. One way to create nodes in the knowledge graph is by using columns from a table. The columns can be used to create nodes with a specific label in the KG, and the columns are mapped to the node's properties. For instance, if a user is interested in the manufacturer of the products, they can create nodes with the label Manufacturer" using the following statement:

```
for row r in (select manufacturer from FoodProducts)
    Merge (n:Manufacturer {name:r.manufacturer})
```

The Manufacturer nodes have a `name` property that is derived from the `manufacturer` column in the `FoodProducts` relation. The Merge clause is used to ensure that the nodes created are distinct, i.e., there are no two Manufacturer nodes with the same name. Nodes can also be created with multiple columns as properties. For example, ProductName" nodes can be created from both the `product-name` and `manufacturer` columns using the following statement:

```
for row r in (select product-name, manufacturer from FoodProducts)
    Merge (n:ProductName {name:r.product-name, manufacturer:r.manufacturer})
```

Rule 2: Mapping Table Rows to KG Edges. The rows in the table can be mapped to edges in the KG as both relational tuples and node edges depict relationships. Each row will be mapped to an edge in the KG. For instance, one can create Product nodes using certain columns, such as "id," as attributes and create Manufacturer nodes using rule 1. Then, the Product nodes can be linked to the Manufacturer nodes using the statement:

```
for row r in (select id, manufacturer from FoodProducts)
    MATCH (p:Product {id: r.id})
    MATCH (m:Manufacturer {name: r.manufacturer})
```

As the Product and Manufacturer nodes are pre-built, the Match clause is used to match the corresponding nodes with the given property values. If the edge does not exist before, it creates an edge between the two matched nodes.

Different mapping rules can be applied based on the user's requirements. Figure 2 shows various mappings from the same data. Different mapping rules can generate KGs with different space costs and different source constraints, which will be explained in detail in Sect. 5.

In the first example, Product, ProductName, Manufacturer and Category nodes are created using columns `id`, `productname`, `manufacturer` and `category` respectively using rule 1. Edges between nodes are created using rule 2. The total number of nodes will be $disc(id)+disc(product-name)+disc(manufacturer)+disc(category)$ where $disc(\cdot)$ is the distinct count of values in the column $\cdot$. It

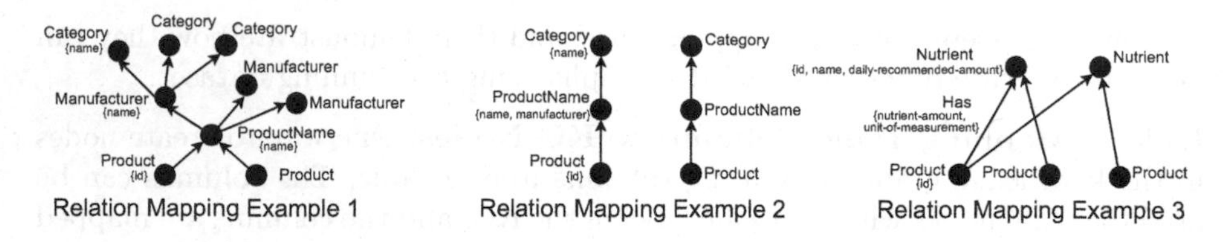

Fig. 2. Different mapping examples for the relational source in the running example.

saves space compared to other mapping rules, however, it loses some source constraints which will be introduced in Sect. 5. A Product node can only link to one node with a certain label, for example, it can only connect to one ProductName (or Manufacturer) node because `id`is the primary key of the relation. For the other relationships, they are m to n mappings. For example, different Product nodes can connect to the same ProductName node, and different ProductName nodes can connect to the same Manufacturer node, and the same ProductName may connect to different Manufacturer nodes. There are other ways to do the mapping. As example 2 shows, the columns `product-name` and `manufacturer` are used together to create the ProductName nodes. Since `product-name` and `manufacturer` columns can determine a product, ProductName nodes have a one-to-one mapping relationship with other types of nodes.

Rule 3: Mapping Multiple Table Columns to KG Nodes. There may be multiple columns, which could be from different tables, that refer to the same entities and can be mapped to the same nodes. Such columns will be used together to create nodes with the same label, while nodes from different columns can have their own associated labels. For instance, consider the `ingredients` column in the `FoodProdcuts` table and the `ingredient-name` column in the `Ingredients` table; these two columns can be mapped to the same nodes. The two columns are used together to create the `AllIngredient` nodes, and the ingredients from the `FoodProdcuts` table have another label `ProductIngredient`, and those from the `Ingredients` table have an `Ingredient` label.

```
for row r in (select unnest(ingredients) as in from FoodProdcuts)
    MERGE (i:AllIngredient{name:r.in}) SET i:ProductIngredient
for row r in (select ingredient-name as in from Ingredients)
    MERGE (i:AllIngredient{name:r.in}) SET i:Ingredient            (1)
```

In this statement, the two `for` loops are used to traverse both tables and create nodes for the ingredients. The first loop handles the `ingredients` column in the `FoodProdcuts` table and creates nodes with the label `ProductIngredient`. The second loop handles the `ingredient-name` column in the `Ingredients` table and creates nodes with the label `Ingredient`. Both loops create nodes with the label `AllIngredient` by using the `MERGE` clause to either match an existing node with the same name or create a new one with the given name. The `SET` clause sets the appropriate label for each node created by the two loops.

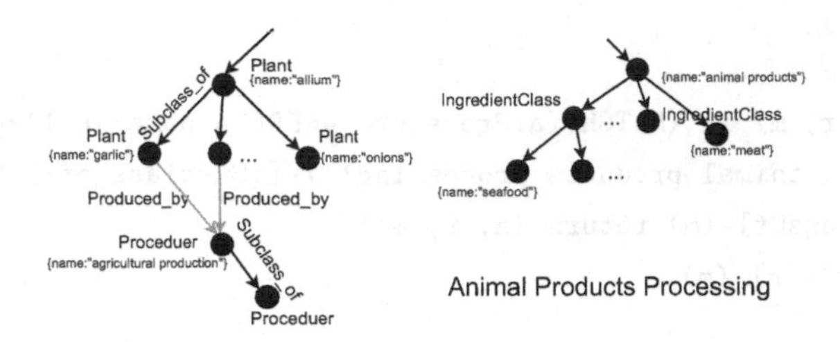

Fig. 3. Mapping example for the graph source in the running example.

Mapping From Query Results. More complex SQL queries including joins between tables can be applied before applying any mapping rules introduced above. The SQL queries is inside the <source-query> syntax, then any mapping rule introduced before can be applied on the query result. For example, in Fig. 2 Example 3, the `Nutrient` and `Product` nodes come from information in the `Nutrient` and `FoodProducts` tables respectively, and the edges between them are mapped from the `NProduct` table. To create the KG, one can write a query to join the three tables and then apply the mapping rules:

```
for row r in (select p.id as pid, n.nid as nid, name as name,
daily-recommended-amount, nutrient-amount, unit-of-measurement
from Nutrient n, FoodProdcuts p, NProduct np
where n.nid = np.nid, p.id = np.pid)
    MERGE (p:Product{id:r.pid})
    MERGE (n:Nutrient{id:r.nid, name:r.name, daily-recommended-amount:
   r.daily-recommended-amount})
    MERGE (p) - [:Has{nutrient-amount:r.nutrient-amount,
    unit-of-measurement:r.unit-of-measurement}] -> (n)
```

4.2 Mapping from Graph Source

When mapping from the graph source to the KG, the <source-query>, which can also be written in Cypher, is used to match the part of the graph data that the user wants to keep in the target KG. Users can also specify some customized rules using Cypher SET clauses to create new edges. Figure 3 shows an example of mapping the ontology graph to the target KG for the running dataset. As stated in the dataset section, the production procedures ontology graph is large and contains unrelated information about food production, and users are only interested in the relevant information about "agricultural production" or "animal products processing." This can be achieved by the following statement:

```
for edge(n, r, m) in (MATCH (a:Procedure WHERE a.name in ['agricultural
production', 'animal products processing'])-[:Subclass_of*]->(n),
(n)-[r:subClassOf]-(m) return (n, r, m))
    MERGE (n)-[r]-(m)
```

Besides, each Plant node should be connected to the agricultural production Procedure node, which can be achieved by the following statement:

```
for node n in (MATCH (n:Plant) return n)
    MATCH (m:Procedure{name:'agricultural production'})
    MERGE (n)-[:Produced_by]->m
```

4.3 Mapping from Multiple Sources

Data from different sources can be linked based on specific rules. For example, in the running example (as shown in Fig. 4), a product from the FoodProducts table can be linked to a Plant node created from the ontology graph if the product contains an ingredient from the plant. Similarly, an Ingredient node from the relations can be linked to an IngredientClass node from the ingredient class ontology using the `subclass_of` edge, if that ingredient is a sub-class of the ingredient class. To create the KG, mapping rules are applied to each single source separately. Then, the cross-source relationships are established by executing a Cypher query in the target KG as the <source-query>,

```
for nodes (n, m) in KG:(MATCH (n:Product)->(x:Ingredient), (m:Plant)
where x.name=m.name return n, m)
    MERGE n-[:Derived_from]->m
for nodes (n, m) in KG:(MATCH (n:Ingredient), (x:IngredientClass)
-[:Subclass_of]->(m:IngredientClass) return n, m)
    MERGE n-[:Subclass_of]->m
```

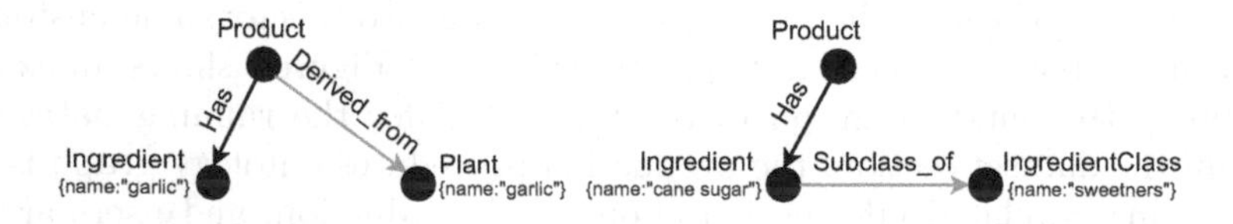

Fig. 4. Mapping examples for multiple source in the running example.

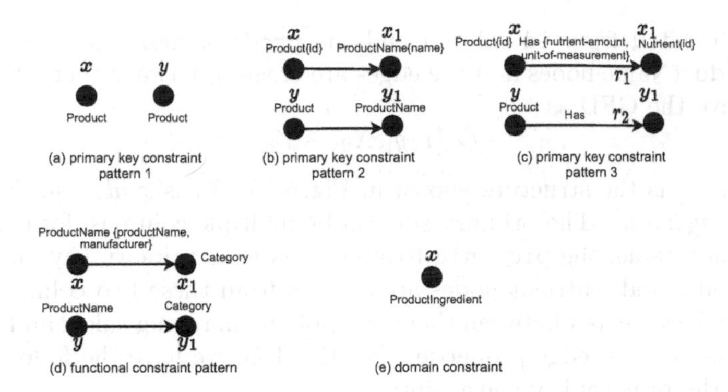

Fig. 5. Source constraints example.

5 Constraints

Consistency checking is important to ensure that a knowledge graph is reliable and accurate. To achieve this, we define three types of constraints. The first type is **source constraints**, which refer to the constraints specified by the data sources used to build the knowledge graph. The target KG must comply with these constraints to ensure consistency. The second type is **general rules**, rules that a good knowledge graph should follow. These rules ensure that the knowledge graph is structured in a coherent and meaningful way. The third type is **user-specified constraints**, which are constraints that are specified by the user to ensure that the knowledge graph meets their specific requirements. To check if a knowledge graph complies with these constraints, we translate these different constraints to our extended graph dependencies which will be used for consistency checking. By checking if the knowledge graph meets these constraints, we can evaluate if it is consistent and of good quality.

5.1 Source Constraints

Constraints from the data sources can be kept in the target KG based on what mapping rules are applied. We mainly consider relational constraints and show how to translate them on the source relation to the graph functional dependency on the KG given the mapping rules applied.

Primary Key Constraint. In relational databases, a primary key constraint is a constraint that ensures that each row in a table has a unique identifier or key value. Figure 5(a)-(c) illustrates the primary key constraint on KG.

For the `FoodProducts` table, the column `id` is a primary key. In Fig. 5(a), the Product nodes are created from the columns from this table including `id` column, thus we have the following graph functional dependency

$$Q_a[x, y], X_1 \rightarrow Y_1$$

where $Q_a[x, y]$ is the structure shown in Fig. 5(a), X_1 is $x.id = y.id$ and Y_1 is $x.A = y.A$ where A is any node property of Product. In Fig. 5(b), Product nodes are created from `id` column and the `product-name` column is used to create ProductName nodes and the edges are created between them, then in the KG, we have the GFD as

$$Q_b[x, y], X_2 \rightarrow Y_2$$

where $Q_b[x, y]$ is the structure shown in Fig. 5(b), X_2 is $x.id = y.id$ and Y_2 is $x_1.name = y_1.name$. The primary key can be multiple columns, for example, in the `NProduct` table, the `pid, nid` together serves as primary key, in Fig. 5(c), Product nodes and Nutrient nodes are created from these two columns respectively and edges created between them by applying mapping rule 2 and the other two columns serve as edge properties. For this KG, we have the following GFD to express the primary key constraint:

$$Q_c[x, x_1, y, y_1, r, r_1], X_3 \rightarrow Y_3$$

where $Q_c[x, x_1, y, y_1, r, r_1]$ is the structure in (c), X_3 is $x.id = y.id, x_1.id = y_1.id$ and Y_3 is $r.nutrient\text{-}amount = r_1.nutrient\text{-}amount, r.unit\text{-}of\text{-}measurement = r_1.unit\text{-}of\text{-}measurement$.

Functional Dependencies. A functional dependency (FD) on relations states that the value of a set of attributes determines the value of another set of attributes. A FD of relation with schema $R(U)$ is an expression of the form $R : X \rightarrow Y$ where $X \subseteq U$ and $Y \subseteq U$.

For example, in the `FoodProducts` table, the `productname` together with `manufacturer` columns determine the category of the product. In Fig. 5 (d), the Product nodes are created from `productname` and `manufacturer`, and they are connected to the Category nodes. We have the following GFD equivalent to the relational FD:

$$Q_d[x, x_1, y, y_1], X_4 \rightarrow Y_4$$

where Q_d is the topological structure in (d), X_4 is $x.productName = y.productName$, $x.manufaturer = y.manufacturer$ and Y_4 is $x_1.category = y1.category$.

Inclusion Dependencies. An inclusion dependency (IND) on pairs of relations of schemas $R(U)$ and $S(V)$ (with R and S not necessarily distinct) is an expression of the form $R[X] \subseteq S[Y]$ where $X \subseteq U$ and $Y \subseteq V$. There is an inclusion dependency between `FoodProducts` and `Ingredients` table as stated in 1. Suppose that these two columns are mapped to the target KG using mapping statement (1), then we have the following GFD: $Q_e[x], X_5 \rightarrow Y_5$ where $Q_e[x]$ is shown in Figure (e) which matches any node with *ProductIngredient* label, X_5 is $\emptyset$ and Y_5 is $x.label = Ingredient$.

5.2 General Rule Constrains

To ensure a high-quality knowledge graph, there are several general constraints that should be satisfied. These constraints are illustrated in Fig. 6.

- The graph should not contain any isolated nodes, as shown in Fig. 6 (a).

- All nodes created from the ontology data source, except for the root nodes, should have a subClassOf parent, as shown in Fig. 6 (b).
- The graph should display edge-label acyclicity i.e., if you just consider a single edge label, the graph will be acyclic (symmetric edge labels are implicit), as shown in Fig. 6 (c).

To provide a uniform way to express different types of constraints, we propose an extension to the graph functional dependency called graph dependency (GD). In this extension, we support node/edge existence statements, allowing us to express constraints on the existence of nodes and edges in the knowledge graph. As stated in Sect. 2, in GFD, X and Y are two sets of literals of $\bar{x}$, and a literal of $\bar{x}$ can either be a constant literal or variable literal. We extended the form of X and Y. For X, it is extended to support the existence statement which says that there exists a node/edge in the knowledge graph with certain properties value or labels which can be constant or related to $\bar{x}$. For example, X can be $\exists node\ n \in KG, n.label = L, n.A = c, n.A' = x.A'$ where $x \in \bar{x}$. For Y, it is extended to support connection between nodes defined in X and nodes in $\bar{x}$, for example, y can be $n \longrightarrow x$ which states that there is an edge from n (node defined in X) to x. With extended GDs, the general constraints can be expressed as follows.

- $Q_a[x], X_1 \rightarrow Y_1$, X_1 is $\exists node\ y \in KG$, Y_1 is `x -> y`.
- $Q_b[x], X_2 \rightarrow Y_2$, X_2 is $\exists node\ y \in KG, y.label = x.label$, Y_2 is `x-[:Subclass_of]->y`.
- $Q_c[x], X_3 \rightarrow Y_3\ X_3 = \emptyset, Y_3 = false$.

where Q_a, Q_b, Q_c match the topological structure depicted in black color in Fig. 6 (a) - (c) respectively.

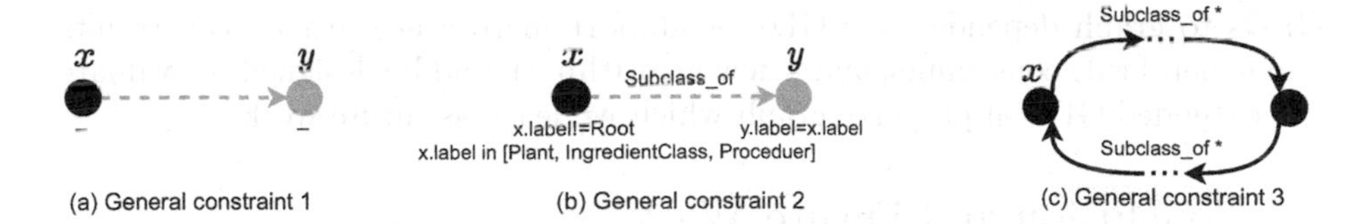

Fig. 6. General constraints example.

5.3 Constraints from Users

Users can specify constraints on the knowledge graph directly using graph dependency expressions to check if the created graph satisfy their specific requirements. We show some examples in Fig. 7, and explain them as follows:

- Any Plant node should be connected to the Procedure node with name "Agricultural Process".

- A Product node which has "garlic" as ingredient should be connected with the Plant node whose name is garlic by the `Derived_from` edge.
- If a product has ingredient which is a subclass of meat or seafood, then the product should be non-vegetarian.

Users can specify these constraints in the extended GD expressions as follows:

- $Q_a[x, y], X_1 \rightarrow Y_1$, X_1 is $\exists edge\ r \in KG, r.label = Produced\_by$, Y_1 is `(x)-[r]->(y)`;
- $Q_b[x, y], X_2 \rightarrow Y_2$, X_2 is $\exists edge\ r \in KG, r.label = Derived\_from$, Y_2 is `(x)-[r]->(y)`;
- $Q_c[x], X_3 \rightarrow Y_3$, $X_3 = \emptyset$, Y_3 is `x.type='non-vegetarian'`.

where Q_a, Q_b, Q_c match the topological structure depicted in black color in Fig. 7 (a) - (c) respectively.

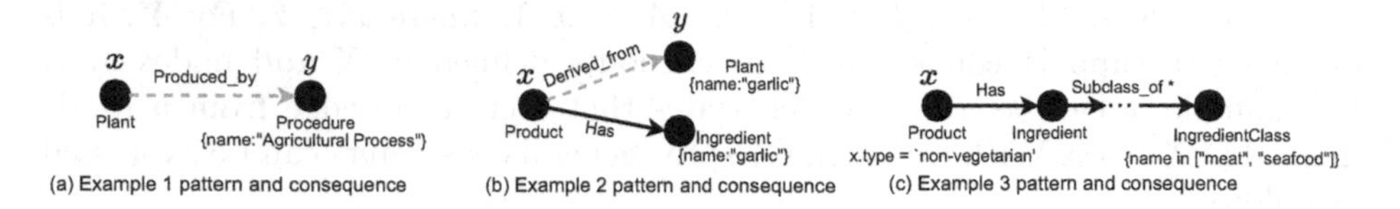

Fig. 7. User-specified constraints example.

5.4 Evaluation of Constraints

There is prior work such as [8] which proposes algorithm to evaluate GFDs on property graph, and the prior algorithm can be directly introduced to evaluate the GFDs derived from the source constraints. However, we extended the original GFDs to graph dependencies (GD) to support more user-specified constraints and general rule constraints, and a new algorithm should be designed to evaluate the extended GDs on property graph which we leave as future work.

6 Conclusion and Future Work

In this paper, we have presented mapping rules for efficiently mapping data from different sources, namely relational and graph data, to a target knowledge graph. We have also defined three types of constraints, derived from source data, user specifications, and common rules of a good knowledge graph, and translated them into unified expressions in the form of GFDs and extended GDs to evaluate the quality, correctness, and consistency of an existing knowledge graph.

Future work includes developing an algorithm to efficiently evaluate the GDs in large-scale knowledge graphs and automating the process of generating equivalent GFDs based on the constraints on the data sources and mapping rules used to create the KG. These contributions will facilitate the development and maintenance of high-quality knowledge graphs.

References

1. Alexe, B., Hernández, M., Popa, L., Tan, W.C.: Mapmerge: correlating independent schema mappings. VLDB J. **21**, 191–211 (2012)
2. Angles, R., Thakkar, H., Tomaszuk, D.: Mapping RDF databases to property graph databases. IEEE Access **8**, 86091–86110 (2020)
3. Asprino, L., Daga, E., Gangemi, A., Mulholland, P.: Knowledge graph construction with a façade: a unified method to access heterogeneous data sources on the web. ACM Trans. Internet Technol. **23**(1), 1–31 (2023)
4. Calvanese, D., et al.: Ontologies and databases: the *DL-Lite* approach. In: Tessaris, S., et al. (eds.) Reasoning Web 2009. LNCS, vol. 5689, pp. 255–356. Springer, Heidelberg (2009). https://doi.org/10.1007/978-3-642-03754-2_7
5. Dong, X., et al.: Knowledge vault: a web-scale approach to probabilistic knowledge fusion. In: Proceedings of the 20th ACM SIGKDD International Conference on Knowledge Discovery and Data Mining, pp. 601–610 (2014)
6. Dooley, D.M., et al.: Foodon: a harmonized food ontology to increase global food traceability, quality control and data integration. NPJ Sci. Food **2**(1), 23 (2018)
7. Fan, W., Hu, C., Liu, X., Lu, P.: Discovering graph functional dependencies. ACM Trans. Database Syst. (TODS) **45**(3), 1–42 (2020)
8. Fan, W., Wu, Y., Xu, J.: Functional dependencies for graphs. In: Proceedings of the 2016 International Conference on Management of Data, pp. 1843–1857 (2016)
9. Mazilu, L., Paton, N.W., Fernandes, A.A., Koehler, M.: Dynamap: schema mapping generation in the wild. In: Proceedings of the 31st International Conference on Scientific and Statistical Database Management, pp. 37–48 (2019)
10. Vrandečić, D., Krötzsch, M.: Wikidata: a free collaborative knowledgebase. Commun. ACM **57**(10), 78–85 (2014)

Using Knowledge Graphs to Model Green Investment Opportunities

Giorgio Grani(✉), Lorenzo Di Rocco, and Umberto Ferraro Petrillo

Sapienza University of Rome, Piazzale Aldo Moro 5, 00185 Rome, Italy
{g.grani,lorenzo.dirocco,umberto.ferraro}@uniroma1.it
https://www.dss.uniroma1.it

Abstract. International institutions are funding renewable energy initiatives to transition to a carbon-neutral economy and combat the effects of climate change. However, the involvement of multiple institutions and the absence of a standardized database of investment opportunities make it difficult for interested parties to discover and consider these opportunities.

In this paper, we propose a standardized modeling approach based on knowledge graphs to represent green change investment opportunities funded by international organizations. Such an approach offers many advantages, including the ability to obtain a comprehensive overview of all available investment opportunities and uncover hidden patterns in the underlying data.

We also report the results of an exploratory analysis of green investments in African countries using traditional graph metrics and community detection algorithms on a knowledge graph created with our model. The results are consistent with the existing literature on this topic.

Keywords: green investment opportunities · knowledge graph · natural language processing

1 Introduction

The dramatic consequences of climate change compel an orderly and steady transition to a carbon-neutral economy based on the adoption of green technologies. To achieve this ambitious goal, many international institutions around the world are funding renewable energy generation, storage, and distribution initiatives. Aside from these noble reasons, interest in these initiatives is moved by geopolitics or by business strategies, like the need to enhance one's reputation or brand image. However, it is difficult to get a comprehensive and easy-to-navigate view of all available investment opportunities in this area, due to the lack of a standardized repository taking into account the plethora of promoting subjects and the large variety of document formats being used.

In this paper, we explore the idea of using a knowledge graph (KG) to represent investment opportunities for the green transition, as implemented by international funding organizations. As a case study, we introduce a KG representing

A. Abelló et al. (Eds.): ADBIS 2023, CCIS 1850, pp. 440–451, 2023.
https://doi.org/10.1007/978-3-031-42941-5_38

the stock of green investments targeting African countries over the past decade, as funded by international organizations such as the African Development Bank and described in a collection of over 2,000 publicly available digital documents. The dataset we used for this purpose comes from the project "using Artificial Intelligence to support GREEn investments for promoting the Ecological Transition" (AIGreet for short) [1]. The project aims to develop a system that automatically detects and analyzes green investment opportunities. This is done using a distributed pipeline that automatically captures and processes all digital documents appearing on a large collection of public agency websites and extracts structured relevant information about investment opportunities.

In the final KG, each funding initiative has been described as a subgraph itself, with nodes and relationships used to describe relevant facts like the type of investment or the provided budget. This led to the creation of a KG, then implemented using the Neo4J technology, consisting of 1,600 nodes and 3,601 edges.

Finally, we conducted an exploratory analysis of green investments in African countries by evaluating our KG using traditional graph metrics, such as node degree distribution, as well as more sophisticated analyses, such as community detection algorithms. The results are consistent with the literature on these topics.

2 Background

In order to mitigate the negative effects of climate change and reduce carbon emissions, the world has become increasingly aware of the need to transition to a greener economy. To reach this goal, there has been in recent years an increasing need for funding to support green projects and initiatives. Development banks and other financial institutions play an important role in this context, as they can leverage their financial resources and their expertise to support projects for the implementation of the green transition. These institutions include multilateral and international development banks, national development banks, commercial banks, private equity firms, as well as other subjects.

Actually, the number of subjects actively engaged in this type of operation is significantly high. According to a study presented in [14], by only considering the total number of public development banks (PDBs) and development finance institutions (DFIs) in the 2021, their number is well over 500 units, with their total assets exceeding $18 trillion. However, it is not easy to give an accurate and comprehensive account of all the subjects involved in this type of operation, both because the definition of development institutions itself is unclear and because the reasons that drive institutions to lend funds for development projects vary widely and are rarely merely philanthropic.

On a side, this translates into the availability of a significant number of financing opportunities, targeting countries from all over the world. On the other side, it becomes problematic for companies, especially those of small and medium size, to keep track of all these opportunities and take advantage of the most interesting ones. It is possible to overcome these problems using web scraping, machine learning, and natural language processing (NLP) techniques. This is the case, for example, of the AIGreet research project [1].

3 Describing Investment Opportunities Using Knowledge Graphs

Knowledge graphs (KG for short) are graph-based data models that capture and represent the semantics of real-world entities, events, concepts, and their corresponding relationships. By doing so, they allow integration in a single framework both structured and unstructured data.

They have become increasingly popular in the last years thanks to their ability to support a wide range of applications (see, e.g. [6]). One of the key advantages of KGs is their ability to simplify the discovery of new insights and knowledge from existing data. By properly choosing a correct representation for the data of interest, it is then possible to analyze the structure of a KG to reveal hidden patterns in the underlying data, identify anomalies and support the data-driven decision-making process.

With respect to the topic addressed in this paper, we note that no approach to modeling green investment opportunities using a KG has yet been proposed. However, there are other contributions of this type on similar topics. For example, the paper in [12] discusses the automatic construction of query-specific knowledge graphs (KGs) for complex research topics in finance. The authors focus on the dataset CODEC, where domain experts create challenging questions and construct long natural language narratives. In [4], the authors propose a knowledge graph-based recommender system that helps identify potential target companies for investment promotion. This is done using a two-tier model based on both local and global knowledge graph reasoning.

In [13], the authors propose a knowledge graph-building approach using semantic technologies to gradually match data from different sources about nationwide enterprises in China. Finally, in [7], the authors present a knowledge graph-based framework for embedding events for quantitative investment. The framework extracts structured events from raw text, constructs a knowledge graph with entities and relationships, and inserts the knowledge graph information into the objective function of a learning model for event embedding.

4 Objective

The main objective of this work is to define a standard representation of the universe of green financing options, as well as the institutions that promote them and other relevant information, based on the use of a knowledge graph (KG).

Given a subject interested in making investment decisions in this area using a data-driven approach, the availability of a KG representation offers several advantages:

- makes it possible to better explain the relationships between different entities thanks to the use of visualization techniques based on the structured nature of a graph.
- simplifies the integration of data from various sources, such as information from international institutions, government agencies, private companies, news agencies, and other open sources.
- allows the discovery of new insights and knowledge from existing data. A proper analysis of the KG is useful to reveal hidden patterns in the underlying data, and identify anomalies.

4.1 A Knowledge Graph Representation for Green Investment Opportunities

International institutions often introduce green investment opportunities to the public through different types of documents such as Requests for Expression of Interest (REOI), General Procurement Notices (GPN), and Specific Procurement Notices (SPN). These institutions then publish further documents such as Invitation for Bids (IFB) or Official Award (Aw) documents to confirm the progress of the procurement process.

In our representation, the publication of each of these documents is modeled as a new *Investment Initiative* carried out by one or more *Funding Organizations*. The initiative itself is directed at one or more *Countries* and may also include an indication of the size of the financial investment. In our case, we only consider investments related to the generation, storage, and distribution of energy from renewable sources. Therefore, each initiative is also associated with one or more *Green Technologies*. Finally, each initiative is also linked to various other relevant information available in the underlying documents, such as the email contacts of those responsible for the process or the URL addresses where new proposals can be submitted.

All these entities are introduced in our KG as labeled nodes linked by edges (see Table 1). As shown in Fig. 1, the labeled nodes in the KG relate to each other by means of edges representing a specific relationship.

5 Case Study: Analyzing Green Investments in African Countries Through the AIGreet Knowledge Graph

In this section, we present the results of an exploratory analysis of green investments in African countries implemented in the last 10 years. This analysis was conducted by examining the knowledge graph resulting from the representation of the AIGreet dataset for green investments using traditional graph metrics, such as the node degree distribution, as well as more sophisticated analyses, such as community detection algorithms.

Table 1. Nodes labels and the corresponding entity types used in our Knowledge Graph to describe green investment initiatives.

Node Label	Entity Description
`Initiative`	Investment initiative aimed at supporting the development of green technologies
`Funding_Org`	Entity that provides financial support to an investment initiative
`Project`	The project to which the investment initiative belongs
`Country`	Entity representing a country targeted by an investment initiative
`Technology`	Green technology being supported by an investment initiative
`Money_Class`	Funding level of an investment initiative
`Phone`	Phone number of a person/organization involved in an investment initiative
`Email`	Email of a person/organization involved in an investment initiative
`Contract`	Type of contract used procuring the investment initiative
`URL`	A URL where information about the investment initiative can be found

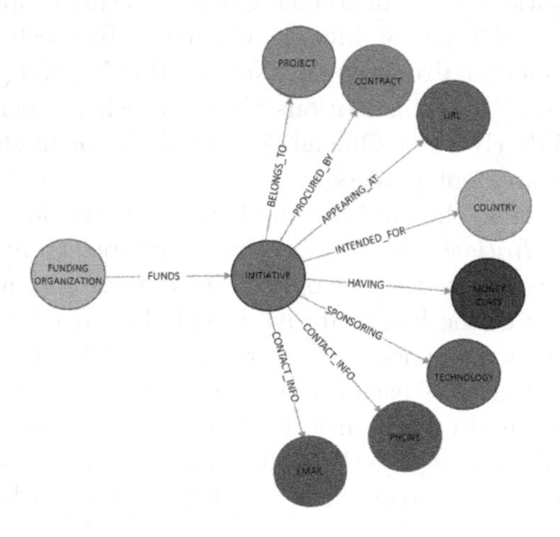

Fig. 1. The semantic schema defining the knowledge graph representation with respect to the entities described in Table 1.

5.1 The AIGreet Green Investments Dataset

The AIGreet dataset is automatically generated through the AIGreet pipeline, collecting the resources from the web and then processing them through machine learning techniques. The AIGreet dataset on green investments presented here comes from the analysis of a collection of over 2,000 digital documents on investment opportunities downloaded from a number of financial institutions (e.g., African Development Bank, European Investment Bank) and news agency websites. These documents were first processed using a text classification module based on deep neural networks available in the AIGreet pipeline to filter documents related to green investment opportunities. This returned 191 documents to be further analyzed. In the next step, these documents were processed using NLP techniques to extract or predict detailed information about the underlying investment opportunities. The resulting graph contains 1,660 nodes connected by 3,601 edges. Specifically, it contains 191 `Initiative` nodes supported by 160 `Funding_Org` nodes. These initiatives include 51 `Country` nodes for investments related to 9 technology nodes.

5.2 Degree Distribution Analysis

The first analysis we performed concerns the degree distribution of our graph. This metric gives us a measure of how many incident edges point to a node, where an incident edge is a connection pointing to the referenced node. The more edges, the higher the degree, i.e., its relevance with respect to the graph.

In our case, this metric is useful to get a quick overview of some relevant properties of the graph, such as identifying the countries that can intercept the most funding initiatives or the most popular green technologies that are funded. Below, we present the various analyses we performed using this metric. In each case, we did not look at the entire graph, but rather at a subgraph that reflects the characteristics we examined.

Assessing the flow of investments targeting each country.
Our goal is to use the degree metric to get an overview of the number of investments targeting each country under study.
In Fig. 2, we give the degree metric for the top 10 countries, computed on a subgraph consisting only of the nodes `Initiative` and `Country`. From the chart, it is clear that investments are concentrated in a small number of countries, with a significant imbalance between them. Values range from a maximum of 27 for Rwanda to a minimum of 1 for Chad, Eritrea, Djibouti, and Cameroon (not shown for clarity). A large number of green investments in Rwanda are a result of the green agricultural growth plan adopted by the Rwandan government in recent years [10]. One strategy in particular aims at diffusing mini-grid solar solutions that provide decentralised sustainable green energy, as explained in [5].

Assessing the popularity of the different green technologies.
Our goal is to get an overview of the most popular green technologies as required by the investments under analysis, by using the degree metric.
We analyzed the degree measure for `Technology` nodes, computed on the subgraph composed of `Initiative` and `Technology` nodes only. The outcoming results show that the most popular green technologies involved in the investments under analysis are *solar* (40.4%, including those based on mini-grid and off-grid solar plants) and `hydro` (26.3%). This indicates a clear investment strategy, taking advantage, on the one hand, of the low installation costs of solar plants and, on the other hand, exploiting the efficiency and durability of hydropower plants. Gas power plants (10.7%), instead, are only dominant in gas-producing nations and their neighbors.

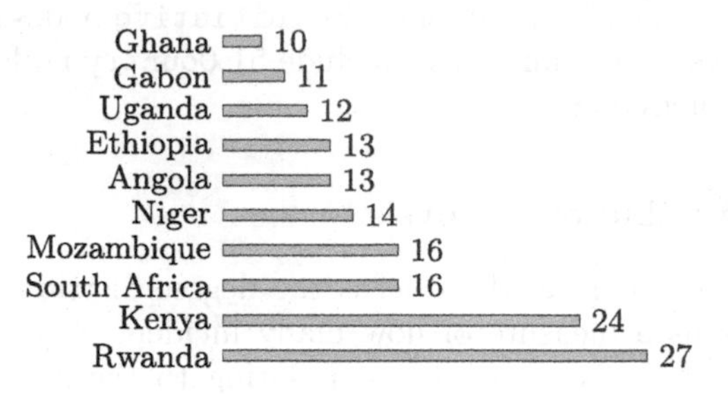

Fig. 2. Degree measure for `Country` nodes in a `Country-Initiative` subgraph. Only `Country` nodes with a degree higher than 9 are reported, over a total of 46 `Country` nodes.

Assessing the overall status of projects.
Different types of documents in the AIGreet dataset may reflect different states for the same project. Therefore, it is of interest to link all of these documents to the project to which they relate in order to develop a general understanding of the status of all projects.
In Fig. 3, we give the degree measure for the type of contracts, organized as a tree and evaluated for the `Contract-Initiative` subgraph. Note that not all documents are mandatory, shortcuts can be taken if the document category allows it. Most of the available documents are of the REOI and GPN type. These contracts are created after a specific grant is awarded to a project and are the basis for building the procurement process. We also note that the number of awarded projects (Aw) is relatively low, meaning many of the projects considered are still to be assigned.
Finally, we note that the Figure does not take into account documents derived from the newspaper articles available in the dataset under consideration, accounting for a 15% of the `Initiative` nodes.

Classifying funding initiatives according to their grants.
In Fig. 4, we classify funding initiatives by the presumed amount in euros of their grants, which is calculated using the degree measure for nodes of type `Money_Class` in the subgraph `Initiative-Money_Class`. From the graph, we see that there is a large gap between very large projects (more than 50mln €, 42.5%) and small projects (less than 100,000€, 25%), with all four remaining classes accounting for no more than the 32.5%.

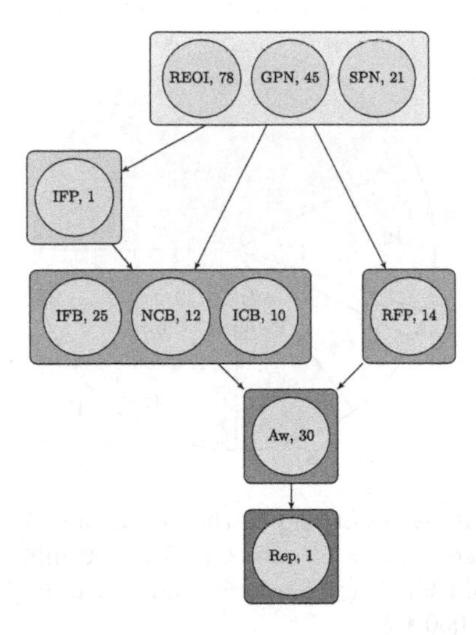

Fig. 3. The contract categories are organized as a tree representing their role in the procurement process. For each node, we also specify its degree metric, which is evaluated in a `:Category-Initiative` subgraph. Labels from left to right, top to bottom: Request for Expression of Interest (REOI), General Procurement Notice (GPN), Specific Procurement Notice (SPN), Invitation for Prequalification (IFP), Invitation for Bids (IFB), National Competitive Bidding (NCB), International Competitive Bidding (ICB), Request for Proposals (RFP), Award (Aw), and Report (Rep).

5.3 Community Detection

The concept of `Initiative` is the key element of our KG. Multiple initiatives may have the same underlying investment opportunity, refer to the same country, or use the same green technology. All of these substructures can be easily found by properly formulating the input queries. However, there is still a higher order of substructures in our KG, which we are interested in and can not be determined by using queries. We are interested in subgraphs revolving around a funding organization, a green technology, or a specific country that have a similar structure.

To find and characterize these subgraphs, we resort to community detection algorithms. These can be used to cluster the nodes of a graph to find groups of entities that are somehow connected. Given the nature of our graph, we chose to use the Louvain [2] algorithm. This is a hierarchical clustering method that maximizes a modularity score measuring how densely connected the nodes in the community are, as opposed to a random network.

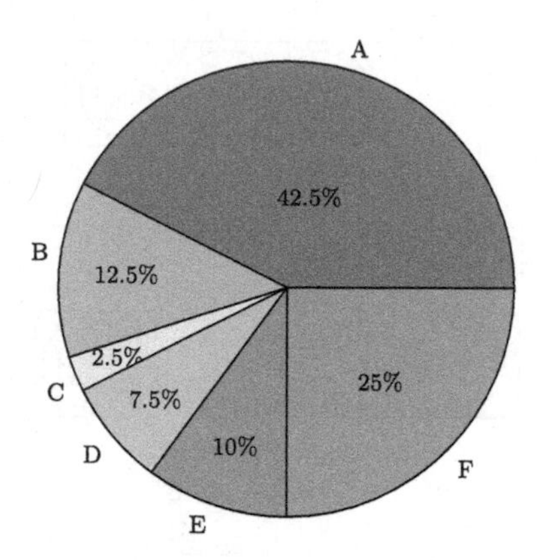

Fig. 4. Pie-chart of initiatives according to the degree metric for `Money_Class` nodes, within the following classes: A ($>$ 50 mln €), B ($\leq$ 50 mln €, and $>$ 10 mln €), C ($\leq$ 10 mln €, and $>$ 5 mln €), D ($\leq$ 5 mln €, and $>$ 1 mln €), E ($\leq$ 1 mln €, and $>$ 100000 €), and F ($\leq$ 100000 €)

Communities can give us different information depending on which subgraph is used to determine them. For this reason, we selected the subgraph consisting of the nodes **`Funding_Org`**, **`Initiative`**, **`Country`**, and **`Technology`**. A community in this subgraph identifies clusters of initiatives where a particular technology is prevalent in a region for a group of funding organizations. This is useful for identifying and strategically targeting areas of investment based on which types of current investors are operating, and in which technology they are putting their financial resources.

Table 2 provides the breakdown of the five most relevant communities resulting from running the execution of the Louvain community detection algorithm on the AIGreet investments KG. In Fig. 5, we project the elements of these communities onto a map of the African continent. In this representation, we indicate for each country the predominant green technology in which the nodes of the initiative are involved. This does not exclude the possibility of diversification but is meant to highlight multinational investment plans. An example is Egypt, where there are several large solar power plants, or South Africa, where interest in renewable energy, in general, has increased recently, as can be seen in [11].

Based on these results, it is interesting to note that there are small subsets of neighboring countries investing in the same green technology. This reflects the multinational nature of green power projects across Africa and the intent to leverage economies of scale by including neighboring countries, as explained in the African Development Bank [8] report. We also found that Saharan and sub-Saharan countries are heavily involved in solar energy initiatives, while hydropower projects involve areas known for their hydrogeological resources. There are two isolated clusters: one relates to thermal power generation and consists of Djibouti, Ethiopia, and Kenya; the other relates to gas and hydrogen power plants and consists of Malawi and Mozambique. This geographic distinction is consistent with the availability of renewable resources reported in [3].

Table 2. Breakdown of the five most relevant communities found using the Louvain community detection algorithm on the AIGreet investments KG, considering only the `Funding_Org`, `Country`, `Initiative`, and `Technology` nodes. In # `Funding_Org`, we indicate the number of `Funding_Org` nodes in the community.

Country	Technology	# Funding_Org	communityId
Malawi Mozambique Rwanda	gas hydrogen	15	358
Djibouti Ethiopia Kenya Mauritius	thermal	10	202
Benin Burundi Gambia Ghana Libya Niger Sierra Leone Sudan Togo Zambia	solar mini-grid off-grid	32	105
Cameroon Gabon Guinea Lesotho Liberia Namibia Sao Tome and Principe Tanzania Uganda	hydro	14	18
Angola Egypt Madagascar Mali Senegal Seychelles South Africa	wind combyned cycle biofuels	15	12

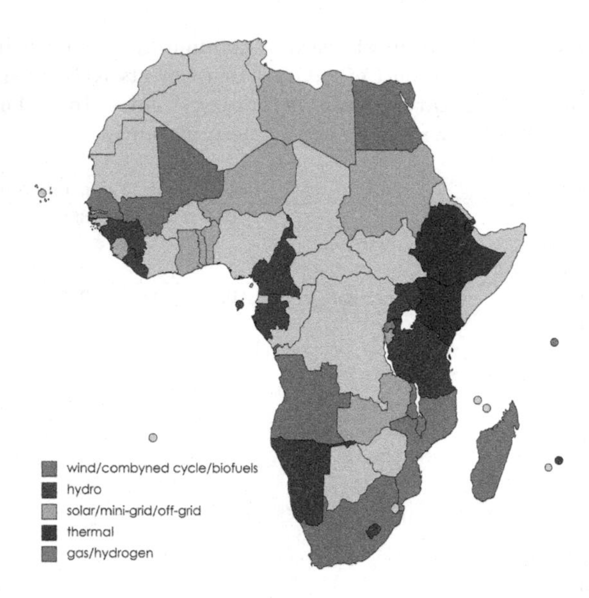

Fig. 5. Projection on the African map of the countries in the communities found using the Louvain community detection algorithm on the AIGreet investments KG. Grey countries are not included in these communities. For each community, we also report in the legend the technology types associated with the included `Initiative` nodes. (Color figure online)

6 Future Directions

As a further direction, we plan to develop an augmented version of our KG including investment opportunities available worldwide, as financed by a larger number of financial institutions, as soon as these data will be available in the framework of the AIGreet project. This will lead to a large KG for green investment opportunities generated through an automatic pipeline. As soon as the size of the KG will increase to this scale, we also plan to carry out our analysis using distributed computing frameworks (as done, e.g., in [9]), so as to keep small the amount of time required by our analysis. Finally, once the information is robust enough, we will include temporal analysis to understand the timelines of projects and investments, determining if there are underlying dynamic structures.

Acknowledgments. This work was partially supported by Università di Roma - La Sapienza Research Project 2021 "Caratterizzazione, sviluppo e sperimentazione di algoritmi efficienti". It was also supported in part by INdAM - GNCS Project 2023 "Approcci computazionali per il supporto alle decisioni nella Medicina di Precisione". It was also supported by the PON Project, DM1062 10/08/2021, "AIGREET: using Artificial Intelligence to support GREEn investments for promoting the Ecological Transition", financed by the "Ministero dell'Università e della Ricerca". Finally, it was supported by an initiative with the industrial partner "Internationalia S.r.L." for the

"Development and design of a system for the automatic detection of green investment documents from freely available resources, based on machine learning and NLP".

References

1. The AIGreet Project. https://www.dss.uniroma1.it/Aigreet (2021)
2. Blondel, V.D., Guillaume, J.-L., Lambiotte, R., Lefebvre, E.: Fast unfolding of communities in large networks. J. Stat. Mech. Theor. Exp. **2008**(10), P10008 (2008)
3. Bouchene, L., Cassim, Z., Engel, H., Jayaram, K., Kendall, A.: Green Africa: a growth and resilience agenda for the continent. Mckinsey Sustain. **28**, 28 (2021)
4. Bu, C., Zhang, J., Yu, X., Wu, L., Wu, X.: Which companies are likely to invest: knowledge-graph-based recommendation for investment promotion. In: 2022 IEEE International Conference on Data Mining (ICDM), pp. 11–20. IEEE (2022)
5. Chang, K.-C., et al.: Standalone and minigrid-connected solar energy systems for rural application in Rwanda: an in situ study. Int. J. Photoenergy **1–22**, 2021 (2021)
6. Chen, X., Jia, S., Xiang, Y.: A review: knowledge reasoning over knowledge graph. Exp. Syst. Appl. **141**, 112948 (2020)
7. Cheng, D., Yang, F., Wang, X., Zhang, Y., Zhang, L.: Knowledge graph-based event embedding framework for financial quantitative investments. In: Proceedings of the 43rd International ACM SIGIR Conference on Research and Development in Information Retrieval, pp. 2221–2230 (2020)
8. Chuku, C.A., Ajayi, V.: Growing Green: Enablers and Barriers for Africa. African Development Bank (2022)
9. Di Rocco, L., Ferraro Petrillo, U., Rombo, S.E.: DIAMIN: a software library for the distributed analysis of large-scale molecular interaction networks. BMC Bioinf. **23**(1), 474 (2022)
10. Hudani, S.E.: The green masterplan: Crisis, state transition and urban transformation in post-genocide Rwanda. Int. J. Urban Reg. Res. **44**(4), 673–690 (2020)
11. Jain, S., Jain, P.: The rise of renewable energy implementation in South Africa. Energy Procedia **143**, 721–726 (2017)
12. Mackie, I., Dalton, J.: Query-specific knowledge graphs for complex finance topics. arXiv preprint arXiv:2211.04142 (2022)
13. Ruan, T., Xue, L., Wang, H., Hu, F., Zhao, L., Ding, J.: Building and exploring an enterprise knowledge graph for investment analysis. In: Groth, P., et al. (eds.) ISWC 2016. LNCS, vol. 9982, pp. 418–436. Springer, Cham (2016). https://doi.org/10.1007/978-3-319-46547-0_35
14. Xu, J., Marodon, R., Ru, X., Ren, X., Wu, X.: What are public development banks and development financing institutions?? qualification criteria, stylized facts and development trends. China Econ. Q. Int. **1**(4), 271–294 (2021)

Knowledge Graphs Embeddings for Link Prediction in the Context of Sustainability

Ylenia Galluzzo[1(✉)] and Francesco Gennusa[2]

[1] Department of Engineering, University of Palermo, Palermo, Italy
ylenia.galluzzo01@unipa.it

[2] Sustainability and Ecological Transition Center, University of Palermo, Palermo, Italy
francesco.gennusa@unipa.it

Abstract. The use of a large amount of heterogeneous, interconnected and quality data is nowadays essential to address challenges related to the development of solutions for "complex systems". In this paper, we propose a framework for building a collaborative platform that can support the System Dynamics experts in identifying new partnerships and innovative solutions. To this end, we use a Knowledge Graph approach that can track the links among stakeholders and store their metadata. It is charge of leveraging ML methodologies and models for data analysis as well as the prediction of connections among stakeholders and potential work tables to facilitate the achievement of the SDGs in specific contexts.

Keywords: Knowledge Graph Embeddings · Link Prediction · Machine Learning · Neo4j · Collaborative Platforms · System Dynamics · Sustainable Development Goals

1 Introduction

The United Nations (UN) have established 17 Sustainable Development Goals (SDGs) as part of the 2030 Agenda for Sustainable Development. These goals aim to address different environmental, social, and economic challenges that the world faces. However, the implementation of these goals can be challenging due to differences in political, social, economic, and cultural contexts across the globe. To address this challenge, localization of the SDGs is critical to achieve the targets set for each goal. Wicked problems [1], such as climate change, poverty, biodiversity imbalances, population dynamics, and access to renewable energy sources, provide tangible examples of how interdependence can limit the effectiveness of policies aimed at addressing a single issue. This implies the need to consider holistically this set of problems and to enhance dynamic approach to policy analysis and stakeholder collaboration to pursue sustainable outcomes [3].

Since the 2030 Agenda is universal, integrative, and transformative, it is crucial to involve the public and private sectors, as well as all stakeholders, through a horizontal integration. Furthermore, vertical alignment between local,

A. Abelló et al. (Eds.): ADBIS 2023, CCIS 1850, pp. 452–464, 2023.
https://doi.org/10.1007/978-3-031-42941-5_39

Table 1. Table of the 17 Sustainable Development Goals (SDGs).

SDGs	SDGs Title
SDG 1	No Poverty
SDG 2	Zero Hunger
SDG 3	Good Health and Well-Being
SDG 4	Quality Education
SDG 5	Gender Equality
SDG 6	Clean Water and Sanitation
SDG 7	Affordable and Clean energy
SDG 8	Decent Work and Economic Growth
SDG 9	Industry, Innovation and Infrastructure
SDG 10	Reduced inequalities
SDG 11	Systainable cities and communities
SDG 12	Responsible consumption and production
SDG 13	Climate action
SDG 14	Life below water
SDG 15	Life on land
SDG 16	Peace, Justice and Strong Institution
SDG 17	Partnership for the goals

regional, national, and international governance levels cannot be overlooked. The combination and coordination of mandates, resources, and capacities of various institutional and societal actors at different governance levels and across sectors is an essential catalyst for sustainable development.

This is particularly important for the localization of Sustainable Development Goals (SDGs), which refers to the process of delineating, executing, and overseeing strategies at the community level to accomplish the global, national, and regional SDGs [2].

In recent years, Knowledge Graphs have been attracting more and more interest in various application contexts (precision medicine [7], biomedical [8], NLP tasks [9–11], recommendation systems [12,13], etc.) as they are able to integrate and relate data from different sources in a single database (e.g. Neo4j). Recently, Knowledge Graphs have been applied in the field of sustainability [14–16]. The research presented here aims to complement existing work by enriching the scientific literature in this field (which is still little researched) and explore a new domain, which is that of relations between stakeholders and working tables, fostering effective collaborations in support of the goals set by the various SDGs of the 2030 Agenda.

In detail, this research aims at enhancing the SDGs localization through the creation of collaborative platforms [6], supported by Knowledge Graphs (KG), which may create shared meaning, understand different perspectives, and foster a common view of the system. The application will be developed to support the expert in System Dynamics. This methodology will be applied to facilitate learning processes in the planning and implementation of policies shared by different stakeholders.

Furthermore, the investigation will be carried out through the utilization of the Horizon Europe Programme Guide 2021–2027 [23], that it is the successor to Horizon 2020 [4]. This guide offers comprehensive instructions on the arrangement, financial plan, and key objectives of the Horizon Europe program. It is significant in terms of promoting cooperation, enhancing the influence of research and innovation, and actively contributing to the development, support, and implementation of EU policies aimed at addressing global challenges.

2 Proposed Approach

In view of the universal, integrated and transformative nature of the 2030 Agenda, effective implementation of Sustainable Development Goals (SDGs) requires the participation of governments across various policy domains and governance levels, in view of collaborative networks with different stakeholders. For this purpose, the System Dynamics will be applied to facilitate the effective implementation of collaborative platforms that are able to combine the policy level of contextual governance with the management level, through the action and interaction of the different organizations involved.

Collaborative platforms are essential for engaging local actors in addressing targets of the SDGs (see Table 1) due to their contribution to the representation of new event not fully explained by current concepts [6]. These platforms facilitate collaboration, create shared understanding, determine each stakeholder's role in localizing the SDGs, and promote the development of strategic resources for implementing collaborative policies in socio-economic, cultural, and ecological transitions. However, they often face challenges in mobilizing resources and expertise to address sustainable development challenges [5].

An inclusive and multi-stakeholder approach to implement the SDGs provides several benefits, including stimulating public participation in implementation programs, providing political support for the sustainable development agenda, and increasing political alignment between stakeholder groups. Moreover, by promoting the convergence of efforts by different development actors, it can ensure the inclusion of marginalized demographic groups and communities, allowing them to have a role in development processes. SDGs localization requires accurate and comprehensive data in support of collaborative platforms (and its facilitators) to assess progress, identify gaps, and inform decision-making. Indeed, data is essential in several ways to support the achievement of the SDGs. Firstly, it enables governments and other stakeholders to establish starting points, monitor progress, and assess the effectiveness of development interventions. Secondly, data is crucial in mobilizing resources for the SDGs by showcasing the need for investment, identifying priorities and monitoring their impact. Lastly, data can also identify new partnerships and innovative solutions that can assist in achieving the SDGs in specific contexts.

2.1 Application Context

With reference to the present research, the application of the System Dynamics framework intends to enhance the comprehension of the causal connections between policies implemented and the results achieved in the short, medium, and long term. Specifically, this approach seeks to account for the tangible and intangible factors that influence the impact of specific policies [3]. Through the facilitation of planning processes within collaborative platforms, the guidance of System Dynamics experts enables the explicit and tacit knowledge of key actors to be capitalized and refined, as well as the information system that supports the planning processes. This is achieved by analyzing cause-and-effect linkages related to both proposed policies and the phenomena that have led to past outcomes and current conditions that key actors aim to address. Therefore, System Dynamics provides fundamental support for the holistic representation of the sustainability of performance concerning the policies adopted and the issues at hand. System Dynamics in support of collaborative platforms can be used to localize SDGs. By understanding the interactions between the specific socioeconomic, environmental, and institutional factors, policymakers and stakeholders can develop targeted strategies that address the unique challenges facing their communities. Including in this process the Horizon Europe program fosters cooperation between different actors, including academia, industry, and society. Indeed, through this program, the EU seeks to address societal challenges, promote excellence in science, and increase competitiveness and sustainability in the global economy. Horizon Europe can promote cooperation among different sectors and stakeholders, such as governments, civil society, and the private sector.

General Use. In this research we describe our Knowledge Graph (KG) data integration framework. Specifically, the proposed framework aims to be in support of the expert of System Dynamics. The following proposed work will improve the productivity of the domain expert, in order to identify, with greater accuracy, potential collaborations between stakeholders. With the aim of improving the prediction of intrinsic connections between stakeholders and between *stakeholders* and *work tables*, we create a Knowledge Graph in a Neo4j graph database. The use of this technology provides a view with transparent access to data from heterogeneous sources, as well as numerous advantages in terms of flexibility and performance. In particular in Fig. 1, we represent the general structure of the framework, that we will describe in detail in every part of it in the following subsection. A Knowledge Graph represents deep information of interconnected data enriched with metadata that can make analyses useful for complex decision-making processes. A Neo4j Knowledge Graph is constructed using typed entities that may have attributes (metadata). Entities are interconnected and connections (arcs) can be weighted.

The first stage of the proposed framework concerns the use of data from the "Horizon Europe Program Guide 2021–2027" [23]. The Horizon EU [23]

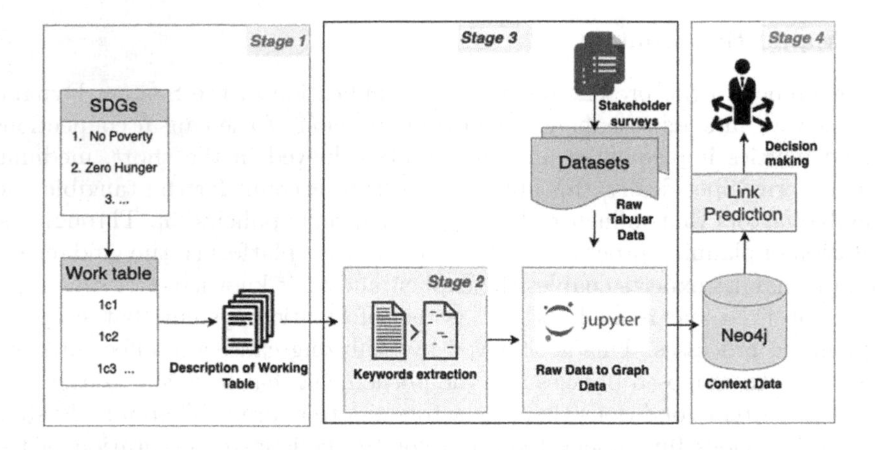

Fig. 1. General Components of the proposed approach.

in particular divides the "Global Challenges and European industrial competitiveness" into six clusters: *"Health"*, *"Culture, Creativity and inclusive society"*, *"Civil security for society, digital, industry and space"*, *"Digital, Industry and Space"*, *"Climate, energy and mobility"*, *"Food, bioeconomy, natural resources, agriculture and the environment"*. The six clusters are complemented by support for the research activities of the Joint Research Centre. Each cluster is composed of several "Destinations", which relate to the impacts expected by the cluster outlined in the strategic plan. Each cluster is also associated with specific SDGs (see Table 1). For example, the activities of the cluster *Health* are directly linked to two United Nations Sustainable Development Goals: SDG 3 (Good Health and Well-Being) and SDG 13 (Climate Action). We used the "Horizon Europe Program Guide 2021–2027" [23] to simulate our work tables for potential stakeholders to work on. The *work tables* are represented by the "Destination". Each "Destination" is represented by a title and a detailed mission description. In our work, we have translated the descriptions of the Destinations presented in the Guide [23] into English language (see Table 2).

Table 2. Example of one Destination with its description and Cluster.

Cluster	Destination Title	Description
HEALTH	Staying healthy in a rapidly changing society	It focuses on the main social challenges outlined at European level: diet and obesity, ageing and demographic change, mental health, digital empowerment in health literacy and personalized prevention. For 2022 it also requires design proposals to improve the availability and use of artificial intelligence (AI) tools to predict the risk of chronic disease occurrence and progression

The *"Descriptions of the working tables"* in Fig. 1 are represented in our framework by the descriptions of the Horizon EU [23] Destinations.

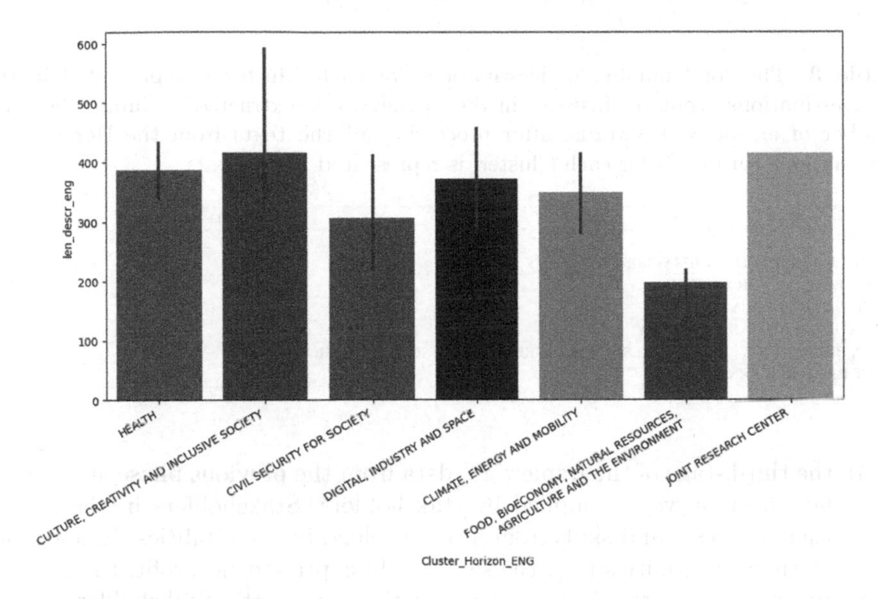

Fig. 2. Length of the texts present on Horizon EU Destinations for each Cluster. The minimum length of the considered texts is 147. The maximum length is 590. The average length of the texts is 329.

In detail, in the Fig. 2, we show the length of the texts present in the several Horizon EU [23] Destinations, for each Cluster. In general, on average the length of the texts considered, from which the keywords are extracted (in the second stage), is equal to 329 characters (an average of 45 words on each text).

The second stage of the proposed framework concerns the use of a *keyword extractor* from the text. In detail, the description of the work table passes under the manual care of a domain expert, who has the task in the first instance of selecting keywords (single and compound) from the descriptions of the Horizon EU Destinations. Subsequently, the keywords are automatically processed via Jupyter Notebook, for a data cleaning phase (e.g. removing duplicate words). Finally, we assess the similarity of the words extracted from the domain expert, converting them into word embeddings. We calculate the vectors representing the strings, using word embeddings pre-trained for the English language. In this work, we use FastText's pre-trained word vector model [17] for the English language. The model has 1 million word vectors trained with subword infomation on Wikipedia 2017, (16B tokens). Next we calculate the cosine similarity between the previously extracted words. If the value exceeds the similarity threshold set to 0.80, then the word is incorporated. The process is thus iterative, and continues until there are no more similarities greater than or equal to the threshold. This step is important because it is necessary to incorporate similar words to improve the quality of the data that will constitute the Knowledge Graph. Moreover, incorrect processing of raw data could significantly impact the generation of new knowledge, as well as go against FAIR principles [18].

Table 3. The total number of Destinations for each Cluster is represented in the "# Destinations" column. Instead, in the "# keywords extracted" column, the total number of extracted keywords, after processing all the texts from the Horizon EU Destinations (in Fig. 2) for each Cluster, is represented.

Cluster Horizon ENG	# Destinations	# keywords extracted
HEALTH	6	33
CULTURE, CREATIVITY AND INCLUSIVE SOCIETY	3	21
CIVIL SECURITY FOR SOCIETY	6	30
DIGITAL, INDUSTRY AND SPACE	6	35
CLIMATE, ENERGY AND MOBILITY	6	30
FOOD, BIOECONOMY, NATURAL RESOURCES,AGRICULTURE AND THE ENVIRONMENT	7	35
JOINT RESEARCH CENTER	1	2

In the third stage of the framework, data from the previous phase are merged with data from surveys compiled by stakeholders. Stakeholders in the survey select their interests and skills from a set of selectable possibilities. In addition, they list their generalities (e.g. the sector public/private/no-profit, the number of employees of the organization of which they are part). Stakeholder data is linked to keywords extracted from the description of the work tables. The graph is generated from raw and tabular data with the specific Jupyter Notebook and Py2Neo library, then is stored in the Neo4j database. In the last stage of the framework, the data stored on the Neo4j graph generates the embedded vectors of the nodes, depending on the appropriate projection of the chosen sub-graph. New potential links are generated between stakeholders, and between *Stakeholders – Destination/work table*. The results of the following phase support the decision maker.

3 Experiments

The framework is based on a labeled property graph (LPG) built on the Neo4j graph data platform. The LPG graph are used to represent complex, connected data structures. This kind of graph are widely used in graph databases, such as Neo4j, as they provide a flexible and efficient way to represent complex data structures and relationships. They differ from RDF graphs because in the LPG graph each node and each arc can have a set of properties and thus an internal structure. Furthermore, LPG is typically queried using Cypher, a query language specifically designed to query a complex graph, perform operations (e.g. aggregation, sorting etc.) efficiently on databases such as Neo4j. Specifically, for this research work, we created a prototype of the framework using *Neo4j Sandbox* (a free, cloud-based instance of Neo4j). We also used the *Neo4j Graph Data Science*

(GDS) [1] library to analyse relationships between data and to improve the prediction of links between entities. The whole graph created consists of 342 nodes (5 entity types that are shown in the Fig. 3) and 6598 links between them.

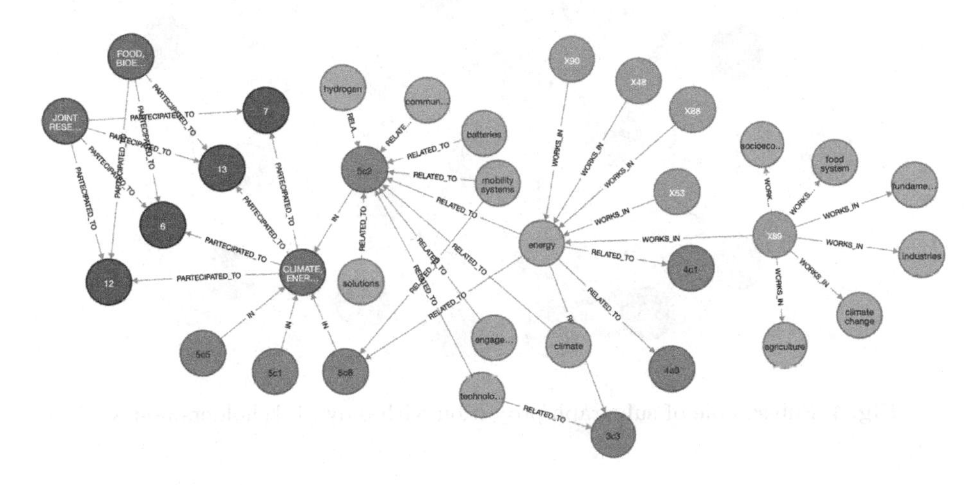

Fig. 3. Sub-sample of association between the 5 types of entity: SDGs (red nodes), Clusters (orange nodes), Destination/Working Tables (blue nodes), Keywords (brown nodes) and Stakeholders (purple nodes).

In addition to the raw data from stakeholder surveys and destination-derived keywords, collaboration weighted undirected links named [CO_WORK] between stakeholders were added based on the counts of shared keywords. A undirect link [WORK_ON] between Stakeholders and Destinations/Working tables was also added. In particular, the links [WORK_ON] were created considering the keywords corresponding to each Stakeholder with the intersection of the keywords corresponding to each Destination (of each Cluster). In this way we create a quick link between Stakeholders and Destinations, so that the connection between these two graph entities could be considered as a target arc for the task of link prediction.

3.1 Results Analysis

To evaluate the advantages of the proposed approach, we test the learning model for solving link prediction on two specific graph projections. First, we train and test the Link Prediction Pipeline on the whole graph. The *Target relationship type* to be predicted in this case is (Stakeholder - [WORK_ON] - Destination). Then, to also evaluate potential collaborations between stakeholders, we test the pipeline on a reduced projection of the graph, which consists only of the Stakeholders entities and the [CO_WORK] relations (in Fig. 4). In this case the *Target relationship type* to be predicted will be [CO_WORK].

[1] https://neo4j.com/docs/graph-data-science/current/.

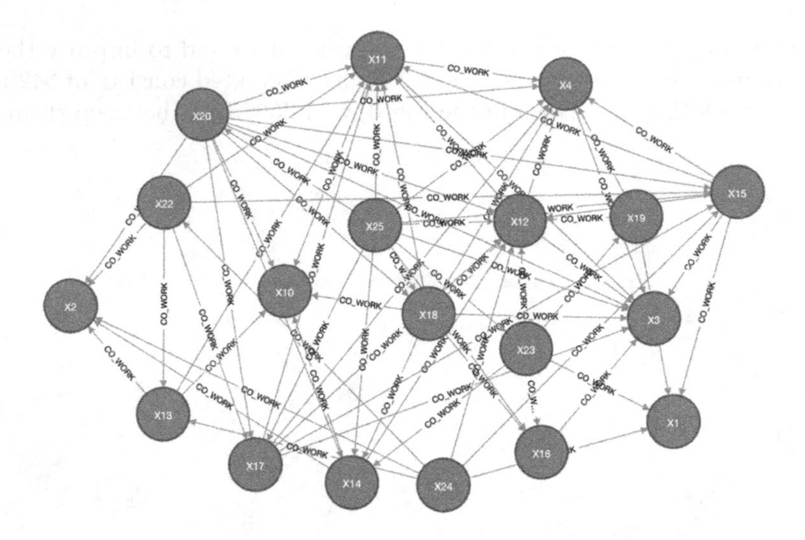

Fig. 4. Sub-sample of sub-graph projection with only Stakeholders nodes.

Link Prediction training pipelines split relationships into sets and add negative relationships to some of these sets. Configuration of the splitting is optional, and default settings are used if not specified. In our case, the configuration of the training phases is set to have a fraction of the graph reserved for training (0.6) and a fraction of the graph reserved for testing (0.4). Internally, the Neo4j pipeline divides positive relationships in the train and test sets. Subsequently, "negative" relationships are added randomly to each set of the graph (train set and test set) which are complementary to the "positive" relationships, and do not overlap with the other. It is important to add the sampled negative relationships to the training pipelines, as it allows you to provide the model with examples of relationships that do not exist in the real graph, helping it learn to discriminate between positive and negative relationships (negativeSamplingRatio, is predefined as 1.0 by default).

Cross-validation with k-folds (validationFolds) is also used to evaluate the performance of the machine learning model on our limited dataset. The number of divisions of the training graph is k=10. In the end, the performance of the models used is evaluated on the average of the results obtained on the k tests. In particular, we compare several basic models (Logistic Regression, Random Forest, MLP) for learning graphs on a link predictions task. Finally, the model with the best AUCPR (Area Under the Precision-Recall Curve metric) results on the test set is extracted and automatically chosen and saved. A model that will average an AUCPR score close to 0.5 indicates that its predictions are not reliable. For the creation of the embeddings of the graph nodes, we use the Fast Random Projection algorithm, or FastRP [19]. It is a random projection algorithm. This technique greatly reduces dimensionality while preserving information on the distance between nodes. In fact, two non-similar nodes do not have to be assigned similar embedding vectors. In the GDS library, the FastRP algorithm is extended from the original algorithm, as the algorithm implemented in GDS considers the properties of the nodes. The resulting embeddings can therefore represent the graph more accurately. We use for the preliminary experiments in this work, a vector of embeddings of size 56 that will be stored as additional properties of each node. Table 4 shows the results of the AUCPR metric on the two projections made on the graph and the two target relationship: [WORK_ON], [CO_WORK].

Table 4. Link Prediction results.

Projection	Target Link	Winning Model	Test Score
Whole Graph	WORK_ON	{maxDepth: 2147483647, minLeafSize: 1, criterion: 'GINI', minSplitSize: 2, numberOfDecisionTrees: 10, methodName: 'RandomForest', numberOfSamplesRatio: 1.0}	**0.94**
Whole Graph	CO_WORK	{maxEpochs: 500, minEpochs: 1, classWeights: [], penalty: 0.0016619220722602157, patience: 1, methodName: 'LogisticRegression', focusWeight: 0.0, batchSize: 100, tolerance: 0.001, learningRate: 0.001}	**0.96**
Stakeholders Graph	CO_WORK	{maxDepth: 2147483647, minLeafSize: 1, criterion: 'GINI', minSplitSize: 2, numberOfDecisionTrees: 10, methodName: 'RandomForest', numberOfSamplesRatio: 1.0}	0.72

The results of both target link predictions are appreciable and satisfactory, but further analyses and experiments should be carried out considering different configuration models for the learning and trying different embedding models (e.g. Node2Vec, HashGNN, GraphSAGE [20–22]) as well as different embedding sizes.

It can be deduced from the preliminary results obtained that the creation of node embeddings for the projection of the entire graph is more suitable for the prediction of the target relations we considered: [CO_WORK] and [WORK_ON]. The training times of the costructed predictive models are shown in Table 5. The experiments were conducted using a MacBook Air with M1 chip, based on Apple Silicon technology with an ARM processor. It is equipped with octa-core CPU, octa-core GPU, a 16-core Neural Engine and 8 GB of unified memory in its hardware.

We note that when we exploit the projection of the entire graph, the training of predictive model for the target relation [CO_WORK] is more time-consuming. This type of result is expected, since we consider a larger number of arcs (there are 3516 stakeholder directed relationships) on which we need to do the training, as well as using the entire graph. The prediction of the target relation

Table 5. Training time of the models.

Projection	Target Link	Training Time
Whole Graph	WORK_ON	4min 30.0s
Whole Graph	CO_WORK	19min 36.9s
Stakeholder Graph	CO_WORK	5min 59.7s

[CO_WORK] is more accurately predicted when we use the entire graph of knowledge.

With the goal of refining the learning of the predictive model of connections between nodes, and avoinding potential cases of overfitting, the Knowledge Graph will also be populated with more data from additional stakeholders in future implementation of the framework.

4 Conclusion

This preliminary version of a collaborative system based on Knowledge Graphs, is an innovative solution for the development of SDGs goals. The use of Knowledge Graphs represents succinctly complex systems such as participatory models, which take into account interrelations between heterogeneous entities. In future versions of the framework, it is proposed to extend the graph of knowledge by integrating data from other sources (e.g. textual documents, geospatial documents etc.), able to enrich the database with additional entities and connections not trivial. We will also work to enrich the Knowledge Graph with information about actual time holdings (dates of past holdings) of stakeholders at specific work tables. This kind of information could refine the suggestions given to the end user of the platform on the stakeholders that are available to participate in a working table with a specific deadline. In conclusion, the future goal is to create an interactive dashboard that can be easily used by the end user (e.g. System Dynamics expert). The ultimate goal is to have a quality and comprehensive knowledge base, useful to scientific, economic, social and cultural progress in all its forms.

Acknowledgements. Part of the research presented here has been funded by the MIUR-PRIN research project "Multicriteria Data Structures and Algorithms: from compressed to learned indexes, and beyond".

References

1. Ansell, C., Torfing, J. (eds.). Public innovation through collaboration and design. Routledge, Milton Park (2014)
2. Perry, B., Diprose, K., Taylor Buck, N., Simon, D.: Localizing the SDGs in England: challenges and value propositions for local government. Front. Sustain. Cities **3**, 74633 (2021)

3. Bianchi, C., Bereciartua, P., Vignieri, V., Cohen, A.: Enhancing urban brownfield regeneration to pursue sustainable community outcomes through dynamic performance governance. Int. J. Public Adm. **44**(2), 100–114 (2021)
4. COM 808. EU Framework programme for research and Innovation-Horizon 2020 (2011)
5. Oosterhof, P.D.: Localizing the sustainable development goals to accelerate implementation of the 2030 agenda for sustainable development (2018)
6. Ansell, C., Gash, A.: Collaborative platforms as a governance strategy. J. Public Adm. Res. Theor. **28**(1), 16–32 (2018)
7. Chandak, P., Huang, K., Zitnik, M.: Building a knowledge graph to enable precision medicine. Sci. Data **10**(1), 67 (2023)
8. Galluzzo, Y.: A review: biological insights on knowledge graphs. In: Chiusano, S., et al. (ed.) New Trends in Database and Information Systems: ADBIS 2022 Short Papers, Doctoral Consortium and Workshops: DOING, K-GALS, MADEISD, MegaData, SWODCH, Turin, Italy, 5–8 September 2022, Proceedings, pp. 388-399. Springer, Cham, August 2022. https://doi.org/10.1007/978-3-031-15743-1_36
9. Schneider, P., Schopf, T., Vladika, J., Galkin, M., Simperl, E., Matthes, F.: A decade of Knowledge Graphs in natural language processing: a survey. arXiv preprint arXiv:2210.00105 (2022)
10. Chakraborty, N., Lukovnikov, D., Maheshwari, G., Trivedi, P., Lehmann, J., Fischer, A.: Introduction to neural network-based question answering over knowledge graphs. Wiley Interdisc. Rev. Data Min. Knowl. Discovery **11**(3), e1389 (2021)
11. Al-Moslmi, T., Ocaña, M.G., Opdahl, A.L., Veres, C.: Named entity extraction for knowledge graphs: a literature overview. IEEE Access **8**, 32862–32881 (2020)
12. Shao, B., Li, X., Bian, G.: A survey of research hotspots and frontier trends of recommendation systems from the perspective of knowledge graph. Exp. Syst. Appl. **165**, 113764 (2021)
13. Guo, Q., et al.: A survey on Knowledge Graph-based recommender systems. IEEE Trans. Knowl. Data Eng. **34**(8), 3549–3568 (2020)
14. Fotopoulou, E., et al.: SustainGraph: a knowledge graph for tracking the progress and the interlinking among the sustainable development goals' targets. Front. Environ. Sci. **10**, 2175 (2022)
15. Sun, Y., Liu, H., Gao, Y., Zheng, M.: Research on the policy analysis of sustainable energy based on policy knowledge graph technology–a case study in china. Systems **11**(2), 102 (2023)
16. Kalaycı, T.E., Bricelj, B., Lah, M., Pichler, F., Scharrer, M.K., Rubeša-Zrim, J.: A Knowledge Graph-based data integration framework applied to battery data management. Sustainability **13**(3), 1583 (2021)
17. Mikolov, T., Grave, E., Bojanowski, P., Puhrsch, C., Joulin, A.: Advances in pre-training distributed word representations. arXiv preprint arXiv:1712.09405 (2017)
18. Wilkinson, M.D., et al.: The FAIR Guiding Principles for scientific data management and stewardship. Sci. Data **3**(1), 1–9 (2016)
19. Chen, H., Sultan, S. F., Tian, Y., Chen, M., Skiena, S.: Fast and accurate network embeddings via very sparse random projection. In: Proceedings of the 28th ACM International Conference on Information and Knowledge Management, pp. 399–408, November 2019
20. Grover, A., Leskovec, J.: node2vec: Scalable feature learning for networks. In: Proceedings of the 22nd ACM SIGKDD International Conference on Knowledge Discovery and Data Mining, pp. 855–864, August 2016
21. Tan, Q., et al.: Learning to hash with graph neural networks for recommender systems. In: Proceedings of The Web Conference 2020, pp. 1988–1998, April 2020

22. Hamilton, W., Ying, Z., Leskovec, J.: Inductive representation learning on large graphs. In: Advances in Neural Information Processing Systems, vol. 30 (2017)
23. Horizon Europe. La guida, APRE (2021)

An ArchiMate-Based Thematic Knowledge Graph for Low-Code Software Development Domain

Karlis Rokis(✉) and Marite Kirikova

Institute of Applied Computer Systems, Riga Technical University, 6A Kipsalas Street, Riga 1048, Latvia

rokis.karlis@gmail.com, marite.kirikova@rtu.lv

Abstract. Knowledge graphs can be considered at their entity type and entity levels. The entity type level is called also an ontology level. This paper illustrates, how ArchiMate modelling language can be applied as the ontology level of a knowledge graph for representing knowledge that concerns an enterprise. In particular, the ArchiMate-based knowledge graph is proposed for representing knowledge about low-code software development with the purpose to make this knowledge available for enterprise management. Navigating the knowledge graph enables stakeholders to exploit knowledge about different aspects of low-code development and promote the use of the low-code development approach in their enterprises. The knowledge graph is constructed and used by applying the capabilities of the Archi tool and its Visualiser.

Keywords: Low-Code Development · ArchiMate · Knowledge Graph

1 Introduction

Low-code development is an emerging approach based on visual software development, reusability of components, and automation and reducing the amount of manual coding. Gartner predicts that 65% of development activity will be related to low-code applications by 2024. Low-code development can be considered a tool-based approach [1]. There are particular software development and delivery principles and peculiarities related to this approach [2]. Multiple aspects influence the successful use of low-code development within organisations and the realisation of its potential benefits. Thus, organisations need to identify what relevant principles, capabilities, and other aspects influence the realisation of the low-code development potential considering the business and IT objectives of the enterprise. However, the question arises, of how the knowledge about low-code development can be represented so that it would be easily accessible and handy for the enterprise management to make decisions about the activities and investments in promoting low-code development in their companies.

The goal of this paper is to illustrate how ArchiMate-based [3]–[6] knowledge graph can be used for amalgamating and using low-code development related knowledge.

A. Abelló et al. (Eds.): ADBIS 2023, CCIS 1850, pp. 465–476, 2023.
https://doi.org/10.1007/978-3-031-42941-5_40

The paper is organised as follows: Sect. 2 describes the research method used. Section 3 briefly overviews currently available knowledge on low-code development and points to the ArchiMate concepts used for representing this knowledge. In Sect. 4, we illustrate, how this knowledge has been amalgamated in the ArchiMate-based knowledge graph. In Sect. 5, the practical application of the knowledge graph is illustrated. Section 6 provides conclusions regarding the applicability of the ArchiMate language as an ontology layer of a knowledge graph.

2 Research Approach

The paper focuses on the aim of helping organisations to identify the main aspects of low-code development to realise its potential benefits to reach business objectives. The target audience for the results of this research is specialists from organisations that intend to adopt or are currently applying low-code development.

The work was split into three phases, which follow design science principles described in [7], namely problem investigation, design, and validation.

The initial phase was problem investigation, which is constituted of problem exploration and conceptualisation. It was performed with the help of a systematic literature review and thematic synthesis to understand the aspects related to low-code development (the details of this phase are out of the scope of this paper).

Then the "Design" phase followed. There the knowledge on low-code development was represented in the ArchiMate-based knowledge graph. Afterwards, the obtained knowledge representation was validated in a company.

The literature review has been selected as the primary method for amalgamating knowledge on low-code development (Sect. 3). A keyword search was used as an initial step for the literature selection [8]. Then the backward reference search was also applied, as suggested by [8]. To summarise, amalgamate, and compare gathered information on the defined research questions, the concepts of the research synthesis method, called thematic synthesis, were applied. In this method, recurring themes from multiple studies are identified, interpreted, explained, and at the end, conclusions are delivered [9].

Based on the results of the first stage, the low-code development knowledge was amalgamated in the knowledge graph, the ontology level of which corresponds to the selected elements of the well-known enterprise architecture representation language ArchiMate (discussed in detail in Sect. 4). The graph was constructed to guide enterprises in recognising the low-code development-related aspects that might be useful for realising low-code development potential and fulfilling their business objectives. The Knowledge Graph was validated with the help of the chief operating officer in a small and medium-sized enterprise (SME) that uses low-code platforms for smart home device application development and internal application development (more details are available in Sect. 5).

3 Knowledge About Low-Code Development

It is considered that the term "low-code" originated back in 2014 when it was defined in a Forrester Research paper [2]. Since then, multiple definitions have been delivered in the current research literature. For instance, in [10], the author describes it as a derivate of fourth-generation programming (4GL) in a combination with the Rapid Application Development (RAD) principles. In [11], low-code development is described as a paradigm that minimises hand-coding and uses visual programming with a graphical interface and model-driven design. It also enables software development for practitioners with various backgrounds and experiences. The paper also establishes the relation to the Agile methodology. From this follows that it would be relevant to consider development methodology principles when defining low-code development. Somewhat similar aspects in defining low-code development are described in [12].

Low-code development is strongly related to low-code development platforms (LCDPs) [13]. Therefore, low-code development can be considered a tool-based approach [1]. LCDPs are cloud- or on-premise-based. They provide visual tools, predefined components, and customisation and configuration options for low-code development [14, 15]. Low-code development platforms combine multiple traditional and low-code development specific components and therefore include multiple features – supported functionalities and services. Several scientific literature sources [15]–[19] have introduced feature diagrams and lists to compare the commonalities and variabilities of the low-code software development platforms. These platforms may include the following features: Requirement modelling support [15, 18]; Visual development tools [12, 15, 16, 20]; Reusability support [1, 15, 18]; Data source specification mechanisms [16, 17]; Interoperability support [15, 16]; Business logic specification mechanisms [13, 18]; Development automation features [10, 13, 15]; Collaborative development support [18]; Artificial intelligence features [16]; Testing and verification support [15]; Deployment support features [1, 15]; Security support [15]; Lifecycle-management features [11, 15], Analysis environment [1, 15, 18]; Scalability [15]; and other features. To represent LCDPs in the knowledge graph, the *Application Component* element at the *Application layer* of ArchiMate was chosen and populated with the *Application Functions* corresponding to the above-mentioned features.

Any software development methodology can be applied for low-code development, but organisations frequently use fast delivery practices [2, 21]. Typical low-code software development stages are described in papers [11] and [15]: Requirements and feasibility analysis; Data modelling; Definition of the user interface; Implementation of business logic and workflows; Integration of external services; Testing and deployment; Customer feedback; and additional features. Researchers suggest that the Agile methodology conforms to low-code software development [11, 22], as it promotes an iterative, frequent delivery approach with continuous stakeholder involvement. In the Citizen Development Live Cycle model SDLC [23], three paths are possible: fast track (application built completely by citizen developers); assisted path (citizen developers engagement with IT); and IT delivery path (application delivered by IT). Various peer-reviewed sources mention the use of RAD within low-code development (e.g., [10, 24], and [25]). RAD follows an iterative approach, and it seems to work well with low-code development regarding fast reviews, feedback, and changing requirements support, but it may not be ideal for projects with long development times, large teams, or a lack of highly skilled developers [26, 27]. The ArchiMate *Business layer* elements, such as *Process*, *Actor*, *Data object,* and *Event* were used to represent the Low-code development life cycle in the knowledge graph.

There are multiple LCDP segments (general-purpose, process app, database app, request-handling, and mobile app platforms), which enable low-code development across multiple application areas. According to Bucaioni A. et al. [28], 21 application areas have been identified in peer-reviewed and grey studies. The most common are web, mobile, enterprise services, business processes, and IoT. Other areas include healthcare, education, databases, request handling, recommender systems, manufacturing, industrial training, domain-specific language (DSL) engineering, social media, process, marketing, desktop, blockchain, automotive, AI, and aeronautics [28]. Additionally, [12] mentions e-commerce and Extract-Transform-Load applications. IT professionals from various organisations have reported such usage scenarios of low-code development as employee, customer, or partner-facing portals, web-based and mobile applications, replacing legacy systems, and extensions to existing systems [29].

Many research papers point to the benefits of low-code development. For instance: Acceleration of the development cycle [1, 11, 12, 15, 17, 19, 30]–[32]; Involvement of citizen developers [1, 11, 12, 15, 17, 22, 30]; Decreased costs [1, 12, 15, 17, 31]; Increased responsiveness to business and market demands [1, 11, 17]; Lowered maintenance effort [11, 15, 17]; Improved collaboration among the development team and business [12]; Promoted digital innovation [33]; and Mitigation of shadow IT [33]. The *Outcome* element of the *Motivational extension* of the Archimate was used to represent the benefits of low-code development.

The challenges of low-code development were analyzed in our previous publication [14], where 23 low-code development challenges were described, and their possible mitigation approaches were overviewed according to the Agile development phases. Some of these challenges are vendor lock-in, selection of development platforms, and debugging. Low-code development usually includes changes in software development and delivery principles – methods, practices, and approaches. These changes can be intentionally led by an organisation, or can appear as side effects due to platform adoption [2]. To get benefits and meet the challenges the following low-code development principles were identified in the related work: Select the right low-code platform [1, 2, 11, 12, 15, 17]–[19]; Comprehend and master the platform [17, 31, 34]; Embrace visual application development and utilisation of predefined components, elements and templates [12, 15, 22, 31, 34, 35]; Enhance reusability [15, 16, 22, 34]; Embrace platform automation capabilities [1, 15, 16, 22, 31, 34]; Extend the functionality with additional customisation when required [12, 15, 16, 20]; Empower citizen developers and establish a fusion teams approach [14, 15]; Promote IT-business collaboration [2, 10, 11, 35]; Address the knowledge gap of citizen developers [15]; Establish governance [1, 2]; Establish and follow an iterative development lifecycle [1, 11, 15, 35]; Embrace the test-and-learn culture for innovation [2, 20, 32]; and Support changing requirements [17, 32]. These principles were represented as *Principle* elements of ArchiMate *Motivational Extension.*

This section demonstrated a text-based amalgamation of knowledge about low-code development. In the next section, we will demonstrate, how this knowledge and more can be represented in the ArchiMate-based knowledge graph.

4 Amalgamating Low-Code Development Knowledge in ArchiMate-Based Knowledge Graph

The representation of knowledge relevant to low-code development promotion in enterprises was obtained in the following three steps: (1) Mapping the low-code development principles, benefits, and business objectives; (2) Creating a capability view of the low-code development in the ArchiMate language; and (3) Merging the mapping obtained in the first step and the capabilities view obtained in the second step.

The knowledge graph was constructed in the ArchiMate language [3] because it allows a clear and concise representation of various enterprise activities and relationships. Also, low-code development is one of the enterprise activities, and, consequently, such representation can be seamlessly integrated into broader enterprise architecture, ensuring alignment with business strategy and objectives. The relationships, in the model, were identified based on several sub-mappings (obtained by the thematic synthesis method [9]) between the concepts discussed in Sect. 3. The detailed discussion of these sub-mappings is out of the scope of this paper.

The relationships between low-code development principles, benefits, and business objectives were identified based on the literature review and a comprehensive list of business imperatives available in [4]. The imperatives were modelled as *Goal* elements of the ArchiMate *Motivational extension* as the goal element describes an enterprise's intent, direction, or desired end state [5]. Additionally, the *Influence* relationships between benefits themselves were also identified. Then the *Influence* relationships between low-code development principles and benefits were established (Fig. 1).

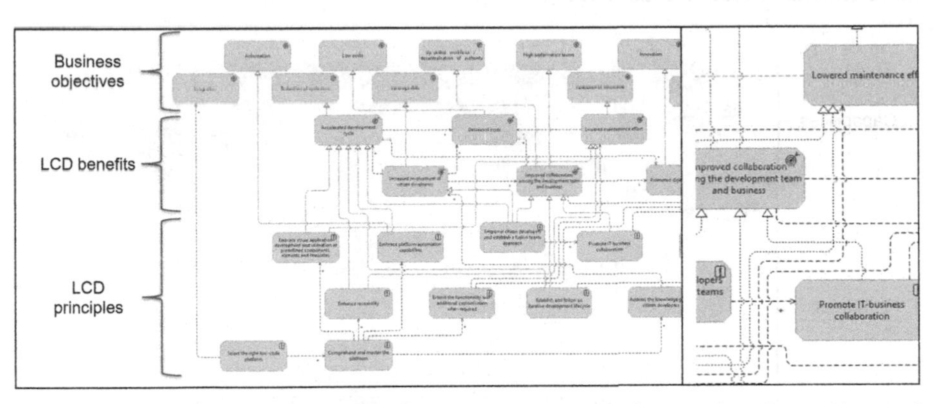

Fig. 1. Relationships between business objectives, low-code development benefits, and low-code development principles.

Then, based on the literature review, the capabilities that represent the abilities of organisations, persons, or systems [3] needed to follow the low-code development principles were derived and represented as ArchiMate's *Strategy layer* elements *Capability*. The capabilities were related to principles with *Realisation* relationships (Fig. 2).

Further, the ArchiMate capabilities planning view was used [5]. This representation was established through the following steps: (i) *Identifying resources*. As a first step,

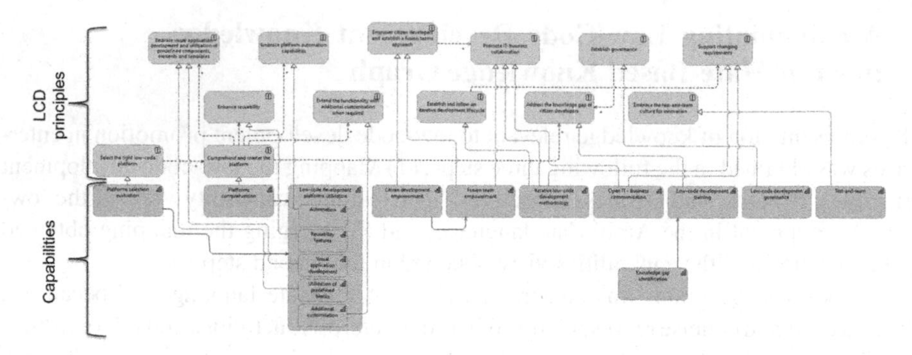

Fig. 2. Low-code development principles and capabilities.

resources were identified based on the literature review. Resources are individual or enterprise-owned tangible, intangible, and human assets [3], and they are modelled with corresponding ArchiMate *Resources* elements at the *Strategy layer*. (ii) *Designing the low-code development lifecycle model.* Amalgamating aspects from the literature review, a low-code development lifecycle model using *Business layer* elements of the ArchiMate modelling language was established. (iii) *Designing a model of a low-code development platform.* Low-code development is a tool-based approach realised on low-code development platforms amalgamating various features as explained in Sect. 3. Amalgamating identified capabilities, resources, a low-code development lifecycle, and the model of LCDP, the knowledge was represented as a capabilities planning view of the ArchiMate modeling language (Fig. 3).

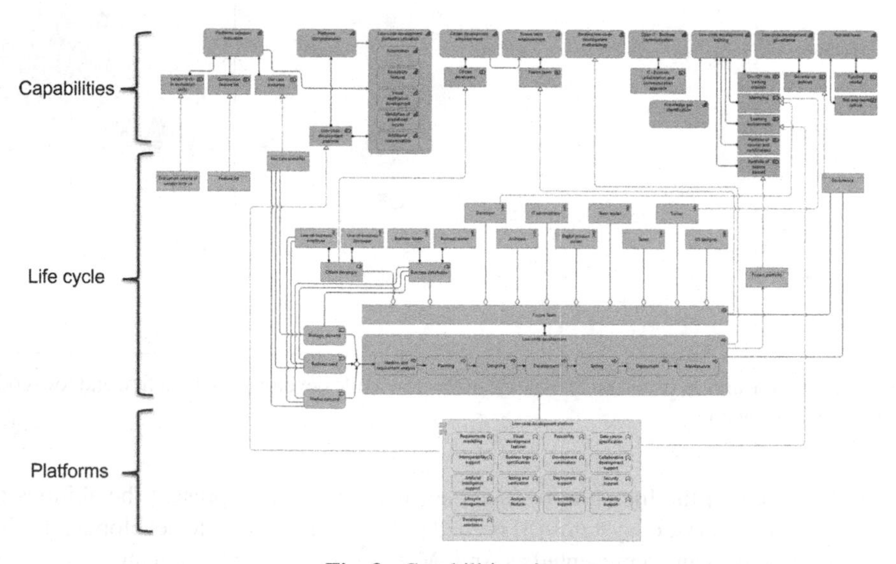

Fig. 3. Capabilities view.

By merging the models discussed above, the low-code development and business objectives knowledge graph, as an ArchiMate model, was obtained. ArchiMate's standardized notation is user-friendly and easy to implement for organisations. It integrates well into enterprise architecture and can be expanded to include more specific practices, actions, and elements related to capabilities, increasing its practical application within organisations. The meta-structure of the model is displayed in Fig. 4. The model can be downloaded from GitHub [36]. The knowledge graph was created in an open-source modelling tool Archi, which provides easy overviewing and browsing features. Relevant details of the elements of the knowledge graph were recorded in the tools "Documentation" field.

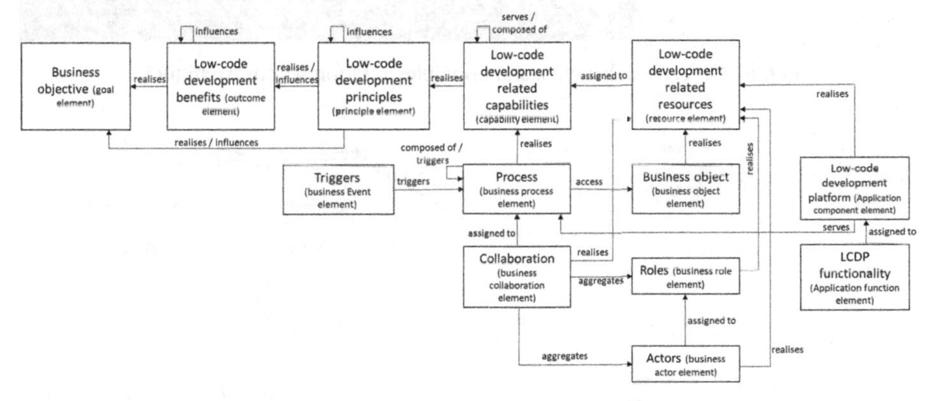

Fig. 4. Meta-structure of the model.

The "Visualiser" feature of the tool allows for tracing element relationships and filtering by depth or by relationship or element type enhancing practical model applicability. In-built drill-down buttons such as "Go into", "Back", and "Home" also help with navigation through the elements of the model [6]. Figure 5 shows an example with the selected business objective "Leverage skills". Related low-code benefits (as outcome elements) can be identified using the Archi Visualiser feature. As shown in Fig. 5, the related benefit is "Increased involvement of citizen developers," and two principles are related to the benefit through "realisation" or "influence" relationships. Principles related to the outcome element through the "realisation" relationship are most important for the realisation of the benefit. In the example, this is the "Empower citizen developers and establish a fusion teams approach" principle.

Additionally, other principles might affect the achievement of a particular benefit and are linked through influence relationships. In the reviewed case, it is the "Address the knowledge gap of citizen developers" principle. These additional impacting principles can be identified as shown in Fig. 6. In Visualiser, particular principles can be viewed, identifying related capabilities and resources. Going further into a particular resource or capability, users can identify related business or application layer elements with relevant descriptions. Returning to the outcome element, users can review other related benefits that influence the realisation of the considered outcome.

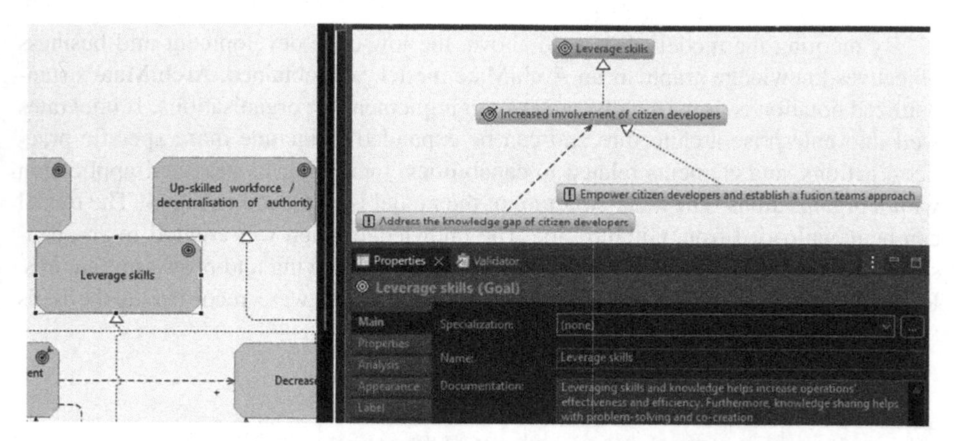

Fig. 5. Objectives, benefits, and principles relation within Archi tool.

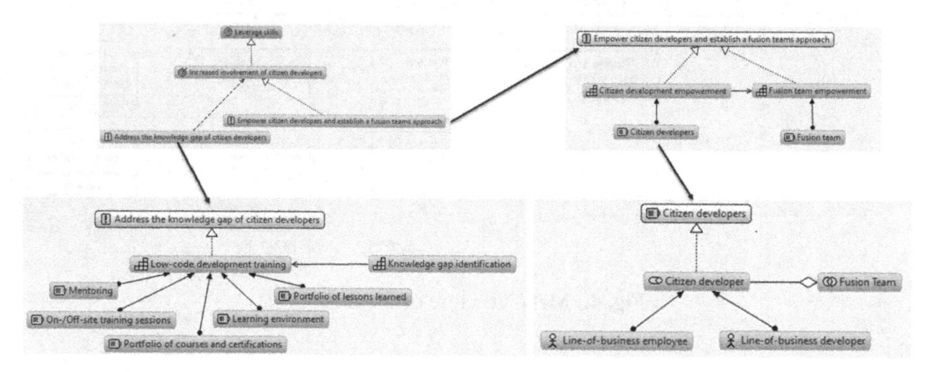

Fig. 6. Navigating the model with Visualizer.

While any navigation route can be beneficial in exploring the knowledge amalgamated in the knowledge graph, for practical use of the graph the sequence represented in Fig. 7 is suggested. This sequence was used when testing the usability of the graph with an enterprise representative as discussed in Sect. 5. The graph representations in Fig. 1–3 and Fig. 5–6 are given at the instance level, while Fig. 7 uses the concepts at the ontology level of the graph, which corresponds to the subset of ArchiMate language elements represented in Fig. 4.

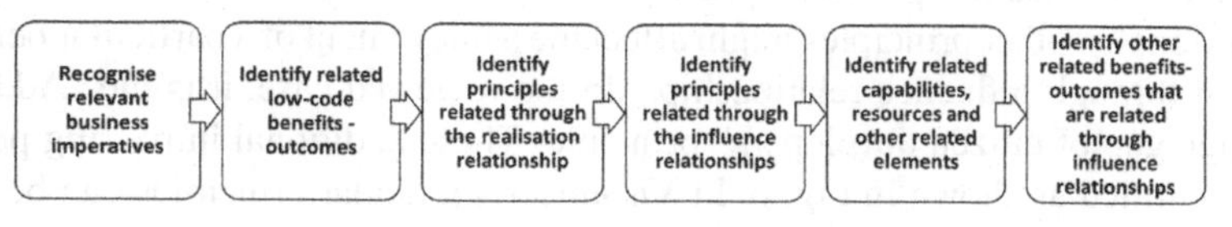

Fig. 7. Suggested navigation sequence.

5 Practical Application of the Proposed Knowledge Graph

The knowledge graph was validated in Company X, an SME whose operating direction is related to developing, importing, and distributing small electronics products. The company currently employs two low-code development platforms – one for a customer-facing mobile application related to smart home devices and another for internal business needs. The focus of validation was on the latter platform, for which the designed knowledge graph was considered. The IT Team realises low-code development on the PowerApps platform and there is no established culture of citizen development within the organisation. However, recent development activities showed close collaboration between IT and business departments, resulting in iterative development and faster feedback from business users.

In the knowledge graph, four identified business imperatives (objectives) led to eight low-code development principles and related capabilities in the validation process. Figure 8 illustrates how the "Reliability of information" objective was explored by the company.

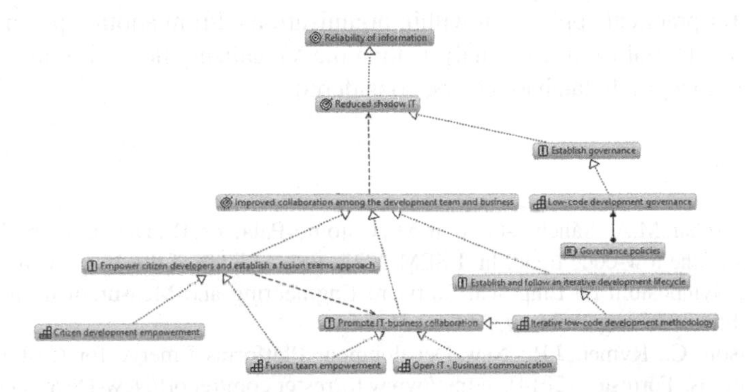

Fig. 8. Knowledge graph's validation: exploring a business objective.

The representative (the chief operating officer) from Company X found the list of identified low-code development aspects to be valuable for evaluating potential focus areas to improve the organisation's ability to reach IT and business objectives. Reviewing the whole model, the representative agreed with the relevance of the objectives, benefits, principles, and other elements included in the model and the established relationships among them. The representative acknowledged that the model helped to identify low-code development aspects that should be evaluated to achieve specific benefits and enhance the achievement of business objectives.

6 Conclusion

In this research in progress, we experimented with ArchiMate language as an ontology level of a knowledge graph. A subset of the ArchiMate language elements where used to create a model (knowledge graph) that represents knowledge on low-code development.

The graph itself was created at the instance level. The results show that the knowledge representation using the ArchiMate-based knowledge graph is applicable for practical use. With its standardised, clear, and concise notation, the ArchiMate-based knowledge graph is easy to implement and is user-friendly for stakeholders.

At the same time, the mentioned benefits are available not just because of the notation. The tool support is very important, as without the visualization possibilities, the stakeholders could easily get lost in a vast number of elements and links in the model (knowledge graph at the instance level). The role of the tool is important also for the change management of the knowledge graph. The model, presented in this paper amalgamates knowledge that was available by 2022; and the factor of subjectivity could not be fully excluded by the methods used in this research. When new knowledge becomes available, or some nodes or links prove to be unnecessary or become not valid, the changes in the model can be relatively easily introduced by the stakeholders. The use of an open-source tool, also, allows for enterprises to customize the model according to their needs.

For future work, the model could be extended, describing capabilities and other related elements with more specific practices, actions, and other details, increasing its usefulness for practical application within organisations. From another perspective, the model has a potential for deeper analysis than the Visualizer allows. For this, an export of the model to a graph database can be considered.

References

1. Rafi, S., Akbar, M.A., Sánchez-Gordón, M., Colomo-Palacios, R.: Devops practitioners' perceptions of the low-code trend. In: ESEM 2022: Proceedings of the 16th ACM/IEEE International Symposium on Empirical Software Engineering and Measurement, pp. 301–306 (2022). https://doi.org/10.1145/3544902.3546635
2. Richardson, C., Rymer, J.R.: New Development Platforms Emerge for Customer-Facing Applications. Forrester (2014). https://www.forrester.com/report/New-Development-Platforms-Emerge-For-CustomerFacing-Applications/RES113411
3. The Open Group. ArchiMate® 3.1 Specification. Homepage (2019). https://pubs.opengroup.org/architecture/archimate3-doc/toc.html
4. Patterson, M.: A structured approach to strategic alignment between business and information technology objectives. South Afr. J. Bus. Manag. **51**(1) (2020). https://doi.org/10.4102/sajbm.v51i1.365
5. Hosiaisluoma, E.: ArchiMate Cookbook Patterns & Examples. Homepage (2019). https://www.hosiaisluoma.fi/blog/
6. Beauvoir, P., Sarrodie, J.-B.: Archi User Guide (2022)
7. Wieringa, R.J.: Design Science Methodology Roel J. Wieringa for Information Systems and Software Engineering. Springer, Heidelberg (2014). https://doi.org/10.1007/978-3-662-43839-8
8. Levy, Y., Ellis, T.J.: A systems approach to conduct an effective literature review in support of information systems research. Inf. Sci. **9**, 181–211(2006). https://doi.org/10.28945/479
9. Cruzes, D.S., Dybå, T.: Recommended steps for thematic synthesis in software engineering. In: International Symposium on Empirical Software Engineering and Measurement, pp. 275–284 (2011). https://doi.org/10.1109/esem.2011.36
10. Waszkowski, R.: Low-code platform for automating business processes in manufacturing. IFAC-PapersOnLine **52**(10), 376–381 (2019). https://doi.org/10.1016/j.ifacol.2019.10.060

11. al Alamin, M.A., Malakar, S., Uddin, G., Afroz, S., Haider, T., bin, Iqbal, A.: An empirical study of developer discussions on low-code software development challenges. In: Proceedings – 2021 IEEE/ACM 18th International Conference on Mining Software Repositories, MSR 2021, pp. 46–57 (2021). https://doi.org/10.1109/MSR52588.2021.00018
12. Luo, Y., Liang, P., Wang, C., Shahin, M., Zhan, J.: Characteristics and challenges of low-code development: the practitioners perspective. In: International Symposium on Empirical Software Engineering and Measurement, pp. 1–11 (2021). https://doi.org/10.1145/3475716.3475782
13. Di Ruscio, D., Kolovos, D., de Lara, J., Pierantonio, A., Tisi, M., Wimmer, M.: Low-code development and model-driven engineering: two sides of the same coin? Softw. Syst. Model. **21**(2), 437–446 (2022). https://doi.org/10.1007/s10270-021-00970-2
14. Rokis, K., Kirikova, M.: Challenges of low-code/no-code software development: a literature review. In: Nazaruka, E., Sandkuhl, K., Seigerroth, E., (eds.), Perspectives in Business Informatics Research. BIR 2022. Lecture Notes in Business Information Processing. Springer, Cham (2022). https://doi.org/10.1007/978-3-031-16947-2_1
15. Sahay, A., Indamutsa, A., di Ruscio, D., Pierantonio, A.: Supporting the understanding and comparison of low-code development platforms. In: Proceedings – 46th Euromicro Conference on Software Engineering and Advanced Applications, SEAA 2020, pp. 171–178 (2020). https://doi.org/10.1109/SEAA51224.2020.00036
16. Bock, A.C., Frank, U.: In search of the essence of low-code: an exploratory study of seven development platforms. In: Companion Proceedings – 24th International Conference on Model-Driven Engineering Languages and Systems, MODELS-C 2021, pp. 57–66 (2021). https://doi.org/10.1109/MODELS-C53483.2021.00016
17. Bock, A.C., Frank, U.: Low-code platform. Bus. Inf. Syst. Eng. **63**(6), 733–740 (2021). https://doi.org/10.1007/s12599-021-00726-8
18. Ihirwe, F., di Ruscio, D., Mazzini, S., Pierini, P., Pierantonio, A.: Low-code engineering for internet of things: a state of research. In: Proceedings – 23rd ACM/IEEE International Conference on Model Driven Engineering Languages and Systems, MODELS-C 2020 – Companion Proceedings, pp. 522–529 (2020). https://doi.org/10.1145/3417990.3420208
19. Oteyo, I.N., Pupo, A.L.S., Zaman, J., Kimani, S., de Meuter, W., Boix, E.G.: Building smart agriculture applications using low-code tools: the case for DisCoPar. In: IEEE AFRICON Conference, pp. 1-6 (2021). https://doi.org/10.1109/AFRICON51333.2021.9570936
20. Richardson, C., Rymer, J.R.: Vendor Landscape: The Fractured, Fertile Terrain of Low-Code Application Platforms the Landscape Reflects a Market in Its Formative Years. Forrester (2016). https://www.forrester.com/report/Vendor-Landscape-The-Fractured-Fertile-Terrain-Of-LowCode-Application-Platforms/RES122549
21. da Cruz, M.A.A., de Paula, H.T.L., Caputo, B.P.G., Mafra, S.B., Lorenz, P., Rodrigues, J.J.P.C.: OLP – a restful open low-code platform. Future Internet **13**(10), 249 (2021). https://doi.org/10.3390/fi13100249
22. Khorram, F., Mottu, J.M., Sunyé, G.: Challenges & opportunities in low-code testing. In: Proceedings – 23rd ACM/IEEE International Conference on Model Driven Engineering Languages and Systems, MODELS-C 2020 – Companion Proceedings, pp. 490–499 (2020). https://doi.org/10.1145/3417990.3420204
23. Project Management Institute. Citizen development: the handbook for creators and change makers (2021)
24. di Sipio, C., di Ruscio, D., Nguyen, P.T.: Democratising the development of recommender systems by means of low-code platforms. In: Proceedings – 23rd ACM/IEEE International Conference on Model Driven Engineering Languages and Systems, MODELS-C 2020 – Companion Proceedings, pp. 471–479 (2020). https://doi.org/10.1145/3417990.3420202

25. Jacinto, A., Lourenço, M., Ferreira, C.: Test mocks for low-code applications built with OutSystems. In: Proceedings – 23rd ACM/IEEE International Conference on Model Driven Engineering Languages and Systems, MODELS-C 2020 – Companion Proceedings, pp. 530–534 (2020). https://doi.org/10.1145/3417990.3420209
26. OutSystems. What Is Rapid Application Development? Homepage. https://www.outsystems.com/glossary/what-is-rapid-application-development/. Accessed 11 Jan 2022
27. Kissflow Inc. Rapid Application Development (RAD) Model: An Ultimate Guide for App Developers in 2022. Homepage. https://kissflow.com/application-development/rad/rapid-application-development/
28. Bucaioni, A., Cicchetti, A., Ciccozzi, F.: Modelling in low-code development: a multi-vocal systematic review. Softw. Syst. Model. **21**, 1959–1981 (2022). https://doi.org/10.1007/s10270-021-00964-0
29. OutSystems. The State of Application Development. Homepage. https://www.outsystems.com/1/state-app-development-trends/
30. Lethbridge, T.C.: Low-code is often high-code, so we must design low-code platforms to enable proper software engineering. In: Margaria, T., Steffen, B. (eds) Leveraging Applications of Formal Methods, Verification and Validation. ISoLA 2021. LNCS, vol. 13036, pp. 202–212. Springer, Cham (2021). https://doi.org/10.1007/978-3-030-89159-6_14
31. Martins, R., Caldeira, F., Sa, F., Abbasi, M., Martins, P.: An overview on how to develop a low-codeapplication using OutSystems. In: 2020 International Conference on Smart Technologies in Computing, Electrical and Electronics (ICSTCEE), Bengaluru, India, 2020, pp. 395–401 (2020). https://doi.org/10.1109/ICSTCEE49637.2020.9277404
32. Krishnaraj, N., Vidhya, R., Shankar, R., Shruthi, N.: Comparative study on various low code business process management platforms. In: 5th International Conference on Inventive Computation Technologies, ICICT 2022 – Proceedings, pp. 591–596 (2022). https://doi.org/10.1109/ICICT54344.2022.9850581
33. Sanchis, R., García-Perales, Ó., Fraile, F., Poler, R.: Low-code as enabler of digital transformation in manufacturing industry. Appl. Sci. **10**(1), 12 (2020). https://doi.org/10.3390/app10010012
34. Bexiga, M., Garbatov, S., Seco, J.C.: Closing the gap between designers and developers in a low code ecosystem. In: Proceedings – 23rd ACM/IEEE International Conference on Model Driven Engineering Languages and Systems, MODELS-C 2020 – Companion Proceedings, pp. 413–422 (2020). https://doi.org/10.1145/3417990.3420195
35. Saay, S., Margaria, T.: Model-driven-design of NREn bridging application: case study AfgREN. In: Proceedings – 2020 IEEE 44th Annual Computers, Software, and Applications Conference, COMPSAC 2020, pp. 1522–1527 (2020). https://doi.org/10.1109/COMPSAC48688.2020.00-39
36. Rokis, K.: ArchiMate-based low-code development model. https://github.com/KarlisRokis/LCD-and-Business-Objectives-Relation-Model

A Knowledge Graph to Analyze Clinical Patient Data

Mariella Bonomo[1,2(✉)], Francesco Ippolito[1], and Salvatore Morfea[2]

[1] Department of Mathematics and Computer Science, University of Palermo, Palermo, Italy
{mariella.bonomo,francesco.ippolito07}@community.unipa.com
[2] Kazaam Lab S.r.l., Palermo, Italy
info@kazaamlab.com

Abstract. Knowledge graphs may be successfully applied to represent different types of relationships among different types of subjects. Here, we propose a Knowledge Graph model to represent patient data coming from clinical folders information. The main aim is to provide suitable classifications of patients, in order to allow a deeper understanding of possible (side) effects that the same treatment may cause on different patients. We have implemented our model using the Neo4J NoSQL database and we present some preliminary analysis, as an example of how our model can be usefully exploited.

Keywords: Clinical Folders · Patient Data · Graph Analysis

1 Introduction

The complexity of biological systems often makes difficult the full understanding of occurrence and progress of complex diseases. It has been recognized that combining clinical and molecular data can provide a deeper understanding of individuals' disease phenotypes, and reveals candidate markers of prognosis and/or treatment [15]. Network and graphs have emerged as powerful models to represent and study the relationships among different biological players, such as for example the physical interaction between proteins involved in the same processes (e.g., [1,9,14]), or the gene co-expression used to discriminate ill from health populations [6]. Interestingly, biological graphs have been applied also for the study of human phenotypes [11], phenotype stratification [2], and for the analysis of genotype-phenotype effects in complex diseases [12].

Due to the fact that biological data increase exponentially, novel approaches based on big data technologies in order to allow their analysis in the distributed started to be proposed (e.g., [7,8]), even with references to graph based analysis [5]. On the other hand, often the type of nodes and edges that need to be included in the analysis may present a sort of heterogeneity that is difficult to manage. As an example, phenotype characteristics of patients should be combined with genotype features, and clinical data on diagnosis and therapies need to be taken into account as well.

A. Abelló et al. (Eds.): ADBIS 2023, CCIS 1850, pp. 477–484, 2023.
https://doi.org/10.1007/978-3-031-42941-5_41

Knowledge graphs are useful models allowing to represent different types of associations among different types of subjects. In the last few years, they have started to be successfully applied also in the biological context [3,10,15]. Here, we propose a Knowledge Graph model to represent patient data coming from clinical folders information. The main aim is to provide suitable classifications of patients, in order to allow a deeper understanding of possible (side) effects that the same treatment may cause on different patients.

We have implemented our model using the Neo4J NoSQL database [16] and we present an example of possible types of analysis that can be performed on it.

2 The Proposed Knowledge Graph

The Proposed Knowledge Graph (KG) represents patient data retrieved from clinical folders information, according to the following schema. *Patient* contains the information and demographic data, *treatment* contains patient's conditions or diagnoses, *Allergy* contains patient's allergy data, *disease* contains patient's disease data. Below the proposed KG schema is described throught the corresponding set of predicates, indicated as tuples of the form ⟨*subject1* - relationship - *subject2*⟩, while Figure 1 illustrates it graphically.

⟨*patient* - follows - *treatment*⟩
⟨*patient* - is affected by - *disease*⟩
⟨*disease* - is associated to - *treatment*⟩
⟨*patient* - has developed - *allergy*⟩
⟨*allergy* - is associated to - *treatment*⟩

Table 1 shows the list of attributes associated to each subject. Attributes on the *Patient* subject represent patient demographics as city, state, country, zip and other information like birth date and address. *Disease* subject is represented by attributes related to the disease and the patient.
The diagnostic process of a disease determines the attributes of *Treatments.* Firstly, a patient has a health problem/disease/allergy, he chooses to engage with the health care system. Consequently, the patient seeks health care, tests a process of information gathering, determining a working diagnosis as start and stop of treatment assigned to the diseases and/or allergy about patients. Finally *Allergy* subject contains information on the begin and/or stop of occurrence of an allergy on the patient, different types of the reaction, severity and disease linked to the patient, treatment associated to it. Most allergies can't be cured, but some treatments are connected to it. We extracted the patients with an allergy persistent in which the attribute stop will be null.

3 Implementation and Analysis

Input Datasets. The medical data is built from [17]. We have 21,795 nodes and total number of edges that represent the relationships between these nodes

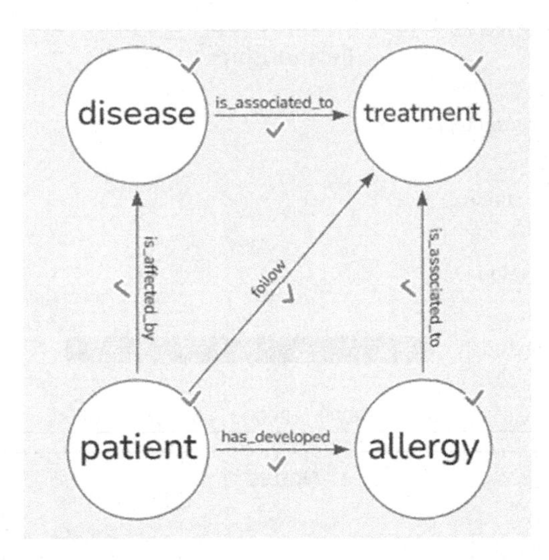

Fig. 1. Representation of a segment of the schema of KG.

Table 1. The attributes assigned to subjects in the previous schema.

Subject	Attributes
Patient	id, birthDate,deathDate, drivers, first, last, marital, race, ethnicity, gender, bithPlace, address, city, state, country, zip
Treatment	start, stop, patient, encounter, code, description
Disease	id, start, stop, patient, encounter, code, description, reasonCode, reasonDescription
Allergy	start, stop, patient, encounter, code, system, description, type, category, reaction, description, type, category, reaction1, description1, severity

are 22, 490. In Fig.2 we show the distribution of the nodes and relationships. The medical data consist of 597 patients, 20, 955 diseases, 97 different types of allergies and 146 treatments for the diagnosis.
In Fig. 3 we show the portion of significative Knowledge graph contructed, where the patients that follow *encounter for symptom*, they are affected by *cystitis and recurrent urinary tract infection, localized primary osteoarthritis of the hand* , in general we have 184 different patients with these characteristics.

Big Data Technologies. The database to properly store and analyze the proposed KG is built using Neo4j [13], an enterprise-strength graph database that combines native graph storage, advanced security, scalable speed-optimized architecture, and associated Cypher Query Language that is a declarative graph query language aiming to be intuitive and human-readable. Nodes, relationships, and properties are described using ascii-art.

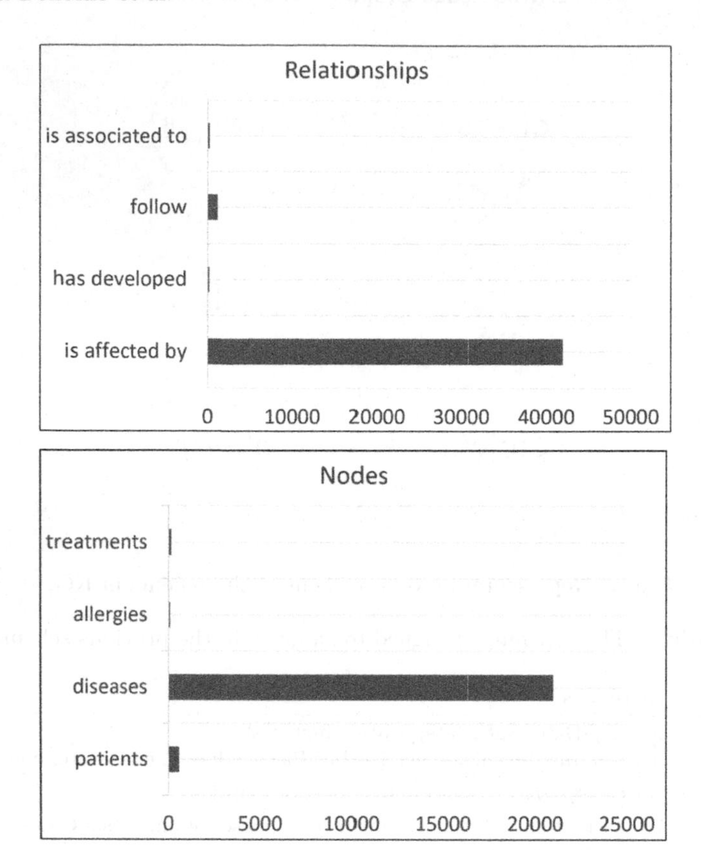

Fig. 2. The Distributions of nodes and relationships on neo4j.

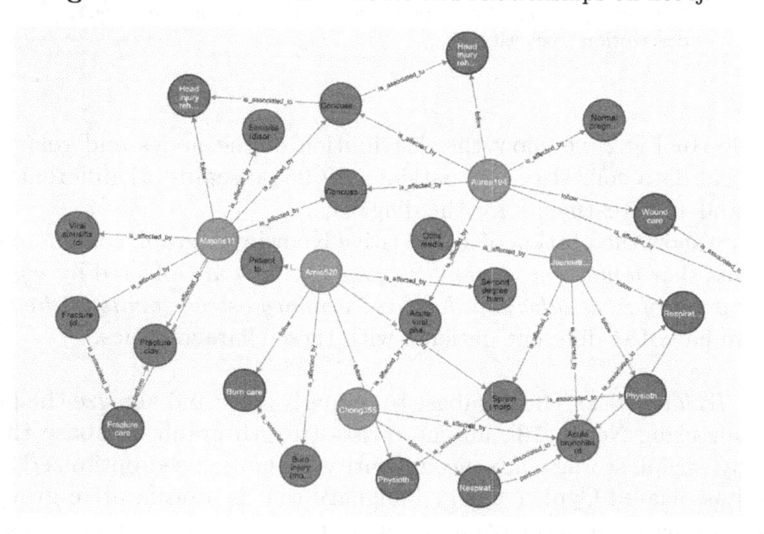

Fig. 3. A small portion of the constructed Knowledge Graph. We show the diseases with blue color, the patients with violet color and the treatments with orange color.

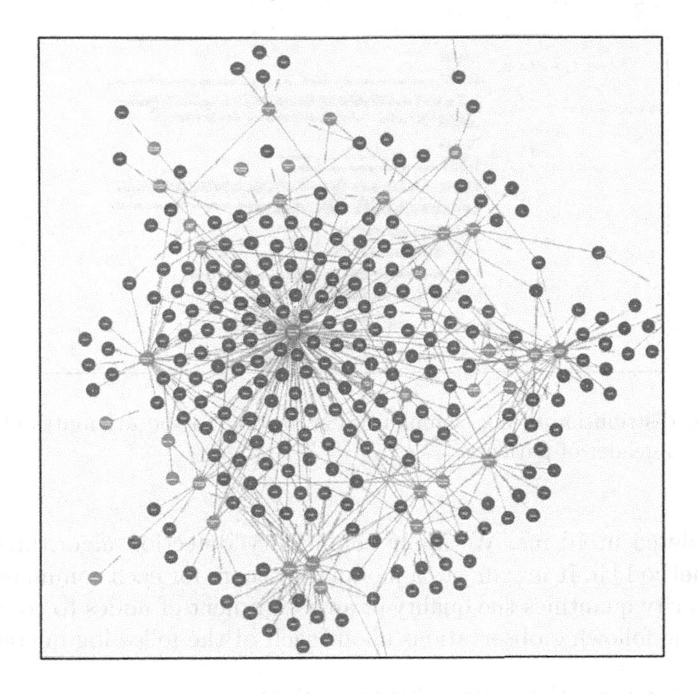

Fig. 4. Example of query q1: a community of patients grouped by common diseases.

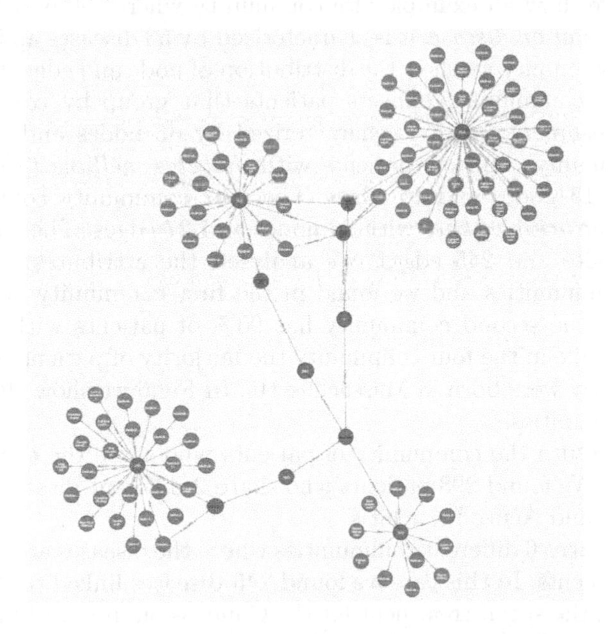

Fig. 5. Example of query q3: the community of diseases grouped by treatments.

Preliminary Analysis. In this section we show the preliminary analysis, where we analysed many queries that have for both their result count and their execution

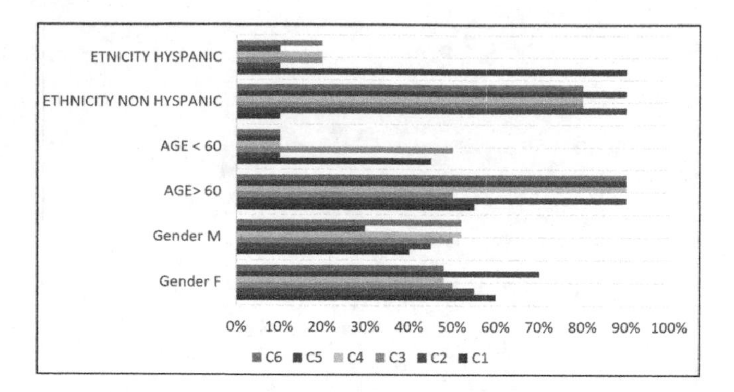

Fig. 6. The distribution of six communities according to the attributes of ethnicity, birthdate and gender of patients.

time completed in 10 ms. We apply community detection algorithms such as Louvain method [4]. It maximizes a modularity score for each community, where the modularity quantifies the quality of an assignment of nodes to communities. We make the following observations about each of the following queries:

- *Query q1:* returns the communities of patients grouped by common diseases. In Fig. 4 we show an example of a community where the patients are linked to *facial laceration disease*, it is characterized by 53 diseases and 257 patients. For each community we show the distribution of node and edge linked to them. The second community connects patients that group by *concussion injury of brain (disorder)* and it is characterized by 43 nodes and 44 edges. The third community connects patients with *diabetes mellitus type 2 (disorder)* and it has 15 nodes and 16 edges. The four community connects patients with *Osteoarthritis of knee* with 26 nodes and 27 edges. The five community has 271 nodes and 245 edges. We analysed the attributes of the patients of these communities and we found in the first community 60% female sex of patients, the second community has 90% of patients with non hyspanic ethnicity; while in the four community the majority of patients was born over 1960 and they were born in Massachusetts. In Fig.6 we show the distribution of the communities.
- *Query q2:* return the community of patients who share the common number of diseases. We found 298 patients who share 2 different diseases: Acute viral pharyngitis and Acute bronchitis.
- *Query 3:* returns 6 different communities where the diseases are linked for each set of treatments. In this case we found 245 diseases linked to 26 treatments, for example the specif treatment for the Concussion injury of brain and Concussion with no loss of consciousness is emergency urgent room admission. In Fig. 5 we show the portion of this graph.

4 Conclusion

We have proposed a Knowledge Graph based model implemented upon the Neo4J NoSQL database. We have presented an example of possible types of analysis that can be performed on it, with the main aim of providing suitable classifications of patients, in order to allow a deeper understanding of possible (side) effects that the same treatment may cause on different patients.

In the future, we plan to extend the proposed Knowledge Graph model by integrating new predicates that involve for example physical interactions between molecular components [1] and genotype-phenotype effects [12], in order to enrich the analysis which can be performed and to provide insights on the occurrence and progress of complex diseases.

Funding

Part of the research presented here has been funded by the research projects: "Multicriteria Data Structures and Algorithms: from compressed to learned indexes, and beyond" (2017WR7SHH) funded by MIUR-PRIN; "Modelling and analysis of big knowledge graphs for web and medical problem solving" (2022, E55F22000270001) and "Computational approaches for decision support in precision medicine" (2023, E53C22001930001), both funded by INdAM GNCS; "AMABILE - Amarelli BIg data and bLockchain Enterprise platform" (CUP: B76G20000880005) funded by the Italian Ministry of Economic Development.

References

1. Ahn, Y., Bagrow, J., Lehmann, S.: Link communities reveal multiscale complexity in networks. Nature **466**, 761–764 (2010)
2. Ben-Hamo, R., Gidoni, M., Efroni, S.: PhenoNet: identification of key networks associated with disease phenotype. Bioinformatics **30**(17), 2399–2405 (2014)
3. Bonomo, M.: Knowledge extraction from biological and social graphs. In: Chiusano, S., et al. (eds.) New Trends in Database and Information Systems - ADBIS 2022 Short Papers, Doctoral Consortium and Workshops: DOING, K-GALS, MADEISD, MegaData, SWODCH, Turin, Italy, September 5–8, 2022, Proceedings. Communications in Computer and Information Science, vol. 1652, pp. 648–656. Springer (2022). https://doi.org/10.1007/978-3-031-15743-1_60
4. De Meo, P., Ferrara, E., Fiumara, G., Provetti, A.: Generalized louvain method for community detection in large networks. In: 2011 11th International Conference on Intelligent Systems Design and Applications, pp. 88–93. IEEE (2011)
5. Di Rocco, L., Ferraro Petrillo, U., Rombo, S.E.: DIAMIN: a software library for the distributed analysis of large-scale molecular interaction networks. BMC Bioinf. **23**(1), 474 (2022)
6. Fassetti, F., Rombo, S.E., Serrao, C.: Discriminative pattern discovery for the characterization of different network populations. Bioinformatics **39**(4) (2023)
7. Ferraro Petrillo, U., Palini, F., Cattaneo, G., Giancarlo, R.: Alignment-free genomic analysis via a big data spark platform. Bioinformatics **37**(12), 1658–1665 (2021)

8. Ferraro Petrillo, U., Sorella, M., Cattaneo, G., Giancarlo, R., Rombo, S.E.: Analyzing big datasets of genomic sequences: fast and scalable collection of k-mer statistics. BMC Bioinf. **20-S**(4), 138:1–138:14 (2019)
9. Fionda, V., Palopoli, L., Panni, S., Rombo, S.E.: Protein-protein interaction network querying by a "Focus and Zoom" approach. In: Elloumi, M., Küng, J., Linial, M., Murphy, R.F., Schneider, K., Toma, C. (eds.) BIRD 2008. CCIS, vol. 13, pp. 331–346. Springer, Heidelberg (2008). https://doi.org/10.1007/978-3-540-70600-7_25
10. Galluzzo, Y.: A review: Biological insights on knowledge graphs. In: Chiusano, S., et al. (eds.) New Trends in Database and Information Systems - ADBIS 2022 Short Papers, Doctoral Consortium and Workshops: DOING, K-GALS, MADEISD, MegaData, SWODCH, Turin, Italy, September 5–8, 2022, Proceedings. Communications in Computer and Information Science, vol. 1652, pp. 388–399. Springer (2022). https://doi.org/10.1007/978-3-031-15743-1_36
11. Hidalgo, C.A., Blumm, N., Barabási, A., Christakis, N.A.: A dynamic network approach for the study of human phenotypes. PLoS Comput. Biol. **5**(4) (2009)
12. Kim, Y., Cho, D., Przytycka, T.M.: Understanding genotype-phenotype effects in cancer via network approaches. PLoS Comput. Biol. **12**(3) (2016)
13. Miller, J.J.: Graph database applications and concepts with neo4j. In: Proceedings of the southern association for information systems conference, Atlanta, GA, USA. vol. 2324 (2013)
14. Pizzuti, C., Rombo, S.E.: An evolutionary restricted neighborhood search clustering approach for PPI networks. Neurocomputing **145**, 53–61 (2014)
15. Santos, A., Colaço, A., Nielsen, A., et al.: A knowledge graph to interpret clinical proteomics data. Nat. Biotech. **40**, 692–702 (2022)
16. Van Bruggen, R.: Learning Neo4j. Packt Publishing Ltd, Birmingham (2014)
17. Walonoski, J., et al.: Synthea: An approach, method, and software mechanism for generating synthetic patients and the synthetic electronic health care record. J. Am. Med. Inf. Assoc. **25**(3), 230–238 (2018)

MADEISD: 5th Workshop on Modern Approaches in Data Engineering, Information System Design

Towards Automatic Conceptual Database Design Based on Heterogeneous Source Artifacts

Goran Banjac(✉), Drazen Brdjanin, and Danijela Banjac

Faculty of Electrical Engineering, University of Banja Luka, Patre 5, 78000 Banja Luka, Bosnia and Herzegovina
{goran.banjac,drazen.brdjanin,danijela.banjac}@etf.unibl.org

Abstract. The paper presents an early prototype of the tool named DBomnia – the first online web-based tool enabling automatic derivation of conceptual database models from heterogeneous source artifacts (business process models and textual specifications). DBomnia employs other pre-existing tools to derive conceptual models from sources of the same type and then integrates those models, whereby the main challenge is related to the integration of conceptual models that are automatically generated and can be considered unreliable. The initial results show that the implemented tool derives models that are more precise and complete compared to the models derived from sources of the same type.

Keywords: AMADEOS · Conceptual schema · Schema integration · Schema matching · TexToData · UML class diagram · Uncertain schema

1 Introduction

The database design process undergoes several typical steps [14], whereby the first and the most important step is conceptual design. The result of the conceptual database design is the *conceptual database model* (CDM), which is platform independent and describes the target database on a high level of abstraction. The result of all subsequent steps, including the database specification for the specific target database management system, can be straightforwardly derived starting from the CDM. This is the main reason why researchers have been significantly interested in the topic of automated CDM design.

Automated CDM design is a research topic since the 1980 s [11], and since then a plethora of papers have been published in the field. Although different types of artifacts (models, textual specifications, recorded speech, etc.) have been introduced as a source in the automated CDM design, the existing tools enable CDM synthesis based on sources of one single type only. Since there is still no tool enabling fully automatic generation of the complete target CDM, we started to investigate the possibilities of increasing the completeness and correctness of automatically generated CDMs by deriving them from heterogeneous source artifacts (compared to the CDMs derived from sources of one single type). Our research goal is to define an approach and implement a tool enabling automatic CDM derivation from a set of heterogeneous source artifacts.

A. Abelló et al. (Eds.): ADBIS 2023, CCIS 1850, pp. 487–498, 2023.
https://doi.org/10.1007/978-3-031-42941-5_42

This paper presents an early prototype of a tool named DBomnia which enables automatic CDM synthesis based on two different types of source artifacts: *business process models* (BPMs) and *textual specifications.* DBomnia employs pre-existing tools to generate CDMs based on sources of the same type and then integrates those CDMs into a single unified CDM. While the state of the art in the schema integration field is the integration of *reliable* schemas (i.e. schemas derived from reliable sources), our task is to integrate *unreliable* schemas (i.e. schemas automatically generated by different tools based on different source artifacts) which are not 100% complete nor 100% correct.

The paper is structured as follows. After this introduction, the second section presents the related work. The third section presents an illustrative example that motivates our research. The fourth section presents the approach and still open issues. The fifth section presents the implemented tool, while the sixth section illustrates its usage. The final section concludes the paper.

2 Related Work

This section presents the related work. Firstly, we provide an overview of the existing approaches and tools to (semi-)automatic CDM design, followed by an overview of the existing approaches to schema matching and integration.

2.1 (Semi-)automatic CDM Design

The existing approaches and tools to (semi-)automatic CDM design derive the target model from the source artifacts of the same type, and can be classified as: *Text-based*, *Model-based*, *Form-based*, and *Speech-based.*

Text-based approaches constitute the oldest and most dominant category. These approaches and tools derive CDMs from textual specifications that are typically unstructured and represented in some *natural language* (NL). They can be further classified (as suggested in [32]) as: (1) *Linguistics-based*, (2) *Pattern-based*, (3) *Case-based*, (4) *Ontology-based*, and (5) *Multiple approaches.*

Most text-based approaches and tools are *linguistics-based.* They use *natural language processing* (NLP) techniques to convert NL text into the CDM. The development of these approaches started with Chen's eleven rules [11] for the translation of English text into the corresponding CDM, which have been further enhanced and extended in [16,24,25]. The most important *linguistics-based* tools are: ER-Converter [24], CM-Builder [15], and LIDA [25]. The main representatives of other categories are: *pattern-based* APSARA [27], *case-based* CABSYDD [13], *ontology-based* OMDDE [30], and HBT [32] belonging to the *multiple approaches.*

The existing text-based tools typically support one single source NL (mainly English) and do not provide multilingual support. Only TexToData [10] enables automatic CDM derivation from textual specifications in different source NLs, even with very complex morphology (e.g. Slavic languages). In our approach presented in this paper, we employ TexToData as one of the generators of uncertain CDMs in the multi-sourced automatic CDM synthesis.

Model-based approaches and tools emerged as an alternative to those that are text-based, in order to avoid their shortcomings mainly related to the modest effectiveness for languages with complex morphology. The existing approaches and tools take source models that can be represented by a number of different notations, which can be classified (according to [6]) as: (1) *process-oriented* (e.g. BPMN), (2) *function-oriented* (e.g. Data Flow Diagram), (3) *communication-oriented* (e.g. Sequence Diagram), and (4) *goal-oriented* (e.g. TROPOS).

There is still no model-based approach nor tool enabling automatic derivation of the complete target CDM from a source model (regardless of the notation). Only a few papers present a set of formal rules for automatic CDM derivation (e.g. [5,31]), while the majority give only guidelines and informal rules that do not enable automatic CDM derivation. Most of the proposed tools are actually transformation programs (e.g. [18,28]) specified in some model-to-model transformation language (e.g. ATL [17]), while only a small number of papers present real CASE tools for automatic model-driven CDM synthesis (e.g. [8,23]).

There is only one single online model-driven tool named AMADEOS [8], which enables automatic derivation of an initial CDM from a set of BPMs. The most recent AMADEOS release [29] supports the whole BPM-driven forward database engineering process by using the standard UML notation in all stages. In our approach presented in this paper, we employ AMADEOS as one of the generators of uncertain CDMs in the multi-sourced automatic CDM synthesis.

Form-based approaches take *collections of forms* as the source, whereby the most important tools are EDDS [12] and IIS*Case [19].

Speech-based approaches constitute the smallest category – in the existing literature, there is only one paper [9] presenting the SpeeD tool that is able to generate CDM from recorded English speech.

In comparison to the existing approaches that consider only sources of the same type, in this paper we consider automatic CDM derivation from heterogeneous source artifacts. In comparison to the existing tools which enable (semi-) automatic CDM derivation from sources of the same type, in this paper we present a prototype named DBomnia that enables automatic CDM derivation from two different types of source artifacts: textual specifications and BPMs.

2.2 Schema Matching and Integration

In the essence of our research objective is the problem of *uncertain* schema integration, i.e. integration of CDMs that are automatically generated by the tools that do not generate 100% correct nor 100% complete models.

Schema integration is a research topic since the 1980 s. It has been heavily investigated and is part of the research to this day. The term schema integration, introduced in [2], is defined as the activity of integrating database schemas. It is a generic term that is used in two contexts. In the database design context it means to produce a global conceptual schema of the proposed database, and in the distributed database management context it means to produce a global schema that represents a virtual view of all databases in such an environment.

Schema integration implies the creation of a unified representation of the schemas from different sources in order to combine data from those sources. It consists of two core parts: *schema matching* and *schema merging*. Schema matching is the process of discovering mappings (correspondences) between different schemas, while schema merging is the process of the unification of different schemas based on the discovered mappings [21].

In the 2000 s researchers started to investigate the topic of generic model management. In [4] authors introduce the term *model management* and propose to generalize and integrate model management operations (such as matching and merging) to support generic model management. The term *model* is used to refer to any structured representation of the data (e.g. relational schema, XML, etc.). After that, numerous papers emerged focusing on generic model management operations, such as generic matching (e.g. [20]) and generic merging (e.g. [26]).

Although some papers deal with the uncertainty that is inherent in the schema matching step [21], the state of the art is the integration of the source schemas that can be described as *reliable*. Reliable source schemas are the product of reliable sources (e.g. when integrating schemas of the databases already in use). In this paper, we deal with the integration of the models with reduced reliability, because tools for automated CDM derivation do not generate 100% correct nor 100% complete models.

3 Motivating Example

This section presents an example that motivates our current research.

As already stated, the existing tools for automated CDM design are not able to automatically generate CDMs that are 100% correct and 100% complete, but our research objective is to try to maximize the correctness and completeness of the CDM by integrating CDMs derived from different sources. In our case, we use CDMs automatically derived from a set of BPMs by AMADEOS[1], and CDMs automatically derived from textual specifications by TexToData[2].

In order to illustrate AMADEOS' capabilities, we prepared a sample source set of BPMs (Fig. 1), which represent two main processes in the Online Library – *Book Borrowing* and *Book Returning*. In the first process: a member creates a borrowing request for a copy of some book edition; a librarian registers the borrowing and issues a book. In the second process: the member returns borrowed book, and the librarian registers returned book.

Figure 2 shows the CDM derived from the sample set of BPMs. From the perspective of a database designer of the target CDM: (1) all classes and seven associations could be evaluated as correctly generated; (2) three associations could be evaluated as surplus (*Librarian-RegisterReturnedBook-Book*, *Librarian-ReceiveMessage_BookRequest-BookRequest*, *Member-SendMessage_Book-Book*), and (3) two associations are missing in the generated CDM (*BookEdition↔Book*, and *Librarian↔Book_Borrowed* that represents the returning of the book).

[1] http://m-lab.etf.unibl.org:8080/amadeos.
[2] http://m-lab.etf.unibl.org:8080/Textodata.

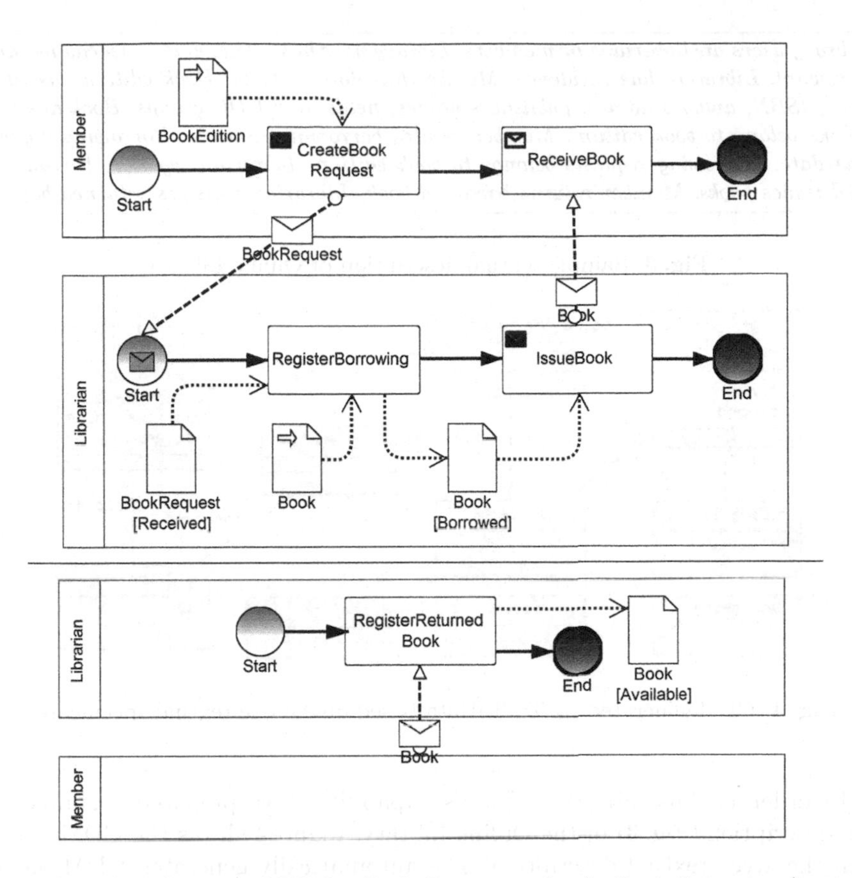

Fig. 1. Sample collection of BPMs: *Book Borrowing* (top), *Book Returning* (bottom)

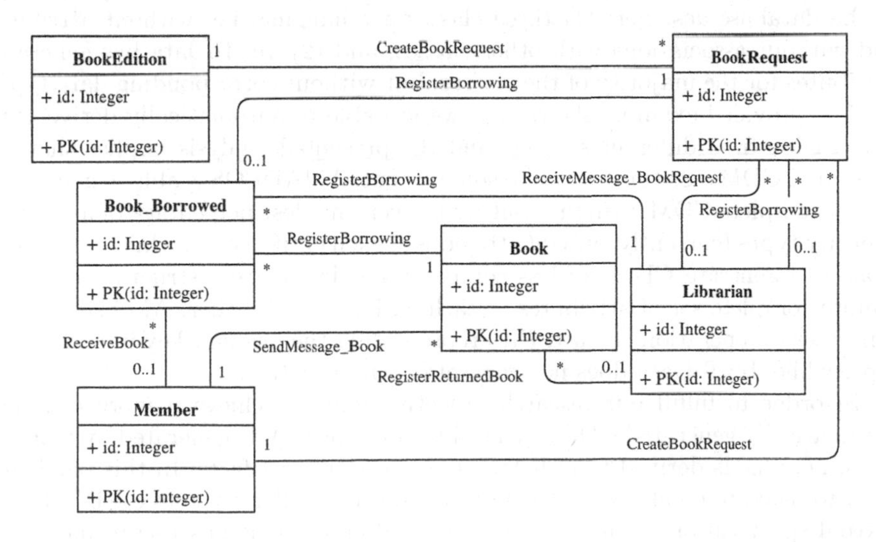

Fig. 2. CDM generated by AMADEOS based on sample collection of BPMs

Library users are librarians or members. Library user has name, email, username, and password. Librarian has residence. Member has date of birth. Book edition has title, year, ISBN, authors names, publishers names, fields, and UDC groups. Book has tag. Books belong to book edition. Member creates borrowing requests. Borrowing request has date. Borrowing requests belongs to book edition. Librarian registers borrowings and issues books. Member returns borrowed book. Librarian registers returned book.

Fig. 3. Sample textual description of Online Library

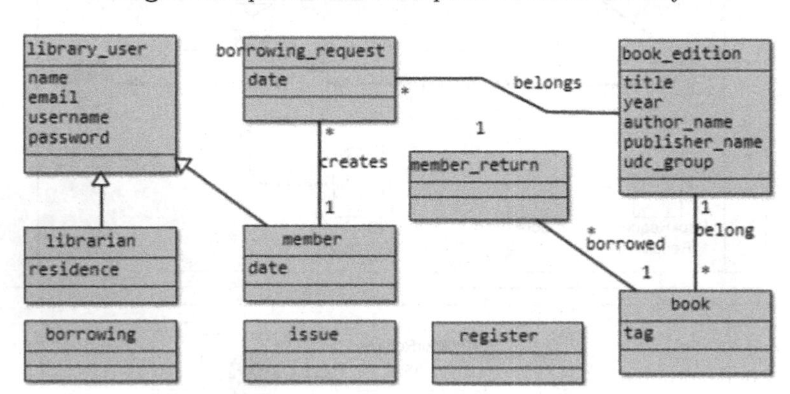

Fig. 4. CDM generated by TexToData based on sample textual specification

In order to illustrate TexToData's capabilities, we prepared a simple textual description (Fig. 3) of the Online Library. Figure 4 shows the CDM derived from the given textual description. The automatically generated CDM has ten classes, two generalizations, and only four associations. From the perspective of the database designer: (1) three classes are hanging, i.e. without attributes and without associations with other classes, and (2) TexToData has generated attributes for the majority of the classes, but without corresponding data types.

The provided example shows that we are able to automatically derive different CDMs from different sources, and the presented analysis implies that the generated CDMs are (un)certain to some extent. AMADEOS is able to generate a highly complete CDM structure but with a very modest percentage of attributes in entity types (currently, only *id* attribute in each entity type), while TexToData is able to generate CDMs of less complete and less correct structure but with a more complete set of attributes in each entity type. Further, AMADEOS also generates an operation (named *PK*) representing the primary key in each entity type, while TexToData does not generate such operations.

In order to fulfill our research objective and to achieve a more complete automatically generated CDM, we need to combine CDMs generated by different tools, i.e. CDMs derived from heterogeneous source artifacts. In this particular case, we need to combine CDMs derived from two different sources (BPMs and textual specifications), but in a more general case – it is possible to have even more different sources (e.g. recorded speech, ontologies, etc.).

4 Approach and Open Issues

Out task is to integrate multiple (unreliable) CDMs into a single unified CDM. Currently, we have only two types of source artifacts (BPMs and textual specifications) so we focus on the integration of two CDMs – CDM derived from BPMs by AMADEOS and CDM derived from textual specifications by TexToData.

The motivating example, as well as conducted experiments [7], suggest that CDM generated by AMADEOS (further referred to as CDM A) is expected to have a more complete and more precise structure than CDM generated by TexToData (further referred to as CDM T). This suggests that, in general, CDM A is more reliable than CDM T. With this assumption, we take CDM A as the starting point. This is in line with [1], where the order of the schemas is the input in the integration process. This also corresponds to the enterprise view (cf. [22]) whose existence is assumed as the input to the integration process. Since CDM A is the starting point, it needs to be mapped against CDM T, and CDM T needs to be merged into CDM A which represents the final CDM.

Both CDMs are in the package of the same name (ICM_CD), so we consider these packages as an implicit match, and classes (entity types) are the first elements that we explicitly match.

The first step in the schema integration process is the schema mapping. Over the years, many schema mapping techniques have been introduced [3]. One of the categories of mapping techniques includes *schema-based* and *instance-based* techniques. We can not use any of the instance-based techniques because we deal with conceptual models, which means that there are no data instances that can be used in the matching process. In order to achieve more precise matching, we apply *hybrid schema-based* matching, whereby we apply a combination of *linguistic matching* and *structure-based matching*. Since the problem of automatic recognition and resolution of semantic anomalies (e.g. synonyms, homonyms, abbreviations, ignored words, etc.) remains to be solved, we plan to introduce auxiliary information usage (e.g. thesaurus). Also, we will take into account typographical errors, e.g. by using distance matching such as Levenshtein distance. *Constraint-based matching* will also be part of the future improvement (e.g. data types can be considered, but first both tools should be improved).

The motivating example shows that (currently) structure-based matching is not a realistic option to match classes (classes in the CDM generated by AMADEOS are lacking attributes, while TexToData generates a lesser number of associations in CDM), so we decided to use linguistic matching where we tokenize class names and try to find the best match for each class in both models. Structure-based matching for classes will be considered in the future.

After finding matched classes in input CDMs, it is necessary to decide what to do with the unmatched classes in both models. At the moment, we consider the class to be relevant in the final CDM if it has attributes or associations with other already matched classes. Such (relevant) classes in CDM A are simply kept in the model, while non-relevant classes are removed. For each relevant class in CDM T, a new class is created in CDM A, and these two classes (class in CDM T and newly created class in CDM A) are considered as a match.

The next step, after the class matching, is the class merging step. Since AMADEOS generates only *id* attribute for each class, then attributes from the classes in CDM T are simply added to the corresponding classes in CDM A.

In the next step, we deal with associations. Although AMADEOS generates associations with lower precision than classes, CDM A is still much more precise and much more complete than CDM T regarding associations. In order to increase the completeness of the associations in the generated CDM, at the moment we try to add associations from CDM T that are missing in CDM A. Both AMADEOS and TexToData generate only binary associations. Therefore, for two classes *Xa* and *Ya* from CDM A that have a match in CDM T (*Xt* and *Yt*, respectively), we compare associations in CDM A between *Xa* and *Ya*, with the associations from CDM T between *Xt* and *Yt* (structure-based matching). We consider such two associations from CDM A and CDM T a match if their cardinalities (upper values of the respective member ends) are equal. We only compare the cardinalities because CDM A is more precise than CDM T regarding participation constraints (lower values of the respective member ends). At the moment, merging of matched associations means only keeping the respective associations unchanged in CDM A. Also, unmatched associations from CDM A are kept unchanged, while unmatched associations from CDM T are added to CDM A. Our future work will include improvements to both tools, as well as improvement of the integration approach (e.g. hybrid technique).

In the last step, we deal with generalizations. Since AMADEOS does not generate generalizations, generalizations from CDM T are added to CDM A. Creating a generalization between a subtype and a supertype forces the removal of the *id* attribute and the corresponding *PK* operation from the subtype.

Finally, the modified CDM A represents the final CDM.

5 Implemented Tool

This section presents the implemented tool named DBomnia[3]. DBomnia is the first online web-based tool providing the functionality of CDM derivation from heterogeneous source artifacts. Currently, DBomnia enables automatic CDM derivation from two different types of source artifacts: textual specifications and collections of BPMs. Figure 5 shows the architecture of the DBomnia tool, while Fig. 6 shows a screenshot of the tool in action.

The process of the CDM synthesis consists of four steps: (1) generation of the CDM from the source collection of BPMs, (2) generation of the CDM from the input textual specification, (3) integration of the generated CDMs, and (4) generation of the diagram layout for the integrated CDM.

In the first step, the source collection of BPMs is sent to the AMADEOS. AMADEOS generates the corresponding CDM and responds with the JSON object which contains generated CDM, execution status, etc. In the second step, the input textual specification is sent to TexToData, which generates the corresponding CDM. TexToData also responds with the JSON object containing generated CDM, error messages (if any), etc.

[3] http://m-lab.etf.unibl.org:8080/dbomnia.

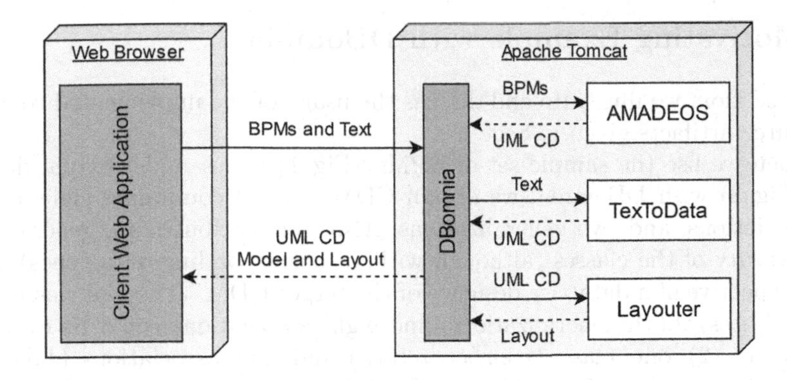

Fig. 5. DBomnia architecture

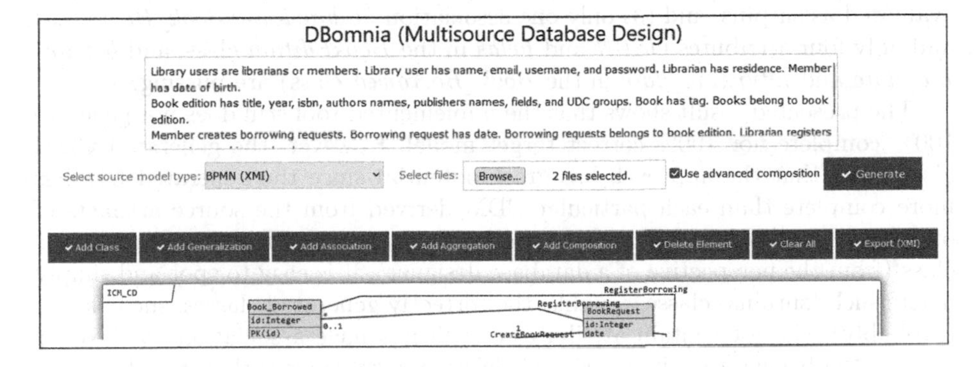

Fig. 6. Screenshot of DBomnia in action

In the third step, DBomnia integrates the CDMs generated in the first two steps, by applying the approach described in the previous section. Then, in the fourth step, it sends the integrated CDM to the Layouter service, which generates the corresponding diagram layout. Layouter is the pre-existing service (also used in TexToData) that provides the functionality of generating a diagram layout for the input UML class diagram (or in our case, CDM represented by the UML class diagram). Layouter responds with the file containing the generated layout.

Finally, DBomnia prepares and returns the response – the JSON object containing the integrated CDM, diagram layout, and status information.

The client web application allows users to input a textual specification and upload a collection of source BPMs. Upon the user's request (click on the button Generate), all source artifacts are sent to DBomnia. When the client application receives the JSON response, it visualizes[4] the class diagram in the browser and enables further manual improvements.

[4] Implementation is based on the jsUML2 library (http://www.jrromero.net/tool_jsUML2.html).

6 Motivating Example with DBomnia

In this section, we illustrate and discuss the usage of the implemented tool with the source artifacts given in Sect. 3.

When we use the sample set of BPMs (Fig. 1) and sample textual description (Fig. 3) with DBomnia, we obtain CDM (Fig. 7) containing eight classes, 12 associations, and two generalizations. Also, the attributes are generated for the majority of the classes (although without corresponding data types). From the perspective of a database designer of the target CDM: (1) seven classes (and all attributes), both generalizations, and eight associations could be evaluated as correct; (2) one class (*member_return*) and four associations (*Librarian-RegisterReturnedBook-Book*, *Member-SendMessage_Book-Book*, *Book-borrowed-member_return*, *Librarian-ReceiveMessage_BookRequest-BookRequest*) could be evaluated as surplus, and (3) only one association (*Librarian*↔*Book_Borrowed*) and only four attributes (*ISBN* and *fields* in the *BookEdition* class, and *borrowing_date* and *returning_date* in the *Book_Borrowed* class) are missing.

The presented result shows that the implemented tool still does not generate 100% complete nor 100% correct target model. However, the generated CDM also shows that the approach has great potential since the generated CDM is more complete than each particular CDM derived from the source artifacts of one single type. Although the automatically generated CDM has one surplus class (from the perspective of a database designer), it is easy to spot and simply delete such (surplus) classes – unlike the correctly generated classes, such classes (probably) will not contain attributes or will not have associations with other classes. We believe that it is easier to delete a surplus class than to add a new class when it's missing in the generated CDM.

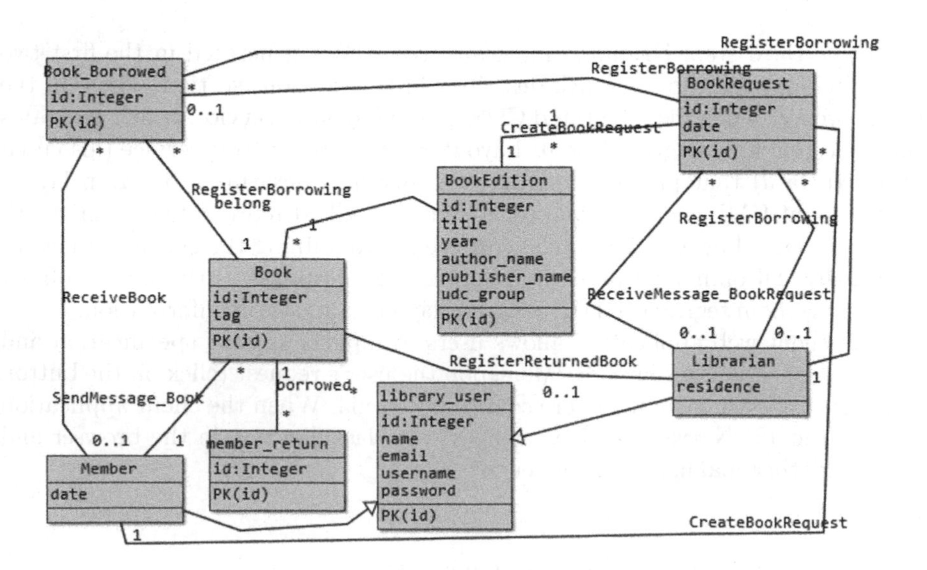

Fig. 7. CDM of Online Library generated by DBomnia

7 Conclusion

In this paper, we presented DBomnia – the first online web-based tool enabling automatic CDM derivation from a heterogeneous set of source artifacts, whereby the currently supported source artifacts are BPMs and textual specifications. DBomnia employs other tools (AMADEOS and TexToData) to generate CDMs from specific source artifacts (AMADEOS derives CDM from BPMs, while TexToData derives CDM from textual specifications) and then integrates the generated CDMs into a single unified CDM. The employed tools generate CDMs that are not 100% complete nor 100% correct, which makes them models with reduced reliability and this constitutes the main challenge.

Even though the presented tool represents an early prototype of the system, the initial results show that CDMs generated by DBomnia are more precise and more complete than CDMs generated by the pre-existing tools. These initial results, although based on a relatively simple example, are the motivation for our future work that will include further improvements of the approach and tool, further improvement of the specific CDM generators, the inclusion of other types of source artifacts, and thorough validation and verification.

References

1. Batini, C., Lenzerini, M.: A methodology for data schema integration in the entity-relationship model. IEEE Trans. Softw. Eng. **SE-10**(6), 650–664 (1984)
2. Batini, C., Lenzerini, M., Navathe, S.B.: A comparative analysis of methodologies for database schema integration. ACM Comput. Surv. **18**(4), 323–364 (1986)
3. Bernstein, P., Madhavan, J., Rahm, E.: Generic schema matching, ten years later. Proc. VLDB Endow. **4**(11), 695–701 (2011)
4. Bernstein, P.A., Halevy, A.Y., Pottinger, R.A.: A vision for management of complex models **29**(4), 55–63 (2000)
5. Brdjanin, D., Maric, S.: An approach to automated conceptual database design based on the UML activity diagram. Comput. Sci. Inf. Syst. **9**(1), 249–283 (2012)
6. Brdjanin, D., Maric, S.: Model-driven techniques for data model synthesis. Electronics **17**(2), 130–136 (2013)
7. Brdjanin, D., Banjac, G., Banjac, D., Maric, S.: An experiment in model-driven conceptual database design. Softw. Syst. Model. **18**(3), 1859–1883 (2019)
8. Brdjanin, D., Vukotic, A., Banjac, D., Banjac, G., Maric, S.: Automatic derivation of the initial conceptual database model from a set of business process models. Comput. Sci. Inf. Syst. **19**(1), 455–493 (2022)
9. Brdjanin, D., Banjac, G., Babic, N., Golubovic, N.: Towards the speech-driven database design. In: Proceedings of TELFOR 2022, pp. 1–4. IEEE (2022)
10. Brdjanin, D., et al.: Towards an online multilingual tool for automated conceptual database design. In: Braubach, L., et al. (eds.) Intelligent Distributed Computing XV, vol. 1089, pp. 144–153. Springer, Cham (2023). https://doi.org/10.1007/978-3-031-29104-3_16
11. Chen, P.: English sentence structure and entity-relationship diagrams. Inf. Sci. **29**(2–3), 127–149 (1983)

12. Choobineh, J., Mannino, M., Nunamaker, J., Konsynsky, B.: An expert database design system based on analysis of forms. IEEE Trans. Softw. Eng. **14**(2), 242–253 (1988)
13. Choobineh, J., Lo, A.W.: CABSYDD: Case-based system for database design. J. Manage. Inf. Syst. **21**(3), 281–314 (2004)
14. Date, C.: An Introduction to Database Systems, 8th edn. Addison-Wesley, Boston (2003)
15. Harmain, H., Gaizauskas, R.: CM-builder: a natural language-based CASE tool for object-oriented analysis. Autom. Softw. Eng. **10**(2), 157–181 (2003)
16. Hartmann, S., Link, S.: English sentence structures and EER modeling. In: Proceedings of the 4th Asia-Pacific Conference on Conceptual Modelling, vol. 67, pp. 27–35 (2007)
17. Jouault, F., Allilaire, F., Bezivin, J., Kurtev, I.: ATL: a model transformation tool. Sci. Comput. Program. **72**(1–2), 31–39 (2008)
18. Kriouile, A., Addamssiri, N., Gadi, T.: An MDA method for automatic transformation of models from CIM to PIM. Am. J. Softw. Eng. Appl. **4**(1), 1–14 (2015)
19. Lukovic, I., Mogin, P., Pavicevic, J., Ristic, S.: An approach to developing complex database schemas using form types. Softw. Pract. Exp. **37**(15), 1621–1656 (2007)
20. Madhavan, J., Bernstein, P., Rahm, E.: Generic schema matching with cupid. In: Proceedings of VLDB 2001, pp. 49–58. Morgan Kaufmann (2001)
21. Magnani, M., Rizopoulos, N., Mc.Brien, P., Montesi, D.: Schema integration based on uncertain semantic mappings. In: Delcambre, L., Kop, C., Mayr, H.C., Mylopoulos, J., Pastor, O. (eds.) ER 2005. LNCS, vol. 3716, pp. 31–46. Springer, Heidelberg (2005). https://doi.org/10.1007/11568322_3
22. Navathe, S., Gadgil, S.: A methodology for view integration in logical database design. In: Eigth International Conference on Very Large Data Bases, pp. 142–164. Morgan Kaufmann (1982)
23. Nikiforova, O., Gusarovs, K., Gorbiks, O., Pavlova, N.: BrainTool: a tool for generation of the UML class diagrams. In: Proceedings of ICSEA 2012, pp. 60–69. IARIA (2012)
24. Omar, N., Hanna, P., McKevitt, P.: Heuristics-based entity-relationship modelling through natural language processing. In: Proceedings of AICS 2004, pp. 302–313 (2004)
25. Overmyer, S.P., Benoit, L., Owen, R.: Conceptual modeling through linguistic analysis using LIDA. In: Proceedings of ICSE 2001, pp. 401–410. IEEE (2001)
26. Pottinger, R., Bernstein, P.: Merging models based on given correspondences, pp. 862–873. VLDB 2003, VLDB Endowment (2003)
27. Purao, S.: APSARA: a tool to automate system design via intelligent pattern retrieval and synthesis. SIGMIS Database **29**(4), 45–57 (1998)
28. Rodriguez, A., Garcia-Rodriguez de Guzman, I., Fernandez-Medina, E., Piattini, M.: Semi-formal transformation of secure business processes into analysis class and use case models: an MDA approach. Inf. Softw. Technol. **52**(9), 945–971 (2010)
29. Spasic, Z., Vukotic, A., Brdjanin, D., Banjac, D., Banjac, G.: UML-based forward database engineering. In: Proceedings of INFOTEH 2023, pp. 1–6. IEEE (2023)
30. Sugumaran, V., Storey, V.C.: Ontologies for conceptual modeling: their creation, use, and management. Data Knowl. Eng. **42**(3), 251–271 (2002)
31. Tan, H.B.K., Yang, Y., Blan, L.: Systematic transformation of functional analysis model in object oriented design and implementation. IEEE Trans. Softw. Eng. **32**(2), 111–135 (2006)
32. Thonggoom, O.: Semi-automatic conceptual data modelling using entity and relationship instance repositories. PhD Thesis, Drexel University (2011)

Lightweight Aspect-Oriented Software Product Lines with Automated Product Derivation

Jakub Perdek(✉) and Valentino Vranić

Institute of Informatics, Information Systems and Software Engineering
Faculty of Informatics and Information Technologies, Slovak University of Technology in Bratislava, Bratislava, Slovakia
{xperdek,vranic}@stuba.sk

Abstract. Aspect-oriented software product lines are not a new idea, but their application is facing two obstacles: establishing software product lines is challenging and aspect-oriented programming is not that widely accepted. In this paper, we address exactly these two obstacles by an approach to establishing lightweight aspect-oriented software product lines with automated product derivation. This is particularly relevant for data preprocessing systems, which are typically custom-built with respect to the data and its structure. They may involve data cleaning, reduction, profiling, validation, etc., which may have variant implementations and may be composed in different settings. The approach is simple and accessible because developers decide about variation points directly in the code without any assumption on development process and applied management. Also, it allows for variability management by making the code readable, configurable, and adaptable mainly to scripts and code fragments in a modular and concise way. The use of annotations helps preserve feature models in code. We presented the approach on the battleship game and data preprocessing pipeline product lines, which include the configuration of features, product derivation mechanism based on annotations applied by the user on certain classes and methods, implementation of features according to the feature model, and the possibility to generate all given software product derivations.

1 Introduction

Software product lines (SPL) are an efficient approach to software reuse. The secret of their success lies in limiting the effort for reuse to a set of related software systems within one organization (sometimes denoted as a family). Such systems have a lot of common features, but they also have some variable features. These have to be implemented in such a way that would enable them to vary. Consequently, a design of a software product line starts by mapping the features it should cover. This requires some kind of feature modeling to be employed, which is not necessarily the academic FODA-like notation [3, 15].

It is obvious that variable features are best implemented in a pluggable fashion, but this is not easy to achieve in traditional object-oriented programming

A. Abelló et al. (Eds.): ADBIS 2023, CCIS 1850, pp. 499–510, 2023.
https://doi.org/10.1007/978-3-031-42941-5_43

because their implementation tends to crosscut many other features [25]. Aspect-oriented programming addresses exactly the issue of crosscutting concerns. The best known aspect-oriented programming language is AspectJ, which is an extension to Java, but many other programming languages exhibit aspect-oriented features, e.g., Python, JavaScript, etc. [9,10].

Aspect-oriented software product lines are not a new idea, but their potential application is facing two obstacles: establishing software product lines is challenging and aspect-oriented programming is not that widely accepted. In this paper, we address exactly these two obstacles by an approach to establishing lightweight aspect-oriented software product lines with automated product derivation. This is particularly relevant for data preprocessing systems, which are typically custom-built with respect to the data and its structure. They may involve data cleaning, reduction, profiling, validation, etc., which may have variant implementations and may be composed in different settings.

The rest of the paper is organized as follows. Section 2 explains the position of aspect-oriented programming in software product lines. Section 3 provides an overview of establishing a software product line for a battleship game. Section 4 introduces annotations and expressions which help to choose content to copy into the resulting project presented on the game derivation based on actual configuration settings. Section 5 presents evaluation and discussion. Section 6 compares the approach proposed in this paper to related work. Section 7 concludes the paper.

2 Software Product Lines and Aspect-Oriented Modularization

Variability is the main part of software product development and products are built by resolving it in a way that can build customer specific products [8]. Software products often emerge from the success of the market with different needs that can be provided by actual knowledge determined by features, relationships among them and between them, and software artifacts that provide the implementation of these features. These known actual knowledge sources are used for systematic reuse introduced by software product line engineering [22]. Evolving products then depend on their real-world applications which should provide a flexible way how to apply constantly changing needs. Given features may affect several places in models and code which can cause problems during its adaptations with other already included features. These places where changes occur [14] are called variation points and represent possible ways how to model variability. Interaction of features can cause a need to generate tailored software artifacts or software artifacts for their next modification for the final application. It is easier to configure the process by applying a change to code at the places where components are generated [25].

Aspect-oriented model driven software product line development models variability on model level and aspects to implement these models, primarily their crosscutting features [23]. Models provide more abstract views of features to be separated and be more effectively managed such as traced according to customer requirements, in comparison to their configuration on code level [23].

The main problems of AspectJ are the necessity to maintain certain conventions about names of methods and classes [17], as well as to divide the application into appropriate parts for changing their behavior and introduce hook methods only to hang up aspect on them [16]. Other issues can arise with certain types of applications. For example, during the refactoring of the database system, there was necessary to use privileged aspects to access non-public variables, but this violates encapsulation [16]. Aspect behavior can modify and use these variables. Aspects often require mentioned supervision on the design of the final solution otherwise the solution will be less maintainable.

3 Establishing a Software Product Line

Our approach assumes that an initial software product exists and that it exhibits good object-oriented modularization. We rely on this, so that we can introduce variable features using aspect-oriented programming. After this step aspects implemented by AspectJ are inseparable parts of the resulting products and product line code. We map both common and variable features by a feature model, an example of which can be found in the next section.

We will explain our approach on an implementation of the battleship game product line. We adopted and adapted the basic game from a publicly available resource.[1] We refactored it significantly to improve its object-oriented modularization.

After changing the visibility of variables, adding a configuration file for the base state of an application, and moving appropriate content to newly created classes, we needed to design configurable features, for which we relied on known aspect-oriented practices [8,16]. We did this using AspectJ, a relatively stable aspect-oriented extension of Java.[2].

We based variable features on AspectJ because this way we can additively integrate concerns where each is implemented as a separate aspect. Thanks to presented modularity the solution is more extendable. For example, the feature for setting player names requires adding another method before creating the functionality to set names to players. Without aspects, only condition statements will be used instead. There is also no possibility to add a player name variable to the Player class only for this case. But, all required functionality such as new variables and methods can be added into segregated aspects. It is also possible to configure or exclude them according to the configuration which can be loaded when the application starts or changing values at runtime. By using aspects, we only need to specify a pointcut that provides the mapping of the certain location where players are created to the executed method. In this aspect method, the program loads names which are typed by players from the input. The whole functionality is in one aspect.

[1] https://github.com/juletx/BattleshipFeatureIDE.

[2] https://www.eclipse.org/aspectj/.

We found that aspects well suited the configuration of settings in the solution by directly putting conditions from the configuration to manage the optional inclusion of given concerns. Different implementations of the same aspect can be used to apply specific configurations reflecting the customer's needs. The configuration is applied only by replacing the parameters of overloaded methods with the loaded values from the configuration file before calling these methods.

4 Product Derivation

Since aspects in AspectJ become active simply if they are included in the compilation, the product derivation in our approach is realized by simply copying the corresponding aspects to the project folder. For non-aspect features, the classes that implement them are copied, of course. This process is managed by annotating certain code parts and specifying the condition inside expressions that are inserted into them. Feature inclusion must follow the constraints set by the feature model in terms of variability if the modeling of the variability is based on feature models. Furthermore, a feature can be included only if its parent is included, even if it is mandatory. Other models can be incorporated accordingly by changing annotations or variables to meet given restrictions.

4.1 Annotating Variation Points

The comprehensive and readable rules which will contain the mapping for the features are needed. We come up with the idea of creating annotations with expressions that allow specifying a condition when a feature should be copied and when not. Each annotation starts with the comment characters (//) followed by an identifier for the given annotation type. The last part consists of an expression in JSON format. It contains variables with assigned values, especially operators as reserved ones (AND or OR) to evaluate the truthfulness of grouped variables in a certain way.

Mapping of some variables with given values to the feature model can be seen in Fig. 1. Each rule is prescribed and handled by a given annotation type and thus changing its purpose needs to be applied in the derivator. Derivation happens in design time by evaluating all these expressions inside the project folder. The result of this operation has an impact on whether this prescribed functionality will be applied or not. For example, if the variable named playerNames mapped to the feature with the same name is set to true and all expressions where its included are evaluated positively, then all necessary (annotated) code parts for this feature will be copied in the final solution. If an expression in the rule is empty, then it is evaluated as true. If a file contains no annotations or all rules have their annotations evaluated as false, then the file will not be copied into the final solution.

In order to copy only modular code fragments and marginally their crosscutting functionality, these annotations are used:

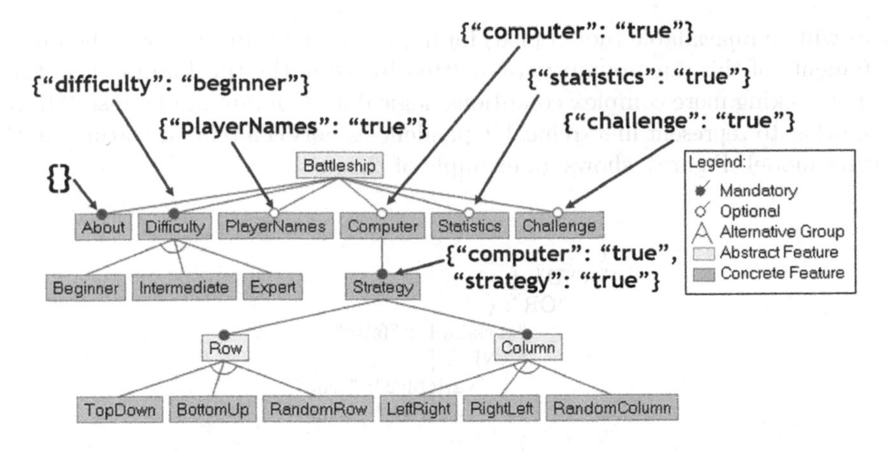

Fig. 1. The battleship game feature model with derivation rules.

//@{} This annotation type is used to annotate classes, interfaces, or aspects to copy them along with their source file. An additional check can be added to the program that one of the keywords related to the used programming language is located after annotation.

//#{} This annotation type is suitable to include or exclude given methods. If the file should be copied, then at least one condition in the annotation should evaluate to true. This annotation type should not be mixed with the first one because nested annotations are not supported.

//%{} By using this annotation type, some import statements (each represented as one line of code) are included or excluded according to including or excluding a given method from the file. Consequently, it should be used restrictively with the second annotation.

Making a product ready is only a matter of splitting features into modular parts and making clones of selected code parts according to the product specification. Effective decomposition into separate files allows copying whole files rather than code fragments. The aspects woven into the code are essential here. If fragmentation would cause these modular parts to lose their main responsibilities, then the latter two annotations are used. The inner content then can be managed effectively. Annotated methods can make other dependencies by importing classes that will not be part of the final product. Their import statements are, thus, managed similarly. The third annotation type solves these problems by including or excluding only one line of code.

A typical example is the functionality for setting names (the PlayerNames feature) which is affecting the Computer and Player classes displayed in the feature model in Fig. 1. Methods that are setting names are implemented in a separate aspect and are annotated with the second annotation. Both of them contain the playerNames variable in their expressions. Except for the first annotation, the second one contains also the computerPlayer variable which prevents problems if the functionality which simulates an opponent (the Computer feature) will be unavailable (not copied) for a given variant. In this case, the import statements of this functionality are omitted by using the third annotation type.

For making more complex conditions, logical *and* and *or* can be inserted into each other to represent hierarchical dependencies between related features in the feature model. Figure 2 shows an example of this.

```
{
"AND": {
    "OR": {
        "variable1": "false",
        "AND": {
            "variable2": "true",
            "variable3": "true"
        }
    },
    "variable4": "true"
}
```

Fig. 2. A complex derivation rule.

4.2 Product Derivation

The product derivation can start by launching the derivator after variable code parts are annotated according to the proposed mapping of the variability model, especially the feature model. This happens in design time by configuring these mapped values. Each one is set to the one that fits the resulting product in this step. No other action is needed from developers, except evaluating errors caused by improperly chosen or configured expressions in place of variation points.

Figure 3 shows the classes involved in product derivation. The DerivationManager class manages product derivation. The ProjectCopier class is used to copy an empty project and the FileCopy class manages the reading and writing of a given file as a stream. During copying, it is necessary to determine whether a file should be copied or not. For this purpose, the program needs to find annotations and evaluate the expression they contain, which is performed by the DerivationAnnotationManager and corresponding annotation specific classes. This search is repeated after any annotation is found until the end of the file. The DerivationVariableProcessor class helps the DerivationAnnotationManager class to recursively evaluate expressions associated with the annotation.

The derivator is available and open for further customization.[3]

[3] https://github.com/jperdek/productLineAnalysisGame.

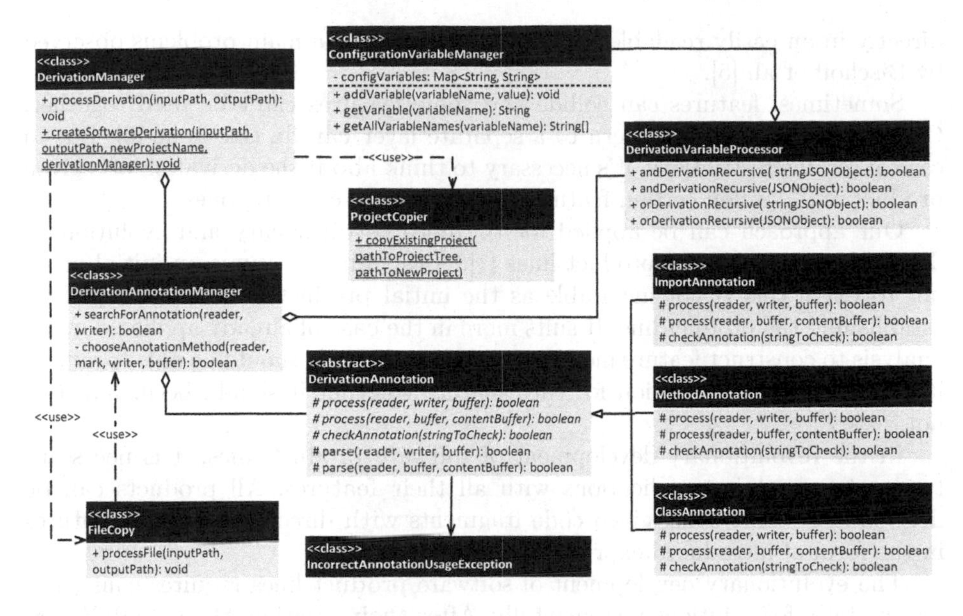

Fig. 3. Product derivator

5 Discussion

Aspects help to move the logic of crosscutting concerns into modular units and enhance functionality without modification of existing code. Well-modularized crosscutting concerns can be easily reused in the form of modules such as in HealthWatcher [11]. Typical examples are large parts of interacting code for logging, exception handling, and third-party functionality which is moved to aspects. According to these observations, we designed a mechanism to easily copy certain parts of a given project and let the project be functional and possibly prepared for the next development phase.

Applying the functionality of AspectJ to given join points can be restricted by adding conditions to their pointcuts. Whole functionality is thus available in resulting products. Many of these conditions remain redundant, except in the case when they need to be selected and changed dynamically at runtime. Our annotations are language-independent and are removed from all derived products, thus they are selected before compilation in design time. Also, if features affect each other, then modularity cannot be preserved with aspects without decomposing this functionality further.

Appropriate domain analysis is required to design common and variable features of a given system in order to understand which products are possible to generate. For this purpose, we used feature models. In reality, many of these models can exist, but only a few of them support the changing needs of the customers [4]. We intended to find a mechanism to preserve feature models in code and make the integration process easier by letting developers manage it directly in an easily readable way as a response to the main problems observed by Bischoff et al. [5].

Sometimes, features can collide and some features can't be used together. Often, adding another concern to a separate layer can fix this issue [6]. But in case of serious restrictions, it's necessary to think about the derivation of a given product and develop certain features without those colliding ones.

Our approach can be applied for the both revolutionary and evolutionary development of software product lines [7]. Although we assume an initial product to exist, this is not inevitable as the initial product can be developed by establishing the product line. It suits more in the case of already applied domain analysis to construct feature models. According to it, we should be able to explicitly formulate an expression for each feature whether it should be included or not.

In the revolutionary development of software product lines, it is necessary to develop modular applications with all their features. All products can be derived after annotating given code fragments with direct mapping to features in the form of introduced expressions.

The evolutionary development of software product lines requires constructing modules for solutions incrementally. After their creation, these modules are annotated, and included expressions are properly configured to manage their integration. In the end, the derivation mechanism omits not annotated code fragments.

6 Related Work

The most similar to our approach is the frame technology, which is also capable of creating easily adaptable components with a language-independent mechanism [18]. However, the code is mixed with tags and it is not possible to compile it. The tag hierarchy can be complex and the tag endings in it can be difficult to pair. On the contrary, expressions in our approach can be recursively evaluated. Unlike the tags in the frame technology, they are in the form of annotations adaptive preserve feature models directly in the place where they occur in a much more modular way.

Another similar work [12] introduced mapping between given models based on a template model as a representation of each available product. Our expressions are not only presence conditions, but rather, as JSON representations, provide space to include more information for further configuration related to domain requirements. Not only removing unwanted features from the feature model but also adding new functionality apart from it, is then possible. This makes our approach not restricted to negative variability. Another significant difference is in the hierarchic organization of variables under used operators which is more comprehensive for the hierarchic structure of feature models. Additionally, the mechanism for evaluating expressions can be configured according to the expected key and value pair.

Plastic partial components were introduced to manage variability inside components [19]. They are just component fragments that are mainly reused by applying invasive composition techniques [1] when aspects are used to manage their selection in the final product. In our approach, annotated parts can evolve together as an already instantiated product or as a derived application on its own.

The variability on the code level can be also handled by pure::variants [20], which provides also many tools, especially for tracing. In our approach, expressions have the same role as the configuration rules, but differ with used JSON format and also are recursively evaluated to directly adapt to the nature of feature models in code. The used format allows processing them by using distributed computing. Also, other functionality relevant to other roles associated with annotated variation points can be configured there. In pure::variants, endings are used, similar to tag endings in the frame technology or conditional compilation. Annotated code is thus less readable. Opposite to this, we want to force developers to annotate only aspects and classes as the primary types of modules, and methods, if necessary.

In our approach, developers make all decisions about variability directly in code, but these are in pure::variants configured and visualized by plugins and tools. Their configuration through used tools is restricted to programmed functions, which do not allow immediately adding developers' own variables with special semantics for changed functionality of the derivator. Such cases can be optional functionality to pretest variants by creating additional logs, letting variants generate data for analysis, adding integration with their tool, or adapting certain business functionality related to the variation point without directly polluting the code.

The key activities for the individual product derivation, such as product configuration, requirements engineering, additional development, integration, deployment, and, finally, product line evolution [21], in our approach are managed not much differently than in the most used derivation approaches such as DOPLER or Pro-PD. Mostly, developers introduce variation points that are configured mainly according to the feature model which is discussed with the customer or chosen for the given domain. As in DOPLER and Pro-PD, the customer requirements are fulfilled iteratively also in our approach. The collaboration with the customer depends then on the project management methodology being applied, but even more on variability modeling. The focus on product configuration should minimize product specific development as in Pro-PD. This benefit is not similar to DOPLER which is more collaborative than our approach and Pro-PD. This is recognizable in situations when unsatisfied user requirements are perceived as product specific implementations [21].

In our approach, it is not necessary to create a specific language for product derivation as it is in FAST [2]. It involves less planning for the product derivation. Activities, products, and roles are not described as in PuLSE-I [26]. The development process is not limited to iterative fulfilling of requirements as in COVAMOF [13].

7 Conclusions and Future Work

In this paper, we introduced an approach to establishing lightweight aspect-oriented software product lines with automated product derivation which is simple and accessible because developers decide about variation points directly in the code without any assumption on the development process and applied management. Also, it allows for variability management by making the code readable, configurable, and adaptable mainly to scripts and code fragments in a modular and concise way. The use of annotations helps preserve feature models in code through their ability to store information about the hierarchy of features.

The information extracted from variation points in the form of JSON format can be processed by known big data systems, such as Hadoop, Pig, or Hive to analyze dependencies and relations between used variables.

We presented the approach on the battleship game and data preprocessing product lines. These examples include the configuration of features, product derivation mechanism based on annotations applied by the user on certain classes and methods, implementation of features according to the feature model, and the possibility to generate all given software product derivations. We need only three types of basic annotations to mark the variety of code fragments for the next separation. We generated all 48 possible derivations and tested their functionality. It's also possible to run one of the supported product instances directly using the base project by setting configuration values in the configuration file. Developers are also allowed to change provided simple derivator functionality to promptly handle demands on data processing.

The approach proposed in this paper can be applied to efficiently handle the variability associated with the customization of data preprocessing mechanisms, derivation of adjusted products which are ready for their next validation, or processing expressions that contain knowledge about variation points by using big data analysis techniques.

We intend to explore this more thoroughly in our future work mainly focused on the ways how to create, process, and evaluate data that contains a lot of variability. Fractals are a good source of it. We will effectively derive and validate a variety of them according to metrics that support the aesthetic perception of users or use other assumptions. The knowledge from variation points will be used during the prediction phase. A small modification of the derivation mechanism is necessary to adapt it to web languages, mainly those where the conditional compilation is unavailable, such as JavaScript or TypeScript.

Derived products from the introduced product line can serve as the basis for another one to implement contradictory features efficiently. They can be quickly improved to support the extraction of such features by introducing or changing the annotations.

We are also exploring how our approach could be applied to manage the diversity of services featured in a research setting of developing rurAllure,[4] an innovative pilgrimage support system [24].

---

[4] https://rurallure.eu/.

Acknowledgements. The work reported here was supported by from the European Union's Horizon 2020 research and innovation program under grant agreement No. 101004887 (rurAllure), the Operational Program Integrated Infrastructure for the project: Support of Research Activities of Excellence Laboratories STU in Bratislava, project no. 313021BXZ1, co-financed by the European Regional Development Fund (ERDF), and the Operational Programme Integrated Infrastructure for the project: Research in the SANET network and possibilities of its further use and development (ITMS code: 313011W988), co-funded by the ERDF, and by the Slovak Research and Development Agency under the contract No. APVV-15-0508.

References

1. Aßmann, U.: Invasive Software Composition. Springer-Verlag, Berlin (2003)
2. Bayer, J., Gacek, C., Muthig, D., Widen, T.: PuLSE-I: deriving instances from a product line infrastructure. In: Proceedings of 7th IEEE International Conference and Workshop on the Engineering of Computer-Based Systems, ECBS 2000 (2000)
3. Berger, T., et al.: A survey of variability modeling in industrial practice. In: Proceedings of 7th International Workshop on Variability Modelling of Software-Intensive Systems, VaMoS 2013. ACM, Pisa (2013)
4. Beuche, D., Dalgarno, M.: Software product line engineering with feature models (2006). https://www.pure-systems.com/fileadmin/downloads/pure-variants/tutorials/SPLWithFeatureModelling.pdf
5. Bischoff, V., Farias, K., Gonçales, L.J., Victória Barbosa, J.L.: Integration of feature models: a systematic mapping study. Inf. Softw. Technol. **105**, 209–225 (2019)
6. Blair, L., Pang, J.: Aspect-oriented solutions to feature interaction concerns using AspectJ (2003)
7. Bosch, J.: Design and Use of Software Architectures: Adopting and Evolving a Product-Line Approach. Addison-Wesley, Boston (2000)
8. Botterweck, G., Lee, K., Thiel, S.: Automating product derivation in software product line engineering. In: Proceedings of Software Engineering 2009. LNI P-143, Gesellschaft für Informatik e.V. (2009)
9. Bystrický, M., Vranić, V.: Preserving use case flows in source code. In: Proceedings of 4th Eastern European Regional Conference on the Engineering of Computer Based Systems, ECBS-EERC 2015. IEEE, Brno (2015)
10. Bálik, J., Vranić, V.: Symmetric aspect-orientation: some practical consequences. In: Proceedings of Proceedings of International Workshop on Next Generation Modularity Approaches for Requirements and Architecture, NEMARA 2012, at AOSD 2012. ACM, Potsdam (2012)
11. Cherait, H., Bounour, N.: History-based approach for detecting modularity defects in aspect oriented software. Informatica **39**(2), 187–194 (2015)
12. Czarnecki, K., Antkiewicz, M.: Mapping features to models: a template approach based on superimposed variants. In: Glück, R., Lowry, M. (eds.) GPCE 2005. LNCS, vol. 3676, pp. 422–437. Springer, Heidelberg (2005). https://doi.org/10.1007/11561347_28
13. Deelstra, S., Sinnema, M., Bosch, J.: Product derivation in software product families: a case study. J. Syst. Softw. **74**(2), 173–194 (2005)
14. Jacobson, I., Griss, M., Jonsson, P.: Software Reuse: Architecture, Process and Organization for Business Success. Addison-Wesley, Boston (1997)

15. Kang, K.C., Cohen, S.G., Hess, J.A., Novak, W.E., Peterson, A.S.: Feature-oriented domain analysis (FODA): a feasibility study. Technical Report CMU/SEI-90-TR-21, Software Engineering Institute, Carnegie Mellon University, Pittsburgh, USA (1990)
16. Kastner, C., Apel, S., Batory, D.: A case study implementing features using AspectJ. In: Proceedings of 11th International Software Product Line Conference, SPLC 2007. IEEE, Kyoto (2007)
17. Laddad, R.: AspectJ in Action: Practical Aspect-Oriented Programming. Manning, Shelter Island (2003)
18. Loughran, N., Rashid, A.: Framed aspects: supporting variability and configurability for AOP. In: Bosch, J., Krueger, C. (eds.) ICSR 2004. LNCS, vol. 3107, pp. 127–140. Springer, Heidelberg (2004). https://doi.org/10.1007/978-3-540-27799-6_11
19. Perez, J., Diaz, J., Costa-Soria, C., Garbajosa, J.: Plastic partial components: a solution to support variability in architectural components. In: 2009 Joint Working IEEE/IFIP Conference on Software Architecture & European Conference on Software Architecture. IEEE, Cambridge (2009)
20. pure::systems: PLE & code–managing variability in source code (2020). https://youtu.be/RlUYjWhJFkM
21. Rabiser, R., O'Leary, P., Richardson, I.: Key activities for product derivation in software product lines. J. Syst. Softw. **84**(2), 285–300 (2011)
22. Reinhartz-Berger, I., Sturm, A., Clark, T., Cohen, S., Bettin, J. (eds.): Domain Engineering, Product Lines, Languages, and Conceptual Models. Springer, Cham (2013)
23. Voelter, M., Groher, I.: Product line implementation using aspect-oriented and model-driven software development. In: 11th International Software Product Line Conference (SPLC 2007). IEEE, Kyoto (2007)
24. Vranić, V., Lang, J., López Nores, M., Pazos Arias, J.J., Solano, J., Laseca, G.: Use case modeling in a research setting of developing an innovative pilgrimage support system. Universal Access in the Information Society (2023). Accepted
25. Vranić, V., Táborský, R.: Features as transformations: a generative approach to software development. Comput. Sci. Inf. Syst. (ComSIS) **13**(3), 759–778 (2016)
26. Weiss, D.M., Lai, C.T.R.: Software Product-Line Engineering: A Family-Based Software Development Process. Addison-Wesley, Boston (1999)

Comparison of Selected Neural Network Models Used for Automatic Liver Tumor Segmentation

Dominik Kwiatkowski and Tomasz Dziubich(✉)

Computer Vision and Artificial Intelligence Laboratory, Department of Computer Architecture, Faculty of Electronics, Telecommunications and Informatics, Gdańsk University of Technology, Gdańsk, Poland
{dominik.kwiatkowski,tomasz.dziubich}@eti.pg.edu.pl

Abstract. Automatic and accurate segmentation of liver tumors is crucial for the diagnosis and treatment of hepatocellular carcinoma or metastases. However, the task remains challenging due to imprecise boundaries and significant variations in the shape, size, and location of tumors. The present study focuses on tumor segmentation as a more critical aspect from a medical perspective, compared to liver parenchyma segmentation, which is the focus of most authors in publications. In this paper, four state-of-the-art models were trained and used to compare with UNet in terms of accuracy. Two of them (namely, based on polar coordinates and Visual Image Transformer (ViT)) were adopted for the specified task. Dice similarity measure is used for the comparison. A unified baseline environment and preprocessing parameters were used. Experiments on the public LiTS dataset demonstrate that the proposed ViT based network can accurately segment liver tumors from CT images in an end-to-end manner, and it outperforms many existing methods (tumour segmentation accuracy 56%, liver parenchyma 94% Dice). The average Dice similarity measure for the considered images was found to be 75%. The obtained results seem to be clinically relevant.

Keywords: Liver tumour segmentation · transformer · convolutional network

1 Introduction

According to the World Health Organisation (WHO), liver cancer is the sixth most common cancer worldwide, accounting for around 830,000 new cases and 800,000 deaths annually. A similar situation occurs in Europe, although the prevalence varies between different regions and countries. According to the latest available data, in 2020, liver cancer accounted for 57,000 new cases and 51,000 deaths in the European region. The highest mortality rates are observed in Eastern and Central Europe (Bulgaria, Romania, and Hungary). The lowest incidence rates are found in Northern Europe (Iceland, Norway, and Sweden)

https://cvlab.eti.pg.gda.pl/.

A. Abelló et al. (Eds.): ADBIS 2023, CCIS 1850, pp. 511–522, 2023.
https://doi.org/10.1007/978-3-031-42941-5_44

[1]. In recent years, Computed Tomography (CT) has become the most widely used biomedical imaging modality in the diagnosis of liver tumors. Liver tumor segmentation plays an essential role in the diagnosis and treatment of hepatocellular carcinoma or metastasis. Usually, liver and tumor segmentations are obtained from radiologists through manual delineation. Labeling volumetric CT images in a slice-by-slice manner is tedious and poorly reproducible.

Automated or semi-automated segmentation techniques would greatly improve accuracy and efficiency. Despite numerous liver segmentation algorithms being suggested by researchers, they have not been able to meet the clinical requirements of being efficient and accurate. This is because every automatic segmentation method has its own restrictions and the intricate structures of the liver, individual variations, and the similarity on the grayscale of the surrounding tissues make it difficult to develop automatic liver segmentation algorithms. It is important to underscore that the task of segmenting tumors is a notably more challenging undertaking when compared to the segmentation of parenchymal tissue, primarily due to the considerable degree of variability and intricacy that tumor regions exhibit within medical imaging data, in contrast to the relatively uniform characteristics of parenchyma.

Based on the literature review (using the IEEE Explore database and the search phrase "liver segmentation" for the period of 2020–2022, as of July 1, 2022), a total of 192 articles were obtained, of which 60 were rejected because they were not related to the topic. 30 works were related to a different modality (or multimodality), while 102 articles were related to CT, of which 23 were related to liver parenchyma segmentation, 15 to tumor segmentation, and 50 to both liver parenchyma and tumor segmentation. The remaining articles covered other organs (such as the gallbladder) or hepatic vessels. This is in line with the practical trend in medicine of evaluating the volume of liver parenchyma and tumors.

This paper presents a comparison study between automatic methods for liver tumor segmentation from contrast-enhanced CT images: Unet++, QUANet, Video Image Transformer (VitSeg) and Polar Image Transformations (PIT), with the reference point being the UNet model, which is commonly used as a baseline. Two of the models, PIT and VitSeg, were adapted to the specific segmentation task at hand. In our paper, PIT is introduced for liver tumor on CT image segmentation. In [2] the authors employed a segmentation approach for the identification of large intestine polyps, liver parenchyma, skin lesions, and anatomical regions of the heart. Notably, tumor segmentation was excluded from the scope of their investigation. Additionally, we introduce an innovative utilization of the Transformer model for tumor segmentation. In contrast to [4], which implemented the Image Transformer model in a skip connection path, our proposed approach utilizes the Vision Transformer (VitSeg) model directly, without the inclusion into a convolutional neural network (CNN).

The remainder of the paper is organized as follows. The related works are provided in Sect. 2. This section presents the methods that were used in the experiments in the following section. The proposed evaluation method is detailed in Sect. 3. The results are presented and discussed in Sect. 4 and the conclusions are drawn in Sect. 5.

2 Related Works

In the past decade, multiple liver and tumor segmentation methods have been developed for CT examination. These approaches can be roughly categorized into two classes: model-based methods and deep learning-based methods. In turn, approaches in the first group are divided into gray level based and structure based, and we can include the following in them: contour based segmentation [12], fuzzy c-mean clustering (FCM) [22], greedy snakes algorithm [21], adaptive thresholding [7], watershed transform applied on gradient magnitude [23], confidence connected region growing algorithm [14], gaussian mixture models [5,11], fast marching method [17], random forest [13], graph cuts and restricted regions of shape constrains [10], etc. The vast majority of these methods are semi-automatic in nature and has been tested on smaller datasets. Their main limitations include low solution generality and long computational time. Due to the remarkable achievements of deep learning in the domain of natural image processing, deep convolutional neural networks have exhibited impressive results in medical image segmentation. Ronneberger et al. [16] firstly used an end-to-end fully convolutional neural network (UNet) to segment medical images. As a result, scientists have commenced using semantic segmentation techniques to segment the liver and tumor in CT images. More detailed reviews of some methods based on the machine learning approach have been proposed in [8] and [15]. In the following subsection, we characterize the models used in our experiments.

2.1 UNet++

One of the first topologies that have emerged in recent years and are based on UNet is UNet++, proposed in the work [24]. This network was designed to address two problems of the traditional UNet. Firstly, in some applications, four levels of the auto-encoder are not necessary. Secondly, the authors identified a restrictive skip connection scheme that prevents effective feature fusion.

The main difference is the design of additional internal blocks that allow for efficient model pruning and improve the performance of the network. This is because in some cases, especially with small datasets, a smaller architecture proves to be more efficient. Thus, by having information from architectures of different heights, the network can evolve better results without testing each height. It can be said that the network learns the importance of information from a specific height for a given problem.

The second improvement is in the fusion function, which collects information from the layer below, after up-sampling, and from all layers to the left. According to the authors, the average Dice metric result is 2.7% higher than the base UNet, which makes this model less precise. It is worth noting that the authors provide results for several different segmentation problems, but in the case of the liver, they rely only on the liver mask.

2.2 QUA-net

QUA-net, proposed in [9], addresses two problems by making changes to the vanilla model (UNet). The first, solved with the Quartet Attention block, is to make better use of and combine information provided in different channels. Attention modules proposed by other authors focus on spatial information or dependencies between channels, but ignore the dependencies between channels in spatial information, as implemented in QUA-net. The second problem is to transfer information from the encoder to the decoder. The authors note that a simple link layer blurs the feature vector and loses detailed information like edge position or fine cancer cell texture, so they propose a much more complicated long-short- skip connection. In turn, the problem of information loss is addressed by additional convolutional layers to better select the relevant feature vector. The improvements made compared to the U–net base model result in a gain of 2.15% Dice for parenchyma and 3.74% for tumors.

2.3 Polar Image Transformations

The authors of the paper [2] proposed a model that receives as input an image in the polar coordinate domain, instead of color channels. This approach is effective for segmenting elliptical shapes because, after the transformation, if the center point is the center of the segmented space, it becomes linear. The method uses a pipeline of activities: localization and segmentation.

The authors proposed two different ways of extracting the transformed image. The first involves a network for segmentation in Cartesian coordinates, then calculating the center of mass of the segmented module. This center becomes the zero point against which segmentation is again performed. The final obtained mask is transformed into Cartesian coordinates. This solution is particularly useful when applied to existing methods. Another advantage of this approach is the ease of use of transfer learning.

The second proposed path is the focal point predictor. Instead of trying to segment the whole image, the network only focuses on the focal point. Based on empirical experience, the authors decided to use 8 connected 'Hourglass' blocks. The authors designed their methods exclusively for liver parenchyma segmentation. The optimization used yields about 3.5% of Dice metric gain for the different networks when using the two-segmentation approach and about 2% for the center-determining network approach.

2.4 Vision Transformer

The VitSeg (Vision Transformer segmenter) architecture is a type of neural network model designed for processing visual data, such as images. The VitSeg model is based on the Visual Image Transformer architecture [6], which in turn is based on the Transformer architecture developed for natural language processing tasks [20]. The VitSeg (Vision Transformer) model was proposed in [18].

The VitSeg architecture consists of a sequence of layers that process image patches, which are smaller rectangular regions of an image. The image patches are linearly projected to a lower dimension and then passed through a series of Transformer blocks. Each Transformer block contains a multi-head attention mechanism and a feedforward neural network layer. The attention mechanism allows the network to selectively focus on important features of the image, while the feedforward layer applies non-linear transformations to these features.

After the Transformer blocks, the resulting feature vectors are passed through a final classification layer that predicts the class label of the image. To improve performance, the VitSeg model is often pre-trained on a large dataset using a self-supervised learning approach, where the model learns to predict the relative positions of image patches within the same image, without requiring explicit class labels. Overall, the VitSeg architecture represents a powerful and effective approach for processing visual data using deep learning techniques.

The primary benefit of transformers resides in their superior speed in comparison to RNN and LSTM networks. Moreover, a contributory factor to their increasing prevalence is their significantly enhanced performance relative to other algorithms.

Transformers suffer from several disadvantages, such as their demanding computational requirements for training, necessitating a truly vast quantity of data to be trained on, and being relatively nascent models, we have yet to fully comprehend their hyperparameters. Additionally, certain examples have revealed that transformers may underperform on hierarchical data.

3 Method and Materials

Providing a succinct analysis of all the methodologies identified in the literature review is deemed unfeasible. Therefore, the decision was made to select four models for comparison with the baseline method (UNet) in a standardized testbed. Our motivation was to use the best method in a diagnostic support system for comparative research in oncology. The basis of this choice was that the authors made the source code (models implemented in the PyTorch framework) publicly available, and the proposed method achieved similar results to modern methods (i.e. 95% Dice for parenchyma, if applicable). We decided to evaluate four models: Unet++, QUA-net, PIT, and VitSeg, which are described in the previous section. In case of PIT, it was adapted to be used in parenchyma and tumor segmentation. Center point was calculated basing on parenchyma mask. Then segmentation on both parenchyma and tumor was conducted on polar base images.

For ViTSeg it was adopted to be used in medical imaging. Dataset specific parameters such as number of channels, resolution, number of classes were set. Also, basic hyperparameters were changed to find proper configuration.

Our study was conducted for multi-class segmentation (tumor and parenchyma). The focus of the experiment is the segmentation of the tumor and, to a lesser extent, the parenchyma. To date, a model based on PIT and VitSeg- has not been used in the segmentation of liver tumors.

3.1 Data Set

We test the proposed method on a publicly available dataset. The LiTS dataset [3] includes 131 abdominal CT scans that were enhanced with contrast (venous phase) and were taken from six different clinical sites using varying scanners and protocols. This has resulted in a wide range of differences in terms of spatial resolution and field of view of the scans. The data and masks were annotated by radiologists of varying experience. The dimensions of the voxel are [0:60–0:98, 0:60–0:98, 0:45–5:0] mm, and most scans are pathological, with tumors of varying sizes, metastases, and cysts. All axial slices in the scans have a size of 512×512, but the number of slices differs greatly, ranging from 42 to 1026.

3.2 Data Preprocessing

Medical image volume data preprocessing was carried out in a slice-wise fashion. First, the Hounsfield unit values were windowed in the range $[-200, 250]$ to enhance the contrast between the liver and the surrounding organs and tissues, to exclude irrelevant organs and tissues.

Next, we normalized values to range from $[-1, 1]$. Additionally, we discard all slices that consist only of background mask, to improve training time. For polar images, at the start of training all central points were calculated using the backbone network(UNet). Then, the polar transformation was applied before feeding data to the network. Note, that for validation and test runs it was transformed back to Cartesian coordinates before Dice and loss calculations.

3.3 Selected Metric

There are several metrics for evaluating the segmentation and localization of the medical image. Each of these metrics is based on a statistical interpretation of the prediction with respect to the truth of the ground. In [19], the authors have presented 20 metrics from different categories. The authors contend that the most suitable binary metrics are the Intersection over Union (IoU), also known as the Jaccard index, and Dice similarity coefficient. We decided to choose Dice coefficient for the segmentation problem that is defined in Eq. (1).

$$Dice = \frac{2TP}{2TP + FP + FN} \tag{1}$$

where: TP – true positives, FP – false positives, FN - false negatives. The Dice metric is more adept at handling the issue of unbalanced classes. In instances where one class exhibits a substantially smaller pixel count relative to the other, the Jaccard metric may yield a value approximating zero, thereby failing to accurately reflect the extent of overlap between these classes.

3.4 Training

The study employed a four-fold cross-validation approach to conduct the experiments. All networks are trained with Adam optimizer with a learning rate equal to 1e-05 for VitSeg and 1e-04 for other models. We use a mix of Dice loss and binary cross-entropy loss with equal weight. The early stop mechanism was used for all runs containing the validation dataset. ReduceLROnPlateau was used with patience=3, factor=0.1, threshold=0.0000001. Note, it was only applied to those sets, that contained the validation subset. All networks used the same methodology and splits, except one described above. Framework can be found at https://github.com/DominikKwiatkowski/tomography.

4 Results and Discussion

In Table 1 the results of the comparison of UNet, UNet++, QUAnet, PIT, and VitSeg in the context of a liver tumor segmentation evaluation are presented. It can be seen that under the evaluation of four indicators, the proposed VitSeg outperforms tumor liver CT image segmentation in five methods, increasing the Dice coefficient by 4.1% over UNet++.

Table 1. Segmentation accuracy in multiclass experiment.

Model	Overall		Parenchyma		Tumor	
	Dice	Std. dev	Dice	Std. dev	Dice	Std. dev
UNet	0.721	0.062	0.947	0.008	0.495	0.112
UNet++	0.735	0.062	**0.952**	**0.008**	0.519	0.112
QUANet	0.720	0.051	0.950	0.005	0.490	0.101
PIT	0.718	0.053	0.944	0.008	0.492	0.102
VitSeg	**0.751**	**0.050**	0.940	0.007	**0.560**	**0.010**

We randomly extracted a case of CT data from the test set and selected one of the slices. Visualization and comparing the algorithmic segmentation results of the slice and the ground labels are shown on the comparison chart in Fig. 4.

4.1 Discussion

Experiments conducted on the LiTS dataset show that Transformer can focus on the target area of the whole image. In most cases, ViTSeg tends to create discontinuous masking compared to UNet like networks. Also, it tends to create smaller masks compared to the ground truth as shown in Fig. 4.1, being conservative in prediction. One can observe a thin boundary (with a width 1–2 pixels) separating the mass region (tumor) from surrounding tissue. Also, it brings outstanding performance in tumor segmentation compared with previous methods.

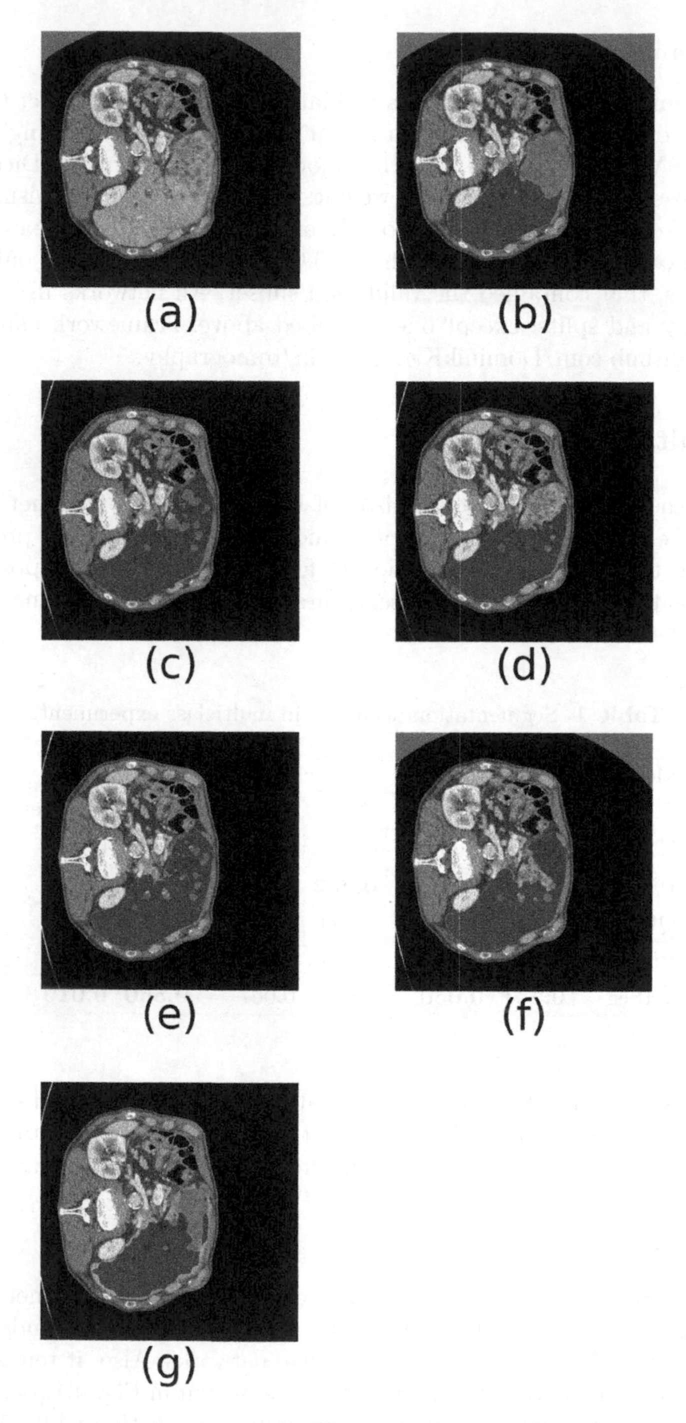

Fig. 1. Inference results for all models. The parenchyma is shown in a red region, and the tumor is in the green one. (a) source image (b) ground truth (c) UNet (d) UNet++ (e) QUANet (f) PIT (g) VitSeg. (Color figure online)

Since tumor is more significant task in real environment, it may improve overall knowledge in this topic.

Differently, UNet like networks tend to create a bigger mask than it is in the ground truth. As a result, UNet based networks are performing narrowly better in liver parenchyma segmentation (95.2% UNet++ vs. 94% VitSeg). The VitSeg network is particularly beneficial for segmentation of the objects with fuzzy boundaries and small targets. This is likely due to the larger capacity of the model (UNet - 7 852 035 parameters, UNet++ - 9 162 819, VitSeg - 36 812 398, respectively) (Fig. 1).

In the case of PIT model, the obtained results are characterized by slightly lower performance compared to the reference model (UNet, ca. 1%), however, in some cases showed a significant increase in performance. As a result, it may be beneficial to test this method for certain models.

UNet++, as a member of convolutional neural network architectures, appears to exhibit superior performance (ca. 1%) in accurately segmentation on the LiTS dataset when compared to other CNN methods.

From experiments, Unet improvements like QUANet and UNet++ does not bring much improvement. Transformer architecture seems to have bigger potential in medical segmentation and can lead to much better results with their scalability (Fig. 2).

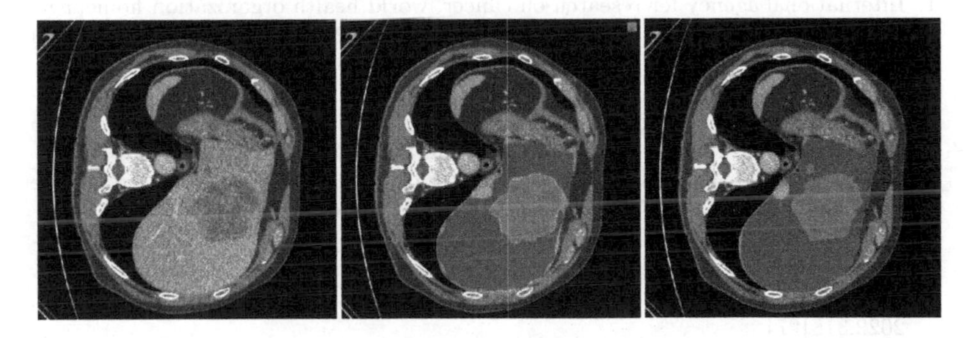

Fig. 2. The tumor margin effect. The parenchymal region is discontinuous from the tumor region. The parenchyma is shown in a red region, and the tumor is in the green one. The source image is on the left, the prediction result using the VitSeg in the middle, and the ground truth is on the right side. (Color figure online)

5 Future Works

In this paper, the segmenter based on a Visual Image Transformer is adopted and applied to liver tumor segmentation. After evaluating five automatic methods, the experiments conducted on the LiTS dataset showed that VitSeg performs competitively in liver CT image segmentation. The improvement is duc to the combination of dense skip connections and attention mechanisms. The experiments also demonstrated that VitSeg can increase the weight of the target region while inhibiting the background region that is not relevant to the segmentation task. Potential benefits can be achieved by utilizing an ensemble of VIT and UNet++ models for parenchyma and tumor segmentation.

Expansion of the model to include segmentation of a greater number of pathological classes is planned.

After evaluating the capabilities of VitSeg in three-dimensional CT image analysis, its implementation is planned as part of the STOS project in the Tri-City Academic Supercomputer Center (Poland).

Acknowledgements. The paper has been supported by founds of Department of Computer Architecture Faculty of Electronics, Telecommunications and Informatics, Gdańsk University of Technology and "The establishment of the STOS (Smart and Transdisciplinary Knowledge Services) Competence Centre in Gdańsk in the field of R&D infrastructure" project (RPPM.01.02.00-22-0001/17). The project is co-financed by the European Regional Development Fund under the Regional Operational Program of the Pomeranian Voivodeship for the years 2014–2020

References

1. International agency for research on cancer, world health organization homepage. https://gco.iarc.fr/today/online-analysis-map?v=2020&mode=cancer. Accessed 4 Apr 2023
2. Benčević, M., Galić, I., Habijan, M., Babin, D.: Training on polar image transformations improves biomedical image segmentation. IEEE Access **9**, 133365–133375 (2021). https://doi.org/10.1109/ACCESS.2021.3116265
3. Bilic, P., et al.: The liver tumor segmentation benchmark (LiTS). Med. Image Anal. **84**, 102680 (2023). https://doi.org/10.1016/j.media.2022.102680
4. Di, S., Zhao, Y.Q., Liao, M., Zhang, F., Li, X.: TD-Net: a hybrid end-to-end network for automatic liver tumor segmentation from CT images. IEEE J. Biomed. Health Inform. **27**(3), 1163–1172 (2023). https://doi.org/10.1109/JBHI.2022.3181974
5. Dong, C., et al.: A knowledge-based interactive liver segmentation using random walks. In: 2015 12th International Conference on Fuzzy Systems and Knowledge Discovery (FSKD), pp. 1731–1736 (2015). https://doi.org/10.1109/FSKD.2015.7382208
6. Dosovitskiy, A., et al.: An image is worth 16x16 words: transformers for image recognition at scale (2021)
7. Farzaneh, N., et al.: Atlas based 3D liver segmentation using adaptive thresholding and superpixel approaches. In: 2017 IEEE International Conference on Acoustics, Speech and Signal Processing (ICASSP), pp. 1093–1097 (2017). https://doi.org/10.1109/ICASSP.2017.7952325
8. Fernández, J.G., Fortunati, V., Mehrkanoon, S.: Exploring automatic liver tumor segmentation using deep learning. In: 2021 International Joint Conference on Neural Networks (IJCNN), pp. 1–8 (2021). https://doi.org/10.1109/IJCNN52387.2021.9533649

9. Hong, L., Wang, R., Lei, T., Du, X., Wan, Y.: QAU-Net: quartet attention U-Net for liver and liver-tumor segmentation. In: 2021 IEEE International Conference on Multimedia and Expo (ICME), pp. 1–6 (2021). https://doi.org/10.1109/ICME51207.2021.9428427
10. Kitrungrotsakul, T., Han, X.H., Chen, Y.W.: Liver segmentation using superpixel-based graph cuts and restricted regions of shape constrains. In: 2015 IEEE International Conference on Image Processing (ICIP), pp. 3368–3371 (2015). https://doi.org/10.1109/ICIP.2015.7351428
11. Li, C., Li, A., Wang, X., Feng, D., Eberl, S., Fulham, M.: A new statistical and Dirichlet integral framework applied to liver segmentation from volumetric CT images. In: 2014 13th International Conference on Control Automation Robotics & Vision (ICARCV), pp. 642–647 (2014). https://doi.org/10.1109/ICARCV.2014.7064379
12. Lim, S.J., Jeong, Y.Y., Ho, Y.S.: Automatic liver segmentation for volume measurement in CT images. J. Vis. Commun. Image Represent. **17**(4), 860–875 (2006). https://doi.org/10.1016/j.jvcir.2005.07.001, https://www.sciencedirect.com/science/article/pii/S1047320305000702
13. Norajitra, T., Maier-Hein, K.H.: 3D statistical shape models incorporating landmark-wise random regression forests for omni-directional landmark detection. IEEE Trans. Med. Imaging **36**(1), 155–168 (2017). https://doi.org/10.1109/TMI.2016.2600502
14. Qiao, S., Xia, Y., Zhi, J., Xie, X., Ye, Q.: Automatic liver segmentation method based on improved region growing algorithm. In: 2020 IEEE 4th Information Technology, Networking, Electronic and Automation Control Conference (ITNEC), vol. 1, pp. 644–650 (2020). https://doi.org/10.1109/ITNEC48623.2020.9085126
15. Rela, M., Suryakari, N.R., Reddy, P.R.: Liver tumor segmentation and classification: a systematic review. In: 2020 IEEE-HYDCON, pp. 1–6 (2020). https://doi.org/10.1109/HYDCON48903.2020.9242757
16. Ronneberger, O., Fischer, P., Brox, T.: U-Net: convolutional networks for biomedical image segmentation (2015). https://doi.org/10.48550/ARXIV.1505.04597
17. Song, X., Cheng, M., Wang, B., Huang, S., Huang, X.: Automatic liver segmentation from CT images using adaptive fast marching method. In: 2013 Seventh International Conference on Image and Graphics, pp. 897–900 (2013). https://doi.org/10.1109/ICIG.2013.181
18. Strudel, R., Garcia, R., Laptev, I., Schmid, C.: Segmenter: transformer for semantic segmentation (2021)
19. Taha, A.A., Hanbury, A.: Metrics for evaluating 3D medical image segmentation: analysis, selection, and tool. BMC Med. Imaging **15**(1), 29 (2015). https://doi.org/10.1186/s12880-015-0068-x
20. Vaswani, A., et al.: Attention is all you need (2017). https://doi.org/10.48550/arXiv.1706.03762
21. Wang, W., Ma, L., Yang, L.: Liver contour extraction using modified snake with morphological multiscale gradients. In: 2008 International Conference on Computer Science and Software Engineering, vol. 6, pp. 117–120 (2008). https://doi.org/10.1109/CSSE.2008.447
22. Yuan, Z., Wang, Y., Yang, J., Liu, Y.: A novel automatic liver segmentation technique for MR images. In: 2010 3rd International Congress on Image and Signal Processing, vol. 3, pp. 1282–1286 (2010). https://doi.org/10.1109/CISP.2010.5647676

23. Zhanpeng, H., Qi, Z., Shizhong, J., Guohua, C.: Medical image segmentation based on the watersheds and regions merging. In: 2016 3rd International Conference on Information Science and Control Engineering (ICISCE), pp. 1011–1014 (2016). https://doi.org/10.1109/ICISCE.2016.218
24. Zhou, Z., Siddiquee, M.M.R., Tajbakhsh, N., Liang, J.: UNet++: redesigning skip connections to exploit multiscale features in image segmentation. IEEE Trans. Med. Imaging **39**(6), 1856–1867 (2020). https://doi.org/10.1109/TMI.2019.2959609

Intelligent Index Tuning Using Reinforcement Learning

Michał Matczak(✉) and Tomasz Czochański

Faculty of Electronics, Telecommunications and Informatics, Gdańsk University of Technology, Gdańsk, Poland
michalmatczak22@gmail.com

Abstract. Index tuning in databases is a critical task that can significantly impact database performance. However, the process of manually configuring indexes is often time-consuming and can be inefficient. In this study, we investigate the process of creating indexes in a database using reinforcement learning. Our research aims to develop an agent that can learn to make optimal decisions for configuring indexes in a chosen database. The paper also discusses an evaluation method to measure database performance. The adopted performance test provides necessary documentation, database schema (on which experiments will be performed) and auxiliary tools such as data generator. This benchmark evaluates a selected database management system in terms of loading, querying and processing power of multiple query streams at once. It is a comprehensive test which results, calculated on measured queries time, will be used in the reinforcement learning algorithm. Our results demonstrate that used index technique requires repeatable benchmark with stable environment and high compute power, which cause cost and time demand. The replication package for this paper is available at GitHub: https://github.com/Chotom/rl-db-indexing.

Keywords: Index tuning · Databases · Reinforcement learning · Benchmark

1 Introduction

Database indexing is a problem most often solved by human analysis and manual changes. To better understand the problem and improve the efficiency of database indexing, researchers have developed various automated tools and techniques that can analyze the database and make index selections based on specific criteria, such as [2, 10]. The index itself is a structure or function designed to make data retrieval more efficient. Many times, it is a tree-based structure or a structure that uses a hash function, explained in [13, 14]. However, adding an index is costly and fraught with increased time of inserting, deleting and updating records. In addition, it increases the usage of a disk space. Therefore, the selection of indexes should be used only when there is a need for it because their excess can cause negative effects in the performance of the database. The motivation for this article is to address the challenges faced in index tuning. As the complexity of queries used in the database increases, the difficulty of selecting optimal indexes does as well.

A. Abelló et al. (Eds.): ADBIS 2023, CCIS 1850, pp. 523–534, 2023.
https://doi.org/10.1007/978-3-031-42941-5_45

In this study, we have used TPC-H [16] as our benchmark to evaluate the performance of the index tuning process. TPC-H is a widely used benchmark in the industry for evaluating the performance of relational databases. It consists of a suite of business-oriented ad hoc queries [6] and concurrent data modifications. This benchmark illustrates decision support systems that examine large volumes of data, execute queries with a high degree of complexity, and give answers to critical business questions. It is a crucial component in our environment, as it provides a standardized and objective way to determine reward in a reinforcement learning algorithm. Although, based on our experience and example TPC-H report [15], a single benchmark run is time-consuming, which is problematic in our case as we need to run multiple times.

Authors' contribution in this paper focuses on several aspects of utility of reinforcement learning with usage of TPC-H benchmark on real-life database management system. Through our research and experimentation, we were able to demonstrate the effectiveness of using reinforcement learning algorithms to automate the process of setting indexes in a database in a simulated environment. Additionally, we were able to identify the challenges that come with using this approach in practice, such as dealing with large standard deviation of benchmark score, the complexity of evaluating the database states or time costly model training. Our research also discusses results from other papers and highlights the importance of considering the limitations and challenges that come with a reinforcement learning approach.

The rest of the paper is organized as follows: Sect. 2 discusses the current and related work on this topic that has been referenced in this article. Section 3 provides a background on the benchmark, environment, and agent used in this study. Section 4 presents the solutions used in this project and provides detailed information on the reinforcement learning algorithm. In Sect. 5, we present the conducted experiments. Section 6 presents a discussion of the results and a comparison of these results with other approaches. Finally, in Sect. 7, we draw conclusions about the method and benchmark used in this study.

2 Related Work and Contribution

2.1 Smart-IX

Our work builds upon the work of Licks et al. [8]. Their algorithm is similar to the one used by us, however it is unclear what were the state, action features. Later, Licks et al. [9] extended the algorithm by introducing a neural network approximator, which is a better idea in most cases. However, as we later showed, linear features approximator seems sufficient for the discussed task. The Smart-IX results differ significantly than ours. We suspect that it is caused by differences in implementation of TPC-H Benchmark. In our opinion, the benchmark in Smart-IX is wrong, at least in one aspect. According to the official TPC-H documentation, each of 22 queries from query stream needs to be executed in the order specified in the documentation. On contrary, queries during query stream in Smart-IX implementation are always run in the same order. In addition, the procedure of inserting the data to database raises doubts as well. The fact that the DBGen error explained at the end of Sect. 3.1 did not affect their TPC-H results is suspicious as well. The reason why the authors implemented their solution in such a way is not explained in their paper. Fortunately, in their paper there is the final index configuration found by their algorithm, which allowed us to compare the results.

2.2 The Case for Automatic Database Administration using Deep Reinforcement Learning

This work [3] explores the use of deep reinforcement learning to administer a database management system (DBMS). It introduces the concept of NoDBA, which applies deep reinforcement learning to automatically tune the system and make recommendations for index selection based on specific workloads. The study highlights the potential of this approach in simplifying DBMS configuration and maintenance tasks. It based on TPC-H tables in PostgreSQL, which was used for workload analysis. The authors used different approach than us and instead of using whole benchmark, they used own queries to optimize indexes. This means that workload measure disregards inserting and deleted rows from tables. Contrary to this approach, we wanted to check whether it is possible to evaluate the index configuration with the usage of complicated benchmark.

3 Background

3.1 TPC-H

TPC-H is a benchmark for evaluating the performance of relational database management systems. This benchmark measures the performance QphH@Size, which is a metric that indicate the number of queries that the system can execute per hour, given a specific database size. This metric gives us a clear and objective way to compare the performance of different indexing configurations and to evaluate the effectiveness of our algorithm. TPC-H consists of a set of SQL queries and data generation tools (called DBGen and QGen) that are designed to simulate a decision support system, with a focus on ad-hoc queries [6] and concurrent modifications to the data.

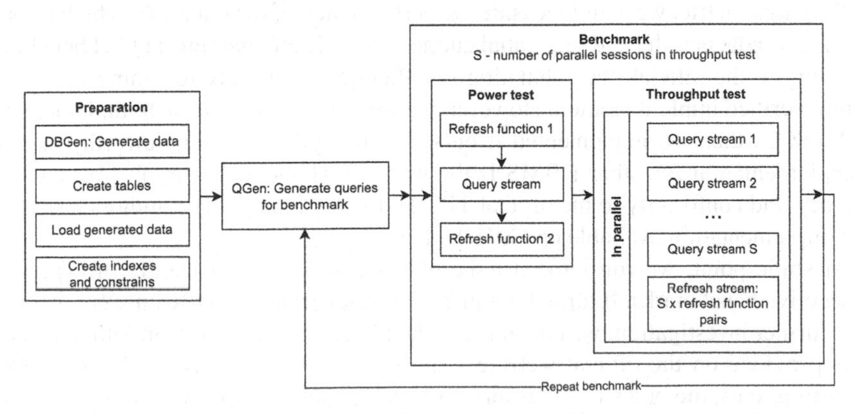

Fig. 1. TPC-H Benchmark structure according to [4] adapted for database indexing problem.

According to Fig. 1 the benchmark includes:

- Refresh function 1 - a sequence of transactions with the database, in which multiple records are inserted.
- Refresh function 2 - a sequence of transactions with the database, in which a certain number of records are deleted. These are the entries already in the table and differ from the data inserted in the refresh function 1.
- Query stream – a sequence of 22 consecutive queries sent to the database in a specific order. Each query stream is numbered, enabling us to determine the order of the queries as specified in the specification [16].

After running both Power test and Throughput test, we calculate metrics with following formulas:

$$Power@Size = \frac{3600 * SF}{\sqrt[24]{\prod_{i=1}^{22} QI(i,0) * \prod_{j=1}^{2} RI(j,0)}}$$

$$Throughput@Size = \frac{S * 22}{T_s} * 3600 * SF$$

where:

- SF – scale factor (size of database in GB)
- S – number of query streams
- QI(i, 0) – time of execution of i-th query in query stream
- RI(j, 0) – time of execution of j-th refresh function
- T_s – time of execution of throughput test

After calculating both metrics, we can finally calculate the final metric, which is considered the benchmark result:

$$QphH@Size = \sqrt{Power@Size * Throughput@Size}$$

With that metric, we could measure the performance of the database, which tests the ability to handle parallel and sequential queries. Since implementing TPC-H benchmark according to the official documentation is difficult, we strongly recommend trying to acquire verified implementation. However, we were unable to find such implementation for MySQL, thus we implemented it ourselves in Python. One noteworthy example of implementation for other DBMS is the work of Thanopoulou et al. [4], due to its accuracy and conformity to the official TPC-H documentation. The authors have made their implementation available on GitHub [12].

At some point, we run into an issue of database's size constantly growing after consecutive benchmarks. It turned out that the ids generated by DBGen were incorrect. After further investigation, we encountered the file called 'BUGS' from official TPC-H Tools provided on the official website – the issue was known as 'Problem #00062'. According to us, the bug still exists and does affect each scale factor. To overcome this issue, we had to manually modify the generated ids. This issue was worth to mention here in case of results reproducibility.

4 Methodology

In this section, we are going to define the crucial elements of our solution for this problem. The approach utilizes Q-learning [5] algorithm, which is a reinforcement learning algorithm used to solve Markov decision processes (MDPs) [11]. Therefore, the concept of agent and environment will be introduced and the problem will be formally defined as a MDP, before final explanation of the algorithm.

4.1 Agent and Environment

In reinforcement learning, an agent interacts with an environment to achieve a certain goal. The agent is responsible for making decisions and taking actions, while the environment provides the agent with rewards. The agent's goal is to learn a policy that maximizes the cumulative reward over time. The environment can be a simulated or real-world system and can be complex and dynamic. In order to effectively learn, the agent must be able to perceive the state of the environment, and the environment must be able to provide the agent with meaningful feedback in the form of rewards (Fig. 2).

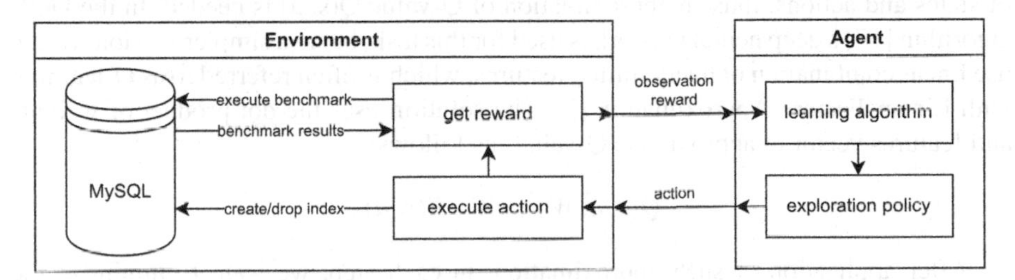

Fig. 2. Agent-Environment communication

Based on the action selected by the agent, a step is executed. In this case, it is the deletion or creation of an index on the chosen column in the table. The database is updated, and then the environment evaluates changes – the TPC-H benchmark is called, which returns a measurement interpreted as a reward. After the step is completed, the environment returns the observation along with the reward to the agent. The agent works by using a policy, which is a function that maps observation of the environment to action to be taken. The policy is updated over time using feedback in the form of rewards, and methods like Q-learning [5] in our case.

4.2 Markov Decision Process

To define this problem as MDP [11], state space, action space, reward and state-transition probabilities need to be addressed. State space, which is current database state can be defined as follows. For each of n columns that can be indexed, we introduce a binary feature (either 1 or 0), which describes whether in the current database state the index on the corresponding column is created or not. State can thus be encoded with an n-element vector of binary numbers. We get $2n + 1$ possible states. Action space is defined similarly – taking an action means either creating or removing an index on a selected column. This gives us 2n possible actions, which can be encoded by a natural number from 0 to 2n–1. The reward for moving from state s to state s' by taking action a is simply a TPC-H Benchmark result on a database with indexes configured as encoded by s'. State-transition probabilities are not used in this setup, since by choosing action, we know exactly in which state we end up.

4.3 Algorithm

To approach this problem, we chose the Q-learning algorithm [5], because we believed it was capable of correctly modeling the relations between the index configuration and TPC-H benchmark results, based on a later described intuition. The algorithms is based on the following Bellman optimality equation:

$$Q(s, a) = \mathbb{E}\left[r_{t+1} + \gamma \max_{a' \in A} Q(s', a') \right]$$

Classical tabular implementation of Q-learning is not possible due to the large amount of states and actions, thus an approximation of Q-value Q(s, a) is needed. In the DQN algorithm [17] a deep neural network is used for this task, but in a simpler version we can use linear combination of hand-crafted features, which is often referred to as Q-learning with Linear Feature Approximation [7]. The solution uses the dot product of weights and features vector to approximate Q-values as follows:

$$\hat{Q}(s, a, W) = W \cdot F(s, a)$$

After application of such approximation, in each step, we want to minimize the following quadratic error between predicted and target Q value:

$$J(W) = \left(Q(s, a) - \hat{Q}(s, a, W) \right)^2$$

where for target Q(s, a) value, we take the following:

$$Q(s, a) = r + \gamma \max_{a' \in A} \hat{Q}(s', a', W)$$

Using this equation requires selection of state, action features, that differentiate the term Q(s', a'), allowing us to select the maximum Q-value based on action a'. An often-used approach, recommended by [1] is to create a different set of weights for each possible action. This solution, however, does not scale well for large action spaces, because the number of weights increases linearly with the number of possible actions and the need for broad exploration is arising. For these reasons, the number of iterations required to achieve a satisfactory level of convergence increases drastically. The way we propose to handle this issue is to introduce one weight for each possible action. Thus, vector F(s, a) represents a combination of n state features defined earlier and 2n features, of which one is 1, if a is the corresponding action, and the rest is 0. The intuition of this approach could be explained by each possible index receiving a weight by which it is being rated. When this value is positive, the introduction of this index is considered beneficial by the model, otherwise it is not. Additionally, the action, which is strictly connected to the next state, receives a value by which it can be rated as well.

For stability and performance reasons, we use a technique called experience replay [18]. In each step, a transition tuple, consisting of state, action, reward and next state is added to the buffer. We then randomly sample the buffer in order to acquire a batch, denoted by the letter B, of tuples s, a, r, s', which we use to perform gradient descent to minimize the error function, which after applying experience replay can be redefined as the following:

$$J(W) = \frac{1}{|B|} \sum_{s,a,r,s' \in B} \left(Q(s,a) - \hat{Q}(s,a,W) \right)^2$$

After final substitution, we are able to calculate the gradient with respect to W:

$$\nabla_W J(W) = \frac{1}{|B|} \sum_{s,a,r,s' \in B} \frac{1}{2} \left(r + \gamma \max_{a' \in A} W \cdot F(s', a') - W \cdot F(s,a) \right) \cdot F(s,a)$$

5 Experiments

5.1 Training in Simulated Environment

Before running the proper training on actual machine with TPC-H Benchmark, we tested the agent on simulated environment to check, if everything was implemented correctly. The simulated environment we propose reflects the intuition of agent described earlier (see Sect. 4.3). For each theoretical column, a random number is generated, which would resemble the potential contribution of index created on this column. The bigger the value, the more important the index on this column is. The reward function in such an environment is a sum of all values for each column, with generated noise added randomly. Specifically, 1/3 of the values were positive, and 2/3 negative – making a total of 45 values. This means that a random selection of columns would most likely produce negative results, and the agent would have to learn the better strategy.

Table 1. Simulation results.

Index configuration	Positive indexes	Negative indexes	Reward
Best possible	15/15	0/30	837.36
Trained strategy	13/15	5/30	671.19
Fully indexed	15/15	30/30	−305.82

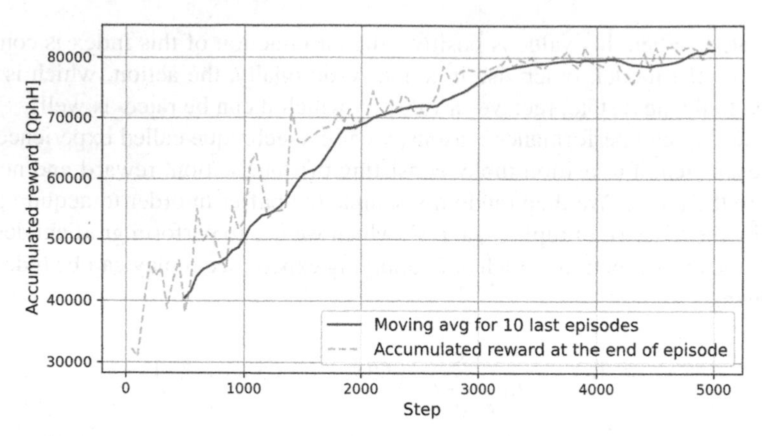

Fig. 3. Sum of rewards throughout training for simulated environment.

5.2 Training in Database Environment

As presented in the article, the final environment will be a MySQL database with implemented TPC-H benchmark. The limited resources allowed for tests to be conducted only on a 100 MB data. Limitations were caused, among others, by the complexity of the benchmark itself, which duration significantly prolonged the learning process, but also by the available hardware resources. The machine specification on which the MySQL server was run was Intel Core i3–8100, 3.60 GHz CPU, 16 GB RAM, Toshiba KXG60ZNV256G SSD Disk. The agent was run on another computer in the same local network. With this configuration, the duration of a training lasted over 72 h, hence conducting tests on a larger amount of data exceeded our possibilities. The number of episodes was set to 120, and the number of steps for each episode was set to 50, the discount factor $\gamma = 0.8$. The number of experience replay's samples is 30. These hyperparameters give us the best balance between good results and the costs of training.

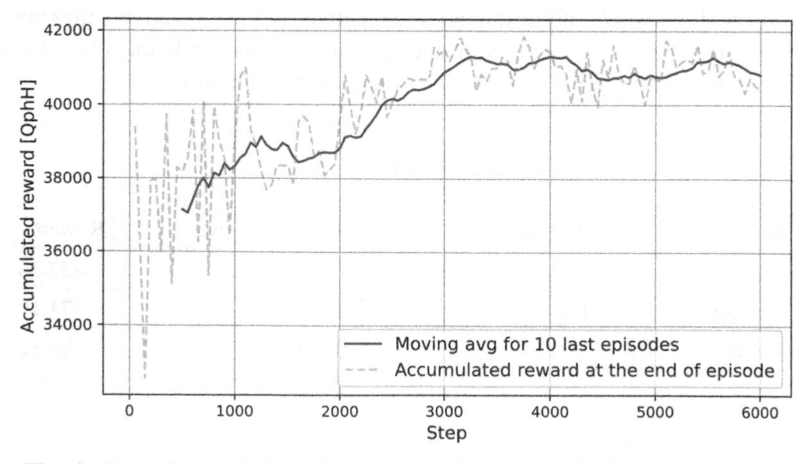

Fig. 4. Sum of rewards throughout training for 100 MB database environments

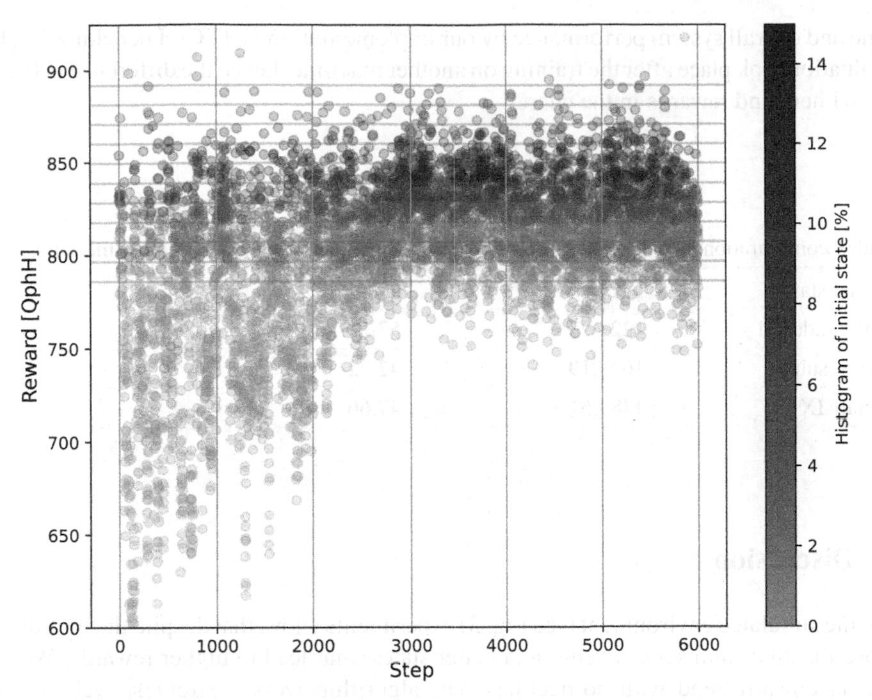

Fig. 5. Reward throughout training in 100 MB database environment with overlaid histogram of benchmark scores in initial state.

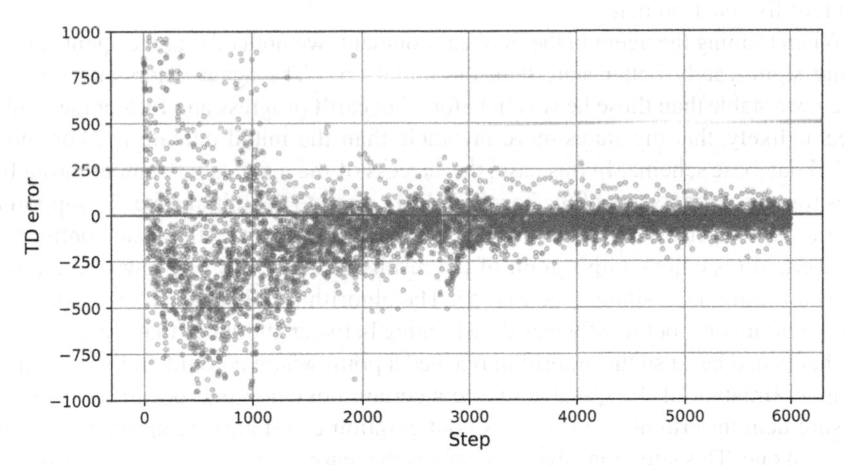

Fig. 6. Temporal difference error for each step performed by the agent in 100 MB database environment.

5.3 Indexes Configuration Evaluation

Apart from reinforcement learning experiments, different indexes configurations have been checked as well. Individual configurations were tested in terms of query response time and overall system performance by our implementation of TPC-H benchmark. This evaluation took place after the training on another machine, hence the difference between QphH here and rewards in the chart.

Table 2. Index configuration evaluation

Index configuration	QphH (mean of 10)	Standard deviation for QphH measures
Initial state	1721.16	47.96
Fully indexed	1221.77	53.52
Our results	1690.19	42.22
Smart-IX [8]	1489.51	47.66

6 Discussion

For the simulated environment (see Fig. 3) experiments show, that despite the introduced noise the algorithm selects better and better states, that lead to higher rewards. We can see an upward trend with no declines. The algorithm works extremely well in such environment (see results in Table 1). It is worth to mention, that the environment was designed to implement the same principles as the agent, which may be different in the final real-live environment.

When training the agent in the final environment, we noticed that the agent is unable to find significantly better state than the initial one. The agent enters states that are more rewardable than those he was in before, but can't progress any further (see Fig. 4). It seems likely, that the states more favorable than the initial one are not common in TPC-H database schema. In this case, the success of the agent is movement through the states for which the results of the benchmark are close to the initial state. To support this hypothesis, we performed numerous TPC-H benchmarks for initial state configuration and overlaid (as color) a histogram of the results onto the chart of rewards that agent received during the training (see Fig. 5). The algorithm failed to achieve states better than the initial one, but mostly was transitioning between the states that were close to it. This happened because the algorithm reached a point where it could no longer improve the approximation of the Q-value due to a certain deviation that occurred. With large measurement uncertainties each index configuration could have been rated worse than the initial one. This situation adversely affects the learning process of the algorithm (see Fig. 6). Performed experiments also suggest that TPC-H benchmark is constructed in a way that does not take any benefit from the introduction of additional indexes. The ad hoc queries introduced by benchmark are too complex for this problem. Finding an optimal solution, better than initial state, in this case may not be possible.

The algorithm in comparison with DQN lacks the ability to discover complicated relationship between indexes. For example, if two indexes p and q work negatively alone, but positively together. However, because of significantly lower number of parameters, it takes much faster to train and, since we lacked the computational power, we couldn't test the performance of DQN.

The results of our algorithm showed higher efficiency than index configuration recommended by the solution described in Smart-IX article (see Table 2). This is probably due to the significant differences in benchmark implementation. This supposition is also confirmed by the results they made available in the paper. According to them, measurements carried out on full indexing gave much better results than the initial condition. In our case, benchmark returns worst results in such scenario. Since TPC-H benchmark tests performance of queries, inserts and deletes, it is expected that countless indexing will slow down insert and delete operations, which will negatively affect TPC-H benchmark results.

7 Conclusion

In summary, the paper presents a complex analysis of index tuning and an attempt to adapt this problem to a machine learning algorithm. Performed experiments confirm that the algorithm works, however, TPC-H as a selected database schema and a way of evaluating index configuration seems improper. We recommend using this algorithm only for specific databases for which we can define a function to evaluate the current index configuration.

The TPC-H benchmark is burdened with high costs and long evaluation, combining this with multiple training steps leads to a very long agent training time, which is impractical. What is more, as the database is a live system and fixed compute power cannot be guaranteed, the benchmark results may differ between evaluations in the same states. This brings another problem that benchmark results are burdened with a large standard deviation. So large, in fact, that the potential gain from creating an index is smaller than the deviation, which makes the training process difficult. In a simulated environment where these problems were not present, the agent demonstrated that the algorithm itself could work with such defined assumptions.

In summary, the environment and the agent work according to the intended purposes and functionally meet the expectations of their application. Unfortunately, due to its very long training time in real-live scenarios, we claim that costs would be higher than potential profit from using it.

References

1. Geramifard, A., Walsh, T.J., Tellex, S., Chowdhary, G., Roy, N., How, J.P.: A tutorial on linear function approximators for dynamic programming and reinforcement learning. Found. Trends® Mach. Learn. **6**(4), 375–451 (2013). https://doi.org/10.1561/2200000042
2. Nanda, A.: Automatic indexing with Oracle Database. Oracle (2021). https://www.oracle.com/news/connect/oracle-database-automatic-indexing.html. Accessed 17 Apr 2023
3. Sharma, A., Schuhknecht, F., Dittrich, J.: The case for automatic database administration using deep reinforcement learning (2018). arXiv preprint. https://arxiv.org/pdf/1801.05643.pdf

4. Thanopoulou, A., Carreira, P., Galhardas, H.: Benchmarking with TPC-H on off-the-shelf hardware. In: 14th International Conference on Enterprise Information Systems. Springer, Wroclaw pp. 205–208 (2012). https://doi.org/10.5220/0004004402050208
5. Watkins, C.J., Dayan, P.: Q-learning. Mach. Learn. **8**, 279–292 (1992). https://doi.org/10.1007/BF00992698
6. Chu, E., Baid, A., Chai, X., Doan, A., Naughton, J.: Combining keyword search and forms for Ad Hoc querying of databases. In: Proceedings of the 2009 ACM SIGMOD International Conference on Management of Data. Association for Computing Machinery, pp. 349–360 (2009). https://doi.org/10.1145/1559845.1559883
7. Melo, F.S., Ribeiro, M.I.: Q-Learning with linear function approximation. In: Bshouty, N.H., Gentile, C. (eds.) COLT 2007. LNCS (LNAI), vol. 4539, pp. 308–322. Springer, Heidelberg (2007). https://doi.org/10.1007/978-3-540-72927-3_23
8. Paludo Licks, G., Colleoni Couto, J., Miehe, P., De Paris, R., Dubugras Ruiz, D., Me-neguzzi, F.: SMARTIX: a database indexing agent based on reinforcement learning. Appl. Intell. **50**, 2575–2588 (2020). https://doi.org/10.1007/s10489-020-01674-8
9. Paludo Licks, G., Meneguzzi, F.: Automated Database Indexing using Model-free Reinforcement Learning (2020). https://doi.org/10.48550/arXiv.2007.14244
10. Kvet, M., Matiaško, K.: Analysis of current trends in relational database indexing. In: 2020 International Conference on Smart Systems and Technologies (SST), Osijek, Croatia, pp. 109–114 (2020). https://doi.org/10.1109/SST49455.2020.9264034
11. Van Otterlo, M., Wiering, M.: Reinforcement learning and Markov decision processes Reinforcement Learn. State-of-the-art, 3–42 (2012). https://doi.org/10.1007/978-3-642-27645-3_1
12. Iftikhar, S., Rosner, F.: Data-Science-Platform/tpch-pgsql Github Repository (2018). https://github.com/Data-Science-Platform/tpch-pgsql. Accessed 19 Apr 2023
13. Mostafa, S.A.: A case study on b-tree database indexing technique. J. Soft Comput. Data Min. 1.1, 27–35 (2020). https://doi.org/10.30880/jscdm.2020.01.01.004
14. Kraska, T., Beutel, A., Chi, E.H., Dean, J., Polyzotis, N.: The case for learned index structures. In: Proceedings of the 2018 International Conference on Management of Data, pp. 489–504 (2018). https://doi.org/10.1145/3183713.3196909
15. Transaction Processing Performance Council (TPC). Dell PowerEdge R6525 using Exasol 7.1 (2021). https://www.tpc.org/results/individual_results/dell/dell~tpch~10000~dell_poweredge_r6525~es~2021-05-26~v04.pdf. Accessed 20 Apr 2023
16. Transaction Processing Performance Council (TPC), (2021), TPC BENCHMARK H (Decision Support) Standard Specification, Revision 3.0.0. https://www.tpc.org/tpc_documents_current_versions/pdf/tpc-h_v3.0.0.pdf. Accessed 17 Apr 2023
17. Mnih, V., et al.: Playing atari with deep reinforcement learning (2013). arXiv preprint arXiv: 1312.5602, https://doi.org/10.48550/arXiv.1312.5602
18. Fedus, W., et al.: Revisiting fundamentals of experience replay. In: International Conference on Machine Learning, pp. 3061–3071, PMLR (2020). https://doi.org/10.48550/arXiv.2007.06700

Automatic Indexing for MongoDB

Lucia de Espona Pernas[1], Anton Vichalkovski[1], William Steingartner[2], and Ela Pustulka[1(✉)]

[1] School of Business, University of Applied Sciences and Arts Northwestern Switzerland FHNW, Riggenbachstrasse 16, 4600 Olten, Switzerland
elzbieta.pustulka@fhnw.ch

[2] Faculty of Electrical Engineering and Informatics, Technical University of Košice, Letná 1/9, 04200 Košice, Slovak Republic

Abstract. We present a new method for automated index suggestion for MongoDB, based solely on the queries (called aggregation pipelines), without requiring data or usage information. The solution handles complex aggregations and is suitable for both cloud and standalone databases. We validated the algorithm on TPC-H and showed that all suggested indexes were used. We report on the performance and provide hints for further development of an automated method of index selection. Our algorithm is, to the best of our knowledge, the first query-based solution for automated indexing in MongoDB.

Keywords: DB Indexing · DB Performance · MongoDB · NoSQL

1 Introduction

Indexing helps optimise database (DB) performance by minimising the number of disk accesses required in query processing. Relational DBs offer a variety of index structures [17]. An index is optimised for systems that read data and should not degrade over time [8,9]. A typical indexing scenario assumes that we have a working system containing data where performance is monitored to identify the bottlenecks that could be removed by indexing or other techniques. In a traditional business scenario, a DB administrator (DBA) tunes the system. Our approach differs, as we assume that we are creating a new DB system with no DBA, i.e. autonomous database [7,10]. The system contains no data, so no statistics are available, and it has to be tuned based only on the queries which are known.

This solution has been designed for Enablerr [1]: an innovative Enterprise Resource Planning (ERP) for small and medium enterprises built on top of MongoDB [13]. Earlier steps were presented in [15], and a versioning solution in [3,4]. Our novel approach assumes an optimisation of forthcoming workloads.

The work [5] describes how to tune MongoDB in a traditional indexing approach which relies on data presence. Our solution fits the Enablerr architecture, to produce a performance optimiser, ahead of system use, either in the cloud or standalone, as justified by some usage scenarios.

A. Abelló et al. (Eds.): ADBIS 2023, CCIS 1850, pp. 535–543, 2023.
https://doi.org/10.1007/978-3-031-42941-5_46

2 Related Work

Our work is related to database cracking (DBC) [6,19] and, more narrowly, index selection [12]. DBC improves query performance by reorganising data storage and reindexing on the fly. As data change, it samples data at query time, creates additional data structures, and uses those to optimise performance. Index data structures have been surveyed by Vitter [17]. Index optimisation and merging with cost models are discussed for instance in [2]. UDO [18] uses machine learning to select the best indexes, however it requires data and an initial set of indexes. MongoDB has a Performance Advisor which can only be used after the data have been queried and not beforehand. None of the work we have seen so far proposes indexes when data statistics are not available.

3 The Indexing Solution

MongoDB v6 defines indexes at the collection level and supports indexes on any field, sub-field, or as a compound index. Queries are simpler, while aggregations are more powerful and consist of multiple stages that can do grouping or unwinding of embedded objects. Our method suggests indexes for aggregations. We transform queries into aggregations automatically, which makes our solution applicable to both queries and aggregations.

Our Auto Index method takes a set of aggregations or queries as JSON and produces a JSON file containing candidate indexes that could potentially improve query performance. The method consists of three steps shown in Fig. 1:

1. Generation: generate the indexes for each pipeline stage and each aggregation in the set.
2. Merging: merge the indexes created in multiple steps for each aggregation.
3. Cleaning: remove duplicate indexes.

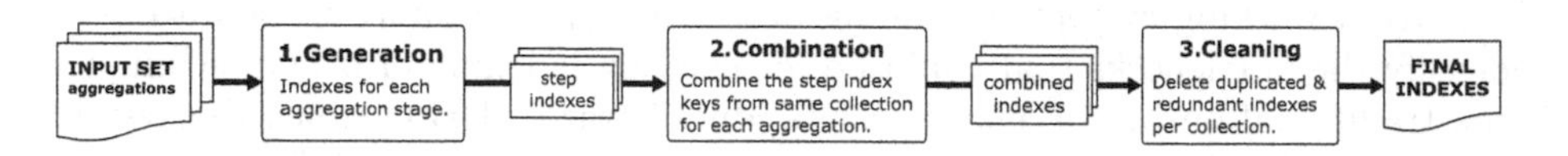

Fig. 1. Auto-Index Processing Stages

Index generation produces indexes for the *$match*, *$lookup*, *$group* and *$sort* stages. It takes into account for each of these operators the previous stages as, for example, no indexes will be used after an unwind. Indexes on the *_id* (unless it is a compound field) are ignored, as they exist by default. We filter out indexes on fields that appear to be arrays based on the aggregation operators used (for example *$size*, exclusive for arrays), as such indexes tend to be extremely large and worsen performance, as we will show later. We support field aliases, complex expressions, logical operators and comparisons.

Index merging takes all the indexes generated for a single aggregation and combines them. The indexes from the match stages are merged and combined with the group and sort indexes. The individual match stage indexes are included in the result set when the configuration option *minimize* is *false*, otherwise only the combinations from the multiple stages are returned.

Cleaning takes all the indexes generated for all aggregations and removes duplicates. If the configuration option *minimize* is *true*, we remove the indexes contained in other indexes. For an empty DB, the configuration *minimize=true* produces an initial set of indexes as a cost function is not applicable. If the aim is to perform an optimisation with existing queries and data, the configuration *minimize=false* will produce all possible indexes to select using a cost function or other optimisation method.

The algorithm is implemented in NodeJS[1] and as an NPM package[2].

4 Evaluation

Our evaluation used TPC-H [14,16] which has eight tables with 22 queries and simulates business behaviour and measures the response time of single queries and the total time. We followed a TPCH-H MongoDB adaptation [11] and performed the experiment with three schemas, see Fig. 2. S1 is a single collection with line items inlined under orders and orders under customers. S2 has three collections, analogously to the original TPC-H. Line items have a bi-directional relation to orders, and orders a one-way reference to customers. S3 has two collections where line items are inlined under orders. DB Collection sizes[3] are shown in Table 1.

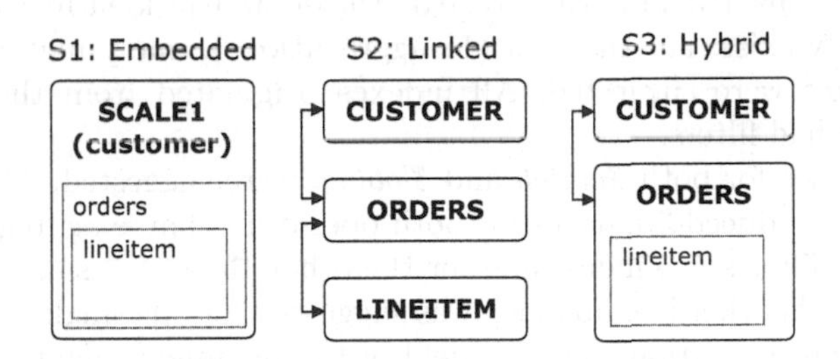

Fig. 2. TPC-H Schema translated to document stores in three schemas: S1, S2 and S3.

Local experiments were executed on a dedicated virtual Linux server with Ubuntu Jammy 22.04, 16GB RAM, 4 virtual CPUs and a 20GB SSD. A standalone MongoDB (6.0.4. Community) was locally installed in the server with default configuration.

[1] https://github.com/pier4all/mongodb-auto-index.
[2] https://www.npmjs.com/package/mongodb-auto-index.
[3] https://github.com/antw0n/universal-database-optimizer-evaluation.

Table 1. Data collection statistics

Schema	Collection	Documents	Document Size	Total Size
S1	scale1	15000	16070B	78.85 MB
S2	customer	15000	267B	1.77 MB
	lineitem	600600	355B	63.50 MB
	orders	150000	234B	13.05 MB
S3	customer	15000	267B	1.77 MB
	orders-lineitem	150000	1580B	88.88 MB

We successfully checked the validity of the indexes suggested, that is, that they are actually used by the DB when running the queries. We also tested performance for each aggregation set and for each single aggregation. The software to reproduce the results is available online[4].

4.1 Results: Index Validity Experiment

We first checked if the suggested indexes were used. The parameter *minimize* was set to false to obtain the maximum number of indexes possible. Table 2 lists the indexes generated for each query set.

In S1, only indexes for *$match* were suggested. The *pqe* query set (4 aggregations), generated one index and the *tpch* set with 11 aggregations generated 4 indexes. Two aggregations had an *$unwind* at the beginning that changes the document root and makes any other index on that collection unusable. Other aggregations had a filtering stage on the document identifier where default indexes already exist. For the *tpch* (11 aggregations), many had early unwinds and no indexes were suggested. All indexes originated from the other three queries which had filters.

In S2, indexes for both *$match* and *$lookup* were suggested. The *pqe* set (10 aggregations) produced 3 indexes for both operators. Three aggregations generated the same filter stage index, and for the other three, the same lookup index was suggested. Another lookup index was suggested for one single query referencing the lineitems collection by a sub-field of the document identifier containing the order key. The rest of 3 aggregations contained unwind stages or filters by the default identifier. Six indexes were suggested for the *tpch* (8 aggregations), two of them for *$lookup* stages and the rest for filters.

In S3, in the *pqe* (4 aggregations), only one index from a *$lookup* stage was suggested for three aggregations. The remaining aggregation started with a *$lookup* but referenced the document identifier. The *tpch* (7 aggregations) produced 4 indexes for *$match* and *$lookup* from five of the aggregations. The two remaining aggregations did not generate any index as they presented an *$unwind* before any potentially optimisable stage.

[4] https://github.com/pier4all/auto-index.

Table 2. TPC-H Suggested Indexes

Set	Stage	Index Name	Collection	Index Key
S1-pqe	match	acctbal	scale1	c_acctbal:1
S1-tpch	match	acctbal	scale1	c_acctbal:1
	match	mktsg	scale1	c_mktsegment:1
	match	mktsg_shipd	scale1	c_mktsegment:1, c_orders.o_lineitems.l_shipdate:1
S2-pqe	match	acctbal	customer	c_acctbal:1
	lookup	orderk	lineitem	_id.l_orderkey:1
	lookup	custk	orders	o_custkey:1
S2-tpch	match	acctbal	customer	c_acctbal:1
	match	mktsg	customer	c_mktsegment:1
	lookup	custk	orders	o_custkey:1
	match	orderd	orders	o_orderdate:1
	lookup	orderk	lineitem	_id.l_orderkey:1
	match	shipd	lineitem	l_shipdate:1
S3-pqe	lookup	custk	orders-lineitem	o_custkey:1
S3-tpch	match	acctbal	customer	c_acctbal:1
	match	mktsg	customer	c_mktsegment:1
	lookup	custk	orders-lineitem	o_custkey:1
	match	orderd	orders-lineitem	o_orderdate:1

Two methods were used to assess index usage:

- **Query Explain**: is the default provided by MongoDB to get the execution plan. However, this does not report index usage on a lookup.
- **Index Stats**: returns the number of times an index was used including lookups.

All the indexes our method proposed were used, which confirms our indexing method was correct.

4.2 Results: Performance Experiment

After index validation, we evaluated performance with data shown in Table 1. Although all the suggested indexes were used by MongoDB, they did not always improve query performance, due to specific data related features that cannot be extracted from the input aggregations. Figure 4 shows the performance at aggregation set level. All the aggregation sets for S2 and S3 were improved by the use of indexes, while for both S1 aggregation sets the index had a negative influence (Fig. 4).

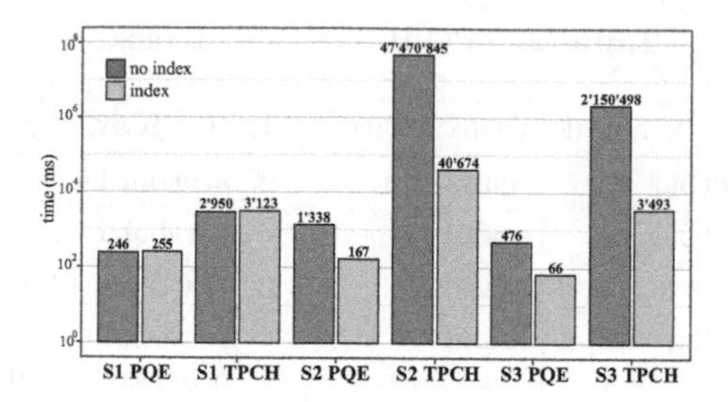

Fig. 3. Auto-Index Performance by query set with all indexes

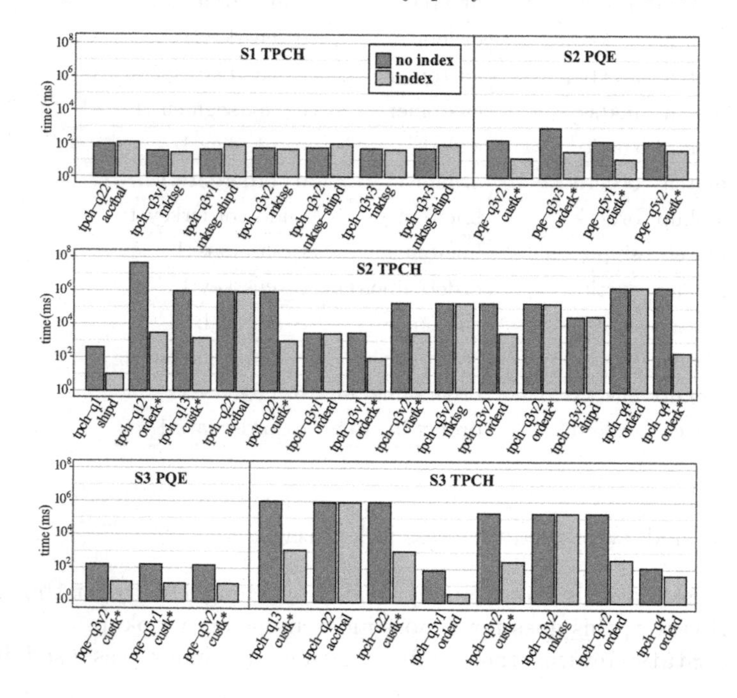

Fig. 4. Auto-Index Performance in a single query – single index setting. The indexes correspond to *$match* except those from a *$lookup*, marked with a star. Only the cases where the index was used by a non-zero execution time aggregation are shown.

We also measured the performance of each query with and without each index proposed for the set it belongs to. Figure 4 shows the performance of each index-aggregation combination for all schemas and query sets, including information about the originating aggregation stage (a star at the end of the index name indicates a *$lookup* index, while the rest were suggested for filters).

S1 is not a realistic design as the embedded array grows over time as customers place new orders, and generates large documents that affect negatively the overall performance. *mktsg-shipd* index worsened the performance of all the aggregations that used it. As shown in Table 2, the second component of this index represents the shipping date of one of the multiple items inside one of the orders with almost 120 K entries, while the collection has 15 K documents and only 3 K are returned. A collection scan is more efficient.

acctbal index suggested for tpch-q22 S1 increased the execution time, due to the low filter specificity retrieving 90% of the collection. Index based fetching has to check around 13 K keys and then bring them to the next aggregation steps, which is slower than a collection scan (15 K documents) [15]. *mktsg* index from the filtering stage of the S1 tpch set aggregations did not cause significant improvement in performance due to the low cardinality of the field. In addition, execution time of the filtering step is very low compared to the rest of the aggregation. For similar reasons, *mktsg* and *acctbal* indexes on S2 and S3 barely improved the performance.

custk index on the customer reference in S2 orders was generated from a lookup. The index strongly reduced the execution times of all the aggregations that used it. A similar improvement was observed for S3 when indexing the same field on the orders-lineitem collection with reductions up to two orders of magnitude, used in three queries in each set. Another lookup index, *orderk*, produced a significant improvement for S2.

orderd index in collection orders in S2 is generated from a filter and made tpch-q3v2 run 60 times faster. Query performance on the same set tpch-q3v1 and tpch-q4 was not affected by the index, as both queries have a filter at the start of the aggregation and this step is very fast without an index. All the three aggregations using this index on the same field in collection orders-lineitem in S3 tpch set improved query performance.

Query sensitivity is a crucial factor in the performance of index *shipd* of the lineitem collection in S2. This index improves tpch-q1 performance 30 fold (filter returns less than 0.2% of the documents) but does not improve tpch-q3v3 (filter returns more than 90% of the documents).

5 Conclusions

We have developed an index suggestion algorithm based on queries that works in the absence of data. We tested it and showed that all the suggested indexes were used by MongoDB.

The *$match* stage performance was often improved by the suggested indexes, although the *$lookup* generated indexes had the higher impact as they are usually very expensive. The *$lookups* can be avoided by modifying the schema, which is not always possible. As the results demonstrate, an index on the collection targeted by the lookup can have a huge impact.

The factors that determine index performance are related to the schema and data, and therefore cannot be derived from the aggregations. It is possible that the same query with the same index on different data results in different performance if the cardinalities or even the values of the document fields change.

The data related factors seen in our experiment were first the number of index entries relative to the collection: indexes on large cardinality embedded arrays should be avoided. Second the query specificity: in the filtering stage, if a large proportion of the documents pass the filter, the index will be detrimental to performance.

Automatic indexes suggested by our method can be used as an initial set to be refined by a second analysis on existing data. An appropriate selection or ranking method for the indexes will need access to real data to take into account both the schema and the actual values.

Acknowledgements. We acknowledge funding from www.innosuisse.ch, Grant No. 44824.1 IP-ICT and support from project 030TUKE-4/2023 from the Cultural and Education Grant Agency of the Slovak Ministry of Education. We thank Saverio Damiani and Anton Lorvi for their work and Prof. Christian Dolfus (HSLU) for providing the server. We have no conflicts of interest to disclose.

References

1. pier4all AG (Ltd.): enablerr - the revolutionary business solution (2021). https://www.enablerr.ch/en/
2. Chaudhuri, S., Narasayya, V.: Index merging. In: ICDE, pp. 296–303. IEEE (1999)
3. de Espona Pernas, L., Pustulka, E.: Document versioning for MongoDB. In: Chiusano, S., et al. (eds.) New Trends in Database and Information Systems. Communications in Computer and Information Science, vol. 1652, pp. 512–524. Springer, Cham (2022). https://doi.org/10.1007/978-3-031-15743-1_47
4. de Espona Pernas, L., Pustulka, E.: MongoDB data versioning performance: local versus Atlas. In: DataPlat. CEUR-WS.org (2023)
5. Harrison, G., Harrison, M.: MongoDB Performance Tuning: Optimizing MongoDB Databases and their Applications. Apress Berkeley, New York (2021)
6. Kersten, M.L., Manegold, S.: Cracking the database store. In: CIDR 2005, pp. 213–224 (2005). https://www.cidrdb.org/
7. Kossmann J., Halfpap S., J.M.S.R.: Magic mirror in my hand, which is the best in the land? An experimental evaluation of index selection algorithms. In: Proceedings of the VLDB, pp. 2382–2395. ACM (2020). https://doi.org/10.14778/3407790.3407832
8. Kvet, M.: Database index balancing strategy. In: 29th Conference of Open Innovations Association (FRUCT), pp. 214–221 (2021)
9. Kvet, M.: Relational data index consolidation. In: 28th Conference of Open Innovations Association (FRUCT), pp. 215–221 (2021)
10. Li, G., et al.: openGauss: an autonomous database system. Proc. VLDB **14**(12), 3028–3042 (2021)
11. Llano-Ríos, T.F., Khalefa, M., Badia, A.: Experimental comparison of relational and NoSQL document systems: the case of decision support. In: Nambiar, R., Poess, M. (eds.) TPCTC 2020. LNCS, vol. 12752, pp. 58–74. Springer, Cham (2021). https://doi.org/10.1007/978-3-030-84924-5_5
12. Lum, V.Y.: On the selection of secondary indexes. In: Proceedings of the 1974 Annual ACM Conference - Volume 2, ACM '74, p. 736. ACM (1974)
13. MongoDB, I.: MongoDB (2021). https://www.mongodb.com
14. Poess, M., Floyd, C.: New TPC benchmarks for decision support and web commerce. SIGMOD Rec. **29**(4), 64–71 (2000)

15. Pustulka, E., von Arx, S., de Espona, L.: Building a NoSQL ERP. In: Yang, X.S., Sherratt, S., Dey, N., Joshi, A. (eds.) Proceedings of Seventh International Congress on Information and Communication Technology. Lecture Notes in Networks and Systems, vol. 448, pp. 671–680. Springer, Singapore (2023). https://doi.org/10.1007/978-981-19-1610-6_59
16. TPC: TPC BenchmarkTM H Standard Specification Revision 3.0.0. TPC (2022). https://www.tpc.org/tpc_documents_current_versions/pdf/tpc-h_v3.0.0.pdf
17. Vitter, J.S.: External memory algorithms and data structures: dealing with massive data. ACM Comput. Surv. **33**(2), 209–271 (2001)
18. Wang, J., Trummer, I., Basu, D.: Demonstrating udo: A unified approach for optimizing transaction code, physical design, and system parameters via reinforcement learning. In: Proceedings of the International Conference on Management of Data, pp. 2794–2797 (2021)
19. Zardbani, F., Afshani, P., Karras, P.: Revisiting the theory and practice of database cracking. In: EDBT 2020, pp. 415–418. OpenProceedings.org (2020)

From High-Level Language to Abstract Machine Code: An Interactive Compiler and Emulation Tool for Teaching Structural Operational Semantics

William Steingartner(✉) and Igor Sivý

Faculty of Electrical Engineering and Informatics, Technical University of Košice, Letná 1/9, 04200 Košice, Slovak Republic
william.steingartner@tuke.sk, igor.sivy@student.tuke.sk

Abstract. This paper examines the development of an abstract machine compiler specifically for the Structural Operational Semantics (SOS) framework. This compiler serves as a formal framework for programming language specification and analysis. The paper focuses on key aspects of the development process, such as translating high-level language constructs into low-level machine instructions and implementing memory models. The ultimate objective of this work is to provide valuable insights into the design and implementation of an abstract machine compiler and emulator, with potential applications in language design and verification. Additionally, the tools developed can be applied in educational settings to aid in the teaching process.

Keywords: abstract machine · compiler · emulation tool · formal methods · structural operational semantics · university didactics · visualization

1 Introduction

When designing a programming language, it is crucial to express program meaning clearly and unambiguously. Formal semantic methods provide a means of achieving this. There are several widely used semantic methods, including structural operational semantics, which emphasizes program execution steps, and abstract language implementation, which helps verify language correctness [8]. To aid in the teaching of such methods, software tools are often used in computer science education [3]. These tools can process language input and allow

This work was supported by national KEGA project 030TUKE-4/2023 – "Application of new principles in the education of IT specialists in the field of formal languages and compilers", granted by the Cultural and Education Grant Agency of the Slovak Ministry of Education, and by the Initiative project "Semantics-Based Rapid Prototyping of Domain-Specific Languages" under the bilateral program "Aktion Österreich - Slowakei, Wissenschafts- und Erziehungskooperation".

A. Abelló et al. (Eds.): ADBIS 2023, CCIS 1850, pp. 544–551, 2023.
https://doi.org/10.1007/978-3-031-42941-5_47

students to analyze outputs, making them ideal for teaching semantic methods. By visualizing the semantic method, students can better understand the underlying concepts.

One important advantage of using abstract machines is that they can provide a basis for provable correct implementation of programs. By specifying the behavior of a program using an abstract machine, it is possible to reason about the correctness of the program without having to consider the details of its physical implementation. This is particularly useful when designing and implementing programming languages, where the correctness of the language implementation is of critical importance.

Overall, the use of abstract machines can play a crucial *rôle* in provable correct implementation of programs and programming languages. By providing a precise and abstract model of computation, abstract machines can help to ensure that the behavior of programs is well-defined and can be reasoned about in a rigorous and systematic way. This can lead to more reliable and efficient implementations of programs and programming languages.

The aim of this work is to create such a teaching tool for abstract language implementation intended for the course Semantics of Programming Languages. The resulting software will be in the form of a visualization environment for this semantic method, which will allow students of the subject to compile the input program and visualize its execution in a graphical interface.

The structure of the paper is as follows. In Sect. 2, we review some works related to the educational abstract and virtual machines. A short overview of an abstract machine for operational semantics is presented in Sect. 3. In Sect. 4, we describe the environment for a visualization of abstract machine in details. The Sect. 5 concludes our paper.

2 Educational Abstract and Virtual Machines

Numerous authors employ visualization software or develop their own programs to enhance the educational experience for students. Their aim is to help students better comprehend the course material by bridging the gap between theoretical concepts and practical applications. By doing so, these efforts result in a more lucid and explicit explanation of the course's procedures and principles.

An informative and in-depth survey of various variations of abstract machines and their use in computer science is given by the authors in the work [1]. This work is also a starting point for further research and development in the field of abstract machines as formal mathematical models.

Emustudio [12] is a software tool that functions as an emulator of multiple virtual architectures including the RAM machine, and has proven effective in teaching data structures and algorithms. Another comparable architecture is discussed in [4].

The Computron tool [5] provides a virtual architecture for executing programs written in a language similar to Jane and compiled into bytecode, making it useful for instruction in compiler construction. Similarly, the focus of the subject was on developing a compiler for this architecture.

A noteworthy software that focuses on algorithm modeling and defining their behavior in first-order logic has been created at the Research Institute for Symbolic Computation at JKU Linz by W. Schreiner [10]. Known as RISCAL (RISC Algorithm Language), this language and its accompanying software system allows users to construct theories in first-order logic, articulate algorithms using a high-level language, and specify their behavior through formal constraints. The software leverages an innovative approach to semantic evaluation, which is grounded in an implementation of the denotational semantics of the RISCAL language.

VCOCO, a visualization tool utilized for teaching compilers, offers an excellent example. It generates LL(1) visible compilers with a visual element added to the compilers created by COCO/R, an earlier compiler generator developed by H. Mössenböck [7], making it an efficient tool for learning. COCO/R is responsible for constructing analyzers, and VCOCO produces visual representations that students can interact with to enhance their understanding of compilers.

It is worth noting that there is a considerable amount of literature on the subject of this article. Although there are many compelling and relevant works that could be discussed, we have chosen to focus on a selected few in this paper due to space limitations.

3 Short Overview of Abstract Machine for Operational Semantics

The term "abstract machine" refers to a mathematical model that formally describes programs according to semantic specifications [1]. One definition of abstract machine for operational semantics was presented by Nielson and Nielson [8], which follows the standard definition of abstract machine for structural operational semantics, utilizing two memory abstractions: linear memory and the set of states.

The abstract machine language is a set of instructions structured like an assembler. These instructions follow an abstract syntax defined as follows:

$$\begin{aligned} \mathit{instr} ::= {} & \texttt{PUSH-}n \mid \texttt{ADD} \mid \texttt{SUB} \mid \texttt{MULT} \mid \texttt{TRUE} \mid \texttt{FALSE} \mid \texttt{EQ} \mid \texttt{LE} \mid \texttt{AND} \mid \texttt{NEG} \mid \\ & \texttt{FETCH-}x \mid \texttt{STORE-}x \mid \texttt{EMPTYOP} \mid \texttt{BRANCH}(c, c) \mid \texttt{LOOP}(c, c), \end{aligned}$$

$$c ::= \varepsilon \mid \mathit{instr} : c.$$

In the second version of the language (with another memory model), the instructions $\texttt{FETCH-}x/\texttt{STORE-}x$ are replaced with the instructions $\texttt{GET-}n/\texttt{PUT-}n$ (x represents a variable, n stands for a numeral). Full specification of the abstract machine can be found in [8].

An abstract machine semantics presents a high-level model of how a computer might go about running a program. Historically, abstract machines were the first way of specifying operational semantics, introduced by Peter Landin in the highly influential 1964 paper "The Mechanical Evaluation of Expressions" [6]. While they were the first type of operational semantics, some researchers have considered them too low-level. Structural operational semantics (introduced by Plotkin, [9]) and reduction semantics (introduced by Felleisen, [2]) were subsequently developed to address this.

To make it easier for students studying the theoretical foundations of the semantics of programming languages to understand and practically learn the basics of the given semantic method, as well as to enable them to try it out on their own, we have designed a tool enabling the emulation of an abstract machine, including a compiler from and to the language of an abstract machine. This tool is presented in the next section.

4 An Environment for a Visualization of Abstract Implementation of a Language

Our visualization tool has multiple tasks. It is a web-based tool that processes different inputs, performs compilation, visualizes program execution, and exports results. It includes a command-line compiler for various output formats and program interpretation. The tool is divided into three packages: editor, compiler, and visualization. In the next subsections, we present their structure and communication.

The visualization tool [11] is a web application divided into three modules. These modules are separate web apps launched together via Docker. They communicate using the HTTP protocol, with each module defining its interface to handle requests from other modules. To facilitate this communication, each module includes a basic HTTP server using the Django web framework. In the following subsections, we provide a detailed description of the design and implementation of each module.

4.1 Editor Module

Editor for providing the input source is a main part of user interface. User is allowed to upload input sources from the local storage or to write them manually. Supported are sources written in Jane and in the abstract machine language, allowing to use also the basic or extended syntax of instructions. The editor utilizes Ace for syntax highlighting, with user-defined regular expressions. For example, the partial code of the definitions looks like this:

```
keywords = ("while|do|repeat|until|for|to|if|then|else|skip");
builtinConstants = ("tt|ff");
```

To simplify the user's work, the editor recognizes selected keywords in both LaTeXand HTML syntax and converts them directly in the source form into the appropriate symbols (e.g. \neg, \lnot, ¬ are changed to ¬, etc.).

4.2 Compiler Module

The compiler module is a web app that receives HTTP requests, performs compilation, and returns results or errors. The compiler is a program implemented in C++ that translates a given source to an executable (visualizable) form that is executed upon the server request.

At the input of the compiler is the source code in the form of a text string. The output is a structure that stores the bytecode itself in the form of an array of bytes and an array of variable names that we need to keep for use in visualization. Common phases in the compilation process facilitate this by supporting both languages in lexical analysis and translating them into a shared intermediate code. The remaining parts of the compiler operate on this intermediate code. The compiler also includes a simple semantic analysis, during which expression types, operand types, l-values, etc. are checked.

Our compiler has a specific feature: lexical analysis is not a separate phase of compilation. Rather than creating tokens upfront, the compiler process dynamically requests tokens from the lexical analyzer as needed.

Code generation takes place immediately the moment it is possible. For expressions, this applies only after the tree structure of the entire expression has been created. The tree is subsequently traversed recursively and, according to the type of expressions in the nodes of the tree, individual instructions are generated, which the expressions represent, and the values with which the instructions work. Unlike expressions, when translating commands, the code can be generated immediately, since we do not need any auxiliary structure here. However, the translation is somewhat more complicated, because when translating the conditional command and the loop commands, we need to calculate the lengths of the conditions and the lengths of the command bodies due to the jump instructions.

As an example, Fig. 1 shows a compiled code for the statement `while` $x \leq 10$ `do` $x := x + 1$.

```
0000 WHILE 9 12
0005 PUSH-10
0010 GET-0 (FETCH-x)
0013 LE
0014 BRANCH_IF_FALSE 21
0017 PUSH-1
0022 GET-0 (FETCH-x)
0025 ADD
0026 PUT-0 (STORE-x)
0029 LOOP 24
0032 BRANCH 4
0035 EMPTYOP
```

Fig. 1. A bytecode for the loop statement `while` $x \leq 10$ `do` $x := x + 1$

When translating abstract machine language source code into bytecode, the same lexical analysis is used as when translating from Jane. Syntactic recursive descent is also used here, and thus the same system of reporting errors during translation is also used here. The language is also translated in one pass with source code as input and bytecode as output. However, the grammar of the abstract machine language is much simpler and we do not need to build any auxiliary data structures during translation.

As an extra funtionality, the program allows also reverse compilation form the bytecode to abstract machine code and to Jane language. Technical details are specified in [11].

4.3 Visualizer Module

The Visualiser module enables users to visualize the execution of the input program using a web-based graphical interface. Users can set the initial state, step through program execution, or run the entire program. They can also save the visualization result in text or LATEXformat.

The visualization is based on a virtual machine that executes bytecode instructions received after translation. It visually represents the execution of both the program and the abstract machine by displaying the state configurations on the web interface. The virtual machine, integrated into the graphical interface, is stack-based and emulates the behavior of an abstract machine. It includes a stack and an array for memory storage, with memory accessed using indexes in instructions such as `FETCH`, `STORE`, `PUT`, and `GET`.

To visualize the execution of abstract machine code, a table is used, containing a textual representation of each instruction. The table is constructed gradually, similar to translating bytecode into abstract machine language. It is represented by an object, where each element's key is the instruction's address, and the value is an object storing the text string with abstract machine code and the address of the next instruction. For example, the table for the code $x := 1$, is depicted in Fig. 2.

```
{
    0:  { instruction: "PUSH-1", nextIp: 5 },
    5:  { instruction: "STORE-x", nextIp: 8 },
    8:  { instruction: "e", nextIp: 9 }
}
```

Fig. 2. Code table in visualizer for a simple assignment

Once the instruction table is constructed, creating the visualization is relatively straightforward. During code execution, the table data is referenced to generate text strings for the visualization. The visualization consists of two parts. The first part shows the abstract machine's transition, including the executed instruction, stack content, and current status. This part is updated after each instruction. The second part displays the status, which is updated after executing `STORE` or `PUT` instructions. It lists program variables, their current values, and indexes of current and previous states. Additionally, the statement is saved as a text string in two formats in the background.

An example of visualization part in the application is in Fig. 3.

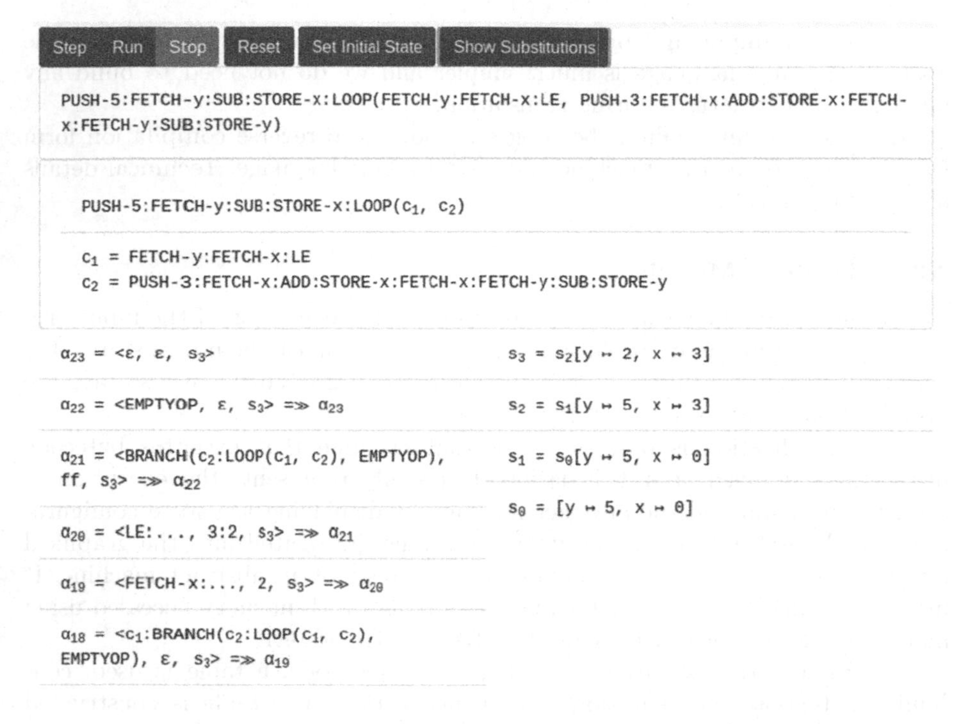

Fig. 3. Window containing a visualization

4.4 User Experience Using the Tool

The presented tool is currently deployed and used at the authors' home university. Its use during face-to-face and online teaching has been received with a positive response. So far, the mentioned semantic method has been taught and practiced using traditional elements – chalk and blackboard, which is sufficiently illustrative, but on the other hand, relatively lengthy when writing more complicated abstract machine code. Visualization of calculations using software is faster and enables checking the solution and preventing errors. The mentioned tool is used during lectures and laboratory exercises (at the time of the lectured topic) and also during independent (home) preparation of students. Overall, such a tool for students is received positively (the survey was conducted in the form of a questionnaire).

5 Conclusion

In this article, we focused on developing a software tool that visualizes a semantic method for an abstract (toy) programming language. Our research followed similar approaches used in visualizing other semantic methods. The main objective was to create a comprehensive environment that allows users to input their source code in a higher-level language (such as the abstract language Jane) or in the language of the abstract machine for operational semantics. The program processes this user input by compiling it into bytecode, which then facilitates the creation of visual steps demonstrating the calculation of an abstract machine. This interactive process enables users to follow the calculations and serves as a teaching aid for the subject of Semantics of Programming Languages.

Potential future enhancements for this application include extending the Jane language with additional constructs, expanding the instruction set for the abstract machine, and incorporating support for visualization in various formats, such as images.

The software tool is part of a comprehensive package designed to support the teaching of formal semantics. It can be effectively utilized during both face-to-face and distance learning scenarios, as well as for independent student preparation and experimentation with the semantic method.

References

1. Diehl, S., Hartel, P., Sestoft, P.: Abstract machines for programming language implementation. Futur. Gener. Comput. Syst. **16**(7), 739–751 (2000)
2. Felleisen, M., Friedman, D.P.: Control operators, the SECD-machine, and the λ-calculus. In: Wirsing, M. (ed.) Formal Description of Programming Concepts - III: Proceedings of the IFIP TC 2/WG 2.2 Working Conference on Formal Description of Programming Concepts - III, Ebberup, Denmark, 25–28 August 1986, pp. 193–222. North-Holland (1987)
3. Haleem, A., Javaid, M., Qadri, M.A., Suman, R.: Understanding the role of digital technologies in education: a review. Sustain. Oper. Comput. **3**, 275–285 (2022)
4. Jakubco, P., Simonak, S.: Utilizing GPGPU in computer emulation. J. Inf. Organ. Sci. **36**(1), 39–53 (2012)
5. Kollár, J.: Computron VM: identification of expert knowledge in virtual computer architecture development. In: CSE 2012 : International Scientific Conference on Computer Science and Engineering, pp. 87–94 (2012)
6. Landin, P.J.: The mechanical evaluation of expressions. Comput. J. **6**(4), 308–320 (1964). https://doi.org/10.1093/comjnl/6.4.308
7. Mössenböck, H.: The compiler generator Coco/R user manual (2010). https://ssw.jku.at/Research/Projects/Coco/Doc/UserManual.pdf
8. Nielson, H.R., Nielson, F.: Semantics with Applications: An Appetizer (Undergraduate Topics in Computer Science). Springer-Verlag, Berlin (2007)
9. Plotkin, G.: A structural approach to operational semantics. J. Log. Algebr. Program. **60–61**, 17–139 (2004). https://doi.org/10.1016/j.jlap.2004.05.001
10. Schreiner, W.: Thinking Programs, Logical Modeling and Reasoning About Languages, Data Computations, and Executions. Springer Nature Switzerland AG, Basel (2021)
11. Sivý, I.: An environment for a visualization of abstract implementation of a language. Technical report, Technical University of Košice, Slovakia (2022)
12. Šipoš, M., Šimoňák, S.: RASP abstract machine emulator - extending the emustudio platform. Acta Electrotechnica et Informatica **17**, 33–41 (2017)

Decentralised Solutions for Preserving Privacy in Group Recommender Systems

Marina Paldauf(✉)

Faculty of Mathematics, Natural Sciences and Information Technology, University of Primorska, Glagoljaška ulica 8, 6000 Koper, Slovenia
marina.paldauf@famnit.upr.si

Abstract. Group Recommender Systems (GRS) combine large amounts of data from various user behaviour signals (likes, views, purchases) and contextual information to provide groups of users with accurate suggestions (e.g. rating prediction, rankings). To handle those large amounts of data, GRS can be extended to use distributed processing and storage solutions (e.g. MapReduce-like algorithms and NoSQL databases). As such, privacy has always been a core issue since most recommendation algorithms rely on user behaviour signals and contextual information that may contain sensitive information. However, existing work in this domain mostly distributes data processing tasks without addressing privacy, and the solutions that address privacy for GRS (e.g. k-anonymisation and local differential privacy) remain centralised. In this paper, we identify and analyse privacy concerns in GRS and provide guidelines on how decentralised techniques can be used to address them.

Keywords: Group recommender systems · Decentralisation · Privacy

1 Introduction

Artificial Intelligence (AI), and more specifically, Machine Learning (ML) techniques [3], have been at the core of Recommender Systems (RSs) to support decision-making. RSs can target a group –i.e. Group RS (GRS) – or an individual –i.e. Single-user RS (SRS) across various domains such as tourism, entertainment, or medicine. RSs rely on some level of users' personal information[1] to provide recommendations that are tailored to users' interests [13], with user-centric RSs, such as collaborative filtering use [17] more personal information than the other RSs. Therefore, the reliance on sensitive personal data requires privacy management to protect users' right to keep their personal information and relationships private [17].

In the recent years, high-profile data breaches, such as Cambridge Analytica in 2018 [6], have highlighted the need for privacy through efficient and transparent data governance to guarantee the availability, usability, integrity and security

The author would like to thank Marko Tkalčič and Michael Mrissa for their help in elaborating the ideas developed in this paper.

[1] Information that can be used to differentiate or trace the identification of a person.

A. Abelló et al. (Eds.): ADBIS 2023, CCIS 1850, pp. 552–560, 2023.
https://doi.org/10.1007/978-3-031-42941-5_48

of data [12]. While several privacy protection methods have been applied to SRSs, including encryption [8], anonymity [18], differential privacy [18], and decentralisation [14], only a few studies propose k-anonymisation and Local Differential Privacy (LDP) [1,19] solutions to the privacy problem in GRSs. Moreover, these techniques rely on a centralised architecture that requires transferring data to a central repository that, as a consequence, becomes a Single-Point-of-Failure (SPOF), more likely to be subject to diverse attacks and, therefore, more vulnerable.

Given (i) the lack of privacy support for GRSs and (ii) the availability of distributed computing solutions, an intuitive approach to privacy protection in GRSs is to take advantage of distributed computing to avoid transferring sensitive data to a central place. Recent technological advances, such as the Internet of Things (IoT) [15], Edge Computing (EC) [5], Blockchain Technology (BT) [20], and AI [10] can help maintain user privacy by providing distributed support to process the data directly where it is produced, thus removing the need for data transfers and eliminating the SPOF problem. However, to our knowledge, there is currently no Distributed GRS (DGRS) addressing privacy. Thus, we focus in this paper on exploring the potential of DGRSs in addressing privacy by analysing existing work in the related research domains and provide guidelines for developing privacy-aware DGRSs.

2 State-of-the Art

This section provides an overview of relevant recent studies related to privacy and decentralisation aspects in RSs. Diagram in Fig. 1 presents the interconnections of subject areas - network architecture, RSs and the systems handling privacy-sensitive data.

2.1 Privacy in Recommender Systems

Privacy[2] is closely related to the security and trust of RS. Security measures protect data and enable privacy, and a system that ensures privacy is one aspect that builds users' trust in the system. Privacy risks emerge when data are collected and shared without the user's explicit consent, such as data leaks and de-anonymisation attempts [18], or when user's personal information can be inferred from the data [17]. For example, in restaurant selection GRSs, external attackers can deduct information such as geographical location based on group recommendation and historical information [19].

Furthermore, extending informed consent to the indirect processing of user preferences is also challenging. For example, collaborative filtering techniques used by RSs can construct a user model based on the data gathered on other users' interactions, which suggests that it might not be possible to completely protect users from the inferences that the system may be able to make about

[2] Privacy is the right to regulate or keep personal information secret [7].

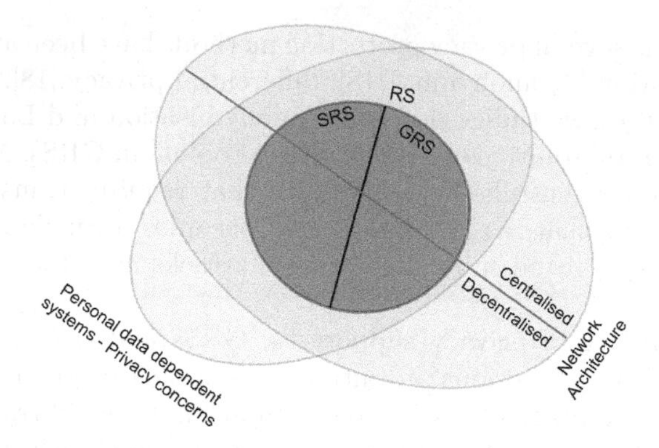

Fig. 1. Regarding network architecture, systems are centralised and decentralised. RSs are split into Single-user RSs (SRS) and Group RSs (GRS), systems requiring personal data and subject to privacy concerns.

them [17]. For example, Himeur (2021) [9], by analysing new versions of RSs, including Big Data and explainable RSs identified that modern RSs' characteristics emphasise security and privacy concerns, as users' data sensitivity and susceptibility to unlawful access are increasing. Additionally, modern technologies, such as BT, IoT, EC, and federated learning (FL) [22] have opened up new research perspectives for privacy-enabling RSs that could help provide potential solutions to decentralisation issues in GRSs. We detail such perspectives in Sect. 3.

2.2 Group Recommender Systems

Communal decision-making is required for ongoing socialisation and individual interaction. GRSs aim to promote decisions that benefit and satisfy all group members to the greatest extent feasible, unlike SRSs, which focus on satisfying one individual only. GRSs rely on the aggregation of individual members' preferences in a group profile or individual recommendations into a group recommendation [16], which increases the level of complexity and privacy issues since the intuitive way to proceed is to collect all the data in one place and analyse it there, leading to the SPOF problem.

Despite the use and benefits of GRSs, research has mainly focused on SRSs addressing privacy concerns [21], with some exceptions. For example, Wang et al. (2020) [19] used perturbation, inter-group random transmission[3] and a median aggregation strategy to maintain the recommendation's accuracy while hiding historical data and preferences from curious parties and malicious users. Ahmed et al. (2016) [1] proposed an k-anonymity[4] based algorithm to solve the structural re-identification attack, in which an anonymous user's identity can be discovered

[3] I.e. a random transmission mechanism inside the group to prevent the main server from knowing users' personal preferences.

[4] I.e. masking member's preferences in a pool of large data and masking their identity.

through data for group recommendation. Elmisery et al. (2017) [7] focused on limiting the number of group members with whom the information needs to be shared. Users provide their information and disclosure preferences to the middle layer run at the end-users' devices, which controls the publishing of data and the interest-group discovery process, and as a result their data gets shared only with the members of the chosen interest group.

2.3 Decentralisation in Recommender Systems

In the context of network architecture, centralisation refers to a network with one central node (e.g. a server) or a cluster of highly interconnected nodes connected to all other nodes (i.e. clients). On the other hand, decentralisation is a general term that can be used to describe network architectures that range from decentralised[5] to distributed[6] [4].

Most existing RSs use centralised architecture despite decentralised collection (e.g. on mobile devices) of private user data, including behaviour, context, domain knowledge and metadata about items. Data is gathered by one or multiple organisations/devices and shared [21], which poses a security and privacy risk [21]. As a solution, Friedman et al. (2015) [8] used encryption for data privacy and security. However, data still needed to be transferred, resulting in the risk of data breaches and leakage.

FL is an emerging decentralised solution that transfers only intermediate results, thereby reducing the risk of data breaches while sharing the knowledge [22]. By deploying FL to RS, the Federated RS (FRS) adopts the decentralised architecture [21]. Actors contain their data privately while collaboratively training the model. Kalloori et al. (2021) [14] assessed the accuracy of single-user FRS and its benefits to all stakeholders by employing the FRS to a number of stakeholders with information about users and user's data at each stakeholder could not be shared with other stakeholders in the structure. The study found that local models had performed worse than federated ones, indicating that FL can help stakeholders exploit each other user's preferences without exchanging user information.

Nevertheless, to the best of our knowledge, only one study [2] has focused on a DGRS that employed a dimension reduction technique and clustering to detect groups of similar users and distributed a group recommendation algorithm that maintained high accuracy and low computation time. The algorithm was applied to a Big Data environment, specifically to a replicated data storage (i.e. NoSQL Cassandra database[7]), which does not remove a SPOF for privacy.

[5] I.e. a combination of multiple centralised networks.

[6] I.e. a network where every node has approximately the same number of connections to other nodes, and there is no hierarchy between nodes.

[7] https://cassandra.apache.org/doc/latest/cassandra/architecture/overview.html.

3 Discussion and Architecture Overview

Existing privacy solutions for GRSs that rely on the perturbation of personal user information presents two main drawbacks. First, if the perturbation is not efficient enough, the user information can be inferred, and second, when the protection is efficient, the recommendations become less accurate. However, our literature review shows that some privacy-preserving distributed solutions [11] could be applied to support the distributed execution of the GRS and address privacy concerns through the use of appropriate encryption methods. Unlike noise addition, privacy-preserving methods, such as FL in SRSs, do not negatively impact the RS' accuracy [14]. FL enables model training using data from remote locations, especially in situations where certain data cannot be transferred due to physical or legal constraints [14]. The solution proposed by Kalloori et al. (2021) [14] relies on distributed raw data and a global model located on the main server, which in turn still represents a SPOF. Therefore, a possible solution is the use of FL technology in GRS, where raw data is kept locally on users' devices while the model is collaboratively trained in a distributed way.

More precisely, our idea is to transfer the indicators that allow the development of the ML model from one peer to another instead of the data itself (see Fig. 2) In this setup, each peer receives a set of instructions to incrementally contribute to enriching the previous model with its own local model. Such exchanges do not involve raw data sharing, and the partially generated models can be protected through appropriate asymmetric cryptography[8] [11].

Based on the above discussion, we present the following initial guidelines for the design of a privacy-addressing DGRS, that integrates a FRS framework [14] and a privacy-preserving decentralised data mining architecture [11]:

1. NODE SETUP: Users' devices form a naturally distributed setup providing data storage and computing power. This setup can be exploited to implement DGRS, with each user's device representing a node in the system.
2. NODE DATA: Each member's data is kept local to prevent any data disclosure, while all users (i.e. clients) have access to the same and all information about the items. Node generation depends on the GRS category[9]:
   (i) if recommendations are generated individually for group members: there each group member's device represents a different node in the system, where each node contains individual recommendations.
   (ii) if recommendations are generated using a group profile combining individual members' preferences: there a randomly chosen group member's device represents a node with a group profile, which is generated using a group modelling strategy, for example, an additive utilitarian [16] strategy in which the ordered ranking of items for a group is the same as the average of cumulative rankings of all members for a particular item.

[8] I.e. each peer is endowed with a pair of public/private key.

[9] GRSs are divided into two categories [16] based on the group preference and recommendation aggregation process.

3. FEDERATED LEARNING: Similarly to existing FRS, we envision a system that keeps user data (i.e. individual recommendations (i) or group preferences (ii)) locally (i.e. in a group setting, local to one node within a group with combined group preferences) while training a global model securely using this data. This approach is similar to Kalloori's example [14], with one crucial difference. In their framework, each stakeholder receives weights from a global server, runs it over their local privately-held training data, calculates the error gradient and updates the parameters to get new model weights. Each stakeholder then returns their updated weights to the global server. Finally, the global server aggregates the model by averaging all the stakeholders' updates. By eliminating the global server's role to kick-start the training process and collecting the updated weights, the SPOF problem is avoided, thus enhancing the system's security and privacy.
4. ONION ROUTING: Hrovatin et al. (2022) [11] onion routing decentralised data processing solution can be utilised to remove the SPOF. It enables secure communications between nodes while maintaining the data locally. A message is built in layers (hence the name onion routing). Each layer contains the model generated from the previous nodes and the address of the next node that possesses the secret key to decipher the next layer of the onion. Upon message reception, each node extracts the data with its private key, updates the model, and encrypts the result with the public key of the next node before sending the message. Therefore, the message always contains the partial model encrypted with the public key of the next node that will receive it, and the message is guaranteed to be processed by each node with its private key.
5. MODEL GENERATION: The initial model may be generated with the items' (i.e. a service, product, or solution) information and randomised members' preferences on one of the existing nodes instead of a global server. The model would then be passed across all actual nodes in a randomised order, updating the model weights based on each node's data (i.e. individual recommendations (i) or group preferences (ii)). The process would finish once all nodes are reached; at this point, the final model is shared with all nodes. The process would finish once all nodes are reached; at this point, the final model is shared with all nodes.
6. ADDITIONAL LAYER OF SECURITY: To add an additional level of security, members' data could be homomorphically encrypted, which enables computations on the data without decrypting it first [21].

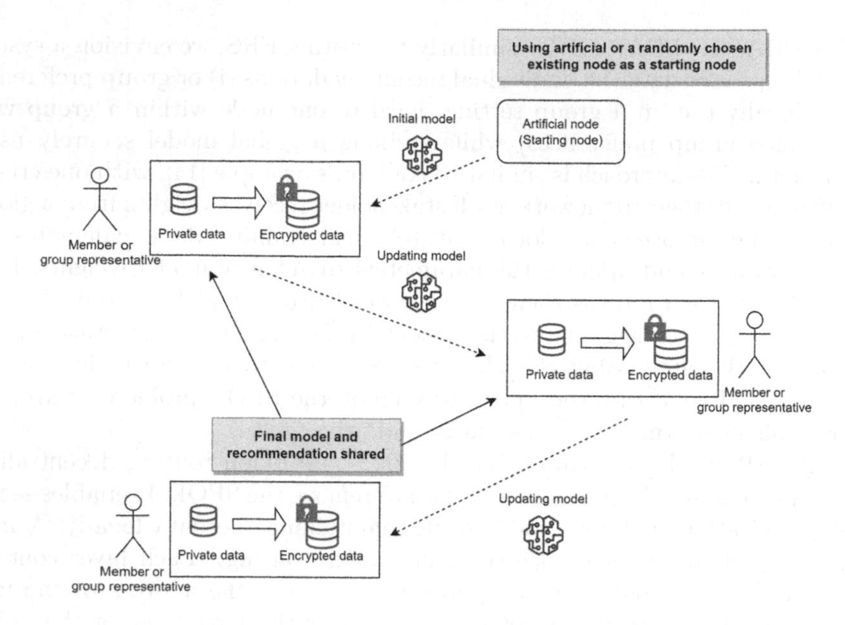

Fig. 2. A decentralised architecture for GRS

4 Conclusion

This paper provides an overview of the latest developments in the field of privacy-preserving GRS. It highlights some research directions towards the design of privacy-aware decentralised GRSs. Extrapolating from existing GRS addressing privacy and distributed privacy solutions in SRS, we present an architecture that makes use of asymmetric encryption and onion routing to support a privacy-preserving DGRS scheme. This article devises the initial concept of our decentralised architecture that balances the individuals' privacy requirements while maximising their utility.

In addition to the main design of the architecture, we provide some discussion of the current state of the art and give some guidelines towards further work in the area of privacy-preserving DGRS solutions that hopefully will ultimately lead to the development of effective and practical solutions for protecting people's privacy in GRS. The proposed design architectures build upon nodes that contain (i) individual recommendations of each group member and (ii) group profile/preferences combining individual members' preferences. Subsequently, FL and onion routing concepts are used to build a DGRS.

To further enhance the privacy of the solution, it should be considered in future work to extend the design with the privacy-preserving generation of individual recommendations of group members or a group modelling strategy used to generate a group profile.

Acknowledgement. The authors gratefully acknowledge the European Commission for funding the InnoRenew CoE project (Grant Agreement #739574) under the Horizon2020 Widespread-Teaming program, the Republic of Slovenia (Investment funding of the Republic of Slovenia and the European Regional Development Fund), and the Slovenian Research Agency ARRS for funding the project J2-2504.

References

1. Ahmed, K.W., Mouri, I.J., Zaman, R., Yeasmin, N.: A privacy preserving personalized group recommendation framework. In: 2016 IEEE 6th International Conference on Advanced Computing (IACC), pp. 594–598. IEEE (2016)
2. Ait Hammou, B., Ait Lahcen, A., Mouline, S.: A distributed group recommendation system based on extreme gradient boosting and big data technologies. Appl. Intell. **49**, 4128–4149 (2019)
3. Biswas, P.K., Liu, S.: A hybrid recommender system for recommending smartphones to prospective customers. Expert Syst. Appl. **208**, 118058 (2022)
4. Bodó, B., Brekke, J.K., Hoepman, J.H.: Decentralisation: a multidisciplinary perspective. Internet Policy Rev. **10**(2), 1–21 (2021)
5. Cao, K., Liu, Y., Meng, G., Sun, Q.: An overview on edge computing research. IEEE Access **8**, 85714–85728 (2020)
6. Confessore, N.: Cambridge Analytica and Facebook: the scandal and the fallout so far (2018). https://www.nytimes.com/2018/04/04/us/politics/cambridge-analytica-scandal-fallout.html. Accessed 30 Jan 2023
7. Elmisery, A.M., Rho, S., Sertovic, M., Boudaoud, K., Seo, S.: Privacy aware group based recommender system in multimedia services. Multimedia Tools Appl. **76**, 26103–26127 (2017)
8. Friedman, A., Knijnenburg, B.P., Vanhecke, K., Martens, L., Berkovsky, S.: Privacy aspects of recommender systems. In: Ricci, F., Rokach, L., Shapira, B. (eds.) Recommender Systems Handbook, pp. 649–688. Springer, Boston (2015). https://doi.org/10.1007/978-1-4899-7637-6_19
9. Himeur, Y., et al.: A survey of recommender systems for energy efficiency in buildings: principles, challenges and prospects. Inf. Fusion **72**, 1–21 (2021)
10. Himeur, Y., Sohail, S.S., Bensaali, F., Amira, A., Alazab, M.: Latest trends of security and privacy in recommender systems: a comprehensive review and future perspectives. Comput. Secur. **118**, 102746 (2022)
11. Hrovatin, N., Tošić, A., Mrissa, M., Kavšek, B.: Privacy-preserving data mining on blockchain-based WSNs. Appl. Sci. **12**(11), 5646 (2022)
12. Janssen, M., Brous, P., Estevez, E., Barbosa, L.S., Janowski, T.: Data governance: organizing data for trustworthy artificial intelligence. Gov. Inf. Q. **37**(3), 101493 (2020)
13. Jeckmans, A.J., Beye, M., Erkin, Z., Hartel, P., Lagendijk, R.L., Tang, Q.: Privacy in recommender systems. Soc. Media Retrieval, 263–281 (2013)
14. Kalloori, S., Klinger, S.: Horizontal cross-silo federated recommender systems. In: RecSys '21, September 27–October 1 2021, Amsterdam, Netherlands (2021)
15. Li, S., Xu, L.D., Zhao, S.: The internet of things: a survey. Inf. Syst. Front. **17**, 243–259 (2015)
16. Masthoff, J.: Group recommender systems: aggregation, satisfaction and group attributes. Recommender Syst. Handb., 743–776 (2015)
17. Milano, S., Taddeo, M., Floridi, L.: Recommender systems and their ethical challenges. AI Soc. **35**, 957–967 (2020)

18. Valdez, A.C., Ziefle, M.: The users' perspective on the privacy-utility trade-offs in health recommender systems. Int. J. Hum. Comput. Stud. **121**, 108–121 (2019)
19. Wang, H., He, K., Niu, B., Yin, L., Li, F.: Achieving privacy-preserving group recommendation with local differential privacy and random transmission. Wirel. Commun. Mob. Comput. **2020**, 1–10 (2020)
20. Yaga, D., Mell, P., Roby, N., Scarfone, K.: Blockchain technology overview. arXiv preprint: arXiv:1906.11078 (2019)
21. Yang, L., Tan, B., Zheng, V.W., Chen, K., Yang, Q.: Federated recommendation systems. In: Yang, Q., Fan, L., Yu, H. (eds.) Federated Learning. LNCS (LNAI), vol. 12500, pp. 225–239. Springer, Cham (2020). https://doi.org/10.1007/978-3-030-63076-8_16
22. Yang, Q., Liu, Y., Chen, T., Tong, Y.: Federated machine learning: concept and applications. ACM Trans. Intell. Syst. Technol. (TIST) **10**(2), 1–19 (2019)

PeRS: 2nd Workshop on Personalization, Recommender Systems

A Recommendation System for Personalized Daily Life Services to Promote Frailty Prevention

Ghassen Frikha[1(✉)], Xavier Lorca[1], Hervé Pingaud[2], Adel Taweel[3], Christophe Bortolaso[4], Katarzyna Borgiel[4], and Elyes Lamine[1,5]

[1] University of Toulouse, IMT Mines Albi, Industrial Engineering Center, Route de Teillet, 81013 Cedex 9 Albi, France
ghassen.frikha@mines-albi.fr
[2] CNRS-LGC, Champollion National University Institute, University of Toulouse, Albi, France
[3] Faculty of Engineering and Technology, Birzeit University, Birzeit, PS, Palestine
[4] Research and Innovation Division, Berger-Levrault, Labège, France
[5] University of Toulouse, ISIS, Champollion National University Institute, Rue Firmin-Oulès, 81104 Castres, France

Abstract. Frailty is a clinical syndrome that commonly occurs in older adults and characterizes an intermediate state between robust health and the loss of autonomy. As such, it is crucial to identify and evaluate frailty to preserve the abilities of older adults and reduce the loss of their functional capabilities. This paper proposes a personalized service recommendation framework designed, to assist senior citizens in selecting appropriate services, to prevent frailty and improve autonomy.

This framework is based on a multidimensional evaluation of the elderly person, considering both user's status information and service-related data. It defines a four steps recommendation process, including personal characteristics identification, needs identification, service types identification, and service identification. To support this, both knowledge-based and rule-based approaches are defined and employed to generate personalized recommendations that align with the individual's specific needs. The effectiveness of the proposal is evaluated using different representative scenarios, of which an example is described in detail that demonstrates the developed recommendation process, algorithms and relevant rules.

Keywords: Recommendation process · Recommender Systems · Elderly · Frailty

1 Introduction

Well-being is essential for successful ageing. Studies link well-being to maintaining functioning, engagement, and life satisfaction in later life [1]. However, ageing can cause a decline in well-being due to resource loss. This decline can lead to problems such as social isolation and depression, further decreasing the person's well-being. This process of functional decline is related to frailty, a concept

A. Abelló et al. (Eds.): ADBIS 2023, CCIS 1850, pp. 563–574, 2023.
https://doi.org/10.1007/978-3-031-42941-5_49

that is related to vulnerability and adverse health outcomes due to physiological decline [2]. Prevention plans aim to slow deterioration through tailored solutions such as physical activity, but selecting suitable options can be difficult due to the broad availability and diversity of services, as well as various factors such as cost, quality, expertise, location, and individual preferences and needs.

In this context, recommender systems provide customized services that cater to users' needs. By generating personalized recommendations, these systems offer the most suitable services that align with the unique needs of elderly individuals, aiming to improve overall well-being and prevent frailty. However, several challenges must be addressed before developing and implementing such systems. Our research aims to define and formalize essential knowledge for making personalized recommendations based on user needs. We identify knowledge for characterizing users and services, determine the necessary representation for embedding it within a suitable software structure, and define mechanisms and algorithms for implementing it in a recommendation system.

To address these challenges, we examined surveys and assessment tools used to evaluate frailty. Key elements were identified for defining frailty and intrinsic capacities, allowing for a better understanding of individual status through survey responses. The proposed recommender system uses this data to identify appropriate services and construct tailored prevention plans based on the specific needs and preferences of the person.

In this paper, we first provide an overview of the well-being of the elderly, with a particular emphasis on the use of recommender systems and their role in promoting healthy ageing and preventing frailty among the elderly population. In the third section, we introduce our proposed recommendation framework, Senselife. In the fourth section, we describe the details of the proposed recommender engine, exploring the algorithms used and the four-step recommendation process. In the fifth section, we present a use case to illustrate the effectiveness of the recommender system. Finally, we conclude with a summary of our key findings and offer perspectives for future research.

2 Background

2.1 Well-Being

The World Health Organization [9] has emphasized the importance of promoting well-being in older adults for healthy ageing and maintaining a high quality of life. Recent research has emphasized the importance of promoting well-being in older adults, as poor quality of life has been linked to frailty. Frailty, which is characterized by a loss of resources in physical, social, and psychological domains [14], can be prevented by promoting interventions like physical activity and social engagement. [5] have found that higher levels of well-being are associated with a lower risk of developing frailty, while frailty is linked to poorer quality of life and well-being [4].

Promoting a complete approach to well-being is essential for preventing frailty and enhancing the overall quality of life for older adults. Therefore, it is necessary to prioritize frailty prevention and well-being promotion in older adults to achieve healthy ageing and prevent or slow down the progression of frailty.

Recommender systems (RS) offer a potential approach to managing frailty and enhancing well-being in older adults. RS deliver personalized recommendations based on user preferences and contextual information [12].

Several RS have been developed to help older adults manage frailty, including Nestore [11], CARE [13] and others [3]. Nestore assesses frailty and recommends services based on individual needs and environment [11]. CARE, on the other hand, uses context-specific recommendations based on sensor data to link well-being subcategories to potential activity recommendations [13]. Other RS focus on specific components such as nutrition [3].

In the context of frailty prevention, RS can evaluate data from numerous sources, such as electronic health records, wearable sensors, and self-reported data, to assess an individual's health condition and situation, and design a preventive plan to their personal requirements and preferences. In the next Sect. 2.2, we will focus on some of the existing systems and compare them to our objectives (Table 1).

2.2 Existing Recommender Systems

We present an overview of existing systems that share similarities with Senselife's functional scope. Senselife is designed around six essential functions, namely:

1. Describing one's environment
2. Expressing one's needs, preferences, desires and satisfaction
3. Frailty Evaluation
4. Benefit from and evaluate services available in one's environment
5. The proposed services fit coherently into one's usual life and schedule
6. Follow-up provided by close ones and healthcare professionals

Table 1. Existing systems comparison

Application	1	2	3	4	5	6
ICOPE [10]	◐	○	●	○	○	●
NESTORE [11]	◐	●	◐	◐	●	◐
CARE [13]	●	◐	◐	○	○	◐
NutelCare [3]	○	◐	○	◐	○	○

○ = Not supported ◐ = Partially supported ● = Fully supported

While ICOPE does not function as a recommender system that suggests services, it is a valuable tool developed by the World Health Organization (WHO)

to assist healthcare professionals in assessing frailty. Although there are other systems available, none of them fully cover the functional scope of Senselife. The system that comes closest is NESTORE, but there are differences in how data is collected and used to generate recommendations. Senselife's approach to data collection and analysis sets it apart from other systems, making it a unique and innovative solution in the field of healthcare technology.

3 Proposed Recommender System

Our proposed recommender system addresses the challenge of managing frailty in older adults by utilizing self-evaluations to provide personalized recommendations tailored to their unique needs and preferences. The system recommends services aimed at enhancing the functional abilities of older adults, matching the supply and demand of services available in the elderly environment.

To achieve this, the Senselife framework comprises three main components: data collection, recommendation generation, and service consumption and usage, as shown in Fig. 1. Defining and utilizing knowledge to create personalized recommendations for each user is a primary challenge. Achieving this requires a deep understanding of users and service characteristics, determining the appropriate format for representing the knowledge, and integrating it seamlessly into the framework's software structure.

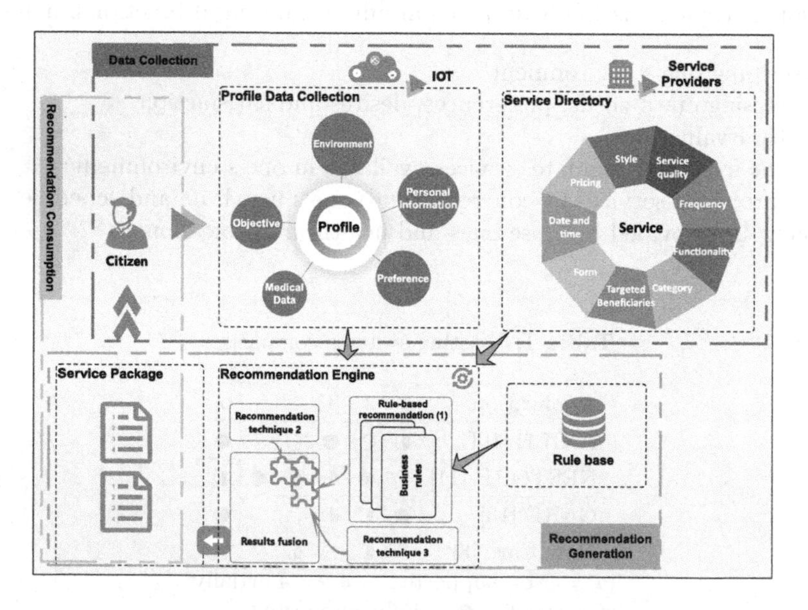

Fig. 1. Senselife framework

3.1 Profile and Service Models

Data collection is the first component of the Senselife framework, and it involves gathering information on both citizens and the services that will be recommended to them. We created two models to help with this process: one for user profiles and another for services. These models assist in identifying and categorizing the data required to generate appropriate suggestions.

User profiling is an important part of Senselife that helps to improve the relevance and quality of recommendations. It involves gathering and evaluating information about target users' preferences and behaviours to better understand their needs and interests. On the other hand, the service model is used to define the key attributes of different offered services and guide the service provider when they introduce new services on the framework.

3.2 The Rules Base

The rule base is an essential element of Senselife's recommendation engine. It contains different types of rules that have different purposes and will be used in various steps of the recommendation process. Figure 2 depicts the typology of rules in our system. It defines six rule types, as follows:

- **Constraint Rule:** This type of rule imposes some constraints on the system. If the constraints are not respected the recommendation can be stopped or modified. An example of constraint rules is the *ConsiderationRules* applied at the beginning of the need identification step of the recommendation process (see Sect. 4.2).
- **Action Rule** (also known as a "condition-action" rule): This type of rule is commonly used in artificial intelligence and expert systems to represent knowledge in the form of logical statements. In the need identification step (Sect. 4.2), there are 3 types of action rules. The first one is *NeedCreation* rule, which is used to create new Needs and add them to the profile if conditions are met. The second one is *PriorityRules*, which can modify needs priority based on the conditions. Finally, *ClassificationRules* are used to classify the identified needs based on Maslow's hierarchy of needs [8].
- **Privacy Rule:** These rules consider the user's privacy preferences and restrict the information that is used to generate recommendations.
- **Association Rule:** These are used to identify patterns of co-occurrence between different person's characteristics and populate the knowledge database by adding new relations. (Section 4.1)
- **Recommendation specific Rule:** According to [7], these are usually called "beyond accuracy" objectives. These rules can define diversity, serendipity, novelty, and coverage of recommender system to enhance its performance.
- **Filtering Rule:** These rules are used to filter the service recommendation. These filters can be contextual or demographic depending on the citizen's preferences and the service provider obligations. These rules are used in the final step of the recommendation process (Sect. 4.4).

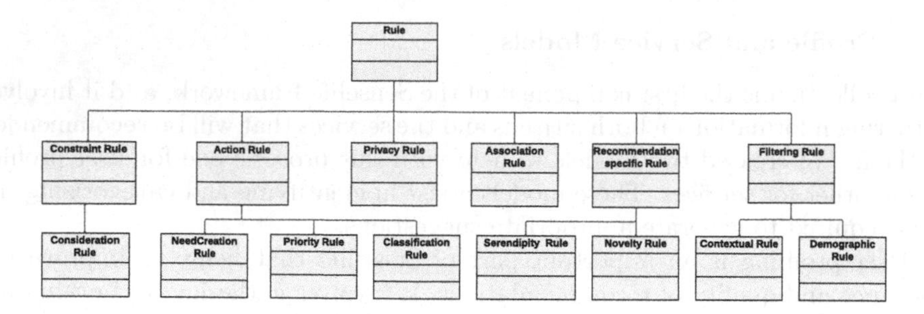

Fig. 2. Rule typology

4 Recommender Engine

For its second component, i.e. recommendation generation, Senselife framework utilizes a recommendation engine to provide personalized services for frailty prevention. Due to the complexity of the citizen and service models, the recommendation engine employs a multi-phase approach to associate the user's needs with appropriate services. This approach involves generating a citizen profile, identifying the user's needs, and selecting suitable service types using knowledge bases. Details regarding these phases will be discussed in the upcoming sub-sections (Fig. 3).

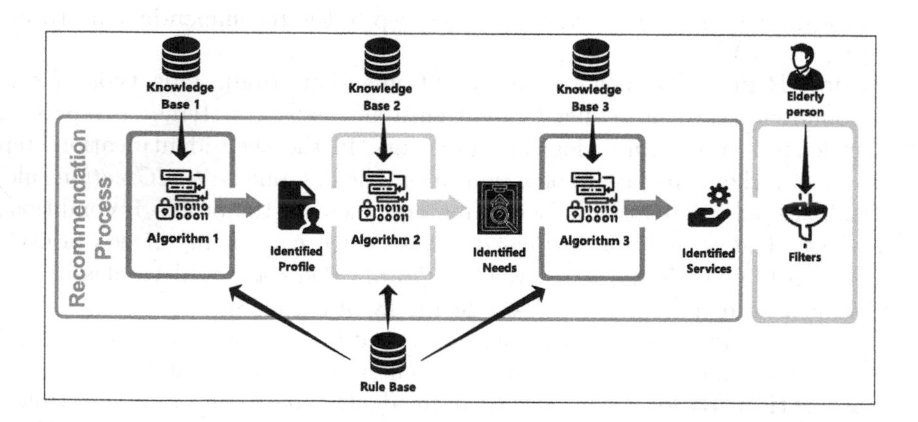

Fig. 3. Recommendation Engine overview

4.1 Personal Characteristics Identification

The first phase of the proposed recommendation system is crucial, it aims to systematically and thoroughly identify the personal characteristics of elderly individuals. However, the identification of these characteristics thoroughly can be challenging. For example, the common method to capture information is to use a survey or a questionnaire, however for detailed information and for the elderly population that can be tedious and often leads to low engagement. To address this issue, we plan to implement an intuitive survey that leverages a knowledge database containing connections between various attributes of an individual. This survey will be dynamic and adapt to the user's responses to previous questions. The knowledge database helps to select the most relevant questions, making the survey more efficient while gathering essential information. For instance, for a person with an ear problem, the system can adapt the questions to focus on the affected characteristics, such as the ability to perform household chores or the quality of sleeping, while it omits questions that are less relevant, such as whether the person has a driver's license.

For a well-connected knowledge, one that identifies common characteristics with prevailing patterns amongst its members, it requires creating a knowledge base with a significant amount of data from real target elderly individuals. Association rules [6] will be used to identify hidden patterns and relationships between different characteristics of the elderly that may not be immediately apparent, leading to the identification of more connected and relevant survey questions.

Algorithm 1. Need Identification Algorithm

Require:
  Input: *profile* : an instance of the class Citizen, containing all the characteristics PC identified in the previous step
Ensure:
  Output: *profile* : the same input instance with the added needs

1: ConsiderationRules(*profile.characteristics*)
2: **for all** need in Knowledge DB **do**
3:   *charPerNeed* ← GroupCharcteristics(*need*, *profile.PC*)
4:   **if** *charPerNeed* NOT *empty* **then**
5:     *profile.need* ← NeedsCreation(*profile*, *charPerNeed*)
6:   **end if**
7: **end for**
8: **for all** need in *profile.needs* **do**
9:   PriorityRules(*profile*, *need*)
10: **end for**
11: **for all** need in *profile.needs* **do**
12:   ClassificationRules(*profile*, *need*)
13: **end for**

4.2 Need Identification

Once the survey data is collected, it will be used to create a citizen profile. Algorithm 1 will then determine if a citizen is eligible for recommendations by applying *ConsiderationRules*, which may include verifying the score of an activity of daily living survey. Once eligibility is confirmed, the generated characteristic-profile helps to identify citizen's needs using both knowledge and rule bases. The initial set of needs and their priorities can be identified in the first step using the knowledge base. In the second step, *PriorityRules* are used to modify the needs priorities, if necessary. The purpose of these rules is to detect associations between needs and characteristics (see Sect. 5 for further explanation).

In the final step of the second phase, *ClassificationRules* use Maslow's hierarchy of needs [8] to assign general priorities to the identified needs. This ensures that we prioritize the most fundamental needs first, and then move on to addressing the less important ones. As a result, the need has two types of priority. The specific priority, which is based on the data provided by the citizen, and a general one, which displays the importance of the need in relation to other needs according to Maslow's hierarchy.

Algorithm 2. Service Types Identification

Require:
  Input: $profile$: an instance of the class Citizen, containing all the characteristics PC and the identified needs in the previous steps
Ensure:
  Output: $profile$: the same input instance with added Service Types

1: **for all** need in $profile.needs$ **do**
2: $serviceTypesPerNeed \leftarrow$ CheckConnectedServiceTypes($profile, need$)
3: $updatedSTPerNeed \leftarrow$ CheckObligatoryPC($serviceTypesPerNeed$)
4: $updatedSTPerNeed2 \leftarrow$ CheckRelatedPriority($updatedSTPerNeed$)
5: $updatedSTPerNeed3 \leftarrow$ CheckRelatedNeeds($updatedSTPerNeed2$)
6: UpdatePofile($profile$, $updatedSTPerNeed3$)
7: **end for**

4.3 Service Types Identification

After identifying the citizen's needs, the service type identification algorithm is executed, an example is depicted in Algorithm 2. Using the knowledge base and the identified needs, the algorithm identifies all related service types. However, there are some specifications for certain service types that need to be taken into consideration. To address this issue, dedicated functions have been implemented.

At the end of this step, a list of service types is identified and proposed to the citizens for confirmation. This list is generated based on the identified needs and the service types that are applicable to the citizen's profile. The citizen can review the list and confirm which service types they are interested in. This process ensures that the citizen receives personalized recommendations that meet their specific needs and preferences.

4.4 Services Recommendation

The final phase of the recommendation process involves scoring the services available in the database using a matching formula 1 that considers citizen's preferences $\boldsymbol{p}$ and service attributes $\boldsymbol{s}$.

$$Sim_{tot}(p, s) = \frac{1}{|I|} \sum_{k \in I} Sim_k \tag{1}$$

The formula matches citizen preferences for services with service attributes, such as cost, location, timing, and rating. These attributes make up the set $\boldsymbol{I}$. To calculate similarity scores between a citizen's profile and a service, the similarity of each attribute must be calculated using formula 2.

$$Sim_k = \frac{1}{1 + \sqrt{\sum_{i=1}^{N} |r_i - r_i'|^2}} \tag{2}$$

For example, this formula calculates the similarity between a citizen's preference for the "Cost" attribute and the real value of "Cost" for a service. If a citizen wants both free and paid services, the cost preference vector in the profile would be $v_{profile} = (r_1, r_2) = (1, 1)$, and the cost preference vector for a paid service would be $v_{service} = (r_1', r_2') = (0, 1)$. In this case $\boldsymbol{N}$ is equal to 2 since the attribute cost has two possible values.

Personalized recommendations are added to the citizen's agenda, ensuring time and location suitability. By considering preferences and service attributes, the system provides customized recommendations that meet the individual's specific needs, as well as increasing satisfaction and quality of life.

5 Use Case

The following use case illustrates the recommendation process. Mr James, a retired 75-year-old, completed the survey, which identified 7 personal characteristics (Fig. 4).

The survey identifies personal characteristics with varying levels that help the system understand the challenges faced by elderly users. For example, assigned levels of 1 indicate the lowest level of difficulty, while values of 0 indicate no issues. Some characteristics are binary, indicating whether or not they exist. For example 1 for living alone characteristic and 0 if not. These levels reflect the difficulty of performing specific activities and will be used by the *NeedCreation* Rules in the future. After identifying personal characteristics, the system uses Algorithm 1 to identify the needs. *ConsiderationRules* are applied to verify the system's ability to help the citizen. For example, one of the rules applied in this use case is the following:

Rule 1 *Profiles with a frailty score (ADL[1]) greater than 4 are ineligible for consideration.*

[1] ADL: Activities of Daily Living survey.
https://www.alz.org/careplanning/downloads/katz-adl.pdf.

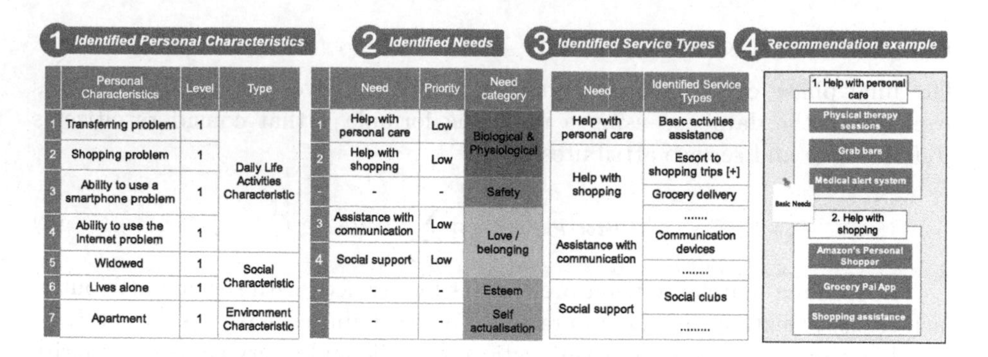

Fig. 4. Recommendation process results in each of the 4 steps

Determining a person's frailty status is a significant role played by the consideration step. The system collects user data by asking questions inspired by frailty surveys. With the help of rules, the system can interpret this data like a healthcare professional. For example, the ADL survey shows that if a person has four or more out of six problems, it indicates severe functional impairment, which is expressed by the rule. James' profile meets all system requirements as defined by the *ConsiderationRules.* Needs are created and assigned priorities using the *NeedCreation* Rules, which are based on the knowledge base. Rule 2 guides the creation of the social support need based on related characteristics. In James' case, two characteristics trigger Social support need with low priority: "widowed" and "living alone".

Rule 2 *Social support need is deemed low priority if one or two related characteristics are present.*

The creation of some other needs is straightforward. For instance, if the "Shopping problem" characteristic is present in the user's profile, the system will automatically generate the need *Help with shopping.* The priority of this need is determined directly by the characteristic level. In this case, rule 3 is triggered:

Rule 3 *Help with shopping need is deemed low priority if the "Shopping problem" characteristic has a level of 1.*

PriorityRules are applied to identified needs in the second phase, revealing relationships between different needs that require special attention. For example, if a person is unable to communicate effectively and living alone, Social support priority might increase to ensure necessary assistance. If someone cannot perform daily activities and communicate properly, Help with personal care priority should increase to ensure appropriate attention and safety.

Rule 4 *A high-priority Assistance with communication need and "living alone" characteristic are factors that warrant an increase in the priority level of a citizen's Social support need.*

In James' case, nothing will be changed since his Assistance with communication need has low priority.

Finally, the *ClassificationRules* will be applied to find out which are the basic needs and those optional based on Maslow's hierarchy of needs.

Rule 5 *Help with personal care and Help with shopping are part of the basic need level*

At the end of the need identification step, the list of needs is added to James' profile (see Fig. 4 part 2). The next algorithm will be responsible for identifying the service types. It will go through the needs sequentially, one by one, respecting their priority order. In James' case, it will start with the need "Help with personal care". For this need, four service types (ST) are identified in the first phase. In the next one, three ST out of four are eliminated by the *CheckRelatedPriority* function. One ST addresses high-priority needs, while two others address medium-priority needs. Only Basic Activities Assistance ST remains, addressing low-priority needs like James' Help with personal care need.

In the second iteration of Algorithm 2, the Help with shopping need is addressed. The iteration involves identifying the Escort to shopping trips ST, which is linked to two needs: Help with shopping and Social support. Consequently, the relevance of the ST is heightened by *CheckRelatedNeed*. This change is denoted by a '+' symbol next to the service type name in Fig. 4.

The third need on the list is Assistance with communication, with four service types identified in the first phase. *CheckObligatoryPC* eliminates two of them (1) sign language interpreters and (2)speech therapy, which require a "Speech problem" characteristic that James does not have. These are some examples of the constraints of service types. They are defined in the knowledge database and utilized by various functions in algorithm 2 to eliminate unnecessary service types that do not align with the citizen's profile.

Finally, the recommender system calculates scores for available services and recommends the top 3 for each need. Recommendations respect the priority of each need and are presented to the citizen. Figure 4 shows an example of a recommendation of the two first needs for James's profile.

6 Conclusion

The paper proposes a personalized recommendation framework that prioritizes the well-being of the elderly and aims to prevent frailty. It defines a four-step recommendation process, which generates personalized recommendations aligned with specific needs. Further, two algorithms were developed for needs identification and service type identification. We assessed the effectiveness of the proposed framework through a series of scenarios and provided examples of rule applications, an example of a use case scenario is described above.

To evaluate the framework using real-world data on a scale, we plan to develop a prototype and conduct usage testing with a group of elderly individuals. The Senselife framework's final component involves monitoring the consumption and use of recommended services, allowing for continuous improvement through user feedback. The prototype's usage data will further enhance the framework's recommendation capabilities, leading to better outcomes for elderly citizens.

References

1. Bowling, A., Dieppe, P.: What is successful ageing and who should define it? BMJ **331**(7531), 1548–1551 (2005)
2. Clegg, A., Young, J., Iliffe, S., Rikkert, M.O., Rockwood, K.: Frailty in elderly people. Lancet **381**(9868), 752–762 (2013)
3. Espin, V., Hurtado, M.V., Noguera, M.: Nutrition for elder care: a nutritional semantic recommender system for the elderly. Expert Syst. **33**(2), 201–210 (2016)
4. Fillit, H., Butler, R.N.: The frailty identity crisis. J. Am. Geriatr. Soc. **57**(2), 348–352 (2009)
5. Furtado, G.E., et al.: Emotional well-being and cognitive function have robust relationship with physical frailty in institutionalized older women. Front. Psychol. **11**, 1568 (2020)
6. Hikmawati, E., Maulidevi, N.U., Surendro, K.: Adaptive rule: a novel framework for recommender system. ICT Express **6**(3), 214–219 (2020)
7. Kaminskas, M., Bridge, D.: Diversity, serendipity, novelty, and coverage: a survey and empirical analysis of beyond-accuracy objectives in recommender systems. ACM Trans. Interact. Intell. Syst. (TiiS) **7**(1), 1–42 (2016)
8. Maslow, A.H.: A theory of human motivation. Psychol. Rev. **50**(4), 370 (1943)
9. World Health Organization. World report on ageing and health. World Health Organization (2015)
10. World Health Organization. Integrated care for older people: guidelines on community-level interventions to manage declines in intrinsic capacity (2017)
11. Orte, S., et al. Dynamic decision support system for Personalised coaching to support active ageing. In: AI* AAL@ AI* IA, pp. 16–36 (2018)
12. Ribeiro, O., Almeida, S., Teixeira, L., Santos, J., Marques, A.: A recommender system for personalized physical activity and nutrition recommendations for older adults with frailty. J. Ambient. Intell. Humaniz. Comput. **10**(2), 635–644 (2019)
13. Rist, T., Seiderer, A., Hammer, S., Mayr, M., Andre, E.: Care-extending a digital picture frame with a recommender mode to enhance well-being of elderly people. In: 2015 9th International Conference on Pervasive Computing Technologies for Healthcare (PervasiveHealth), pp. 112–120. IEEE (2015)
14. Schuurmans, J.E.H.M.: Promoting well-being in frail elderly people: theory and intervention (2004)

Design-Focused Development of a Course Recommender System for Digital Study Planning

Michaela Ochs(✉), Tobias Hirmer, Katherina Past, and Andreas Henrich

Chair of Media Informatics, University of Bamberg, Bamberg, Germany
{Michaela.Ochs,Tobias.Hirmer,Andreas.Henrich}@uni-bamberg.de,
pastkatherina@gmail.com

Abstract. In recent years, universities have been offering a rising number of courses, creating not only greater flexibility of choice, but simultaneously more complexity for students in their course selection. Therefore, this paper seeks to identify key aspects that should be considered when designing a course recommender system (CRS) in the higher education context. To achieve this, students' selection criteria and processes were analyzed and condensed into requirements. Based upon these, a prototype was implemented and evaluated via think-aloud user tests. The results of this study indicate that a multidimensional approach shall be taken to optimize the user experience. As our main contribution, we identify six guidelines for the design of an effective CRS.

Keywords: Study Planning · Course Recommendation · Decision Support · Design-Focused

1 Introduction

Within the field of European higher education, students face multiple options in the process of planning their individual way through their studies. Students select between different courses that contribute to modules, which are thematically specialized and self-contained learning units. Students must complete a specific amount of credits in different groups of modules, which vary depending on their study program. This results in a complex network of connections between modules and courses [6], making study planning a challenging decision problem, theoretically grounded by decision theory [14]. In the following, we will therefore reference common terminology associated with this field. Supporting (student) decisions is an interesting field for recommender systems (RS), where, in particular, the process of course selection is already a popular use case.

This research was conducted as part of the projects "Developing Digital Cultures for Teaching (DiKuLe)" and "Learning from Learners (VoLL-KI)" and was financed by "Stiftung Innovation in der Hochschullehre" as well as "Künstliche Intelligenz in der Hochschulbildung (Artificial Intelligence in Higher Education)".

A. Abelló et al. (Eds.): ADBIS 2023, CCIS 1850, pp. 575–582, 2023.
https://doi.org/10.1007/978-3-031-42941-5_50

CRS for higher education can be characterized by individual conditions that influence their selection and use. Factors such as grades, career prospects, and social considerations can affect the recommendation of university courses [10]. In this specialized research field, Ma et al. [8–10] provide related papers with an explicit focus on the user experience. Initially, patterns in students' course selection were analyzed and based on this, a framework with a matching course recommendation algorithm was envisioned and qualitatively evaluated [10]. The two subsequent papers [8,9] focused on the design and evaluation of an interactive CRS. Here, recommendation technologies and an interactive user interface were combined for flexible course scheduling and user control [8,9].

Building on this previous work, this paper is structured as follows: After introducing the methodological foundation, we present the requirements that were elicited in the qualitative study. A CRS prototype that has undergone a first iteration of user feedback is introduced thereafter and the results of the evaluation are described. To conclude, the paper presents six guidelines for designing a CRS for study planning derived from the requirements. We thereby contribute guidance in designing a user-centered CRS for study planning, focusing on the design rather than algorithmic details, which is the next step to take.

2 Method

This work is guided by the double diamond design process model, which structures the design process of service products [3]. The framework comprises the four phases *Discover*, *Define*, *Develop* and *Deliver* – the first and second of which are referred to as *research phase* and the third and fourth as *design phase* [3]. The double diamond design process model is user-centered and encourages iterative thinking, revision and solutions – which corresponds to the design-focused aspects of this research. It underlies and guides this contribution as visualized in Fig. 1. In the research phase, students' course selection behavior was discovered in a qualitative, explorative study, which resulted in a list of requirements for a CRS. For this purpose, 25 previously collected interviews (among others from [6]), asking students from different faculties (mostly computer and education science) about their process of and problems with study planning, have been re-examined in a complementary secondary analysis [5]. *Thematic analysis* allowed to extract attitudinal and behavioral data [16] based on which *thematic networks* [2] and *affinity diagrams* [13] were further used to structure the results of the previous analysis [16] in order to derive requirements. Following this research phase, a prototypical CRS was designed and tested in the design phase. For this purpose, the *How-Might-We-Method* was employed to translate requirements into questions in order to generate solution ideas [7], which were realized by iterative *interaction prototyping* [12]. The resulting prototype was evaluated in a semi-structured *user test ($n = 6$)* with the *concurrent think aloud method* [4] where users had to perform given tasks while sharing their thoughts. The interviews (Ø 75 min) were analyzed via the method of *structured content analysis* [11]. Next, Chap. 3 describes the results of the research phase and then Chap. 4 introduces the design of the prototype and the results of the evaluation.

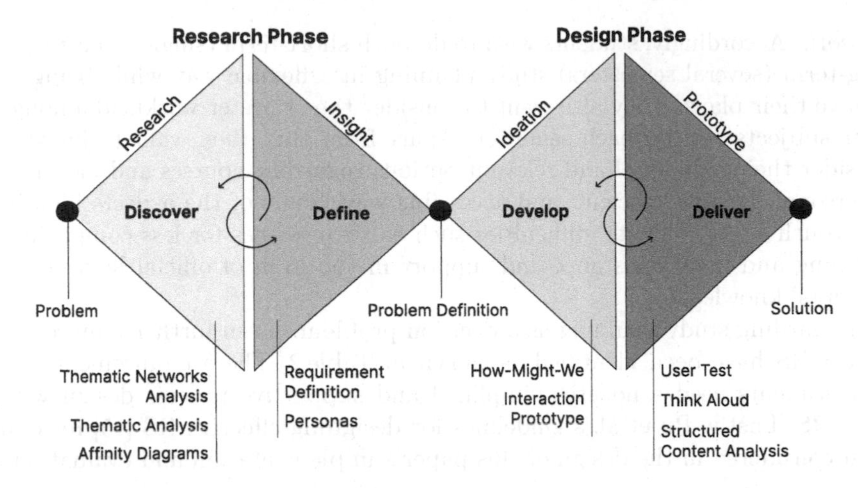

Fig. 1. Methodical Structure According to the Double Diamond Decision Process Model (Adapted from [3])

3 Requirements

This chapter presents the findings of the exploratory study on students' course selection process within the research phase. The requirements presented are based on two thematic networks identified in the analysis: *course selection criteria* and *course selection process*. Table 1 shows eight criteria that are relevant for selecting courses that have been derived from the first theme.

Table 1. List of Course Selection Criteria Which Shall Be Considered (Random Order)

No	Criteria	Examples
1.1	Course Features	Credits, difficulty, mandatory/elective,
1.2	Module Context	Courses needed to complete module,
1.3	Organizer	Chair, person, expectations,
1.4	Time Constraints	Time, day, semester, conflicting courses,
1.5	Organizational Constraints	Course cycle, capacity, online/in-class,
1.6	Social Constraints and Recommendations	Ratings, choices and recommendation of peers,
1.7	Individual Interests and Goals	Credits, focus of study, career,
1.8	Individual Prerequisites and Study History	Prior knowledge, failed courses,

In addition to the consideration of these criteria, further criteria related to the semester and study planning process have been derived from the second thematic network. Accordingly, students wish to do both short-term (single semester) and long-term (several semesters) study planning in a flexible way while being able to save their plans. They also want to consider the semester workload among all their subjects and for each semester. Apart from this, they want to know and consider their individual and relevant options regarding courses and modules in a personal, helpful, efficient, and accessible way. Some of the aspects identified also touch upon students' difficulties such as their wishes for less complexity in planning and more assistance and support in the form of official help or older students' knowledge.

Regarding study planning as a decision problem, seven further requirements for a CRS have been identified as shown in Table 2. These requirements show that students need a holistic, simplified and supportive process design within the CRS. Lastly, Pu et al.'s guidelines for designing effective RS [15] have also been considered in the design of this paper's implementation and evaluation.

Table 2. List of Requirements Regarding the Decision Problem and Process

No	Criteria
2.1	Staying Informed about Short- and Long-Term Consequences of Selections
2.2	Making Decisions under Risk, not Insecurity [14]
2.3	Knowing all Options
2.4	Being Faced with a Given Set of Options
2.5	Being Supported during the Multilevel Course Selection Process
2.6	Being Supported with Individual and Social Decisions
2.7	Being Supported during all Phases of the Decision Process within Semester and Study Planning

4 Design and Evaluation of the CRS

This section describes the final realization of the previously described requirements as a prototypical implementation created with the design tool *Figma*[1], including the insights from the user evaluation from the first iteration of the design process model as introduced in Chap. 2. The results of the evaluation are described thereafter.

Based on the requirements that have been introduced, a knowledge-based and a collaborative recommendation approach were selected for our CRS. On the one hand, the choice of the knowledge-based approach responds to the requirements of personal recommendations and support during the entire course selection and planning process (cf. requirements 2.4, 2.5, 2.7) as it can adapt to specific needs based on user specifications. On updating the specifications, users can observe changes in the composition of recommended items and revise their input, leading to control and trust in the system. On the other hand, the collaborative

[1] https://www.figma.com/.

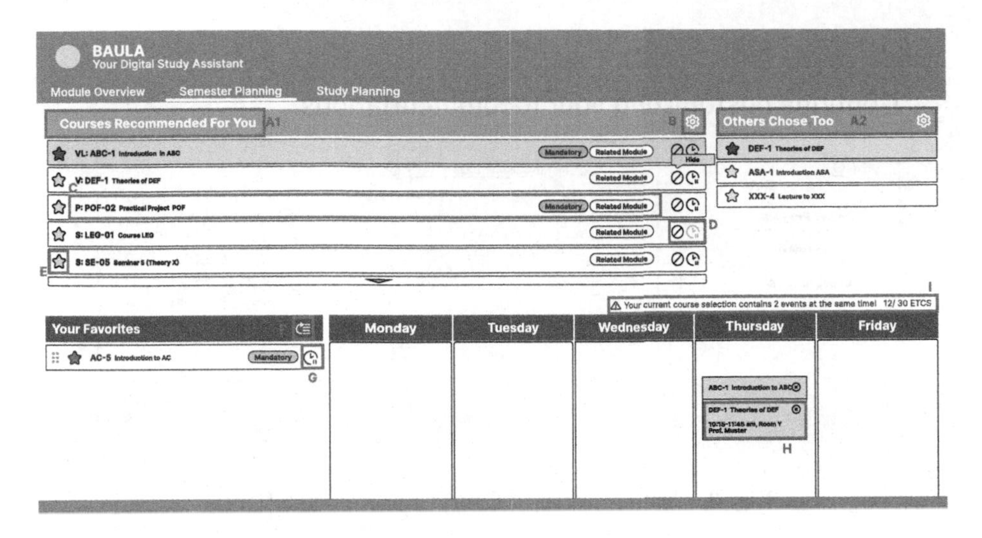

Fig. 2. Semester Planning Page of the CRS

approach was found to be helpful, as it addresses students' wish for multi-source recommendations, i.e. by students with the same or a higher semester count or students within the same degree program (cf. requirement 1.6). Since students do not always remain constant within their interests and evaluations, an item-based approach was chosen over the user-based approach [1], which may be refined by preliminary filtering of the users according to their study program.

With respect to the prototypical realization, the semester planning process illustrated in Fig. 2 constitutes a significant aspect of the CRS. We will thus primarily focus on this process in the subsequent sections to provide a comprehensive demonstration.

Within the knowledge-based system, user input is incorporated via a search-based system that queries students' course preferences [1] (i. e. semester to be planned, personal interests, courses of interest). Thereby, requirements such as 2.1 can be fulfilled as students may be notified that they have the prerequisites for the courses of interest. Students can customize their specifications for course recommendations further according to requirements 1.1, 1.2, and 1.5 - 1.8 within the dialogue depicted in Fig. 3, which opens when clicking on (Fig. 2 - B). A list of recommended courses can be generated based on these specifications (Fig. 2 - A1), the accuracy and comprehensibility of which may be improved by the option of adjusting the specifications anytime. Within the list, relevant information that describes the course is displayed for each item (cf. requirement 1.1; Fig. 2 - C). Clicking on one item displays more detailed information about the course. Certain courses can be hidden (Fig. 2 - D left), moved to a later semester (Fig. 2 - D right) or added to a list of favorites (Fig. 2 - E).

In the context of the the collaborative approach, using subjective ratings may negatively impact the relationship between students and lecturers. Thus, only

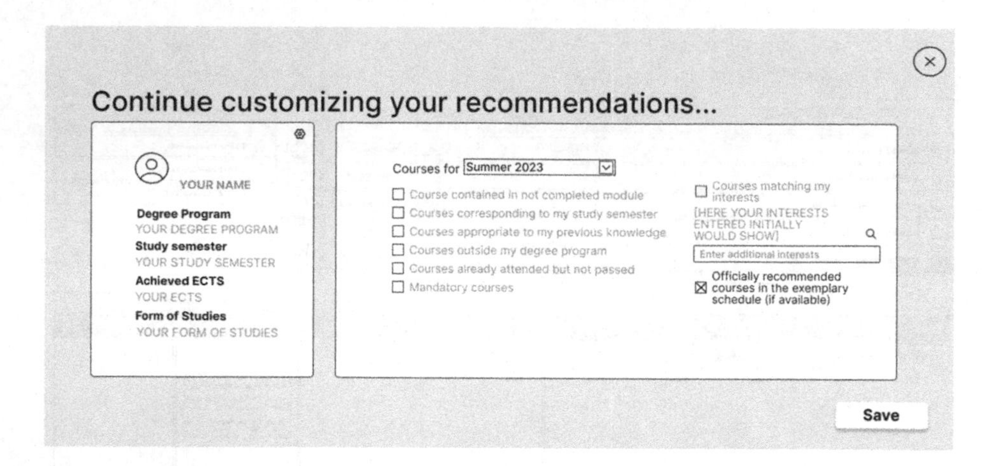

Fig. 3. Customization of Course Recommendations

indirect ratings (*taken* = *recommended, not taken* = *not recommended*) should be used. To circumvent cold-start issues, the rating modalities may be derived automatically from process-generated data that exists in related systems. In our case, the cold start problem for new items was addressed by adding a banner with new courses to the interface of the course selection process that students have to close. Lastly, a major aspect within the collaborative method was to create the possibility for students to adjust in which way other students should be similar to them, i.e. regarding focus of study, degree program or faculty. Based on their input, a list of courses is generated and displayed to the user (Fig. 2 - A2). An equivalent list was used for the long-term study planning component.

Apart from these elements, further options of individualization are provided. Thus, the list of favorites can be sorted according to specific course selection criteria (cf. Chapter 3, Fig. 2 - F). From the list of favorites, courses can be scheduled into the timetable (Fig. 2 - G), which shows courses as calendar items with essential course details (Fig. 2 - H). In case of discrepancies, e.g. in the form of overlaps, differences from individual goals or missing prerequisites, a hint is displayed (Fig. 2 - I).

In the user evaluation, as it was described in Chap. 2, the overall recommendation system and procedures were found to be helpful. Students mentioned that the planning process is easier and more comprehensible and that such a system would motivate them to more detailed planning, while creating a more secure feeling when planning. In addition, the chosen recommendation methods were considered to be useful and in accordance with their needs.

5 Discussion

Based on the previous findings, supported by the user evaluation, this research suggests six guidelines for designing a CRS:

- *The short-term and long-term course selection criteria used in practice should be included in the recommendation generation*, which can help to better match the recommendations to the needs of the user, which might lead to a higher satisfaction and usage rate of the application.
- *The CRS should not only provide course recommendations, but should also accompany the entire planning and decision-making process.* Thereby, the effectiveness and efficiency of the system may be improved, maximizing the benefits for students.
- *The potential (negative) consequences of selecting a recommendation should be made accessible to students*, which can encourage an informed decision, prevent an insufficient user experience and strengthen the trust in the system.
- *The (additional) effort of the use of the recommendation system should be kept as low as possible and be proportional to its use*, which particularly refers to the enquiry of specifications and the terminology used. Efficiency, familiarity and comprehensibility can increase the value for the students.
- *Adaptation to different students and their needs should be possible within the CRS*, so it can help realize individual and self-paced study progress.
- *The visible course information within the CRS is to be carefully sorted and presented according to its relevance*, whereby students get a quick overview of the most important course features.

With regard to these findings, a major limitation in terms of external validity is that the results are not fully generalizable as the sample of students could have been increased to more faculties and the requirements have been elicited within a specific University context. Also, the reliability and objectivity of the results could be improved further using inter-rater reliability and triangulation. Further limitations are related to the prototype, i.e. the *How-Might-We method* is prone to fluctuating results, influencing the reproducibility. To enhance internal validity and mitigate issues of subjectivity, we have focused on methodological precision and reviewed our findings within the team of authors.

6 Conclusion

Based upon exisiting interviews this work has extracted requirements for a CRS within the context of study planning. These were implemented prototypically and evaluated in a user test. As has been shown, a user-centered CRS shall include multidimensional criteria which can be condensed into six guidelines.

Thereby, this research contributes to filling the research gap within the user-centered design of CRS. It is also an extension to the previously more algorithmically driven design of course recommendation algorithms. Moreover, it shows how a user-centered design can be developed and tested from an application context of RS, similar to Ma et al. [8–10]. Further steps are a second iteration regarding the double diamond model, as well as the selection and implementation of a specific recommendation algorithm for this specific context.

References

1. Aggarwal, C.C.: Recommender Systems. Springer, Cham (2016). https://doi.org/10.1007/978-3-319-29659-3
2. Attride-Stirling, J.: Thematic networks: an analytic tool for qualitative research. Qual. Res. **1**(3), 385–405 (2001). https://doi.org/10.1177/146879410100100307
3. Council, D.: A study of the design process: Eleven lessons: Managing design in eleven global brands. https://www.designcouncil.org.uk/our-work/skills-learning/resources/11-lessons-managing-design-global-brands/ (2007), Accessed 19 Apr 2023
4. Ericsson, K.A., Simon, H.A.: How to study thinking in everyday life: contrasting think-aloud protocols with descriptions and explanations of thinking. Mind Culture Activity **5**(3), 178–186 (1998). https://doi.org/10.1207/s15327884mca0503_3
5. Heaton, J.: Secondary analysis of qualitative data: An overview. Hist. Soc. Res. **33**(3), 33–45 (2008)
6. Hirmer, T., Etschmann, J., Henrich, A.: Requirements and Prototypical Implementation of a Study Planning Assistant in CS Programs. In: 2022 International Symposium on Educational Technology (ISET). pp. 281–285 (2022). https://doi.org/10.1109/ISET55194.2022.00066
7. Holderfield, G.: Generating How Might We (HMW) Questions from Insights - Using Design Principles to Innovate and Find Opportunities. https://www.coursera.org/lecture/leadership-design-innovation/generating-how-might-we-hmw-questions-from-insights-JMRYd (2017), Accessed 19 Apr 2023
8. Ma, B., Lu, M., Taniguchi, Y., Konomi, S.: CourseQ: The impact of visual and interactive course recommendation in university environments. Research and Practice in Technology Enhanced Learning 16 (2021). https://doi.org/10.1186/s41039-021-00167-7
9. Ma, B., Lu, M., Taniguchi, Y., Konomi, S.: Exploration and explanation: An interactive course recommendation system for university environments. CEUR Workshop Proceedings 2903 (2021)
10. Ma, B., Taniguchi, Y., Konomi, S.: Course Recommendation for University Environments. In: Proceedings of The 13th International Conference on Educational Data Mining (EDM 2020). pp. 460–466 (2020)
11. Mayring, P., Fenzl, T.: Qualitative Inhaltsanalyse. In: Handbuch Methoden der empirischen Sozialforschung, pp. 633–648. Springer, Wiesbaden (2019). https://doi.org/10.1007/978-3-658-21308-4_42
12. Moser, C.: User Experience Design-Mit Erlebniszentrierter Softwareentwicklung Zu Produkten. Die Begeistern. Springer, Berlin Heidelberg (2012) https://doi.org/10.1007/978-3-642-13363-3
13. Pernice, K.: Affinity Diagramming: Collaboratively Sort UX Findings & Design Ideas. https://www.nngroup.com/articles/affinity-diagram/ (2019), Accessed 19 Apr 2023
14. Pfister, H.R., Jungermann, H., Fischer, K.: Die Psychologie der Entscheidung. Springer, Heidelberg (2017). https://doi.org/10.1007/978-3-662-53038-2
15. Pu, P., Chen, L., Hu, R.: Evaluating recommender systems from the user's perspective: survey of the state of the art. User Model. User-Adap. Interact. **22**(4), 317–355 (2012)
16. Rosala, M.: How to Analyze Qualitative Data from UX Research: Thematic Analysis. https://www.nngroup.com/articles/thematic-analysis/ (2019), Accessed 19 Apr 2023

Systematic Literature Review on Click Through Rate Prediction

Paulina Leszczełowska(✉), Maria Bollin, and Marcin Grabski

Faculty of Electronics, Telecommunications and Informatics, Gdańsk University of Technology, Gdańsk, Poland
paulinaleszczelowska@gmail.com

Abstract. The ability to anticipate whether a user will click on an item is one of the most crucial aspects of operating an e-commerce business, and clickthrough rate prediction is an attempt to provide an answer to this question. Beginning with the simplest multilayer perceptrons and progressing to the most sophisticated attention networks, researchers employ a variety of methods to solve this issue. In this paper, we present the findings of a comprehensive literature review that will assist researchers in getting a head start on developing new solutions. The most prevalent models were variants of the state-of-the-art DeepFM model.

Keywords: Click-through Rate · Recommender Systems · Machine Learning · Deep Learning · DeepFM

1 Introduction

CTR prediction is an essential task in online advertising and recommender systems [6], with the objective of estimating the likelihood that a user will click on an item or advertisement [3]. CTR can be used as a significant indicator of the effectiveness of Internet advertising. The most important aspect of CTR prediction is understanding and extracting the interactive functions hidden behind the user's click behavior. [18]

In recent times, a great number of models that are based on deep learning have been proposed with the intention of capturing high-order feature interactions in terms of CTR prediction [12]. An inexperienced analyst can easily become overwhelmed by the abundance of different approaches to the problem. Therefore, we have decided to compile a systematic literature review that summarizes the available data on various CTR prediction methods.

2 Literature Review Process Design and Execution

In order to determine the focus of our research study, the following four research questions were formulated: (1) How can AI assist in the software development

A. Abelló et al. (Eds.): ADBIS 2023, CCIS 1850, pp. 583–590, 2023.
https://doi.org/10.1007/978-3-031-42941-5_51

process?, (2) Which feature transformations are used in the CTR prediction? (3) Which datasets and data types are used in the CTR prediction? (4) Which metrics are used to assess the performance of obtained models?

The databases that we have chosen were: Scopus and IEEExplore, and the search string used were as follows: ("click through rate prediction" OR "CTR prediction" OR "click prediction"). We decided to only include English-language conference papers published between 2018 and 2023 with DOI, abstract, and full paper available. We excluded any paper that had a form of review. Following the removal of duplicates, we were left with 143 papers for further analysis.

The papers were tagged based on their titles. Each of us assigned a score between 0 (totally irrelevant) and 2 (totally relevant) for a given title. All the papers with a total score of 6 were evaluated further, which totaled 41 articles. The next step involved tagging articles based on their abstracts, and similarly, only those with a total score of at least 6 were considered for further analysis. We concluded this phase with a total of 29 articles.

3 Results

3.1 Datasets

The most commonly used datasets in the articles were Avazu, Criteo, and Taobao. Few datasets included short videos or images of advertisements, such as "Tame-map." The number of articles with author-created datasets was relatively small. The number of articles that utilized each dataset is depicted in Fig. 1.

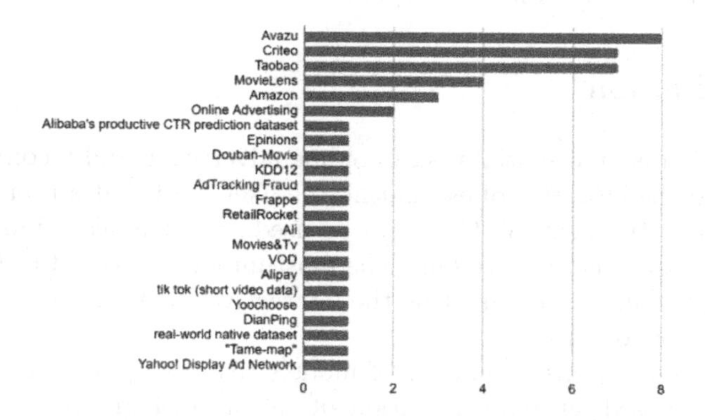

Fig. 1. Number of uses of a given dataset

3.2 Type of Data

Depending on the model chosen by the authors, the data required an appropriate format. The most common data type was feature vector, and it was used in 23 articles. Other utilized data types included text [20], metadata [20], sequence [1,17,22,24], matrix [11], and image [4,20].

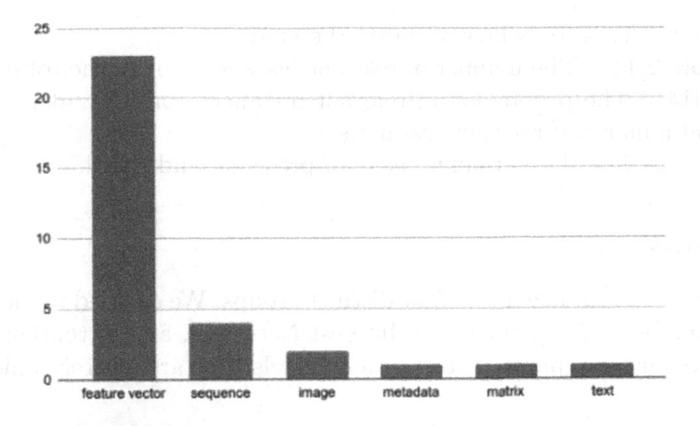

Fig. 2. Number of uses of a given type of data

3.3 Feature Transformation

In a total of 29 articles, 14 feature extraction methods were identified. Among those 14, 10 were mentioned only once. These include: sampling patches from images [20], representing text data as Paragraph Vector [20], dimension conversion layer [10], applying a density matrix [12], multimodal fusion [20], k-means clustering [22], categorical feature pruning [14], boosted decision trees [14], normalization [14], and regularization [5].

The most prevalent feature transformation method was embedding. This was commonly implemented using the Embedding Layer to encode sparse features into low-dimensional vectors [1]. One-hot encoding of the data was the second most commonly used method for transforming features [13].

Normalization and factorization were used twice each. Normalization aims to prevent large features from distorting the feature space by limiting the numerical values to the range 0 to 1 [14]. Two articles used factorization prior to feeding the data into the models. Utilizing inner products, factorization machines and matrix factorization are used to model feature interactions.

3.4 Metrics

The most commonly used metric was AUC (area under the ROC curve) [9]. This metric was mentioned in 26 different articles. Less commonly used metrics include:

- MSE (mean squared error) [4,20]- It represents the prediction error;
- KL-D (Kullback–Leibler divergence) [20] - It indicates the prediction error;
- Pearson's product moment correlation coefficient (R) [20] - A higher R indicates better performance.
- Hit Ration (HR) [22] - Recall metric for top-K recommendations.
- Normalized Discounted Cumulative Gain (NDCG) [20] - Indicates that the score of each recommendation result correlation is accumulated as the score of the entire recommendation list;

- Accuracy [2,17] - Tells how accurate the model is;
- Precision [2,14] - The number of relevant instances out of the total instances;
- Recall [14] - The proportion of relevant instances correctly identified out of the total number of relevant instances.
- F1-score [14,17]- the harmonic mean of precision and recall.

3.5 Models

The models were organized into four distinct groups. We decided to include models based on DeepFM architecture, Interest Networks, and Attention Networks as categories and a separate category for models that are entirely unique.

DeepFM/xDeepFM. DeepFM consists of two input-sharing components: the FM component and the deep component. Deep extracts high-order combination features, while FM extracts low-order features [7]. As commonly known as DeepFM, its upgraded version, called xDeepFM, is used as a base for further development. Extreme Deep Factorization Machine (xDeepFM) integrates the capabilities of convolutional neural networks (CNN) and recurrent neural networks (RNN). We discovered eight solutions based on the DeepFM or xDeepFM network; in the following paragraphs, we will discuss their innovations.

ACDeepFm, the first of the proposed models, incorporates a self-attentive mechanism. It employs the self-attentive mechanism to discover the intrinsic relationships between input features and models the input features' weights in an adaptive manner. A compressed interaction network (CIN) is added to mine the impact of order combinations of features on the prediction results. Information from the self-attentive mechanism module, the deep neural networks module, and the CIN module is fed into the subsequent multilayer perceptron [29].

The second solution introduced the concept of ECANET, which modifies the model embedding layer to dynamically learn the significance of embedding features and build a new deep CTR prediction model [16].

DeepFaFM provides a scalable coding matrix with adjustable factors that can dynamically balance operation pressure based on the actual production environment. It outperforms DeepFm by enhancing the inadequate feature interactions in the operation and enhancing the model's accuracy [28].

The following article presents the xDeepFm model and uses Bayesian optimization. In addition to the fundamental xDeepFm model, Bayesian optimization is employed to determine the optimal hyperparameters [25].

The DeepFm model cannot highlight the influence of the user's behavior, which has a strong correlation with click behavior. Consequently, two behavior-focused algorithms, DeepFM&Habit (Pro) and DeepFm&Habit (LR), are proposed. The algorithms are used to emphasize the significance of individual behavioral patterns for predicting outcomes. The Pro version employs probabilistic prediction, while the LR version employs linear regression [9].

Interest Network. The Deep Interest Network (DIN) calculates in an adaptive manner the representation vector of user interests by taking into account the relevance of historical behaviors when presented with a candidate advertisement. The representation of user interests is dominated by behaviors with greater relevance to the candidate ad, which receive higher weights when activated [27].

The first paper implemented DIN for the Taobao dataset. Changes to DIN's configuration, such as the number of layers and number of neural units per layer, were part of the modifications introduced. Additionally, the author proposed a novel mixed-loss metric for the training of the model in order to achieve more reliable results [2].

Another example of an architecture relying on interest would be a Multi-Interest Self-Supervised (MISS) framework. The main goal of this framework is to improve feature embeddings by employing interest-level contrastive losses. The proposed MISS framework has the capability of being integrated into preexisting CTR prediction models in the form of a "plug-in" component [3].

User Evolving Interest Network (UEIN) was developed to capture dynamic user interests adaptively from user-item sequence interactions by incorporating item category information into user behavior sequences. The proposed model divides the user's past interactions into long- and short-term segments and uses the attention network and MLP to learn the user's interests [22].

Attention Network. Attention Networks (AN) are neural network models that employ the attention mechanism [20]. Two of the found papers ([3,22]) were previously mentioned when discussing Interest Networks, as many of their implementations also incorporate an attention mechanism.

First article demonstrated a multimodal approach utilizing image, metadata, and text data. The proposed AN consists of fully connected layers, a batch normalization layer, and a softmax layer. The model's output is a column vector, with each row's element representing the weight of each feature [20].

The adaptive depth attention (ADA) model automatically discovers the high-order feature interactions of unprocessed data. It consists of a multi-head self-attention neural network that learns feature interactions and a network depth control module that regulates the network depth required for the interaction between different feature fields [10].

Novel Approaches. First, from the solutions, proposed a model that extracts high-impact features using CNN and predicts and classifies them using FM, which can learn the relationship between mutually distinct feature components. In comparison to the single structure model, the CNN-FM hybrid model increases the accuracy [15].

Similitude Attentive Relation Network (SARN) is a model based on the Relation Network (RN) paradigm. The user's behavior is represented as a graph, with nodes corresponding to visited items and edges representing relationships. The model attempts to learn the similarity between items in a semantic space by employing a learnable dot-product operation and combining the item representations and relational information to produce the final relations [1].

The next paper examines the listwise user-specific CTR prediction for the entire recommendation list, as opposed to a single item. A joint prediction framework based on matrix factorization and recurrent neural network (RNN) is proposed [19].

Another study proposed a method called FINT, which employs the Field-aware Interaction layer. It captures high-order feature interactions while retaining low-order field information. A unique feature interaction vector is maintained for each field, containing interaction information about the field and each of the other fields in any order within K. [26]

Dual Adaptive Factorization Network (DAFN) can simultaneously learn element-wise and vector-wise cross features of arbitrary order and make extensive use of the information about the original embedding vectors. It incorporates multiple components, such as Talking-heads Attention, Logarithmic Transformation Layer, and Residual Networks [8].

The following paper introduces an embedding and deep learning-based model FMSDAELR. This model can effectively model multi-field categorical data, extract the complex non linearly correlated relationship, and capture the key features contributing to CTR prediction for advertising [5].

The 3M (Multi-task, Multi-interest, Multi-feature) model considers user interests from the standpoint of item characteristics. The 3M model trains multiple interest vectors for each user and extracts multiple characteristic vectors for each item. A multi-task learning model is used to connect the characteristic vectors with the interest vectors and train them to obtain multiple interest scores [21].

The Extreme Cross Network (XCrossNet) learns explicitly dense and sparse feature interactions. As a feature structure-oriented model, XCrossNet results in a more expressive representation and a more accurate prediction [23].

The final paper in this category introduced a novel model, Convolutional Click Prediction Model (RES-IN), based on a residual unit with an inception module. RES-IN can extract local-global key feature interactions from an input instance containing diverse elements [11].

4 Summary

The majority of researchers choose to employ embedding mechanisms for improved data representation. Due to the vast variety of model architectures, we were able to classify them according to their underlying concepts. The most prominent models were those based on the DeepFM model and its variants. There was an extensive range of datasets used, with many of them being unique to each article. However, three standouts can be distinguished based on occurrence number: Avazu, Criteo, and Taobao. When it comes to the data types used in the studies, we can clearly identify one type—feature vector—that was used in 23 of the 29 papers.

References

1. Deng, H., Wang, Y., Luo, J., Hu, J.: Similitude attentive relation network for click-through rate prediction. In: 2020 International Joint Conference on Neural Networks (IJCNN) (2020)
2. Di, S.: Deep interest network for taobao advertising data click-through rate prediction. In: 2021 International Conference on Communications, Information System and Computer Engineering (CISCE) (2021)
3. Guo, W., et al.: Miss: Multi-interest self-supervised learning framework for click-through rate prediction. In: 2022 IEEE 38th International Conference on Data Engineering (ICDE) (2022)
4. Inoue, D., Matsumoto, S.: Predicting CTR of regional flyer images using CNN. In: 2022 12th International Congress on Advanced Applied Informatics (IIAI-AAI) (2022)
5. Jiang, Z., et al.: A CTR prediction approach for advertising based on embedding model and deep learning. In: 2018 IEEE Intl Conference on Parallel & Distributed Processing with Applications, Ubiquitous Computing & Communications, Big Data & Cloud Computing, Social Computing & Networking, Sustainable Computing & Communications (ISPA/IUCC/BDCloud/SocialCom/SustainCom) (2018)
6. Karpus, A., Raczyńska, M., Przybylek, A.: Things you might not know about the k-nearest neighbors algorithm. In: 11th International Joint Conference on Knowledge Discovery, Knowledge Engineering and Knowledge Management (2019)
7. Li, L.S., Hong, J., Min, S., Xue, Y.: A novel CTR prediction model based on deepfm for taobao data. In: 2021 IEEE International Conference on Artificial Intelligence and Industrial Design (AIID) (2021)
8. Li, X., Wang, Z., Wu, X., Yuan, B., Wang, X.: A dual adaptive factorization network for CTR prediction. 2021 IEEE 23rd International Conference on High Performance Computing & Communications; 7th International Conference on Data Science & Systems; 19th International Conference on Smart City; 7th International Conference on Dependability in Sensor, Cloud & Big Data Systems & Application (HPCC/DSS/SmartCity/DependSys) (2021)
9. Li, Y., Wang, Y., Chen, C., Huang, J.: CTR prediction with user behavior: An augmented method of deep factorization machines. In: 2019 IEEE 14th International Conference on Intelligent Systems and Knowledge Engineering (ISKE) (2019)
10. Liu, S., Chen, D., Shao, J.: Ada: adaptive depth attention model for click - through rate prediction. In: 2021 International Joint Conference on Neural Networks (IJCNN) (2021)
11. Ni, Z., Ma, X., Sun, X., Bian, L.: A click prediction model based on residual unit with inception module. In: PRICAI 2019: Trends in Artificial Intelligence, pp. 393–403 (2019)
12. Niu, T., Hou, Y.: Density matrix based convolutional neural network for click-through rate prediction. 2020 3rd International Conference on Artificial Intelligence and Big Data (ICAIBD) (2020)
13. K. Potdar, T. S., and C. D. A comparative study of categorical variable encoding techniques for neural network classifiers. Int. J. Comput. Appl. **175**(4), 7–9 (2017)
14. Qiu, X., Zuo, Y., Liu, G.: Etcf: An ensemble model for CTR prediction. In: 2018 15th International Conference on Service Systems and Service Management (ICSSSM) (2018)

15. She, X., Wang, S.: Research on advertising click-through rate prediction based on CNN-FM hybrid model. In: 2018 10th International Conference on Intelligent Human-Machine Systems and Cybernetics (IHMSC) (2018)
16. Shi, X., Yang, Y., Tao, C.: Deep interest network for taobao advertising data click-through rate prediction. CTR prediction model considering the importance of embedding vector. In: 2021 IEEE International Conference on Artificial Intelligence and Computer Applications (ICAICA) (2021)
17. Wang, G., Wang, X.: Time-aware multi-layer interest extraction network for click-through rate prediction. In: 2022 Asia Conference on Algorithms, Computing and Machine Learning (CACML) (2022)
18. Wang, P., Sun, M., Wang, Z., Zhou, Y.: A novel CTR prediction based model using xdeepfm network. 2021 IEEE International Conference on Computer Science, Electronic Information Engineering and Intelligent Control Technology (CEI) (2021)
19. Wang, R., Guo, P., Fan, X., Li, B., Zhang, W., Xin, X.: Listwise click-through rate prediction with item-item interactions. In: 2018 IEEE International Conference on Systems, Man, and Cybernetics (SMC) (2018)
20. Xia, B., Wang, X., Yamasaki, T., Aizawa, K., Seshime, H.: Deep neural network-based click-through rate prediction using multimodal features of online banners. In: 2019 IEEE Fifth International Conference on Multimedia Big Data (BigMM) (2019)
21. Xie, Y., Li, M., Lu, K., Shah, S.B., Zheng, X.: Multi-task learning model based on multiple characteristics and multiple interests for CTR prediction. In: 2022 IEEE Conference on Dependable and Secure Computing (DSC) (2022)
22. Xu, J., Shi, X., Qiao, H., Shang, M., He, X., HeQ.: Uein: A user evolving interests network for click-through rate prediction. In: 2021 IEEE 15th International Conference on Big Data Science and Engineering (BigDataSE) (2021)
23. Yu, R., et al.: XCrossNet: Feature Structure-Oriented Learning for Click-Through Rate Prediction. In: Karlapalem, K., et al. (eds.) PAKDD 2021. LNCS (LNAI), vol. 12713, pp. 436–447. Springer, Cham (2021). https://doi.org/10.1007/978-3-030-75765-6_35
24. Yuan, H., He, C.: Click-through rate prediction model based on dynamic graph attention mechanism network. In: 2021 4th International Conference on Robotics, Control and Automation Engineering (RCAE) (2021)
25. Zhang, Y.: CTR prediction model using xdeepfm and bayesian optimization. In: 2021 IEEE International Conference on Computer Science, Artificial Intelligence and Electronic Engineering (CSAIEE) (2021)
26. Zhao, Z., Yang, S., Liu, G., Feng, D., Xu, K.: Fint: Field-aware interaction neural network for click-through rate prediction. ICASSP 2022–2022 IEEE International Conference on Acoustics, Speech and Signal Processing (ICASSP) (2022)
27. Zhou, G., et al.: Deep interest network for click-through rate prediction. In: Proceedings of the 24th ACM SIGKDD International Conference on Knowledge Discovery & Data Mining (2018)
28. Zhou, X., Shi, Y.: Deepfafm :a field-array factorization machine based neural network for CTR prediction. In: 2020 IEEE 4th Information Technology, Networking, Electronic and Automation Control Conference (ITNEC) (2020)
29. Zhu, T., Li, S., Liang, C., Liu, B., Li, X.: Product click-through rate prediction model integrating self-attention mechanism. In: 2021 3rd International Conference on Advances in Computer Technology, Information Science and Communication (CTISC) (2021)

Overcoming the Cold-Start Problem in Recommendation Systems with Ontologies and Knowledge Graphs

Stanislav Kuznetsov(✉) and Pavel Kordík

Faculty of Information Technology, Czech Technical University in Prague, Prague, Czechia
{kuznesta,kordikp}@fit.cvut.cz

Abstract. Many recommendation systems struggle with the cold-start problem, especially in the early stages of a new application, when there are few active users and limited interactions. Traditional approaches like Collaborative Filtering cannot provide enough recommendations, and text-based methods, while helpful, do not offer sufficient context. This paper argues against the idea that the cold-start phase will eventually disappear and proposes a solution to enhance recommendation performance from the start. We propose using Ontologies and Knowledge Graphs to add a semantic layer to text-based methods and improve the recommendation performance in cold-start scenarios. Our approach uses ontologies to generate a knowledge graph that captures item text attributes' implicit and explicit characteristics, extending the item profile with similar semantic keywords. We evaluate our method against state-of-the-art text feature extraction techniques and present the results of our experiments.

Keywords: Semantic-based Recommender Systems · Knowledge Graphs · Ontology · Cold-start Problem · Ontology-based user models

1 Introduction

The main goal of recommender systems is to help users discover items that are highly relevant to their interests, but that they are unaware of. Collaborative filtering algorithms, which recommend items to users based on similar user interactions, have successfully achieved this goal. However, these methods are not effective in cold-start scenarios, where interactions are scarce, and many items do not have any interaction data.

To overcome this issue, we can employ content-based methods, wherein algorithms utilize item attributes or generate additional information about the items. One common approach is to rely on similarities based on identical words in attributes. However, it is important to note that this method is not always reliable and can potentially result in false recommendations.

Our approach tackles this problem by generating more in-depth knowledge of the domain using an ontology-based approach. This approach generates an

A. Abelló et al. (Eds.): ADBIS 2023, CCIS 1850, pp. 591–603, 2023.
https://doi.org/10.1007/978-3-031-42941-5_52

ontological profile for each item and compares them to improve recommendation performance. The knowledge graph helps filter out unnecessary words and improve the information value of the item profile.

To show the benefits of our approach, we compared it with standard methods such as Term frequency Inverse Document Frequency (Tf-Idf) weighting and the Language-Agnostic Sentence Representations (LASER) toolkit, which uses a pre-trained neural network to generate word embeddings. Our ontology-based approach showed better performance in the cold-start recommendation task, as it benefits from semantic understanding and can enrich item profiles from the knowledge base. Our findings were experimentally verified on a movie database using the *top*@N evaluation methodology.

The paper is organized as follows. Section 2 provides the background methods that we use in our research; Sect. 3 describes our approach in detail. In Sect. 4, we present a comprehensive list of related work. Section 5 presents the result of our experiments. Finally, the discussion and conclusion are presented in Sect. 6 and Sect. 7.

2 Background

In most cases, items have textual attributes. We can use these attributes to compute similarities between two items and recommend the most similar items. Below, we present some common approaches for computing the text similarities between items. Collaborative-based recommender systems are not useful in cold-start systems because they cannot find a group of users with similar behaviour, as there are no or few preceding user interactions. Content-based recommender systems perform better because they can utilise item attributes to find similarities.

Cold-Start Problem. The cold-start problem is a well-known challenge in the field of recommender systems, where a new user or item has no or limited interaction history in the system, making it challenging to generate personalized recommendations. According to a study by [1,5,16] the cold-start problem can arise due to various reasons, including a lack of data, a high level of novelty, and a lack of user/item information. The authors noted that addressing this problem is crucial in developing effective recommender systems that can provide personalized recommendations to users.

Similarity Embeddings. In this paper, we suggest adding a semantic layer to improve the accuracy of item recommendations by using ontologies and knowledge graphs (KGs). Specifically, we expand the item profile with semantic keywords to increase the probability of finding similar items. According to Meymandpour and Davis [12], semantic similarity measures reflect the relationship between the meaning of two concepts, entities, terms, sentences, or documents, and are computed based on a set of factors derived from a knowledge representation model, such as thesauri, taxonomies, and ontologies. Various similarity measures have been proposed based on the application context and its knowledge representation model. These measures can be classified into three categories:

- Distance-based models that use the structural representation of the context, such as vector space-based measures such as TF-IDF [17] and cosine similarity, structural similarity measures like SimRank [8], ASCOS [3], and P-Rank [24], Propagated Linked Data Semantic Distance (PLDSD) and SPrank algorithm [14].
- Feature-based models that define concepts or entities as sets of features, such as the Jaccard and Dice coefficients and the Tversky ratio model.
- Statistical methods that consider statistics derived from the underlying context, such as information content (IC)-based similarity measures and statistical.

LASER. stands for Language-Agnostic Sentence Representations toolkit[1], used to accelerate the transfer of natural language processing (NLP) applications to many more languages beyond English [2]. The toolkit works in many languages written in different alphabets. To enable the recommendation of similar items, authors adopt the encoder / decoder approach or sequence-to-sequence processing, which shares the same architecture as neural machine translators.

Algorithm 1. Similarity-Based CB Algorithm

1: **for** $\forall$ Item $i \in I$ **do** ▷ Preprocessing Phase
2: Get textual attributes and prepossess
3: Prepare embedding $\vec{j_i}$
4: **end for**
5: Merge all embedding to matrix $M_{i,\vec{j}}$ where $i \in I, \vec{j} \in \vec{E_i}$
6: Reduce dimensionality of $M_{i,\vec{j}}$
7: For $\forall$ (i_n, i_m) compute $similarityMatrix = \cos(i_n, i_m)$, $i_n \neq i_m$

8: **function** ItemToItem($itemId, n_{sim}, similarityMatrix$)
return The number of n_{sim} most similar items to item=$itemId$ by $similarityMatrix$
9: **end function**
▷ Recommendation

10: **function** Recommendation($UserInterItems, n_{sim}, simMatrix, N$)
11: **for** $\forall$ Item i in UserInterItems **do**
12: retRecomArray add results of $ItemToItem(i, n_{sim}, simMatrix)$
13: **end for**
return retRecomArray[0:N] where N is number of recommended item in $Top@N$
14: **end function**

[1] https://github.com/facebookresearch/LASER.

Similarity-Based Algorithm CB. This algorithm is a similarity-based content-based algorithm. The principle is to use item text attributes as a text corpus and calculate the similarity. The algorithm will recommend movies based on internal similarities. The final recommendation is calculated based on the cosine distance between film embeddings. The schema of the algorithm is in Algorithm 1.

We use the same algorithms for each similarity method. In our experiments, we use methods such as Tf-Idf [9], LASER [2] and our ontology. For each item in the dataset, we compute embedding with a particular method; then we get the matrix $M_{i,\vec{j}}$ where i is a movie and $\vec{j_i}$ is an embedding. From this matrix we computed the cosine similarity for each movie pair. Finally, for each movie, we get the list of most similar movies by cosine similarity. For *top@N* recommendation, we use the *ItemToItem* approach, where for each movie in the user interaction profile, we recommend n_{sim} most similar movies. The final recommendation contains movies that we sort by item score. n_{sim} is a parameter that we can use to control the quality of the recommendation.

3 Our Approach

At the beginning of this section, we summarize our approach in Table 2. Subsequently, we introduce the concept of our ontology and discuss the implicit and explicit word embedding features employed in our study. In Sect. 5, we evaluate the benefits of ontological similarity in comparison to the similarities presented in the previous section through a series of experiments.

In the following text, we discuss the concept of ontologies and the difference between them and knowledge graphs, which is essential for understanding our concept.

Table 3. Summary of our approach

1. We gather the text attributes from all items and create a single text corpus automatically.
2. We will use a Stanford CoreNLP keyword extraction algorithm to enrich the existing keywords dictionary with new keywords. All keywords are in basic form.
3. For creating an KG, we use two types of relations, implicit and explicit
   (a) To generate implicit relations, we use the interaction between words, such as the number of co-occurrences of two keywords throughout the dataset or the average year of occurrence of a keyword. These relations are dataset dependent.
   (b) To generate explicit relations, we use external embedding representations. In our approach, we use the ConceptNet framework, word2vec and GloVe. These relations add dependencies that are typical within language, and these relations are dataset independence.
4. Final KG has nodes representing keywords and edges with implicit and explicit relations and weight parameters. The main weight parameter combines the implicit and explicit weights, chosen through ablation study (see Table 1). To use KG in the ontology model, we set a threshold for main weights as a hyperparameter that can be optimized.
5. For each Item in our dataset, we generate a profile with a list of keywords (nodes) and connections between keywords (edges). We can change the number of keywords in the profile dynamically by adding or removing the ontologically closest keywords.

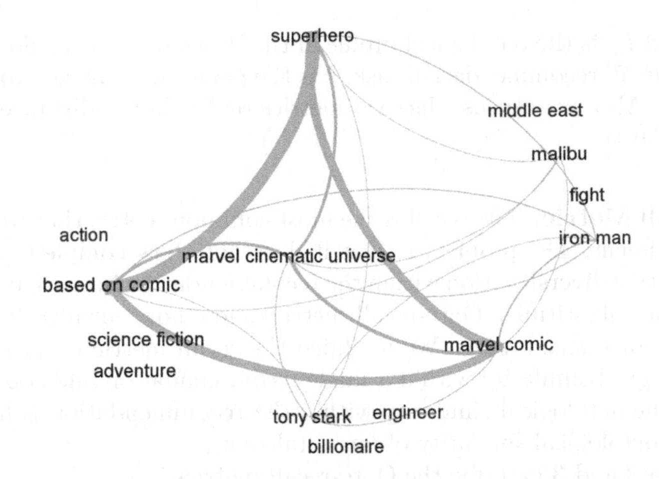

Fig. 1. The basic ontological profile of the Iron Man (2008) movie.

Ontology and Knowledge Graph Architecture. Our approach utilizes ontologies, which are a formal representation of concepts and relationships within a specific domain [6]. We integrate external sources such as ConceptNet, word2vec, and GloVe to generate new knowledge from the graph. A reasoning engine extends Item profiles with related keywords. Our graph has oriented edges with attributes representing ratios between connected nodes, enabling KG space exploration. Text attributes are preprocessed and mapped to the KG to generate an ontology profile, which is converted to embeddings for finding item similarities (see Fig. 1). The approach facilitates profile exploration and dynamic extension by including related nodes above a threshold. Keyword extraction algorithms are utilized to extract keywords from the text corpus for KG construction.

Ontological-Similarity Metric. Our approach to computing similarity between items involves using their respective ontology profiles. This ontological-similarity metric is based on semantic-similarity and uses cosine similarity to compare the two profiles. Each profile is an embedding of the item's text attributes in the ontology, with the size equal to the number of keywords in the ontology. For each keyword in the item's ontology profile, we place a one in its corresponding position and fill other positions with zeros. We use this metric alongside other metrics in our content-based algorithm and as an additional recommendation metric to show the impact of ontological similarity on recommendation.

The formula presents ontological-similarity metric:

$$OntoSim(\overrightarrow{I_n}, \overrightarrow{I_m})^{COS} = \frac{1}{k} \cdot \sum_{i,j=1, i \neq j}^{k} \frac{\overrightarrow{I_i} \cdot \overrightarrow{I_j}}{\|\overrightarrow{I_i}\| \|\overrightarrow{I_j}\|} \quad (1)$$

where $\overrightarrow{I_n}$ and $\overrightarrow{I_m}$ is the ontological profile of the Items inside a single recommendation at *top@N* recommendation task, k is the position of the keyword in items embeddings. Also, we can use Jaccard metrics or Euclidian distance instead of Cosine Similarity.

Onto-Recall Metric. The recall is the most common metric that we use in our offline experiment. The problem is that it does not show complete information about the actual diversification within the recommendation because it only shows the hits of our algorithms. Onto-recall metrics show how our algorithm can hit ontologically diversified items. We modified the recall metric using a filter that tracks ontological similarity within a single recommendation and counts as a hit only when the ontological similarity within the recommendation is higher than the average ontological similarity of our catalogue.

Equations 2 and 3 describe the Onto-recall metrics.

$$OntoTP = \begin{cases} 1, & \text{if } OntoSim(\overrightarrow{I_n}, \overrightarrow{I_m}) \geq OntoSim(\overrightarrow{I_{all}}) \\ 0, & \text{otherwise.} \end{cases} \quad (2)$$

where $\overrightarrow{I_n}$ is the ontological profile of the items inside a single recommendation in the top@N recommendation task and $OntoSim(\overrightarrow{I_{all}})$ is the average ontological similarity between all ontological profiles of our items inside our catalog.

$$Onto - recall = \frac{OntoTP}{OntoTP + FP} \quad (3)$$

where $OntoTP$ is the total number of hits, defined by Eq. 2 and FP are all other recommendations. In the calculation of Onto-recall, we exclusively consider the recommendations that include the control item and have a semantic difference greater than the dataset's average of 6%. This average was determined by calculating the average similarity of all items within our datasets. We use Onto-recall and Catalogue coverage in all the experiments in this paper.

Implicit Word Embedding Features. Our knowledge graph (KG) has nodes (keywords) and edges that connect them based on several metrics. These metrics are simple but effective in helping us understand the relationships between the keywords. One important metric is "common occurrence," which counts how often two keywords appear together in the dataset. Another metric is the average year that a keyword appears in the dataset. By comparing two keywords with the same meaning using this metric, we can identify newer keywords that are replacing older ones.

The KG's implicit embedding is generated from the dataset, which is not ideal because it is biased towards the dataset's characteristics, and it can create connections that do not exist in natural language. Therefore, we need to use explicit embeddings to incorporate unbiased semantic information

Explicit Word Embedding Features. ConceptNet is a framework that helps computers understand the meanings of words used by people. It contains word embeddings like word2vec and GloVe but has advantages such as being multilingual, having useful properties, and being linked with Linked Open Data. It allows the use of the same terms in other dictionaries, extending the amount of information. ConceptNet uses properties and metrics for word similarity and context, including related words, synonyms, and associated terms with similar meanings. We also experiment with word2vec and GloVe embeddings to define relationships between nodes.

4 Related Work

This section provides an overview of the state-of-the-art approaches. Several surveys [4,13,15,18,19] describe the main problems that researchers focus on in the domain of recommendation systems. In all these papers, we find the cold-start problem. Therefore, even if it is not a new problem, there is no single approach to solving or reducing it. The authors of the survey [11] describe the concept of using the Knowledge Graph-Based Recommender system.

KGs have recently gained popularity in generating recommendations as they address issues like data sparsity and cold-start problems. Nodes in a KG represent entities and edges represent relations between entities, enabling mapping of items and their attributes into the KG to understand item relationships. KG-based recommender systems can be categorized into embedding-based, connection-based, and propagation-based methods based on how they use KG information. Our approach is similar to KG embedding-based methods, such as DKN [21], CKE [23], and MKR [22], which are two-stage learning, joint learning, and multi-task learning methods, respectively [7].

Table 1. The Table shows the result of the ablation study we made at the cold-start dataset, where the value is the Onto-recall at top@N recommendation task. Parameter *impl* defines the value of the implicit weight, *expl* defines the value of the explicit weight. Also, we try to use dimension reduction approaches to increase performance; unfortunately, this approach does not improve our performance. We use the combination of implicit and explicit weight inside our model.

Parametrs	Dataset 2–3	Dataset 3–5
impl > 0	0.006	0.012
expl > 0	0.004	0.010
impl > 0 and expl > 0	**0.016**	**0.027**
impl > 0 or expl > 0	0.012	0.020
SVD(impl > 0 and expl > 0)	0.08	0.016
NMF(impl > 0 and expl > 0)	0.08	0.010

5 Offline Experiments

To test the effectiveness of our recommendation system, we need to simulate how users would interact with it. The most common way to do this is by using historical data and hiding some user interactions. We then predict the value of these hidden interactions and compare them to the actual values to see how well the system performs.

We typically use a procedure called "leave-one-out cross-validation" to test the system. This involves testing the system for each user in the system, by hiding one interaction at a time and predicting its value based on the other interactions. To compare different recommendation models, we use two metrics: Onto-recall and coverage. These metrics measure how well the system recommends diverse items that cover a large portion of the catalog. Users prefer more diverse recommendations with greater coverage, so we optimize for both metrics.

Datasets. Our goal in this section is to test our algorithms in a cold-start environment. In a cold-start scenario, the system transitions gradually from a low number of interactions to a sufficient amount. To test this, we created five datasets from the MovieLens 20M dataset, each with the same number of interactions. We removed noisy and low-interaction items before splitting the data. Dataset 2-3 simulates the cold-start problem, while the others simulate the transition to a normal system. We use text corpus to compute Tf-Idf and generate LASER embeddings during data preprocessing. For ontology profile, we use original keywords and extended nodes generated by ontology degree level of two. We evaluate our algorithms using Onto-recall and coverage metrics, which optimize diversity and coverage. We present our new datasets in the Table 2.

Table 2. Cold-user datasets and characteristics: Total interaction inside each dataset, the total number of users, average interaction per user, the total number of movies and the name of the experiment.

Dataset	Total inter.	User count	Average user inter.	Movies count	Experiment
2–3	12975	5252	2.5	2735	Cold-start system
3–5	46054	10763	4	4809	Transition
5–10	194911	26784	7	7637	Transition
10–20	1063130	70170	15	10686	Transition
20–50	1879318	58864	32	12173	"Normal" system

In this experiment, we demonstrate the results of the comparison of Tf-Idf, LASER and ontology-similarity methods. For each method, we prepare one instance of the single CB algorithm presented in Sect. 2. We use the leave-one-out cross-validation technique for the evaluation of the algorithms. The result is presented in the Onto-recall-coverage optimisation space. Also, we use the model capacity parameters described here [20].

Cold-User Recommendation Experiment. We set the hyper-parameter n_{sim} that we use for capacity manipulation. The value of this parameter means how many similar items should return for each user item. When this parameter is one, it uses one item for each item in the user profile. While the parameter is set to ten, the final recommendation will contain only the item similar to the first (last view) user item. All our CB algorithms instances use the same parameter. The complete results of the experiment can be found in Table 3.

Figure 2 shows the comparison of CB algorithms in each dataset. In the result of the experiment, we can see that the ontology-based model is Pareto dominant compared to other CB algorithms. The Tf-Idf and LASER algorithm is Pareto non-dominant compared to ontology-based. The reason is that in our approach, ontology could provide more relevant keywords, and if the film has a small number of nodes, we can increase this number with semantically close words.

6 Discussion

We compared different algorithms for text feature extraction in recommending movies to users with few interactions. Our ontology-based approach performed better than other methods in certain aspects and was comparable to state-of-the-art methods in computing item similarities. We preprocessed the data by removing irrelevant keywords and adding more semantic information to improve the accuracy of our profiles. This approach is particularly useful for datasets with few interactions.

In addition to preprocessing the data, our approach uses an extended movie profile that includes a new keyword that is semantically similar to the original keyword, resulting in more keywords than other methods. This increases the probability of finding the nearest or most similar item. However, for datasets with more user interactions, the advantage of our approach over CF methods diminishes, and the LASER algorithm is closer to our ontology model than Tf-Idf.

Our approach has been successfully applied to a range of domains, starting with education data mining [10]. In this domain, we recommend internships and job offers to students at FIT, CTU[2] by developing an ontology based on the subjects they study, using implicit and explicit information. We update our knowledge graph with student results, such as their marks, and generate a unique ontology profile for each student. To make recommendations, we use a combination of behavioural collaborative filtering (CF) methods based on anonymous student interactions, along with our content-based (CB) algorithm. Our CB algorithm re-ranks the recommendations and increases the scores of items that are most similar to the student's study profile.

[2] https://fit.cvut.cz/en/cooperation/for-students/job-offers/partners-and-sponzors.

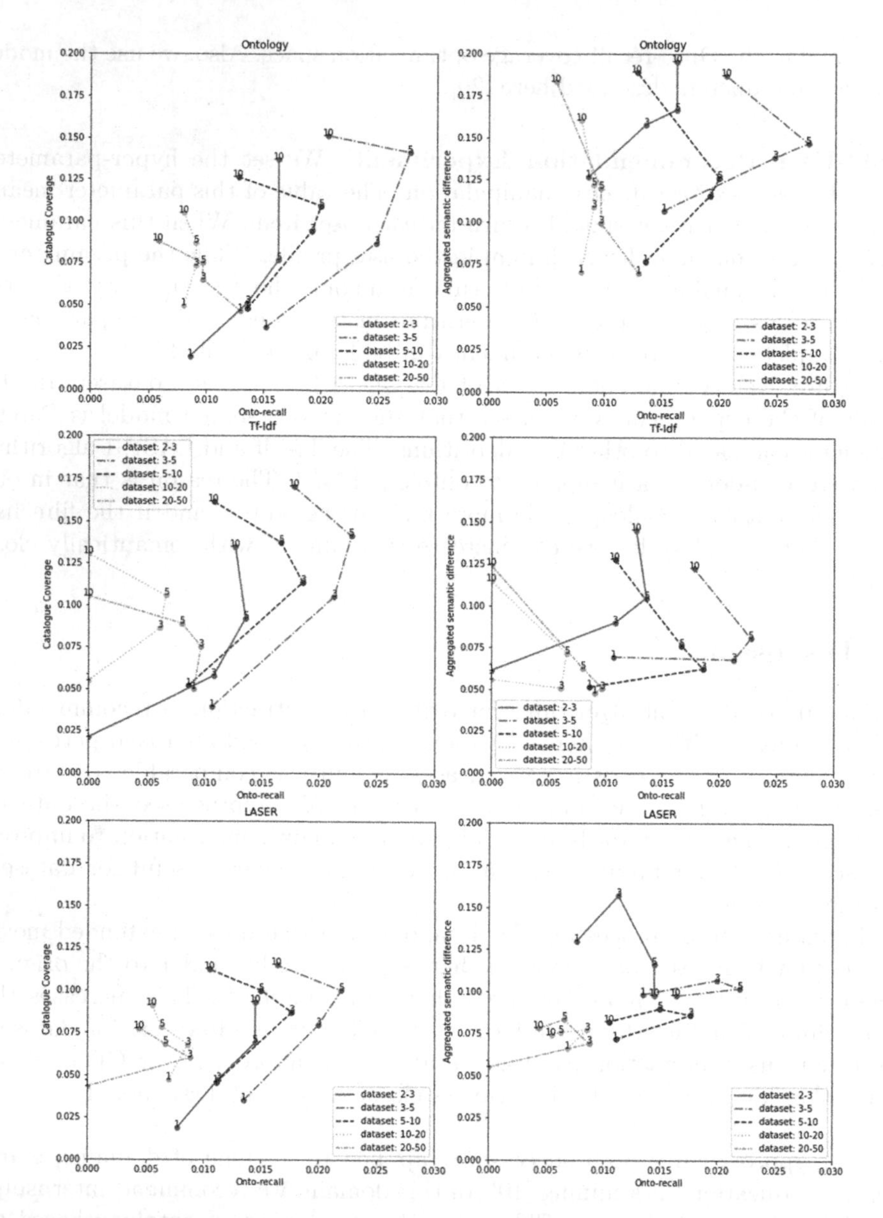

Fig. 2. The Figure displays the results of our experiment for each model on different datasets. Our Onto-model outperforms other models in all datasets, with the best Onto-recall. The optimal dataset is 3–5 for all models, and the more similar items we recommend for a single item, the more semantic similarity grows. Onto-model recommends more semantically diverse items than other models, with an average semantic difference higher than the dataset average. Additionally, the graph lines indicate the Similarity-Based Algorithm, with numbers corresponding to model parameters, and the best results are on datasets 3–5 and 5–10, with the highest Onto-recall and catalogue coverage. Tf-Idf has the best catalogue coverage, but Onto-recall is more critical for recommending relevant items during a cold start.

Table 3. Results of the cold-user recommendation experiment. Where n_{sim} is a hyper-parameter that we use for capacity manipulation. The value of this parameter means how many similar items should return for each user item in the user interaction profile; see Algorithm 1. Onto-recall is the metric described in Sect. 3; Catalogue coverage is the metric that shows the per cent of catalogue we could cover in our recommendations; Ontological-similarity is the metric described in Sect. 3. The best results of the metrics for each dataset are in bold.

Model	Parameter	Onto-recall	Catalogue coverage	Ontological-similarity
Tf-Idf				
Dataset 2–3	$n_{sim} = 5$	0.013	0.09	0.11
Dataset 3–5	$n_{sim} = 5$	0.022	**0.14**	0.09
Dataset 5–10	$n_{sim} = 3$	0.018	**0.11**	0.07
Dataset 10–20	$n_{sim} = 5$	0.006	**0.1**	0.08
Dataset 20–50	$n_{sim} = 3$	0.009	**0.075**	0.05
LASER				
Dataset 2–3	$n_{sim} = 10$	0.014	0.09	0.1
Dataset 3–5	$n_{sim} = 5$	0.022	0.1	0.11
Dataset 5–10	$n_{sim} = 3$	0.017	0.08	0.09
Dataset 10–20	$n_{sim} = 1$	**0.01**	0.09	0.07
Dataset 20–50	$n_{sim} = 1$	0.01	0.04	0.06
Ontology				
Dataset 2–3	$n_{sim} = 10$	**0.016**	**0.13**	**0.2**
Dataset 3–5	$n_{sim} = 5$	**0.027**	**0.14**	**0.15**
Dataset 5–10	$n_{sim} = 5$	**0.02**	0.1	**0.13**
Dataset 10–20	$n_{sim} = 5$	**0.01**	0.08	**0.13**
Dataset 20–50	$n_{sim} = 1$	**0.013**	0.046	**0.07**

In addition, we have implemented our CB algorithms with ontological similarity on the Experts.ai platform[3]. On this platform, we generate an ontology and knowledge graph from various expert academic outcomes, projects, patents, and other sources. Using our approach, we extend search results by adding the most similar keywords to search queries, increasing the probability of finding appropriate experts. We also use our approach to recommend the next similar expert to users who have already seen an expert. This has proven to be highly effective in practice and has led to significant improvements in search accuracy and recommendation quality.

7 Conclusion

This paper proposes an ontology-based approach to solve the cold-start problem in recommendation systems. Our method achieves better results than traditional

[3] https://find.experts.ai/.

content-based methods in MovieLens data, and we demonstrate its potential in other domains.

Our approach extends item profiles with similar ontological keywords to recommend similar items without user interaction, which can be easily applied to other domains since the ontological profile is self-explanatory. Compared to traditional attribute similarity-based CB algorithms, our ontological similarity method increases Onto-recall and catalogue coverage by 5% and 4%, respectively, in cold-start datasets.

Our algorithms have been successfully used in both the education data mining domain, where we recommend internships to students, and in the production version of the Experts.ai platform, where we use our approach to increase the chances of finding appropriate experts by adding more similar keywords to users' search queries and recommending the next most similar experts.

In our future work, we will focus on exploring ontology-based algorithms to enhance recommendation diversity and serendipity in cold-start environments with limited user-item interaction data. By leveraging ontological knowledge, our approach aims to address the challenges associated with recommending in such environments, resulting in recommendations that are more diverse and serendipitous. Our ultimate goal is to develop a universal ontology-based algorithm that integrates collaborative filtering methods to further improve performance.

References

1. Anwar, T., Uma, V., Hussain, M.I., Pantula, M.: Collaborative filtering and KNN based recommendation to overcome cold start and sparsity issues: a comparative analysis. Multimedia Tools Appl. (2022). https://doi.org/10.1007/s11042-021-11883-z
2. Artetxe, M., Schwenk, H.: Massively multilingual sentence embeddings for zero-shot cross-lingual transfer and beyond (2018). https://doi.org/10.48550/arXiv.1812.10464
3. Chen, H.H., Giles, C.L.: ASCOS: an asymmetric network structure context similarity measure. In: Proceedings of the 2013 IEEE/ACM International Conference on Advances in Social Networks Analysis and Mining (2013). https://doi.org/10.1145/2492517.2492539
4. Vinodhini, G., Suban, R., Venil, P.: A state of the art survey on cold start problem in a collaborative filtering system. Int. J. Sci. Technol. Res. **9**, 2606–2612 (2020). IJSTR-0420-33718
5. Goldberg, D., Nichols, D., Oki, B.M., Terry, D.: Using collaborative filtering to weave an information tapestry. Commun. ACM (1992). https://doi.org/10.1145/138859.138867
6. Gruber, T.R.: A translation approach to portable ontology specifications. Knowl. Acquis. (1993). https://doi.org/10.1006/knac.1993.1008
7. Guo, Q., et al.: A survey on knowledge graph-based recommender systems. IEEE Trans. Knowl. Data Eng. (2022). https://doi.org/10.1109/TKDE.2020.3028705
8. Jeh, G., Widom, J.: Simrank: A measure of structural-context similarity. In: Proceedings of the Eighth ACM SIGKDD International Conference on Knowledge Discovery and Data Mining (2002). https://doi.org/10.1145/775047.775126

9. Jurafsky, D., Martin, J.H.: Speech and Language Processing (2nd Edition). Prentice-Hall, Inc. (2009). https://dl.acm.org/doi/book/10.5555/1214993
10. Kuznetsov, S., Kordík, P., Řehořek, T., Dvořák, J., Kroha, P.: Reducing cold start problems in educational recommender systems. In: 2016 International Joint Conference on Neural Networks (IJCNN) (2016). https://doi.org/10.1109/IJCNN.2016.7727600
11. Liu, J., Duan, L.: A survey on knowledge graph-based recommender systems. In: 2021 IEEE 5th Advanced Information Technology, Electronic and Automation Control Conference (IAEAC) (2021). https://doi.org/10.1109/IAEAC50856.2021.9390863
12. Meymandpour, R., Davis, J.G.: A semantic similarity measure for linked data: an information content-based approach. Knowl. Based Syst. (2016). https://doi.org/10.1016/j.knosys.2016.07.012
13. Mishra, N., Chaturvedi, S., Vij, A., Tripathi, S.: Research problems in recommender systems. J. Phys. Conf. Series (2021). https://doi.org/10.1088/1742-6596/1717/1/012002
14. Noia, T.D., Ostuni, V.C., Tomeo, P., Sciascio, E.D.: SPrank: semantic path-based ranking for top-n recommendations using linked open data. ACM Trans. Intell. Syst. Technol. **8**, 1–34 (2016). https://doi.org/10.1145/2899005
15. Patel, K., Patel, H.B.: A state-of-the-art survey on recommendation system and prospective extensions. Comput. Electron. Agric. (2020). https://doi.org/10.1016/j.compag.2020.105779
16. Patro, S.G.K., Mishra, B.K., Panda, S.K., Kumar, R., Long, H.V., Taniar, D.: Cold start aware hybrid recommender system approach for e-commerce users. Soft Comput. (2022). https://doi.org/10.1007/s00500-022-07378-0
17. Sammut, C., Webb, G.I.: Tf-idf. In: Encyclopedia of Machine Learning (2010). https://doi.org/10.1007/978-0-387-30164-8
18. Sun, Z., et al.: Research commentary on recommendations with side information: a survey and research directions. Electron. Commerce Res. Appl. (2019). https://doi.org/10.1016/j.elerap.2019.100879
19. Tarus, J.K., Niu, Z., Mustafa, G.: Knowledge-based recommendation: a review of ontology-based recommender systems for e-learning. Artif. Intell. Rev. (2018). https://doi.org/10.1007/s10462-017-9539-5
20. Řehořek, T.: Manipulating the Capacity of Recommendation Models in Recall-Coverage Optimization. Ph.D. thesis, Faculty of Information Technology, Czech Technical University in Prague (2018). http://hdl.handle.net/10467/81823
21. Wang, H., Zhang, F., Xie, X., Guo, M.: DKN: deep knowledge-aware network for news recommendation. In: Proceedings of the 2018 World Wide Web Conference (2018). https://doi.org/10.1145/3178876.3186175
22. Wang, H., Zhang, F., Zhao, M., Li, W., Xie, X., Guo, M.: Multi-task feature learning for knowledge graph enhanced recommendation. In: The World Wide Web Conference (2019). https://doi.org/10.1145/3308558.3313411
23. Zhang, F., Yuan, N.J., Lian, D., Xie, X., Ma, W.Y.: Collaborative knowledge base embedding for recommender systems. In: Proceedings of the 22nd ACM SIGKDD International Conference on Knowledge Discovery and Data Mining (2016). https://doi.org/10.1145/2939672.2939673
24. Zhao, P., Han, J., Sun, Y.: P-rank: a comprehensive structural similarity measure over information networks. In: Proceedings of the 18th ACM Conference on Information and Knowledge Management (2009). https://doi.org/10.1145/1645953.1646025

Recommender Chatbot as a Tool for Collaborative Business Intelligence in Tourism Domain

Olga Cherednichenko(✉) and Fahad Muhammad

Univ Lyon, Univ_Lyon 2, UR ERIC – 5, 25 Avenue Mendès France, 69676 Bron Cedex, France
olga.cherednichenko@univ-lyon2.fr

Abstract. Nowadays big amounts of data are created by different sources. Big Data allows for expanding the analysis scope beyond the in-house data sources and taking into consideration publicly available data on the web and external sources. Collaborative Business Intelligence goes beyond the organization frame and collecting people in one virtual space in order to give their opinion concerning certain issue results in an opportunity to boost the decision-making process and make it more reasonable. In this research we focus on tourism domain. Tourists often search for information from different sources and compare found facts. This leads to spending significant amounts of time comparing descriptions and photos across different platforms before deciding. To overcome this challenge, the use of virtual assistants has become increasingly common. These virtual assistants can perform various functions beyond just comparing prices, such as comparing user reviews. We introduce a framework for a virtual assistant that incorporates the following crucial elements: a conversational unit, a task identification, data exploration component, and a recommendation model.

Keywords: Collaborative Business Intelligence · Tourism · Data Exploration · Chatbot

1 Introduction

The decision-making process is a multifaceted endeavor that predominantly relies on the information accessible to the decision-maker. In today's contemporary society, we find ourselves amidst an era characterized by vast reservoirs of data across diverse domains. These data repositories hold considerable potential for facilitating decision preparation and execution. Nonetheless, these data are often collected and stored by disparate software systems, utilizing different formats, and presenting varying levels of accessibility and security. Moreover, data can be plagued by issues such as incompleteness, contradictions, and unreliability. To address the challenges associated with processing such extensive data volumes, the expertise of data analysts is sought. Business intelligence (BI) emerges as a valuable aid in deriving meaningful insights and shaping strategic decisions. By scrutinizing historical and current data, BI tools proficiently present outcomes through visually intuitive formats.

A. Abelló et al. (Eds.): ADBIS 2023, CCIS 1850, pp. 604–611, 2023.
https://doi.org/10.1007/978-3-031-42941-5_53

A substantial impediment in harnessing the full potential of accumulated data arises from the lack of direct communication between technical specialists, decision-makers, and business process analysts. To overcome this hurdle, the adoption of a collaborative business intelligence (CBI) approach becomes imperative. However, organizing effective collaboration among potential users, decision-makers, and technical specialists in the realm of data analysis proves to be a complex task. To tackle these challenges, the BI4people project [1] has been established with the purpose of providing solutions.

The core objective of BI4people is to democratize the benefits of Business Intelligence, ensuring its accessibility to the widest possible audience. This is achieved through the implementation of a software-as-a-service model for the data warehousing process, encompassing the integration of diverse and disparate data sources, as well as intuitive analysis and data visualization [2]. The central concept revolves around uniting individuals within a virtual space, fostering their active participation by encouraging the exchange of comments and opinions for the collective benefit. Furthermore, the utilization of others' findings or insights fosters a collaborative approach to Business Intelligence, thus giving rise to what is known as Collaborative BI.

When it comes to planning their travels, tourists often find themselves scouring various sources of information, including Google Maps, feedback on booking sites, and conducting comparisons of the gathered facts. This process demands significant time and effort. Consequently, tourists aspire to make informed decisions based on the experiences of others. However, most tourists do not possess expertise in data analysis or the ability to make rational decisions. Therefore, they require a user-friendly tool that incorporates data visualization capabilities, offering potential solutions and ideas in a comprehensible manner.

2 Background

Data exploration plays a crucial role in the broader Business Intelligence (BI) process, encompassing the collection, identification, and analysis of data to unveil valuable insights and patterns. However, it is important to acknowledge that data analysis can be a daunting task for inexperienced users.

Collaborative BI entails employing collaborative tools and methodologies to bolster Business Intelligence (BI) processes. These tools can encompass a range of technologies, including social networks, brainstorming sessions, chat applications, and more, all aimed at facilitating knowledge sharing and collaboration among business users, analysts, and technical specialists. The underlying concept is that by harnessing the collective knowledge and expertise of a group, superior solutions can be generated, leading to enhanced decision-making compared to working in isolation.

Chatbots play a vital role in supporting users by providing information, searching data, and executing routine tasks. A chatbot is a computer program designed to simulate human conversation, utilizing text or voice interactions with users [3]. These intelligent agents are typically created with specific tasks in mind, aimed at providing information or performing designated functions. Notably, there exist chatbots and Conversational Agents (CAs) that enable users to execute complex data analysis tasks by combining commands within nested conversations [4, 5].

Chatbots offer numerous advantages for Collaborative Business Intelligence (CBI), including: 24/7 availability, personalization, and scalability. These advantages make chatbots invaluable tools within the CBI landscape, facilitating seamless communication, support, and personalized experiences for users engaging in data analysis and decision-making processes.

Thus, the research questions are following.

RQ1. Can a virtual assistant using a conversational interface help simplify the data exploration process for non-experienced users?
RQ2. What features of a virtual assistant are most helpful in assisting users who have limited knowledge of the domain and business intelligence?
RQ3. How can virtual assistants be designed to integrate in BI platform?

3 The State of the Art

In today's world, there exists variety of chatbots and recommender models. A well-known example of recommendation systems was considered when Netflix offered a prize for the recommendation algorithm of predicting movie ratings by users [6]. The prize was awarded in September (2009), after over three years of work.

There are many general purpose chatbots to help users in solving general problems. Ranoliya et al. designs a chatbot for university related FAQs [7]. They implement chatbot using Artificial Intelligence Markup Language (AIML) and Latent Semantic Analysis (LSA). Narducci et al. develop a Telegram chatbot, based on a conversational recommender system, which is able to provide personalized recommendations in the movie domain. They implement critiquing strategies for improving the recommendation accuracy as well [8].

Besides these, there are many specialized Chatbots for applications such as customer care, health care, medical diagnoses, etc. Mathew et al. proposes a chatbot for disease prediction and treatment recommendation to avoid the time-consuming method of visiting hospitals [9]. Similar to this research, Tode et al. developed a chatbot for medical purpose using deep learning to help poor and needy people [10]. Preez et al. proposes an intelligent web-based voice recognition chat bot [11].

Now-a-days, many hotels, travel agencies, and airline companies are adopting Chatbots as a recent technology to help them boost their businesses. Ukpabi et al. explores chatbot adoption in tourism services [12]. Pillai et al. proposes AI-based chatbot for hospitality and tourism by extending the technology adoption model (TAM) with context-specific variables [13]. Nica et al. present foundation stones behind a chatbot for e-tourism that allows people textually communicate with the purpose of booking hotels, planning trips, and asking for interesting sights worth being visit [14]. They demonstrated how model-based reasoning can be used for enhancing user experience during a chat.

Since mobile technology has been widely used, therefore many researchers propose mobile based solution for tourism industry. Abowd et al. develop a mobile application, named Cyberguide, which provide a hand-held context-aware tour guide [15]. Similarly, Horozov et al. also used the location parameter as a key criterion for personalized POI recommendations in mobile environments [16]. Their application works with the physical entities such as restaurants and can be used in the context of restaurant recommender system. The research work of Zheng et al. is also based on user generated GPS traces for learning travel recommendations [17]. Zhiyang et al. develop a user-based collaborative filtering approach for generating tourist attraction recommendations as the list of preference attractions for the tourist [18].

Generally, there exist two major groups of recommendation systems. First group explores properties of the items recommended and named as Content-based systems. Second group, Collaborative filtering, considers similarity measure to recommend items/users and focus on the relationship between users and items. The items which are preferred by similar users are recommended.

4 Methods and Materials

Conversational agents (CAs) are computer programs designed to engage in natural conversations with human users. They can be categorized as either chatbots for informal chatting or task-oriented agents for providing users with specific information related to a task [19, 20]. CAs may use text-based input and output or more complex modalities such as speech. The handling of dialogue is a crucial component of any CA, which can range from simple template-matching systems to complex deep neural networks.

While chatbots and task-oriented agents are the primary categories for classifying CAs, other aspects such as input and output modalities, applicability, and the agent's ability to self-learn could also be considered [21]. Chatbots can be classified into three main groups based on their response generation: pattern-based, information-retrieval, and generative. Task-oriented agents must provide precise and consistent answers to fact-based conversations on specific topics. The general pipeline depicts the various stages of processing as separate components (Fig. 1). The three basic components are natural language understanding (NLU), cognitive processing, and natural language generation (NLG). NLU involves identifying precisely what the user said or meant, while cognitive processing executes the dialogue policy to construct the knowledge required for the response using a knowledge base. NLG involves formulating the answer as a sentence.

In order to investigate the effectiveness of integrating a virtual assistant for data exploration, we have chosen to utilize a scenario focused on the analysis of tourist data. This scenario requires our virtual assistant to be equipped with a set of statistical commands as well as possess text analysis capabilities. The data exploration process entails a systematic examination and analysis of datasets with the goal of gaining insights, identifying patterns, and uncovering valuable information. This process typically involves several key steps, including data collection, data cleaning and preprocessing, descriptive statistics, data visualization, exploratory data analysis (EDA), hypothesis testing, and iterative refinement.

To organize and streamline these steps, we propose dividing them into distinct groups. The first group includes data collection and preprocessing, which involves tasks such as metadata creation and data cleaning. The second group focuses on data exploration, encompassing activities such as descriptive statistics, EDA, and hypothesis formulation. The third group centers around data visualization, allowing for effective representation

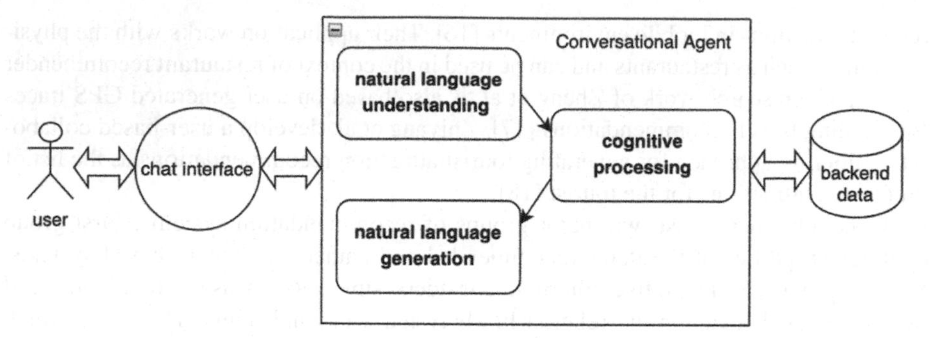

Fig. 1. The basic pipeline for conversational agent.

and interpretation of the data. Lastly, the fourth group involves the collection of feedback, which facilitates the iterative improvement of the virtual assistant's performance. These groupings align with the general framework established in the BI4People project [1]. While we specifically collect and clean touristic data for testing our ideas, it is important to note that the virtual assistant framework is designed to be applicable to datasets from a tourism domain. The domain plays a significant role in terms of vocabulary and the development of predefined behavior models within the assistant.

In various domains, such as data analysis or collaborative chats, users frequently engage in sequential interactions with entities known as "items". These interactions can occur over an extended period, and their characteristics evolve over time. As a result, a network of temporal connections is formed between users and items. Effectively recommending items in real-time and predicting changes in user behavior present significant challenges in these domains, necessitating further exploration and research. Representation learning has emerged as a powerful technique for capturing the evolving properties of users and items by generating low-dimensional embeddings.

The generation of dynamic embeddings poses several fundamental challenges for existing methods. Firstly, most current approaches only generate embeddings for users when they take actions, overlooking the temporal aspect of user behavior. Secondly, items exhibit both stationary properties that remain constant and time-evolving properties. However, existing methods often focus exclusively on either stationary or time-evolving properties, failing to capture the complete dynamics. Additionally, many methods rely on scoring all items for each user to predict user-item interactions, which becomes impractical for datasets with millions of items and users. These limitations highlight the need to address these challenges and develop more accurate and efficient techniques for real-time recommendations and predictions across diverse domains.

In this research, our focus lies in the task of tourist data exploration, where we aim to identify and generalize queries and commands as items. Our ultimate goal is to provide recommendations for non-expert users in the field of data exploration specifically in the tourist domain. As our research is in its early stages, we begin by investigating fundamental aspects to lay the groundwork for developing a recommender assistant framework.

5 Results

The scenario begins with loading the data into a dataframe. To ensure the dataset is ready for analysis, we conduct prior data loading, cleaning, and labeling procedures. An essential step in this process is creating metadata, which allows us to generate a set of initial data research questions that can be suggested to users at the start. Additionally, metadata exploration enables us to construct a trace matrix, aiding in matching available data to users' queries. The next task involves transforming users' queries into commands that invoke relevant methods and return numerical data as answers. It is imperative to provide an explanation of the obtained data, and to accomplish this, we propose the utilization of an abstract semantic tree to identify commands and parameters, along with a prompt generator to convert data into plain text. While data visualization can be considered a separate task, it plays a vital role in presenting the results. Users are given the opportunity to comment or annotate the results, and the recommender unit can collect feedback. Encouraging users to provide feedback becomes the second important task for the virtual assistant.

The communication aspect of the virtual assistant is crucial, as it is responsible for understanding and generating dialogue with the user, maintaining context, and interacting with other software components. Current solutions offer a wide range of features and interfaces for implementing natural language chatting. However, our focus lies in supporting data exploration as the core task of the virtual assistant. Our main idea is to automatically generate data processing commands based on the text of the user's query. We identify two main challenges in this process: building a set of reference queries and implementing a set of basic commands. Consequently, there are two sets of items that users interact with within the system.

We have adopted an approach based on representation learning to tackle our research problem. Representation learning involves automatically acquiring meaningful and compact representations of data, typically in the form of low-dimensional vectors or embeddings. In our case, we create various types of embeddings, including user trajectories that capture user-item interactions, user features, and item features. Each interaction is associated with a feature vector that contains additional information such as textual content. By leveraging these embeddings, we can make predictions about user queries for new users or new interactions. Additionally, we can predict user behavior, enabling us to recommend relevant actions and data analysis approaches. This allows us to provide our users with new questions and different perspectives on the touristic data, enhancing their exploration experience in the domain.

6 Conclusion and Discussion

In order to generalize the discussed features of a virtual personal assistant, we propose the framework that outlines four essential components of a virtual personal assistant:

- Natural language conversation processing: This component is responsible for processing the dialogue between the user and the virtual assistant. The virtual assistant must accurately identify the user's intentions and execute the appropriate dialogue policy to construct a response. It then generates a response in natural language that is easily understandable by the user.

- Task identification: This component involves identifying the services that best match the user's needs based on natural language processing and semantic analysis of the input text. It also extracts relevant keywords for information retrieval, behavior annotation, and matching.
- Recommender model: This component is responsible for personalizing the experience based on the user's preferences, history, and behavior. It employs machine learning and artificial intelligence algorithms to learn from data exploration and user interactions.

Thus, the framework provides a comprehensive framework for developing a virtual assistant software by identifying the main modules necessary for facilitating the software creation.

Acknowledgements. The research study depicted in this paper is funded by the French National Research Agency (ANR), project ANR-19-CE23–0005 BI4people (Business intelligence for the people).

References

1. Business intelligence for the people. https://eric.univ-lyon2.fr/bi4people/index-en.html
2. Muhammad, F., Darmont, J., Favre, C.: The Collaborative Business Intelligence Ontology (CBIOnt). 18e journées Business Intelligence et Big Data (EDA-22), B-18. Clermont-Ferrand, Octobre 2022, RNTI (2022)
3. Power, D., Sharda, R.: Business intelligence and analytics. Wiley (2015). https://doi.org/10.1002/9781118785317.weom070011
4. Google Assistant. https://assistant.google.com/
5. Amazon Alexa Voice AI. https://developer.amazon.com/en-US/alexa
6. Koren, Y.: The BellKor solution to the Netflix grand prize (2009). https://www2.seas.gwu.edu/~simhaweb/champalg/cf/papers/KorenBellKor2009.pdf
7. Ranoliya, B.R., Raghuwanshi, N., Singh, S.: Chatbot for university related FAQs. In: 2017 International Conference on Advances in Computing, Communications and Informatics (ICACCI), pp. 1525–1530. Udupi, India (2017)
8. Narducci, F., de Gemmis, M., Lops, P., Semeraro, G.: Improving the user experience with a conversational recommender system. In: Ghidini, C., Magnini, B., Passerini, A., Traverso, P. (eds.) AI*IA 2018. LNCS (LNAI), vol. 11298, pp. 528–538. Springer, Cham (2018). https://doi.org/10.1007/978-3-030-03840-3_39
9. Mathew, R.B., Varghese, S., Joy S.E., Alex, S.S.: Chatbot for disease prediction and treatment recommendation using machine learning. In: 2019 3rd International Conference on Trends in Electronics and Informatics (ICOEI), pp. 851–856. IEEE (2019). https://doi.org/10.1109/ICOEI.2019.8862707
10. Tode, V., Gadge, H., Madane, S., Kachare, P., Deokar, A.: A chatbot for medical purpose using deep learning. Int. J. Eng. Res. Technol. (IJERT) **10**(5), 441–447 (2021)
11. du Preez, S., Lall, M., Sinha, S.: An intelligent web-based voice chat bot. In: EUROCON 2009, pp. 386–391. IEEE (2009). https://doi.org/10.1109/EURCON.2009.5167660
12. Ukpabi, D.C., Aslam, B., Karjaluoto, H.: Chatbot adoption in tourism services: a conceptual exploration. In: Robots, Artificial Intelligence, and Service Automation in Travel, Tourism and Hospitality, pp. 105–121. Emerald Publishing Limited, Bingley (2019). https://doi.org/10.1108/978-1-78756-687-320191006

13. Pillai, R., Sivathanu, B.: Adoption of AI-based chatbots for hospitality and tourism. Int. J. Contemp. Hosp. Manag. **32**(10), 3199–3226 (2020). https://doi.org/10.1108/IJCHM-04-2020-0259
14. Nica, I., Tazl, O.A., Wotawa, F.: Chatbot-based Tourist Recommendations Using Model-based Reasoning. Configuration Workshop 2018, pp. 25–30. Graz, Austria (2018)
15. Abowd, G.D., Atkeson, C.G., Hong, J., Long, S., Kooper, R., Pinkerton, M.: Cyberguide: a mobile context aware tour guide. Wireless Netw. **3**, 421–433 (1997)
16. Horozov, T, Narasimhan, N., Vasudevan, V.: Using location for personalized POI recommendations in mobile environments. In: International Symposium on Applications and the Internet (SAINT 2006), vol. 129. IEEE Press (2006). https://doi.org/10.1109/SAINT.2006.55
17. Zheng, Y., Xie, X.: Learning travel recommendations from user generated GPS traces. ACM Trans. Intell. Syst. Technol. **2**(1), 1–29, Article 2 (2011)
18. Jia, Z., Yang, Y., Gao, W., Chen, X.: User-based collaborative filtering for tourist attraction recommendations. In: 2015 IEEE International Conference on Computational Intelligence & Communication Technology, pp. 22–25. IEEE (2015). https://doi.org/10.1109/CICT.2015.20
19. Masche, J., Le, N.-T.: A review of technologies for conversational systems. In: Le, N. T., Van Do, T., Nguyen, N.T., Thi, H.A.L. (eds.) ICCSAMA 2017. AISC, vol. 629, pp. 212–225. Springer, Cham (2018). https://doi.org/10.1007/978-3-319-61911-8_19
20. Svenningsson, N., Faraon, M.: Artificial intelligence in conversational agents: a study of factors related to perceived humanness in chatbots. In: Proceedings of the 2019 2nd Artificial Intelligence and Cloud Computing Conference (AICCC 2019), pp. 151–161. Association for Computing Machinery, New York (2020). https://doi.org/10.1145/3375959.3375973
21. Diederich, S., Brendel, A.B., Kolbe, L.M.: Towards a taxonomy of platforms for conversational agent design. In: 14th International Conference on Wirtschaftsinformatik (WI2019), pp. 1100–1114. Siegen, Germany (2019)

Neural Graph Collaborative Filtering: Analysis of Possibilities on Diverse Datasets

Dariusz Kobiela(✉), Jan Groth, Michał Sieczczyński, Rafał Wolniak, and Krzysztof Pastuszak

Faculty of Electronics, Telecommunications and Informatics, Gdańsk University of Technology, Gabriela Narutowicza 11/12, Gdańsk 80-233, Poland

s175656@student.pg.edu.pl

Abstract. This paper continues the work by Wang et al. [17]. Its goal is to verify the robustness of the NGCF (Neural Graph Collaborative Filtering) technique by assessing its ability to generalize across different datasets. To achieve this, we first replicated the experiments conducted by Wang et al. [17] to ensure that their replication package is functional. We received sligthly better results for ndcg@20 and somewhat poorer results for recall@20, which may be due to the randomness. Afterward, we applied their framework to four additional datasets (NYC2014, TOKYO2014, Yelp2022, and MovieLens1M) and compared NGCF with HOP-Rec [18] and MF-BPR [14] as in the original study. Our results confirm that NGCF outperforms other models in terms of ndcg@20. However, when considering recall@20, either HOP-Rec or MF-BPR performed better on the new datasets. This finding suggests that NGCF may have been optimized for the datasets used in the original paper. Furthermore, we analyzed the models' performance using recall@K and ndcg@K, where K was set to 1, 5, 10, and 40. The obtained results support our previous finding. The replication package for this paper can be found in our GitHub repository [1].

Keywords: NGCF · Graph neural network · Collaborative Filtering · Recommendation Systems

1 Introduction

Collaborative filtering (CF) is a type of personalized user preferences recommendation technique based on the assumption that users with a similar rating history will like the same things in the future [7,9]. Graph Neural Network (GNN) are a generalization of Convolutional Neural Networks (CNN) [10]. GNNs can act identically to CNNs and on the same data (e.g. images). In addition, they take into account dependencies between objects and enable adding new ones. They are used to make inferences about objects, about their relationships and interactions. Collaborative filtering technique was embedded together with Graph Neural Network in order to create NGCF - Neural Graph Collaborative Filtering algorithm,

A. Abelló et al. (Eds.): ADBIS 2023, CCIS 1850, pp. 612–619, 2023.
https://doi.org/10.1007/978-3-031-42941-5_54

which exploits the user-item graph structure by propagating embeddings on it. This leads to the expressive modeling of high-order connectivity y in user-item graph, effectively injecting the collaborative signal into the embedding process in an explicit manner. The recommendation effect of NGCF algorithm was checked by several authors, including varations like INGCF (Improved NGCF) [15], Deep NGCF [13], S-NGCF (socially-aware NGCF) [16], FTL-NGCF (federated transfer learning NGCF) [19].

The NGCF algorithm obtains the highest results in terms of recall and ndcg metrics, among the algorithms such as HOP-Rec or Collaborative Memory Network [17]. Prepared datasets contain information about users and corresponding items, thus having a graph character - they consist of objects between which there are relations. Each object is described by a feature vector x, and each object has a decision attribute value y. Performed research shows the NGCF model performance on five selected datasets in terms of precision and ndcg metrics, along with the dataset preparation process. Data preprocessing steps were done via python scripts and are described and shown in the authors' github page [1].

2 Input Data

Six datasets available from the online open sources were selected as the input data for the model. For each dataset, 80% of historical interactions of each user was randomly selected to constitute the training set, and the remaining 20% was used as the test set. 10% of interactions from the training set was then randomly selected as a validation set to tune hyper-parameters. Each observed user-item interaction was treated as a positive instance and was then paired using negative sampling strategy with one negative item that the user did not consume before. Most of the datasets can be downloaded from RecBole webpage [8].

- **Gowalla (dataset size: 6 MB)**: Gowalla is the check-in dataset [11] obtained from Gowalla webpage [2], where users share their locations by checking-in. To ensure the quality of the dataset, a 10-core setting was used [5], i.e., retaining users and items with at least ten interactions.
- **Amazon-book (dataset size: 17 MB)**: Amazon-review [12] is a widely used collection of datasets for product recommendation [4]. From there Amazon-book (Amazon-review/Books) dataset was selected. Similarly, the 10-core setting was applied to ensure that each user and item have at least ten interactions.
- **Yelp (Yelp2018 size: 30 MB, Yelp2022 size: 37 MB)**: Yelp2022 dataset is the latest version of Yelp dataset, which contains 908,915 tips by 1,987,897 users over 1.2 million business attributes like hours, parking, availability, and ambience aggregated check-ins over time for each of the 131,930 businesses. The research performed by Wang et al. [17] used Yelp2018 dataset, which was adapted from the 2018 edition of the Yelp challenge. The local businesses like restaurants and bars are viewed as the items. The same 10-core setting was used in order to ensure data quality.

- **Foursquare dataset (NYC size: 1.1 MB, TKY size: 2.7 MB)**: Foursquare dataset contains check-ins in NYC and Tokyo collected for about 10 months (from 12 April 2012 to 16 February 2013). It contains 227,428 check-ins in New York city and 573,703 check-ins in Tokyo. Each check-in is associated with its time stamp, its GPS coordinates and its semantic meaning (represented by fine-grained venue-categories). This dataset is originally used for studying the spatiotemporal regularity of user activity in LBSNs. The same 10-core setting was used in order to ensure data quality.
- **MovieLens dataset (MovieLens1M size: 4.5 MB)**: MovieLens dataset contains 1,000,209 anonymous ratings of approximately 3,900 movies made by 6,040 MovieLens users who joined MovieLens in year 2000 [3]. It was decided to use MovieLens1M movie ratings, which is recommended for education and development. It is stable benchmark dataset. Each user has at least 20 ratings. Released 2/2003.

Exact statistics about the datasets used can be seen in the Table 1.

Table 1. Statistics of the datasets

Dataset	#Users	#Items	#Interactions	Density	#Train	#Test	Size [MB]
Gowalla	29,858	40,981	1,027,370	0.00084	810,128	217,242	5.72
Yelp2018	226,334	5,261,669	3,511,627	0.000003	2,354,021	1,157,606	29.3
Amazon-Book	52,643	91,599	2,984,108	0.00062	2,380,730	603,378	17.1
Yelp2022	287,103	6,990,280	4,392,104	0.000002	2,942,336	1,449,768	37.0
NYC2014	1083	38,333	227,428	0.00548	158,683	68,745	1.14
TKY2014	2293	61,858	573,703	0.00404	400,532	173,171	2.68
MovieLens1M	6041	3953	1,000,208	0.04188	697,377	302,831	4.56

3 Models and Evaluation Metrics

Wang et al. [17] proposed to integrate the user-item interactions - more specifically the bipartite graph structure - into the embedding process. They developed a new recommendation framework - Neural Graph Collaborative Filtering (NGCF), which exploits the user-item graph structure by propagating embeddings on it. This leads to the expressive modeling of high-order connectivity in user-item graphs, effectively injecting the collaborative signal into the embedding process in an explicit manner. Proposed framework is shown in the Fig. 1. In order to compare the NGCF-3 model with other state-of-the art solutions, it was decided to use HOP-Rec and MF with BPR models, similarly as in the article presented by Wang et al. [17]. HOP-Rec [18] - High-Order Proximity for implicit Recommendation - is a state-of-the-art graph-based model, where the high-order neighbours derived from random walks are exploited to enrich the user-item interaction data. MF with BPR [14] - this is Matrix Factorization optimized by the Bayesian Personalized Ranking (BPR) loss, which exploits the user-item direct interactions only as the target value of interaction function. For each user in the test set, all the items that the user has not interacted with are treated as the negative items. Then each method outputs the user's preference scores over all the items, except the positive ones used in the training set. To evaluate the effectiveness of top-K recommendation and preference ranking, two widely-used evaluation protocols were adopted [6,18]: recall@K and ndcg@K. By default, K = 20. Different K values were also checked and the results are presented in the experiments section. Results reports the average metrics for all users in the test set.

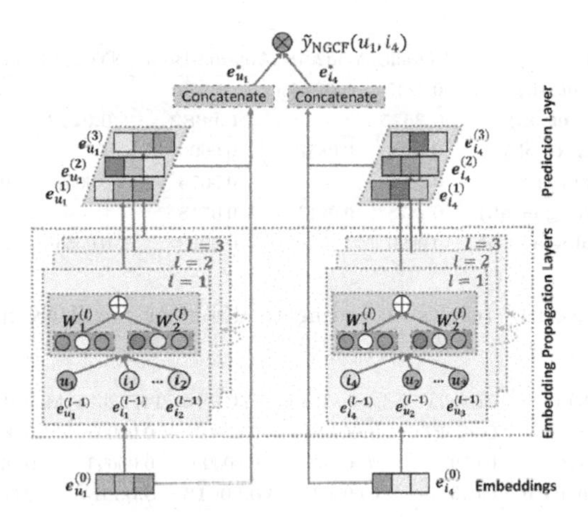

Fig. 1. An illustration of NGCF model architecture (the arrowed lines present the flow of information). The representations of user u_1 (left) and item i_4 (right) are refined with multiple embedding propagation layers, whose outputs are concatenated to make the final prediction.

4 Computational Resources

Model was trained on a single A100 40GB SXM GPU card. Resources used allowed for conducting experiments on 5 out of 6 prepared datasets: Gowalla, Amazon-Book, NYC2014, TKY2014, MovieLens1M. It was not possible to test the models on the Yelp2022 dataset. The size of the Yelp2022 dataset is about 37 MB, which resulted in model size too large to load into the GPU memory. Even reducing the dataset size by using Yelp2018 dataset instead, which has a size of about 29 MB, did not help to curtail the problem. It was clearly visible when compared to other datasets. Amazon-book dataset (about 17 MB size) and TKY2014 dataset (about 3 MB size) required 19 h and an hour of model training, respectively.

Table 2. Comparison of results according to recall@20 metric for different models and datasets

Dataset	Gowalla	Yelp2018	Amazon-Book	NYC2014	TKY2014	MovieLens1M
NGCF (Wang et al.)	**0.1547**	**0.0438**	**0.0344**	-	-	-
NGCF (Kobiela et al.)	0.1440	-	0.0286	0.0012	0.0013	0.0209
HOP-Rec (Wang et al.)	0.1399	0.0388	0.0309	-	-	-
HOP-Rec (Kobiela et al.)	0.1013	-	0.0087	0.0026	**0.0040**	**0.0228**
MF with BPR (Wang et al.)	0.1291	0.0317	0.0250	-	-	-
MF with BPR (Kobiela et al.)	0.1109	-	0.0097	**0.0029**	0.0034	0.0177

Table 3. Comparison of results according to ndcg@20 metric for different models and datasets

Dataset	Gowalla	Yelp2018	Amazon-Book	NYC2014	TKY2014	MovieLens1M
NGCF (Wang et al.)	0.2237	**0.0926**	0.0630	-	-	-
NGCF (Kobiela et al.)	**0.2625**	-	**0.0682**	**0.0212**	**0.0174**	**0.1006**
HOP-Rec (Wang et al.)	0.2128	0.0857	0.0606	-	-	-
HOP-Rec (Kobiela et al.)	0.0885	-	0.0075	0.0052	0.0073	0.0379
MF with BPR (Wang et al.)	0.1878	0.0617	0.0518	-	-	-
MF with BPR (Kobiela et al.)	0.0990	-	0.0078	0.0056	0.0071	0.0241

Table 4. Comparison of results according to different recall@K metric for different models and datasets

recall@1	Gowalla	Amazon-Book	NYC2014	TKY2014	MovieLens1M
NGCF-3	**0.02827**	**0.00219**	0.00007	0.00005	0.00195
HOP-Rec	0.01976	0.00066	0.00009	0.00006	**0.00229**
MF with BPR	0.02319	0.00062	**0.00013**	**0.00014**	0.00152
recall@5	Gowalla	Amazon-Book	NYC2014	TKY2014	MovieLens1M
NGCF-3	**0.07159**	**0.00900**	0.00041	0.00028	0.00709
HOP-Rec	0.04852	0.00304	**0.00079**	0.00093	**0.00776**
MF with BPR	0.05677	0.00294	0.00069	**0.00095**	0.00590
recall@10	Gowalla	Amazon-Book	NYC2014	TKY2014	MovieLens1M
NGCF-3	**0.10198**	**0.01615**	0.00076	0.00066	0.01207
HOP-Rec	0.07142	0.00515	**0.00148**	**0.00207**	**0.01324**
MF with BPR	0.08064	0.00540	**0.00148**	0.00184	0.01046
recall@20	Gowalla	Amazon-Book	NYC2014	TKY2014	MovieLens1M
NGCF-3	**0.14402**	**0.02864**	0.00120	0.00132	0.02089
HOP-Rec	0.10133	0.00872	0.00260	**0.00399**	**0.02276**
MF with BPR	0.11094	0.00971	**0.00292**	0.00339	0.01773
recall@40	Gowalla	Amazon-Book	NYC2014	TKY2014	MovieLens1M
NGCF-3	**0.20288**	**0.04957**	0.00226	0.00247	0.03677
HOP-Rec	0.13955	0.01621	**0.00553**	**0.00726**	**0.03945**
MF with BPR	0.15134	0.01815	0.00527	0.00660	0.03051

Table 5. Comparison of results according to different ndcg@K metric for different models and datasets

ndcg@1	Gowalla	Amazon-Book	NYC2014	TKY2014	MovieLens1M
NGCF	**0.17004**	**0.01748**	**0.00462**	0.00392	0.04371
HOP-Rec	0.12318	0.00671	0.00369	0.00349	**0.05000**
MF with BPR	0.14103	0.00562	0.00369	**0.00741**	0.02583
ndcg@5	Gowalla	Amazon-Book	NYC2014	TKY2014	MovieLens1M
NGCF	**0.19087**	**0.03351**	**0.01090**	**0.00755**	**0.05432**
HOP-Rec	0.08045	0.00607	0.00509	0.00654	0.04181
MF with BPR)	0.09192	0.00552	0.00528	0.00729	0.02394
ndcg@10	Gowalla	Amazon-Book	NYC2014	TKY2014	MovieLens1M
NGCF	**0.22418**	**0.04789**	**0.01608**	**0.01141**	**0.07203**
HOP-Rec	0.08074	0.00625	0.00519	0.00720	0.03827
MF with BPR	0.09150	0.00623	0.00562	0.00745	0.02291
ndcg@20	Gowalla	Amazon-Book	NYC2014	TKY2014	MovieLens1M
NGCF	**0.26249**	**0.06817**	**0.02121**	**0.01739**	**0.10062**
HOP-Rec	0.08847	0.00745	0.00525	0.00730	0.03788
MF with BPR	0.09899	0.00780	0.00560	0.00713	0.02408
ndcg@40	Gowalla	Amazon-Book	NYC2014	TKY2014	MovieLens1M
NGCF	**0.30709**	**0.09668**	**0.03256**	**0.02759**	**0.14893**
HOP-Rec	0.10010	0.01027	0.00592	0.00798	0.04217
MF with BPR	0.11132	0.01104	0.00593	0.00766	0.02830

5 Results

Results obtained by Wang et al. [17] are compared to the results obtained by the authors of this article and can be seen in the Tables 2 and 3. The highest metric value for every dataset is bolded and the cell is coloured in green. Then, the comparison of different K values for both metrics (recall and ndcg) was performed. Obtained results can be seen in the Tables 4 and 5.

6 Conclusions and Future Work

Results of the replication attempts show, that NCGF model performs slightly better or not worse than in the research performed by Wang et al. [17] when using ndcg metric (Table 3). In almost every case, regardless of the chosen K parameter value (either 1, 5, 10, 20 or 40) the highest metric results were obtained for the NGCF model on every dataset (Table 5). However, when looking at the recall metric, the results were different. Results obtained by Wang et al. [17] on datasets Gowalla and Amazon-Book were confirmed (Table 2). Notwithstanding, in the case of new datasets (NYC2014, TKY2014 and MovieLens1M), the model was worse in every case. Regardless of the chosen K parameter values (either 1, 5, 10, 20 or 40), the NGCF model never succeeded to be the best one. In every case, either HOP-Rec model or MF with BPR model achieved higher results (Table 4). Obtained results shows that the model does not perform as good on the new datasets as on the primary ones. Results obtained on the NYC2014, TKY2014 and MovieLens1M shows that NGCF model in the best case is not worse than in the results obtained by Wang et al., but in some cases it is 10-times worse (according to the recall@20 metric, Table 2). Models were compared using both metrics. According to recall@20 metric (Table 2), results achieved by the authors of the article on the Gowalla dataset are almost the same as the one achieved by Wang et al. [17]. Results obtained on Amazon-Book dataset are slightly lower. When looking to ndcg@20 metric (Table 3), results achieved by the authors of the article on the Gowalla dataset are slightly higher on the Amazon-Book dataset and on the Gowalla dataset then the ones achieved by Wang et al. [17]. Results obtained on new datasets (NYC2014, TKY2014 and MovieLens1M) are much lower (even about 10 times lower order of values for NYC2014 and TKY2014) than the ones obtained by Wang et al. Obtained results shows the need to improve the NGCF model by using a broader variety of datasets.

Acknowledgements. Computations were carried out using the computers of Centre of Informatics Tricity Academic Supercomputer & Network (https://task.gda.pl).

References

1. https://github.com/dariuszkobiela/ngcf-neural-graph-collaborative-filtering
2. https://snap.stanford.edu/data/loc gowalla.html: Gowalla dataset
3. Harper, F.M., Konstan., J.A.: The movielens datasets: History and context. ACM Trans. Interact. Intell. Syst.(TiiS) **5**, 4, Article 19 (December 2015), 19 (2015). https://doi.org/10.1145/2827872
4. He, R., McAuley, J.: Ups and downs: Modeling the visual evolution of fashion trends with one-class collaborative filtering, pp. 507–517 (2016). https://doi.org/10.1145/2872427.2883037
5. He, R., McAuley, J.: VBPR: Visual bayesian personalized ranking from implicit feedback, pp. 144–150 (2016)
6. He, X., Liao, L., Zhang, H., Nie, L., Hu, X., Chua, T.S.: Neural collaborative filtering (2017)
7. Herlocker, J.L., Konstan, J.A., Borchers, A., Riedl, J.: An algorithmic framework for performing collaborative filtering, pp. 230–237. SIGIR '99, Association for Computing Machinery, New York, NY, USA (1999). https://doi.org/10.1145/312624.312682
8. https://recbole.io/dataset _list.html: Recbole dataset list
9. Karpus., A., Raczyńska., M., Przybylek., A.: Things you might not know about the k-nearest neighbors algorithm, pp. 539–547. INSTICC, SciTePress (2019). https://doi.org/10.5220/0008365005390547
10. Kipf, T.N., Welling, M.: Semi-supervised classification with graph convolutional networks. In: 5th International Conference on Learning Representations, ICLR 2017, Toulon, France, April 24–26, 2017, Conference Track Proceedings. OpenReview.net (2017). https://openreview.net/forum?id=SJU4ayYgl
11. Liang, D., Charlin, L., McInerney, J., Blei, D.M.: Modeling user exposure in recommendation (2015). https://doi.org/10.48550/ARXIV.1510.07025

12. McAuley, J., Leskovec, J.: Hidden factors and hidden topics: Understanding rating dimensions with review text, pp. 165–172 (10 2013). https://doi.org/10.1145/2507157.2507163
13. Pan, R. Yu, Q., Xiong, H., Liu, Z.: Recommendation algorithm based on deep graph neural network. J. Comput. Appl. http://www.joca.cn/EN/10.11772/j.issn.1001-9081.2022091361
14. Rendle, S., Freudenthaler, C., Gantner, Z., Schmidt-Thieme, L.: Bpr: Bayesian personalized ranking from implicit feedback. In: Proceedings of the 25th Conference on Uncertainty in Artificial Intelligence, UAI 2009 (2012)
15. Sun, W., Chang, K., Zhang, L., Meng, K.: INGCF: An improved recommendation algorithm based on NGCF. In: Lai, Y., Wang, T., Jiang, M., Xu, G., Liang, W., Castiglione, A. (eds.) Algorithms and Architectures for Parallel Processing, pp. 116–129. Springer International Publishing, Cham (2022)
16. Tsai, Y.C., Guan, M., Li, C.T., Cha, M., Li, Y., Wang, Y.: Predicting new adopters via socially-aware neural graph collaborative filtering. In: Tagarelli, A., Tong, H. (eds.) Computational Data and Social Networks, pp. 155–162. Springer International Publishing, Cham (2019)
17. Wang, X., He, X., Wang, M., Feng, F., Chua, T.S.: Neural graph collaborative filtering, pp. 165–174. SIGIR'19, Association for Computing Machinery, New York, NY, USA (2019). https://doi.org/10.1145/3331184.3331267
18. Yang, J.H., Chen, C.M., Wang, C.J., Tsai, M.F.: Hop-rec: High-order proximity for implicit recommendation, pp. 140–144. RecSys '18, Association for Computing Machinery, New York, NY, USA (2018). https://doi.org/10.1145/3240323.3240381
19. Yaqi, L., Shuzhen, F., Lingyu, W., Chong, H., Ruixue, W.: Neural graph collaborative filtering for privacy preservation based on federated transfer learning **40**(6), 729–742 (2022). https://doi.org/10.1108/EL-06-2022-0141

Performance and Reproducibility of BERT4Rec

Aleksandra Gałka(✉), Jan Grubba, and Krzysztof Walentukiewicz

Gdansk University of Technology, 80-233 Gdansk, Poland
ola@galka.pl, grubba.jan@wp.pl, krzysztof.walentukiewicz2@gmail.com

Abstract. Sequential Recommendation with Bidirectional Encoder Representations from Transformer, BERT4Rec, is an efficient and effective model for sequential recommendation, regarded as a state-of-the-art in this field. However, studies of said architecture achieve varying results, causing doubts whether it is possible to consistently reproduce the result with reported training parameters. This study aims to test the performance and reproducibility of a BERT4Rec implementation on MovieLens1M and Netflix Prize datasets. Overall findings suggest that while using a proper implementation, BERT4Rec can still be called a state-of-the-art solution, additional work is needed to increase reproducibility of the original model. Moreover, possible avenues for further improvement are discussed.

Keywords: Recommendation systems · Transformers · Reproducibility

1 Introduction

In recent years, transformer-based [28] models have become the state-of-the-art for a number of natural language processing applications, including recommendation systems. Due to their high performance on various datasets, these models are increasingly being used in a variety of businesses and organizations where accurate forecasts and suggestions are critical components. The most significant and groundbreaking element is the attention mechanism which enable these models to simulate dynamic user preferences [9,14,21,26,31]. Previous recommendation transformer models that used unidirectional self-attention mechanisms used only past context to model future recommendations [9]. As a result, the model's capacity to capture more complex patterns and relationships is limited. To overcome this limitation the BERT4Rec model [26] takes advantage of the bidirectional self-attention mechanism, which enables the model to capture past and future relationships between the input and output sequences and so explains better the non-obvious patterns within the data.

The purpose of this research is to investigate the performance and reproducibility of the BERT4Rec model. Moreover, in previous studies a notable dataset was omitted, namely the Netflix Prize Dataset [1], a valuable resource for researchers in the field of recommendation systems. Therefore, evaluation of

A. Abelló et al. (Eds.): ADBIS 2023, CCIS 1850, pp. 620–628, 2023.
https://doi.org/10.1007/978-3-031-42941-5_55

the performance of the BERT4Rec model on the Netflix Prize Dataset is also the aim of this study. The results from our experiments and analyses are presented below.

2 Background

In this section the works related to our research has been mentioned, including a short overview of general recommendation models, their problem with modeling sequential user preferences, the solution to this problem that is a transformer based architecture and prior usage of BERT4Rec models.

2.1 Traditional Recommendation Models

Traditional recommendation models such as Collaborative Filtering (CF) and Content-based recommendation systems have been widely used in the past for recommendation tasks. [3,6,10–12,23] Collaborative filtering models leverage the user-item interaction data, such as ratings, clicks, or purchases, to learn the latent preferences of users and items. Content-based recommendation systems, on the other hand, use the similarities between users and items based on their attributes to provide recommendations. These systems recommend items that are similar to the ones that a user has previously interacted with. They utilize the information, usually based on expert knowledge, associated with users and items to learn the features that characterize users and items [15,18]. There also exists a wide range of different recommendation models eg. click through rate predictors [17], but they are out of the scope of this paper.

2.2 Transformer-Based Sequential Recommendations Models

While traditional recommendation models have previously been successful, they struggle to account for sequential user preferences that can evolve over time. Markov Chain models were formerly used in sequential recommendation systems [22], but more recent neural network-based models have proven to be more effective [7,16,27,29]. The biggest shortcoming of traditional neural networks used for sequential recommendation was the limited length of sequences, which was resolved by introduction of transformer based architectures capable of utilizing sequences of greater length. Due to this, more sophisticated models, such as SASRec [9], S^3-Rec [33] have emerged. Furthermore transformer-based sequential recommendation models have seen considerable performance increases since the introduction of the BERT language model [2] and its ability to understand the context of words in a sentence.

2.3 BERT4Rec

BERT4Rec leverages the BERT language model to achieve state-of-the-art results on various recommendation tasks. The model employs a Cloze objective

[19] to train on masked items and uses bidirectional self attention to capture the contextual information of words in a sequence. This enables BERT4Rec to model the sequential patterns in user behavior, which are essential for recommendation tasks where the user's preferences may evolve over time. The model has demonstrated superior performance over previous transformer-based and traditional recommendation models on several benchmark datasets.

The original paper of BERT4Rec has been challenged by subsequent papers that questioned its claimed advantages over existing neural and traditional methods [4,30,33]. A systematic review [20] has been conducted to verify the performance of BERT4Rec. The review found that BERT4Rec can indeed outperform SASRec, former state-of-the-art, however required more training time compared to the default configuration in order to replicate originally reported results. Therefore the review noted that some results reported in subsequent papers were based on underfitted versions of BERT4Rec.

The authors have also discovered, that by modifying the parameters or the model architecture some of the implementations, mainly BERT4Rec-VAE [8], DeBERTa4Rec and ALBERT4Rec, even achieved better performance than the original model, while also reducing the training time.

3 Experimental Setup

In this section the procedure conducted to evaluate reproducibility and establish performance of BERT4Rec on the MovieLens1M and Netflix Prize dataset has been presented.

3.1 Datasets Used

MovieLens1M [5]
MovieLens1M (ML-1M) is a dataset that GroupLens Research released in February 2003 via the movielens.org website. It contains 1 million ratings of 3952 movies by 6040 users, who each rated at least 20 movies on a scale of 1 to 5 stars. The dataset consists of quadruplets in the form of: [*UserID*,*MovieID*, *Rating*, *Timestamp*]. Moreover, the dataset provides information about the users' gender, age range and occupation, and the movies title and genre.

Netflix Prize [1]
Netflix Prize Dataset contains 100,480,507 anonymous movie ratings from 480,189 subscribers on 17,770 movie titles. The data is grouped by movie ID. Each subsequent line in the dataset corresponds to a rating from a customer and its date. Additionally, the dataset includes information about the movies such as the title and release year, but no information is provided about the users to protect their privacy. The training set is constructed such that the average user rated over 200 movies, and the average movie was rated by over 5000 users. However, there is a wide variance in the data, with some movies having as few as 3 ratings, while one user rated over 17,000 movies.

3.2 Procedure

In order to reproduce and evaluate the performance of BERT4Rec on the ML-1M and Netflix Prize datasets an implementation has been selected. In [20] authors failed to reproduce the results of the original implementation [26] using the default parameters and reported training time. However they provided their own implementation of BERT4Rec, as well as two other derivative models, that significantly outperform the original implementation. A model combining BERT4Rec with Variational Autoencoder [13] - BERT4Rec-VAE-Pytorch [8] has also been shown to exceed the performance of the original BERT4Rec implementation, while also taking less time to train. The aforementioned problems with the original implementation and poor performance of RecBole [32] was encountered in this study as well. Furthermore, we were unable to launch [20], therefore an implementation of BERT4Rec-VAE-Pytorch by Furyton was selected [25].

For training and evaluation phases all the best practises were followed. During evaluation the leave-one-out method has been carried out - for each user the last item was held out as the test data, and the second to last as validation. For evaluation each ground truth in the test set, has been paired with 100 popularity-sampled negative items - items that the user has not interacted with.

In order to test the reproducibility of BERT4Rec, original and well-performing models from [20] have been chosen as a comparison. The parameters set for the training on ML-1M dataset are shown in the Table 1. All the other hyperparameters (Adam, learning rate $1e-4$) have been set as in the original paper.

Table 1. Parameters of the models

Implementation	Original	SysReview	BERT4Rec-VAE	Furyton
Sequence length	200	50	100	50
Training epochs	200			
Item masking probability	0.2	0.2	0.15	0.2
Embedding size	64	64	256	64
Transformer blocks	2			
Attention heads	2	2	4	2

For the training purposes the Netflix Prize dataset had to be rearranged. The original data had ratings from customers and dates grouped by *MovieID*. The new data format had four elements in each row: [*UserID*,*MovieID*, *Rating*, *Timestamp*] sorted by *UserID* and *Timestamp*. Following this, the rating and timestamp were deleted, in order to prepare the data for the model input. The same was done with the ML-1M dataset.

3.3 Metrics

In order to check the evaluation results of the experiments the following metrics were used. The higher the result, the better performance.

NDCG@10 Normalized Discounted Cumulative Gain at rank 10 is a measure of the effectiveness of a ranked search algorithm. It is a variation of the traditional Discounted Cumulative Gain and is a way to measure the relevance of the returned documents, by considering both the order of the documents and the relevance scores.

RECALL@10 Recall at rank 10 is a ratio between the number of relevant documents in the top 10 results and all relevant documents in the entire dataset.

4 Results

During the reproduction the following results presented in the Table 2 have been achieved. The comparision was made with the original [26], systematic review [20] and BERT4Rec-VAE-Pytorch [8] implementation.

Table 2. The results of different implementations of BERT4Rec on ML-1M dataset. In brackets the difference noticed with the Furyton implementation.

MovieLens-1M dataset		
Model\ Metric	NDCG@10	Recall@10
Original	0.4818 (−9.30%)	0.6970 (−2.75%)
SysReview	0.4602 (−5.04%)	0.6865 (−1.27%)
BERT4Rec-VAE	0.4533 (−3.60%)	0.6698 (+1.19%)
Furyton	0.4370	0.6778

Subsequently, the results achieved on the Netflix Prize dataset are presented in the Table 3. As there was no previous evaluation of transformer based models on this dataset, the comparison was made with two notable autoencoder-based models recommender models - Recommender VAE (RecVAE) and The Embarrassingly Shallow Autoencoder (EASE) [24]. Additionally, as not all the metrics used in this study were reported in [24], the NDCG@10, MRR@10 and Recall@10 are presented only for BERT4Rec to compare with the results of ML-1M.

The plots of a validation metric, NDCG@10, during training are presented in Fig. 1. and Fig. 2.

Table 3. The results of different implementations on the Netflix Prize dataset. In brackets the improvement noticed with the Furyton implementation.

Netflix Prize Dataset					
Model\ Metric	Recall@20	Recall@50	NDCG@100	Recall@10	NDCG@10
RecVAE	0.361 (+110%)	0.452 (+97%)	0.394 (+31%)	-	-
EASE	0.31 (+109%)	0.445 (+100%)	0.393 (+31%)	-	-
Furyton	0.758	0.891	0.515	0.657	0.446

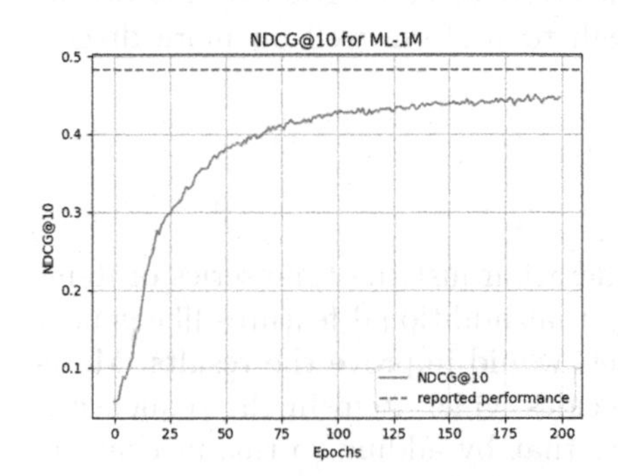

Fig. 1. NDCG@10 metric values over 200 epochs for the ML-1M dataset

Fig. 2. NDCG@10 metric values over 200 epochs for the Netflix Prize dataset

5 Discussion and Conclusions

To evaluate the performance of the BERT4Rec model on the Netflix Prize dataset, firstly, the reproducibility study of BERT4Rec implementation on ML-1M dataset was performed, to compare the results of the chosen implementation to the other, most significant, mentioned in [20].

All of the results vary by not more than 10%. One notable fact is that, the reported levels of the original implementation were achieved by training the model 30 times longer, as per [20]. When taken into consideration, this leads to the conclusion that the replication was successful. It is also worth noticing, that slight modification in the BERT4Rec-VAE-Pytorch model architecture, presented in Table 1, result in insignificant changes in reported metrics, which shows, that it is both reliable and robust.

Furthermore, the Netflix Prize dataset results presented in the Table 3 confirm that transformer solutions, especially BERT4Rec, significantly outperform previously used models.

Even though the size of Netflix Prize dataset is about 100 times larger, the results obtained on the Netflix Prize dataset were comparable to those on the ML-1M dataset. Since the characteristics of the Netflix Prize dataset ensures a wide variance of the data, it more accurately represents users' real preferences. Based on that, the results achieved are even more impressive.

Considering the learning curves on 200 epochs of NDCG@10 parameter, the curve for ML-1M [Fig. 1] grows quite fast at the beginning of the training only to slow down in the later phases, nonetheless keeping the growth trend. However NDCG@10 value for Netflix Prize [Fig. 2] increases rapidly at first and then after passing the 25th epoch start oscillating, increasing and decreasing value of the metric in small intervals (+/- 0.01 from the trend), although still increasing overall value of NCDG@10.

Although it needs more research, preliminary conclusions can be made, that BERT4Rec is a very flexible model, that not only achieves good results on the benchmark datasets, but also robust enough to perform well on more diverse data.

6 Future Work

To date BERT4Rec model takes into consideration just the time series of items for each user. However there is a possibility that additional features like gender or age, that are available in ML-1M dataset, would improve the results. Moreover, BERT4Recs' attention mechanism decides which item in the sequence is more important than the other. We suspect that by adding to this mechanism a weighted sequence based on the latest user's choice, the performance could improve as well.

Acknowledgements. Special thanks to Shiguang Wu from Shandong University for sharing his implementation of BERT4Rec model.

References

1. Bennett, J., Lanning, S., et al.: The netflix prize. In: KDD cup and workshop. vol. 2007, p. 35. New York (2007)
2. Devlin, J., Chang, M.W., Lee, K., Toutanova, K.: Bert: Pre-training of deep bidirectional transformers for language understanding. preprint arXiv:1810.04805 (2018)
3. Elahi, M., Ricci, F., Rubens, N.: A survey of active learning in collaborative filtering recommender systems. Comput. Sci. Rev. **20**, 29–50 (2016)
4. Fan, Z., Liu, Z., Zhang, J., Xiong, Y., Zheng, L., Yu, P.S.: Continuous-time sequential recommendation with temporal graph collaborative transformer. In: 30th ACM International Conference on Information and Knowledge Management, pp. 433–442 (2021)
5. Harper, F., Konstan, J.: Movielens data set. ACM Trans. Interact. Intell. Syst. **5**(4), 1–19 (2006)
6. Herlocker, J.L., Konstan, J.A., Terveen, L.G., Riedl, J.: Evaluating collaborative filtering recommender systems. ACM Trans. Inf. Syst. **22**, 5–53 (2004)
7. Hidasi, B., Karatzoglou, A., Baltrunas, L., Tikk, D.: Session-based recommendations with recurrent neural networks. preprint arXiv:1511.06939 (2015)
8. Jaywonchung: Pytorch implementation of bert4rec and netflix vae. https://github.com/jaywonchung/BERT4Rec-VAE-Pytorch
9. Kang, W.C., McAuley, J.: Self-attentive sequential recommendation. In: 2018 IEEE ICDM, pp. 197–206. IEEE (2018)

10. Kannout, E.: Context clustering-based recommender systems. In: 2020 15th FedCSIS, pp. 85–91. IEEE (2020)
11. Karpus, A., Raczynska, M., Przybylek, A.: Things you might not know about the k-nearest neighbors algorithm. In: KDIR, pp. 539–547 (2019)
12. Koren, Y.: Factorization meets the neighborhood: a multifaceted collaborative filtering model. In: 14th ACM SIGKDD, pp. 426–434 (2008)
13. Liang, D., Krishnan, R.G., Hoffman, M.D., Jebara, T.: Variational autoencoders for collaborative filtering (2018)
14. Lin, Z., et al.: A structured self-attentive sentence embedding. preprint arXiv:1703.03130 (2017)
15. Lops, P., De Gemmis, M., Semeraro, G.: Content-based recommender systems: State of the art and trends. Recommender sys. handbook, pp. 73–105 (2011)
16. Ma, C., Kang, P., Liu, X.: Hierarchical gating networks for sequential recommendation. In: 25th ACM SIGKDD, pp. 825–833 (2019)
17. Paulina Leszczełowska, M.B., Grabski, M.: Systematic literature review on click through rate prediction. In: 2nd Workshop on PeRS at ADBIS'23 (2023)
18. Pazzani, M.J., Billsus, D.: Content-based recommendation systems. In: Brusilovsky, P., Kobsa, A., Nejdl, W. (eds.) The Adaptive Web. LNCS, vol. 4321, pp. 325–341. Springer, Heidelberg (2007). https://doi.org/10.1007/978-3-540-72079-9_10
19. Petrov, A., Macdonald, C.: Effective and efficient training for sequential recommendation using recency sampling. In: 16th ACM Conference on Recommender System (2022)
20. Petrov, A., Macdonald, C.: A systematic review and replicability study of bert4rec for sequential recommendation. In: 16th ACM Conference on Recommender System (2022)
21. Radford, A., Narasimhan, K., Salimans, T., Sutskever, I., et al.: Improving language understanding by generative pre-training (2018)
22. Rendle, S., Freudenthaler, C., Schmidt-Thieme, L.: Factorizing personalized markov chains for next-basket recommendation. In: 19th International Conference on World wide web, pp. 811–820 (2010)
23. Sarwar, B.M., Karypis, G., Konstan, J.A., Riedl, J.: Item-based collaborative filtering recommendation algorithms. In: The Web Conf. (2001)
24. Shenbin, I., Alekseev, A., Tutubalina, E., Malykh, V., Nikolenko, S.I.: Recvae: A new variational autoencoder for top-n recommendations with implicit feedback. In: 13th International Conference on Web Search and Data Mining. ACM (jan 2020). https://doi.org/10.1145/3336191.3371831
25. Shiguang, W.: Furyton/recommender-baseline-model: Common used recommend baseline model. https://github.com/Furyton/Recommender-Baseline-Model
26. Sun, F., et al.: Bert4rec: Sequential recommendation with bidirectional encoder representations from transformer. In: 28th ACM International Conference on Information and Knowledge Management (2019)
27. Tang, J., Wang, K.: Personalized top-n sequential recommendation via convolutional sequence embedding. In: 11th ACM International Conference on Web Search and Data Mining, pp. 565–573 (2018)
28. Vaswani, A., et al.: Attention is all you need. In: Advances in Neural Information Processing System, vol. 30 (2017)
29. Yuan, F., Karatzoglou, A., Arapakis, I., Jose, J.M., He, X.: A simple convolutional generative network for next item recommendation. In: 12th ACM International Conference on Web Search and Data Mining, pp. 582–590 (2019)

30. Yue, Z., He, Z., Zeng, H., McAuley, J.: Black-box attacks on sequential recommenders via data-free model extraction. In: 15th ACM Conference on Recommender System (2021)
31. Zang, R., Zuo, M., Zhou, J., Xue, Y., Huang, K.: Sequence and distance aware transformer for recommendation systems. In: 2021 IEEE ICWS, pp. 117–124 (2021)
32. Zhao, W.X., et al.: Recbole 2.0: Towards a more up-to-date recommendation library. In: 31st ACM International Conference on Information & Knowledge Management, pp. 4722–4726 (2022)
33. Zhou, K., et al.: S3-rec: Self-supervised learning for sequential recommendation with mutual information maximization. In: 29th ACM International Conference on Information & Knowledge Management, pp. 1893–1902 (2020)

Doctoral Consortium

Towards Reliable Machine Learning

Simona Nisticò(✉)

DIMES Department, University of Calabria, Rende, Italy
simona.nistico@dimes.unical.it

Abstract. The increasing interest in machine and deep learning applications has highlighted how much powerful is to have notable amounts of data and complex models which can leverage them to solve hard tasks. The reverse of the medal of this complexity is a lack of interpretability for obtained results and an impossibility to inspect data to prevent possible issues like biases or to find new knowledge.

This paper focuses on the Explainable Artificial Intelligence topic, aimed to overcome the highlighted critical issues by providing users and practitioners with the instruments for a more aware use of models and data, giving an overview of some of the open problems and describing some methodologies proposed to face them.

Keywords: Explainable AI · Explainability · Interpretability · Fairness

1 Introduction

Recently, many fields witnessed breakthroughs brought by complex machine and deep learning solutions in the automatic resolution of different tasks. Goals reached by AI solutions in general and particularly by these models have awarded this field with increasing attention by academia as well as industry.

The complexity level characterising them makes very difficult, even for AI expert, to understand the reasons that justify their outcomes. It constitutes a drawback, since their application in contexts in which the resulting decisions have consequences is not possible because they can not give a justification for their results. This aspect lowers user trust, additionally, the possibility to return a justification sometimes represents a requirement [10].

Even data used to build AI models are difficult to inspect both due to their size and richness of available information. This represents an obstacle, not only for the detection of possible threats in the data sets (an example are the possible biases), but also for gaining new knowledge, so it is important to consider data explainability as well.

These motivations lead my PhD project which is devoted to consider explainability under these two perspectives, in order to provide to experts and final users the instruments to respectively verify and better understand AI solutions outcomes.

A. Abelló et al. (Eds.): ADBIS 2023, CCIS 1850, pp. 631–638, 2023.
https://doi.org/10.1007/978-3-031-42941-5_56

2 Related Work

Accordingly to what said in Sect. 1, we will considered the explainability problem in a two-sided structure focused on model and data explainability. In this section an overview of the state-of-art is provided for both the considered problems.

2.1 Model Explainability

Regarding the issue of explaining models outcome, various settings arises from different levels of model's information availability and from the stage in which interpretability problem is considered. Particularly, looking at the second discriminating factor, it is possible to circumscribe two classes of approaches.

The first class collects *Post-hoc explanation* methods, which aim at explaining trained black-box models using only the information provided by their output.

In order to explain the prediction for a certain data instance, some Post-hoc methodologies perturb the input, collect information about how the outcome changes, and then exploit them to estimate the level of importance of each feature. Among them, SHAP [14] is a game-theory inspired method that attempts to enhance interpretability by computing the importance values for each features. While SHAP is applicable to different data types, RISE [17] method focuses on image data. It generates an importance map indicating how salient each pixel is for the model's prediction by probing the model with randomly masked versions of the input image to observe how the output changes.

Other approaches exploit information carried by the gradient, focusing on neural networks explanation. The authors of [24] propose two methodologies to explain ConvNets through visualization leveraging the computation of the gradient of the class score with respect to the input image. Grad-CAM [22] uses the gradients of any target concept, which can be related to classification or other tasks, flowing into the final convolutional layer to produce a coarse localization map which highlights the important regions in the image for predicting the concept. This approach, contrary to the latter presented one, is applicable to any neural network.

Another way to decline explainability is in terms of examples, by providing to users counterfactuals, which are objects that show an example of something similar to the considered sample that brings the model to produce a different outcome, as explanations. Among the methods of this category it is possible to find LORE [9], which learns an interpretable classifier in a neighbourhood of the sample to explain, which is generated by employing a genetic algorithm, using model outcome as labels, and then extracting from the interpretable classifier an explanation consisting of a decision rule and a set of counterfactual. DiCE, which is proposed in [16], is a model agnostic approach for generating counterfactual examples that are both actionable and diverse, diversity requirement allows to increase the richness of information delivered to the user.

Finally, there are methodologies that explain models via local surrogates, which are self-interpretable functions that mimic the decisions of black-box models locally. LIME [18] explains black-box models by converting data object into a domain composed by interpretable features, and then perturbing it in that domain and querying the model to learn a simple model (the local surrogate) by means of this generated dataset. The type of explanations provided changes in Anchor [19] where if-then rules having high precision, called anchors, are created and utilised in order to represent local, sufficient conditions for prediction.

Conversely, the goal changes if we consider *explainability by default*, where the objective is to build models that give by construction an explanation for their results without using external tools. In this context, less work has been made in literature. In addiction to models that can be explained by virtue of their simplicity, one of the earliest attempts has been to develop self-explainability leveraging the attention mechanism that some models use, like in [15] where the visual attention used by the model can be exploited even for explanation.

Another possible approach is the one adopted by ProtoPNet [5] which decomposes the image by finding prototypical parts, and combines evidence from the prototypes to make a final classification. The strategy adopted is to bind the latent space of neural networks to be aligned with concept known from the training data.

In [6] they introduce the Concept Withening (CV) mechanism, which alters a given neural network layer to better understand its computation. When added to a convolution neural network, it decorrelates and normalizes the latent space and, additionally, aligns its axes with known concepts of interest.

Recently, some efforts has been made to built frameworks in which a classifier and an explainer learns jointly. For example, in [20], they propose an end-to-end model training procedure in which the classifier takes as input an image and the explainer's resulting explanation, as well as a three-phase training process and two alternative custom loss functions which aims at promoting sparsity and spatial continuity in explanations.

2.2 Data Explainability

For what concerns data explainability, in this paper, we will focus on the problem of explaining anomalies in a data set. This task, referred as *anomaly explanation*, is relevant since a deeper understanding of outliers helps decision makers to use effectively the information provided and to act consciously.

To find features which characterize most samples outlierness some methods focus only on a search perspective by leveraging features selection approaches, which are widely used to improve classification model performances. One example is the approach presented in [7] which searches the outlier in subspaces of the original feature space in order to find that one in which outliers are well discriminated from regular objects. Also [1], which approach is to finds the top k attributes associated with the largest outlier score for the anomaly in order to explain its outlierness, belongs to this group.

Other methods, belonging to the score-and-search methodologies class, employ measures to rank object features and then selects it. So they give to users, together with the features highlighted as anomalous, a score which gives an insight of features "relevance level". Different criteria are used to score features, in [8] density, which is there computed through a kernel-density estimation, is employed to rank attributes, in [25] authors propose to use the Z-score and the Isolation Path score (iPath) as outlierness measures, another score, proposed in [21], is SiNNE. COIN [13] method exploits the anomalous point neighbourhood by dividing it into clusters, and then training on each of them a simple classification model, combining the learned parameters to compute a score for each attribute.

3 Methods and Results

The PhD project presented in this paper covers the two perspectives introduced in Sect. 1. In this section, which structure follow the same organization of Sect. 2, we present published results.

3.1 Model Explainability

Considering post-hoc explainability problem, two approaches have been propose during the PhD period to tackle this issue. The first born from a deep study of LIME [18] from which it emerged that the generation mechanism of this method can potentially produce objects with artifacts during neighbourhood generation process, problem shared with a lot of others perturbation-based approaches, and limits the explanations to be expressed only in terms of what is present in the data object. So *Semantic-LIME* ($\mathcal{S}$-LIME) [3], a novel approach which exploits unsupervised learning to replace user-provided interpretable components with self-generated semantic features, has been proposed to overcome the above-depicted flaws considering text domain. In our proposal the latent space of a task-dependent Denoising Adversarial Auto-encoder [23] has been employed for neighbourhood generation, then the neighbourhood points are converted into an interpretable representation and explanation is produced following the standard LIME procedure which is now able to consider also features not present in the original data-point. Beyond enlarging the possible explanation types, our proposal address above-highlighted LIME issue related to out-of-distribution samples using latent space sampling instead of random perturbation, but it adds the overhead of having a task-dependent latent space learning phase. To compare the quality of explanations with the ones of the standard LIME algorithm, a measure of significance of the words returned as explanations, based on the model predictions (see [3] for more details), has been proposed. The comparison of the explanations of $\mathcal{S}$-LIME and LIME though this metrics has highlighted how the explanation of $\mathcal{S}$-LIME detects all the most significative words of the original sentence but, at the same time, is able to find also significative words which can potentially change the model outcome, reaching higher significance levels.

The second method proposed for post-hoc explainability adopts an adversarial inspired approach and focuses on image classification context under the assumption to have a model which gives as output the class probability. The method proposed, which is called *Mask Image to Local Explain* (MILE) [2], learns a transformation of the input that produces a novel instance able to confuse the black-box model by maximizing the probability of belonging to a class that differs from the one assigned to the original input instance. Since we exploit the novel instance as an explanation of the black-box decision, we simultaneously require that it minimizes the dissimilarity from the instance to explain.

The proposed architecture is composed by a Mask Generator module and Mask Applier module, which aim respectively at learning the transformation which allows to obtain an object of another class near to the considered instance, and to apply that transformation to the sample in order to obtain the counterfactual. Additive transformations has been considered. The main contribution here is to propose an approach which avoids to have out-of-distribution samples, since it not requires a generation step and which not requires to approximate the model to a local surrogate, which reasonably is less expressive compare to the black-box model.

To train the modules the following loss is used:

$$\mathcal{L}(x, x', \ell') = H(p_f(y|x'); \ell') + \lambda \|x - x'\|_2 \quad (1)$$

where $H(p_f(y|x'); \ell')$ is the cross entropy between the distribution $p_f(y|x')$ of the class probabilities associated to x' by f and the "true" distribution assigning x' to the class ℓ', given by the one hot representation of the class, $\|x - x'\|_2$ is the Mean Squared Error obtained when x is approximated by means of x', and λ is an hyper-parameter representing the trade-off between the two loss contributions.

To test MILE effectiveness a biased model has been trained on a synthetic data set built by taking class 9 and 3 of MNIST [12] and modifying all the training samples of class 3 adding a black rectangle in the bottom left corner. Then MILE has been employed to compute explanations for this model for the validation set composed by modified 9s and normal 3s, showing that its explanation has been able to find the bias learned by the model.

3.2 Data Explainability

Regarding data explainability, outlier explanation problem has been considered where, given a data set DS and an object o known to be an outlier, the goal is to find an *explanation* e for o.

The idea exploited in [2] has been rearranged to fit this context. The explanation e searched is a pair $\langle c, m \rangle$ where c denotes a set of attributes and m denotes a set of values, one for each element in c, such that for o by taking the values m for the attributes c would make o an inlier in DS. From it originates MMOAM method [4], which realizes a deep learning solution using a generative component coupled with an adversarial component, which structure is depicted in Fig. 1. The adversarial module tries to predict if the input data point is anomalous or not.

The information it outputs are then fed into the generative module that, in its turn, tries to single out the set of changes (specified as a set of attribute-values pairs) needed for the outlier data point (also referred to as "sample data point" in the following) to be transformed into an inlier w.r.t. the reference data set.

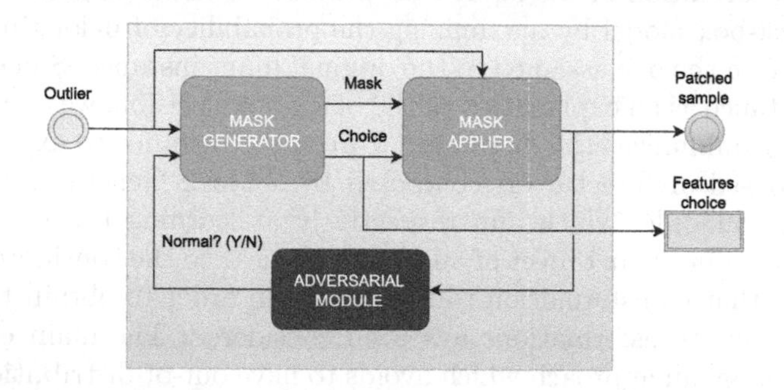

Fig. 1. Proposed architecture. The Mask Generator and the Mask Applier represent the generative module of the MMOAM methodology, the adversarial module acts as an oracle.

In deference to the Occam's razor principle, for an outlier explanation to be significant, the transformation that changes the nature of the sample, transforming it from anomalous to normal, must be *minimal*, which means that the number of features returned as justification of the given data point outlierness must be as small as possible. It has been, at the best of our knowledge, the first methodology to employ neural networks for outlier explanation task and, additionally, the explanation provided points out even the entity of the modification needed to make the object normal, this represents a novelty in the outlier explanation field.

The training of this architecture is led by the following custom loss function:

$$\ell_{\mathcal{G}}(o, X_n) = \alpha_1 \cdot \ell_{focal}(\mathcal{A}(\mathcal{G}(o)), 0) + \alpha_2 \cdot \frac{1}{|X_n|} \sum_{x_n \in X_n} \frac{||\mathcal{G}(o) - x_n||_2^2}{\sqrt{d}} + \\ + \alpha_3 \cdot \ell_{bce}(\mathcal{G}_c(o), 0) + \alpha_4 \cdot \frac{||\mathcal{G}_m(o)||_2^2}{\sqrt{d}}, \quad (2)$$

where $\mathcal{G}(o)$ is the modified version o' of o produced by the generative module, $\mathcal{A}(\mathcal{G}(o))$ is the classification obtained from the adversarial module for o', $\mathcal{G}_m(o)$ and $\mathcal{G}_c(o)$ denote the feature modification vector ϕ_m and the feature selection vector ϕ_c produced by the generative module, respectively, X_n are the data set normal samples, ℓ_{focal} is the focal loss, ℓ_{bce} is the binary cross-entropy, and d is the number of dimensions. Several experiments have been performed that demonstrate the effectiveness of the approach in giving successful transformations and its capacity to outperform the competitor algorithms SOD [11] and COIN [13] in terms of quality of the subspaces highlighted as most outlying for each object.

4 Conclusion

XAI research field is heavily diversified to answer to the needs of AI users and experts. The work of my PhD project has been devoted to explore this variety and to propose solutions for some of the detected open problems. Particularly, regarding post-hoc model explanability, two methods, $\mathcal{S}$-LIME and MILE, have been proposed which follow the objective of addressing some drawback of state-of-art approaches. Moving to data explainability, we proposed MMOAM, which for the first time employs neural networks to face outlier explanation problem and which provides richer and more precise explanations compared to its competitors.

Although a wide range of problems and scenarios have been covered, there are still a lot of challenges to face. An interesting future development would be to deeply explore explainability by default field with a particular focus on anomaly detection which can largely benefits from it due to its more strict requirements for certain applications.

Acknowledgments. I acknowledge the support of the PNRR project FAIR - Future AI Research (PE00000013), Spoke 9 - Green-aware AI, under the NRRP MUR program funded by the NextGenerationEU.

References

1. Angiulli, F., Fassetti, F., Manco, G., Palopoli, L.: Outlying property detection with numerical attributes. Data Min. Knowl. Disc. **31**(1), 134–163 (2017)
2. Angiulli, F., Fassetti, F., Nisticò, S.: Finding local explanations through masking models. In: Yin, H., et al. (eds.) IDEAL 2021. LNCS, vol. 13113, pp. 467–475. Springer, Cham (2021). https://doi.org/10.1007/978-3-030-91608-4_46
3. Angiulli, F., Fassetti, F., Nisticò, S.: Local interpretable classifier explanations with self-generated semantic features. In: Soares, C., Torgo, L. (eds.) DS 2021. LNCS (LNAI), vol. 12986, pp. 401–410. Springer, Cham (2021). https://doi.org/10.1007/978-3-030-88942-5_31
4. Angiulli, F., Fassetti, F., Nisticò, S., Palopoli, L.: Outlier explanation through masking models. In: Advances in Databases and Information Systems: 26th European Conference, ADBIS 2022, Turin, Italy, 5–8 September 2022, Proceedings, pp. 392–406. Springer (2022). https://doi.org/10.1007/978-3-031-15740-0_28
5. Chen, C., Li, O., Tao, D., Barnett, A., Rudin, C., Su, J.K.: This looks like that: deep learning for interpretable image recognition. In: Advances in Neural Information Processing Systems 32 (2019)
6. Chen, Z., Bei, Y., Rudin, C.: Concept whitening for interpretable image recognition. Nat. Mach. Intell. **2**(12), 772–782 (2020)
7. Dang, X.H., Assent, I., Ng, R.T., Zimek, A., Schubert, E.: Discriminative features for identifying and interpreting outliers. In: 2014 IEEE 30th International Conference on Data Engineering, pp. 88–99. IEEE (2014)
8. Duan, L., Tang, G., Pei, J., Bailey, J., Campbell, A., Tang, C.: Mining outlying aspects on numeric data. Data Min. Knowl. Disc. **29**(5), 1116–1151 (2015)
9. Guidotti, R., Monreale, A., Giannotti, F., Pedreschi, D., Ruggieri, S., Turini, F.: Factual and counterfactual explanations for black box decision making. IEEE Intell. Syst. **34**(6), 14–23 (2019)

10. Hamon, R., Junklewitz, H., Malgieri, G., Hert, P.D., Beslay, L., Sanchez, I.: Impossible explanations? beyond explainable ai in the gdpr from a Covid-19 use case scenario.In: FAccT 2021, pp. 549–559. ACM, New York (2021). https://doi.org/10.1145/3442188.3445917
11. Kriegel, H.-P., Kröger, P., Schubert, E., Zimek, A.: Outlier detection in axis-parallel subspaces of high dimensional data. In: Theeramunkong, T., Kijsirikul, B., Cercone, N., Ho, T.-B. (eds.) PAKDD 2009. LNCS (LNAI), vol. 5476, pp. 831–838. Springer, Heidelberg (2009). https://doi.org/10.1007/978-3-642-01307-2_86
12. LeCun, Y.: The mnist database of handwritten digits (1998). http://yann.lecun.com/exdb/mnist/
13. Liu, N., Shin, D., Hu, X.: Contextual outlier interpretation. In: Proceedings of the 27th International Joint Conference on Artificial Intelligence, IJCAI 2018, pp. 2461–2467. AAAI Press (2018)
14. Lundberg, S., Lee, S.I.: A unified approach to interpreting model predictions. arXiv preprint arXiv:1705.07874 (2017)
15. Mnih, V., Heess, N., Graves, A., et al.: Recurrent models of visual attention. In: Advances in Neural Information Processing Systems 27 (2014)
16. Mothilal, R.K., Sharma, A., Tan, C.: Explaining machine learning classifiers through diverse counterfactual explanations. In: Proceedings of the 2020 ACM FAccT, pp. 607–617 (2020)
17. Petsiuk, V., Das, A., Saenko, K.: Rise: randomized input sampling for explanation of black-box models. arXiv preprint arXiv:1806.07421 (2018)
18. Ribeiro, M.T., Singh, S., Guestrin, C.: " Why should i trust you?" explaining the predictions of any classifier. In: Proceedings of the 22nd ACM SIGKDD KDD, pp. 1135–1144 (2016)
19. Ribeiro, M.T., Singh, S., Guestrin, C.: Anchors: high-precision model-agnostic explanations. In: Proceedings of the AAAI Conference on Artificial Intelligence, vol. 32 (2018)
20. Rio-Torto, I., Fernandes, K., Teixeira, L.F.: Understanding the decisions of CNNS: an in-model approach. Pattern Recogn. Lett. **133**, 373–380 (2020)
21. Samariya, D., Aryal, S., Ting, K.M., Ma, J.: A new effective and efficient measure for outlying aspect mining. In: Huang, Z., Beek, W., Wang, H., Zhou, R., Zhang, Y. (eds.) WISE 2020. LNCS, vol. 12343, pp. 463–474. Springer, Cham (2020). https://doi.org/10.1007/978-3-030-62008-0_32
22. Selvaraju, R.R., Cogswell, M., Das, A., Vedantam, R., Parikh, D., Batra, D.: Grad-cam: Visual explanations from deep networks via gradient-based localization. In: Proceedings of the IEEE ICCV, pp. 618–626 (2017)
23. Shen, T., Mueller, J., Barzilay, R., Jaakkola, T.: Educating text autoencoders: latent representation guidance via denoising. In: ICML, pp. 8719–8729. PMLR (2020)
24. Simonyan, K., Vedaldi, A., Zisserman, A.: Visualising image classification models and saliency maps. In: Deep Inside Convolutional Networks (2014)
25. Vinh, N.X., et al.: Discovering outlying aspects in large datasets. Data Min. Knowl. Disc. **30**(6), 1520–1555 (2016)

Tackling the RecSys Side Effects via Deep Learning Approaches

Erica Coppolillo(✉)

University of Calabria, 87036 Arcavacata, CS, Italy
erica.coppolillo@unical.it

Abstract. Digital platforms, such as social media and e-commerce websites, widely make use of Recommender Systems to provide value to users. However, social consequences of such uses and their long-term effects are still unclear and controversial, spanning from influencing users' preferences to perpetrating bias embedded in the data. In this work, we present our current research activities on these phenomena, based in particular on a proper use of ML/DL techniques: we overview the state of the art and then highlight our contribution, focusing on the study of opinions drift and unfairness in recommendations.

Keywords: Machine Learning · Deep Learning · Recommender Systems · Bias · Fairness

1 Introduction and Motivation

Recommender Systems (RecSys) can be defined as "*software tools and techniques providing suggestions for items to be of use to a user. The suggestions relate to various decision-making processes, such as what items to buy, what music to listen to, or what online news to read.*" [29]. Nowadays, also due to a growing amount of available content, they are widely adopted by several digital platforms. On the one hand, they represent a viable way to help users to navigate large volumes of information; on the other hand, they build mechanisms that allow new content to emerge. As a downside, recommendation algorithms have also been blamed of inducing detrimental effects on users (such as echo chambers [25], rabbit holes [21,27], filter bubbles [34], and radicalization [17,22,28]), and of perpetuating bias and unfairness intrinsic in the data (such as popularity bias [1]).

Since there is still no consensus in literature [28], the long-term effects of RecSys are currently under scrutiny and deserve further investigation. In this complex and evolving scenario, Artificial Intelligence and ML/DL techniques can play a key role in analyzing and mitigating the risks due to these potentially noxious phenomena.

In this work, we provide an overview of the above-mentioned problems and discuss current and future research lines. Specifically:

- Sect. 2 surveys the state-of-the-art DL approaches which investigate the effects of RecSys and propose innovative solutions;

A. Abelló et al. (Eds.): ADBIS 2023, CCIS 1850, pp. 639–646, 2023.
https://doi.org/10.1007/978-3-031-42941-5_57

- In Sect. 3 we present our current research line: we first propose a novel methodology for analyzing the opinion drift phenomenon over users, and a further solution to alleviate the presence of bias in data;
- Finally, Sect. 4 concludes the work and illustrate some new research lines.

2 Related Work

2.1 RecSys Influence over Users

We can observe a growing research line studying the long-term effects of RecSys. In particular, user behavior deviation from its natural evolution has been widely studied in the literature, thus still remaining an open challenge.

In [8], the authors claim that recommenders homogenize user behavior without increasing the utility of their suggestions. Recent works observe how people-recommenders can exacerbate social media critical issues[13,16,30], such as polarization, misinformation, and pre-existing inequalities in user communities.

[27] and [21] show that YouTube recommendations lead to the genesis of "*rabbit holes*", i.e., a journey toward increasingly radicalized contents that can exaggerate user bias, belief, and opinion.

Videos on channels belonging to four political leanings are analyzed in [27], who observe that some center channels serve as gateways to push users to far-right ideology. Similarly, [26] provide empirical modeling for analyzing the radicalization phenomenon over conspiracy theory discussions in Reddit, showing how users progressively pass through various radicalization stages.

[25] propose a probabilistic generative model that discovers echo-chambers in a social network by introducing a Generalized Expectation Maximization algorithm, that clusters users according to how information spread among them.

Finally, [9] define a counterfactual model called "organic model", which is used to compare user outcomes influenced by a recommender against the outcomes under the natural evolution of the users' choices. The main finding of this work states that the recommender and organic model dramatically differ from each other, thus highlighting the influence of the recommendation.

On the other hand, several recent works argue the opposite. [23] analyze different politically-oriented communities on YouTube, claiming the recommendation algorithm plays little role in the consumption of radicalizing content. Similarly, authors in [10] show that exposure to alternative and extremist channel videos on YouTube is heavily concentrated among a small group of people with high prior levels of gender and racial resentment, hence suggesting that YouTube's algorithms are not the main reason that sends people toward them.

Given this evident lack of consensus in the literature, we want to contribute to a further investigation.

2.2 Bias in Recommendations

Bias in computer systems can be defined as a *"systematic and unfair discrimination against certain individuals or groups of individuals in favor of others."* [18].

It is a phenomenon that deeply affects the recommendation algorithms in various forms since they are fed with data whose gathering process is observational rather than experimental [11]. One of the forms of bias is the so-called *popularity bias*, which implies an over-exposure of already high-popular items, neglecting niche ones. This pernicious phenomenon reduces not only the personalization (i.e., exacerbating user experience homogenization) but also the fairness and the variety of the suggested items. Long-term consequences triggered by the feedback loop among the user, recommender, and data can be even more detrimental [24,32]. In particular, as mentioned in [3,7], unfair recommendations are concentrated on groups of users interested in long-tail and less popular items.

To mitigate the popularity bias, three possible techniques can be devised:

- **Preprocessing**: A first approach is to deliberately modify the input dataset to train a model favoring low-frequency items [6,15].
- **In-processing**: Another approach is to modify the learning phase of a recommender system by altering its optimization process [12,14,33].
- **Post-processing**: These approaches try to modify the generated recommendation list to boost the exposure of low-popular items, as in [2,4,31].

Our proposal focuses on the first approach, since the in-processing strategy necessarily requires retouching the model, potentially lacking in adaptability, while the risk of the post-processing strategy is to reduce the predictive capability of the chosen recommendation algorithm, thus downgrading the global performance.

3 Ongoing Research

In the following, we describe our current research line: we first present a simulation methodology to study and evaluate the effects of RecSys on the users' preferences, together with novel mitigation strategies; then, we show an effective solution applicable to reduce popularity bias in recommendations.

3.1 Algorithmic Drift

In our research, we consider a typical recommendation scenario in which some top-ranked items [20] are proposed and accepted by the user [5]. Items (that can take the form of media, news, videos, etc.) are tagged as *harmful* or *neutral*, while users are categorized as *non-*, *semi-* or *radicalized*, based on the percentage of harmful interactions in their preference history. Here, we investigate whether and how RecSys can affect and modify the users' preferences, as they could be used to manipulate their opinions and behaviors. In more detail, we devise a simulation methodology for modeling the evolution of the interactions between users and items. In our setting, we assume that such interactions are fully driven by a (black-box) collaborative filtering algorithm.

Given the initial users interactions, our goal is to generate a probabilistic graph G_u, whose nodes include the subset of items for which the user u had at least one interaction during the simulation. In this way, estimating the users' leaning drift built on this graph corresponds to estimating the recommender's influence in the long term. Figure 1 depicts an overview of the proposed simulation methodology.

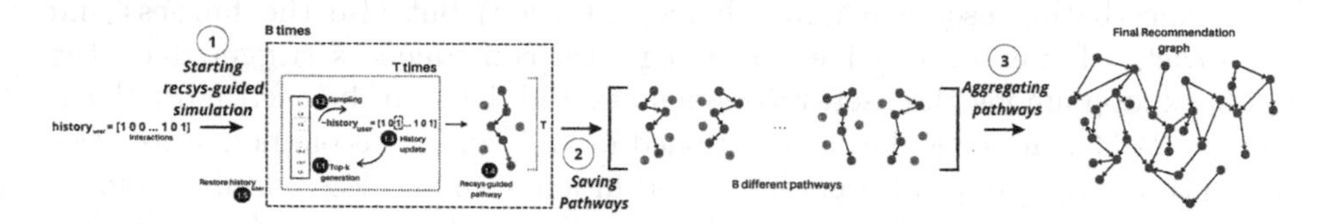

Fig. 1. Overview of the simulation methodology modeled to mimic the users-RecSys interactions in the long-term.

Our final aim is to evaluate how much the recommendation algorithm changes user preferences after a certain number of interactions. In a nutshell, given a recommendation model pre-trained with the initial user interactions (classified as neutral or harmful), we want to estimate if the RecSys will tend to suggest more extreme content to initially unbiased users, and, vice-versa, neutral and harmless content to already radicalized ones. We refer to this tendency as *algorithmic drift*.

To quantify it, we defined a novel graph-based metric to be applied over the probabilistic graph G_u of a user u, by adapting the Random Walk Controversy Score (RWC) proposed by [19] to quantify the radicalization of user opinions in an online social network. In our context, the metric describes whether the user is more inclined to encounter and remain in neutral pathways, or in harmful ones.

Let us consider a user u, and its probabilistic graph G_u. The measure of algorithmic drift $A(G_u)$ can be hence defined as:

$$A(G_u) = Pr(I_h|I_h) \times Pr(I_h|I_n) - Pr(I_n|I_n) \times Pr(I_n|I_h)\,, \tag{1}$$

where $Pr(X|Y)$ is the probability that a random walk ends on an element belonging to the set X, given that the walk starts on an element of the set Y. So defined, $A(G_u) \in [-1, 1]$ represents the probability of moving from harmful to neutral items and remaining in cliques of neutral ones, with respect to the probability of moving from neutral to harmful items and remaining in cliques of harmful ones. The larger the value, the more the user u will consume harmful content in the long run.

In our experiments, we found that all the adopted recommendation algorithms actually drift users' preferences, as Fig. 2 shows, consistently with the most of works previously mentioned. Notably, our setting allows us to point out two further considerations regarding the phenomenon: (*i*) it is more prominent when the portion of semi-radicalized users increases, thus making this subpopulation a sort of *bridge* in enabling the effect; (*ii*) it is not observed when a natural evolution (i.e., the organic model) is assumed.

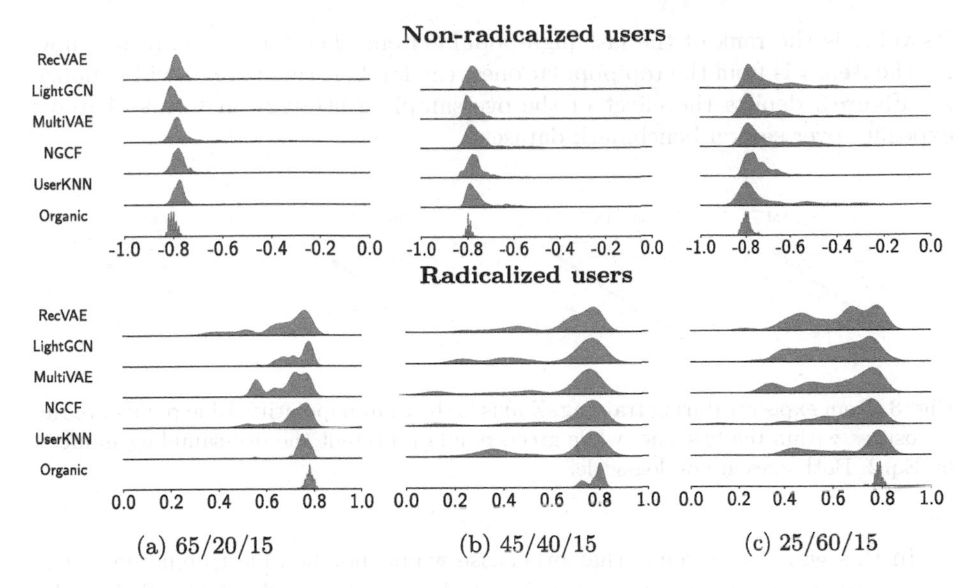

Fig. 2. $A(G_u)$ computed over non-/radicalized users by varying the proportion of the semi-radicalized population (20%, 40%, 60%, respectively).

3.2 Popularity Bias

ML-based systems typically aim for higher accuracy, but a recent research trend focuses on balancing the quality of performances with the ethical or discriminatory effects of ML-based recommendations. We addressed this problem by adopting a solution that considerably boosts the exposure of niche items, without degrading global performances. Notably, our approach is *conservative*, in the sense that can be applied to any DL ranking algorithm based on pairwise comparison, without the need of retouching the model.

Oversampling. The oversampling strategy consists of re-calibrating the positive items' exposure, inversely to their popularity. In other words, rather than sampling, for each occurrence in our dataset, a fixed number n of negative items as in the standard approach, we apply a dynamic sampling scheme. Let n'_i be a term that is inversely proportional to the popularity of the item, defined as:

$$n'_i = n_0 \frac{max(\boldsymbol{\rho})}{\rho_i \, d_i} \,, \tag{2}$$

where n_0 is a constant, $\boldsymbol{\rho}$ is the popularity distribution of all the items, ρ_i is the popularity of the item i, and d_i is the discrete scaling factor that controls the sampling exposure. The latter, which dynamically rebalances the exposures, is defined as:

$$d_i = \frac{r_i}{max(1, h)} + 1 \,, \tag{3}$$

where r_i is the rank of the item i, and h represents the highest rank of the set of items for which we want to preserve a certain number of pairwise comparisons.

Its value is the rank of the last high-popular item. The term $\frac{r_i}{h}$ indicates how far the item i is from the top popular ones: the farther, the more d_i will penalize n_i'. Figure 3 depicts the effect of the oversampling strategy in terms of items exposure over several benchmark datasets.

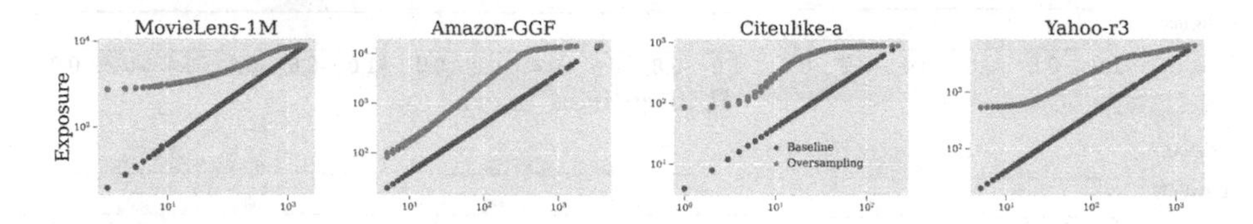

Fig. 3. Item exposure during training. X axis is the item popularity. Blue points are the exposures within the baseline, while green points represent the oversampling induced by Eq. 2. Both axes are on log-scale.

In this way, niche items, that otherwise would not be able to emerge in the recommendation process, receive a boosted exposure, without penalizing the high-popular ones. In our experiments, we observed that this approach leads to better performances in the evaluation phase over low-popular items, with no degradation in terms of global accuracy.

4 Conclusions and Future Works

In this work, we presented our current research activities aimed at analyzing and mitigating the impact of RecSys in changing users' preferences and affecting niche items. Specifically, we presented:

- A simulation methodology that mimics the interactions of users-RecSys in the long-term, allowing us to assess the capability of the recommender in drifting users preferences;
- A conservative approach for mitigating the problem of popularity bias, that can be applied over any DL algorithm based on pairwise comparison.

As future work, we are interested in extending our preliminary studies concerning radicalization phenomena, relying on more sophisticated and realistic simulation approaches. Moreover, in addition to the popularity bias, we aim at exploring different kinds of bias that can emerge in the recommendation process.

References

1. Abdollahpouri, H., Burke, R., Mobasher, B.: Controlling popularity bias in learning-to-rank recommendation. In: RecSys 2017, pp. 42–46. Association for Computing Machinery, New York (2017). https://doi.org/10.1145/3109859.3109912
2. Abdollahpouri, H., Burke, R., Mobasher, B.: Managing popularity bias in recommender systems with personalized re-ranking. In: International Florida Artificial Intelligence Research Society Conference, FLAIRS 2019, pp. 413–418 (2019)
3. Abdollahpouri, H., Mansoury, M., Burke, R., Mobasher, B.: The unfairness of popularity bias in recommendation. In: Workshop on Recommendation in Multi-stakeholder Environments. CEUR Workshop Proceedings 2019, vol. 2440 (2019)
4. Abdollahpouri, H., Mansoury, M., Burke, R., Mobasher, B.: The connection betwe, n popularity bias, calibration, and fairness in recommendation. In: ACM Conference on Recommender Systems, RecSys 2020, pp. 726–731 (2020)
5. Agarwal, A., Wang, X., Li, C., Bendersky, M., Najork, M.: Addressing trust bias for unbiased learning-to-rank. Association for Computing Machinery, New York (2019)
6. Boratto, L., Fenu, G., Marras, M.: Connecting user and item perspectives in popularity debiasing for collaborative recommendation. Inform. Process. Manag. **58**(1), 102387 (2021)
7. Borges, R., Stefanidis, K.: On measuring popularity bias in collaborative filtering data. In: EDBT Workshop on BigVis 2020: Big Data Visual Exploration and Analytics. EDBT/ICDT Workshops (2020)
8. Chaney, A.J.B., Stewart, B.M., Engelhardt, B.E.: How algorithmic confounding in recommendation systems increases homogeneity and decreases utility. In: Proceedings of the 12th ACM Conference on Recommender Systems, pp. 224–232. ACM, Vancouver British Columbia Canada (Sep 2018). https://doi.org/10.1145/3240323.3240370
9. Chang, S., Ugander, J.: To recommend or not? a model-based comparison of item-matching processes. In: Proceedings of the International AAAI Conference on Web and Social Media, vol. 16, pp. 55–66 (2022)
10. Chen, A., Nyhan, B., Reifler, J., Robertson, R.E., Wilson, C.: Subscriptions and external links help drive resentful users to alternative and extremist youtube videos (April 2022)
11. Chen, J., Dong, H., Wang, X., Feng, F., Wang, M., He, X.: Bias and debias in recommender system: a survey and future directions. arXiv preprint arXiv:2010.03240 (2020)
12. Chen, Z., Wu, J., Li, C., Chen, J., Xiao, R., Zhao, B.: Co-training disentangled domain adaptation network for leveraging popularity bias in recommenders. Association for Computing Machinery, New York (2022). https://doi.org/10.1145/3477495.3531952
13. Cinus, F., Minici, M., Monti, C., Bonchi, F.: The effect of people recommenders on echo chambers and polarization. In: Proceedings of the International AAAI Conference on Web and Social Media, vol. 16, pp. 90–101 (2022)
14. Ding, S., et al.: Interpolative Distillation For Unifying Biased and Debiased Recommendation, pp. 40–49. Association for Computing Machinery, New York (2022). https://doi.org/10.1145/3477495.3532002
15. Ekstrand, M.D., .: All the cool kids, how do they fit in?: Popularity and demograet alphic biases in recommender evaluation and effectiveness. In: Conference on Fairness, Accountability, and Transparency, PMLR 2018, pp. 172–186 (2018)

16. Fabbri, F., Croci, M.L., Bonchi, F., Castillo, C.: Exposure inequality in people recommender systems: The long-term effects. In: Proceedings of the International AAAI Conference on Web and Social Media, vol. 16, pp. 194–204 (2022)
17. Fabbri, F., Wang, Y., Bonchi, F., Castillo, C., Mathioudakis, M.: Rewiring what-to-watch-next recommendations to reduce radicalization pathways. Association for Computing Machinery (2022)
18. Friedman, B., Nissenbaum, H.: Bias in computer systems. ACM Trans. Inform. Syst. **14**(3), 330–347 (1996)
19. Garimella, K., Morales, G.D.F., Gionis, A., Mathioudakis, M.: Quantifying controversy in social media (2017)
20. Guan, Z., Cutrell, E.: An eye tracking study of the effect of target rank on web search. Association for Computing Machinery (2007)
21. Haroon, M., Chhabra, A., Liu, X., Mohapatra, P., Shafiq, Z., Wojcieszak, M.: Youtube, the great radicalizer? auditing and mitigating ideological biases in youtube recommendations. arXiv preprint arXiv:2203.10666 (2022)
22. Hosseinmardi, H., Ghasemian, A., Clauset, A., Mobius, M., Rothschild, D.M., Watts, D.J.: Examining the consumption of radical content on youtube. In: Proceedings of the National Academy of Sciences (2021)
23. Hosseinmardi, H., Ghasemian, A., Clauset, A., Mobius, M., Rothschild, D.M., Watts, D.J.: Examining the consumption of radical content on youtube. Proc. Natl. Acad. Sci. **118**(32), e2101967118 (2021). https://doi.org/10.1073/pnas.2101967118
24. Mansoury, M., Abdollahpouri, H., Pechenizkiy, M., Mobasher, B., Burke, R.: Feedback loop and bias amplification in recommender systems. In: ACM International Conference on Information and Knowledge Management, CIKM 2020, pp. 2145–2148 (2020)
25. Minici, M., Cinus, F., Monti, C., Bonchi, F., Manco, G.: Cascade-based echo chamber detection (2022). https://doi.org/10.48550/ARXIV.2208.04620
26. Phadke, S., Samory, M., Mitra, T.: Pathways through conspiracy: the evolution of conspiracy radicalization through engagement in online conspiracy discussions. In: Proceedings of the International AAAI Conference on Web and Social Media, vol. 16, pp. 770–781 (2022)
27. Ribeiro, M.H., Ottoni, R., West, R., Almeida, V.A., Meira Jr, W.: Auditing radicalization pathways on youtube. In: Proceedings of the 2020 Conference on Fairness, Accountability, and Transparency (2020)
28. Ribeiro, M.H., Veselovsky, V., West, R.: The amplification paradox in recommender systems (2023)
29. Ricci, F., Rokach, L., Shapira, B.: Recommender Systems Handbook, vol. 1–35, pp. 1–35 (Oct 2010). https://doi.org/10.1007/978-0-387-85820-3_1
30. Santos, F.P., Lelkes, Y., Levin, S.A.: Link recommendation algorithms and dynamics of polarization in online social networks. Proc. Natl. Acad. Sci. **118**(50), e2102141118 (2021)
31. Steck, H.: Calibrated recommendations. In: ACM Conference on Recommender Systems, RecSys 2018, pp. 154–162 (2018)
32. Tsintzou, V., Pitoura, E., Tsaparas, P.: Bias disparity in recommendation systems. CoRR abs/ arXiv: 1811.01461 (2018)
33. Xv, G., Lin, C., Li, H., Su, J., Ye, W., Chen, Y.: Neutralizing popularity bias in recommendation models. Association for Computing Machinery, New York (2022). https://doi.org/10.1145/3477495.3531907
34. Zhang, H., Zhu, Z., Caverlee, J.: Evolution of filter bubbles and polarization in news recommendation (2023)

Towards a Researcher-in-the-loop Driven Curation Approach for Quantitative and Qualitative Research Methods

Alejandro Adorjan(✉)

Universidad ORT Uruguay Facultad de Ingeniería, Montevideo, Uruguay
adorjan@ort.edu.uy

Abstract. This article describes challenges and initial results concerning the modelling researcher-in-the-loop (RITL) curation for qualitative and quantitative (quanti-qualitative) research methodologies. The guiding research question of the work is *how to model quanti-quali research methods' data and processes to identify the properties and constraints that define them to provide an adapted curation approach?* Our work enhances techniques for exploring and curating data collections and research processes in the context of data-driven methods. We propose a RITL curation approach that models qualitative research processes as data-driven workflows. These workflows manipulate content through collaborative phases defining specific quanti-qualitative research project milestones. The content is related to initial and generated content and the conditions in which the workflows go from one state to another through working groups' decision-making processes. Curated content enables querying and exploring the content to get an insight into the conditions in which research results are produced and allow reproducibility.

Keywords: data curation · processes curation · researcher-in-the-loop · quantitative and qualitative research methods

1 Introduction

Social sciences and humanities studies address problems by combining qualitative and quantitative methods. Qualitative research inquires the understanding a given research problem or topic from the local population's perspective, and systematically uses a predefined set of procedures to answer research questions. Therefore, it collects evidence and produces findings that are not predetermined and applicable beyond the study's immediate boundaries. Quantitative methods help researchers to interpret and better understand the complex reality of a given situation and the implications of quantitative data.

Research methods with (quanti)-qualitative perspectives should curate data, the processes they perform (defining research questions, stating, calibrating and

Supported by the National Agency for Research and Innovation (ANII), Uruguay.
Advisors: Regina Motz, UdelaR, Uruguay; Genoveva Vargas-Solar, CNRS, LIRIS, France.

A. Abelló et al. (Eds.): ADBIS 2023, CCIS 1850, pp. 647–655, 2023.
https://doi.org/10.1007/978-3-031-42941-5_58

validating a theoretical framework, converging to a corpus, defining data collection tools) and the context in which data are produced through processes (interviews, focus group, research team decision-making criteria and consensus protocols). Traditional curation approaches and systems do not provide adapted tools for curating this heterogeneous content. Therefore, our work enhances techniques for exploring and curating data collections and research processes in the context of data-driven methods.

This paper describes challenges and initial results concerning the modelling researcher-in-the-loop (RITL) curation for (quanti)-qualitative research methodologies. The guiding research question of the work is *how to model quanti-qualitative research methods' data and processes to identify the properties and constraints that define them to provide an adapted curation approach?* To answer the question, we propose a Researcher-in-the-loop (RITL) curation approach that models (quanti)-qualitative research processes as data-driven workflows. These workflows manipulate content through collaborative phases that define specific milestones of qualitative research projects. The content is related to initial and generated content and the conditions in which the workflows go from one state to another through working groups' decision-making processes. Curated content enables querying and exploring the content to get an insight into the conditions in which research results are produced and allow reproducibility.

The remainder of the article is organised as follows. Section 2 introduces the background and related work. Section 3 presents the modelling of research methods. Section 4 offers the general lines of the proposed Researcher-in-the-Loop based curation approach. Finally, the expected results and future work lines are presented in Sect. 5.

2 Background and Related Work

Theoretical and methodological frameworks are the basis for sustaining (quali)-qualitative research projects. Qualitative Research explores human aspects and new phenomena, capturing individuals' thoughts, feelings, or interpretations of meaning and process [6]. Qualitative research explores complex phenomena and it can increase research quality by providing a method of theory building [2].

Curated data collections drive a path in scientific knowledge acquisition, reuse and management. Data Curation is a process that manage, maintain and validate data to ensure continued access [8]. It is a critical component of data science, and an relevant aspect of data work [17]. Many related curation activities, processes and data models have been proposed in the literature [7,11,12]. In our work, data curation is "the process of identifying, systematizing, managing and versioning research data produced along the project stages". Quantitative and qualitative research methodologies usually apply ad-hoc data curation strategies that keep track of the data that describe the tools, techniques, hypothesis, and data harvesting criteria defined a priori by a scientific team. Data sets at an early collection stage are generally not ready for analysis or preservation. Thus, extensive preprocessing, cleaning, transformation, and documentation actions are required to support usability, sharing, and preservation over time [10].

CAQDAS-Computer-Assisted Qualitative Data Analysis Software. CAQDAS is a well-known tool for qualitative research [20]. CAQDAS are tools for qualitative researchers to apply statistical techniques and machine learning algorithms. ATLAS.ti [1], Dedoose [4], MAXQDA [16], NVivo [13] implement the REFI-QDA standard, an interoperability exchange format. CAQDAS [3] researchers and practitioners can perform annotation, labelling, querying, audio and video transcription, pattern discovery, and report generation. Furthermore, CAQDAS tools allow the creation of field notes, thematic coding, search for connections, memos (thoughtful comments) creation, contextual analysis, frequency analysis, word location and data analysis presentation in different reporting formats [5]. The REFI-QDA (Rotterdam Exchange Format Initiative)[1] the standard allows the exchange of qualitative data to enable reuse in QDAS [9]. QDA software such as ATLAS.ti [1], Dedoose [4], MAXQDA [16], NVivo [13], QDAMiner [14], Quirkos [15] and Transana [18] adopt REFI-QDA standard.

Discussion. Research infrastructures are intended to provide access to scientific data and resources needed for problem-solving. Research repositories are the standard for publishing data collections to the research communities. Scientific data, including models, should be created using the FAIR[2](Findable, Accessible, Interoperable and Reusable) principles. For example, Dataverse[3] is an open-source data repository software installed in dozens of institutions globally to support public community repositories or institutional research data repositories [19]. FAIR principles provides guidelines for scientific infrastructure supporting the reuse of data [19]. However, these guidelines systems do not fully provide a framework for systematically exploring, curating, and reusing research artefacts.

3 Modelling Qualitative Research Phases

We propose a model for qualitative research process as a set of phases with a spiral life cycle (see Fig. 1). In these phases (problem statement, data acquisition, data management, and data analysis) artefacts such as interviews, surveys, codebooks, field diaries, algorithmic and mathematical tools are produced and modified continuously. Modifications on artefacts can generate new ones and versions of existing ones. The curation in this context must also extract, collect and maintain information about the conditions in which new content and versions are created. For example, instruments of interview design, transcription analysis, coding criteria, and thematic analysis undergo a dynamic change in their elaboration. Moreover, methodological strategies can change, reformulating the instrument, re-calibrating it in another direction, and impacting the data, metadata, and dependencies in the next iteration. Consensus within the research team allows for the rigorous analysis of coding and thematic analysis criteria throughout the process. Notice that the previous concept highly contrasts with

[1] https://www.qdasoftware.org.
[2] https://www.go-fair.org/fair-principles/.
[3] https://dataverse.org.

quantitative methods, in which the hypotheses are generally invariant once the problem is established.

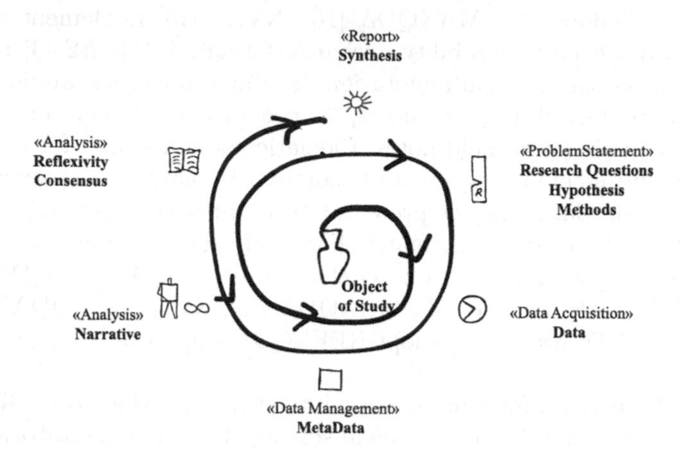

Fig. 1. Spiral Qualitative Research Phases.

The research addresses the following research question: *how to model qualitative research methods' data and processes to identify the properties and constraints that define them to provide an adapted curation approach?* The main challenge of the current work is to model the theoretical framework, phases, interactions between the phases and constant evolution of the artefacts that changes along the research project.

4 Researcher-in-the-Loop Driven Curation for Hybrid Research Methods

We propose a Researcher-in-the-Loop (RITL) workflow for modelling (quanti)-qualitative research methods. The model has five phases: problem statement, data acquisition, data management, data analysis and report. In these phases, research artefacts are produced and versioned, guided by consensus-based decision-making by the research team. The workflow is associated with a curation approach that collects (meta)-data of all artefacts produced during its life cycle using a Versioning Control system strategy. This configuration management strategy reproduces the phases and consensus processes for "validating", reusing and modifying artefacts.

This RITL-based curation approach promotes the exploration and reproducibility of a (quanti)-qualitative research process, exposing the content production and versioning operations that keep track of the evolution of the artefacts until new knowledge and conclusions are produced.

Spiral Research Process and Research-in-the-Loop. Fig. 1 illustrates the spiral research process for each phase of a (quanti)-qualitative research approach: problem statement, data acquisition, data management, and data analysis.

- *Problem statement* refers to the theory review stage, formulation of research questions, the definition of methodologies, and construction of the theoretical framework.
- *Data acquisition* is the phase devoted to data collection, exploration, cleaning, and reliability verification.
- *Data management* refers to metadata generation, evaluation, and contextualization.
- *Analysis* consists of a round of experiments and measurements, incorporating debates and reflections on the results of previous stages.
- *Reporting* is a phase devoted to visualization, evaluation, writing process, and final publication of scientific work.

Researchers conduct qualitative activities in different stages, such as exploratory interviews, weekly reflection seminars, manual generation of analytical coding, and analytical coding of the entire process. Every phase and its stages adopt a spiral process where artefacts are produced through research actions done within fieldwork activities, which use digital solutions such as algorithms, Mathematical models and text processing tools, always with researchers' intervention. This intervention is what we call Researcher-in-the-Loop (RITL).

(Quanti)-Qualitative Methods Content Model. Q2MCM is based on the notion of artefact[4] to model the diverse content that must be curated for (quanti)-qualitative research methods. Q2MCM artefacts are documents containing:

- data collection "tools" like surveys, interviews, codebooks, field diaries, etc.
- descriptions of the design criteria of these tools for choosing focal groups, people applying the tools, objectives, and analysis protocols
- the documents produced of the iteration phases like theoretical framework, research questions, bibliography, corpora, comments on the documents, quantitative and categorical analysis results
- the information describing the context of the (quanti)-qualitative research, including the description of the research team, the conditions in which decision-making is done (agreements, consensus) along the research phases and the provenance of results and agreements.

[4] The ACM https://www.acm.org/publications/policies/artifact-review-and-badging-current defines artefact as "a digital object that was either created by the authors to be used as part of the study or generated by the experiment itself.".

Qualitative Content Production Cycles Protocol Through Versioning. The spiral production of content happens during the phases/stages of the method with commenting, discussion, revisiting and consensual validation of content. We propose to model these spiral life cycles through a versioning strategy that manages the storage and historicization of content with different "validity" statuses (defined by the researcher(s)), associating it with the conditions and protocol under which researchers specify this status. The principle (see Fig. 2) is that at every "phase" in a project, researchers propose, comment, discuss and validate the content and can adopt a consensus protocol to decide, validate, modify or revisit the content and restart the phase/stage or continue and/or define to the next phase/stage the project.

Curation Approach for (Quanti)-Qualitative Research Methods. The curation approach's objective is to extract and collect meta-data about the artefacts, the phases/cycles, and the context of a project (research team, focal groups). These meta-data are organised into versions that reproduce the spiral life cycle defined by the research method. The versioning strategies (status and operators) are adapted to the characteristics of the (quanti)-qualitative research methods. Our Q2MCM models the different types of meta-data to be automatically or manually collected.

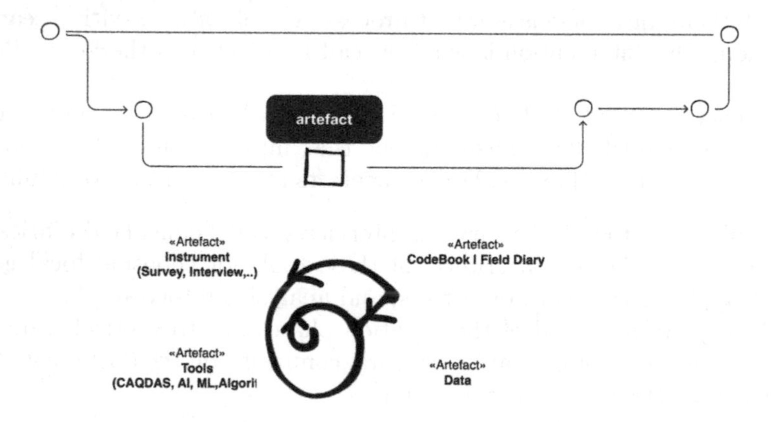

Fig. 2. Artefacts versioning principle.

Throughout the research process development, researchers provide feedback on the process phase that the research team defines internally. Analytical notes or descriptive tags comment on the artefact, and new textual entries can be added in the field diary or new code category in a codebook for thematic analysis. Comments, tags, analytical notes and codebook entries associated with an artefact with their provenance are meta-data produced manually that curate the artefact. Finally, a new release version can be identified when the project has evolved to a stable version declared by researchers and gathered by the versioning strategy.

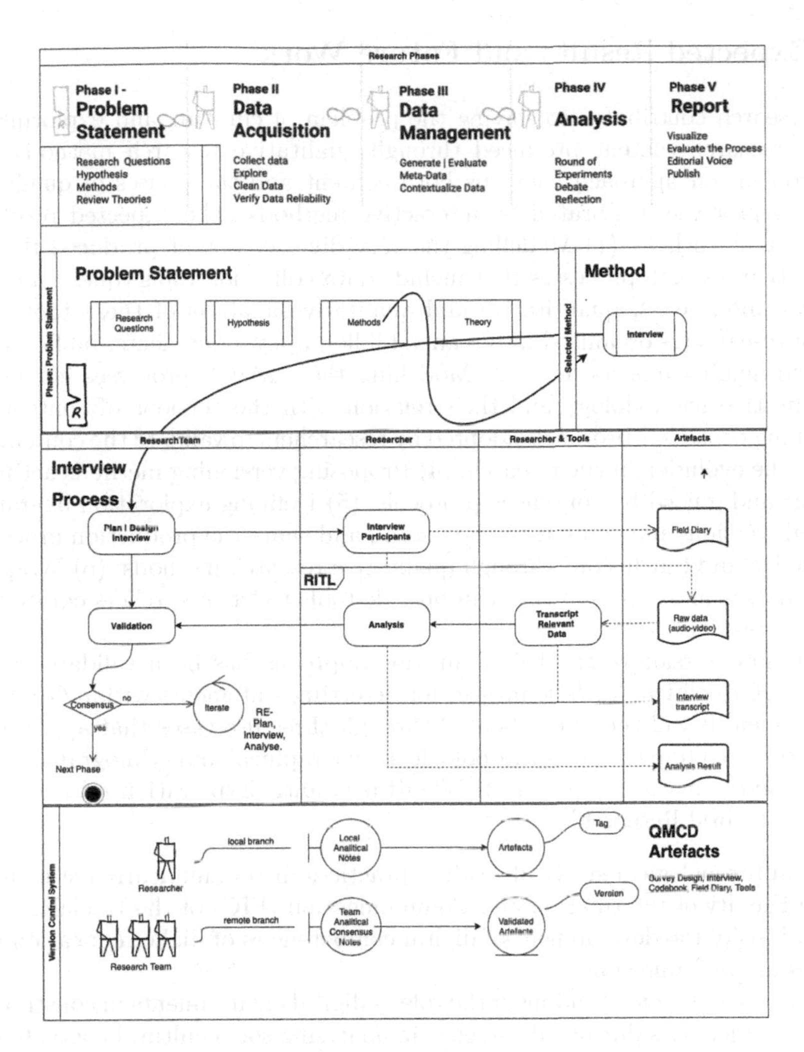

Fig. 3. RITL driven curation system for (quanti)-qualitative research methodologies.

RITL Driven Curation System. The system allows researchers to curate the content produced during the phases of a project (see Fig. 2). It provides tools for uploading digital content, creating associated meta-data (comments, feedback, tags), and interacting asynchronously with other researchers commenting and completing their meta-data. It also provides tools for maintaining the versioning of the content along the different phases of a project and its validation, and also the consensus process that makes the phases evolve from one to another (e.g., branching, validating, committing). The system allows the exploration and analysis of the projects' curated content. Exploration and analysis of curated content will enable them to make informed decisions and have insight into their research practice.

5 Expected Results and Future Work

Our research contributes to solving the problem of curating and exploring scientific research content produced through qualitative research methods. The curation in our approach concerns both content and the phases of qualitative research processes calibrated by interactive methods. The expected results of the research include: (1) Modelling the very diverse content produced through qualitative research processes that include data collection tools (questionnaires, surveys, interviews), quantitative and qualitative analysis of these tools, profiles of researchers organised into teams, bibliography, documents, bibliography, decision-making reports, etc. (2) Modelling the research processes composing a qualitative methodology and their relation with the versions of content. (3) Modelling consensus protocols adopted by researchers to validate the content and decide the evolution of the research. (4) Proposing versioning methods acting on content and guided by consensus protocols. (5) Defining exploration operations through versions and content to understand and share the production process of knowledge and conclusions through qualitative research methods. (6) Wrapping these models into a system that can provide tools to let researchers curate their research results.

The first version of the RITL curation approach has been validated in the context of a political, artistic project for detecting statements within Grafitti in Montevideo. It will then be validated through three use cases that apply qualitative research to perform anthropological, geographical and ethnographic data-driven studies in the context of MENTOR (seMantic Exploration aNd curaTion of Open Hybrid Research):

1. An anthropological study of reading practices in communication students of the Faculty of Information and Communication (FIC) of the UdelaR.
2. Analysis of the development of digital competencies of higher education abilities in pandemic times.
3. Analysis and understanding of the role of digital environments in constructing gender identities during adolescence in analysing socio-cultural contexts with populations.

Acknowledgments. This work is part of MENTOR proyect FMV-1-2021-1-167914, supported by the National Agency for Research and Innovation (ANII), in Uruguay.

References

1. ATLAS.ti: ATLAS.ti. https://atlasti.com (Accessed April 2023)
2. Charmaz, K., Thornberg, R.: The pursuit of quality in grounded theory. Qual. Res. Psychol. **18**(3), 305–327 (2021)
3. Chen, N., Drouhard, M., Kocielnik, R., Suh, J., Aragon, C.: Using machine learning to support qualitative coding in social science: shifting the focus to ambiguity. ACM Trans. Interactive Intell. Syst. **8**(2), 1–20 (2018)
4. Dedoose: Dedoose. https://www.dedoose.com/ (Accessed April 2023)

5. Evers, J.C.: Current issues in qualitative data analysis software (qdas): a user and developer perspective. Qual. Report **23**(13), 61–73 (2018)
6. Given, L.M.: The Sage encyclopedia of qualitative research methods. Sage publications (2008)
7. Han, L., Chen, T., Demartini, G., Indulska, M., Sadiq, S.: A data-driven analysis of behaviors in data curation processes. ACM Trans. Inform. Syst. **41**(3), 1–35 (2023)
8. Hourcle, J.A., King, T.A.: Vocabulary for Virtual Observatories and Data Systems. In: AGU Fall Meeting Abstracts, vol. 2010, pp. IN11B-1076 (Dec 2010)
9. Karcher, S., Kirilova, D.D., Pagé, C., Weber, N.: How data curation enables epistemically responsible reuse of qualitative data. Qualit. Report **26**(6), 1996–2010 (2021)
10. Lafia, S., Thomer, A., Bleckley, D., Akmon, D., Hemphill, L.: Leveraging machine learning to detect data curation activities. In: 2021 IEEE 17th International Conference on eScience (eScience), pp. 149–158. IEEE (2021)
11. Lee, D., Stvilia, B.: Practices of research data curation in institutional repositories: A qualitative view from repository staff. PloS One **12**(3) (2017)
12. Martín, L., Sánchez, L., Lanza, J., Sotres, P.: Development and evaluation of artificial intelligence techniques for IOT data quality assessment and curation. Internet of Things **22**, 100779 (2023)
13. NVivo: Nvivo. https://www.qsrinternational.com/ (Accessed April 2023)
14. QDAMiner: Qdaminer. https://provalisresearch.com/products/qualitative-data-analysis-software/ (Accessed April 2023)
15. Quirkos: Quirkos. https://www.quirkos.com (Accessed April 2023)
16. Software, V.: Maxqda. http://maxqda.com (Accessed April 2023)
17. Thomer, A., et al.: The craft and coordination of data curation: complicating "workflow" views of data science. arXiv preprint arXiv:2202.04560 (2022)
18. Transana: Transana. https://www.transana.com (Accessed April 2023)
19. Wilkinson, M.D., et al.: The fair guiding principles for scientific data management and stewardship. Scientific Data **3**(1), 1–9 (2016)
20. Woods, M., Macklin, R., Lewis, G.K.: Researcher reflexivity: exploring the impacts of caqdas use. Int. J. Soc. Res. Methodol. **19**(4), 385–403 (2016)

Deep Learning Techniques for Television Broadcast Recognition

Federico Candela(✉)

DIIES Department, University Mediterranea of Reggio Calabria, 89124 Reggio Calabria, Italy
federico.candela@unirc.it

Abstract. With the advent of the digital age, the amount of video content available has grown exponentially, especially in the context of television programs, Deep learning has proven to be an important technology capable of addressing the challenges associated with the classification of television programs. One of the main problems in video classification is the complexity of the data itself, moreover, video classification of TV programs by deep learning requires a large amount of training data. This paper proposes several deep learning techniques and the advancement of research in this area for TV program classification on long-form video.

Keywords: Video Classification · Deep Learning · TV-Channels

1 Introduction

With the advent of the digital age, the amount of available video content has grown exponentially, especially in the context of television programs. The amount of audiovisual data available has necessitated the adoption of advanced methods for analyzing and organizing this content. In this context, deep learning has emerged as an important technology that can address the challenges associated with the classification of television programs. For example, content providers can use this technology to index their video archives, facilitating program search and retrieval. Television broadcasters can use video classification to monitor the circulation and popularity of their content. In addition, viewers can benefit from personalized recommendations based on their interests and preferences. One of the main problems with video classification is the complexity of the data itself. Videos are composed of a sequence of related frames, thus creating a highly dimensional stream of information. Managing and analyzing this large amount of data requires significant computational resources, resulting in scalability issues and long run times. The more channel-specific the training data, the greater the probability of program recognition. A major problem arises when one wants to generalize the process for multiple TV stations. For

University Mediterranea of Reggio Calabria, Co.Re.Com.Calabria, AGCOM (Communications Guarantee Authority).

A. Abelló et al. (Eds.): ADBIS 2023, CCIS 1850, pp. 656–667, 2023.
https://doi.org/10.1007/978-3-031-42941-5_59

example, the sequence of news headlines on one channel is different from that of another channel, which poses significant problems for training a deep neural network in terms of updating the network in case new headline sequences change and are introduced by a station. In addition, the number of fps of which a title sequence of a program is composed turns out to contain little meaningful data for training, unless one wants to incorporate many more title sequences from various programs into a single class, which, however, poses the problem previously described, namely, updating by a program broadcaster. It is an open challenge for audiovisual traffic management and monitoring agencies, such as the Communications Guarantee Authority (AGCOM), to be able to automate the program classification process.

L'AGCOM assigns Regional Communications Committees (Co.Re.Com.) to monitor local audiovisual broadcasts and report any problems. These tasks include identifying problems such as exceeding the time limit for a program, airing unauthorized commercials, broadcasting programs or scenes intended only for adult audiences, identifying debates and classifying them by type, identifying the opening and closing title sequence of a program, and more according to a nationally regulated table [1]. The research developed so far aims to make a strong scientific contribution in this area, solving the previously mentioned problems and developing a new methodology adaptable to the entire television world. The following paper proposes a series of innovative methodologies, SSIM with ResNet50 enables the accurate classification of television acronyms and general contents of a program, the development of this method led to the integration of an LSTM network for storing the entire program stream. Then, through a shot boundary detection technique, we moved on to work directly on scene changes that are significant for commercials and the opening and closing title sequence of a program, classifying both on the optical stream through the Transformers network and on the stacked frames through CNN. The following paper is structured as follows: Sect. 2 gives an overview of the state of the art, Sect. 3 discusses the methodologies applied so far, Sect. 4 discusses the experimental results obtained, and Sect. 5 summarizes the conclusions.

2 State of the Art

Video classification using deep learning is a dynamic and constantly evolving research field capable of analyzing and solving problems ranging from everyday use cases to high-risk domains. The emergence of high-quality publicly available datasets such as ImageNet [2], UCF-101 [3], Youtube8M [4], and many others, have contributed to the tremendous growth of video image classification by deep learning in addition to the significant implementation of GPUs that has enabled the considerable acceleration of computational processes. The best methodologies used for video classification by deep learning depend on the context and specific objectives of the problem. However, there are some common techniques and deep learning architectures that are often employed for video classification such as: Convolutional Neural Networks (CNNs) whose origins date back to the

Necognitron [5] use convolutional filters to extract spatial features over time from each frame of video, and then ultimately fully connected layers for classification. The effectiveness of such networks has led to further developments over time by modifying the architecture and improving performance, C3D (Convolutional 3D) for example, proposed by Du Tran, et al. [6] uses three-dimensional convolutions for extracting spatiotemporal features from videos. I3D (Infilated 3D ConvNet) [7], on the other hand, introduced by Joao Carreira et al. combines the pre-trained weights of a 2D convolutional neural network (e.g., ResNet or Inception) with three-dimensional convolutions for video recognition. The Two-Stream Networks, a family of architectures proposed by Karen Simonyan et al. [8] initially proposed the Two-Stream CNN model, which combines an input stream for static (spatial) images with a stream for optical (temporal) streams. In [9] Lee et al. describe an approach using deep learning techniques, specifically a convolutional neural network (CNN), to extract relevant features from television programs. These features are then used to predict online video advertising inventory, enabling advertisers and broadcasters to optimize their advertising planning. For predicting online video advertising inventory based on television programs. The objective of the study was to develop a predictive model that can estimate advertising inventory for online videos based on television programs, However, it has limitations in terms of inventory per time period in a program unit, and the scope of application to online video channels related to television programs is limited. Narducci et al. [10] present a comparison of machine learning-based approaches for TV program retrieval and classification. It examines different machine learning techniques and features used and evaluates their performance on a specific dataset. The personalization scenario considered in the study has a low number of available training examples. This may pose a challenge for machine learning algorithms, as it may be more difficult to obtain reliable results with a limited number of training data. Gomes et al. [11] propose the effectiveness of using deep neural networks for IMDb score prediction of television series, focusing on the specific case of the "Arrow" series. The authors also analyze the challenges and considerations of applying deep learning techniques to this specific prediction problem. However, the study focuses on the specific case of the TV series "Arrow." However, the model's ability to predict IMDb score may vary when applied to other television series with different characteristics. This may limit the generalizability of the study's findings to other television series. Moreover, the state of the art has seen methodologies that originated for purposes other than video recognition such as Transformer [12] neural networks, which have become popular in natural language processing but have also been applied to video classification, employed for this task. In this case, video frames are treated as sequences of tokens, and attention operations are applied to capture spatial and temporal relationships between frames. These developments have made it possible to apply video recognition by deep learning to various problems such as object identification [13], motion detection [14], video surveillance systems [15], and human action recognition [16].

3 Methodology

This section introduces the methods so far proposed for classifying long-form videos from TV channels.

3.1 SSIM

The Similarity Structure Index Measure (SSIM) [17] is a metric used to evaluate the quality of images or videos. It is designed to evaluate the structural similarity between two images, taking into account three main components, brightness, contrast, and texture structure. The SSIM index is a function of these three components:

$$SSIM(x,y) = [l(x,y)^{\alpha} \cdot c(x,y)^{\beta} \cdot s(x,y)^{\gamma}]$$

$\gamma > 0, \beta > 0, \alpha > 0$ are used to modify the three relief components. To simplify the expression. Using the parameter $\alpha = \beta = \gamma = 1$.

SSIM with ResNet50 CNN. The architecture proposed in [18] is based on the use of an image comparison system using the Similarity Index Measure Structure (SSIM) with ResNet50 [19]. The use of SSIM technique was applied to avoid updating the training data of ResNet50 in the case of the insertion of a new program, and thus a new TV title sequence, by the broadcaster, making it easier to extract a single image for comparison rather than retraining ResNet50 as shown in Fig. 1. ResNet50 performs the function of classifying general example cases, pre-trained on ImageNet, the model is structured into a base model that is frozen during new training via backpropagation, and the head model that contains FullyConnected layers that unlike the base will receive training via backpropagation. Such a network is designed to operate on stacked frames where each frame of the video under consideration is compared with standard test images. Test images were selected from the same channels depicting the opening and closing title sequence of a program. Consider some test images representing a particular moment of detection of the program title sequence of dimensions $N \cdot N$ equal to those of the frames in the video. Consider some test images: $\sum_{i=1}^{n} x_i$ and video frames per second $\sum_{i=1}^{n} y_i$, we want to show a specific point that can be detected. We measure the similarity index between test images and video fps by obtaining a floating-point value: $SSIM(x_i, y_i) = f_n$. By setting a value t we calculate:

$$f_n > t$$

By setting t to a high value, it is possible to accurately recognize when an image of interest represents a program's title sequence. Once the image of interest is identified, ResNet50 begins classification on general content depicting human actions such as playing a sport.

The advantage of this methodology is that it is possible to accurately recognize the opening and closing instants of television programs through their title

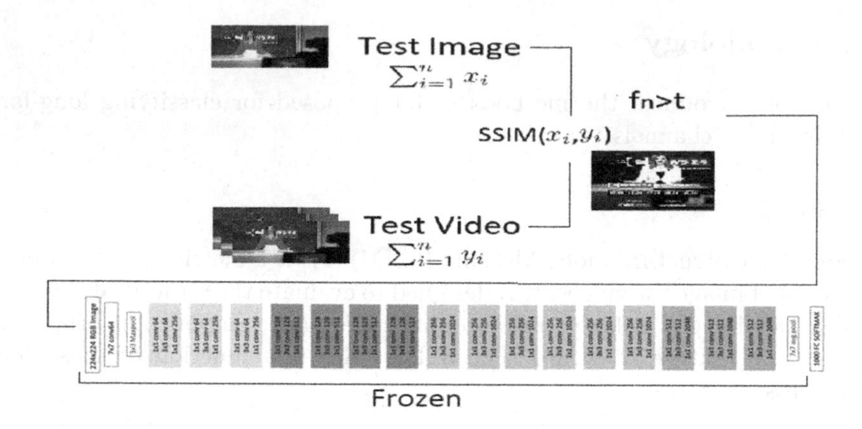

Fig. 1. SSIM with CNN Architecture.

sequences. In addition, by considering test images of title sequences for comparison with video through SSIM, it is possible to avoid creating new channel-specific datasets on the opening and closing title sequences, update ResNet50 training data in case a channel updates the title sequences of a program, it will be necessary to extract only two images of the closing and opening title sequence of the program. Training ResNet50 alone on the opening and closing titles of the program is also inconvenient since these are usually only a few seconds in duration and many frames turn out to be identical, this poses the problem of training data sparsity that cannot even be solved using the data augmentation technique. While effective, the proposed method has many limitations; the test images of the opening and closing titles must be the same size as the video which makes it unsuitable for different video formats.

LSTM Integration. Long Short-Term Memory (LSTM) [20] networks are a type of recurrent neural network (RNN) that have been introduced to overcome the limitations of traditional neural networks in modeling long-term data sequences, in this case allowing the optical stream provided by the SSIM and CNN framework to be worked on as training data for the network. This process allows the program to be stored by assigning it a higher-level classification tag that can be Tv Sports News. LSTMs have a special structure called a "memory cell" that allows them to maintain and update information over time. This memory cell can store relevant information for extended periods without degradation or discarding. Therefore, LSTMs are capable of retaining and recalling knowledge over the long term, even when past information is distant in time.

The recurrent network computes the sequence of the hidden vector:

$$h = (h_1, h_2, ..., h_n)$$

and the sequence of the output value:

$$y = (y_1, y_2, ..., y_n)$$

from an input sequence:

$$x = (x_1, x_2, ..., x_n)$$

and:

$$h_t = \sigma(W_{ih}x_t + W_{hh}h_{t-1} + b_h)$$

W denotes the weight matrices, b the polarization vectors, and σ the activation function of the hidden layer. The LSTM unit consists of three ports, an input port (i_t), governs the flow of new information to be added to the memory cell state. A forgetting port (f_t), governs which part of the previous state of the memory cell is to be forgotten or discarded and an output port (o_t), determines how much of the memory cell's state is to be output as output.

The advantage of this methodology is the approach of working directly on the optical flow, achieving higher accuracies than SSIM with ResNet50 working on individual frames. The methodology appears to be effective for recognizing the entire broadcast within a stream. In addition, through the proposed method it is possible to perform predictions on sports transmissions of different channels used for training even though it does not accurately identify the opening and closing instants of the broadcast. Although effective, the proposed methodology has limitations in terms of fixed video length, as LSTM networks require a fixed length of the input sequence, otherwise, segmentation techniques must be used. This limitation poses the classification for recordings whose content must already be known and thus know that there is a sports news program within the stream to be analyzed.

3.2 ShotBoundary Detection

The proposed approach aims to divide a video sequence into smaller subsequences and then segment the whole video. By analyzing scene changes, the video is divided into sub-sequences such as $x1, x2, x3$, where each sub-sequence represents an independent semantic event that will be used as input to the neural network.

To preprocess the video, the first step is to use the background and object subtraction technique to identify the boundaries of the shots. This process generates a binary mask $M(i, j)$ shown in Fig. 2 that indicates the area of the image occupied by the foreground object. The mask is created by applying an adaptive threshold to the difference map between the current image $I(i, j)$ and the background model $B(i, j)$. Specifically, the adaptive threshold is determined based on the standard deviation μ of the difference map and is set as the mean σ plus a constant multiplied k by the standard deviation verifying condition:

$$M(i,j) = \begin{cases} 1 \text{ if } |I(i,j) - B(i,j)| > \mu + k \cdot \sigma \\ 0 \text{ oterwhise} \end{cases}$$

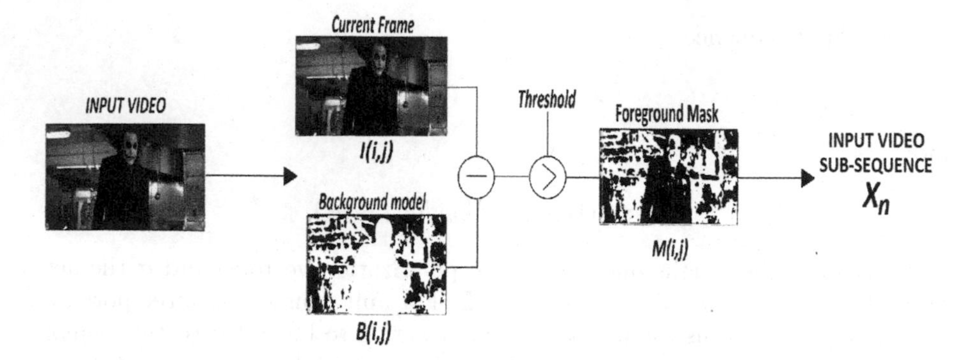

Fig. 2. Shot Boundary Detection.

ShotBoundary with ResNet50. The implementation of frame boundary detection with ResNet50 performs the function of classifying scene changes of video sub-sequences, which are significant for identifying the beginning or end of a TV program. In this case, the network is trained on images including landscapes, studio reports, games, sales, commercials, advertisements, and theme songs. This implementation has training data that allows the approach to be generalized to multiple television stations. The advantage of this method is that it allows us to classify a type of program within a television broadcast; it is adaptable to multiple stations, but it is limited by the fact that the proposed methodology does not capture the beginning and ending moments of a broadcast, and the predominant scene changes within a program may be inaccurately unclassified because ResNet50 works on classifying individual frames and not the entire optical stream.

ShotBoundary with Transformers. The proposed framework works on optical flow. Transformer networks take advantage of the global attention mechanism, which enables them to take into account the long-distance connections between individual video frames. This is especially helpful when there is a chance that non-local distribution of classification-relevant information in the video. On the other hand, CNN may struggle to recognize long-distance links because they are built to discover local patterns. Given a video $\sum_{i=1}^{n} Y_i$ where Y_i denotes the total number of fps in a video, with shot boundary detection using the background/object subtraction technique, we divide it into sub-sequences of duration dependent on the new scene change. The sub-sequences will be the input to the Transformer that will classify each scene change. Deep Transformer networks are based on stacking multiple layers of Transformer modules. Each module consists of two main sub-modules: the encoder and the decoder. The encoder processes the input and is divided into sequences of tokens, which are transformed into vectors by an embedding operation. These vectors represent the characteristics of the tokens. The input vectors are passed through a series of multiple attention blocks. Each multiple attention block calculates importance weights for each token based on its relationship to other tokens in the sequence, generating a weighted context for each token. After multiple attention, a feed-forward layer is applied to transform the intermediate contextual representations in a nonlinear manner. The multiple attention and feed-forward blocks are stacked multiple times to allow the encoder to learn increasingly complex representations of the data. The output of the last encoder block represents the final representation of the input.

The output of the encoder is used as the input to the decoder. The decoder initially, is given a sequence start token that is processed by an embedding operation. The decoder uses a combination of multiple attention and cross-attention to generate the output step by step. Multiple attention allows the decoder to focus on different parts of the contextual input, while cross-attention considers both the output generated up to that point and the output of the last encoder block. After attention, a feed-forward layer is applied to transform the intermediate representations non-linearly. The process of attention and feed-forward is repeated several times in the decoder. The output of the last decoder block is converted into a probability distribution over the possible output tokens, using a softmax function.

Working directly on video streams and not on impolite frames has great advantages for television program classification. In particular, predictions of classification scenes as advertisements turn out to be very accurate. Limitations arise in the choice of training data, particularly in the identification of program sequence titles that may be compromised and classified incorrectly or as advertisements. Therefore, the merging of classes described below was decided for the creation of the dataset.

4 Experimental Results

All experimental tests were done on a dedicated server bearing the following characteristics: Intel(R) Xeon(R) Gold 6126 CPU. @ 2.6 64 KiB BIOS, 64 GiB DIMM DDR4 System Memory, 2 x GV100GL [Tesla V100 PCIe 32 GB]. The suggested frameworks were developed with Python and the Keras package together with Tensorflow. The video classification test was conducted on the same channels. We considered the three local channels: LaC, Rtv, and Esperia Tv. We examined 2 24-h videos for each channel. Various datasets were considered for the experimental phase; in particular, the tests conducted were done on the following datasets:

SSIM with ResNet50. Two databases are examined in the experiment: one for SSIM[32] with test images representing the beginning and end of TV theme songs. The fps of the test video have the same format as the test images. The second dataset for CNN was obtained from [33]. It contains five classes of images: basketball (495 images), soccer (799 images), formula1 (638 images), MotoGP (679 images), and tennis (718 images). The latter has been divided into 80% and 20% for training and testing, respectively.

SSIM with ResNet50+LSTM. Data training was carried out by whole-program recognition using the SSIM with ResNet50 technique, in an attempt to generalize the recognition of programs from multiple television channels, through a post-processing script. By selecting the program start and end intervals on file.csv (renamed according to the video examined) produced by the SSIM with ResNet50 result, it was possible to intercept the data stream of sports news programs. Specifically, the following were selected for training data creation: 2 test images, one for the opening title sequence and one for the closing title sequence of the sports newscasts of LACTV, ESPERIATV, and REGGIO TV. ResNet50 was trained on the datasets mentioned in the previous paragraph. The datasets created include TV SPORTS NEWS LACTV (53 videos), TV SPORTS NEWS ESPERIATV (43 videos), and TV SPORTS NEWS RTV (48 videos).

ShotBoundaryDetection with ResNet50. ResNet training was done on specially created data such as Geo documentaries (826), Religious events (769), Game shows (525), Talk shows (685), Sales (470), commercial_advertising_television_signs (560).

ShotBoundaryDetection with Transformer. Transformers training was done on specially created data from the same TV stations considered for the test: Geo documentaries(152), Religious events(132), Gameshow (145), Talk shows (129), Sales(191), TV news(123), and football, basketball, and tennis merged into one sports folder(308).

4.1 Performance Evaluation Metrics

The following metrics are used to measure the performance of the proposed framework by indicating with P a positive condition, with N a negative condition, then:

ResNet50

- TP indicates the number of correctly identified frames.
- FP indicates the number of frames classified differently.
- TN frames not correctly classified.
- FN denotes not classified frames or anomalies.

LSTM and Transformer

- TP indicates the number of correctly identified scenes.
- FP indicates the number of differently identified scenes.
- TN denotes incorrectly identified scenes.
- FN denotes unidentified scenes or anomalies.

The performance of the framework was thoroughly evaluated using:

$$Accuracy : \frac{TP + TN}{TP + TN + FP + FN}$$

Table 1 shows the results of the various frameworks in terms of accuracy on the training of 120 epochs.

Table 1. Network accuracy

METHODOLOGY	ACCURACY
SSIM with ResNet50	95%
SSIM with ResNet50+LSTM	94%
Shot Boundary Detection with ResNet50	94%
Shot Boundary Detection with Transformer	88%

4.2 Evaluation of the Proposed Methods

The proposed methodologies while analyzing their pros and cons have achieved excellent results. A study showing results compared with the proposed methodologies can be seen in Table 2. In [21] a structure based on deep learning with CNN and MLP for BBC TV programs in nine genres that correspond to BBC gender categories, 93.7% results were achieved. In Table 2 we also compare the proposed methods with the state of the art of video classification, the proposed examples most relevant to the study used UCF-101, which is significant for human actions within television programs, furthermore, the extent of the datasets used for experimentation shows how the described techniques can be implemented for other purposes as well.

Table 2. Network accuracy

METHODOLOGY	*ACCURACY*
SSIM with ResNet50	95.0%
SSIM with ResNet50+LSTM	94.3%
Shot Boundary Detection with ResNet50	94.0%
Shot Boundary Detection with Transformer	88.0%
CNN MLP [21]	93.7%
LSTM with 30 Frame Unroll (Optical Flow + Image Frames) [22]	88.6
Two-Stream CNN (Optical Flow + Image Frames, Averaging) [23]	86.9
Res3D [24]	85.8

5 Conclusion

The following work demonstrates the continuous improvement and search for innovative solutions to solve problems on television broadcast recognition, seeking a solution that can be generalized to multiple broadcasters. The results obtained so far show that the use of deep learning for television program classification has reached very high levels of accuracy. Through deep learning, the models can automatically learn the discriminating characteristics of television programs, enabling them to make accurate and consistent predictions about a wide variety of content. As can be seen, there is a strong dependence on training data. The goal posed is to be able to generalize the whole process adaptable not only to local or national broadcasting horizons but to the audiovisual world, this goal could pose new frontiers for self-learning developments.

References

1. https://www.agcom.it/documents/10179/539063/Allegato+12- . Accessed 13 Nov 2008
2. Deng, J., Dong, W., Socher, R., Li, L.-J., Li, K., Fei-Fei, L.: ImageNet: a large-scale hierarchical image database. In: IEEE Conference on Computer Vision and Pattern Recognition 2009, pp. 248–255 (2009). https://doi.org/10.1109/CVPR.2009.5206848
3. Soomro, K., Zamir, A.R., Shah, M.: UCF101: a dataset of 101 human actions classes from videos in the wild. arXiv preprint arXiv:1212.0402 (2012)
4. Li, F., et al.: Temporal modeling approaches for large-scale Youtube-8m video understanding. https://arxiv.org/abs/1707.04555. Accessed Jul 2017
5. Fukushima, K.: Neocognitron: a hierarchical neural network capable of visual pattern recognition. Neural Netw. **1**(2), 119–130 (1988)
6. Tran, D., et al.: Learning spatiotemporal features with 3d convolutional networks. In: Proceedings of the IEEE International Conference on Computer Vision (2015
7. Carreira, J., Zisserman, A.: Quo vadis, action recognition? A new model and the kinetics dataset.. In: Proceedings of the IEEE Conference on Computer Vision and Pattern Recognition (2017)
8. Simonyan, K., Zisserman, A.: Two-stream convolutional networks for action recognition in videos. In: Advances in Neural Information Processing Systems, vol. 27 (2014)
9. Lee, S.-H., Yoon, S.-H., Kim, H.-W.: Prediction of online video advertising inventory based on TV programs: a deep learning approach. IEEE Access **9**, 22516–22527 (2021)
10. Narducci, F., et al.: TV-program retrieval and classification: a comparison of approaches based on machine learning. Inf. Syst. Front. **20**, 1157–1171 (2018)
11. Gomes, A.L., et al.: Predicting IMDb rating of TV series with deep learning: the case of arrow. In: XVIII Brazilian Symposium on Information Systems (2022)
12. Vaswani, A., et al.: Attention is all you need. In: Advances in Neural Information Processing Systems, vol. 30 (2017)
13. Ouyang, W., Zeng, X., Wang, X., et al.: DeepID-net: object detection with deformable part based convolutional neural networks. IEEE Trans. Pattern Anal. Mach. Intell. **39**(7), 1320–1334 (2017)
14. Doulamis, N., Voulodimos, A.: FAST-MDL: fast adaptive supervised training of multi-layered deep learning models for consistent object tracking and classification. In: 2016 IEEE International Conference on Imaging Systems and Techniques (IST). IEEE (2016)
15. Ramachandra, B., Jones, M., Vatsavai, R.: Learning a distance function with a Siamese network to localize anomalies in videos. In: Proceedings of the IEEE/CVF Winter Conference on Applications of Computer Vision, pp. 2598–2607 (2020)

16. Jaouedi, N., Boujnah, N., Bouhlel, M.S.: A new hybrid deep learning model for human action recognition. J. King Saud Univ. Comput. Inf. Sci. **32**(4), 447–453 (2020)
17. Wang, Z., Bovik, A.C., Sheikh, H.R., Simoncelli, E.P.: Image quality assessment: from error visibility to structural similarity. IEEE Trans. Image Process. **13**(4), 600–612 (2004)
18. Candela, F., Morabito, F.C., Zagaria, C.F.: Television programs classification via deep learning approach using SSMI-CNN. In: Mahmud, M., Ieracitano, C., Kaiser, M.S., Mammone, N., Morabito, F.C. (eds.) AII 2022. CCIS, vol. 1724, pp. 293–307. Springer, Cham (2022). https://doi.org/10.1007/978-3-031-24801-6_21
19. Krishna, S.T., Kalluri, H.K.: Deep learning and transfer learning approaches for image classification. Int. J. Recent Technol. Eng. (IJRTE) **7**(5S4), 427–432 (2019)
20. Hochreiter, S., Schmidhuber, J.: Long short-term memory. Neural Comput. **9**(8), 1735–1780 (1997). https://doi.org/10.1162/neco.1997.9.8.1735
21. Pham, L., et al.: An audio-based deep learning framework for BBC television programme classification. In: 2021 29th European Signal Processing Conference (EUSIPCO). IEEE (2021)
22. Ng, J.Y.-H., et al.: Beyond short snippets: deep networks for video classification. In: Proceedings of the IEEE Conference on Computer Vision and Pattern Recognition (2015)
23. Karpathy, G.T., Shetty, S., Leung, T., Sukthankar, R., Fei-Fei, L.: Large-scale video classification with convolutional neural networks. In: Proceedings of CVPR, Columbus, Ohio, USA, pp. 1725–1732 (2014)
24. Tran, D., Ray, J., Shou, Z., Chang, S.-F., Paluri, M.: Convnet architecture search for spatiotemporal feature learning. arXiv:1708.05038 (2017)

Process Mining Solutions for Public Administration

Simona Fioretto[(✉)]

University of Naples Federico II, 08544 Naples, Italy
simona.fioretto@unina.it

Abstract. Artificial Intelligence and the digital transformation of Public Administration are two of the main topics of current years. Process management in Public Administration is crucial to ensure the execution and the effectiveness of back-end and front-end e-services. While Artificial Intelligence is quite common throughout various sectors, the application of AI in Public Administration is still under development and continuous improvement. The use and application of AI techniques could improve the digital transformation and the effectiveness of public administration; among them Process Mining techniques could help in the discovery and management of business processes. The goal of this paper is to show the current AI application for PA, and the process discovery technique which could practically improve PA effectiveness.

Keywords: Artificial Intelligence · Process mining · Public Administration

1 Introduction

The digital transformation process of Public Administration in European countries is not yet complete. The efficiency of PA services is the most important growth factor for a country's economy: citizens need the services of public administration in many cases, from the completion of everyday tasks to the opening of new projects or jobs. Clearly, a faulty PA service can have a strong impact on the lives of citizens and the economic growth of the country. In this context can be helpful to highlight the definition of E-government given by [5] which is "he application of ICT within the public administration to optimize its internal and external functions. Also known as e-gov, digital government, online government or in a certain context transformations government, it refers to the use of Internet technology as a platform for exchanging information, delivering services and transacting with citizens, businesses, and other arms of government. E-Government can be used by the legislature, judiciary, or administration, in order to improve internal efficiency, the delivery of public services, or processes of democratic governance. The primary delivery models are Government-to-Citizen or Government-to-Customer (G2C), Government-to-Business (G2B), Government-to-Government (G2G) and Government-to- Employees (G2E)."

Supported by University of Naples Federico II.

A. Abelló et al. (Eds.): ADBIS 2023, CCIS 1850, pp. 668–675, 2023.
https://doi.org/10.1007/978-3-031-42941-5_60

The integration of AI technologies into service delivery models could bring great benefits and public value to citizens, depending on how they are used [6]. For these reasons, Europe is paying close attention to the digitalisation of public administration and is focusing its efforts and investments on the integration of innovative solutions such as artificial intelligence in this sector. In fact, the evaluation of the 2023 E-Government Development Index, which is the weighted average of three independent component indices: the Online Services Index (OSI), the Telecommunications Infrastructure Index (TII) and the Human Capital Index (HCI) [5], shows that the European Union is a leader in the implementation of e-government applications.

The delivery of quality government services depends on the appropriate and efficient execution of processes. As highlighted, AI tools can be used to improve the quality of government services and Process Mining is an area of AI that uses data to obtain the model of the real process execution in a given environment. Process Mining can be helpful in modelling the real behaviour of a process and identifying the associated bottlenecks and their causes in the real work activities. Using the information contained in event logs, PM is able to extract the processes, analyse their performance and improve their efficiency. PA has many information systems where data and information is stored, from which it is possible to extract valuable information for improving PA work activities, discovering processes and supporting PA digital transformation.

2 Research Objectives

The purpose of this section is to provide an overview of the gaps and problems on which the research of the PhD is focused. The context of this research is based on a national project dealing with the improvement of public administrations and is currently in the early stages, being the first year of research. After an initial finding based on practical experience in a public administration and a review of European objectives and projects related to the introduction of AI tools and innovative solutions for the digitisation process of PA, the specific area of process management was identified as one of the first problems causing slowness, delays and errors in the PA workflow. By studying and improving PA processes, it is possible to find the most critical tasks and activities related to PA problems. Process mining can actually help in this research by discovering, exploring and changing PA processes.

The AI solution can be integrated to support public administrations in the internal management of business processes and thus in organisational management. In particular, it will be used for the most effective implementation of services. In fact, the management of business processes can be improved through the use of process mining, which can support the discovery of the process and its associated bottlenecks and repetitive activities.

PM can answer some questions about the PA:

- What are the actual business processes of the organisation?
- Are there some actions that can be removed or even not repeated?
- Is every action in the process necessary?

- Can the time taken to complete the process be reduced?

The process mining applications will be used at different stages of application and maturity of the research project:

- In the initial phase, the goal is to extract the processes and then use process discovery;
- in a second phase, when you know the processes that are performed and those that should be realised, the goal is to use conformance checking applications.
- In the future, it is hoped to be able to carry out process enhancement.

In conclusion, the main goal of this thesis is to find out how to implement Process Mining in the Public Administration domain.

3 Related Work

This section focuses on the analysis of public administration digital transformation and AI applications, as well as current process mining applications in public administration.

3.1 Artificial Intelligence for Public Administration

This section aims to provide a brief overview of the context of AI in public administration. The application of artificial intelligence in e-Government services is a relatively new field that lacks description. Some authors have given an overview of these applications, trying to systematise them in different categories. In [11] the organisation of the current AI studies in public administration is made, distinguishing 5 categories:

1. AI public service: deals with AI applications aimed at improving the time and quality of services offered;
2. Working and social environment influenced by AI: includes the study of the impact of AI on the social environment;
3. AI for public order and law: these two areas can be supported by AI applications for predictive models to minimise damage and casualties from natural disasters and to support surveillance by public authorities;
4. AI for ethics: assessing benefits and threats to society;
5. AI for government policy: assessing laws and policies that AI might consider and how they might adapt to AI applications.

The authors [4] as part of the landscaping task of the AI-Watch of the European Commission started beginning of 2019 give an overview of the current implementations of Artificial Intelligence in public services. After a systematic review they conclude that most of the AI-technologies currently in use are a form of Natural Language Processing technology, such as Chatbots or Speech Recognition (29%), followed by Pattern recognition (25%) and Image processing (20%), so all AI technologies that want to have an automatic recognition in order to provide more accurate predictions. Indeed regarding the typology of policy sector, most of the AI-adoptions in Europe seem to take place in the “General Public Services” domain rather than in other policy sectors.

3.2 Process Mining in Public Administration

The application of Process Mining in Public Administration is not very common in literature. Also [1] highlight the importance of Process Mining applications in PA, showing with real application in municipalities that conformance checking is relevant as compliance in public administration particularly due to many regulations which have legal deadlines consequently having temporal constraints.

Regarding the applications of PM in PA, [8] highlights that process mining could not only contribute to the optimization of administrative processes in the public sector, but could also take a more citizen-centered viewpoint including considering the customer-specific parts of the process and the interactions between authorities that are relevant to the customer. Similarly, the customer experience in financial transactions be optimized by process mining.

An increasing area of application of PM in PA is related to health sector. In [7] the authors through a literature review show PM application to Hospital Information Systems. Other applications are related to the extraction of business processes from documents in which there are policies. These are called policy based extraction, and tend to use Natural Language Processing techniques for process discovery from documents and policies. [3] proposes a new kind of extraction, moving from event logs to policies, calling it Policy-based Process Mining (PBPM), to automatically extract process information from policy documents in text; an important role and step forward was done by [2] who extracts the process from a text without making assumptions on text structure, furthermore makes a proper anaphora resolution to identify concepts that are referenced using pronouns and articles.

4 Method

In this section Process Mining is shown and analysed in order to give a brief introduction to the technique.

4.1 Process Mining Technique

The term Process Mining was introduced by Wil Van der Aalst in a research proposal. Process mining is a field relatively new, which can be viewed as an intersection between data science and process science where "process science" refers to the discipline that combines knowledge from information technology and knowledge from management science to improve and run operational processes [9]. The PM idea is based on knowledge extraction from event logs present in today's information systems; however, this data, which are stored in unstructured form, often need laboriously extraction.

Event logs can be used to conduct three types of process mining:

- **discovery** technique takes an event log and produces a model without using any a-priori information;

- **conformance** technique compares an existing process with an event log of the same process;
- **enhancement** technique improves an existing process model using information about the actual process recorded in event log, making changes to the a-priori model.

To talk about process mining we must identify the case and the activity. In the following context, a "case" refers to a specific instance or occurrence of a process. It represents a unique execution through the process. A case consists of a series of events or tasks that run in a specific order and are associated with a common identifier. The activity represents a distinct step within a process.

The minimal requirements for process mining are that any event can be related to both a **case** and an **activity** and that events within a case are ordered, because without ordering information it is impossible to discover causal dependencies in process models [9].

4.2 Process Discovery

The aim of this subsection is to give an overview and brief introduction to process discovery.

A Brief Introduction to Process Discovery. Discovery task and control-flow perspective are known as process discovery. Discovery brings as input an event log and gives as output a model which represents a process in different notations (Petri Net, BPMN, EPC or UML). The aim of process discovery is to extract a process and map this process from event logs. Given the definitions of event logs in [9] it is possible to sum up some concepts of process discovery.

Given an event log, an algorithm of process discovery is a function γ that maps this event log into a process model. The general process discovery problem is defined as follow [9]:

Definition 1. *Let L be an event log, a process discovery algorithm is a function that maps L onto a process model such that the model is "representative" for the behavior seen in the event log. The challenge is to find such an algorithm.*

The problem is to find this algorithm. This definition is quite general, so to make it more concrete [9] make assumptions on the target format and on the event log and reformulate this definition in a specific process problem discovery, but it is not the purpose of this paper to go into the detail of the formulation.

Function γ defines a so-called "Play-In" technique, which takes an input with the aim of constructing a model. Play-In is often referred to as inference.

An example of the mentioned γ function is the α -algorithm. Given a simple event log it produces a Petri net that could replay the log. The α -algorithm was one of the first process discovery algorithms that could adequately deal with concurrency, however, the α -algorithm has problems with noise, infrequent/incomplete behavior, and complex routing constructs [9] so it is not a real practical application of process discovery.

Typical characteristics of process discovery algorithms are representational bias, the class of process models that can be discovered, ability to deal with noise, completeness assumptions and approach used. Regarding the approach there are five families of approaches: Direct Algorithmic Approaches, Two-Phase Approaches, Divide-and-Conquer Approaches, Computational Intelligence Approaches, Partial Approaches.

Existing framework make it possible to use different tools on the same data set and to compare the mining results. Among them the most widely shared is ProM [10].

The process discovery requires that the model discovered should be "representative" for the behavior seen in the event log. The above requirement is the so-called "fitness" requirement [9]. There must exists a trade-off among the following four quality criteria:

- fitness: the discovered model should allow for the behavior seen in the event log;
- precision: the discovered model should not allow for behavior completely unrelated to what was seen in the event log;
- generalization: the discovered model should generalize the example behavior seen in the event log;
- simplicity: the discovered model should be as simple as possible.

Based on the above criteria and the objective of the research, appropriate evaluation metrics are chosen.

5 Future Work

This section explores a possible future direction for the research project.

The current state of the art in the use of process mining in public administration lacks of several real-world applications. E-services need to be improved for primary delivery models (G2C, G2B, G2G, G2E). Process Mining could be used to improve such services, from process discovery to conformance checking and process enhancement. There are some open questions that are still object of observation for this research and that must be explored and studied yet in order to do process mining.

- evaluate availability and type of real event log;
- evaluate the type of algorithm to be used for process discovery;
- alternatively use simulated processes;
- assessing the impact of the lack of interoperability among PA systems on the application of PM;

The future work could be summarised as follow:

- working on event log;
- choosing the discovery algorithm;
- choosing the right evaluation metrics;

– implement and simulate process mining.

In addition, since for conformance checking there is the need to check if a process is respective to imposed policies, it could also be interesting to use both policy-based process discovery and process discovery for process mining.

6 Conclusion

The aim of this paper is to show the application of Artificial Intelligence in public administration and process mining techniques in the same field. The need to improve e-government services, also by integrating AI solutions, pays attention to some AI solutions that can accelerate the digital transformation process. Then, PM is presented as one of the possible techniques to achieve the aforementioned goals. Finally, through the structure of a future work, the paper tries to show the still open questions and the possible future research directions.

Acknowledgment. This publication article was produced while attending the PhD programme in Information and Communication Technology at the University of Naples, Cycle XXXVIII, with the support of a scholarship financed by the Ministerial Decree no. 351 of 9th April 2022, based on the NRRP - funded by the European Union - NextGenerationEU - Mission 4 "Education and Research", Component 1 "Enhancement of the offer of educational services: from nurseries to universities" - Investment 3.4 "Advanced teaching and university skills" OR Investment 4.1 "Extension of the number of research doctorates and innovative doctorates for public administration and cultural heritage".

References

1. van der Aalst, W.M.: Process discovery: capturing the invisible. IEEE Comput. Intell. Mag. **5**(1), 28–41 (2010)
2. Friedrich, F., Mendling, J., Puhlmann, F.: Process model generation from natural language text. In: Mouratidis, H., Rolland, C. (eds.) CAiSE 2011. LNCS, vol. 6741, pp. 482–496. Springer, Heidelberg (2011). https://doi.org/10.1007/978-3-642-21640-4_36
3. Li, J., Wang, H.J., Zhang, Z., Zhao, J.L.: A policy-based process mining framework: mining business policy texts for discovering process models. IseB **8**, 169–188 (2010)
4. Misuraca, G., van Noordt, C., Boukli, A.: The use of AI in public services: results from a preliminary mapping across the EU. In: Proceedings of the 13th International Conference on Theory and Practice of Electronic Governance, pp. 90–99 (2020)
5. Nations, U.: United nation e-government survey 2022 (2022). https://publicadministration.un.org/egovkb/en-us/Reports/UN-E-Government-Survey-2022
6. van Noordt, C., Misuraca, G.: Artificial intelligence for the public sector: results of landscaping the use of AI in government across the European union. Gov. Inf. Q. **39**(3), 101714 (2022)
7. Rojas, E., Munoz-Gama, J., Sepúlveda, M., Capurro, D.: Process mining in healthcare: a literature review. J. Biomed. Inform. **61**, 224–236 (2016)

8. Thiede, M., Fuerstenau, D., Bezerra Barquet, A.P.: How is process mining technology used by organizations? A systematic literature review of empirical studies. Bus. Process. Manag. J. **24**(4), 900–922 (2018)
9. Van Der Aalst, W.: Process Mining: Data Science in Action, vol. 2. Springer, Heidelberg (2016). https://doi.org/10.1007/978-3-662-49851-4
10. van Dongen, B.F., de Medeiros, A.K.A., Verbeek, H.M.W., Weijters, A.J.M.M., van der Aalst, W.M.P.: The ProM framework: a new era in process mining tool support. In: Ciardo, G., Darondeau, P. (eds.) The prom framework: a new era in process mining tool support. LNCS, vol. 3536, pp. 444–454. Springer, Heidelberg (2005). https://doi.org/10.1007/11494744_25
11. Wirtz, B.W., Weyerer, J.C., Geyer, C.: Artificial intelligence and the public sector-applications and challenges. Int. J. Publ. Adm. **42**(7), 596–615 (2019)

Intelligent Technologies for Urban Progress: Exploring the Role of AI and Advanced Telecommunications in Smart City Evolution

Enea Vincenzo Napolitano(✉)

Department of Electrical Engineering and Information Technology, University of Naples Federico II, Naples, Italy
eneavincenzo.napolitano@unina.it

Abstract. The rapid urbanization and associated challenges necessitate the development of smart cities that leverage intelligent technologies, such as artificial intelligence (AI) and advanced telecommunications, to drive urban progress. In this paper the author describes one of the possible developments of his PhD research, the study aims to propose a framework for integrating AI and advanced telecommunications into smart city infrastructure to enhance urban services, optimize decision-making processes, and improve the overall quality of life. By examining the benefits, challenges, and future prospects, this research provides insights into the potential of intelligent technologies in shaping the cities of tomorrow. Ultimately, this paper contributes to the understanding of how AI and advanced telecommunications can drive urban progress and foster the sustainable evolution of smart cities.

Keywords: Smart City · Artificial Intelligence · Advanced Telecommunications Network

1 Introduction

With the rapid acceleration of urbanisation, cities around the world are facing unprecedented challenges in areas such as traffic management, waste management, energy use and security. These challenges, characterised by their complexity and multi-dimensionality, cannot be effectively addressed by traditional urban planning and management methods alone. The emergence of the concept of 'smart cities', where advanced technology is seamlessly integrated into urban infrastructure, has the potential to revolutionise urban planning, living and governance. In this context, advances in technologies such as Artificial Intelligence (AI) and advanced telecommunications mark a new era in urban progress. This paper aims to explore this potential, focusing on the possible development trajectories of the author's doctoral research.

As a key driver of the Fourth Industrial Revolution, AI has the transformative power to fundamentally reshape urban life. Advanced telecommunications, as

A. Abelló et al. (Eds.): ADBIS 2023, CCIS 1850, pp. 676–683, 2023.
https://doi.org/10.1007/978-3-031-42941-5_61

the backbone of digital connectivity, amplifies these effects by enabling real-time data transfer and seamless connectivity. However, the integration of these technologies into urban infrastructure is still in its early stages and presents numerous challenges that require careful consideration.

The aim of this paper is to contribute to the understanding of how AI and advanced telecommunications can catalyse urban progress and promote the sustainable development of smart cities. The goal goes beyond simply improving efficiency and quality of life to creating urban landscapes that are resilient, adaptable and sustainable in the face of changing demographic and climatic conditions.

By exploring the potential of AI and advanced telecommunications in smart city development, this research aims to shed light on innovative approaches and strategies that can shape future urban environments. By addressing the challenges and harnessing the opportunities presented by these technologies, cities can strive for sustainable, inclusive and resilient urban progress. Through the author's doctoral research, this paper aims to contribute valuable insights and propose pathways towards the realisation of intelligent, connected and sustainable smart cities.

2 Smart City

The term 'smart city' encompasses a broad and evolving concept involving the integration of advanced technologies and data-driven solutions to improve the quality of life, sustainability and efficiency of urban areas [11]. Smart cities make use of digital technologies such as the Internet of Things (IoT) [22], artificial intelligence [14] and advanced analytics to optimise the use of resources, improve infrastructure and enhance the delivery of public services. These technologies enable the collection, analysis and use of real-time data to inform decision-making and improve the overall functioning of cities [3].

The following are the main urban advances of smart cities: Sustainability, energy efficiency, smart mobility and efficient resource management.

Sustainability is a fundamental pillar of smart cities. It involves the adoption of practices that minimise environmental impact [23], promote social inclusion and ensure economic prosperity. Smart cities strive to achieve environmental sustainability by integrating renewable energy sources, implementing energy efficient technologies and adopting sustainable urban planning strategies [21]. In addition, social sustainability focuses on creating inclusive communities that prioritise equal access to resources, services and opportunities for all residents.

Energy efficiency plays a critical role in smart cities due to the growing demand for energy and the need to mitigate climate change [27]. Smart cities use advanced technologies to optimise energy consumption in buildings, transport systems and public infrastructure. This includes smart grid systems that enable efficient distribution and management of energy, smart meters that monitor energy usage, and intelligent building systems that optimise energy consumption based on occupancy and demand [2]. By promoting energy efficiency, smart cities can reduce greenhouse gas emissions, lower energy costs and improve energy resilience.

Intelligent mobility is another key feature of smart cities, focusing on transforming transport systems to be more efficient, accessible and sustainable [12]. Smart cities use technologies such as intelligent transport systems, connected vehicles and shared mobility platforms to improve traffic flow, enhance public transport services and promote active and alternative modes of transport. Intelligent mobility solutions enable real-time monitoring of traffic conditions, optimise routes and promote seamless integration between different modes of transport [15]. The aim is to reduce congestion, emissions and travel times, while improving the overall transport experience for residents.

Efficient resource management is essential for the long-term sustainability of smart cities. It involves the effective and sustainable use of resources such as water, waste and land. Smart cities use technologies such as sensors [28], data analytics [17], and automation to monitor and manage resource consumption, optimise waste management systems, and enable efficient use of water and land. By adopting smart resource management practices, smart cities can reduce waste, conserve natural resources, and minimise environmental impact.

By integrating these key aspects, smart cities can create sustainable, resilient and liveable urban environments that improve the well-being and quality of life of their residents. The integration of advanced technologies, data-driven solutions and sustainable practices paves the way for the evolution of smart cities, fostering innovation, efficiency and a better future for urban communities.

3 AI and Advanced Telecommunications in Smart Cities

The integration of AI and advanced telecommunications has emerged as a transformative force in shaping the development of smart cities, driving urban progress and improving the quality of life for citizens.

3.1 AI in Smart Cities

AI technologies play a critical role in enabling smart cities to harness the vast amounts of data generated from multiple sources such as sensors, devices and urban systems [4]. By applying machine learning algorithms and data analytics techniques [24], cities can extract valuable insights and patterns from these data sets. This enables informed decision-making and resource optimisation in key areas such as urban planning, traffic management, energy consumption and emergency response. AI-driven data analytics enable cities to proactively address challenges and improve the overall efficiency and effectiveness of city services.

Within smart cities, AI contributes to the development of intelligent infrastructure [6]. By using AI algorithms, cities can optimise the operation and maintenance of critical infrastructure systems. For example, in the energy sector,

AI enables the prediction of energy demand patterns [29], facilitating the efficient distribution and management of electricity through smart grids. AI is also improving transport systems [25] by enabling real-time traffic monitoring, intelligent routing and adaptive traffic management. In addition, AI-driven solutions enable efficient waste management [5] by optimising waste collection routes and schedules based on real-time data analysis.

AI technologies also have significant potential to revolutionise governance processes [19] in smart cities. Through automation and machine learning, cities can streamline administrative tasks, improve service delivery and enhance public safety. AI-powered chatbots and virtual assistants provide personalised and efficient support to citizens, enabling 24/7 access to information and services [8]. Data-driven decision-making is further facilitated by AI, providing policymakers with predictive analytics and modelling capabilities. This enables cities to implement more effective urban management strategies and address the complex challenges faced by urban environments.

3.2 Advanced Telecommunications in Smart Cities

Advanced telecommunications infrastructure, including technologies such as 5G networks and fibre broadband, play a critical role in supporting the seamless communication and connectivity required in smart cities [26]. These advanced networks enable real-time data exchange and support the Internet of Things ecosystem within smart cities. With improved connectivity, cities can deploy a wide range of applications and services, including smart grids, intelligent transport systems, remote healthcare and real-time monitoring systems.

In addition, advanced telecommunications support the deployment of sensor networks and IoT devices [20], which are integral to the collection and transmission of real-time data in smart cities. These sensors collect information on air quality, noise levels, traffic patterns, waste management and more. The data collected from these sources can be analysed using AI algorithms, enabling cities to gain valuable insights into city operations, make data-driven decisions and optimise resource allocation [26]. The integration of advanced telecommunications with sensor networks and IoT devices increases the overall intelligence and efficiency of smart city systems. In addition to infrastructure and data exchange, advanced telecommunications technologies foster greater citizen engagement in smart cities. Digital platforms, mobile applications and smart city portals enable citizens to access real-time information, participate in decision-making processes and provide feedback to city authorities. These platforms facilitate the delivery of e-governance services, smart citizen services and citizen-centric applications [1]. By leveraging advanced telecommunications, cities can create inclusive environments where citizens are empowered to contribute to urban development, making cities more responsive and citizen-centric. Overall, the integration of AI and advanced telecommunications brings significant advances to smart cities, enhancing their capabilities in decision-making, infrastructure optimisation, connectivity, and citizen engagement. However, challenges related to privacy, security and equitable access need to be addressed to ensure the responsible and inclusive use of these technologies in smart city contexts.

4 Future Directions and Research Opportunities

As the concept of smart cities continues to evolve and mature, the field offers a wealth of future directions and research opportunities. This section explores some of these areas and highlights the potential for further research and innovation. As AI technologies advance, there are promising opportunities to explore new applications in smart cities. These include the development of AI-based systems for intelligent decision making, predictive analytics and urban simulation. Future research can focus on improving the capabilities of AI algorithms, exploring novel techniques such as deep learning and reinforcement learning, and integrating AI with other emerging technologies to address complex urban challenges. In addition, exploring ethical considerations, transparency and accountability in AI applications in smart cities is crucial to ensure responsible and unbiased decision-making.

Edge Computing and IoT in Smart Cities. The proliferation of IoT devices and the exponential growth of data generated by them pose challenges in terms of data storage, processing and privacy. Edge computing, which brings computing and data storage closer to the source of data generation, offers a promising solution for smart cities [10]. Future research can explore the potential of edge computing architectures, investigate resource optimisation and data management strategies, and develop secure and scalable frameworks for efficient real-time data processing and analysis.

Privacy and Security in Smart Cities. As smart cities rely heavily on data collection and analysis, ensuring the privacy and security of sensitive information becomes critical [9]. Future research can focus on developing robust frameworks and protocols to protect privacy, ensure secure communication between devices and systems, and address cybersecurity risks [9]. In addition, research into privacy-enhancing technologies, such as differential privacy and secure data exchange protocols, can facilitate responsible data use while preserving citizens' privacy rights.

Citizen Engagement and Social Inclusion. Ensuring that technology-driven solutions in smart cities are inclusive and benefit all members of the community is paramount. Future research can focus on understanding and addressing challenges related to the digital divide and social inequalities within smart cities [16]. This includes exploring ways to increase citizen engagement, promoting participatory decision-making processes, and developing inclusive digital platforms and services that meet the diverse needs of the population [16]. In addition, research efforts can explore the social and cultural impacts of smart city technologies, fostering a better understanding of their impact on community dynamics and social cohesion.

Policy and Governance for Smart Cities. The development and implementation of effective policies and governance frameworks [19] are essential for the successful deployment of smart city technologies. Future research can explore the policy and regulatory landscape and investigate best practices for governance, privacy and data management in smart cities [13]. In addition, studies can focus on assessing the economic, social and environmental impacts of smart city initiatives, helping policy makers to make informed decisions and design sustainable urban policies [18].

Integration of Renewable Energy Sources. With an increasing focus on sustainability and reducing carbon emissions, the integration of renewable energy sources into smart cities is an exciting avenue of research. Future studies can explore innovative solutions for incorporating technologies such as solar, wind and geothermal energy into the energy infrastructure of smart cities [7]. This includes developing smart grid systems that efficiently manage the generation, distribution and consumption of renewable energy, exploring energy storage solutions and assessing the economic and environmental benefits of such integration.

In conclusion, there are many future directions and research opportunities in the field of smart cities. Advances in AI, IoT, renewable energy integration, privacy and security, citizen engagement and governance have the potential to shape the future of urban environments. By exploring these areas and addressing the associated challenges, researchers can contribute to the development of smarter, more sustainable and inclusive cities that improve the quality of life for their inhabitants. Continued research and innovation is essential to realise the full potential of smart cities and to meet the evolving needs of urban communities.

5 Conclusion

This paper has explored the role of AI and advanced telecommunications in the development of smart cities. As a first year Ph.D. student, the research presented in this paper is at an early stage and no concrete results have yet been obtained. However, the paper has provided an overview of the potential impact of AI and advanced telecommunications in shaping the future of smart cities.

The concept of smart cities is constantly evolving, driven by the integration of advanced technologies and data-driven solutions. In exploring future directions and research opportunities, several areas have emerged as promising avenues for further investigation. These areas include advanced AI applications in smart cities, edge computing and IoT, renewable energy integration, privacy and security considerations, citizen engagement and social inclusion, and policy and governance frameworks. By addressing these areas, researchers can contribute to the development of smarter, more sustainable and inclusive cities.

As a first year Ph.D. student, the research journey is just beginning, the research presented in this paper serves as a starting point and lays the foundation for future studies in the field of smart cities.

In summary, this paper has shed light on the potential of AI and advanced telecommunications in the context of smart cities, highlighting the need for further research to unlock their full potential. As research continues, it is expected that concrete results and new insights will emerge, contributing to the advancement of the field. By addressing the challenges and exploring the opportunities presented by AI and advanced telecommunications, researchers can contribute to the development of innovative solutions and policies that will shape the future of smart cities, making them more sustainable, efficient and liveable for their inhabitants.

Acknowledgments. Author's PhD is funded via *PNRR-Partenariato Esteso PE14-RESearch and innovation on future Telecommunications systems and networks (RESTART).*

References

1. Aguiar, A., Portugal, P.: 5g in smart cities. INESC TEC Sci. Soc. **1**(3) (2021)
2. Akcin, M., Kaygusuz, A., Karabiber, A., Alagoz, S., Alagoz, B.B., Keles, C.: Opportunities for energy efficiency in smart cities. In: 2016 4th International Istanbul Smart Grid Congress and Fair (ICSG), pp. 1–5. IEEE (2016)
3. Al Nuaimi, E., Al Neyadi, H., Mohamed, N., Al-Jaroodi, J.: Applications of big data to smart cities. J. Internet Serv. Appl. **6**(1), 1–15 (2015)
4. Allam, Z., Dhunny, Z.A.: On big data, artificial intelligence and smart cities. Cities **89**, 80–91 (2019)
5. Anagnostopoulos, T., et al.: Challenges and opportunities of waste management in IoT-enabled smart cities: a survey. IEEE Trans. Sustain. Comput. **2**(3), 275–289 (2017)
6. Aoun, C.: The smart city cornerstone: urban efficiency. Schneider Electric White Paper **1**, 1–13 (2013)
7. Atasoy, T., Akınç, H.E., Erçin, Ö.: An analysis on smart grid applications and grid integration of renewable energy systems in smart cities. In: 2015 International Conference on Renewable Energy Research and Applications (ICRERA), pp. 547–550. IEEE (2015)
8. Borda, A., Bowen, J.P.: Smart cities and digital culture: models of innovation. In: Museums and Digital Culture: New Perspectives and Research, pp. 523–549 (2019)
9. Braun, T., Fung, B.C., Iqbal, F., Shah, B.: Security and privacy challenges in smart cities. Sustain. Urban Areas **39**, 499–507 (2018)
10. Khan, L.U., Yaqoob, I., Tran, N.H., Kazmi, S.A., Dang, T.N., Hong, C.S.: Edge-computing-enabled smart cities: a comprehensive survey. IEEE Internet Things J. **7**(10), 10200–10232 (2020)
11. Komninos, N., Mora, L.: Exploring the big picture of smart city research. Scienze Regionali **17**(1), 15–38 (2018)
12. Lewicki, W., Stankiewicz, B., Olejarz-Wahba, A.A.: The role of intelligent transport systems in the development of the idea of smart city. In: Sierpiński, G. (ed.) TSTP 2019. AISC, vol. 1091, pp. 26–36. Springer, Cham (2020). https://doi.org/10.1007/978-3-030-35543-2_3
13. Lopes, N.V.: Smart governance: a key factor for smart cities implementation. In: 2017 IEEE International Conference on Smart Grid and Smart Cities (ICSGSC), pp. 277–282. IEEE (2017)
14. Mahdavinejad, M.S., Rezvan, M., Barekatain, M., Adibi, P., Barnaghi, P., Sheth, A.P.: Machine learning for internet of things data analysis: a survey. Digit. Commun. Netw. **4**(3), 161–175 (2018)
15. Maldonado Silveira Alonso Munhoz, P.A., et al.: Smart mobility: the main drivers for increasing the intelligence of urban mobility. Sustainability **12**(24), 10675 (2020)

16. Malek, J.A., Lim, S.B., Yigitcanlar, T.: Social inclusion indicators for building citizen-centric smart cities: a systematic literature review. Sustainability **13**(1), 376 (2021)
17. Ogawa, K., Kanai, K., Nakamura, K., Kanemitsu, H., Katto, J., Nakazato, H.: IoT device virtualization for efficient resource utilization in smart city IoT platform. In: 2019 IEEE International Conference on Pervasive Computing and Communications Workshops (PerCom Workshops), pp. 419–422. IEEE (2019)
18. Paskaleva, K.A.: Enabling the smart city: the progress of city e-governance in Europe. Int. J. Innov. Region. Dev. **1**(4), 405–422 (2009)
19. Pereira, G.V., Parycek, P., Falco, E., Kleinhans, R.: Smart governance in the context of smart cities: a literature review. Inf. Polity **23**(2), 143–162 (2018)
20. Poncha, L.J., Abdelhamid, S., Alturjman, S., Ever, E., Al-Turjman, F.: 5G in a convergent internet of things era: an overview. In: 2018 IEEE International Conference on Communications Workshops (ICC Workshops), pp. 1–6. IEEE (2018)
21. Ramirez Lopez, L.J., Grijalba Castro, A.I.: Sustainability and resilience in smart city planning: a review. Sustainability **13**(1), 181 (2020)
22. Talari, S., Shafie-Khah, M., Siano, P., Loia, V., Tommasetti, A., Catalão, J.P.: A review of smart cities based on the internet of things concept. Energies **10**(4), 421 (2017)
23. Toli, A.M., Murtagh, N.: The concept of sustainability in smart city definitions. Front. Built Environ. **6**, 77 (2020)
24. Ullah, Z., Al-Turjman, F., Mostarda, L., Gagliardi, R.: Applications of artificial intelligence and machine learning in smart cities. Comput. Commun. **154**, 313–323 (2020)
25. Xiong, Z., Sheng, H., Rong, W., Cooper, D.E.: Intelligent transportation systems for smart cities: a progress review. Sci. China Inf. Sci. **55**, 2908–2914 (2012)
26. Yang, C., et al.: Using 5g in smart cities: a systematic mapping study. Intell. Syst. Appl. 200065 (2022)
27. Yu, Y., Zhang, N.: Does smart city policy improve energy efficiency? Evidence from a quasi-natural experiment in china. J. Clean. Prod. **229**, 501–512 (2019)
28. Zahoor, S., Javaid, S., Javaid, N., Ashraf, M., Ishmanov, F., Afzal, M.K.: Cloud-fog-based smart grid model for efficient resource management. Sustainability **10**(6), 2079 (2018)
29. Zekić-Sušac, M., Mitrović, S., Has, A.: Machine learning based system for managing energy efficiency of public sector as an approach towards smart cities. Int. J. Inf. Manag. **58**, 102074 (2021)

Automatic Discovery of Zones of Interests with Maritime Trajectory Mining

Omar Ghannou(✉)

Aix-Marseille University, CNRS, LIS, Marseille, France
omar.ghannou@univ-amu.fr

Abstract. In this paper, we describe a method and an algorithm for trajectory mining to identify specific trajectory points called zones of interest (ZOI). Our method ingests both raw GPS data and contextual ones (ZOI candidates neighborhood) in order to classify ZOI. The used data is a real GPS-based trajectories (e.g., a container shipment with maritime and terrestrial data) that are collected from thousands of commercial ships within the TNTM French project (https://www.pole-mer-bretagne-atlantique.com/en/maritime-ports-infrastructure-and-transport/project/tntm).

Keywords: Trajectory Data · Maritime · Terrestrial · Zone Of Interest · ZOI · Mobility Analysis · Data Mining · Clustering · Stops detection · DBScan

Introduction

This paper describes the preliminary findings of the Ph.D. thesis work conducted as part of the TNTM project. This work aims to mine trajectories to identify specific stop points that constitutes ZOI, in contributing to the overall TNTM goal, i.e., addressing environmental concerns such as reducing CO_2 emissions.

The Mobility analysis refers to analyzing the movement to figure out the patterns of mobile objects to optimize their behavior and performance. It involves many techniques to model these movements and make decisions based on observed or predicted data. Mobility analysis is widely used in AI applications such as object tracking, autonomous navigation, swarm robotics, etc. It aims to improve mobility aspects for multiple purposes, such as ensuring effective tracking and determining optimal routes for agents, considering obstacles avoidance, energy consumption optimization, and addressing environmental concerns such as CO_2 emissions.

Mobility analysis can offer a significant contribute in the transportation and logistics domains; as this thesis is carried out within the TNTM logistic National French project (Transformation Numérique du Transport Maritime). our concern is the identification and classification of Zones-of-interest (ZOI) in a contextual approach that relies on clustering and the detection of stops in trajectories, and it combines GPS-data and surrounding context for a reliable identification and classification.

A. Abelló et al. (Eds.): ADBIS 2023, CCIS 1850, pp. 684–692, 2023.
https://doi.org/10.1007/978-3-031-42941-5_62

The rest of the paper is organized as follows: Sect. 1, and Sect. 2 reviews some related work. Section 3 describes our approach, together with some preliminary results. Finally, we concludes the paper and highlights some future directions.

1 Problem Statement

A Zone of Interest (ZOI) refers to a geographical area of particular interest for studying mobility patterns, transportation systems, or urban planning. Analyzing mobility through ZOI can provide valuable insights into transportation patterns, travel behavior, congestion, and other factors. ZOI discovery is a relatively idle domain where the proposed works in the literature are very limited, which is why in this work as a ZOI problematic there are various challenges or considerations we address to get a reliable identification: As first, defining the ZOI, is the challenge of defining the boundaries of the Zone of Interest accurately. Also, data availability is another challenge as the analysis relies on diverse data sources. There also exist data quality issues, such as accuracy, completeness, or consistency. An additional challenge is that the analysis should consider both spatial and temporal dimensions. Without forgetting an essential need and challenge, which is the detection of the transportation mode, where understanding the different modes of transportation through the ZOI is crucial for any analysis. Last but not least, an objective of this project is the predictability of multiple metrics, such as the next ZOI, next mode of transportation, arrival time, etc.

This paper addresses one of the mentioned challenges: the detection and definition of ZOI based on trajectory data, in considering both the spatial and temporal dimensions. To this end, we propose a clustering method to find stops in trajectories, hence defining the ZOI, as explained in detail in the third section. In the next section, we review some related work in trajectory mining, focusing on clustering which is the basis of our method.

2 Related Works

Trajectory mining or trajectory data mining is an interdisciplinary research area that focuses on analyzing and extracting knowledge from trajectory data which consists of spatio-temporal information collected from moving objects, such as GPS logs, mobile phone signals, and surveillance cameras. Trajectory mining has been widely applied in various domains, such as transportation, urban planning, social network analysis, and environmental monitoring, etc. Recently, *Deep Learning* techniques have been applied to trajectory Mining, as they are able to capture complex spatio-temporal patterns and dependencies. As example, [3] proposed the usage of Recurrent Neural Networks (RNNs) model to predict the next locations in a vehicle's trajectory, given previous locations. Convolutional Neural Networks (CNNs) models also have been used for trajectory or next point in trajectory prediction, some of them are combined with LSTM. Deep learning also can be found in trajectory clustering or anomaly detection. Such features are of great interest and can be used as a basis to compare our algorithms. *Privacy Preservation* is an important aspect to consider in the collection of trajectory data, and such case applies to our dataset as we show later; Several studies have been proposed to cover this issue, among them: local preferential anonymity (LPA), trajectory anonymization, differential privacy, and homomorphic encryption.

As part of TNTM project which has among its aims the environmental axis, we find many of environmental monitoring aspects rely on trajectory data mining, as air pollution monitoring, noise pollution monitoring, and wildlife tracking. Several studies have proposed methods to analyze trajectory data for environmental monitoring; [4] proposes a tree-based approach that makes use of sequence analysis methods to extract spatio-temporal patterns in animal trajectories. In this environmental axis we can find also other applications such trajectory-based anomaly detection, trajectory clustering, and trajectory hotspot detection. Before embarking on an in-depth analysis, it is imperative to establish a technical definition of the concept of a trajectory.

Trajectories are considered as a subset of spatio-temporal data with the particular property of displaying the evolution in the spatial dimensions by time dimension [9]. Many definitions were proposed to trajectories, who do not articulate the origin of trajectories, while other definitions are explicitly considering trajectories to be representations of moving objects. In the modeling of object's movements there are a few alternatives to consider regarding representation of time and space properties. Literature provides at least two principal, fundamentally different, definitions on what trajectories are. One of the overall definitions concerns a continuous map from time into the 2D/3D plane, so that the position of the object can be calculated at any time point.

Definition 1. *(Trajectory) A trajectory is a continuous mapping from R to R^2. formally :*

$$\begin{aligned} \tau : \mathbb{R} &\rightarrow \mathbb{R}^2 \\ t &\mapsto (x, y) \end{aligned}$$

However, in the alternative definition the movement abstraction differs from the movements notion as continuous. This discrete trajectory is instead a polyline in three-dimensional space. At each specific, recorded, timestamp the object is represented as a point, for which only the position, but not the extent, is relevant. This is the most utilized definition of trajectories (Fig. 1).

Definition 2. *(Trajectory) A trajectory T is an ordered list of spatio-temporal samples $p_1, p_2, p_3, ..., p_n$. Each $p_i = (x_i, y_i, t_i)$, where x_i, y_i are the spatial coordinates of the sampled point and t_i is the timestamp at which the position of the point is sampled, with $t_1 < t_2 < t_3 < ... < t_n$.*

Trajectory clustering research has attracted attention due to its ability to address multiple issues, such as urban planning or pattern recognition. Many algorithms in this domain have been proposed so far, among which the

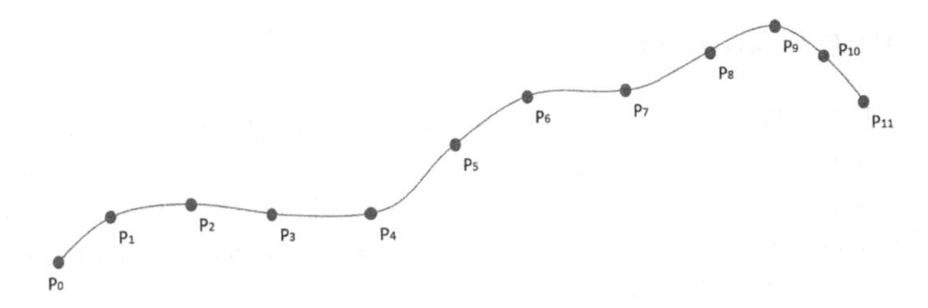

Fig. 1. Trajectory example according to the 2nd definition

renown Density-based Spatial Clustering of Applications with Noise (DBScan) [5]. DBScan effectively clusters trajectories based on their density, offering the advantage of noise detection. This adaptability to density patterns, trajectory structures, and noisy data contribute to its popularity and applicability in the domain.

In [8], a DBScan variant is proposed: clustering rules are defined for trajectories rather than points (i.e. neighborhood, distance, etc.). This variant also performs sub-trajectories clustering which is important for detecting common sub-trajectories among two or more trajectories that are different. Another work of interest [7],improves the previous variant in providing clustering by region and by trajectory. A framework proposed in [10] presents a hierarchical density-based spatial trajectory clustering that uses Hausdorff distance as the similarity measure where the distance between each two trajectories is represented by a single value in a similarity matrix. Other researchers used DBScan on trajectories to get stops rather than clustering the whole trajectories or sub trajectories which is beneficial in some domains, where stop area detection or identification is needed. Our work focuses on the usage of density-based clustering in order to get the potential stop areas or ZOI; We can find also some proposed works that cover the stops detection as in [2], where the T-DBScan is proposed as a spatiotemporal variant of DBSCAN for Stop detection.

Trajectory data quality is also an issue, and for example, in the TNTM context, AIS maritime data suffer from quality issues such as timestamp desynchronization. One idea suggested in [1] is to use a novel approach named Context-enhanced Trajectory Reconstruction to improve Call-detail-Records based data quality to ensure the completeness of human trajectories.

Finally, several studies have proposed methods to incorporate contextual information into trajectory mining which is known as contextual trajectory mining methods, such as context-aware trajectory clustering, context-aware trajectory prediction, and context-aware zone-of-interest detection. [6] proposes a method that considers additional information from virtual reality to the dataset to predict pedestrian trajectory using a multi-input network of Long Short-Term Memory. In the TNTM context, we focus on these aspects of context-aware prediction to get additional information for ZOI classification purposes.

3 Proposed Approach

Let us recall TNTM project objectives which are both economic and environmental optimization. In this context, one of the first objectives is understanding ships' trajectories in extracting mobility patterns. Therefore this thesis addresses multiple issues as discussed before, but the work described in this paper focuses on ZOI detection and identification.

As our main topic is ZOI discovery or ZOI defining, which means defining the boundaries of a geographical area with a particular interest and building a contextual view of the area, thus defines a ZOI; To develop this framework we need to identify the stops being made by a moving object (i.e. containers, ships, vehicles, etc.) within a trajectory. This is the challenge as we need a reliable algorithm to extract the stops from a trajectory.

DBScan is suitable for density clustering, and many applications rely on it to extract the stops clusters within a trajectory. Still, in multiple scenarios, DBScan clusters do not always represent the real stops (i.e., we have false positives). In the case of round trips, an object goes through many places at least twice, which makes that many points are likely to be identified as stops, which is not the case. That is why considering the spatial dimensions only in this clustering, even with adapted parameters to the trajectory type, increases the false positives.

In this work, we propose a method that uses contextual *overground* information (i.e. spatial context) and an algorithm that we call STC-DBScan; a spatiotemporal variant of DBScan, which uses density to find the clusters, but it also considers the temporal dimension. With this approach, we avoid the issues created by the round trips and even the slowdowns considered as stops by defining a constraint on the time values. Figure 2 shows the flow of the proposed method.

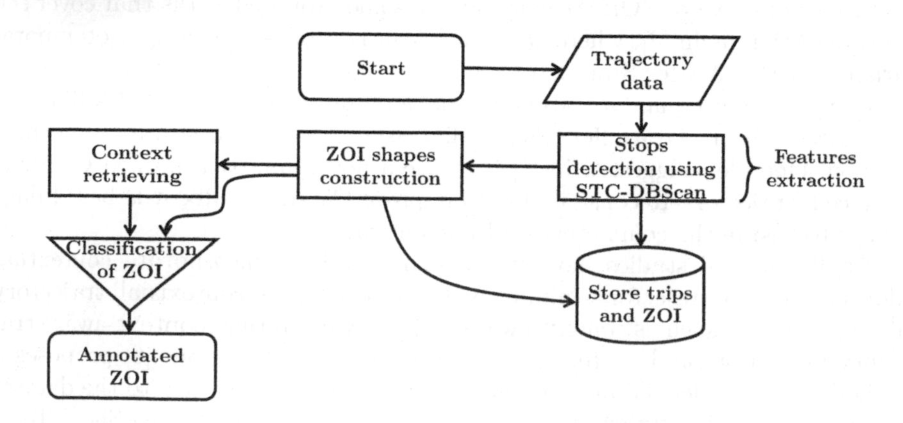

Fig. 2. The Framework of ZOI discovery

Once the stops are identified, our algorithm infer the shapes of the different ZOI based on the stops contained in all the clusters. We use a density based algorithm and a convex hull shaping algorithm to generate the ZOI shapes from the extracted stops. Figure 3 presents some examples of such ZOI shapes. Finally, the last phase of our approach is to classify the ZOI based on the retrieved information from the overground. The proposed algorithm allows the detection of ZOI and their identification (i.e. classification based on the interest). This method is still under development. Finally, Besides its ability to identify stops, STC-DBSCAN can extract trips between stops, see Algorithm 1, which is useful for some of the objectives of the TNTM project such as mobility mean detection.

Fig. 3. ZOI Shapes extracted from the *Porto taxi dataset*

The last phase of this method is the ZOI context retrieving and classification. In our approach, we opt to use OpenStreetMap (OSM) as the source of information. Currently the classification part is under development. Our experiments mostly used TNTM data (confidential) but we also used the public Porto Taxi dataset. Figure 4 shows an example of context retrieving based on ZOI shape, using our algorithm (under development).

Algorithm 1: Spatiotemporal Clustering DBScan Algorithm

```
Data: T : trajectory points, t1 : spatial threshold, t2 : temporal threshold
Result: ST : Trips, S : Stops
P_pre ← T[0];
P_curr ← T[0];
for ( idx = 1; idx < Size(T); idx = idx + 1 ) {
    P_pre ← P_curr; P_curr ← T[idx];
    In_Move ← Distance(P_pre, P_curr) >= t1;
    if In_Move is True then
        StopIndices.Add(idx − 1)  ▷ if marked stop, then mark it in movement;
    else
        StopIndices.Add(idx − 1)  ▷ if marked movement, then mark it in stop;
    end
}
for ( idx = 0; idx < Size(T); idx = idx + 1 ) {
    if T[idx] is in a Stop Area then
        Define temp_trip and add T[idx];
        if T[idx] is last index in a stop area then
            Define bbox of point;
            if BboxDiagDistance(T[idx]) < spatialThreshold and
             cumulatedTimeOnZone > t2 then
                Add to current trip first point of temp_trip;
                ST ← current_trip; and start new current trip;
            else
                Add the whole temp_trip to current trip;
            end
        end
    else
        Add T[idx] to current trip;
        if idx = last index then
            ST ← current_trip;
        end
    end
}
```

(a) ZOI Satellite Picture

(b) Context of ZOI

Fig. 4. ZOI context retrieving algorithm

Conclusion and Future Direction

In this work, we described a method for the automatic discovery of ZOI and stops extraction from maritime and terrestrial trajectories using the proposed STC-DBScan as a spatiotemporal algorithm. Currently, beside improving the method and the results, we are working on its validation, thus we intend to present a comparative study for the stops extraction algorithm. However, with only a handful of similar ZOI identification methods according to our knowledge, we will try to present also a comparative study for the whole method or validate it using relevant machine learning validation methods. This ongoing work is the first step of the Ph.D. thesis, and we plan to pursue in addressing other TNTM issues, for instance, defining precisely the ZOI shape, improving the context retrieving, prediction of transportation mode, waiting time, ETA (expected arrival time) in ZOI, and intermediate ZOI in future works.

Acknowledgment. Thanks to the team members, Etienne Thuillier and Omar Boucelma for their contributions in this work.

References

1. Chen, G., Viana, A.C., Fiore, M., Sarraute, C.: Complete trajectory reconstruction from sparse mobile phone data. EPJ Data Sci. **8**(1), 1–24 (2019). https://doi.org/10.1140/epjds/s13688-019-0206-8
2. Chen, W., Ji, M., Wang, J.: T-dbscan: a spatiotemporal density clustering for GPS trajectory segmentation. Int. J. Online Eng. (iJOE) **10**, 19 (2014). https://doi.org/10.3991/ijoe.v10i6.3881
3. Choi, S., Yeo, H., Kim, J.: Network-wide vehicle trajectory prediction in urban traffic networks using deep learning. Transport. Res. Rec. J. Transport. Res. Board **2672**, 036119811879473 (2018). https://doi.org/10.1177/0361198118794735

4. De Groeve, J., et al.: Extracting spatio-temporal patterns in animal trajectories: an ecological application of sequence analysis methods. Methods Ecol. Evol. **7**(3), 369–379 (2016). https://doi.org/10.1111/2041-210X.12453, https://besjournals.onlinelibrary.wiley.com/doi/abs/10.1111/2041-210X.12453
5. Ester, M., Kriegel, H.P., Sander, J., Xu, X.: A density-based algorithm for discovering clusters in large spatial databases with noise. In: Proceedings of the Second International Conference on Knowledge Discovery and Data Mining. KDD'96, pp. 226–231. AAAI Press (1996)
6. Kalatian, A., Farooq, B.: A context-aware pedestrian trajectory prediction framework for automated vehicles. Transport. Res. Part C Emerg. Technol. **134**, 103453 (2022). https://doi.org/10.1016/j.trc.2021.103453, https://www.sciencedirect.com/science/article/pii/S0968090X21004423
7. Lee, J.G., Han, J., Li, X., Gonzalez, H.: Traclass: trajectory classification using hierarchical region-based and trajectory-based clustering. Proc. VLDB Endow. **1**(1), 1081–1094 (2008). https://doi.org/10.14778/1453856.1453972
8. Lee, J.G., Han, J., Whang, K.Y.: Trajectory clustering: a partition-and-group framework. In: Proceedings of the 2007 ACM SIGMOD International Conference on Management of Data. SIGMOD '07, pp. 593–604. Association for Computing Machinery, New York, NY, USA (2007). https://doi.org/10.1145/1247480.1247546
9. Wagner, R., de Macedo, J.A.F., Raffaetà, A., Renso, C., Roncato, A., Trasarti, R.: Mob-warehouse: a semantic approach for mobility analysis with a trajectory data warehouse. In: Parsons, J., Chiu, D. (eds.) ER 2013. LNCS, vol. 8697, pp. 127–136. Springer, Cham (2014). https://doi.org/10.1007/978-3-319-14139-8_15
10. Wang, L., Chen, P., Chen, L., Mou, J.: Ship AIS trajectory clustering: an hdbscan-based approach. J. Marine Sci. Eng. **9**(6) (2021). https://doi.org/10.3390/jmse9060566, https://www.mdpi.com/2077-1312/9/6/566

Author Index

A. Abelló et al. (Eds.): ADBIS 2023, CCIS 1850, pp. 693–695, 2023.
https://doi.org/10.1007/978-3-031-42941-5

Printed in the United States
by Baker & Taylor Publisher Services

Printed in the United States
by Baker & Taylor Publisher Services